Precalculus
Mathematics for Calculus

SIXTH EDITION

James Stewart
McMaster University and University of Toronto

Lothar Redlin
The Pennsylvania State University

Saleem Watson
California State University, Long Beach

Prepared by

Andrew Bulman-Fleming

BROOKS/COLE
CENGAGE Learning

Australia • Brazil • Japan • Korea • Mexico • Singapore • Spain • United Kingdom • United States

For product information and technology assistance,
contact us at
**Cengage Learning Customer & Sales Support,
1-800-354-9706**

For permission to use material from this text or product,
submit all requests online at
www.cengage.com/permissions
Further permissions questions can be emailed to
permissionrequest@cengage.com

ISBN-13: 978-0-8400-6879-8
ISBN-10: 0-8400-6879-4

Brooks/Cole
20 Davis Drive
Belmont, CA 94002-3098
USA

Cengage Learning is a leading provider of
customized learning solutions with office locations
around the globe, including Singapore, the United
Kingdom, Australia, Mexico, Brazil, and Japan.
Locate your local office at:
www.cengage.com/global

Cengage Learning products are represented in
Canada by Nelson Education, Ltd.

To learn more about Brooks/Cole, visit
www.cengage.com/brookscole

Purchase any of our products at your local
college store or at our preferred online store
www.cengagebrain.com

Printed in the United States of America
1 2 3 4 5 6 7 15 14 13 12 11

CONTENTS

CHAPTER 6   TRIGONOMETRIC FUNCTIONS: RIGHT TRIANGLE APPROACH   227

CHAPTER 7   ANALYTIC TRIGONOMETRY   245

CHAPTER 8   POLAR COORDINATES AND PARAMETRIC EQUATIONS   269

1 FUNDAMENTALS

1.1 REAL NUMBERS

1. (a) The natural numbers are $\{1, 2, 3, \ldots\}$.

(b) The numbers $\{\ldots, -3, -2, -1, 0\}$ are integers but not natural numbers.

(c) Any irreducible fraction $\dfrac{p}{q}$ with $q \neq 1$ is rational but is not an integer. Examples: $\frac{3}{2}, -\frac{5}{12}, \frac{1729}{23}$.

(d) Any number which cannot be expressed as a ratio $\dfrac{p}{q}$ of two integers is irrational. Examples are $\sqrt{2}$, $\sqrt{3}$, π, and e.

3. The set of numbers between but not including 2 and 7 can be written as $\{x \mid 2 < x < 7\}$, and $(2, 7)$ in interval notation.

5. (a) Natural number: 50

(b) Integers: $0, -10, 50$

(c) Rational numbers: $0, -10, 50, \frac{22}{7}, 0.538, 1.2\overline{3}, -\frac{1}{3}$

(d) Irrational numbers: $\sqrt{7}, \sqrt[3]{2}$

7. Commutative Property for addition

9. Associative Property for addition

11. Distributive Property

13. Commutative Property for multiplication

15. $x + 3 = 3 + x$

17. $4(A + B) = 4A + 4B$

19. $3(x + y) = 3x + 3y$

21. $4(2m) = (4 \cdot 2)m = 8m$

23. $-\frac{5}{2}(2x - 4y) = -\frac{5}{2}(2x) + \frac{5}{2}(4y) = -5x + 10y$

25. (a) $\frac{3}{10} + \frac{4}{15} = \frac{9}{30} + \frac{8}{30} = \frac{17}{30}$

(b) $\frac{1}{4} + \frac{1}{5} = \frac{5}{20} + \frac{4}{20} = \frac{9}{20}$

27. (a) $\frac{2}{3}\left(6 - \frac{3}{2}\right) = \frac{2}{3} \cdot 6 - \frac{2}{3} \cdot \frac{3}{2} = 4 - 1 = 3$

(b) $0.25\left(\frac{8}{9} + \frac{1}{2}\right) = \frac{1}{4}\left(\frac{16}{18} + \frac{9}{18}\right) = \frac{1}{4} \cdot \frac{25}{18} = \frac{25}{72}$

29. (a) $\dfrac{\frac{2}{3}}{\frac{2}{3}} - \dfrac{\frac{2}{3}}{2} = 2 \cdot \frac{3}{2} - \frac{2}{3} \cdot \frac{1}{2} = 3 - \frac{1}{3} = \frac{9}{3} - \frac{1}{3} = \frac{8}{3}$

(b) $\dfrac{\frac{1}{12}}{\frac{1}{8} - \frac{1}{9}} = \dfrac{\frac{1}{12}}{\frac{1}{8} - \frac{1}{9}} \cdot \frac{72}{72} = \frac{6}{9 - 8} = \frac{6}{1} = 6$

31. (a) $2 \cdot 3 = 6$ and $2 \cdot \frac{7}{2} = 7$, so $3 < \frac{7}{2}$

(b) $-6 > -7$

(c) $3.5 = \frac{7}{2}$

33. (a) False

(b) True

35. (a) False   **(b)** True

37. (a) $x > 0$ **(b)** $t < 4$

(c) $a \geq \pi$ **(d)** $-5 < x < \frac{1}{3}$

(e) $|p - 3| \leq 5$

39. (a) $A \cup B = \{1, 2, 3, 4, 5, 6, 7, 8\}$

(b) $A \cap B = \{2, 4, 6\}$

41. (a) $A \cup C = \{1, 2, 3, 4, 5, 6, 7, 8, 9, 10\}$

(b) $A \cap C = \{7\}$

43. (a) $B \cup C = \{x \mid x \le 5\}$

(b) $B \cap C = \{x \mid -1 < x < 4\}$

45. $(-3, 0) = \{x \mid -3 < x < 0\}$

47. $[2, 8) = \{x \mid 2 \le x < 8\}$

49. $[2, \infty) = \{x \mid x \ge 2\}$

51. $x \le 1 \Leftrightarrow x \in (-\infty, 1]$

53. $-2 < x \le 1 \Leftrightarrow x \in (-2, 1]$

55. $x > -1 \Leftrightarrow x \in (-1, \infty)$

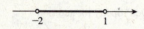

57. (a) $[-3, 5]$ **(b)** $(-3, 5]$

59. $(-2, 0) \cup (-1, 1) = (-2, 1)$

61. $[-4, 6] \cap [0, 8) = [0, 6]$

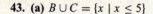

63. $(-\infty, -4) \cup (4, \infty)$

65. (a) $|100| = 100$

(b) $|-73| = 73$

67. (a) $||-6| - |-4|| = |6 - 4| = |2| = 2$

(b) $\frac{-1}{|-1|} = \frac{-1}{1} = -1$

69. (a) $|(-2) \cdot 6| = |-12| = 12$

(b) $\left| \left(-\frac{1}{3}\right)(-15) \right| = |5| = 5$

71. $|(-2) - 3| = |-5| = 5$

73. (a) $|17 - 2| = 15$

(b) $|21 - (-3)| = |21 + 3| = |24| = 24$

(c) $\left| -\frac{3}{10} - \frac{11}{8} \right| = \left| -\frac{12}{40} - \frac{55}{40} \right| = \left| -\frac{67}{40} \right| = \frac{67}{40}$

75. (a) Let $x = 0.777\ldots$. So $10x = 7.7777\ldots \Leftrightarrow x = 0.7777\ldots \Leftrightarrow 9x = 7$. Thus, $x = \frac{7}{9}$.

(b) Let $x = 0.2888\ldots$. So $100x = 28.8888\ldots \Leftrightarrow 10x = 2.8888\ldots \Leftrightarrow 90x = 26$. Thus, $x = \frac{26}{90} = \frac{13}{45}$.

(c) Let $x = 0.575757\ldots$. So $100x = 57.5757\ldots \Leftrightarrow x = 0.5757\ldots \Leftrightarrow 99x = 57$. Thus, $x = \frac{57}{99} = \frac{19}{33}$.

77. Distributive Property

79. (a) When $L = 60$, $x = 8$, and $y = 6$, we have $L + 2(x + y) = 60 + 2(8 + 6) = 60 + 28 = 88$. Because $88 \le 108$ the post office will accept this package.

When $L = 48$, $x = 24$, and $y = 24$, we have $L + 2(x + y) = 48 + 2(24 + 24) = 48 + 96 = 144$, and since $144 \not\le 108$, the post office will *not* accept this package.

(b) If $x = y = 9$, then $L + 2(9 + 9) \le 108 \Leftrightarrow L + 36 \le 108 \Leftrightarrow L \le 72$. So the length can be as long as 72 in. = 6 ft.

81. Let $x = \dfrac{m_1}{n_1}$ and $y = \dfrac{m_2}{n_2}$ be rational numbers. Then $x + y = \dfrac{m_1}{n_1} + \dfrac{m_2}{n_2} = \dfrac{m_1 n_2 + m_2 n_1}{n_1 n_2}$,

$x - y = \dfrac{m_1}{n_1} - \dfrac{m_2}{n_2} = \dfrac{m_1 n_2 - m_2 n_1}{n_1 n_2}$, and $x \cdot y = \dfrac{m_1}{n_1} \cdot \dfrac{m_2}{n_2} = \dfrac{m_1 m_2}{n_1 n_2}$. This shows that the sum, difference, and product

of two rational numbers are again rational numbers. However the product of two irrational numbers is not necessarily

irrational; for example, $\sqrt{2} \cdot \sqrt{2} = 2$, which is rational. Also, the sum of two irrational numbers is not necessarily irrational;

for example, $\sqrt{2} + \left(-\sqrt{2}\right) = 0$ which is rational.

83.

x	1	2	10	100	1000
$\dfrac{1}{x}$	1	$\dfrac{1}{2}$	$\dfrac{1}{10}$	$\dfrac{1}{100}$	$\dfrac{1}{1000}$

As x gets large, the fraction $1/x$ gets small. Mathematically, we say that $1/x$ goes to zero.

x	1	0.5	0.1	0.01	0.001
$\dfrac{1}{x}$	1	$\dfrac{1}{0.5} = 2$	$\dfrac{1}{0.1} = 10$	$\dfrac{1}{0.01} = 100$	$\dfrac{1}{0.001} = 1000$

As x gets small, the fraction $1/x$ gets large. Mathematically, we say that $1/x$ goes to infinity.

85. (a) Subtraction is not commutative. For example, $5 - 1 \neq 1 - 5$.

(b) Division is not commutative. For example, $5 \div 1 \neq 1 \div 5$.

1.2 EXPONENTS AND RADICALS

1. (a) Using exponential notation we can write the product $5 \cdot 5 \cdot 5 \cdot 5 \cdot 5 \cdot 5$ as 5^6.

(b) In the expression 3^4, the number 3 is called the *base* and the number 4 is called the *exponent*.

3. (a) Using exponential notation we can write $\sqrt[3]{5}$ as $5^{1/3}$.

(b) Using radicals we can write $5^{1/2}$ as $\sqrt{5}$.

(c) No. $\sqrt{5^2} = \left(5^2\right)^{1/2} = 5^{2(1/2)} = 5$ and $\left(\sqrt{5}\right)^2 = \left(5^{1/2}\right)^2 = 5^{(1/2)2} = 5$.

5. Because the denominator is of the form $\sqrt{a}$, we multiply numerator and denominator by $\sqrt{a}$: $\dfrac{1}{\sqrt{3}} = \dfrac{1}{\sqrt{3}} \cdot \dfrac{\sqrt{3}}{\sqrt{3}} = \dfrac{\sqrt{3}}{3}$.

7. $\dfrac{1}{\sqrt{5}} = 5^{-1/2}$

9. $4^{2/3} = \sqrt[3]{4^2} = \sqrt[3]{16}$

11. $\sqrt[5]{5^3} = 5^{3/5}$

13. $a^{2/5} = \sqrt[5]{a^2}$

15. (a) $-3^2 = -9$

(b) $(-3)^2 = 9$

(c) $\left(\dfrac{1}{3}\right)^4 (-3)^2 = \left(3^{-4}\right)\left(3^2\right) = 3^{-4+2} = 3^{-2} = \dfrac{1}{9}$

17. (a) $\left(\dfrac{5}{3}\right)^0 2^{-1} = 1 \cdot 2^{-1} = \dfrac{1}{2}$

(b) $\dfrac{2^{-3}}{3^0} = \dfrac{2^{-3}}{1} = \dfrac{1}{8}$

(c) $\left(\dfrac{1}{4}\right)^{-2} = (4)^2 = 16$

19. (a) $\sqrt{16} = \sqrt{4^2} = 4$

(b) $\sqrt[4]{16} = \sqrt[4]{2^4} = 2$

(c) $\sqrt[4]{\dfrac{1}{16}} = \sqrt[4]{\left(\dfrac{1}{2}\right)^4} = \dfrac{1}{2}$

21. (a) $\sqrt{\dfrac{4}{9}} = \sqrt{\left(\dfrac{2}{3}\right)^2} = \dfrac{2}{3}$

(b) $\sqrt[4]{256} = \sqrt[4]{4^4} = 4$

(c) $\sqrt[6]{\dfrac{1}{64}} = \sqrt[6]{\dfrac{1}{2^6}} = \dfrac{\sqrt[6]{1}}{\sqrt[6]{2^6}} = \dfrac{1}{2}$

23. (a) $\left(\dfrac{4}{9}\right)^{-1/2} = \left(\dfrac{2^2}{3^2}\right)^{-1/2} = \dfrac{2^{-1}}{3^{-1}} = \dfrac{3}{2}$

25. When $x = 3$, $y = 4$, $z = -1$ we have
$$\sqrt{x^2 + y^2} = \sqrt{3^2 + 4^2} = \sqrt{9 + 16} = \sqrt{25} = 5.$$

(b) $(-32)^{2/5} = \left[(-2)^5\right]^{2/5} = (-2)^2 = 4$

(c) $-32^{2/5} = -\left[\left(2^5\right)^{2/5}\right] = -2^2 = -4$

27. When $x = 3$, $y = 4$, $z = -1$ we have

$(9x)^{2/3} + (2y)^{2/3} + z^{2/3} = (9 \cdot 3)^{2/3} + (2 \cdot 4)^{2/3} + (-1)^{2/3} = \left(3^3\right)^{2/3} + \left(2^3\right)^{2/3} + (1)^{1/3} = 3^2 + 2^2 + 1 = 9 + 4 + 1 = 14$

29. $\sqrt{32} + \sqrt{18} = \sqrt{16 \cdot 2} + \sqrt{9 \cdot 2} = \sqrt{4^2 \cdot 2} + \sqrt{3^2 \cdot 2} = 4\sqrt{2} + 3\sqrt{2} = 7\sqrt{2}$

31. $\sqrt[5]{96} + \sqrt[5]{3} = \sqrt[5]{32 \cdot 3} + \sqrt[5]{3} = \sqrt[5]{2^5 \cdot 3} + \sqrt[5]{3} = 2\sqrt[5]{3} + \sqrt[5]{3} = 3\sqrt[5]{3}$

33. $\sqrt{16x} + \sqrt{x^5} = \sqrt{4^2 \cdot x} + \sqrt{x^4 \cdot x} = 4\sqrt{x} + x^2\sqrt{x} = \left(x^2 + 4\right)\sqrt{x}$

35. (a) $x^8 x^2 = x^{8+2} = x^{10}$

37. (a) $\dfrac{y^{10}y^0}{y^7} = y^{10+0-7} = y^3$

(b) $\left(3y^2\right)\left(4y^5\right) = 3\,(4)\,y^{2+5} = 12y^7$

(b) $\dfrac{x^6}{x^{10}} = x^{6-10} = x^{-4} = \dfrac{1}{x^4}$

(c) $x^2 x^{-6} = x^{2-6} = x^{-4} = \dfrac{1}{x^4}$

(c) $\dfrac{a^9 a^{-2}}{a} = a^{9-2-1} = a^6$

39. (a) $\left(a^2 a^4\right)^3 = \left(a^6\right)^3 = a^{6 \cdot 3} = a^{18}$

41. (a) $\left(4x^2 y^4\right)\left(2x^5 y\right) = 4 \cdot 2x^{2+5}y^{4+1} = 8x^7 y^5$

(b) $\left(\dfrac{a^2}{4}\right)^3 = \dfrac{a^{2 \cdot 3}}{4^3} = \dfrac{a^6}{64}$

(b) $\left(8a^2 z\right)\left(\tfrac{1}{2}a^3 z^4\right) = 8 \cdot \tfrac{1}{2}a^{2+3}z^{1+4} = 4a^5 z^5$

(c) $(3z)^2 \left(6z^2\right)^{-3} = 3^2 z^2 \cdot$

$(3 \cdot 2)^{-3}\left(z^2\right)^{-3} = 3^{-1}2^{-3}z^2 z^{-6} = \dfrac{1}{24z^4}$

43. (a) $\left(5x^2 y^3\right)\left(3x^2 y^5\right)^4 = 5 \cdot 3^4 x^{2+2 \cdot 4} y^{3+5 \cdot 4}$

45. (a) $\dfrac{6y^3 z}{2yz^2} = \dfrac{6}{2}y^{3-1}z^{1-2} = \dfrac{3y^2}{z}$

$= 405x^{10}y^{23}$

(b) $\left(2a^3 b^2\right)^2 \left(5a^2 b^5\right)^3 = 2^2 \cdot 5^3 a^{3 \cdot 2 + 2 \cdot 3} b^{2 \cdot 2 + 5 \cdot 3}$

(b) $\dfrac{\left(xy^2 z^3\right)^4}{\left(x^2 y^2 z\right)^3} = x^{4-2 \cdot 3}y^{2 \cdot 4 - 2 \cdot 3}z^{3 \cdot 4 - 3} = \dfrac{y^2 z^9}{x^2}$

$= 500a^{12}b^{19}$

47. (a) $\left(\dfrac{a^2}{b}\right)^5 \left(\dfrac{a^3 b^2}{c^3}\right)^3 = a^{2 \cdot 5 + 3 \cdot 3}b^{-5+2 \cdot 3}c^{-3 \cdot 3} = \dfrac{a^{19}b}{c^9}$

(b) $\dfrac{\left(u^{-1}v^2\right)^2}{\left(u^3 v^{-2}\right)^3} = u^{-1 \cdot 2 - 3 \cdot 3}v^{2 \cdot 2 - (-2)3} = \dfrac{v^{10}}{u^{11}}$

49. (a) $\dfrac{8a^3 b^{-4}}{2a^{-5}b^5} = 4a^{3-(-5)}b^{-4-5} = \dfrac{4a^8}{b^9}$

(b) $\left(\dfrac{y}{5x^{-2}}\right)^{-3} = 5^{-1(-3)}x^{-(-2)(-3)}y^{-3} = \dfrac{125}{x^6 y^3}$

51. (a) $\left(\dfrac{3a}{b^3}\right)^{-1} = 3^{-1}a^{-1}b^{-3(-1)} = \dfrac{b^3}{3a}$

(b) $\left(\dfrac{q^{-1}r^{-1}s^{-2}}{r^{-5}sq^{-8}}\right)^{-1} = \dfrac{r^{-5}sq^{-8}}{q^{-1}r^{-1}s^{-2}} = q^{-8-(-1)}r^{-5-(-1)}s^{1-(-2)} = \dfrac{s^3}{q^7 r^4}$

53. $\sqrt[4]{x^4} = |x|$

55. $\sqrt[4]{16x^8} = \sqrt[4]{2^4 x^8} = 2x^2$

57. $\sqrt[6]{64a^6 b^7} = \sqrt[6]{2^6 \cdot a^6 \cdot b^6 \cdot b} = 2ab\sqrt[6]{b}$

59. $\sqrt[3]{\sqrt{64x^6}} = \left(8\left|x^3\right|\right)^{1/3} = 2\,|x|$

61. (a) $x^{3/4}x^{5/4} = x^{3/4+5/4} = x^2$

   (b) $y^{2/3}y^{4/3} = y^{2/3+4/3} = y^2$

63. (a) $\dfrac{w^{4/3}w^{2/3}}{w^{1/3}} = w^{4/3+2/3-1/3} = w^{5/3}$

   (b) $\dfrac{s^{5/2}\left(2s^{5/4}\right)^2}{s^{1/2}} = 2^2 s^{(5/2)+(5/4)2-1/2} = 4s^{9/2}$

65. (a) $\left(8a^6b^{3/2}\right)^{2/3} = 8^{2/3}a^{6(2/3)}b^{(3/2)(2/3)} = 4a^4b$

   (b) $\left(4a^6b^8\right)^{3/2} = 4^{3/2}a^{6(3/2)}b^{8(3/2)} = 8a^9b^{12}$

67. (a) $\dfrac{\left(8s^3t^3\right)^{2/3}}{\left(s^4t^{-8}\right)^{1/4}} = 8^{2/3}s^{3(2/3)-4(1/4)}t^{3(2/3)-(-8)(1/4)} = 4st^4$

   (b) $\dfrac{\left(32y^{-5}z^{10}\right)^{1/5}}{\left(64y^6z^{-12}\right)^{-1/6}} = 32^{1/5}64^{-(-1/6)}y^{-5(1/5)-6(-1/6)}z^{10(1/5)-(-12)(-1/6)} = 4$

69. (a) $\left(\dfrac{x^{-2/3}}{y^{1/2}}\right)\left(\dfrac{x^{-2}}{y^{-3}}\right)^{1/6} = x^{-2/3+-2(1/6)}y^{-1/2-(-3)(1/6)} = \dfrac{1}{x}$

   (b) $\left(\dfrac{4y^3z^{2/3}}{x^{1/2}}\right)^2\left(\dfrac{x^{-3}y^6}{8z^4}\right)^{1/3} = 4^2 8^{-1/3}x^{-1/2(2)-3(1/3)}y^{3(2)+6(1/3)}z^{2/3(2)-4(1/3)} = \dfrac{8y^8}{x^2}$

71. (a) $\sqrt[6]{y^5}\sqrt[3]{y^2} = y^{5/6}\cdot y^{2/3} = y^{5/6+2/3} = y^{3/2}$

   (b) $\left(5\sqrt[3]{x}\right)\left(2\sqrt[4]{x}\right) = 5\cdot 2x^{1/3+1/4} = 10x^{7/12}$

73. (a) $\sqrt{4st^3}\sqrt[6]{s^3t^2} = 4^{1/2}s^{1/2+3/6}t^{3/2+2/6} = 2st^{11/6}$

   (b) $\dfrac{\sqrt[4]{x^7}}{\sqrt[4]{x^3}} = x^{7/4-3/4} = x$

75. (a) $\sqrt[3]{y\sqrt{y}} = \left(y^{1+1/2}\right)^{1/3} = y^{(3/2)(1/3)} = y^{1/2}$

   (b) $\sqrt{\dfrac{16u^3v}{uv^5}} = \sqrt{\dfrac{16u^2}{v^4}} = \dfrac{4u}{v^2}$

77. (a) $69{,}300{,}000 = 6.93 \times 10^7$

   (b) $7{,}200{,}000{,}000{,}000 = 7.2 \times 10^{12}$

   (c) $0.000028536 = 2.8536 \times 10^{-5}$

   (d) $0.0001213 = 1.213 \times 10^{-4}$

79. (a) $3.19 \times 10^5 = 319{,}000$

   (b) $2.721 \times 10^8 = 272{,}100{,}000$

   (c) $2.670 \times 10^{-8} = 0.00000002670$

   (d) $9.999 \times 10^{-9} = 0.000000009999$

81. (a) $5{,}900{,}000{,}000{,}000 \text{ mi} = 5.9 \times 10^{12} \text{ mi}$

   (b) $0.0000000000004 \text{ cm} = 4 \times 10^{-13} \text{ cm}$

   (c) 33 billion billion molecules $= 33 \times 10^9 \times 10^9 = 3.3 \times 10^{19}$ molecules

83. $\left(7.2 \times 10^{-9}\right)\left(1.806 \times 10^{-12}\right) = 7.2 \times 1.806 \times 10^{-9} \times 10^{-12} \approx 13.0 \times 10^{-21} = 1.3 \times 10^{-20}$

85. $\dfrac{1.295643 \times 10^9}{\left(3.610 \times 10^{-17}\right)\left(2.511 \times 10^6\right)} = \dfrac{1.295643}{3.610 \times 2.511} \times 10^{9+17-6} \approx 0.1429 \times 10^{19} = 1.429 \times 10^{19}$

87. $\dfrac{(0.0000162)(0.01582)}{(594621000)(0.0058)} = \dfrac{\left(1.62 \times 10^{-5}\right)\left(1.582 \times 10^{-2}\right)}{\left(5.94621 \times 10^8\right)\left(5.8 \times 10^{-3}\right)} = \dfrac{1.62 \times 1.582}{5.94621 \times 5.8} \times 10^{-5-2-8+3}0.074 \times 10^{-12}$

   $= 7.4 \times 10^{-14}$

89. (a) $\dfrac{1}{\sqrt{10}} = \dfrac{1}{\sqrt{10}} \cdot \dfrac{\sqrt{10}}{\sqrt{10}} = \dfrac{\sqrt{10}}{10}$

(b) $\sqrt{\dfrac{2}{x}} = \sqrt{\dfrac{2}{x}} \cdot \sqrt{\dfrac{x}{x}} = \dfrac{\sqrt{2x}}{x}$

(c) $\sqrt{\dfrac{x}{3}} = \sqrt{\dfrac{x}{3}} \cdot \sqrt{\dfrac{3}{3}} = \dfrac{\sqrt{3x}}{3}$

91. (a) $\dfrac{2}{\sqrt[3]{x}} = \dfrac{2}{\sqrt[3]{x}} \cdot \dfrac{\sqrt[3]{x^2}}{\sqrt[3]{x^2}} = \dfrac{2\sqrt[3]{x^2}}{x}$

(b) $\dfrac{1}{\sqrt[4]{y^3}} = \dfrac{1}{\sqrt[4]{y^3}} \cdot \dfrac{\sqrt[4]{y}}{\sqrt[4]{y}} = \dfrac{\sqrt[4]{y}}{y}$

(c) $\dfrac{x}{y^{2/5}} = \dfrac{x}{\sqrt[5]{y^2}} \cdot \dfrac{\sqrt[5]{y^3}}{\sqrt[5]{y^3}} = \dfrac{x\sqrt[5]{y^3}}{y}$

93. (a) b^5 is negative since a negative number raised to an odd power is negative.

(b) b^{10} is positive since a negative number raised to an even power is positive.

(c) ab^2c^3 we have (positive) (negative)^2 (negative)^3 = (positive) (positive) (negative) which is negative.

(d) Since $b - a$ is negative, $(b - a)^3$ = (negative)^3 which is negative.

(e) Since $b - a$ is negative, $(b - a)^4$ = (negative)^4 which is positive.

(f) $\dfrac{a^3c^3}{b^6c^6} = \dfrac{(\text{positive})^3 (\text{negative})^3}{(\text{negative})^6 (\text{negative})^6} = \dfrac{(\text{positive}) (\text{negative})}{(\text{positive}) (\text{positive})} = \dfrac{\text{negative}}{\text{positive}}$ which is negative.

95. Since one light year is 5.9×10^{12} miles, Centauri is about $4.3 \times 5.9 \times 10^{12} \approx 2.54 \times 10^{13}$ miles away or 25,400,000,000,000 miles away.

97. Volume = (average depth) (area) = $\left(3.7 \times 10^3 \text{ m}\right) \left(3.6 \times 10^{14} \text{ m}^2\right) \left(\dfrac{10^3 \text{ liters}}{\text{m}^3}\right) \approx 1.33 \times 10^{21}$ liters

99. The number of molecules is equal to

$$(\text{volume}) \cdot \left(\dfrac{\text{liters}}{\text{m}^3}\right) \cdot \left(\dfrac{\text{molecules}}{22.4 \text{ liters}}\right) = (5 \cdot 10 \cdot 3) \cdot \left(10^3\right) \cdot \left(\dfrac{6.02 \times 10^{23}}{22.4}\right) \approx 4.03 \times 10^{27}$$

101. (a) Using $f = 0.4$ and substituting $d = 65$, we obtain $s = \sqrt{30fd} = \sqrt{30 \times 0.4 \times 65} \approx 28$ mi/h.

(b) Using $f = 0.5$ and substituting $s = 50$, we find d. This gives $s = \sqrt{30fd} \Leftrightarrow 50 = \sqrt{30 \cdot (0.5)d} \Leftrightarrow 50 = \sqrt{15d} \Leftrightarrow 2500 = 15d \Leftrightarrow d = \frac{500}{3} \approx 167$ feet.

103. Since $10^6 = 10^3 \cdot 10^3$ it would take 1000 days ≈ 2.74 years to spend the million dollars.

Since $10^9 = 10^3 \cdot 10^6$ it would take $10^6 = 1,000,000$ days ≈ 2739.72 years to spend the billion dollars.

105. (a)

n	1	2	5	10	100
$2^{1/n}$	$2^{1/1} = 2$	$2^{1/2} = 1.414$	$2^{1/5} = 1.149$	$2^{1/10} = 1.072$	$2^{1/100} = 1.007$

So when n gets large, $2^{1/n}$ decreases to 1.

(b)

n	1	2	5	10	100
$\left(\frac{1}{2}\right)^{1/n}$	$\left(\frac{1}{2}\right)^{1/1} = 0.5$	$\left(\frac{1}{2}\right)^{1/2} = 0.707$	$\left(\frac{1}{2}\right)^{1/5} = 0.871$	$\left(\frac{1}{2}\right)^{1/10} = 0.933$	$\left(\frac{1}{2}\right)^{1/100} = 0.993$

So when n gets large, $\left(\frac{1}{2}\right)^{1/n}$ increases to 1.

1.3   ALGEBRAIC EXPRESSIONS

1. The polynomial $2x^5 + 6x^4 + 4x^3$ has three terms: $2x^5$, $6x^4$, and $4x^3$. The factor $2x^3$ is common to each term, so $2x^5 + 6x^4 + 4x^3 = 2x^3 \left(x^2 + 3x + 2\right)$. [In fact, the polynomial can be factored further as $2x^3 (x + 2) (x + 1)$.]

3. The Special Product Formula for the "square of a sum" is $(A + B)^2 = A^2 + 2AB + B^2$. So
$(2x + 3)^2 = (2x)^2 + 2(2x)(3) + 3^2 = 4x^2 + 12x + 9$.

5. The Special Factoring Formula for the "difference of squares" is $A^2 - B^2 = (A - B)(A + B)$. So
$4x^2 - 25 = (2x - 5)(2x + 5)$.

7. Type: trinomial. Terms: x^2, $-3x$, and 7. Degree: 2.

9. Type: monomial. Terms: -8. Degree: 0.

11. Type: four-term polynomial. Terms: x, $-x^2$, x^3, and $-x^4$. Degree: 4.

13. $(12x - 7) - (5x - 12) = 12x - 7 - 5x + 12 = 7x + 5$

15. $\left(3x^2 + x + 1\right) + \left(2x^2 - 3x - 5\right) = 5x^2 - 2x - 4$

17. $\left(x^3 + 6x^2 - 4x + 7\right) - \left(3x^2 + 2x - 4\right) = x^3 + 6x^2 - 4x + 7 - 3x^2 - 2x + 4 = x^3 + 3x^2 - 6x + 11$

19. $8(2x + 5) - 7(x - 9) = 16x + 40 - 7x + 63 = 9x + 103$

21. $2(2 - 5t) + t^2(t - 1) - \left(t^4 - 1\right) = 4 - 10t + t^3 - t^2 - t^4 + 1 = -t^4 + t^3 - t^2 - 10t + 5$

23. $(3t - 2)(7t - 4) = 21t^2 - 12t - 14t + 8 = 21t^2 - 26t + 8$ **25.** $(3x + 5)(2x - 1) = 6x^2 + 10x - 3x - 5 = 6x^2 + 7x - 5$

27. $(x + 3y)(2x - y) = 2x^2 + 5xy - 3y^2$ **29.** $(3x + 4)^2 = (3x)^2 + 4^2 + 2(3x)(4) = 9x^2 + 24x + 16$

31. $(2u + v)^2 = 4u^2 + 4uv + v^2$ **33.** $(2x + 3y)^2 = 4x^2 + 12xy + 9y^2$

35. $(x + 5)(x - 5) = x^2 - 5^2 = x^2 - 25$ **37.** $(3x - 4)(3x + 4) = (3x)^2 - 4^2 = 9x^2 - 16$

39. $\left(\sqrt{x} + 2\right)\left(\sqrt{x} - 2\right) = x - 4$

41. $(y + 2)^3 = y^3 + 3y^2(2) + 3y\left(2^2\right) + 2^3 = y^3 + 6y^2 + 12y + 8$

43. $(1 - 2r)^3 = 1^3 - 3\left(1^2\right)(2r) + 3(1)(2r)^2 - (2r)^3 = -8r^3 + 12r^2 - 6r + 1$

45. $(x + 2)\left(x^2 + 2x + 3\right) = x^3 + 2x^2 + 3x + 2x^2 + 4x + 6 = x^3 + 4x^2 + 7x + 6$

47. $(2x - 5)\left(x^2 - x + 1\right) = 2x^3 - 2x^2 + 2x - 5x^2 + 5x - 5 = 2x^3 - 7x^2 + 7x - 5$

49. $\sqrt{x}\left(x - \sqrt{x}\right) = x\sqrt{x} - \left(\sqrt{x}\right)^2 = x\sqrt{x} - x$ **51.** $y^{1/3}\left(y^{2/3} + y^{5/3}\right) = y^{1/3 + 2/3} + y^{1/3 + 5/3} = y^2 + y$

53. $\left(x^2 - a^2\right)\left(x^2 + a^2\right) = x^4 - a^4$ **55.** $\left(\sqrt{a} - b\right)\left(\sqrt{a} + b\right) = a - b^2$

57. $\left((x - 1) + x^2\right)\left((x - 1) - x^2\right) = (x - 1)^2 - \left(x^2\right)^2 = x^2 - 2x + 1 - x^4 = -x^4 + x^2 - 2x + 1$

59. $(2x + y - 3)(2x + y + 3) = (2x + y)^2 - 3^2 = 4x^2 + 4xy + y^2 - 9$

61. $-2x^3 + 16x = -2x\left(x^2 - 8\right)$ **63.** $y(y - 6) + 9(y - 6) = (y - 6)(y + 9)$

65. $2x^2y - 6xy^2 + 3xy = xy(2x - 6y + 3)$ **67.** $x^2 + 2x - 3 = (x - 1)(x + 3)$

69. $8x^2 - 14x - 15 = (2x - 5)(4x + 3)$ **71.** $3x^2 - 16x + 5 = (3x - 1)(x - 5)$

73. $(3x + 2)^2 + 8(3x + 2) + 12 = [(3x + 2) + 2][(3x + 2) + 6] = (3x + 4)(3x + 8)$

75. $9a^2 - 16 = (3a)^2 - 4^2 = (3a - 4)(3a + 4)$

77. $27x^3 + y^3 = (3x)^3 + y^3 = (3x + y)\left[(3x)^2 + 3xy + y^2\right] = (3x + y)\left(9x^2 - 3xy + y^2\right)$

79. $8s^3 - 125t^3 = (2s)^3 - (5t)^3 = (2s - 5t)\left[(2s)^2 + (2s)(5t) + (5t)^2\right] = (2s - 5t)\left(4s^2 + 10st + 25t^2\right)$

81. $x^2 + 12x + 36 = x^2 + 2(6x) + 6^2 = (x + 6)^2$

83. $x^3 + 4x^2 + x + 4 = x^2(x+4) + 1(x+4) = (x+4)\left(x^2 + 1\right)$

85. $2x^3 + x^2 - 6x - 3 = x^2(2x+1) - 3(2x+1) = (2x+1)\left(x^2 - 3\right)$. This can be further factored as

$(2x+1)\left(x - \sqrt{3}\right)\left(x - \sqrt{3}\right)$.

87. $x^3 + x^2 + x + 1 = x^2(x+1) + 1(x+1) = (x+1)\left(x^2 + 1\right)$

89. $x^{5/2} - x^{1/2} = x^{1/2}\left(x^2 - 1\right) = \sqrt{x}\,(x-1)(x+1)$

91. Start by factoring out the power of x with the smallest exponent, that is, $x^{-3/2}$. So

$$x^{-3/2} + 2x^{-1/2} + x^{1/2} = x^{-3/2}\left(1 + 2x + x^2\right) = \frac{(1+x)^2}{x^{3/2}}.$$

93. Start by factoring out the power of $\left(x^2 + 1\right)$ with the smallest exponent, that is, $\left(x^2 + 1\right)^{-1/2}$.

Thus, $\left(x^2 + 1\right)^{1/2} + 2\left(x^2 + 1\right)^{-1/2} = \left(x^2 + 1\right)^{-1/2}\left[\left(x^2 + 1\right) + 2\right] = \dfrac{x^2 + 3}{\sqrt{x^2 + 1}}$.

95. $12x^3 + 18x = 6x\left(2x^2 + 3\right)$ **97.** $x^2 - 2x - 8 = (x-4)(x+2)$

99. $2x^2 + 5x + 3 = (2x+3)(x+1)$ **101.** $9x^2 - 36x - 45 = 9\left(x^2 - 4x - 5\right) = 9(x-5)(x+1)$

103. $49 - 4y^2 = (7 - 2y)(7 + 2y)$ **105.** $t^2 - 6t + 9 = (t-3)^2$

107. $4x^2 + 4xy + y^2 = (2x + y)^2$

109. $(a+b)^2 - (a-b)^2 = [(a+b) - (a-b)][(a+b) + (a-b)] = (2b)(2a) = 4ab$

111. $x^2\left(x^2 - 1\right) - 9\left(x^2 - 1\right) = \left(x^2 - 1\right)\left(x^2 - 9\right) = (x-1)(x+1)(x-3)(x+3)$

113. $8x^3 - 125 = (2x)^3 - 5^3 = (2x-5)\left[(2x)^2 + (2x)(5) + 5^2\right] = (2x-5)\left(4x^2 + 10x + 25\right)$

115. $x^3 + 2x^2 + x = x\left(x^2 + 2x + 1\right) = x(x+1)^2$

117. $x^4y^3 - x^2y^5 = x^2y^3\left(x^2 - y^2\right) = x^2y^3(x+y)(x-y)$

119. $2x^3 + 4x^2 + x + 2 = \left(2x^3 + 4x^2\right) + (x+2) = 2x^2(x+2) + (1)(x+2) = (x+2)\left(2x^2 + 1\right)$ (factor by grouping)

121. $(x-1)(x+2)^2 - (x-1)^2(x+2) = (x-1)(x+2)[(x+2) - (x-1)] = 3(x-1)(x+2)$

123. Start by factoring $y^2 - 7y + 10$, and then substitute $a^2 + 1$ for y. This gives

$\left(a^2 + 1\right)^2 - 7\left(a^2 + 1\right) + 10 = \left[\left(a^2 + 1\right) - 2\right]\left[\left(a^2 + 1\right) - 5\right] = \left(a^2 - 1\right)\left(a^2 - 4\right) = (a-1)(a+1)(a-2)(a+2)$

125. $5\left(x^2 + 4\right)^4(2x)(x-2)^4 + \left(x^2 + 4\right)^5(4)(x-2)^3 = 2\left(x^2 + 4\right)^4(x-2)^3\left[(5)(x)(x-2) + \left(x^2 + 4\right)(2)\right]$

$= 2\left(x^2 + 4\right)^4(x-2)^3\left(5x^2 - 10x + 2x^2 + 8\right) = 2\left(x^2 + 4\right)^4(x-2)^3\left(7x^2 - 10x + 8\right)$

127. $\left(x^2 + 3\right)^{-1/3} - \frac{2}{3}x^2\left(x^2 + 3\right)^{-4/3} = \left(x^2 + 3\right)^{-4/3}\left[\left(x^2 + 3\right) - \frac{2}{3}x^2\right] = \left(x^2 + 3\right)^{-4/3}\left(\frac{1}{3}x^2 + 3\right) = \dfrac{\frac{1}{3}x^2 + 3}{\left(x^2 + 3\right)^{4/3}}$

129. (a) $\frac{1}{2}\left[(a+b)^2 - \left(a^2 + b^2\right)\right] = \frac{1}{2}\left[a^2 + 2ab + b^2 - a^2 - b^2\right] = \frac{1}{2}(2ab) = ab$.

    (b) $\left(a^2 + b^2\right)^2 - \left(a^2 - b^2\right)^2 = \left[\left(a^2 + b^2\right) - \left(a^2 - b^2\right)\right]\left[\left(a^2 + b^2\right) + \left(a^2 - b^2\right)\right]$

$= \left(a^2 + b^2 - a^2 + b^2\right)\left(a^2 + b^2 + a^2 - b^2\right) = \left(2b^2\right)\left(2a^2\right) = 4a^2b^2$

(c) LHS $= \left(a^2 + b^2\right)\left(c^2 + d^2\right) = a^2c^2 + a^2d^2 + b^2c^2 + b^2d^2$.

RHS $= (ac + bd)^2 + (ad - bc)^2 = a^2c^2 + 2abcd + b^2d^2 + a^2d^2 - 2abcd + b^2c^2 = a^2c^2 + a^2d^2 + b^2c^2 + b^2d^2$.

So LHS = RHS, that is, $\left(a^2 + b^2\right)\left(c^2 + d^2\right) = (ac + bd)^2 + (ad - bc)^2$.

(d) $4a^2c^2 - \left(c^2 - b^2 + a^2\right)^2 = (2ac)^2 - \left(c^2 - b^2 + a^2\right)$

$= \left[(2ac) - \left(c^2 - b^2 + a^2\right)\right]\left[(2ac) + \left(c^2 - b^2 + a^2\right)\right]$ (difference of squares)

$= \left(2ac - c^2 + b^2 - a^2\right)\left(2ac + c^2 - b^2 + a^2\right)$

$= \left[b^2 - \left(c^2 - 2ac + a^2\right)\right]\left[\left(c^2 + 2ac + a^2\right) - b^2\right]$ (regrouping)

$= \left[b^2 - (c-a)^2\right]\left[(c+a)^2 - b^2\right]$ (perfect squares)

$= [b - (c-a)][b + (c-a)][(c+a) - b][(c+a) + b]$ (each factor is a difference of squares)

$= (b - c + a)(b + c - a)(c + a - b)(c + a + b)$

$= (a + b - c)(-a + b + c)(a - b + c)(a + b + c)$

(e) $x^4 + 3x^2 + 4 = \left(x^4 + 4x^2 + 4\right) - x^2 = \left(x^2 + 2\right)^2 - x^2 = \left[\left(x^2 + 2\right) - x\right]\left[\left(x^2 + 2\right) + x\right]$

$= \left(x^2 - x + 2\right)\left(x^2 + x + 2\right)$

131. The volume of the shell is the difference between the volumes of the outside cylinder (with radius R) and the inside cylinder (with radius r). Thus $V = \pi R^2 h - \pi r^2 h = \pi \left(R^2 - r^2\right) h = \pi (R - r)(R + r) h = 2\pi \cdot \dfrac{R + r}{2} \cdot h \cdot (R - r)$. The average radius is $\dfrac{R + r}{2}$ and $2\pi \cdot \dfrac{R + r}{2}$ is the average circumference (length of the rectangular box), h is the height, and $R - r$ is the thickness of the rectangular box. Thus $V = \pi R^2 h - \pi r^2 h = 2\pi \cdot \dfrac{R + r}{2} \cdot h \cdot (R - r) = 2\pi \cdot$ (average radius) $\cdot$ (height) $\cdot$ (thickness)

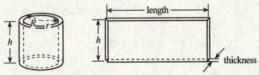

133. (a) The degree of the product is the sum of the degrees.

(b) The degree of a sum is at most the largest of the degrees — it could be smaller than either. For example, the degree of $\left(x^3\right) + \left(-x^3 + x\right) = x$ is 1.

135. (a) $A^4 - B^4 = \left(A^2 - B^2\right)\left(A^2 + B^2\right) = (A - B)(A + B)\left(A^2 + B^2\right)$

$A^6 - B^6 \quad = \left(A^3 - B^3\right)\left(A^3 + B^3\right)$ (difference of squares)

$= (A - B)\left(A^2 + AB + B^2\right)(A + B)\left(A^2 - AB + B^2\right)$ (difference and sum of cubes)

(b) $12^4 - 7^4 = 20{,}736 - 2{,}401 = 18{,}335$; $12^6 - 7^6 = 2{,}985{,}984 - 117{,}649 = 2{,}868{,}335$

(c) $18{,}335 = 12^4 - 7^4 = (12 - 7)(12 + 7)\left(12^2 + 7^2\right) = 5(19)(144 + 49) = 5(19)(193)$

$2{,}868{,}335 \quad = 12^6 - 7^6 = (12 - 7)(12 + 7)\left[12^2 + 12(7) + 7^2\right]\left[12^2 - 12(7) + 7^2\right]$

$= 5(19)(144 + 84 + 49)(144 - 84 + 49) = 5(19)(277)(109)$

137. (a) $x^4 + x^2 - 2 = \left(x^2 - 1\right)\left(x^2 + 2\right) = (x - 1)(x + 1)\left(x^2 + 2\right)$

(b) $x^4 + 2x^2 + 9 = \left(x^4 + 6x^2 + 9\right) - 4x^2 = \left(x^2 + 3\right)^2 - (2x)^2 = \left[\left(x^2 + 3\right) - 2x\right]\left[\left(x^2 + 3\right) + 2x\right]$

$$= \left(x^2 - 2x + 3\right)\left(x^2 + 2x + 3\right)$$

(c) $x^4 + 4x^2 + 16 = \left(x^4 + 8x^2 + 16\right) - 4x^2 = \left(x^2 + 4\right)^2 - (2x)^2$

$$= \left[\left(x^2 + 4\right) - 2x\right]\left[\left(x^2 + 4\right) + 2x\right] = \left(x^2 - 2x + 4\right)\left(x^2 + 2x + 4\right)$$

(d) $x^4 + 2x^2 + 1 = \left(x^2 + 1\right)^2$

1.4   RATIONAL EXPRESSIONS

1. (a) $\dfrac{3x}{x^2 - 1}$ is a rational expression.

(b) $\dfrac{\sqrt{x + 1}}{2x + 3}$ is not a rational expression. A rational expression must be a polynomial divided by a polynomial, and the numerator of the expression is $\sqrt{x + 1}$, which is not a polynomial.

(c) $\dfrac{x(x^2 - 1)}{x + 3} = \dfrac{x^3 - x}{x + 3}$ is a rational expression.

3. To multiply two rational expressions we multiply their *numerators* together and multiply their *denominators* together. So $\dfrac{2}{x + 1} \cdot \dfrac{x}{x + 3}$ is the same as $\dfrac{2 \cdot x}{(x + 1) \cdot (x + 3)} = \dfrac{2x}{x^2 + 4x + 3}$.

5. The domain of $4x^2 - 10x + 3$ is all real numbers.

7. Since $x - 4 \neq 0$ we have $x \neq 4$. Domain: $\{x \mid x \neq 4\}$

9. Since $x + 3 \geq 0$, $x \geq -3$. Domain; $\{x \mid x \geq -3\}$

11. $x^2 - x - 2 = (x + 1)(x - 2) \neq 0 \Leftrightarrow x \neq -1$ or 2, so the domain is $\{x \mid x \neq -1, 2\}$.

13. $\dfrac{3(x + 2)(x - 1)}{6(x - 1)^2} = \dfrac{3(x - 1) \cdot (x + 2)}{3(x - 1) \cdot 2(x - 1)} = \dfrac{x + 2}{2(x - 1)}$

15. $\dfrac{x - 2}{x^2 - 4} = \dfrac{x - 2}{(x - 2)(x + 2)} = \dfrac{1}{x + 2}$

17. $\dfrac{x^2 + 6x + 8}{x^2 + 5x + 4} = \dfrac{(x + 2)(x + 4)}{(x + 1)(x + 4)} = \dfrac{x + 2}{x + 1}$

19. $\dfrac{y^2 + y}{y^2 - 1} = \dfrac{y(y + 1)}{(y - 1)(y + 1)} = \dfrac{y}{y - 1}$

21. $\dfrac{2x^3 - x^2 - 6x}{2x^2 - 7x + 6} = \dfrac{x\left(2x^2 - x - 6\right)}{(2x - 3)(x - 2)} = \dfrac{x(2x + 3)(x - 2)}{(2x - 3)(x - 2)} = \dfrac{x(2x + 3)}{2x - 3}$

23. $\dfrac{4x}{x^2 - 4} \cdot \dfrac{x + 2}{16x} = \dfrac{4x}{(x - 2)(x + 2)} \cdot \dfrac{x + 2}{16x} = \dfrac{1}{4(x - 2)}$

25. $\dfrac{x^2 - 2x - 15}{x^2 - 9} \cdot \dfrac{x + 3}{x - 5} = \dfrac{(x - 5)(x + 3)}{(x - 3)(x + 3)} \cdot \dfrac{x + 3}{x - 5} = \dfrac{x + 3}{x - 3}$

27. $\dfrac{t - 3}{t^2 + 9} \cdot \dfrac{t + 3}{t^2 - 9} = \dfrac{(t - 3)(t + 3)}{(t^2 + 9)(t - 3)(t + 3)} = \dfrac{1}{t^2 + 9}$

29. $\dfrac{x^2 + 7x + 12}{x^2 + 3x + 2} \cdot \dfrac{x^2 + 5x + 6}{x^2 + 6x + 9} = \dfrac{(x + 3)(x + 4)}{(x + 1)(x + 2)} \cdot \dfrac{(x + 2)(x + 3)}{(x + 3)(x + 3)} = \dfrac{x + 4}{x + 1}$

31. $\dfrac{x + 3}{4x^2 - 9} \div \dfrac{x^2 + 7x + 12}{2x^2 + 7x - 15} = \dfrac{x + 3}{4x^2 - 9} \cdot \dfrac{2x^2 + 7x - 15}{x^2 + 7x + 12} = \dfrac{x + 3}{(2x - 3)(2x + 3)} \cdot \dfrac{(x + 5)(2x - 3)}{(x + 3)(x + 4)} = \dfrac{x + 5}{(2x + 3)(x + 4)}$

33. $\dfrac{2x^2+3x+1}{x^2+2x-15} \div \dfrac{x^2+6x+5}{2x^2-7x+3} = \dfrac{2x^2+3x+1}{x^2+2x-15} \cdot \dfrac{2x^2-7x+3}{x^2+6x+5} = \dfrac{(2x+1)(x+1)}{(x-3)(x+5)} \cdot \dfrac{(2x-1)(x-3)}{(x+1)(x+5)}$

$$= \dfrac{(2x+1)(2x-1)}{(x+5)(x+5)} = \dfrac{(2x+1)(2x-1)}{(x+5)^2}$$

35. $\dfrac{\dfrac{x^3}{x+1}}{\dfrac{x}{x^2+2x+1}} = \dfrac{x^3}{x+1} \cdot \dfrac{x^2+2x+1}{x} = \dfrac{x^3(x+1)(x+1)}{(x+1)x} = x^2(x+1)$

37. $\dfrac{x/y}{z} = \dfrac{x}{y} \cdot \dfrac{1}{z} = \dfrac{x}{yz}$

39. $2 + \dfrac{x}{x+3} = \dfrac{2(x+3)}{x+3} + \dfrac{x}{x+3} = \dfrac{2x+6+x}{x+3} = \dfrac{3x+6}{x+3} = \dfrac{3(x+2)}{(x+3)}$

41. $\dfrac{1}{x+5} + \dfrac{2}{x-3} = \dfrac{x-3}{(x+5)(x-3)} + \dfrac{2(x+5)}{(x+5)(x-3)} = \dfrac{x-3+2x+10}{(x+5)(x-3)} = \dfrac{3x+7}{(x+5)(x-3)}$

43. $\dfrac{1}{x+1} - \dfrac{1}{x+2} = \dfrac{x+2}{(x+1)(x+2)} + \dfrac{-(x+1)}{(x+1)(x+2)} = \dfrac{x+2-x-1}{(x+1)(x+2)} = \dfrac{1}{(x+1)(x+2)}$

45. $\dfrac{x}{(x+1)^2} + \dfrac{2}{x+1} = \dfrac{x}{(x+1)^2} + \dfrac{2(x+1)}{(x+1)(x+1)} = \dfrac{x+2x+2}{(x+1)^2} = \dfrac{3x+2}{(x+1)^2}$

47. $u + 1 + \dfrac{u}{u+1} = \dfrac{(u+1)(u+1)}{u+1} + \dfrac{u}{u+1} = \dfrac{u^2+2u+1+u}{u+1} = \dfrac{u^2+3u+1}{u+1}$

49. $\dfrac{1}{x^2} + \dfrac{1}{x^2+x} = \dfrac{1}{x^2} + \dfrac{1}{x(x+1)} = \dfrac{x+1}{x^2(x+1)} + \dfrac{x}{x^2(x+1)} = \dfrac{2x+1}{x^2(x+1)}$

51. $\dfrac{2}{x+3} - \dfrac{1}{x^2+7x+12} = \dfrac{2}{x+3} - \dfrac{1}{(x+3)(x+4)} = \dfrac{2(x+4)}{(x+3)(x+4)} + \dfrac{-1}{(x+3)(x+4)}$

$$= \dfrac{2x+8-1}{(x+3)(x+4)} = \dfrac{2x+7}{(x+3)(x+4)}$$

53. $\dfrac{1}{x+3} + \dfrac{1}{x^2-9} = \dfrac{1}{x+3} + \dfrac{1}{(x-3)(x+3)} = \dfrac{x-3}{(x-3)(x+3)} + \dfrac{1}{(x-3)(x+3)} = \dfrac{x-2}{(x-3)(x+3)}$

55. $\dfrac{2}{x} + \dfrac{3}{x-1} - \dfrac{4}{x^2-x} = \dfrac{2}{x} + \dfrac{3}{x-1} - \dfrac{4}{x(x-1)} = \dfrac{2(x-1)}{x(x-1)} + \dfrac{3x}{x(x-1)} + \dfrac{-4}{x(x-1)} = \dfrac{2x-2+3x-4}{x(x-1)} = \dfrac{5x-6}{x(x-1)}$

57. $\dfrac{1}{x^2+3x+2} - \dfrac{1}{x^2-2x-3} = \dfrac{1}{(x+2)(x+1)} - \dfrac{1}{(x-3)(x+1)}$

$$= \dfrac{x-3}{(x-3)(x+2)(x+1)} + \dfrac{-(x+2)}{(x-3)(x+2)(x+1)} = \dfrac{x-3-x-2}{(x-3)(x+2)(x+1)} = \dfrac{-5}{(x-3)(x+2)(x+1)}$$

59. $\dfrac{x+\dfrac{1}{x+2}}{x-\dfrac{1}{x+2}} = \dfrac{x(x+2)+1}{x(x+2)-1} = \dfrac{x^2+2x+1}{x^2+2x-1} = \dfrac{(x+1)^2}{x^2+2x-1}$

61. $\dfrac{\dfrac{x+2}{x-1} - \dfrac{x-3}{x-2}}{x+2} = \dfrac{(x+2)(x-2)-(x-3)(x-1)}{(x-2)(x-1)(x+2)} = \dfrac{x^2-4-\left(x^2-4x+3\right)}{(x-2)(x-1)(x+2)} = \dfrac{4x-7}{(x-2)(x-1)(x+2)}$

63. $\dfrac{\dfrac{x}{y}-\dfrac{y}{x}}{\dfrac{1}{x^2}-\dfrac{1}{y^2}} = \dfrac{\dfrac{x^2-y^2}{xy}}{\dfrac{y^2-x^2}{x^2y^2}} = \dfrac{x^2-y^2}{xy}\cdot\dfrac{x^2y^2}{y^2-x^2} = \dfrac{xy}{-1} = -xy.$ An alternative method is to multiply the

numerator and denominator by the common denominator of both the numerator and denominator, in this case x^2y^2:

$$\dfrac{\dfrac{x}{y}-\dfrac{y}{x}}{\dfrac{1}{x^2}-\dfrac{1}{y^2}} = \dfrac{\left(\dfrac{x}{y}-\dfrac{y}{x}\right)}{\left(\dfrac{1}{x^2}-\dfrac{1}{y^2}\right)}\cdot\dfrac{x^2y^2}{x^2y^2} = \dfrac{x^3y-xy^3}{y^2-x^2} = \dfrac{xy\left(x^2-y^2\right)}{y^2-x^2} = -xy.$$

65. $\dfrac{x^{-2}-y^{-2}}{x^{-1}+y^{-1}} = \dfrac{\dfrac{1}{x^2}-\dfrac{1}{y^2}}{\dfrac{1}{x}+\dfrac{1}{y}} = \dfrac{\dfrac{y^2}{x^2y^2}-\dfrac{x^2}{x^2y^2}}{\dfrac{y}{xy}+\dfrac{x}{xy}} = \dfrac{y^2-x^2}{x^2y^2}\cdot\dfrac{xy}{y+x} = \dfrac{(y-x)(y+x)xy}{x^2y^2(y+x)} = \dfrac{y-x}{xy}$

Alternatively, $\dfrac{x^{-2}-y^{-2}}{x^{-1}+y^{-1}} = \dfrac{\left(\dfrac{1}{x^2}-\dfrac{1}{y^2}\right)}{\left(\dfrac{1}{x}+\dfrac{1}{y}\right)}\cdot\dfrac{x^2y^2}{x^2y^2} = \dfrac{y^2-x^2}{xy^2+x^2y} = \dfrac{(y-x)(y+x)}{xy(y+x)} = \dfrac{y-x}{xy}.$

67. $1-\dfrac{1}{1-\dfrac{1}{x}} = 1-\dfrac{x}{x-1} = \dfrac{x-1-x}{x-1} = \dfrac{1}{1-x}$

69. $\dfrac{\dfrac{1}{1+x+h}-\dfrac{1}{1+x}}{h} = \dfrac{(1+x)-(1+x+h)}{h(1+x)(1+x+h)} = -\dfrac{1}{(1+x)(1+x+h)}$

71. $\dfrac{\dfrac{1}{(x+h)^2}-\dfrac{1}{x^2}}{h} = \dfrac{x^2-(x+h)^2}{hx^2(x+h)^2} = \dfrac{x^2-\left(x^2+2xh+h^2\right)}{hx^2(x+h)^2} = -\dfrac{2x+h}{x^2(x+h)^2}$

73. $\sqrt{1+\left(\dfrac{x}{\sqrt{1-x^2}}\right)^2} = \sqrt{1+\dfrac{x^2}{1-x^2}} = \sqrt{\dfrac{1-x^2}{1-x^2}+\dfrac{x^2}{1-x^2}} = \sqrt{\dfrac{1}{1-x^2}} = \dfrac{1}{\sqrt{1-x^2}}$

75. $\dfrac{3(x+2)^2(x-3)^2-(x+2)^3(2)(x-3)}{(x-3)^4} = \dfrac{(x+2)^2(x-3)[3(x-3)-(x+2)(2)]}{(x-3)^4}$

$\qquad\qquad = \dfrac{(x+2)^2(3x-9-2x-4)}{(x-3)^3} = \dfrac{(x+2)^2(x-13)}{(x-3)^3}$

77. $\dfrac{2(1+x)^{1/2}-x(1+x)^{-1/2}}{1+x} = \dfrac{(1+x)^{-1/2}[2(1+x)-x]}{1+x} = \dfrac{x+2}{(1+x)^{3/2}}$

79. $\dfrac{3(1+x)^{1/3}-x(1+x)^{-2/3}}{(1+x)^{2/3}} = \dfrac{(1+x)^{-2/3}[3(1+x)-x]}{(1+x)^{2/3}} = \dfrac{2x+3}{(1+x)^{4/3}}$

81. $\dfrac{1}{2-\sqrt{3}} = \dfrac{1}{2-\sqrt{3}}\cdot\dfrac{2+\sqrt{3}}{2+\sqrt{3}} = \dfrac{2+\sqrt{3}}{4-3} = \dfrac{2+\sqrt{3}}{1} = 2+\sqrt{3}$

83. $\dfrac{2}{\sqrt{2}+\sqrt{7}} = \dfrac{2}{\sqrt{2}+\sqrt{7}}\cdot\dfrac{\sqrt{2}-\sqrt{7}}{\sqrt{2}-\sqrt{7}} = \dfrac{2\left(\sqrt{2}-\sqrt{7}\right)}{2-7} = \dfrac{2\left(\sqrt{2}-\sqrt{7}\right)}{-5} = \dfrac{2\left(\sqrt{7}-\sqrt{2}\right)}{5}$

85. $\dfrac{y}{\sqrt{3}+\sqrt{y}} = \dfrac{y}{\sqrt{3}+\sqrt{y}}\cdot\dfrac{\sqrt{3}-\sqrt{y}}{\sqrt{3}-\sqrt{y}} = \dfrac{y\left(\sqrt{3}-\sqrt{y}\right)}{3-y} = \dfrac{y\sqrt{3}-y\sqrt{y}}{3-y}$

87. $\dfrac{1-\sqrt{5}}{3} = \dfrac{1-\sqrt{5}}{3}\cdot\dfrac{1+\sqrt{5}}{1+\sqrt{5}} = \dfrac{1-5}{3\left(1+\sqrt{5}\right)} = \dfrac{-4}{3\left(1+\sqrt{5}\right)}$

89. $\dfrac{\sqrt{r}+\sqrt{2}}{5} = \dfrac{\sqrt{r}+\sqrt{2}}{5} \cdot \dfrac{\sqrt{r}-\sqrt{2}}{\sqrt{r}-\sqrt{2}} = \dfrac{r-2}{5\left(\sqrt{r}-\sqrt{2}\right)}$

91. $\sqrt{x^2+1}-x = \dfrac{\sqrt{x^2+1}-x}{1} \cdot \dfrac{\sqrt{x^2+1}+x}{\sqrt{x^2+1}+x} = \dfrac{x^2+1-x^2}{\sqrt{x^2+1}+x} = \dfrac{1}{\sqrt{x^2+1}+x}$

93. $\dfrac{16+a}{16} = \dfrac{16}{16}+\dfrac{a}{16} = 1+\dfrac{a}{16}$, so the statement is true.

95. This statement is false. For example, take $x = 2$, then LHS $= \dfrac{2}{4+x} = \dfrac{2}{4+2} = \dfrac{2}{6} = \dfrac{1}{3}$, while

RHS $= \dfrac{1}{2}+\dfrac{2}{x} = \dfrac{1}{2}+\dfrac{2}{2} = \dfrac{3}{2}$, and $\dfrac{1}{3} \neq \dfrac{3}{2}$.

97. This statement is false. For example, take $x = 0$ and $y = 1$. Then substituting into the left side we obtain

LHS $= \dfrac{x}{x+y} = \dfrac{0}{0+1} = 0$, while the right side yields RHS $= \dfrac{1}{1+y} = \dfrac{1}{1+1} = \dfrac{1}{2}$, and $0 \neq \dfrac{1}{2}$.

99. This statement is true: $\dfrac{-a}{b} = (-a)\left(\dfrac{1}{b}\right) = (-1)(a)\left(\dfrac{1}{b}\right) = (-1)\left(\dfrac{a}{b}\right) = -\dfrac{a}{b}$.

101. (a) $R = \dfrac{1}{\dfrac{1}{R_1}+\dfrac{1}{R_2}} = \dfrac{1}{\dfrac{1}{R_1}+\dfrac{1}{R_2}} \cdot \dfrac{R_1 R_2}{R_1 R_2} = \dfrac{R_1 R_2}{R_2 + R_1}$

(b) Substituting $R_1 = 10$ ohms and $R_2 = 20$ ohms gives $R = \dfrac{(10)(20)}{(20)+(10)} = \dfrac{200}{30} \approx 6.7$ ohms.

103.

x	2.80	2.90	2.95	2.99	2.999	3	3.001	3.01	3.05	3.10	3.20
$\dfrac{x^2-9}{x-3}$	5.80	5.90	5.95	5.99	5.999	?	6.001	6.01	6.05	6.10	6.20

From the table, we see that the expression $\dfrac{x^2-9}{x-3}$ approaches 6 as x approaches 3. We simplify the expression:

$\dfrac{x^2-9}{x-3} = \dfrac{(x-3)(x+3)}{x-3} = x+3$, $x \neq 3$. Clearly as x approaches 3, $x+3$ approaches 6. This explains the result in the table.

105. Answers will vary.

Algebraic Error	Counterexample
$\dfrac{1}{a}+\dfrac{1}{b} \neq \dfrac{1}{a+b}$	$\dfrac{1}{2}+\dfrac{1}{2} \neq \dfrac{1}{2+2}$
$(a+b)^2 \neq a^2+b^2$	$(1+3)^2 \neq 1^2+3^2$
$\sqrt{a^2+b^2} \neq a+b$	$\sqrt{5^2+12^2} \neq 5+12$
$\dfrac{a+b}{a} \neq b$	$\dfrac{2+6}{2} \neq 6$
$\left(a^3+b^3\right)^{1/3} \neq a+b$	$\left(2^3+2^3\right)^{1/3} \neq 2+2$
$\dfrac{a^m}{a^n} \neq a^{m/n}$	$\dfrac{3^5}{3^2} \neq 3^{5/2}$
$a^{-1/n} \neq \dfrac{1}{a^n}$	$64^{-1/3} \neq \dfrac{1}{64^3}$

1.5   EQUATIONS

1. **(a)** This is true: If $a = b$, then $a + x = b + x$.

   (b) This is false, because the number could be zero. However, it is true that multiplying each side of an equation by a *nonzero* number always gives an equivalent equation.

   (c) This is false. For example, $-5 = 5$ is false, but $(-5)^2 = 5^2$ is true.

3. **(a)** The solutions of the equation $x^2 (x - 4) = 0$ are $x = 0$ and $x = 4$.

   (b) To solve the equation $x^3 - 4x^2 = 0$ we *factor* the left-hand side: $x^2 (x - 4) = 0$, as above.

5. The equation $(x + 1)^2 - 5(x + 1) + 6 = 0$ is of *quadratic* type. To solve the equation we set $W = x + 1$. The resulting quadratic equation is $W^2 - 5W + 6 = 0 \Rightarrow (W - 3)(W - 2) = 0 \Rightarrow W = 2$ or $W = 3 \Rightarrow x + 1 = 2$ or $x + 1 = 3 \Rightarrow x = 1$ or $x = 2$. You can verify that these are both solutions to the original equation.

7. **(a)** When $x = -2$, LHS $= 4(-2) + 7 = -8 + 7 = -1$ and RHS $= 9(-2) - 3 = -18 - 3 = -21$. Since LHS $\neq$ RHS, $x = -2$ is not a solution.

   (b) When $x = 2$, LHS $= 4(-2) + 7 = 8 + 7 = 15$ and RHS $= 9(2) - 3 = 18 - 3 = 15$. Since LHS $=$ RHS, $x = 2$ is a solution.

9. **(a)** When $x = 2$, LHS $= \dfrac{1}{2} - \dfrac{1}{2 - 4} = \dfrac{1}{2} - \dfrac{1}{-2} = \dfrac{1}{2} + \dfrac{1}{2} = 1$ and RHS $= 1$. Since LHS $=$ RHS, $x = 2$ is a solution.

   (b) When $x = 4$ the expression $\dfrac{1}{4 - 4}$ is not defined, so $x = 4$ is not a solution.

11. $2x + 7 = 31 \Leftrightarrow 2x = 24 \Leftrightarrow x = 12$

13. $\frac{1}{2}x - 8 = 1 \Leftrightarrow \frac{1}{2}x = 9 \Leftrightarrow x = 18$

15. $-7w = 15 - 2w \Leftrightarrow -5w = 15 \Leftrightarrow w = -3$

17. $\frac{1}{2}y - 2 = \frac{1}{3}y \Leftrightarrow 3y - 12 = 2y$ (multiply both sides by the LCD, 6) $\Leftrightarrow y = 12$

19. $2(1 - x) = 3(1 + 2x) + 5 \Leftrightarrow 2 - 2x = 3 + 6x + 5 \Leftrightarrow 2 - 2x = 8 + 6x \Leftrightarrow -6 = 8x \Leftrightarrow x = -\frac{3}{4}$

21. $x - \frac{1}{3}x - \frac{1}{2}x - 5 = 0 \Leftrightarrow 6x - 2x - 3x - 30 = 0$ (multiply both sides by the LCD, 6) $\Leftrightarrow x = 30$

23. $\dfrac{1}{x} = \dfrac{4}{3x} + 1 \Rightarrow 3 = 4 + 3x$ (multiply both sides by the LCD, $3x$) $\Leftrightarrow -1 = 3x \Leftrightarrow x = -\frac{1}{3}$

25. $\dfrac{3}{x + 1} - \dfrac{1}{2} = \dfrac{1}{3x + 3} \Rightarrow 3(6) - (3x + 3) = 2$ [multiply both sides by the LCD, $6(x + 1)$] $\Leftrightarrow 18 - 3x - 3 = 2 \Leftrightarrow$
    $-3x + 15 = 2 \Leftrightarrow -3x = -13 \Leftrightarrow x = \frac{13}{3}$

27. $(t - 4)^2 = (t + 4)^2 + 32 \Leftrightarrow t^2 - 8t + 16 = t^2 + 8t + 16 + 32 \Leftrightarrow -16t = 32 \Leftrightarrow t = -2$

29. $PV = nRT \Leftrightarrow R = \dfrac{PV}{nT}$

31. $P = 2l + 2w \Leftrightarrow 2w = P - 2l \Leftrightarrow w = \dfrac{P - 2l}{2}$

33. $\dfrac{ax + b}{cx + d} = 2 \Leftrightarrow ax + b = 2(cx + d) \Leftrightarrow ax + b = 2cx + 2d \Leftrightarrow ax - 2cx = 2d - b \Leftrightarrow (a - 2c)x = 2d - b \Leftrightarrow x = \dfrac{2d - b}{a - 2c}$

35. $a^2 x + (a - 1) = (a + 1)x \Leftrightarrow a^2 x - (a + 1)x = -(a - 1) \Leftrightarrow \left(a^2 - (a + 1)\right)x = -a + 1 \Leftrightarrow \left(a^2 - a - 1\right)x = -a + 1$
    $\Leftrightarrow x = \dfrac{-a + 1}{a^2 - a - 1}$

37. $V = \frac{1}{3}\pi r^2 h \Leftrightarrow r^2 = \dfrac{3V}{\pi h} \Rightarrow r = \pm\sqrt{\dfrac{3V}{\pi h}}$

39. $a^2 + b^2 = c^2 \Leftrightarrow b^2 = c^2 - a^2 \Rightarrow b = \pm\sqrt{c^2 - a^2}$

41. $h = \frac{1}{2}gt^2 + v_0t \Leftrightarrow \frac{1}{2}gt^2 + v_0t - h = 0$. Using the quadratic formula,

$$t = \frac{-(v_0) \pm \sqrt{(v_0)^2 - 4\left(\frac{1}{2}g\right)(-h)}}{2\left(\frac{1}{2}g\right)} = \frac{-v_0 \pm \sqrt{v_0^2 + 2gh}}{g}.$$

43. $x^2 + x - 12 = 0 \Leftrightarrow (x-3)(x+4) = 0 \Leftrightarrow x - 3 = 0$ or $x + 4 = 0$. Thus, $x = 3$ or $x = -4$.

45. $x^2 - 7x + 12 = 0 \Leftrightarrow (x-4)(x-3) = 0 \Leftrightarrow x - 4 = 0$ or $x - 3 = 0$. Thus, $x = 4$ or $x = 3$.

47. $4x^2 - 4x - 15 = 0 \Leftrightarrow (2x+3)(2x-5) = 0 \Leftrightarrow 2x + 3 = 0$ or $2x - 5 = 0$. Thus, $x = -\frac{3}{2}$ or $x = \frac{5}{2}$.

49. $3x^2 + 5x = 2 \Leftrightarrow 3x^2 + 5x - 2 = 0 \Leftrightarrow (3x-1)(x+2) = 0 \Leftrightarrow 3x - 1 = 0$ or $x + 2 = 0$. Thus, $x = \frac{1}{3}$ or $x = -2$.

51. $2x^2 = 8 \Leftrightarrow x^2 = 4 \Leftrightarrow x = \pm 2$.

53. $(3x+2)^2 = 10 \Leftrightarrow 3x + 2 = \pm\sqrt{10} \Leftrightarrow 3x = -2 \pm \sqrt{10} \Leftrightarrow x = -\frac{2}{3} \pm \frac{\sqrt{10}}{3}$.

55. $x^2 + 2x - 5 = 0 \Leftrightarrow x^2 + 2x = 5 \Leftrightarrow x^2 + 2x + 1 = 5 + 1 \Leftrightarrow (x+1)^2 = 6 \Rightarrow x + 1 = \pm\sqrt{6} \Leftrightarrow x = -1 \pm \sqrt{6}$.

57. $x^2 - 6x - 11 = 0 \Leftrightarrow x^2 - 6x = 11 \Leftrightarrow x^2 - 6x + 9 = 11 + 9 \Leftrightarrow (x-3)^2 = 20 \Rightarrow x - 3 = \pm 2\sqrt{5} \Leftrightarrow x = 3 \pm 2\sqrt{5}$.

59. $2x^2 + 8x + 1 = 0 \Leftrightarrow x^2 + 4x + \frac{1}{2} = 0 \Leftrightarrow x^2 + 4x = -\frac{1}{2} \Leftrightarrow x^2 + 4x + 4 = -\frac{1}{2} + 4 \Leftrightarrow (x+2)^2 = \frac{7}{2} \Rightarrow x + 2 = \pm\sqrt{\frac{7}{2}}$

$\Leftrightarrow x = -2 \pm \frac{\sqrt{14}}{2}$.

61. $4x^2 - x = 0 \Leftrightarrow x^2 - \frac{1}{4}x = 0 \Leftrightarrow x^2 - \frac{1}{4}x + \frac{1}{64} = \frac{1}{64} \Leftrightarrow \left(x - \frac{1}{8}\right)^2 = \frac{1}{64} \Rightarrow x - \frac{1}{8} = \pm\frac{1}{8} \Leftrightarrow x = \frac{1}{8} \pm \frac{1}{8}$, so $x = \frac{1}{8} - \frac{1}{8} = 0$

or $x = \frac{1}{8} + \frac{1}{8} = \frac{1}{4}$.

63. $x^2 - 2x - 15 = 0 \Leftrightarrow (x+3)(x-5) = 0 \Leftrightarrow x + 3 = 0$ or $x - 5 = 0$. Thus, $x = -3$ or $x = 5$.

65. $x^2 - 7x + 10 = 0 \Leftrightarrow (x-5)(x-2) = 0 \Leftrightarrow x - 5 = 0$ or $x - 2 = 0$. Thus, $x = 5$ or $x = 2$.

67. $2x^2 + x - 3 = 0 \Leftrightarrow (x-1)(2x+3) = 0 \Leftrightarrow x - 1 = 0$ or $2x + 3 = 0$. If $x - 1 = 0$, then $x = 1$; if $2x + 3 = 0$, then

$x = -\frac{3}{2}$.

69. $3x^2 + 6x - 5 = 0 \Leftrightarrow x^2 + 2x - \frac{5}{3} = 0 \Leftrightarrow x^2 + 2x = \frac{5}{3} \Leftrightarrow x^2 + 2x + 1 = \frac{5}{3} + 1 \Leftrightarrow (x+1)^2 = \frac{8}{3} \Rightarrow$

$x + 1 = \pm\sqrt{\frac{8}{3}} \Leftrightarrow x = -1 \pm \frac{2\sqrt{6}}{3}$.

71. $z^2 - \frac{3}{2}z + \frac{9}{16} = 0 \Rightarrow \left(z - \frac{3}{4}\right)^2 = 0 \Leftrightarrow z = \frac{3}{4}$.

73. $4x^2 + 16x - 9 = 0 \Leftrightarrow (2x-1)(2x+9) = 0 \Leftrightarrow 2x - 1 = 0$ or $2x + 9 = 0$. If $2x - 1 = 0$, then $x = \frac{1}{2}$; if $2x + 9 = 0$, then

$x = -\frac{9}{2}$.

75. $w^2 = 3(w-1) \Leftrightarrow w^2 - 3w + 3 = 0 \Rightarrow w = \frac{-(-3) \pm \sqrt{(-3)^2 - 4(1)(3)}}{2(1)} = \frac{3 \pm \sqrt{9-12}}{2} = \frac{3 \pm \sqrt{-3}}{2}$. Since the

discriminant is less than 0, the equation has no real solution.

77. $10y^2 - 16y + 5 = 0 \Rightarrow$

$$x = \frac{-b \pm \sqrt{b^2 - 4ac}}{2a} = \frac{-(-16) \pm \sqrt{(-16)^2 - 4(10)(5)}}{2(10)} = \frac{16 \pm \sqrt{256 - 200}}{20} = \frac{16 \pm \sqrt{56}}{20} = \frac{8 \pm \sqrt{14}}{10}.$$

79. $D = b^2 - 4ac = (-6)^2 - 4(1)(1) = 32$. Since D is positive, this equation has two real solutions.

81. $D = b^2 - 4ac = (2.20)^2 - 4(1)(1.21) = 4.84 - 4.84 = 0$. Since $D = 0$, this equation has one real solution.

83. $D = b^2 - 4ac = (5)^2 - 4(4)\left(\frac{13}{8}\right) = 25 - 26 = -1$. Since D is negative, this equation has no real solution.

85. $\dfrac{1}{x-1} + \dfrac{1}{x+2} = \dfrac{5}{4} \Leftrightarrow 4(x-1)(x+2)\left(\dfrac{1}{x-1} + \dfrac{1}{x+2}\right) = 4(x-1)(x+2)\left(\dfrac{5}{4}\right) \Leftrightarrow$

$4(x+2) + 4(x-1) = 5(x-1)(x+2) \Leftrightarrow 4x+8+4x-4 = 5x^2+5x-10 \Leftrightarrow 5x^2-3x-14 = 0 \Leftrightarrow$

$(5x+7)(x-2) = 0$. If $5x+7 = 0$, then $x = -\frac{7}{5}$; if $x-2 = 0$, then $x = 2$. The solutions are $-\frac{7}{5}$ and 2.

87. $\dfrac{x^2}{x+100} = 50 \Rightarrow x^2 = 50(x+100) = 50x+5000 \Leftrightarrow x^2-50x-5000 = 0 \Leftrightarrow (x-100)(x+50) = 0 \Leftrightarrow x-100 = 0$

or $x+50 = 0$. Thus $x = 100$ or $x = -50$. The solutions are 100 and -50.

89. $\dfrac{x+5}{x-2} = \dfrac{5}{x+2} + \dfrac{28}{x^2-4} \Rightarrow (x+2)(x+5) = 5(x-2) + 28 \Leftrightarrow x^2+7x+10 = 5x-10+28 \Leftrightarrow x^2+2x-8 = 0 \Leftrightarrow$

$(x-2)(x+4) = 0 \Leftrightarrow x-2 = 0$ or $x+4 = 0 \Leftrightarrow x = 2$ or $x = -4$. However, $x = 2$ is inadmissible since we can't divide by 0 in the original equation, so the only solution is -4.

91. $\sqrt{2x+1} + 1 = x \Leftrightarrow \sqrt{2x+1} = x-1 \Rightarrow 2x+1 = (x-1)^2 \Leftrightarrow 2x+1 = x^2-2x+1 \Leftrightarrow 0 = x^2-4x = x(x-4)$.

Potential solutions are $x = 0$ and $x-4 \Leftrightarrow x = 4$. These are only potential solutions since squaring is not a reversible

operation. We must check each potential solution in the original equation. Checking $x = 0$: $\sqrt{2(0)+1} + 1 \overset{?}{=} (0)$,

$\sqrt{1} + 1 \overset{?}{=} 0$, NO! Checking $x = 4$: $\sqrt{2(4)+1} + 1 \overset{?}{=} (4), \sqrt{9} + 1 \overset{?}{=} 4, 3+1 \overset{?}{=} 4$, Yes. The only solution is $x = 4$.

93. $2x + \sqrt{x+1} = 8 \Leftrightarrow \sqrt{x+1} = 8-2x \Rightarrow x+1 = (8-2x)^2 \Leftrightarrow x+1 = 64-32x+4x^2 \Leftrightarrow$

$0 = 4x^2-33x+63 = (4x-21)(x-3)$. Potential solutions are $x = \frac{21}{4}$ and $x = 3$. Substituting each of these solutions

into the original equation, we see that $x = 3$ is a solution, but $x = \frac{21}{4}$ is not. Thus 3 is the only solution.

95. Let $w = x^2$. Then $x^4-13x^2+40 = \left(x^2\right)^2 - 13x^2 + 40 = 0$ becomes $w^2-13w+40 = 0 \Leftrightarrow (w-5)(w-8) = 0$.

So $w-5 = 0 \Leftrightarrow w = 5$, and $w-8 = 0 \Leftrightarrow w = 8$. When $w = 5$, we have $x^2 = 5 \Rightarrow x = \pm\sqrt{5}$. When $w = 8$, we have

$x^2 = 8 \Rightarrow x = \pm\sqrt{8} = \pm2\sqrt{2}$. The solutions are $\pm\sqrt{5}$ and $\pm2\sqrt{2}$.

97. $2x^4+4x^2+1 = 0$. The LHS is the sum of two nonnegative numbers and a positive number, so $2x^4+4x^2+1 \geq 1 \neq 0$.

This equation has no real solution.

99. Let $u = x^{2/3}$. Then $0 = x^{4/3} - 5x^{2/3} + 6$ becomes $u^2 - 5u + 6 = 0 \Leftrightarrow (u-3)(u-2) = 0 \Leftrightarrow u-3 = 0$ or $u-2 = 0$. If

$u-3 = 0$, then $x^{2/3} - 3 = 0 \Leftrightarrow x^{2/3} = 3 \Rightarrow x = \pm3^{3/2}$. If $u-2 = 0$, then $x^{2/3} - 2 = 0 \Leftrightarrow x^{2/3} = 2 \Rightarrow x = \pm2^{3/2}$.

The solutions are $\pm3^{3/2}$ and $\pm2^{3/2}$; that is, $x = \pm2\sqrt{2}$ and $x = \pm3\sqrt{3}$.

101. Let $W = (x+1)^{1/2}$. Then $4(x+1)^{1/2} - 5(x+1)^{3/2} + (x+1)^{5/2} = 0 \Leftrightarrow 4W - 5W^3 + W^5 = 0 \Leftrightarrow$

$W\left(W^4 - 5W^2 + 4\right) = 0 \Leftrightarrow W\left(W^2 - 4\right)\left(W^2 - 1\right) = 0 \Leftrightarrow W(W-2)(W+2)(W-1)(W+1) = 0 \Leftrightarrow W = -2,$

$-1, 0, 1,$ or 2. If $W < 0$, then $W = \sqrt{x+1}$ has no solution. If $W = 0$, then $x = -1$; if $W = 1$, then $\sqrt{x+1} = 1 \Leftrightarrow x = 1$,

and if $W = 2$, then $\sqrt{x+1} = 2 \Leftrightarrow x = 3$. The solutions are $-1, 0,$ and 3.

103. Let $u = x^{1/6}$. (We choose the exponent $\frac{1}{6}$ because the LCD of 2, 3, and 6 is 6.) Then $x^{1/2} - 3x^{1/3} = 3x^{1/6} - 9 \Leftrightarrow$

$x^{3/6} - 3x^{2/6} = 3x^{1/6} - 9 \Leftrightarrow u^3 - 3u^2 = 3u - 9 \Leftrightarrow 0 = u^3 - 3u^2 - 3u + 9 = u^2(u-3) - 3(u-3) = (u-3)\left(u^2-3\right)$.

So $u-3 = 0$ or $u^2 - 3 = 0$. If $u-3 = 0$, then $x^{1/6} - 3 = 0 \Leftrightarrow x^{1/6} = 3 \Leftrightarrow x = 3^6 = 729$. If $u^2 - 3 = 0$, then

$x^{1/3} - 3 = 0 \Leftrightarrow x^{1/3} = 3 \Leftrightarrow x = 3^3 = 27$. The solutions are 729 and 27.

105. $|3x+5| = 1$. So $3x+5 = 1 \Leftrightarrow 3x = -4 \Leftrightarrow x = -\frac{4}{3}$ or $3x+5 = -1 \Leftrightarrow 3x = -6 \Leftrightarrow x = -2$.

107. $|x-4| = 0.01$. So $x-4 = 0.01 \Leftrightarrow x = 4.01$ or $x-4 = -0.01 \Leftrightarrow x = 3.99$.

109. Using $h_0 = 288$, we solve $0 = -16t^2 + 288$, for $t \geq 0$. So $0 = -16t^2 + 288 \Leftrightarrow 16t^2 = 288 \Leftrightarrow t^2 = 18 \Rightarrow$

$t = \pm\sqrt{18} = \pm3\sqrt{2}$. Thus it takes $3\sqrt{2} \approx 4.24$ seconds for the ball the hit the ground.

111. We are given $v_o = 40$ ft/s.

   (a) Setting $h = 24$, we have $24 = -16t^2 + 40t \Leftrightarrow 16t^2 - 40t + 24 = 0 \Leftrightarrow 8\left(2t^2 - 5t + 3\right) = 0 \Leftrightarrow 8(2t-3)(t-1) = 0$

      $\Leftrightarrow t = 1$ or $t = 1\frac{1}{2}$. Therefore, the ball reaches 24 feet in 1 second (ascending) and again after $1\frac{1}{2}$ seconds (descending).

(b) Setting $h = 48$, we have $48 = -16t^2 + 40t \Leftrightarrow 16t^2 - 40t + 48 = 0 \Leftrightarrow 2t^2 - 5t + 6 = 0 \Leftrightarrow$

$t = \dfrac{5 \pm \sqrt{25 - 48}}{4} = \dfrac{5 \pm \sqrt{-23}}{4}$. However, since the discriminant $D < 0$, there are no real solutions, and hence the ball never reaches a height of 48 feet.

(c) The greatest height h is reached only once. So $h = -16t^2 + 40t \Leftrightarrow 16t^2 - 40t + h = 0$ has only one solution. Thus $D = (-40)^2 - 4(16)(h) = 0 \Leftrightarrow 1600 - 64h = 0 \Leftrightarrow h = 25$. So the greatest height reached by the ball is 25 feet.

(d) Setting $h = 25$, we have $25 = -16t^2 + 40t \Leftrightarrow 16t^2 - 40t + 25 = 0 \Leftrightarrow (4t - 5)^2 = 0 \Leftrightarrow t = 1\frac{1}{4}$. Thus the ball reaches the highest point of its path after $1\frac{1}{4}$ seconds.

(e) Setting $h = 0$ (ground level), we have $0 = -16t^2 + 40t \Leftrightarrow 2t^2 - 5t = 0 \Leftrightarrow t(2t - 5) = 0 \Leftrightarrow t = 0$ (start) or $t = 2\frac{1}{2}$. So the ball hits the ground in $2\frac{1}{2}$ s.

113. (a) The shrinkage factor when $w = 250$ is $S = \dfrac{0.032(250) - 2.5}{10{,}000} = \dfrac{8 - 2.5}{10{,}000} = 0.00055$. So the beam shrinks $0.00055 \times 12.025 \approx 0.007$ m, so when it dries it will be $12.025 - 0.007 = 12.018$ m long.

(b) Substituting $S = 0.00050$ we get $0.00050 = \dfrac{0.032w - 2.5}{10{,}000} \Leftrightarrow 5 = 0.032w - 2.5 \Leftrightarrow 7.5 = 0.032w \Leftrightarrow$

$w = \dfrac{7.5}{0.032} \approx 234.375$. So the water content should be 234.375 kg/m^3.

115. (a) The fish population on January 1, 2002 corresponds to $t = 0$, so $F = 1000\left(30 + 17(0) - (0)^2\right) = 30{,}000$. To find when the population will again reach this value, we set $F = 30{,}000$, giving

$30000 = 1000\left(30 + 17t - t^2\right) \Leftrightarrow 30000 = 30000 + 17000t - 1000t^2 \Leftrightarrow 0 = 17000t - 1000t^2 = 1000t(17 - t) \Leftrightarrow t = 0$ or $t = 17$. Thus the fish population will again be the same 17 years later, that is, on January 1, 2019.

(b) Setting $F = 0$, we have $0 = 1000\left(30 + 17t - t^2\right) \Leftrightarrow t^2 - 17t - 30 = 0 \Leftrightarrow$

$t = \dfrac{17 \pm \sqrt{289 + 120}}{-2} = \dfrac{17 \pm \sqrt{409}}{-2} = \dfrac{17 \pm 20.22}{2}$. Thus $t \approx -1.612$ or $t \approx 18.612$. Since $t < 0$ is inadmissible, it follows that the fish in the lake will have died out 18.612 years after January 1, 2002, that is on August 12, 2020.

117. Setting $P = 1250$ and solving for x, we have $1250 = \frac{1}{10}x(300 - x) = 30x - \frac{1}{10}x^2 \Leftrightarrow \frac{1}{10}x^2 - 30x + 1250 = 0$. Using the

quadratic formula, $x = \dfrac{-(-30) \pm \sqrt{(-30)^2 - 4\left(\frac{1}{10}\right)(1250)}}{2\left(\frac{1}{10}\right)} = \dfrac{30 \pm \sqrt{900 - 500}}{0.2} = \dfrac{30 \pm 20}{0.2}$. Thus $x = \dfrac{30 - 20}{0.2} = 50$

or $x = \dfrac{30 + 20}{0.2} = 250$. Since he must have $0 \le x \le 200$, he should make 50 ovens per week.

119. Since the total time is 3 s, we have $3 = \dfrac{\sqrt{d}}{4} + \dfrac{d}{1090}$. Letting $w = \sqrt{d}$, we have $3 = \frac{1}{4}w + \frac{1}{1090}w^2 \Leftrightarrow \frac{1}{1090}w^2 + \frac{1}{4}w - 3 = 0$

$\Leftrightarrow 2w^2 + 545w - 6540 = 0 \Rightarrow w = \dfrac{-545 \pm 591.054}{4}$. Since $w \ge 0$, we have $\sqrt{d} = w \approx 11.51$, so $d = 132.56$. The well is 132.6 ft deep.

121. When we multiplied by x, we introduced $x = 0$ as a solution. When we divided by $x - 1$, we are really dividing by 0, since $x = 1 \Leftrightarrow x - 1 = 0$.

123. $x^2 - 9x + 20 = (x-4)(x-5) = 0$, so $x = 4$ or $x = 5$. The roots are 4 and 5. The product is $4 \cdot 5 = 20$,
and the sum is $4 + 5 = 9$. $x^2 - 2x - 8 = (x-4)(x+2) = 0$, so $x = 4$ or $x = -2$. The roots are 4 and -2.
The product is $4 \cdot (-2) = -8$, and the sum is $4 + (-2) = 2$. $x^2 + 4x + 2 = 0$, so using the quadratic formula,
$$x = \frac{-4 \pm \sqrt{4^2 - 4(1)(2)}}{2(1)} = \frac{-4 \pm \sqrt{8}}{2} = \frac{-4 \pm 2\sqrt{2}}{2} = -2 \pm \sqrt{2}.$$ The roots are $-2 - \sqrt{2}$ and $-2 + \sqrt{2}$. The product
is $\left(-2 - \sqrt{2}\right) \cdot \left(-2 + \sqrt{2}\right) = 4 - 2 = 2$, and the sum is $\left(-2 - \sqrt{2}\right) + \left(-2 + \sqrt{2}\right) = -4$. In general, if $x = r_1$ and
$x = r_2$ are roots, then $x^2 + bx + c = (x - r_1)(x - r_2) = x^2 - r_1 x - r_2 x + r_1 r_2 = x^2 - (r_1 + r_2)x + r_1 r_2$. Equating the
coefficients, we get $c = r_1 r_2$ and $b = -(r_1 + r_2)$.

1.6   MODELING WITH EQUATIONS

1. An equation modeling a real-world situation can be used to help us understand a real-world problem using mathematical
methods. We translate real-world ideas into the language of algebra to construct our model, and translate our mathematical
results back into real-world ideas in order to interpret our findings.

3. (a) A square of side x has area $A = x^2$.

   (b) A rectangle of length l and width w has area $A = lw$.

   (c) A circle of radius r has area $A = \pi r^2$.

5. A painter paints a wall in x hours, so the fraction of the wall she paints in one hour is $\dfrac{1 \text{ wall}}{x \text{ hours}} = \dfrac{1}{x}$.

7. If n is the first integer, then $n + 1$ is the middle integer, and $n + 2$ is the third integer. So the sum of the three consecutive
integers is $n + (n+1) + (n+2) = 3n + 3$.

9. If s is the third test score, then since the other test scores are 78 and 82, the average of the three test scores is
$\dfrac{78 + 82 + s}{3} = \dfrac{160 + s}{3}$.

11. If x dollars are invested at $2\frac{1}{2}\%$ simple interest, then the first year you will receive $0.025x$ dollars in interest.

13. Since w is the width of the rectangle, the length is three times the width, or $3w$. Then
area $=$ length $\times$ width $= 3w \times w = 3w^2$ ft^2.

15. Since distance $=$ rate $\times$ time we have distance $= s \times (45 \text{ min}) \dfrac{1 \text{ h}}{60 \text{ min}} = \frac{3}{4}s$ mi.

17. If x is the quantity of pure water added, the mixture will contain 25 oz of salt and $3 + x$ gallons of water. Thus the
concentration is $\dfrac{25}{3 + x}$.

19. If d is the number of days and m the number of miles, then the cost of a rental is $C = 65d + 0.20m$. In this case, $d = 3$
and $C = 275$, so we solve for m: $275 = 65 \cdot 3 + 0.20m \Leftrightarrow 275 = 195 + 0.2m \Leftrightarrow 0.2m = 80 \Leftrightarrow m = \dfrac{80}{0.2} = 400$. Thus,
Michael drove 400 miles.

21. Let m be the amount invested at $4\frac{1}{2}\%$. Then $12{,}000 - m$ is the amount invested at 4%.
Since the total interest is equal to the interest earned at $4\frac{1}{2}\%$ plus the interest earned at 4%, we have
$$525 = 0.045m + 0.04(12{,}000 - m) \Leftrightarrow 525 = 0.045m + 480 - 0.04m \Leftrightarrow 45 = 0.005m \Leftrightarrow m = \frac{45}{0.005} = 9000.$$ Thus
$\$9000$ is invested at $4\frac{1}{2}\%$, and $\$12{,}000 - 9000 = \$3000$ is invested at 4%.

23. Using the formula $I = Prt$ and solving for r, we get $262.50 = 3500 \cdot r \cdot 1 \Leftrightarrow r = \dfrac{262.5}{3500} = 0.075$ or 7.5%.

25. Let x be her monthly salary. Since her annual salary $= 12 \times$ (monthly salary) $+$ (Christmas bonus) we have
$97{,}300 = 12x + 8{,}500 \Leftrightarrow 88{,}800 = 12x \Leftrightarrow x \approx 7{,}400$. Her monthly salary is $\$7{,}400$.

27. Let h be the amount that Craig inherits. So $(x + 22,000)$ is the amount that he invests and doubles. Thus $2(x + 22,000) = 134,000 \Leftrightarrow 2x + 44,000 = 134,000 \Leftrightarrow 2x = 90,000 \Leftrightarrow x = 45,000$. So Craig inherits $45,000.

29. Let x be the hours the assistant worked. Then $2x$ is the hours the plumber worked. Since the labor charge is equal to the plumber's labor plus the assistant's labor, we have $4025 = 45(2x) + 25x \Leftrightarrow 4025 = 90x + 25x \Leftrightarrow 4025 = 115x \Leftrightarrow x = \frac{4025}{115} = 35$. Thus the assistant works for 35 hours, and the plumber works for $2 \times 35 = 70$ hours.

31. All ages are in terms of the daughter's age 7 years ago. Let y be age of the daughter 7 years ago. Then $11y$ is the age of the movie star 7 years ago. Today, the daughter is $y + 7$, and the movie star is $11y + 7$. But the movie star is also 4 times his daughter's age today. So $4(y + 7) = 11y + 7 \Leftrightarrow 4y + 28 = 11y + 7 \Leftrightarrow 21 = 7y \Leftrightarrow y = 3$. Thus the movie star's age today is $11(3) + 7 = 40$ years.

33. Let p be the number of pennies. Then p is the number of nickels and p is the number of dimes. So the value of the coins in the purse is the value of the pennies plus the value of the nickels plus the value of the dimes. Thus $1.44 = 0.01p + 0.05p + 0.10p \Leftrightarrow 1.44 = 0.16p \Leftrightarrow p = \frac{1.44}{0.16} = 9$. So the purse contains 9 pennies, 9 nickels, and 9 dimes.

35. Let l be the length of the garden. Since area $=$ width $\cdot$ length, we obtain the equation $1125 = 25l \Leftrightarrow l = \frac{1125}{25} = 45$ ft. So the garden is 45 feet long.

37. Let x be the length of a side of the square plot. As shown in the figure, area of the plot $=$ area of the building $+$ area of the parking lot. Thus, $x^2 = 60(40) + 12,000 = 2,400 + 12,000 = 14,400 \Rightarrow x = \pm 120$. So the plot of land measures 120 feet by 120 feet.

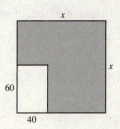

39. Let w be the width of the garden in feet. Then the length is $w + 10$. Thus $875 = w(w + 10) \Leftrightarrow w^2 + 10w - 875 = 0 \Leftrightarrow (w + 35)(w - 25) = 0$. So $w + 35 = 0$ in which case $w = -35$, which is not possible, or $w - 25 = 0$ and so $w = 25$. Thus the width is 25 feet and the length is 35 feet.

41. Let w be the width of the garden in feet. We use the perimeter to express the length l of the garden in terms of width. Since the perimeter is twice the width plus twice the length, we have $200 = 2w + 2l \Leftrightarrow 2l = 200 - 2w \Leftrightarrow l = 100 - w$. Using the formula for area, we have $2400 = w(100 - w) = 100w - w^2 \Leftrightarrow w^2 - 100w + 2400 = 0 \Leftrightarrow (w - 40)(w - 60) = 0$. So $w - 40 = 0 \Leftrightarrow w = 40$, or $w - 60 = 0 \Leftrightarrow w = 60$. If $w = 40$, then $l = 100 - 40 = 60$. And if $w = 60$, then $l = 100 - 60 = 40$. So the length is 60 feet and the width is 40 feet.

43. Let l be the length of the lot in feet. Then the length of the diagonal is $l + 10$. We apply the Pythagorean Theorem with the hypotenuse as the diagonal. So $l^2 + 50^2 = (l + 10)^2 \Leftrightarrow l^2 + 2500 = l^2 + 20l + 100 \Leftrightarrow 20l = 2400 \Leftrightarrow l = 120$. Thus the length of the lot is 120 feet.

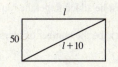

45. (a) First we write a formula for the area of the figure in terms of x. Region A has dimensions 10 cm and x cm and region B has dimensions 6 cm and x cm. So the shaded region has area $(10 \cdot x) + (6 \cdot x) = 16x$ cm^2. We are given that this is equal to 144 cm^2, so $144 = 16x \Leftrightarrow x = \frac{144}{16} = 9$ cm.

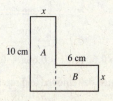

(b) First we write a formula for the area of the figure in terms of x. Region A has dimensions 14 in. and x in. and region B has dimensions $(13 + x)$ in. and x in. So the area of the figure is $(14 \cdot x) + [(13 + x)x] = 14x + 13x + x^2 = x^2 + 27x$. We are given that this is equal to 160 in^2, so $160 = x^2 + 27x \Leftrightarrow x^2 + 27x - 160 = 0$ $\Leftrightarrow (x + 32)(x - 5) \Leftrightarrow x = -32$ or $x = 5$. x must be positive, so $x = 5$ in.

47. Let x be the width of the strip. Then the length of the mat is $20 + 2x$, and the width of the mat is $15 + 2x$. Now the perimeter is twice the length plus twice the width, so $102 = 2(20 + 2x) + 2(15 + 2x) \Leftrightarrow 102 = 40 + 4x + 30 + 4x \Leftrightarrow 102 = 70 + 8x \Leftrightarrow 32 = 8x \Leftrightarrow x = 4$. Thus the strip of mat is 4 inches wide.

49. Let h be the height the ladder reaches (in feet). Using the Pythagorean Theorem we have $\left(7\frac{1}{2}\right)^2 + h^2 = \left(19\frac{1}{2}\right)^2 \Leftrightarrow$ $\left(\frac{15}{2}\right)^2 + h^2 = \left(\frac{39}{4}\right)^2 \Leftrightarrow h^2 = \left(\frac{39}{4}\right)^2 - \left(\frac{15}{2}\right)^2 = \frac{1521}{4} - \frac{225}{4} = \frac{1296}{4} = 324$. So $h = \sqrt{324} = 18$.

51. Let x be the length of the man's shadow, in meters. Using similar triangles, $\dfrac{10 + x}{6} = \dfrac{x}{2} \Leftrightarrow 20 + 2x = 6x \Leftrightarrow 4x = 20 \Leftrightarrow$ $x = 5$. Thus the man's shadow is 5 meters long.

53. Let x be the amount (in mL) of 60% acid solution to be used. Then $300 - x$ mL of 30% solution would have to be used to yield a total of 300 mL of solution.

	60% acid	30% acid	Mixture
mL	x	$300 - x$	300
Rate (% acid)	0.60	0.30	0.50
Value	$0.60x$	$0.30(300 - x)$	$0.50(300)$

Thus the total amount of pure acid used is $0.60x + 0.30(300 - x) = 0.50(300) \Leftrightarrow 0.3x + 90 = 150 \Leftrightarrow x = \dfrac{60}{0.3} = 200$. So 200 mL of 60% acid solution must be mixed with 100 mL of 30% solution to get 300 mL of 50% acid solution.

55. Let x be the number of grams of silver added. The weight of the rings is 5×18 g $= 90$ g.

	5 rings	Pure silver	Mixture
Grams	90	x	$90 + x$
Rate (% gold)	0.90	0	0.75
Value	0.90(90)	$0x$	$0.75(90 + x)$

So $0.90(90) + 0x = 0.75(90 + x) \Leftrightarrow 81 = 67.5 + 0.75x \Leftrightarrow 0.75x = 13.5 \Leftrightarrow x = \frac{13.5}{0.75} = 18$. Thus 18 grams of silver must be added to get the required mixture.

57. Let x be the number of liters of coolant removed and replaced by water.

	60% antifreeze	60% antifreeze (removed)	Water	Mixture
Liters	3.6	x	x	3.6
Rate (% antifreeze)	0.60	0.60	0	0.50
Value	0.60(3.6)	$-0.60x$	$0x$	0.50(3.6)

so $0.60(3.6) - 0.60x + 0x = 0.50(3.6) \Leftrightarrow 2.16 - 0.6x = 1.8 \Leftrightarrow -0.6x = -0.36 \Leftrightarrow x = \dfrac{-0.36}{-0.6} = 0.6$. Thus 0.6 liters must be removed and replaced by water.

59. Let c be the concentration of fruit juice in the cheaper brand. The new mixture that Jill makes will consist of 650 mL of the original fruit punch and 100 mL of the cheaper fruit punch.

	Original Fruit Punch	Cheaper Fruit Punch	Mixture
mL	650	100	750
Concentration	0.50	c	0.48
Juice	$0.50 \cdot 650$	$100c$	$0.48 \cdot 750$

So $0.50 \cdot 650 + 100c = 0.48 \cdot 750 \Leftrightarrow 325 + 100c = 360 \Leftrightarrow 100c = 35 \Leftrightarrow c = 0.35$. Thus the cheaper brand is only 35% fruit juice.

61. Let t be the time in minutes it would take Candy and Tim if they work together. Candy delivers the papers at a rate of $\frac{1}{70}$ of the job per minute, while Tim delivers the paper at a rate of $\frac{1}{80}$ of the job per minute. The sum of the fractions of the job that each can do individually in one minute equals the fraction of the job they can do working together. So we have $\frac{1}{t} = \frac{1}{70} + \frac{1}{80} \Leftrightarrow 560 = 8t + 7t \Leftrightarrow 560 = 15t \Leftrightarrow t = 37\frac{1}{3}$ minutes. Since $\frac{1}{3}$ of a minute is 20 seconds, it would take them 37 minutes 20 seconds if they worked together.

63. Let t be the time, in hours, it takes Karen to paint a house alone. Then working together, Karen and Betty can paint a house in $\frac{2}{3}t$ hours. The sum of their individual rates equals their rate working together, so $\frac{1}{t} + \frac{1}{6} = \frac{1}{\frac{2}{3}t} \Leftrightarrow \frac{1}{t} + \frac{1}{6} = \frac{3}{2t} \Leftrightarrow 6 + t = 9 \Leftrightarrow t = 3$. Thus it would take Karen 3 hours to paint a house alone.

65. Let t be the time, in hours it takes Irene to wash all the windows. Then it takes Henry $t + \frac{3}{2}$ hours to wash all the windows, and the sum of the fraction of the job per hour they can do individually equals the fraction of the job they can do together. Since 1 hour 48 minutes $= 1 + \frac{48}{60} = 1 + \frac{4}{5} = \frac{9}{5}$, we have $\frac{1}{t} + \frac{1}{t + \frac{3}{2}} = \frac{1}{\frac{9}{5}} \Leftrightarrow$

$\frac{1}{t} + \frac{2}{2t + 3} = \frac{5}{9} \Rightarrow 9(2t + 3) + 2(9t) = 5t(2t + 3) \Leftrightarrow 18t + 27 + 18t = 10t^2 + 15t \Leftrightarrow 10t^2 - 21t - 27 = 0$

$\Leftrightarrow t = \frac{-(-21) \pm \sqrt{(-21)^2 - 4(10)(-27)}}{2(10)} = \frac{21 \pm \sqrt{441 + 1080}}{20} = \frac{21 \pm 39}{20}$. So $t = \frac{21 - 39}{20} = -\frac{9}{10}$

or $t = \frac{21 + 39}{20} = 3$. Since $t < 0$ is impossible, all the windows are washed by Irene alone in 3 hours and by Henry alone in $3 + \frac{3}{2} = 4\frac{1}{2}$ hours.

67. Let t be the time in hours that Wendy spent on the train. Then $\frac{11}{2} - t$ is the time in hours that Wendy spent on the bus. We construct a table:

	Rate	Time	Distance
By train	40	t	$40t$
By bus	60	$\frac{11}{2} - t$	$60\left(\frac{11}{2} - t\right)$

The total distance traveled is the sum of the distances traveled by bus and by train, so $300 = 40t + 60\left(\frac{11}{2} - t\right) \Leftrightarrow$

$300 = 40t + 330 - 60t \Leftrightarrow -30 = -20t \Leftrightarrow t = \frac{-30}{-20} = 1.5$ hours. So the time spent on the train is $5.5 - 1.5 = 4$ hours.

69. Let r be the speed of the plane from Montreal to Los Angeles. Then $r + 0.20r = 1.20r$ is the speed of the plane from Los Angeles to Montreal.

	Rate	Time	Distance
Montreal to L.A.	r	$\dfrac{2500}{r}$	2500
L.A. to Montreal	$1.2r$	$\dfrac{2500}{1.2r}$	2500

The total time is the sum of the times each way, so $9\frac{1}{6} = \dfrac{2500}{r} + \dfrac{2500}{1.2r} \Leftrightarrow \dfrac{55}{6} = \dfrac{2500}{r} + \dfrac{2500}{1.2r} \Leftrightarrow$

$55 \cdot 1.2r = 2500 \cdot 6 \cdot 1.2 + 2500 \cdot 6 \Leftrightarrow 66r = 18{,}000 + 15{,}000 \Leftrightarrow 66r = 33{,}000 \Leftrightarrow r = \frac{33{,}000}{66} = 500$. Thus the plane flew at a speed of 500 mi/h on the trip from Montreal to Los Angeles.

71. Let x be the rate, in mi/h, at which the salesman drove between Ajax and Barrington.

Cities	Distance	Rate	Time
Ajax $\to$ Barrington	120	x	$\dfrac{120}{x}$
Barrington $\to$ Collins	150	$x + 10$	$\dfrac{150}{x + 10}$

We have used the equation time $= \dfrac{\text{distance}}{\text{rate}}$ to fill in the "Time" column of the table. Since the second part of the trip

took 6 minutes (or $\frac{1}{10}$ hour) more than the first, we can use the time column to get the equation $\dfrac{120}{x} + \dfrac{1}{10} = \dfrac{150}{x + 10} \Rightarrow$

$120 (10) (x + 10) + x (x + 10) = 150 (10x) \Leftrightarrow 1200x + 12{,}000 + x^2 + 10x = 1500x \Leftrightarrow x^2 - 290x + 12{,}000 = 0 \Leftrightarrow$

$x = \dfrac{-(-290) \pm \sqrt{(-290)^2 - 4(1)(12{,}000)}}{2} = \dfrac{290 \pm \sqrt{84{,}100 - 48{,}000}}{2} = \dfrac{290 \pm \sqrt{36{,}100}}{2} = \dfrac{290 \pm 190}{2} = 145 \pm 95$. Hence, the salesman drove either 50 mi/h or 240 mi/h between Ajax and Barrington. (The first choice seems more likely!)

73. Let r be the rowing rate in km/h of the crew in still water. Then their rate upstream was $r - 3$ km/h, and their rate downstream was $r + 3$ km/h.

	Distance	Rate	Time
Upstream	6	$r - 3$	$\dfrac{6}{r - 3}$
Downstream	6	$r + 3$	$\dfrac{6}{r + 3}$

Since the time to row upstream plus the time to row downstream was 2 hours 40 minutes $= \frac{8}{3}$ hour, we get the equation

$\dfrac{6}{r - 3} + \dfrac{6}{r + 3} = \dfrac{8}{3} \Leftrightarrow 6 (3) (r + 3) + 6 (3) (r - 3) = 8 (r - 3) (r + 3) \Leftrightarrow 18r + 54 + 18r - 54 = 8r^2 - 72 \Leftrightarrow$

$0 = 8r^2 - 36r - 72 = 4 \left(2r^2 - 9r - 18\right) = 4 (2r + 3) (r - 6)$. Since $2r + 3 = 0 \Leftrightarrow r = -\frac{3}{2}$ is impossible, the solution is

$r - 6 = 0 \Leftrightarrow r = 6$. So the rate of the rowing crew in still water is 6 km/h.

75. Let x be the distance from the fulcrum to where the mother sits. Then substituting the known values into the formula given, we have $100 (8) = 125x \Leftrightarrow 800 = 125x \Leftrightarrow x = 6.4$. So the mother should sit 6.4 feet from the fulcrum.

77. We have that the volume is 180 ft³, so $x (x - 4) (x + 9) = 180 \Leftrightarrow x^3 + 5x^2 - 36x = 180 \Leftrightarrow x^3 + 5x^2 - 36x - 180 = 0$

$\Leftrightarrow x^2 (x + 5) - 36 (x + 5) = 0 \Leftrightarrow (x + 5) \left(x^2 - 36\right) = 0 \Leftrightarrow (x + 5) (x + 6) (x - 6) = 0 \Rightarrow x = 6$ is the only positive solution. So the box is 2 feet by 6 feet by 15 feet.

79. Let x be the length of one side of the cardboard, so we start with a piece of cardboard x by x. When 4 inches are removed from each side, the base of the box is $x - 8$ by $x - 8$. Since the volume is 100 in^3, we get $4(x-8)^2 = 100 \Leftrightarrow$ $x^2 - 16x + 64 = 25 \Leftrightarrow x^2 - 16x + 39 = 0 \Leftrightarrow (x-3)(x-13) = 0.$ So $x = 3$ or $x = 13$. But $x = 3$ is not possible, since then the length of the base would be $3 - 8 = -5$, and all lengths must be positive. Thus $x = 13$, and the piece of cardboard is 13 inches by 13 inches.

81. Let r be the radius of the tank, in feet. The volume of the spherical tank is $\frac{4}{3}\pi r^3$ and is also $750 \times 0.1337 = 100.275$. So $\frac{4}{3}\pi r^3 = 100.275 \Leftrightarrow r^3 = 23.938 \Leftrightarrow r = 2.88$ feet.

83. Let x be the length, in miles, of the abandoned road to be used. Then the length of the abandoned road not used is $40 - x$, and the length of the new road is $\sqrt{10^2 + (40 - x)^2}$ miles, by the Pythagorean Theorem. Since the cost of the road is cost per mile $\times$ number of miles, we have $100{,}000x + 200{,}000\sqrt{x^2 - 80x + 1700} = 6{,}800{,}000$ $\Leftrightarrow 2\sqrt{x^2 - 80x + 1700} = 68 - x$. Squaring both sides, we get $4x^2 - 320x + 6800 = 4624 - 136x + x^2 \Leftrightarrow$ $3x^2 - 184x + 2176 = 0 \Leftrightarrow x = \frac{184 \pm \sqrt{33856 - 26112}}{6} = \frac{184 \pm 88}{6} \Leftrightarrow x = \frac{136}{3}$ or $x = 16$. Since $45\frac{1}{3}$ is longer than the existing road, 16 miles of the abandoned road should be used. A completely new road would have length $\sqrt{10^2 + 40^2}$ (let $x = 0$) and would cost $\sqrt{1700} \times 200{,}000 \approx 8.3$ million dollars. So no, it would not be cheaper.

85. Let x be the height of the pile in feet. Then the diameter is $3x$ and the radius is $\frac{3}{2}x$ feet. Since the volume of the cone is 1000 ft^3, we have $\frac{\pi}{3}\left(\frac{3x}{2}\right)^2 x = 1000 \Leftrightarrow \frac{3\pi x^3}{4} = 1000 \Leftrightarrow x^3 = \frac{4000}{3\pi} \Leftrightarrow x = \sqrt[3]{\frac{4000}{3\pi}} \approx 7.52$ feet.

87. Let h be the height in feet of the structure. The structure is composed of a right cylinder with radius 10 and height $\frac{2}{3}h$ and a cone with base radius 10 and height $\frac{1}{3}h$. Using the formulas for the volume of a cylinder and that of a cone, we obtain the equation $1400\pi = \pi(10)^2\left(\frac{2}{3}h\right) + \frac{1}{3}\pi(10)^2\left(\frac{1}{3}h\right) \Leftrightarrow 1400\pi = \frac{200\pi}{3}h + \frac{100\pi}{9}h \Leftrightarrow 126 = 6h + h$ (multiply both sides by $\frac{9}{100\pi}$) $\Leftrightarrow 126 = 7h \Leftrightarrow h = 18$. Thus the height of the structure is 18 feet.

89. Let h be the height of the break, in feet. Then the portion of the bamboo above the break is $10 - h$. Applying the Pythagorean Theorem, we obtain $h^2 + 3^2 = (10-h)^2 \Leftrightarrow h^2 + 9 = 100 - 20h + h^2 \Leftrightarrow -91 = -20h \Leftrightarrow$ $h = \frac{91}{20} = 4.55$. Thus the break is 4.55 ft above the ground.

91. Let x equal the original length of the reed in cubits. Then $x - 1$ is the piece that fits 60 times along the length of the field, that is, the length is $60(x-1)$. The width is $30x$. Then converting cubits to ninda, we have $375 = 60(x-1) \cdot 30x \cdot \frac{1}{12^2} = \frac{25}{2}x(x-1) \Leftrightarrow 30 = x^2 - x \Leftrightarrow x^2 - x - 30 = 0 \Leftrightarrow (x-6)(x+5) = 0.$ So $x = 6$ or $x = -5$. Since x must be positive, the original length of the reed is 6 cubits.

1.7   INEQUALITIES

1. (a) If $x < 5$, then $x - 3 < 5 - 3 \Rightarrow x - 3 < 2$.

   (b) If $x \leq 5$, then $3 \cdot x \leq 3 \cdot 5 \Rightarrow 3x \leq 15$.

   (c) If $x \geq 2$, then $-3 \cdot x \leq -3 \cdot 2 \Rightarrow -3x \leq -6$.

   (d) If $x < -2$, then $-x > 2$.

3. (a) The solution of the inequality $|x| \leq 3$ is the interval $[-3, 3]$.

   (b) The solution of the inequality $|x| \geq 3$ is a union of two intervals $(-\infty, -3] \cup [3, \infty)$.

5. $x = -2$: $3 - 2(-2) \overset{?}{\le} \frac{1}{2}$. No, $7 \not\le \frac{1}{2}$. $x = -1$: $3 - 2(-1) \overset{?}{\le} \frac{1}{2}$. No, $6 \not\le \frac{1}{2}$. $x = 0$: $3 - 2(0) \overset{?}{\le} \frac{1}{2}$. No, $3 \not\le \frac{1}{2}$.

$x = \frac{1}{2}$: $3 - 2\left(\frac{1}{2}\right) \overset{?}{\le} \frac{1}{2}$. No, $2 \not\le \frac{1}{2}$. $x = 1$: $3 - 2(1) \overset{?}{\le} \frac{1}{2}$. No, $1 \not\le \frac{1}{2}$.

$x = \sqrt{2}$: $3 - 2\left(\sqrt{2}\right) \overset{?}{\le} \frac{1}{2}$. Yes, $3 - 2\sqrt{2} \le \frac{1}{2}$. $x = 2$: $3 - 2(2) \overset{?}{\le} \frac{1}{2}$. Yes, $-1 \le \frac{1}{2}$.

$x = 4$: $3 - 2(4) \overset{?}{\le} \frac{1}{2}$. Yes, $5 \le \frac{1}{2}$.

The elements $\sqrt{2}$, 2, and 4 all satisfy the inequality.

7. $x = -2$: $1 \overset{?}{<} 2(-2) - 4 \overset{?}{\le} 7$. No, since $2(-2) - 4 = -8$ and $1 \not< -8$.

$x = -1$: $1 \overset{?}{<} 2(-1) - 4 \overset{?}{\le} 7$. No, since $2(-1) - 4 = -6$ and $1 \not< -6$.

$x = 0$: $1 \overset{?}{<} 2(0) - 4 \overset{?}{\le} 7$. No, since $2(0) - 4 = -4$ and $1 \not< -4$.

$x = \frac{1}{2}$: $1 \overset{?}{<} 2\left(\frac{1}{2}\right) - 4 \overset{?}{\le} 7$. No, since $2\left(\frac{1}{2}\right) - 4 = -3$ and $1 \not< -3$.

$x = 1$: $1 \overset{?}{<} 2(1) - 4 \overset{?}{\le} 7$. No, since $2(1) - 4 = -2$ and $1 \not< -2$.

$x = \sqrt{2}$: $1 \overset{?}{<} 2\left(\sqrt{2}\right) - 4 \overset{?}{\le} 7$. No, since $2\sqrt{2} - 4 < 0$ and $1 \not< 0$.

$x = 2$: $1 \overset{?}{<} 2(2) - 4 \overset{?}{\le} 7$. No, since $2(1) - 4 = -2$ and $1 \not< -2$.

$x = 4$: $1 \overset{?}{<} 2(4) - 4 \overset{?}{\le} 7$. Yes, $2(4) - 4 = 4$ and $1 < 4 \le 7$.

Only 4 satisfies the inequality.

9. $x = -2$: $\frac{1}{(-2)} \overset{?}{\le} \frac{1}{2}$. Yes, $-\frac{1}{2} \le \frac{1}{2}$.

$x = -1$: $\frac{1}{(-1)} \overset{?}{\le} \frac{1}{2}$. Yes, $-1 \le \frac{1}{2}$. $x = 0$: $\frac{1}{0} \overset{?}{\le} \frac{1}{2}$. No, $\frac{1}{0}$ is not defined.

$x = \frac{1}{2}$: $\frac{1}{1/2} \overset{?}{\le} \frac{1}{2}$. No, $2 \not\le \frac{1}{2}$. $x = 1$: $\frac{1}{1} \overset{?}{\le} \frac{1}{2}$. No, $1 \not\le \frac{1}{2}$.

$x = \sqrt{2}$: $\frac{1}{\sqrt{2}} \overset{?}{\le} \frac{1}{2}$. No, $2 \not\le \sqrt{2}$. $x = 2$: $\frac{1}{2} \overset{?}{\le} \frac{1}{2}$. Yes, $\frac{1}{2} \le \frac{1}{2}$.

$x = 4$: $\frac{1}{4} \overset{?}{\le} \frac{1}{2}$. Yes. The elements -2, -1, 2, and 4 all satisfy the inequality.

11. $2x \le 7 \Leftrightarrow x \le \frac{7}{2}$. Interval: $\left(-\infty, \frac{7}{2}\right]$.

Graph:

13. $2x - 5 > 3 \Leftrightarrow 2x > 8 \Leftrightarrow x > 4$

Interval: $(4, \infty)$.   Graph:

15. $7 - x \ge 5 \Leftrightarrow -x \ge -2 \Leftrightarrow x \le 2$

Interval: $(-\infty, 2]$.   Graph:

17. $2x + 1 < 0 \Leftrightarrow 2x < -1 \Leftrightarrow x < -\frac{1}{2}$

Interval: $\left(-\infty, -\frac{1}{2}\right)$.   Graph:

19. $3x + 11 \le 6x + 8 \Leftrightarrow 3 \le 3x \Leftrightarrow 1 \le x$

Interval: $[1, \infty)$.   Graph:

21. $\frac{1}{2}x - \frac{2}{3} > 2 \Leftrightarrow \frac{1}{2}x > \frac{8}{3} \Leftrightarrow x > \frac{16}{3}$

Interval: $\left(\frac{16}{3}, \infty\right)$.   Graph:

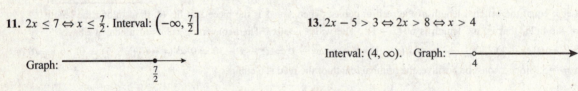

23. $\frac{1}{3}x + 2 < \frac{1}{6}x - 1 \Leftrightarrow \frac{1}{6}x < -3 \Leftrightarrow x < -18$

Interval:

$(-\infty, -18)$. Graph: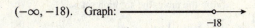

25. $4 - 3x \leq -(1 + 8x) \Leftrightarrow 4 - 3x \leq -1 - 8x \Leftrightarrow 5x \leq -5$
$\Leftrightarrow x \leq -1$

Interval: $(-\infty, -1]$. Graph:

27. $2 \leq x + 5 < 4 \Leftrightarrow -3 \leq x < -1$

Interval: $[-3, -1)$. Graph:

29. $-1 < 2x - 5 < 7 \Leftrightarrow 4 < 2x < 12 \Leftrightarrow 2 < x < 6$

Interval: $(2, 6)$. Graph:

31. $-2 < 8 - 2x \leq -1 \Leftrightarrow -10 < -2x \leq -9 \Leftrightarrow 5 > x \geq \frac{9}{2}$
$\Leftrightarrow \frac{9}{2} \leq x < 5$

Interval: $\left[\frac{9}{2}, 5\right)$. Graph:

33. $\frac{1}{6} < \frac{2x - 13}{12} \leq \frac{2}{3} \Leftrightarrow 2 < 2x - 13 \leq 8$ (multiply each

expression by 12) $\Leftrightarrow 15 < 2x \leq 21 \Leftrightarrow \frac{15}{2} < x \leq \frac{21}{2}$.

Interval: $\left(\frac{15}{2}, \frac{21}{2}\right]$. Graph:

35. $(x + 2)(x - 3) < 0$. The expression on the left of the inequality changes sign where $x = -2$ and where $x = 3$. Thus we must check the intervals in the following table.

Interval	$(-\infty, -2)$	$(-2, 3)$	$(3, \infty)$
Sign of $x + 2$	$-$	$+$	$+$
Sign of $x - 3$	$-$	$-$	$+$
Sign of $(x + 2)(x - 3)$	$+$	$-$	$+$

From the table, the solution set is
$\{x \mid -2 < x < 3\}$. Interval: $(-2, 3)$.

Graph:

37. $x(2x + 7) \geq 0$. The expression on the left of the inequality changes sign where $x = 0$ and where $x = -\frac{7}{2}$. Thus we must check the intervals in the following table.

Interval	$\left(-\infty, -\frac{7}{2}\right)$	$\left(-\frac{7}{2}, 0\right)$	$(0, \infty)$
Sign of x	$-$	$-$	$+$
Sign of $2x + 7$	$-$	$+$	$+$
Sign of $x(2x + 7)$	$+$	$-$	$+$

From the table, the solution set is
$\left\{x \mid x \leq -\frac{7}{2} \text{ or } 0 \leq x\right\}$.

Interval: $\left(-\infty, -\frac{7}{2}\right] \cup [0, \infty)$.

Graph:

39. $x^2 - 3x - 18 \leq 0 \Leftrightarrow (x + 3)(x - 6) \leq 0$. The expression on the left of the inequality changes sign where $x = 6$ and where $x = -3$. Thus we must check the intervals in the following table.

Interval	$(-\infty, -3)$	$(-3, 6)$	$(6, \infty)$
Sign of $x + 3$	$-$	$+$	$+$
Sign of $x - 6$	$-$	$-$	$+$
Sign of $(x + 3)(x - 6)$	$+$	$-$	$+$

From the table, the solution set is
$\{x \mid -3 \leq x \leq 6\}$. Interval: $[-3, 6]$.

Graph:

41. $2x^2 + x \geq 1 \Leftrightarrow 2x^2 + x - 1 \geq 0 \Leftrightarrow (x+1)(2x-1) \geq 0$. The expression on the left of the inequality changes sign where $x = -1$ and where $x = \frac{1}{2}$. Thus we must check the intervals in the following table.

Interval	$(-\infty, -1)$	$\left(-1, \frac{1}{2}\right)$	$\left(\frac{1}{2}, \infty\right)$
Sign of $x + 1$	$-$	$+$	$+$
Sign of $2x - 1$	$-$	$-$	$+$
Sign of $(x+1)(2x-1)$	$+$	$-$	$+$

From the table, the solution set is $\left\{x \mid x \leq -1 \text{ or } \frac{1}{2} \leq x\right\}$.

Interval: $(-\infty, -1] \cup \left[\frac{1}{2}, \infty\right)$.

Graph:

43. $3x^2 - 3x < 2x^2 + 4 \Leftrightarrow x^2 - 3x - 4 < 0 \Leftrightarrow (x+1)(x-4) < 0$. The expression on the left of the inequality changes sign where $x = -1$ and where $x = 4$. Thus we must check the intervals in the following table.

Interval	$(-\infty, -1)$	$(-1, 4)$	$(4, \infty)$
Sign of $x + 1$	$-$	$+$	$+$
Sign of $x - 4$	$-$	$-$	$+$
Sign of $(x+1)(x-4)$	$+$	$-$	$+$

From the table, the solution set is $\{x \mid -1 < x < 4\}$. Interval: $(-1, 4)$.

Graph:

45. $x^2 > 3(x+6) \Leftrightarrow x^2 - 3x - 18 > 0 \Leftrightarrow (x+3)(x-6) > 0$. The expression on the left of the inequality changes sign where $x = 6$ and where $x = -3$. Thus we must check the intervals in the following table.

Interval	$(-\infty, -3)$	$(-3, 6)$	$(6, \infty)$
Sign of $x + 3$	$-$	$+$	$+$
Sign of $x - 6$	$-$	$-$	$+$
Sign of $(x+3)(x-6)$	$+$	$-$	$+$

From the table, the solution set is $\{x \mid x < -3 \text{ or } 6 < x\}$.

Interval: $(-\infty, -3) \cup (6, \infty)$.

Graph:

47. $x^2 < 4 \Leftrightarrow x^2 - 4 < 0 \Leftrightarrow (x+2)(x-2) < 0$. The expression on the left of the inequality changes sign where $x = -2$ and where $x = 2$. Thus we must check the intervals in the following table.

Interval	$(-\infty, -2)$	$(-2, 2)$	$(2, \infty)$
Sign of $x + 2$	$-$	$+$	$+$
Sign of $x - 2$	$-$	$-$	$+$
Sign of $(x+2)(x-2)$	$+$	$-$	$+$

From the table, the solution set is $\{x \mid -2 < x < 2\}$. Interval: $(-2, 2)$.

Graph:

49. $(x+2)(x-1)(x-3) \leq 0$. The expression on the left of the inequality changes sign when $x = -2$, $x = 1$, and $x = 3$. Thus we must check the intervals in the following table.

Interval	$(-\infty, -2)$	$(-2, 1)$	$(1, 3)$	$(3, \infty)$
Sign of $x + 2$	$-$	$+$	$+$	$+$
Sign of $x - 1$	$-$	$-$	$+$	$+$
Sign of $x - 3$	$-$	$-$	$-$	$+$
Sign of $(x+2)(x-1)(x-3)$	$-$	$+$	$-$	$+$

From the table, the solution set is $\{x \mid x \leq -2 \text{ or } 1 \leq x \leq 3\}$.   Interval: $(-\infty, -2] \cup [1, 3]$.

Graph:

51. $(x - 4)(x + 2)^2 < 0$. Note that $(x + 2)^2 > 0$ for all $x \neq -2$, so the expression on the left of the original inequality changes sign only when $x = 4$. We check the intervals in the following table.

Interval	$(-\infty, -2)$	$(-2, 4)$	$(4, \infty)$
Sign of $x - 4$	$-$	$-$	$+$
Sign of $(x + 2)^2$	$+$	$+$	$+$
Sign of $(x - 4)(x + 2)^2$	$-$	$-$	$+$

From the table, the solution set is $\{x \mid x \neq -2 \text{ and } x < 4\}$. We exclude the endpoint -2 since the original expression cannot be 0. Interval: $(-\infty, -2) \cup (-2, 4)$.

Graph:

53. $(x - 2)^2 (x - 3)(x + 1) \leq 0$. Note that $(x - 2)^2 \geq 0$ for all x, so the expression on the left of the original inequality changes sign only when $x = -1$ and $x = 3$. We check the intervals in the following table.

Interval	$(-\infty, -1)$	$(-1, 2)$	$(2, 3)$	$(3, \infty)$
Sign of $(x - 2)^2$	$+$	$+$	$+$	$+$
Sign of $x - 3$	$-$	$-$	$-$	$+$
Sign of $x + 1$	$-$	$+$	$+$	$+$
Sign of $(x - 2)^2 (x - 3)(x + 1)$	$+$	$-$	$-$	$+$

From the table, the solution set is $\{x \mid -1 \leq x \leq 3\}$. Interval: $[-1, 3]$. Graph:

55. $x^3 - 4x > 0 \Leftrightarrow x\left(x^2 - 4\right) > 0 \Leftrightarrow x(x + 2)(x - 2) > 0$. The expression on the left of the inequality changes sign where $x = 0$, $x = -2$ and where $x = 4$. Thus we must check the intervals in the following table.

Interval	$(-\infty, -2)$	$(-2, 0)$	$(0, 2)$	$(2, \infty)$
Sign of x	$-$	$-$	$+$	$+$
Sign of $x + 2$	$-$	$+$	$+$	$+$
Sign of $x - 2$	$-$	$-$	$-$	$+$
Sign of $x(x + 2)(x - 2)$	$-$	$+$	$-$	$+$

From the table, the solution set is $\{x \mid -2 < x < 0 \text{ or } x > 2\}$. Interval: $(-2, 0) \cup (2, \infty)$. Graph:

57. $\dfrac{x - 3}{x + 1} \geq 0$. The expression on the left of the inequality changes sign where $x = -1$ and where $x = 3$. Thus we must check the intervals in the following table.

Interval	$(-\infty, -1)$	$(-1, 3)$	$(3, \infty)$
Sign of $x + 1$	$-$	$+$	$+$
Sign of $x - 3$	$-$	$-$	$+$
Sign of $\dfrac{x - 3}{x + 1}$	$+$	$-$	$+$

From the table, the solution set is $\{x \mid x < -1 \text{ or } x \leq 3\}$. Since the denominator cannot equal 0 we must have $x \neq -1$. Interval: $(-\infty, -1) \cup [3, \infty)$.

Graph:

59. $\dfrac{4x}{2x+3} > 2 \Leftrightarrow \dfrac{4x}{2x+3} - 2 > 0 \Leftrightarrow \dfrac{4x}{2x+3} - \dfrac{2(2x+3)}{2x+3} > 0 \Leftrightarrow \dfrac{-6}{2x+3} > 0$. The expression on the left of the inequality changes sign where $x = -\frac{3}{2}$. Thus we must check the intervals in the following table.

Interval	$\left(-\infty, -\frac{3}{2}\right)$	$\left(-\frac{3}{2}, \infty\right)$
Sign of -6	$-$	$-$
Sign of $2x+3$	$-$	$+$
Sign of $\dfrac{-6}{2x+3}$	$+$	$-$

From the table, the solution set is $\left\{x \mid x < -\frac{3}{2}\right\}$.

Interval: $\left(-\infty, -\frac{3}{2}\right)$.

Graph:

$-\frac{3}{2}$

61. $\dfrac{2x+1}{x-5} \le 3 \Leftrightarrow \dfrac{2x+1}{x-5} - 3 \le 0 \Leftrightarrow \dfrac{2x+1}{x-5} - \dfrac{3(x-5)}{x-5} \le 0 \Leftrightarrow \dfrac{-x+16}{x-5} \le 0$. The expression on the left of the inequality changes sign where $x = 16$ and where $x = 5$. Thus we must check the intervals in the following table.

Interval	$(-\infty, 5)$	$(5, 16)$	$(16, \infty)$
Sign of $-x+16$	$+$	$+$	$-$
Sign of $x-5$	$-$	$+$	$+$
Sign of $\dfrac{-x+16}{x-5}$	$-$	$+$	$-$

From the table, the solution set is $\{x \mid x < 5 \text{ or } x \ge 16\}$. Since the denominator cannot equal 0, we must have $x \ne 5$.

Interval: $(-\infty, 5) \cup [16, \infty)$.

Graph:

$5 \qquad 16$

63. $\dfrac{4}{x} < x \Leftrightarrow \dfrac{4}{x} - x < 0 \Leftrightarrow \dfrac{4}{x} - \dfrac{x \cdot x}{x} < 0 \Leftrightarrow \dfrac{4-x^2}{x} < 0 \Leftrightarrow \dfrac{(2-x)(2+x)}{x} < 0$. The expression on the left of the inequality changes sign where $x = 0$, where $x = -2$, and where $x = 2$. Thus we must check the intervals in the following table.

Interval	$(-\infty, -2)$	$(-2, 0)$	$(0, 2)$	$(2, \infty)$
Sign of $2+x$	$-$	$+$	$+$	$+$
Sign of x	$-$	$-$	$+$	$+$
Sign of $2-x$	$+$	$+$	$+$	$-$
Sign of $\dfrac{(2-x)(2+x)}{x}$	$+$	$-$	$+$	$-$

From the table, the solution set is $\{x \mid -2 < x < 0 \text{ or } 2 < x\}$. Interval: $(-2, 0) \cup (2, \infty)$. Graph:

$-2 \qquad 0 \qquad 2$

65. $1 + \dfrac{2}{x+1} \le \dfrac{2}{x} \Leftrightarrow 1 + \dfrac{2}{x+1} - \dfrac{2}{x} \le 0 \Leftrightarrow \dfrac{x(x+1)}{x(x+1)} + \dfrac{2x}{x(x+1)} - \dfrac{2(x+1)}{x(x+1)} \le 0 \Leftrightarrow \dfrac{x^2+x+2x-2x-2}{x(x+1)} \le 0 \Leftrightarrow$

$\dfrac{x^2+x-2}{x(x+1)} \le 0 \Leftrightarrow \dfrac{(x+2)(x-1)}{x(x+1)} \le 0$. The expression on the left of the inequality changes sign where $x = -2$, where $x = -1$, where $x = 0$, and where $x = 1$. Thus we must check the intervals in the following table.

Interval	$(-\infty, -2)$	$(-2, -1)$	$(-1, 0)$	$(0, 1)$	$(1, \infty)$
Sign of $x + 2$	$-$	$+$	$+$	$+$	$+$
Sign of $x - 1$	$-$	$-$	$-$	$-$	$+$
Sign of x	$-$	$-$	$-$	$+$	$+$
Sign of $x + 1$	$-$	$-$	$+$	$+$	$+$
Sign of $\dfrac{(x+2)(x-1)}{x(x+1)}$	$+$	$-$	$+$	$-$	$+$

Since $x = -1$ and $x = 0$ yield undefined expressions, we cannot include them in the solution. From the table, the solution

set is $\{x \mid -2 \le x < -1 \text{ or } 0 < x \le 1\}$. Interval: $[-2, -1) \cup (0, 1]$. Graph:

67. $\dfrac{6}{x-1} - \dfrac{6}{x} \ge 1 \Leftrightarrow \dfrac{6}{x-1} - \dfrac{6}{x} - 1 \ge 0 \Leftrightarrow \dfrac{6x}{x(x-1)} - \dfrac{6(x-1)}{x(x-1)} - \dfrac{x(x-1)}{x(x-1)} \ge 0 \Leftrightarrow$

$\dfrac{6x - 6x + 6 - x^2 + x}{x(x-1)} \ge 0 \Leftrightarrow \dfrac{-x^2 + x + 6}{x(x-1)} \ge 0 \Leftrightarrow \dfrac{(-x+3)(x+2)}{x(x-1)} \ge 0$. The

expression on the left of the inequality changes sign where $x = 3$, where $x = -2$, where $x = 0$, and where $x = 1$. Thus we must check the intervals in the following table.

Interval	$(-\infty, -2)$	$(-2, 0)$	$(0, 1)$	$(1, 3)$	$(3, \infty)$
Sign of $-x + 3$	$+$	$+$	$+$	$+$	$-$
Sign of $x + 2$	$-$	$+$	$+$	$+$	$+$
Sign of x	$-$	$-$	$+$	$+$	$+$
Sign of $x - 1$	$-$	$-$	$-$	$+$	$+$
Sign of $\dfrac{(-x+3)(x+2)}{x(x-1)}$	$-$	$+$	$-$	$+$	$-$

From the table, the solution set is $\{x \mid -2 \le x < 0 \text{ or } 1 < x \le 3\}$. The points $x = 0$ and $x = 1$ are excluded from the

solution set because they make the denominator zero. Interval: $[-2, 0) \cup (1, 3]$. Graph:

69. $\dfrac{x+2}{x+3} < \dfrac{x-1}{x-2} \Leftrightarrow \dfrac{x+2}{x+3} - \dfrac{x-1}{x-2} < 0 \Leftrightarrow \dfrac{(x+2)(x-2)}{(x+3)(x-2)} - \dfrac{(x-1)(x+3)}{(x-2)(x+3)} < 0 \Leftrightarrow$

$\dfrac{x^2 - 4 - x^2 - 2x + 3}{(x+3)(x-2)} < 0 \Leftrightarrow \dfrac{-2x - 1}{(x+3)(x-2)} < 0$. The expression on the left of the inequality

changes sign where $x = -\frac{1}{2}$, where $x = -3$, and where $x = 2$. Thus we must check the intervals in the following table.

Interval	$(-\infty, -3)$	$\left(-3, -\frac{1}{2}\right)$	$\left(-\frac{1}{2}, 2\right)$	$(2, \infty)$
Sign of $-2x - 1$	$+$	$+$	$-$	$-$
Sign of $x + 3$	$-$	$+$	$+$	$+$
Sign of $x - 2$	$-$	$-$	$-$	$+$
Sign of $\dfrac{-2x-1}{(x+3)(x-2)}$	$+$	$-$	$+$	$-$

From the table, the solution set is $\left\{x \mid -3 < x < -\frac{1}{2} \text{ or } 2 < x\right\}$. Interval: $\left(-3, -\frac{1}{2}\right) \cup (2, \infty)$.

Graph:
$$\underset{-3 \qquad -\frac{1}{2} \qquad 2}{\circ\qquad\circ\qquad\circ\longrightarrow}$$

71. $x^4 > x^2 \Leftrightarrow x^4 - x^2 > 0 \Leftrightarrow x^2\left(x^2 - 1\right) > 0 \Leftrightarrow x^2(x - 1)(x + 1) > 0$. The expression on the left of the inequality changes sign where $x = 0$, where $x = 1$, and where $x = -1$. Thus we must check the intervals in the following table.

Interval	$(-\infty, -1)$	$(-1, 0)$	$(0, 1)$	$(1, \infty)$
Sign of x^2	+	+	+	+
Sign of $x - 1$	−	−	−	+
Sign of $x + 1$	−	+	+	+
Sign of $x^2(x - 1)(x + 1)$	+	−	−	+

From the table, the solution set is $\{x \mid x < -1 \text{ or } 1 < x\}$. Interval: $(-\infty, -1) \cup (1, \infty)$. Graph:
$$\underset{-1 \qquad\qquad 1}{\circ\qquad\qquad\circ\longrightarrow}$$

73. $|x| \le 4 \Leftrightarrow -4 \le x \le 4$. Interval: $[-4, 4]$.

Graph:
$$\underset{-4 \qquad\qquad 4}{\bullet\!\!-\!\!\!-\!\!\!-\!\!\bullet\longrightarrow}$$
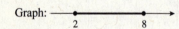

75. $|2x| > 7$ is equivalent to $2x > 7 \Leftrightarrow x > \frac{7}{2}$; or $2x < 7 \Leftrightarrow$
$x < -\frac{7}{2}$.

Interval: $\left(-\infty, -\frac{7}{2}\right) \cup \left(\frac{7}{2}, \infty\right)$.

Graph:
$$\underset{-\frac{7}{2} \qquad\qquad \frac{7}{2}}{\circ\qquad\qquad\circ\longrightarrow}$$

77. $|x - 5| \le 3 \Leftrightarrow -3 \le x - 5 \le 3 \Leftrightarrow 2 \le x \le 8$.
Interval: $[2, 8]$.

Graph:
$$\underset{2 \qquad\qquad 8}{\bullet\!\!-\!\!\!-\!\!\!-\!\!\bullet\longrightarrow}$$
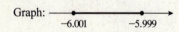

79. $|2x - 3| \le 0.4 \Leftrightarrow -0.4 \le 2x - 3 \le 0.4 \Leftrightarrow$
$2.6 \le 2x \le 3.4 \Leftrightarrow 1.3 \le x \le 1.7$.
Interval: $[1.3, 1.7]$.

Graph:
$$\underset{1.3 \qquad\qquad 1.7}{\bullet\!\!-\!\!\!-\!\!\!-\!\!\bullet\longrightarrow}$$

81. $|3x - 2| \ge 5 \Leftrightarrow \left|x - \frac{2}{3}\right| \ge \frac{5}{3}$, so either $x - \frac{2}{3} \ge \frac{5}{3} \Leftrightarrow$
$x \ge \frac{7}{3}$ or $x - \frac{2}{3} \le -\frac{5}{3} \Leftrightarrow x \le -1$.
Interval: $(-\infty, -1] \cup \left[\frac{7}{3}, \infty\right)$.

Graph:
$$\underset{-1 \qquad\qquad \frac{7}{3}}{\bullet\qquad\qquad\bullet\longrightarrow}$$

83. $\left|\dfrac{x - 2}{3}\right| < 2 \Leftrightarrow -2 < \dfrac{x - 2}{3} < 2 \Leftrightarrow -6 < x - 2 < 6 \Leftrightarrow$
$-4 < x < 8$.
Interval: $(-4, 8)$.

Graph:
$$\underset{-4 \qquad\qquad 8}{\circ\qquad\qquad\circ\longrightarrow}$$

85. $|x + 6| < 0.001 \Leftrightarrow -0.001 < x + 6 < 0.001 \Leftrightarrow$
$-6.001 < x < -5.999$.
Interval: $(-6.001, -5.999)$.

Graph:
$$\underset{-6.001 \qquad\qquad -5.999}{\bullet\!\!-\!\!\!-\!\!\!-\!\!\bullet\longrightarrow}$$
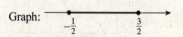

87. $8 - |2x - 1| \ge 6 \Leftrightarrow -|2x - 1| \ge -2 \Leftrightarrow |2x - 1| \le 2 \Leftrightarrow$
$-2 \le 2x - 1 \le 2 \Leftrightarrow -1 \le 2x \le 3 \Leftrightarrow -\frac{1}{2} \le x \le \frac{3}{2}$.
Interval: $\left[-\frac{1}{2}, \frac{3}{2}\right]$.

Graph:
$$\underset{-\frac{1}{2} \qquad\qquad \frac{3}{2}}{\bullet\!\!-\!\!\!-\!\!\!-\!\!\bullet\longrightarrow}$$

89. $|x| < 3$ **91.** $|x - 7| \ge 5$ **93.** $|x| \le 2$ **95.** $|x| > 3$ **97.** $|x - 1| \le 3$

99. For $\sqrt{16 - 9x^2}$ to be defined as a real number we must have $16 - 9x^2 \geq 0 \Leftrightarrow (4 - 3x)(4 + 3x) \geq 0$. The expression in the inequality changes sign at $x = \frac{4}{3}$ and $x = -\frac{4}{3}$.

Interval	$\left(-\infty, -\frac{4}{3}\right)$	$\left(-\frac{4}{3}, \frac{4}{3}\right)$	$\left(\frac{4}{3}, \infty\right)$
Sign of $4 - 3x$	$+$	$+$	$-$
Sign of $4 + 3x$	$-$	$+$	$+$
Sign of $(4 - 3x)(4 + 3x)$	$-$	$+$	$-$

Thus $-\frac{4}{3} \leq x \leq \frac{4}{3}$.

101. For $\left(\dfrac{1}{x^2 - 5x - 14}\right)^{1/2}$ to be defined as a real number we must have $x^2 - 5x - 14 > 0 \Leftrightarrow (x - 7)(x + 2) > 0$. The expression in the inequality changes sign at $x = 7$ and $x = -2$.

Interval	$(-\infty, -2)$	$(-2, 7)$	$(7, \infty)$
Sign of $x - 7$	$-$	$-$	$+$
Sign of $x + 2$	$-$	$+$	$+$
Sign of $(x - 7)(x + 2)$	$+$	$-$	$+$

Thus $x < -2$ or $7 < x$, and the solution set is $(-\infty, -2) \cup (7, \infty)$.

103. (a) $a(bx - c) \geq bc$ (where $a, b, c > 0$) $\Leftrightarrow bx - c \geq \dfrac{bc}{a} \Leftrightarrow bx \geq \dfrac{bc}{a} + c \Leftrightarrow x \geq \dfrac{1}{b}\left(\dfrac{bc}{a} + c\right) = \dfrac{c}{a} + \dfrac{c}{b} \Leftrightarrow x \geq \dfrac{c}{a} + \dfrac{c}{b}$.

(b) We have $a \leq bx + c < 2a$, where $a, b, c > 0 \Leftrightarrow a - c \leq bx < 2a - c \Leftrightarrow \dfrac{a - c}{b} \leq x < \dfrac{2a - c}{b}$.

105. Inserting the relationship $C = \frac{5}{9}(F - 32)$, we have $20 \leq C \leq 30 \Leftrightarrow 20 \leq \frac{5}{9}(F - 32) \leq 30 \Leftrightarrow 36 \leq F - 32 \leq 54 \Leftrightarrow 68 \leq F \leq 86$.

107. Let x be the average number of miles driven per day. Each day the cost of Plan A is $30 + 0.10x$, and the cost of Plan B is 50. Plan B saves money when $50 < 30 + 0.10x \Leftrightarrow 20 < 0.1x \Leftrightarrow 200 < x$. So Plan B saves money when you average more than 200 miles a day.

109. We need to solve $6400 \leq 0.35m + 2200 \leq 7100$ for m. So $6400 \leq 0.35m + 2200 \leq 7100 \Leftrightarrow 4200 \leq 0.35m \leq 4900 \Leftrightarrow 12{,}000 \leq m \leq 14{,}000$. She plans on driving between 12,000 and 14,000 miles.

111. (a) Let x be the number of \$3 increases. Then the number of seats sold is $120 - x$. So $P = 200 + 3x \Leftrightarrow 3x = P - 200 \Leftrightarrow x = \frac{1}{3}(P - 200)$. Substituting for x we have that the number of seats sold is $120 - x = 120 - \frac{1}{3}(P - 200) = -\frac{1}{3}P + \frac{560}{3}$.

(b) $90 \leq -\frac{1}{3}P + \frac{560}{3} \leq 115 \Leftrightarrow 270 \leq 360 - P + 200 \leq 345 \Leftrightarrow 270 \leq -P + 560 \leq 345 \Leftrightarrow -290 \leq -P \leq -215 \Leftrightarrow 290 \geq P \geq 215$. Putting this into standard order, we have $215 \leq P \leq 290$. So the ticket prices are between \$215 and \$290.

113. $0.0004 \leq \dfrac{4{,}000{,}000}{d^2} \leq 0.01$. Since $d^2 \geq 0$ and $d \neq 0$, we can multiply each expression by d^2 to obtain $0.0004d^2 \leq 4{,}000{,}000 \leq 0.01d^2$. Solving each pair, we have $0.0004d^2 \leq 4{,}000{,}000 \Leftrightarrow d^2 \leq 10{,}000{,}000{,}000 \Rightarrow d \leq 100{,}000$ (recall that d represents distance, so it is always nonnegative). Solving $4{,}000{,}000 \leq 0.01d^2 \Leftrightarrow 400{,}000{,}000 \leq d^2 \Rightarrow 20{,}000 \leq d$. Putting these together, we have $20{,}000 \leq d \leq 100{,}000$.

115. $128 + 16t - 16t^2 \geq 32 \Leftrightarrow -16t^2 + 16t + 96 \geq 0 \Leftrightarrow -16\left(t^2 - t - 6\right) \geq 0 \Leftrightarrow -16(t - 3)(t + 2) \geq 0$. The expression on the left of the inequality changes sign at $x = -2$, at $t = 3$, and at $t = -2$. However, $t \geq 0$, so the only endpoint is $t = 3$.

Interval	$(0, 3)$	$(3, \infty)$
Sign of -16	$-$	$-$
Sign of $t - 3$	$-$	$+$
Sign of $t + 2$	$+$	$+$
Sign of $-16(t - 3)(t + 2)$	$+$	$-$

So $0 \leq t \leq 3$.

117. $240 \geq v + \dfrac{v^2}{20} \Leftrightarrow \frac{1}{20}v^2 + v - 240 \leq 0 \Leftrightarrow \left(\frac{1}{20}v - 3\right)(v + 80) \leq 0$. The expression in the inequality changes sign at $v = 60$ and $v = -80$. However, since v represents the speed, we must have $v \geq 0$.

Interval	$(0, 60)$	$(60, \infty)$
Sign of $\frac{1}{20}v - 3$	$-$	$+$
Sign of $v + 80$	$+$	$+$
Sign of $\left(\frac{1}{20}v - 3\right)(v + 80)$	$-$	$+$

So Kerry must drive between 0 and 60 mi/h.

119. Let x be the length of the garden and w its width. Using the fact that the perimeter is 120 ft, we must have $2x + 2w = 120$ $\Leftrightarrow w = 60 - x$. Now since the area must be at least 800 ft^2, we have $800 < x(60 - x) \Leftrightarrow 800 < 60x - x^2 \Leftrightarrow x^2 - 60x + 800 < 0 \Leftrightarrow (x - 20)(x - 40) < 0$. The expression in the inequality changes sign at $x = 20$ and $x = 40$. However, since x represents length, we must have $x > 0$.

Interval	$(0, 20)$	$(20, 40)$	$(40, \infty)$
Sign of $x - 20$	$-$	$+$	$+$
Sign of $x - 40$	$-$	$-$	$+$
Sign of $(x - 20)(x - 40)$	$+$	$-$	$+$

The length of the garden should be between 20 and 40 feet.

121. $\left|\dfrac{h - 68.2}{2.9}\right| \leq 2 \Leftrightarrow -2 \leq \dfrac{h - 68.2}{2.9} \leq 2 \Leftrightarrow -5.8 \leq h - 68.2 \leq 5.8 \Leftrightarrow 62.4 \leq h \leq 74.0$. Thus 95% of the adult males are between 62.4 in. and 74.0 in.

123. The rule we want to apply here is "$a < b \Rightarrow ac < bc$ if $c > 0$ and $a < b \Rightarrow ac > bc$ if $c < 0$". Thus we cannot simply multiply by x, since we don't yet know if x is positive or negative, so in solving $1 < \dfrac{3}{x}$, we must consider two cases.

Case 1: $x > 0$ Multiplying both sides by x, we have $x < 3$. Together with our initial condition, we have $0 < x < 3$.

Case 2: $x < 0$ Multiplying both sides by x, we have $x > 3$. But $x < 0$ and $x > 3$ have no elements in common, so this gives no additional solution. Hence, the only solutions are $0 < x < 3$.

1.8   COORDINATE GEOMETRY

1. The point that is 3 units to the right of the y-axis and 5 units below the x-axis has coordinates $(3, -5)$.

3. The point midway between (a, b) and (c, d) is $\left(\dfrac{a+c}{2}, \dfrac{b+d}{2}\right)$. So the point midway between $(1, 2)$ and $(7, 10)$ is

$\left(\dfrac{1+7}{2}, \dfrac{2+10}{2}\right) = \left(\dfrac{8}{2}, \dfrac{12}{2}\right) = (4, 6)$.

5. (a) To find the x-intercept(s) of the graph of an equation we set y equal to 0 in the equation and solve for x: $2(0) = x + 1$
$\Leftrightarrow x = -1$, so the x-intercept of $2y = x + 1$ is -1.

(b) To find the y-intercept(s) of the graph of an equation we set x equal to 0 in the equation and solve for y: $2y = 0 + 1 \Leftrightarrow$
$y = \frac{1}{2}$, so the y-intercept of $2y = x + 1$ is $\frac{1}{2}$.

7.

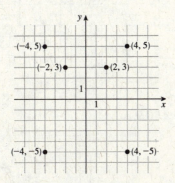

9. The two points are $(0, 2)$ and $(3, 0)$.

(a) $d = \sqrt{(3-0)^2 + (0-(-2))^2} = \sqrt{3^2 + 2^2} = \sqrt{9+4} = \sqrt{13}$

(b) midpoint: $\left(\dfrac{3+0}{2}, \dfrac{0+2}{2}\right) = \left(\dfrac{3}{2}, 1\right)$

11. The two points are $(-3, 3)$ and $(5, -3)$.

(a) $d = \sqrt{(-3-5)^2 + (3-(-3))^2} = \sqrt{(-8)^2 + 6^2} = \sqrt{64+36} = \sqrt{100} = 10$

(b) midpoint: $\left(\dfrac{-3+5}{2}, \dfrac{3+(-3)}{2}\right) = (1, 0)$

13. (a)

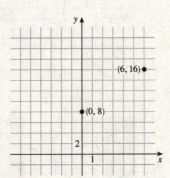

(b) $d = \sqrt{(0-6)^2 + (8-16)^2}$

$= \sqrt{(-6)^2 + (-8)^2} = \sqrt{100} = 10$

(c) midpoint: $\left(\dfrac{0+6}{2}, \dfrac{8+16}{2}\right) = (3, 12)$

15. (a)

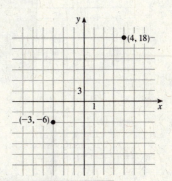

(b) $d = \sqrt{(-3-4)^2 + (-6-18)^2}$

$= \sqrt{(-7)^2 + (-24)^2} = \sqrt{49+576} = \sqrt{625} = 25$

(c) midpoint: $\left(\dfrac{-3+4}{2}, \dfrac{-6+18}{2}\right) = \left(\dfrac{1}{2}, 6\right)$

17. (a)

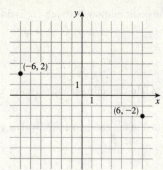

(b) $d = \sqrt{(6 - (-6))^2 + (-2 - 2)^2} = \sqrt{12^2 + (-4)^2}$

$= \sqrt{144 + 16} = \sqrt{160} = 4\sqrt{10}$

(c) midpoint: $\left(\dfrac{6 - 6}{2}, \dfrac{-2 + 2}{2} \right) = (0, 0)$

19. $d(A, B) = \sqrt{(1 - 5)^2 + (3 - 3)^2} = \sqrt{(-4)^2} = 4.$

$d(A, C) = \sqrt{(1 - 1)^2 + (3 - (-3))^2} = \sqrt{(6)^2} = 6.$ So the area is $4 \cdot 6 = 24.$

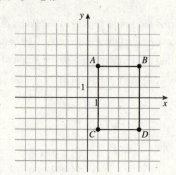

21. From the graph, the quadrilateral $ABCD$ has a pair of parallel sides, so $ABCD$ is a trapezoid. The area is $\left(\dfrac{b_1 + b_2}{2} \right) h$. From the graph we see that

$b_1 = d(A, B) = \sqrt{(1 - 5)^2 + (0 - 0)^2} = \sqrt{4^2} = 4;$

$b_2 = d(C, D) = \sqrt{(4 - 2)^2 + (3 - 3)^2} = \sqrt{2^2} = 2;$ and h is the difference in

y-coordinates is $|3 - 0| = 3$. Thus the area of the trapezoid is $\left(\dfrac{4 + 2}{2} \right) 3 = 9.$

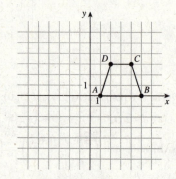

23.

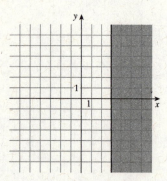

25.

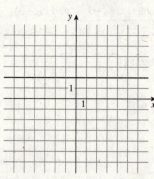

27.

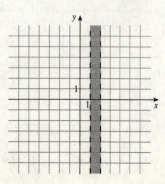

29.

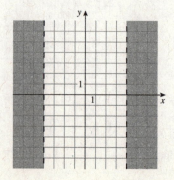

31.

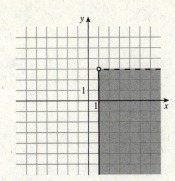

33. $d\,(0, A) = \sqrt{(6-0)^2 + (7-0)^2} = \sqrt{6^2 + 7^2} = \sqrt{36+49} = \sqrt{85}.$

$d\,(0, B) = \sqrt{(-5-0)^2 + (8-0)^2} = \sqrt{(-5)^2 + 8^2} = \sqrt{25+64} = \sqrt{89}.$

Thus point $A\,(6, 7)$ is closer to the origin.

35. $d\,(P, R) = \sqrt{(-1-3)^2 + (-1-1)^2} = \sqrt{(-4)^2 + (-2)^2} = \sqrt{16+4} = \sqrt{20} = 2\sqrt{5}.$

$d\,(Q, R) = \sqrt{(-1-(-1))^2 + (-1-3)^2} = \sqrt{0 + (-4)^2} = \sqrt{16} = 4.$ Thus point $Q\,(-1, 3)$ is closer to point R.

37. Since we do not know which pair are isosceles, we find the length of all three sides.

$d\,(A, B) = \sqrt{(-3-0)^2 + (-1-2)^2} = \sqrt{(-3)^2 + (-3)^2} = \sqrt{9+9} = \sqrt{18} = 3\sqrt{2}.$

$d\,(C, B) = \sqrt{(-3-(-4))^2 + (-1-3)^2} = \sqrt{1^2 + (-4)^2} = \sqrt{1+16} = \sqrt{17}.$

$d\,(A, C) = \sqrt{(0-(-4))^2 + (2-3)^2} = \sqrt{4^2 + (-1)^2} = \sqrt{16+1} = \sqrt{17}.$ So sides AC and CB have the same length.

39. (a) Here we have $A = (2, 2)$, $B = (3, -1)$, and $C = (-3, -3)$. So

$d\,(A, B) = \sqrt{(3-2)^2 + (-1-2)^2} = \sqrt{1^2 + (-3)^2} = \sqrt{1+9} = \sqrt{10};$

$d\,(C, B) = \sqrt{(3-(-3))^2 + (-1-(-3))^2} = \sqrt{6^2 + 2^2} = \sqrt{36+4} = \sqrt{40} = 2\sqrt{10};$

$d\,(A, C) = \sqrt{(-3-2)^2 + (-3-2)^2} = \sqrt{(-5)^2 + (-5)^2} = \sqrt{25+25} = \sqrt{50} = 5\sqrt{2}.$

Since $[d\,(A, B)]^2 + [d\,(C, B)]^2 = [d\,(A, C)]^2$, we conclude that the triangle is a right triangle.

(b) The area of the triangle is $\frac{1}{2} \cdot d\,(C, B) \cdot d\,(A, B) = \frac{1}{2} \cdot \sqrt{10} \cdot 2\sqrt{10} = 10.$

41. We show that all sides are the same length (its a rhombus) and then show that the diagonals are equal. Here we have
$A = (-2, 9)$, $B = (4, 6)$, $C = (1, 0)$, and $D = (-5, 3)$. So

$d\,(A, B) = \sqrt{(4-(-2))^2 + (6-9)^2} = \sqrt{6^2 + (-3)^2} = \sqrt{36+9} = \sqrt{45};$

$d\,(B, C) = \sqrt{(1-4)^2 + (0-6)^2} = \sqrt{(-3)^2 + (-6)^2} = \sqrt{9+36} = \sqrt{45};$

$d\,(C, D) = \sqrt{(-5-1)^2 + (3-0)^2} = \sqrt{(-6)^2 + (-3)^2} = \sqrt{36+9} = \sqrt{45};$

$d\,(D, A) = \sqrt{(-2-(-5))^2 + (9-3)^2} = \sqrt{3^2 + 6^2} = \sqrt{9+36} = \sqrt{45}.$ So the points form a

rhombus. Also $d\,(A, C) = \sqrt{(1-(-2))^2 + (0-9)^2} = \sqrt{3^2 + (-9)^2} = \sqrt{9+81} = \sqrt{90} = 3\sqrt{10},$

and $d\,(B, D) = \sqrt{(-5-4)^2 + (3-6)^2} = \sqrt{(-9)^2 + (-3)^2} = \sqrt{81+9} = \sqrt{90} = 3\sqrt{10}.$ Since the diagonals are equal,
the rhombus is a square.

43. Let $P = (0, y)$ be such a point. Setting the distances equal we get

$$\sqrt{(0-5)^2 + (y - (-5))^2} = \sqrt{(0-1)^2 + (y-1)^2} \Leftrightarrow$$

$$\sqrt{25 + y^2 + 10y + 25} = \sqrt{1 + y^2 - 2y + 1} \Rightarrow y^2 + 10y + 50 = y^2 - 2y + 2 \Leftrightarrow 12y = -48 \Leftrightarrow y = -4.$$ Thus, the point
is $P = (0, -4)$. Check:

$$\sqrt{(0-5)^2 + (-4 - (-5))^2} = \sqrt{(-5)^2 + 1^2} = \sqrt{25 + 1} = \sqrt{26};$$

$$\sqrt{(0-1)^2 + (-4-1)^2} = \sqrt{(-1)^2 + (-5)^2} = \sqrt{25+1} = \sqrt{26}.$$

45. As indicated by Example 3, we must find a point $S(x_1, y_1)$ such that the midpoints
of PR and of QS are the same. Thus

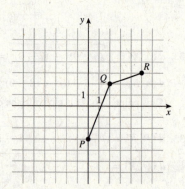

$$\left(\frac{4 + (-1)}{2}, \frac{2 + (-4)}{2} \right) = \left(\frac{x_1 + 1}{2}, \frac{y_1 + 1}{2} \right).$$ Setting the x-coordinates equal,

we get $\dfrac{4 + (-1)}{2} = \dfrac{x_1 + 1}{2} \Leftrightarrow 4 - 1 = x_1 + 1 \Leftrightarrow x_1 = 2.$ Setting the

y-coordinates equal, we get $\dfrac{2 + (-4)}{2} = \dfrac{y_1 + 1}{2} \Leftrightarrow 2 - 4 = y_1 + 1 \Leftrightarrow y_1 = -3.$

Thus $S = (2, -3)$.

47. (a)

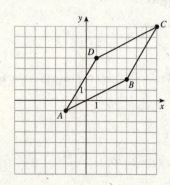

(b) The midpoint of AC is $\left(\dfrac{-2 + 7}{2}, \dfrac{-1 + 7}{2} \right) = \left(\dfrac{5}{2}, 3 \right)$, the midpoint

of BD is $\left(\dfrac{4 + 1}{2}, \dfrac{2 + 4}{2} \right) = \left(\dfrac{5}{2}, 3 \right)$.

(c) Since the they have the same midpoint, we conclude that the
diagonals bisect each other.

49. $(0, 0): 0 - 2(0) - 1 \stackrel{?}{=} 0 \Leftrightarrow -1 \stackrel{?}{=} 0.$ No.

$(1, 0): 1 - 2(0) - 1 \stackrel{?}{=} 0 \Leftrightarrow -1 + 1 \stackrel{?}{=} 0.$ Yes.

$(-1, -1): (-1) - 2(-1) - 1 \stackrel{?}{=} 0 \Leftrightarrow -1 + 2 - 1 \stackrel{?}{=} 0.$ Yes.

So $(1, 0)$ and $(-1, -1)$ are points on the graph of this equation.

51. $(0, -2): (0)^2 + (0)(-2) + (-2)^2 \stackrel{?}{=} 4 \Leftrightarrow 0 + 0 + 4 \stackrel{?}{=} 4.$ Yes.

$(1, -2): (1)^2 + (1)(-2) + (-2)^2 \stackrel{?}{=} 4 \Leftrightarrow 1 - 2 + 4 \stackrel{?}{=} 4.$ No.

$(2, -2): (2)^2 + (2)(-2) + (-2)^2 \stackrel{?}{=} 4 \Leftrightarrow 4 - 4 + 4 \stackrel{?}{=} 4.$ Yes.

So $(0, -2)$ and $(2, -2)$ are points on the graph of this equation.

53. To find x-intercepts, set $y = 0$. This gives $0 = 4x - x^2 \Leftrightarrow 0 = x(4 - x) \Leftrightarrow 0 = x$ or $x = 4$, so the x-intercept are 0 and 4.

To find y-intercepts, set $x = 0$. This gives $y = 4(0) - 0^2 \Leftrightarrow y = 0$, so the y-intercept is 0.

55. To find x-intercepts, set $y = 0$. This gives $x^4 + 0^2 - x(0) = 16 \Leftrightarrow x^4 = 16 \Leftrightarrow x = \pm 2$. So the x-intercept are -2 and 2.

To find y-intercepts, set $x = 0$. This gives $0^4 + y^2 - (0)y = 16 \Leftrightarrow y^2 = 16 \Leftrightarrow y = \pm 4$. So the y-intercept are -4 and 4.

57. $y = -x + 4$

x	y
-4	8
-2	6
0	4
1	3
2	2
3	1
4	0

When $y = 0$ we get $x = 4$. So the x-intercept is 4, and $x = 0 \Rightarrow$ $y = 4$, so the y-intercept is 4.

x-axis symmetry: $(-y) = -x + 4 \Leftrightarrow y = x - 4$, which is not the same as $y = -x + 4$, so the graph is not symmetric with respect to the x-axis.

y-axis symmetry: $y = -(-x) + 4 \Leftrightarrow y = x + 4$, which is not the same as $y = -x + 4$, so the graph is not symmetric with respect to the y-axis.

Origin symmetry: $(-y) = -(-x) + 4 \Leftrightarrow y = -x - 4$, which is not the same as $y = -x + 4$, so the graph is not symmetric with respect to the origin.

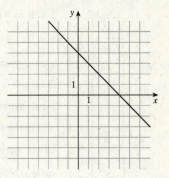

59. $2x - y = 6$

x	y
-1	-8
0	-6
1	-4
2	-2
3	0
4	2
5	4

When $y = 0$ we get $2x = 6$ So the x-intercept is 3. When $x = 0$ we get $-y = 6$ so the y-intercept is -6.

x-axis symmetry: $2x - (-y) = 6 \Leftrightarrow 2x + y = 6$, which is not the same, so the graph is not symmetric with respect to the x-axis.

y-axis symmetry: $2(-x) - y = 6 \Leftrightarrow 2x + y = -6$, so the graph is not symmetric with respect to the y-axis.

Origin symmetry: $2(-x) - (-y) = 6 \Leftrightarrow -2x + y = 6$, which not he same, so the graph is not symmetric with respect to the origin.

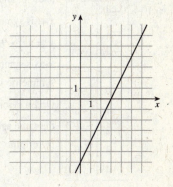

61. $y = 1 - x^2$

x	y
-3	-8
-2	-3
-1	0
0	1
1	0
2	-3
3	-8

$y = 0 \Rightarrow 0 = 1 - x^2 \Leftrightarrow x^2 = 1 \Rightarrow x = \pm 1$, so the x-intercepts are 1 and -1, and $x = 0 \Rightarrow y = 1 - (0)^2 = 1$, so the y-intercept is 1.

x-axis symmetry: $(-y) = 1 - x^2 \Leftrightarrow -y = 1 - x^2$, which is not the same as $y = 1 - x^2$, so the graph is not symmetric with respect to the x-axis.

y-axis symmetry: $y = 1 - (-x)^2 \Leftrightarrow y = 1 - x^2$, so the graph is symmetric with respect to the y-axis.

Origin symmetry: $(-y) = 1 - (-x)^2 \Leftrightarrow -y = 1 - x^2$ which is not the same as $y = 1 - x^2$. The graph is not symmetric with respect to the origin.

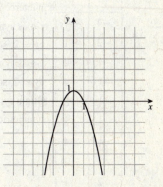

63. $4y = x^2 \Leftrightarrow y = \frac{1}{4}x^2$

x	y
-6	9
-4	4
-2	1
0	0
2	1
4	4
3	2

$y = 0 \Rightarrow 0 = \frac{1}{4}x^2 \Leftrightarrow x^2 = 0 \Rightarrow x = 0$, so the x-intercept is 0, and

$x = 0 \Rightarrow y = \frac{1}{4}(0)^2 = 0$, so the y-intercept is 0.

x-axis symmetry: $(-y) = \frac{1}{4}x^2$, which is not the same as $y = \frac{1}{4}x^2$, so the graph is not symmetric with respect to the x-axis.

y-axis symmetry: $y = \frac{1}{4}(-x)^2 \Leftrightarrow y = \frac{1}{4}x^2$, so the graph is symmetric with respect to the y-axis.

Origin symmetry: $(-y) = \frac{1}{4}(-x)^2 \Leftrightarrow -y = \frac{1}{4}x^2$, which is not the same as $y = \frac{1}{4}x^2$, so the graph is not symmetric with respect to the origin.

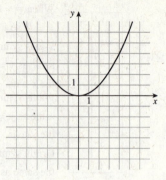

65. $y = x^2 - 9$

x	y
-4	7
-3	0
-2	-5
-1	-8
0	-9
1	-8
2	-5
3	0
4	7

$y = 0 \Rightarrow 0 = x^2 - 9 \Leftrightarrow x^2 = 9 \Rightarrow x = \pm 3$, so the x-intercepts are 3 and -3, and $x = 0 \Rightarrow y = (0)^2 - 9 = -9$, so the y-intercept is -9.

x-axis symmetry: $(-y) = x^2 - 9$, which is not the same as $y = x^2 - 9$, so the graph is not symmetric with respect to the x-axis.

y-axis symmetry: $y = (-x)^2 - 9 \Leftrightarrow y = x^2 - 9$, so the graph is symmetric with respect to the y-axis.

Origin symmetry: $(-y) = (-x)^2 - 9 \Leftrightarrow -y = x^2 - 9$, which is not the same as $y = x^2 - 9$, so the graph is not symmetric with respect to the origin.

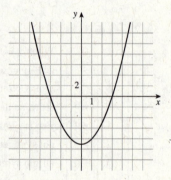

67. $xy = 2 \Leftrightarrow y = \dfrac{2}{x}$

x	y
-4	$-\frac{1}{2}$
-2	-1
-1	-2
$-\frac{1}{2}$	-4
$-\frac{1}{4}$	-8
$\frac{1}{4}$	8
$\frac{1}{2}$	4
1	2
2	1
4	$\frac{1}{2}$

$y = 0$ or $x = 0 \Rightarrow 0 = 2$, which is impossible, so this equation has no x-intercept and no y-intercept.

x-axis symmetry: $x(-y) = 2 \Leftrightarrow -xy = 2$, which is not the same as $xy = 2$, so the graph is not symmetric with respect to the x-axis.

y-axis symmetry: $(-x)y = 2 \Leftrightarrow -xy = 2$, which is not the same as $xy = 2$, so the graph is not symmetric with respect to the y-axis.

Origin symmetry: $(-x)(-y) = 2 \Leftrightarrow xy = 2$, so the graph is symmetric with respect to the origin.

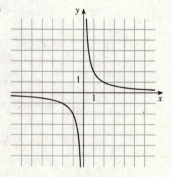

69. $y = \sqrt{4 - x^2}$. Since the radicand (the inside of the square root) cannot be negative, we must have $4 - x^2 \geq 0 \Leftrightarrow x^2 \leq 4 \Leftrightarrow$ $|x| \leq 2$.

x	y
-2	0
-1	$\sqrt{3}$
0	4
1	$\sqrt{3}$
2	0

$y = 0 \Rightarrow 0 = \sqrt{4 - x^2} \Leftrightarrow 4 - x^2 = 0 \Leftrightarrow x^2 = 4 \Rightarrow x = \pm 2$, so the x-intercept are -2 and 2, and $x = 0 \Rightarrow y = \sqrt{4 - (0)^2} = \sqrt{4} = 2$, so the y-intercept is 2. Since $y \geq 0$, the graph is not symmetric with respect to the x-axis.

y-axis symmetry: $y = \sqrt{4 - (-x)^2} = \sqrt{4 - x^2}$, so the graph is symmetric with respect to the y-axis. Also, since $y \geq 0$ the graph is not symmetric with respect to the origin.

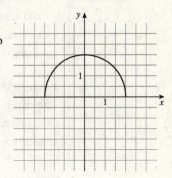

71. Solve for x in terms of y: $x + y^2 = 4 \Leftrightarrow x = 4 - y^2$

x	y
-12	-4
-5	-3
0	-2
3	-1
4	0
3	1
0	2
-5	3
-12	4

$y = 0 \Rightarrow x + 0^2 = 4 \Leftrightarrow x = 4$, so the x-intercept is 4, and $x = 0 \Rightarrow$ $0 + y^2 = 4 \Rightarrow y = \pm 2$, so the y-intercepts are -2 and 2.

x-axis symmetry: $x + (-y)^2 = 4 \Leftrightarrow x + y^2 = 4$, so the graph is symmetric with respect to the x-axis.

y-axis symmetry: $(-x) + y^2 = 4 \Leftrightarrow -x + y^2 = 4$, which is not the same, so the graph is not symmetric with respect to the y-axis.

Origin symmetry: $(-x) + (-y)^2 = 4 \Leftrightarrow -x + y^2 = 4$, which is not the same as $x + y^2 = 4$, so the graph is not symmetric with respect to the origin.

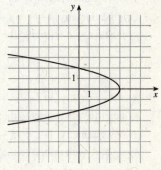

73. $y = 16 - x^4$

x	y
-3	-65
-2	0
-1	15
0	16
1	15
2	0
3	-65

$y = 0 \Rightarrow 0 = 16 - x^4 \Rightarrow x^4 = 16 \Rightarrow x^2 = 4 \Rightarrow x = \pm 2$, so the x-intercepts are ± 2, and so $x = 0 \Rightarrow y = 16 - 0^4 = 16$, so the y-intercept is 16.

x-axis symmetry: $(-y) = 16 - x^4 \Leftrightarrow y = -16 + x^4$, which is not the same as $y = 16 - x^4$, so the graph is not symmetric with respect to the x-axis. y-axis symmetry: $y = 16 - (-x)^4 = 16 - x^4$, so the graph is symmetric with respect to the y-axis.

Origin symmetry: $(-y) = 16 - (-x)^4 \Leftrightarrow -y = 16 - x^4$, which is not the same as $y = 16 - x^4$, so the graph is not symmetric with respect to the origin.

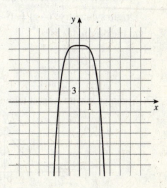

75. $y = 4 - |x|$

x	y
-6	-2
-4	0
-2	2
0	4
2	2
4	0
6	-2

$y = 0 \Rightarrow 0 = 4 - |x| \Leftrightarrow |x| = 4 \Rightarrow x = \pm 4$, so the x-intercepts are -4 and 4, and $x = 0 \Rightarrow y = 4 - |0| = 4$, so the y-intercept is 4.

x-axis symmetry: $(-y) = 4 - |x| \Leftrightarrow y = -4 + |x|$, which is not the same as $y = 4 - |x|$, so the graph is not symmetric with respect to the x-axis.

y-axis symmetry: $y = 4 - |-x| = 4 - |x|$, so the graph is symmetric with respect to the y-axis.

Origin symmetry: $(-y) = 4 - |-x| \Leftrightarrow y = -4 + |x|$, which is not the same as $y = 4 - |x|$, so the graph is not symmetric with respect to the origin.

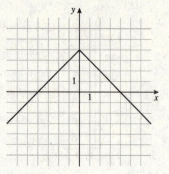

77. x-axis symmetry: $(-y) = x^4 + x^2 \Leftrightarrow y = -x^4 - x^2$, which is not the same as $y = x^4 + x^2$, so the graph is not symmetric with respect to the x-axis.

y-axis symmetry: $y = (-x)^4 + (-x)^2 = x^4 + x^2$, so the graph is symmetric with respect to the y-axis.

Origin symmetry: $(-y) = (-x)^4 + (-x)^2 \Leftrightarrow -y = x^4 + x^2$, which is not the same as $y = x^4 + x^2$, so the graph is not symmetric with respect to the origin.

79. x-axis symmetry: $x^2(-y)^2 + x(-y) = 1 \Leftrightarrow x^2 y^2 - xy = 1$, which is not the same as $x^2 y^2 + xy = 1$, so the graph is not symmetric with respect to the x-axis.

y-axis symmetry: $(-x)^2 y^2 + (-x)y = 1 \Leftrightarrow x^2 y^2 - xy = 1$, which is not the same as $x^2 y^2 + xy = 1$, so the graph is not symmetric with respect to the y-axis.

Origin symmetry: $(-x)^2(-y)^2 + (-x)(-y) = 1 \Leftrightarrow x^2 y^2 + xy = 1$, so the graph is symmetric with respect to the origin.

81. x-axis symmetry: $(-y) = x^3 + 10x \Leftrightarrow y = -x^3 - 10x$, which is not the same as $y = x^3 + 10x$, so the graph is not symmetric with respect to the x-axis.

y-axis symmetry: $y = (-x)^3 + 10(-x) \Leftrightarrow y = -x^3 - 10x$, which is not the same as $y = x^3 + 10x$, so the graph is not symmetric with respect to the y-axis.

Origin symmetry: $(-y) = (-x)^3 + 10(-x) \Leftrightarrow -y = -x^3 - 10x \Leftrightarrow y = x^3 + 10x$, so the graph is symmetric with respect to the origin.

83. Symmetric with respect to the y-axis.

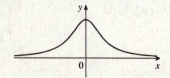

85. Symmetric with respect to the origin.

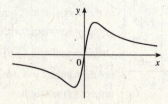

87. $x^2 + y^2 = 9$ has center $(0, 0)$ and radius 3.

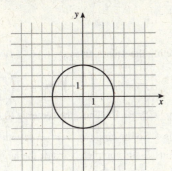

89. $(x - 3)^2 + y^2 = 16$ has center $(3, 0)$ and radius 4.

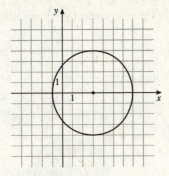

91. $(x + 3)^2 + (y - 4)^2 = 25$ has center $(-3, 4)$ and radius 5.

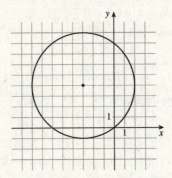

93. Using $h = 2$, $k = -1$, and $r = 3$, we get $(x - 2)^2 + (y - (-1))^2 = 3^2 \Leftrightarrow (x - 2)^2 + (y + 1)^2 = 9$.

95. The equation of a circle centered at the origin is $x^2 + y^2 = r^2$. Using the point $(4, 7)$ we solve for r^2. This gives $(4)^2 + (7)^2 = r^2 \Leftrightarrow 16 + 49 = 65 = r^2$. Thus, the equation of the circle is $x^2 + y^2 = 65$.

97. The center is at the midpoint of the line segment, which is $\left(\dfrac{-1+5}{2}, \dfrac{1+9}{2} \right) = (2, 5)$. The radius is one half the diameter, so $r = \frac{1}{2}\sqrt{(-1-5)^2 + (1-9)^2} = \frac{1}{2}\sqrt{36+64} = \frac{1}{2}\sqrt{100} = 5$. Thus, the equation of the circle is $(x - 2)^2 + (y - 5)^2 = 5^2$ or $(x - 2)^2 + (y - 5)^2 = 25$.

99. Since the circle is tangent to the x-axis, it must contain the point $(7, 0)$, so the radius is the change in the y-coordinates. That is, $r = |-3 - 0| = 3$. So the equation of the circle is $(x - 7)^2 + (y - (-3))^2 = 3^2$, which is $(x - 7)^2 + (y + 3)^2 = 9$.

101. From the figure, the center of the circle is at $(-2, 2)$. The radius is the change in the y-coordinates, so $r = |2 - 0| = 2$. Thus the equation of the circle is $(x - (-2))^2 + (y - 2)^2 = 2^2$, which is $(x + 2)^2 + (y - 2)^2 = 4$.

103. Completing the square gives $x^2 + y^2 - 4x + 10y + 13 = 0 \Leftrightarrow x^2 - 4x + \left(\dfrac{-4}{2} \right)^2 + y^2 + 10y + \left(\dfrac{10}{2} \right)^2 = -13 + \left(\dfrac{4}{2} \right)^2 + \left(\dfrac{10}{2} \right)^2$
$\Leftrightarrow x^2 - 4x + 4 + y^2 + 10y + 25 = -13 + 4 + 25 \Leftrightarrow (x - 2)^2 + (y + 5)^2 = 16$.
Thus, the center is $(2, -5)$, and the radius is 4.

105. Completing the square gives $x^2 + y^2 - \frac{1}{2}x + \frac{1}{2}y = \frac{1}{8} \Leftrightarrow x^2 - \frac{1}{2}x + \left(\dfrac{-1/2}{2} \right)^2 + y^2 + \frac{1}{2}y + \left(\dfrac{1/2}{2} \right)^2 = \frac{1}{8} + \left(\dfrac{-1/2}{2} \right)^2 + \left(\dfrac{1/2}{2} \right)^2$
$\Leftrightarrow x^2 - \frac{1}{2}x + \frac{1}{16} + y^2 + \frac{1}{2}y + \frac{1}{16} = \frac{1}{8} + \frac{1}{16} + \frac{1}{16} = \frac{2}{8} = \frac{1}{4} \Leftrightarrow \left(x - \frac{1}{4} \right)^2 + \left(y + \frac{1}{4} \right)^2 = \frac{1}{4}$. Thus, the circle has center $\left(\frac{1}{4}, -\frac{1}{4} \right)$ and radius $\frac{1}{2}$.

107. Completing the square gives $2x^2 + 2y^2 - 3x = 0 \Leftrightarrow x^2 + y^2 - \frac{3}{2}x = 0 \Leftrightarrow \left(x - \frac{3}{4}\right)^2 + y^2 = \left(\frac{3}{4}\right)^2 = \frac{9}{16}$.

Thus, the circle has center $\left(\frac{3}{4}, 0\right)$ and radius $\frac{3}{4}$.

109. $\left\{(x, y) \mid x^2 + y^2 \le 1\right\}$. This is the set of points inside

(and on) the circle $x^2 + y^2 = 1$.

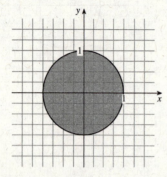

111. Completing the square gives $x^2 + y^2 - 4y - 12 = 0$

$\Leftrightarrow x^2 + y^2 - 4y + \left(\frac{-4}{2}\right)^2 = 12 + \left(\frac{-4}{2}\right)^2 \Leftrightarrow$

$x^2 + (y - 2)^2 = 16$. Thus, the center is $(0, 2)$, and the

radius is 4. So the circle $x^2 + y^2 = 4$, with center $(0, 0)$

and radius 2, sits completely inside the larger circle. Thus,

the area is $\pi 4^2 - \pi 2^2 = 16\pi - 4\pi = 12\pi$.

113. (a) $d(A, B) = \sqrt{3^2 + 4^2} = \sqrt{25} = 5$.

(b) We want the distances from $C = (4, 2)$ to $D = (11, 26)$. The walking distance is

$|4 - 11| + |2 - 26| = 7 + 24 = 31$ blocks. Straight-line distance is

$\sqrt{(4 - 11)^2 + (2 - 26)^2} = \sqrt{7^2 + 24^2} = \sqrt{625} = 25$ blocks.

(c) The two points are on the same avenue or the same street.

115. (a) Closest: 2 Mm. Farthest: 8 Mm.

(b) When $y = 2$ we have $\dfrac{(x - 3)^2}{25} + \dfrac{2^2}{16} = 1 \Leftrightarrow \dfrac{(x - 3)^2}{25} + \dfrac{1}{4} = 1 \Leftrightarrow \dfrac{(x - 3)^2}{25} = \dfrac{3}{4} \Leftrightarrow (x - 3)^2 = \dfrac{75}{4}$. Taking the square

root of both sides we get $x - 3 = \pm\sqrt{\dfrac{75}{4}} = \pm\dfrac{5\sqrt{3}}{2} \Leftrightarrow x = 3 \pm \dfrac{5\sqrt{3}}{2}$. So $x = 3 - \dfrac{5\sqrt{3}}{2} \approx -1.33$ or $x = 3 + \dfrac{5\sqrt{3}}{2} \approx 7.33$.

The distance from $(-1.33, 2)$ to the center $(0, 0)$ is $d = \sqrt{(-1.33 - 0)^2 + (2 - 0)^2} = \sqrt{5.7689} \approx 2.40$ Mm. The

distance from $(7.33, 2)$ to the center $(0, 0)$ is $d = \sqrt{(7.33 - 0)^2 + (2 - 0)^2} = \sqrt{57.7307} \approx 7.60$ Mm.

117. (a) The point $(3, 7)$ is reflected to the point $(-3, 7)$.

(b) The point (a, b) is reflected to the point $(-a, b)$.

(c) Since the point $(-a, b)$ is the reflection of (a, b), the point $(-4, -1)$ is the reflection of $(4, -1)$.

(d) $A = (3, 3)$, so $A' = (-3, 3)$; $B = (6, 1)$, so $B' = (-6, 1)$; and $C = (1, -4)$, so $C' = (-1, -4)$.

119. We need to find a point $S(x_1, y_1)$ such that $PQRS$ is a parallelogram. As

indicated by Example 3, this will be the case if the diagonals PR and QS bisect

each other. So the midpoints of PR and QS are the same. Thus

$\left(\dfrac{0 + 5}{2}, \dfrac{-3 + 3}{2}\right) = \left(\dfrac{x_1 + 2}{2}, \dfrac{y_1 + 2}{2}\right)$. Setting the x-coordinates equal, we get

$\dfrac{0 + 5}{2} = \dfrac{x_1 + 2}{2} \Leftrightarrow 0 + 5 = x_1 + 2 \Leftrightarrow x_1 = 3$.

Setting the y-coordinates equal, we get $\dfrac{-3 + 3}{2} = \dfrac{y_1 + 2}{2} \Leftrightarrow -3 + 3 = y_1 + 2 \Leftrightarrow$

$y_1 = -2$. Thus $S = (3, -2)$.

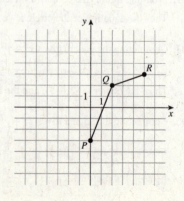

121. (a) (i) $(x - 2)^2 + (y - 1)^2 = 9$, the center is at $(2, 1)$, and the radius is 3. $(x - 6)^2 + (y - 4)^2 = 16$, the center is at $(6, 4)$, and the radius is 4. The distance between centers is

$$\sqrt{(2 - 6)^2 + (1 - 4)^2} = \sqrt{(-4)^2 + (-3)^2} = \sqrt{16 + 9} = \sqrt{25} = 5. \text{ Since } 5 < 3 + 4, \text{ these}$$

circles intersect.

(ii) $x^2 + (y - 2)^2 = 4$, the center is at $(0, 2)$, and the radius is 2. $(x - 5)^2 + (y - 14)^2 = 9$, the center is at $(5, 14)$, and the radius is 3. The distance between centers is

$$\sqrt{(0 - 5)^2 + (2 - 14)^2} = \sqrt{(-5)^2 + (-12)^2} = \sqrt{25 + 144} = \sqrt{169} = 13. \text{ Since } 13 > 2 + 3,$$

these circles do not intersect.

(iii) $(x - 3)^2 + (y + 1)^2 = 1$, the center is at $(3, -1)$, and the radius is 1. $(x - 2)^2 + (y - 2)^2 = 25$, the center is at $(2, 2)$,

and the radius is 5. The distance between centers is $\sqrt{(3 - 2)^2 + (-1 - 2)^2} = \sqrt{1^2 + (-3)^2} = \sqrt{1 + 9} = \sqrt{10}$.

Since $\sqrt{10} < 1 + 5$, these circles intersect.

(b) As shown in the diagram, if two circles intersect, then the centers of the circles and one point of intersection form a triangle. So because in any triangle each side has length less than the sum of the other two, the two circles will intersect only if the distance between their centers, d, is less than or equal to the sum of the radii, r_1 and r_2. That is, the circles will intersect if $d \leq r_1 + r_2$.

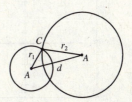

1.9   GRAPHING CALCULATORS; SOLVING EQUATIONS AND INEQUALITIES GRAPHICALLY

1. The solutions of the equation $x^2 - 2x - 3 = 0$ are the x-intercepts of the graph of $y = x^2 - 2x - 3$.

3. (a) From the graph, it appears that the graph of $y = x^4 - 3x^3 - x^2 + 3x$ has x-intercepts -1, 0, 1, and 3, so the solutions to the equation $x^4 - 3x^3 - x^2 + 3x = 0$ are $x = -1$, $x = 0$, $x = 1$, and $x = 3$.

(b) From the graph, we see that where $-1 \leq x \leq 0$ or $1 \leq x \leq 3$, the graph lies below the x-axis. Thus, the inequality $x^4 - 3x^3 - x^2 + 3x \leq 0$ is satisfied for $\{x \mid -1 \leq x \leq 0 \text{ or } 1 \leq x \leq 3\} = [-1, 0] \cup [1, 3]$.

5. $y = x^4 + 2$

(a) $[-2, 2]$ by $[-2, 2]$

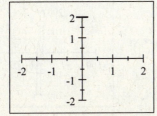

(b) $[0, 4]$ by $[0, 4]$

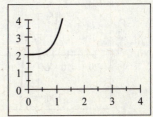

(c) $[-8, 8]$ by $[-4, 40]$

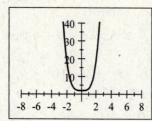

(d) $[-40, 40]$ by $[-80, 800]$

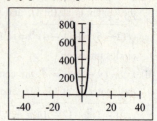

The viewing rectangle in part (c) produces the most appropriate graph of the equation.

7. $y = 100 - x^2$

(a) $[-4, 4]$ by $[-4, 4]$

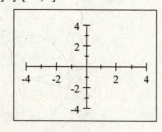

(b) $[-10, 10]$ by $[-10, 10]$

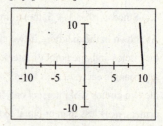

(c) $[-15, 15]$ by $[-30, 110]$

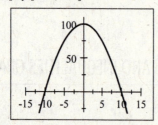

(d) $[-4, 4]$ by $[-30, 110]$

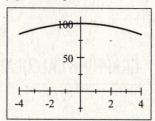

The viewing rectangle in part (c) produces the most appropriate graph of the equation.

9. $y = 10 + 25x - x^3$

(a) $[-4, 4]$ by $[-4, 4]$

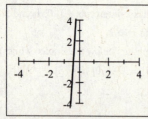

(b) $[-10, 10]$ by $[-10, 10]$

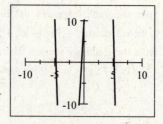

(c) $[-20, 20]$ by $[-100, 100]$

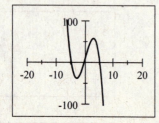

(d) $[-100, 100]$ by $[-200, 200]$

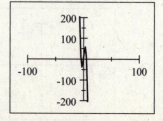

The viewing rectangle in part (c) produces the most appropriate graph of the equation.

11. $y = 100x^2$, $[-2, 2]$ by $[-10, 400]$

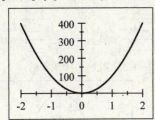

13. $y = 4 + 6x - x^2$, $[-4, 10]$ by $[-10, 20]$

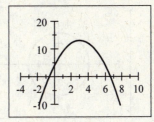

15. $y = \sqrt[4]{256 - x^2}$. We require that $256 - x^2 \geq 0 \Rightarrow$ $-16 \leq x \leq 16$, so we graph $y = \sqrt[4]{256 - x^2}$ in the viewing rectangle $[-20, 20]$ by $[-1, 5]$.

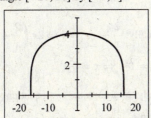

17. $y = 0.01x^3 - x^2 + 5$, $[-50, 150]$ by $[-2000, 2000]$

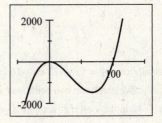

19. $y = x^4 - 4x^3$, $[-4, 6]$ by $[-50, 100]$

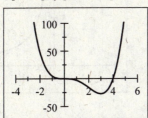

21. $y = 1 + |x - 1|$, $[-3, 5]$ by $[-1, 5]$

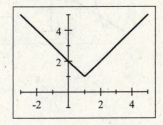

23. Although the graphs of $y = -3x^2 + 6x - \frac{1}{2}$ and $y = \sqrt{7 - \frac{7}{12}x^2}$ appear to intersect in the viewing rectangle $[-4, 4]$ by $[-1, 3]$, there is no point of intersection. You can verify this by zooming in.

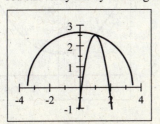

25. The graphs of $y = 6 - 4x - x^2$ and $y = 3x + 18$ appear to have two points of intersection in the viewing rectangle $[-6, 2]$ by $[-5, 20]$. You can verify that $x = -4$ and $x = -3$ are exact solutions.

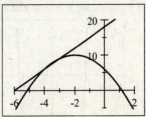

27. $x^2 + y^2 = 9 \Leftrightarrow y^2 = 9 - x^2 \Rightarrow y = \pm\sqrt{9 - x^2}$. So we graph the functions $y_1 = \sqrt{9 - x^2}$ and $y_2 = -\sqrt{9 - x^2}$ in the viewing rectangle $[-6, 6]$ by $[-4, 4]$.

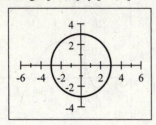

29. $4x^2 + 2y^2 = 1 \Leftrightarrow 2y^2 = 1 - 4x^2 \Leftrightarrow y^2 = \dfrac{1 - 4x^2}{2} \Rightarrow$

$y = \pm\sqrt{\dfrac{1 - 4x^2}{2}}$. So we graph the functions

$y_1 = \sqrt{\dfrac{1 - 4x^2}{2}}$ and $y_2 = -\sqrt{\dfrac{1 - 4x^2}{2}}$ in the viewing

rectangle $[-1.2, 1.2]$ by $[-0.8, 0.8]$.

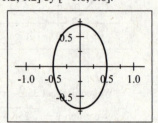

31. Algebraically: $x - 4 = 5x + 12 \Leftrightarrow -16 = 4x \Leftrightarrow x = -4$. Graphically: We graph the two equations $y_1 = x - 4$ and $y_2 = 5x + 12$ in the viewing rectangle $[-6, 4]$ by $[-10, 2]$. Zooming in, we see that the solution is $x = -4$.

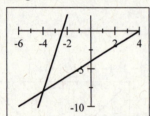

33. Algebraically: $\dfrac{2}{x} + \dfrac{1}{2x} = 7 \Leftrightarrow 2x\left(\dfrac{2}{x} + \dfrac{1}{2x}\right) = 2x\,(7)$

$\Leftrightarrow 4 + 1 = 14x \Leftrightarrow x = \frac{5}{14}$.

Graphically: We graph the two equations $y_1 = \dfrac{2}{x} + \dfrac{1}{2x}$

and $y_2 = 7$ in the viewing rectangle $[-2, 2]$ by $[-2, 8]$. Zooming in, we see that the solution is $x \approx 0.36$.

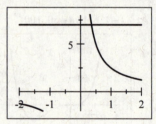

35. Algebraically: $x^2 - 32 = 0 \Leftrightarrow x^2 = 32 \Rightarrow$

$x = \pm\sqrt{32} = \pm 4\sqrt{2}$.

Graphically: We graph the equation $y_1 = x^2 - 32$ and determine where this curve intersects the x-axis. We use the viewing rectangle $[-10, 10]$ by $[-5, 5]$. Zooming in, we see that solutions are $x \approx 5.66$ and $x \approx -5.66$.

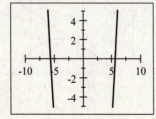

37. Algebraically: $x^2 + 9 = 0 \Leftrightarrow x^2 = -9$, which has no real solution.

Graphically: We graph the equation $y = x^2 + 9$ and see that this curve does not intersect the x-axis. We use the viewing rectangle $[-5, 5]$ by $[-5, 30]$.

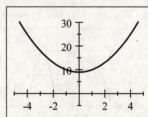

39. Algebraically: $16x^4 = 625 \Leftrightarrow x^4 = \frac{625}{16} \Rightarrow$ $x = \pm\frac{5}{2} = \pm2.5$.

Graphically: We graph the two equations $y_1 = 16x^4$ and $y_2 = 625$ in the viewing rectangle $[-5, 5]$ by $[610, 640]$. Zooming in, we see that solutions are $x = \pm2.5$.

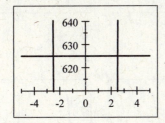

41. Algebraically: $(x - 5)^4 - 80 = 0 \Leftrightarrow (x - 5)^4 = 80 \Rightarrow$ $x - 5 = \pm\sqrt[4]{80} = \pm2\sqrt[4]{5} \Leftrightarrow x = 5 \pm 2\sqrt[4]{5}$.

Graphically: We graph the equation $y_1 = (x - 5)^4 - 80$ and determine where this curve intersects the x-axis. We use the viewing rectangle $[-1, 9]$ by $[-5, 5]$. Zooming in, we see that solutions are $x \approx 2.01$ and $x \approx 7.99$.

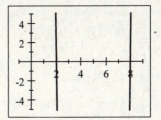

43. We graph $y = x^2 - 7x + 12$ in the viewing rectangle $[0, 6]$ by $[-0.1, 0.1]$. The solutions appear to be exactly $x = 3$ and $x = 4$. [In fact $x^2 - 7x + 12 = (x - 3)(x - 4)$.]

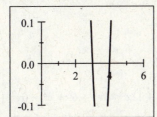

45. We graph $y = x^3 - 6x^2 + 11x - 6$ in the viewing rectangle $[-1, 4]$ by $[-0.1, 0.1]$. The solutions are $x = 1.00$, $x = 2.00$, and $x = 3.00$.

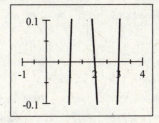

47. We first graph $y = x - \sqrt{x + 1}$ in the viewing rectangle $[-1, 5]$ by $[-0.1, 0.1]$ and find that the solution is near 1.6. Zooming in, we see that solutions is $x \approx 1.62$.

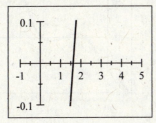

49. We graph $y = x^{1/3} - x$ in the viewing rectangle $[-3, 3]$ by $[-1, 1]$. The solutions are $x = -1$, $x = 0$, and $x = 1$, as can be verified by substitution.

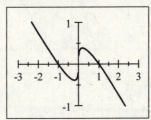

51. We graph $y = \sqrt{2x + 1} + 1$ and $y = x$ in the viewing rectangle $[-1, 6]$ by $[0, 6]$ and see that the only solution to the equation $\sqrt{2x + 1} + 1 = x$ is $x = 4$, which can be verified by substitution.

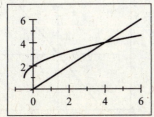

53. We graph $y = 2x^4 + 4x^2 + 1$ in the viewing rectangle $[-2, 2]$ by $[-5, 40]$ and see that the equation $2x^4 + 4x^2 + 1 = 0$ has no solution.

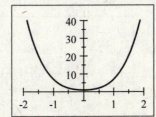

55. $x^3 - 2x^2 - x - 1 = 0$, so we start by graphing the function $y = x^3 - 2x^2 - x - 1$ in the viewing rectangle $[-10, 10]$ by $[-100, 100]$. There appear to be two solutions, one near $x = 0$ and another one between $x = 2$ and $x = 3$. We then use the viewing rectangle $[-1, 5]$ by $[-1, 1]$ and zoom in on the only solution, $x \approx 2.55$.

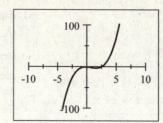 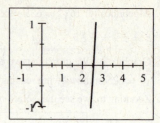

57. $x(x-1)(x+2) = \frac{1}{6}x \Leftrightarrow$

$x(x-1)(x+2) - \frac{1}{6}x = 0$. We start by graphing the function $y = x(x-1)(x+2) - \frac{1}{6}x$ in the viewing rectangle $[-5, 5]$ by $[-10, 10]$. There appear to be three solutions. We then use the viewing rectangle $[-2.5, 2.5]$ by $[-1, 1]$ and zoom into the solutions at $x \approx -2.05$, $x = 0.00$, and $x \approx 1.05$.

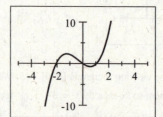 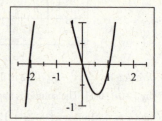

59. We graph $y = x^2$ and $y = 3x + 10$ in the viewing rectangle $[-4, 7]$ by $[-5, 30]$. The solution to the inequality is $[-2, 5]$.

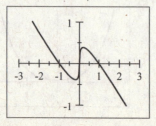

61. Since $x^3 + 11x \leq 6x^2 + 6 \Leftrightarrow x^3 - 6x^2 + 11x - 6 \leq 0$, we graph $y = x^3 - 6x^2 + 11x - 6$ in the viewing rectangle $[0, 5]$ by $[-5, 5]$. The solution set is $(-\infty, 1.0] \cup [2.0, 3.0]$.

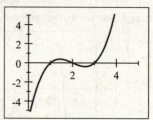

63. Since $x^{1/3} \leq x \Leftrightarrow x^{1/3} - x < 0$, we graph $y = x^{1/3} - x$ in the viewing rectangle $[-3, 3]$ by $[-1, 1]$. From this, we find that the solution set is $(-1, 0) \cup (1, \infty)$.

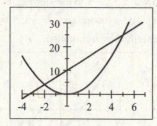

65. Since $(x+1)^2 < (x-1)^2 \Leftrightarrow (x+1)^2 - (x-1)^2 < 0$, we graph $y = (x+1)^2 - (x-1)^2$ in the viewing rectangle $[-2, 2]$ by $[-5, 5]$. The solution set is $(-\infty, 0)$.

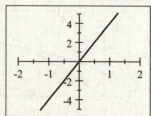

67. We graph the equations $y = 3x^2 - 3x$ and $y = 2x^2 + 4$ in the viewing rectangle $[-2, 6]$ by $[-10, 50]$. We see that the two curves intersect at $x = -1$ and at $x = 4$, and that the first curve is lower than the second for $-1 < x < 4$. Thus, we see that the inequality $3x^2 - 3x < 2x^2 + 4$ has solution set $(-1, 4)$.

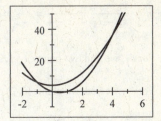

69. We graph the equation $y = (x - 2)^2 (x - 3) (x + 1)$ in the viewing rectangle $[-2, 4]$ by $[-15, 10]$ and see that the inequality $(x - 2)^2 (x - 3) (x + 1) \le 0$ has the solution set $[-1, 3]$.

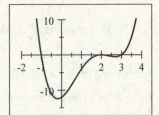

71. As in Example 6, we graph the equation $y = x^3 - 6x^2 + 9x - \sqrt{x}$ in the viewing rectangle $[0, 10]$ by $[-2, 15]$. We see the two solutions found in Example 6 and what appears to be an additional solution at $x = 0$. In the viewing rectangle $[0, 0.05]$ by $[-0.25, 0.25]$, we find yet another solution at $x \approx 0.01$. We can verify that $x = 0$ is an exact solution by substitution.

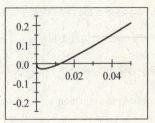

73. (a)

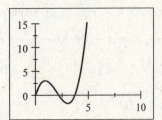

(b) Using a zoom or trace function, we find that $y \ge 10$ for $x \ge 66.7$. We could estimate this since if $x < 100$, then $\left(\frac{x}{5280}\right)^2 \le 0.00036$. So for $x < 100$ we have $\sqrt{1.5x + \left(\frac{x}{5280}\right)^2} \approx \sqrt{1.5x}$. Solving $\sqrt{1.5x} > 10$ we get $1.5 > 100$ or $x > \frac{100}{1.5} = 66.7$ mi.

75. Answers will vary.

77. Calculators perform operations in the following order: exponents are applied before division and division is applied before addition. Therefore, Y_1=x^1/3 is interpreted as $y = \frac{x^1}{3} = \frac{x}{3}$, which is the equation of a line. Likewise, Y_2=x/x+4 is interpreted as $y = \frac{x}{x} + 4 = 1 + 4 = 5$. Instead, enter the following: Y_1=x^(1/3), Y_2=x/(x+4).

1.10 LINES

1. We find the "steepness" or slope of a line passing through two points by dividing the difference in the y-coordinates of these points by the difference in the x-coordinates. So the line passing through the points $(0, 1)$ and $(2, 5)$ has slope $\frac{5 - 1}{2 - 0} = 2$.

3. The point-slope form of the equation of the line with slope 3 passing through the point $(1, 2)$ is $y - 2 = 3(x - 1)$.

5. $m = \dfrac{y_2 - y_1}{x_2 - x_1} = \dfrac{2 - 0}{4 - 0} = \dfrac{2}{4} = \dfrac{1}{2}$

7. $m = \dfrac{y_2 - y_1}{x_2 - x_1} = \dfrac{0 - 2}{-10 - 2} = \dfrac{-2}{-12} = \dfrac{1}{6}$

9. $m = \dfrac{y_2 - y_1}{x_2 - x_1} = \dfrac{4 - 3}{2 - 4} = \dfrac{1}{-2} = -\dfrac{1}{2}$

11. $m = \dfrac{y_2 - y_1}{x_2 - x_1} = \dfrac{6 - (-3)}{-1 - 1} = \dfrac{9}{-2} = -\dfrac{9}{2}$

13. For ℓ_1, we find two points, $(-1, 2)$ and $(0, 0)$ that lie on the line. Thus the slope of ℓ_1 is $m = \dfrac{y_2 - y_1}{x_2 - x_1} = \dfrac{2 - 0}{-1 - 0} = -2$.

For ℓ_2, we find two points $(0, 2)$ and $(2, 3)$. Thus, the slope of ℓ_2 is $m = \dfrac{y_2 - y_1}{x_2 - x_1} = \dfrac{3 - 2}{2 - 0} = \dfrac{1}{2}$. For ℓ_3 we find the points

$(2, -2)$ and $(3, 1)$. Thus, the slope of ℓ_3 is $m = \dfrac{y_2 - y_1}{x_2 - x_1} = \dfrac{1 - (-2)}{3 - 2} = 3$. For ℓ_4, we find the points $(-2, -1)$ and

$(2, -2)$. Thus, the slope of ℓ_4 is $m = \dfrac{y_2 - y_1}{x_2 - x_1} = \dfrac{-2 - (-1)}{2 - (-2)} = \dfrac{-1}{4} = -\dfrac{1}{4}$.

15. First we find two points $(0, 4)$ and $(4, 0)$ that lie on the line. So the slope is $m = \dfrac{0 - 4}{4 - 0} = -1$. Since the y-intercept is 4, the equation of the line is $y = mx + b = -1x + 4$. So $y = -x + 4$, or $x + y - 4 = 0$.

17. We choose the two intercepts as points, $(0, -3)$ and $(2, 0)$. So the slope is $m = \dfrac{0 - (-3)}{2 - 0} = \dfrac{3}{2}$. Since the y-intercept is -3, the equation of the line is $y = mx + b = \dfrac{3}{2}x - 3$, or $3x - 2y - 6 = 0$.

19. Using the equation $y - y_1 = m(x - x_1)$, we get $y - 3 = 5(x - 2) \Leftrightarrow -5x + y = -7 \Leftrightarrow 5x - y - 7 = 0$.

21. Using the equation $y - y_1 = m(x - x_1)$, we get $y - 7 = \dfrac{2}{3}(x - 1) \Leftrightarrow 3y - 21 = 2x - 2 \Leftrightarrow -2x + 3y = 19 \Leftrightarrow 2x - 3y + 19 = 0$.

23. First we find the slope, which is $m = \dfrac{y_2 - y_1}{x_2 - x_1} = \dfrac{6 - 1}{1 - 2} = \dfrac{5}{-1} = -5$. Substituting into $y - y_1 = m(x - x_1)$, we get $y - 6 = -5(x - 1) \Leftrightarrow y - 6 = -5x + 5 \Leftrightarrow 5x + y - 11 = 0$.

25. Using $y = mx + b$, we have $y = 3x + (-2)$ or $3x - y - 2 = 0$.

27. We are given two points, $(1, 0)$ and $(0, -3)$. Thus, the slope is $m = \dfrac{y_2 - y_1}{x_2 - x_1} = \dfrac{-3 - 0}{0 - 1} = \dfrac{-3}{-1} = 3$. Using the y-intercept, we have $y = 3x + (-3)$ or $y = 3x - 3$ or $3x - y - 3 = 0$.

29. Since the equation of a horizontal line passing through (a, b) is $y = b$, the equation of the horizontal line passing through $(4, 5)$ is $y = 5$.

31. Since $x + 2y = 6 \Leftrightarrow 2y = -x + 6 \Leftrightarrow y = -\dfrac{1}{2}x + 3$, the slope of this line is $-\dfrac{1}{2}$. Thus, the line we seek is given by $y - (-6) = -\dfrac{1}{2}(x - 1) \Leftrightarrow 2y + 12 = -x + 1 \Leftrightarrow x + 2y + 11 = 0$.

33. Any line parallel to $x = 5$ will have undefined slope and be of the form $x = a$. Thus the equation of the line is $x = -1$.

35. First find the slope of $2x + 5y + 8 = 0$. This gives $2x + 5y + 8 = 0 \Leftrightarrow 5y = -2x - 8 \Leftrightarrow y = -\dfrac{2}{5}x - \dfrac{8}{5}$. So the slope of the line that is perpendicular to $2x + 5y + 8 = 0$ is $m = -\dfrac{1}{-2/5} = \dfrac{5}{2}$. The equation of the line we seek is $y - (-2) = \dfrac{5}{2}(x - (-1)) \Leftrightarrow 2y + 4 = 5x + 5 \Leftrightarrow 5x - 2y + 1 = 0$.

37. First find the slope of the line passing through $(2, 5)$ and $(-2, 1)$. This gives $m = \dfrac{1 - 5}{-2 - 2} = \dfrac{-4}{-4} = 1$, and so the equation of the line we seek is $y - 7 = 1(x - 1) \Leftrightarrow x - y + 6 = 0$.

39. (a)

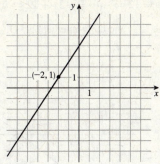

(b) $y - 1 = \frac{3}{2}(x - (-2)) \Leftrightarrow 2y - 2 = 3(x + 2) \Leftrightarrow$
$2y - 2 = 3x + 6 \Leftrightarrow 3x - 2y + 8 = 0.$

41.

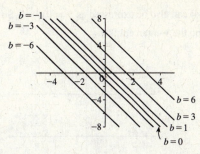

$y = -2x + b$, $b = 0, \pm1, \pm3, \pm6$. They have the same slope, so they are parallel.

43.

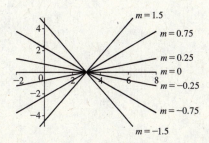

$y = m(x - 3)$, $m = 0, \pm0.25, \pm0.75, \pm1.5$. Each of the lines contains the point $(3, 0)$ because the point $(3, 0)$ satisfies each equation $y = m(x - 3)$. Since $(3, 0)$ is on the x-axis, we could also say that they all have the same x-intercept.

45. $x + y = 3 \Leftrightarrow y = -x + 3$. So the slope is -1, and the y-intercept is 3.

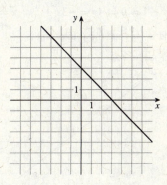

47. $x + 3y = 0 \Leftrightarrow 3y = -x \Leftrightarrow y = -\frac{1}{3}x$. So the slope is $-\frac{1}{3}$, and the y-intercept is 0.

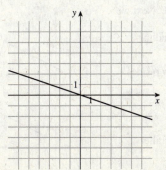

49. $\frac{1}{2}x - \frac{1}{3}y + 1 = 0 \Leftrightarrow -\frac{1}{3}y = -\frac{1}{2}x - 1 \Leftrightarrow y = \frac{3}{2}x + 3.$
So the slope is $\frac{3}{2}$, and the y-intercept is 3.

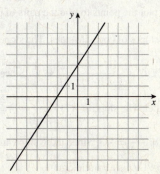

51. $y = 4$ can also be expressed as $y = 0x + 4$. So the slope is 0, and the y-intercept is 4.

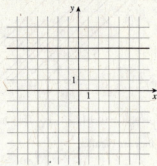

53. $3x - 4y = 12 \Leftrightarrow -4y = -3x + 12 \Leftrightarrow y = \frac{3}{4}x - 3$. So the slope is $\frac{3}{4}$, and the y-intercept is -3.

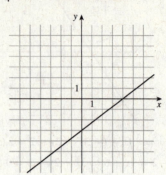

55. $3x + 4y - 1 = 0 \Leftrightarrow 4y = -3x + 1 \Leftrightarrow y = -\frac{3}{4}x + \frac{1}{4}$. So the slope is $-\frac{3}{4}$, and the y-intercept is $\frac{1}{4}$.

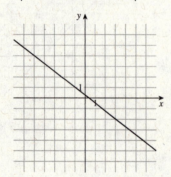

57. We first plot the points to find the pairs of points that determine each side. Next we find the slopes of opposite sides. The slope of AB is $\dfrac{4 - 1}{7 - 1} = \dfrac{3}{6} = \dfrac{1}{2}$, and the slope of DC is $\dfrac{10 - 7}{5 - (-1)} = \dfrac{3}{6} = \dfrac{1}{2}$. Since these slope are equal, these two sides are parallel. The slope of AD is $\dfrac{7 - 1}{-1 - 1} = \dfrac{6}{-2} = -3$, and the slope of BC is $\dfrac{10 - 4}{5 - 7} = \dfrac{6}{-2} = -3$. Since these slope are equal, these two sides are parallel. Hence $ABCD$ is a parallelogram.

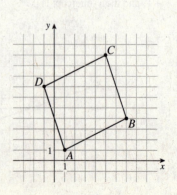

59. We first plot the points to find the pairs of points that determine each side. Next we

find the slopes of opposite sides. The slope of AB is $\dfrac{3-1}{11-1} = \dfrac{2}{10} = \dfrac{1}{5}$ and the

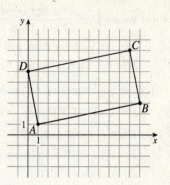

slope of DC is $\dfrac{6-8}{0-10} = \dfrac{-2}{-10} = \dfrac{1}{5}$. Since these slope are equal, these two sides

are parallel. Slope of AD is $\dfrac{6-1}{0-1} = \dfrac{5}{-1} = -5$, and the slope of BC is

$\dfrac{3-8}{11-10} = \dfrac{-5}{1} = -5$. Since these slope are equal, these two sides are parallel.

Since (slope of AB) $\times$ (slope of AD) $= \frac{1}{5} \times (-5) = -1$, the first two sides are

each perpendicular to the second two sides. So the sides form a rectangle.

61. We need the slope and the midpoint of the line AB. The midpoint of AB is $\left(\dfrac{1+7}{2}, \dfrac{4-2}{2} \right) = (4, 1)$, and the slope of

AB is $m = \dfrac{-2-4}{7-1} = \dfrac{-6}{6} = -1$. The slope of the perpendicular bisector will have slope $\dfrac{-1}{m} = \dfrac{-1}{-1} = 1$. Using the

point-slope form, the equation of the perpendicular bisector is $y - 1 = 1(x - 4)$ or $x - y - 3 = 0$.

63. (a) We start with the two points $(a, 0)$ and $(0, b)$. The slope of the line that contains them is $\dfrac{b-0}{0-a} = -\dfrac{b}{a}$. So the equation

of the line containing them is $y = -\dfrac{b}{a} x + b$ (using the slope-intercept form). Dividing by b (since $b \neq 0$) gives

$\dfrac{y}{b} = -\dfrac{x}{a} + 1 \Leftrightarrow \dfrac{x}{a} + \dfrac{y}{b} = 1.$

(b) Setting $a = 6$ and $b = -8$, we get $\dfrac{x}{6} + \dfrac{y}{-8} = 1 \Leftrightarrow 4x - 3y = 24 \Leftrightarrow 4x - 3y - 24 = 0.$

65. Let h be the change in your horizontal distance, in feet. Then $-\dfrac{6}{100} = \dfrac{-1000}{h} \Leftrightarrow h = \dfrac{100,000}{6} \approx 16,667$. So the change

in your horizontal distance is about 16,667 feet.

67. (a) The slope is $0.0417D = 0.0417(200) = 8.34$. It represents the increase in dosage for each one-year increase in the

child's age.

(b) When $a = 0$, $c = 8.34(0 + 1) = 8.34$ mg.

69. (a)

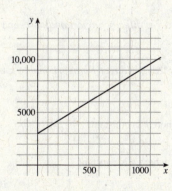

(b) The slope is the cost per toaster oven, \$6. The y-intercept, \$3000, is the monthly fixed cost — the cost that is incurred no matter how many toaster ovens are produced.

71. (a) Using n in place of x and t in place of y, we find that the slope is $\dfrac{t_2 - t_1}{n_2 - n_1} = \dfrac{80 - 70}{168 - 120} = \dfrac{10}{48} = \dfrac{5}{24}$. So the linear

equation is $t - 80 = \frac{5}{24}(n - 168) \Leftrightarrow t - 80 = \frac{5}{24}n - 35 \Leftrightarrow t = \frac{5}{24}n + 45.$

(b) When $n = 150$, the temperature is approximately given by $t = \frac{5}{24}(150) + 45 = 76.25°$ F $\approx 76°$ F.

73. (a) We are given $\dfrac{\text{change in pressure}}{10 \text{ feet change in depth}} = \dfrac{4.34}{10} = 0.434$. Using P for

pressure and d for depth, and using the point $P = 15$ when $d = 0$, we

have $P - 15 = 0.434 (d - 0) \Leftrightarrow P = 0.434d + 15$.

(c) The slope represents the increase in pressure per foot of descent. The
y-intercept represents the pressure at the surface.

(d) When $P = 100$, then $100 = 0.434d + 15 \Leftrightarrow 0.434d = 85 \Leftrightarrow$
$d = 195.9$ ft. Thus the pressure is 100 lb/in^3 at a depth of
approximately 196 ft.

(b)

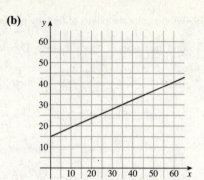

75. (a) Using d in place of x and C in place of y, we find the slope to be
$\dfrac{C_2 - C_1}{d_2 - d_1} = \dfrac{460 - 380}{800 - 480} = \dfrac{80}{320} = \dfrac{1}{4}$. So the linear equation is
$C - 460 = \frac{1}{4} (d - 800) \Leftrightarrow C - 460 = \frac{1}{4}d - 200 \Leftrightarrow C = \frac{1}{4}d + 260$.

(b) Substituting $d = 1500$ we get $C = \frac{1}{4} (1500) + 260 = 635$. Thus, the
cost of driving 1500 miles is \$635.

(d) The y-intercept represents the fixed cost, \$260.

(e) It is a suitable model because you have fixed monthly costs such as
insurance and car payments, as well as costs that occur as you drive,
such as gasoline, oil, tires, etc., and the cost of these for each
additional mile driven is a constant.

(c)

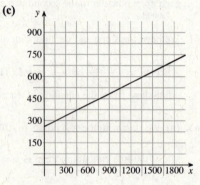

The slope of the line represents the
cost per mile, \$0.25.

77. Slope is the rate of change of one variable per unit change in another variable. So if the slope is positive, then the
temperature is rising. Likewise, if the slope is negative then the temperature is decreasing. If the slope is 0, then the
temperature is not changing.

1.11   MAKING MODELS USING VARIATION

1. If the quantities x and y are related by the equation $y = 3x$ then we say that y is *directly proportional* to x, and the constant
of *proportionality* is 3.

3. If the quantities x, y, and z are related by the equation $z = 3\dfrac{x}{y}$ then we say that z is *directly proportional* to x and *inversely*
proportional to y.

5. $T = kx$, where k is constant.

7. $v = \dfrac{k}{z}$, where k is constant.

9. $y = \dfrac{ks}{t}$, where k is constant.

11. $z = k\sqrt{y}$, where k is constant.

13. $V = klwh$, where k is constant.

15. $R = \dfrac{ki}{Pt}$, where k is constant.

17. Since y is directly proportional to x, $y = kx$. Since $y = 42$ when $x = 6$, we have $42 = k (6) \Leftrightarrow k = 7$. So $y = 7x$.

19. R is inversely proportional to s, so $R = \dfrac{k}{s}$. Since $R = 3$ when $s = 4$, we have $3 = \dfrac{k}{4} \Leftrightarrow k = 12$. So $R = \dfrac{12}{s}$.

21. Since M varies directly as x and inversely as y, $M = \dfrac{kx}{y}$. Since $M = 5$ when $x = 2$ and $y = 6$, we have $5 = \dfrac{k(2)}{6} \Leftrightarrow$

$k = 15$. Therefore $M = \dfrac{15x}{y}$.

23. Since W is inversely proportional to the square of r, $W = \dfrac{k}{r^2}$. Since $W = 10$ when $r = 6$, we have $10 = \dfrac{k}{(6)^2} \Leftrightarrow k = 360$.

So $W = \dfrac{360}{r^2}$.

25. Since C is jointly proportional to l, w, and h, we have $C = klwh$. Since $C = 128$ when $l = w = h = 2$, we have

$128 = k(2)(2)(2) \Leftrightarrow 128 = 8k \Leftrightarrow k = 16$. Therefore, $C = 16lwh$.

27. Since s is inversely proportional to the square root of t, we have $s = \dfrac{k}{\sqrt{t}}$. Since $s = 100$ when $t = 25$, we have $100 = \dfrac{k}{\sqrt{25}}$

$\Leftrightarrow 100 = \dfrac{k}{5} \Leftrightarrow k = 500$. So $s = \dfrac{500}{\sqrt{t}}$.

29. (a) The force F needed is $F = kx$.

   (b) Since $F = 40$ when $x = 5$, we have $40 = k(5) \Leftrightarrow k = 8$.

   (c) From part (b), we have $F = 8x$. Substituting $x = 4$ into $F = 8x$ gives $F = 8(4) = 32$ N.

31. (a) $C = kpm$

   (b) Since $C = 60{,}000$ when $p = 120$ and $m = 4000$, we get $60{,}000 = k(120)(4000) \Leftrightarrow k = \frac{1}{8}$. So $C = \frac{1}{8}pm$.

   (c) Substituting $p = 92$ and $m = 5000$, we get $C = \frac{1}{8}(92)(5000) = \$57{,}500$.

33. (a) $P = ks^3$.

   (b) Since $P = 96$ when $s = 20$, we get $96 = k \cdot 20^3 \Leftrightarrow k = 0.012$. So $P = 0.012s^3$.

   (c) Substituting $x = 30$, we get $P = 0.012 \cdot 30^3 = 324$ watts.

35. $L = \dfrac{k}{d^2}$. Since $L = 70$ when $d = 10$, we have $70 = \dfrac{k}{10^2}$ so $k = 7{,}000$. Thus $L = \dfrac{7{,}000}{d^2}$. When $d = 100$ we get

$L = \dfrac{7{,}000}{100^2} = 0.7$ dB.

37. $P = kAv^3$. If $A = \frac{1}{2}A_0$ and $v = 2v_0$, then $P = k\left(\frac{1}{2}A_0\right)(2v_0)^3 = \frac{1}{2}kA_0\left(8v_0^3\right) = 4kA_0v_0^3$. The power is increased by a

factor of 4.

39. $F = kAs^2$. Since $F = 220$ when $A = 40$ and $s = 5$. Solving for k we have $220 = k(40)(5)^2 \Leftrightarrow 220 = 1000k \Leftrightarrow$

$k = 0.22$. Now when $A = 28$ and $F = 175$ we get $175 = 0.220(28)s^2 \Leftrightarrow 28.4090 = s^2$ so $s = \sqrt{28.4090} = 5.33$ mi/h.

41. (a) $R = \dfrac{kL}{d^2}$

   (b) Since $R = 140$ when $L = 1.2$ and $d = 0.005$, we get $140 = \dfrac{k(1.2)}{(0.005)^2} \Leftrightarrow k = \dfrac{7}{2400} = 0.002916\overline{6}$.

   (c) Substituting $L = 3$ and $d = 0.008$, we have $R = \dfrac{7}{2400} \cdot \dfrac{3}{(0.008)^2} = \dfrac{4375}{32} \approx 137\ \Omega$.

43. (a) For the sun, $E_S = k6000^4$ and for earth $E_E = k300^4$. Thus $\dfrac{E_S}{E_E} = \dfrac{k6000^4}{k300^4} = \left(\dfrac{6000}{300}\right)^4 = 20^4 = 160{,}000$. So the sun

produces $160{,}000$ times the radiation energy per unit area than the Earth.

   (b) The surface area of the sun is $4\pi(435{,}000)^2$ and the surface area of the Earth is $4\pi(3{,}960)^2$. So the sun has

$\dfrac{4\pi(435{,}000)^2}{4\pi(3{,}960)^2} = \left(\dfrac{435{,}000}{3{,}960}\right)^2$ times the surface area of the Earth. Thus the total radiation emitted by the sun is

$160{,}000 \times \left(\dfrac{435{,}000}{3{,}960}\right)^2 = 1{,}930{,}670{,}340$ times the total radiation emitted by the Earth.

45. Let S be the final size of the cabbage, in pounds, let N be the amount of nutrients it receives, in ounces, and let c be the number of other cabbages around it. Then $S = k\dfrac{N}{c}$. When $N = 20$ and $c = 12$, we have $S = 30$, so substituting, we have

$30 = k\frac{20}{12} \Leftrightarrow k = 18$. Thus $S = 18\dfrac{N}{c}$. When $N = 10$ and $c = 5$, the final size is $S = 18\left(\frac{10}{5}\right) = 36$ lb.

47. (a) Since f is inversely proportional to L, we have $f = \dfrac{k}{L}$, where k is a positive constant.

    (b) If we replace L by $2L$ we have $\dfrac{k}{2L} = \frac{1}{2} \cdot \dfrac{k}{L} = \frac{1}{2}f$. So the frequency of the vibration is cut in half.

49. Examples include radioactive decay and exponential growth in biology.

CHAPTER 1 REVIEW

1. Commutative Property for addition.

3. Distributive Property.

5. $[-2, 6) = \{x \mid -2 \le x < 6\}$

7. $x \ge 5 \Leftrightarrow x \in [5, \infty)$

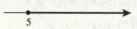

9. $|3 - |-9|| = |3 - 9| = |-6| = 6$

11. $2^{-3} - 3^{-2} = \frac{1}{8} - \frac{1}{9} = \frac{9}{72} - \frac{8}{72} = \frac{1}{72}$

13. $216^{-1/3} = \dfrac{1}{216^{1/3}} = \dfrac{1}{\sqrt[3]{216}} = \frac{1}{6}$

15. $\dfrac{\sqrt{242}}{\sqrt{2}} = \sqrt{\frac{242}{2}} = \sqrt{121} = 11$

17. $2^{1/2}8^{1/2} = \sqrt{2} \cdot \sqrt{8} = \sqrt{16} = 4$

19. $\dfrac{x^2 (2x)^4}{x^3} = \dfrac{x^2 \cdot 16x^4}{x^3} = 16x^{2+4-3} = 16x^3$

21. $\left(3xy^2\right)^3 \left(\frac{2}{3}x^{-1}y\right)^2 = 27x^3y^6 \cdot \frac{4}{9}x^{-2}y^2 = 27 \cdot \frac{4}{9}x^{3-2}y^{6+2} = 12xy^8$

23. $\sqrt[3]{(x^3y)^2\, y^4} = \sqrt[3]{x^6y^4y^2} = \sqrt[3]{x^6y^6} = x^2y^2$

25. $\left(\dfrac{9x^3y}{y^{-3}}\right)^{1/2} = \left(9x^3y^4\right)^{1/2} = 3x^{3/2}y^2$

27. $\dfrac{8r^{1/2}s^{-3}}{2r^{-2}s^4} = 4r^{(1/2)-(-2)}s^{-3-4} = 4r^{5/2}s^{-7} = \dfrac{4r^{5/2}}{s^7}$

29. $78{,}250{,}000{,}000 = 7.825 \times 10^{10}$

31. $\dfrac{ab}{c} \approx \dfrac{(0.00000293)\left(1.582 \times 10^{-14}\right)}{2.8064 \times 10^{12}} = \dfrac{\left(2.93 \times 10^{-6}\right)\left(1.582 \times 10^{-14}\right)}{2.8064 \times 10^{12}} = \dfrac{2.93 \cdot 1.582}{2.8064} \times 10^{-6-14-12}$

       $\approx 1.65 \times 10^{-32}$

33. $12x^2y^4 - 3xy^5 + 9x^3y^2 = 3xy^2\left(4xy^2 - y^3 + 3x^2\right)$

35. $x^2 + 3x - 10 = (x + 5)(x - 2)$

37. $4t^2 - 13t - 12 = (4t + 3)(t - 4)$

39. $25 - 16t^2 = (5 - 4t)(5 + 4t)$

41. $x^6 - 1 = \left(x^3 - 1\right)\left(x^3 + 1\right) = (x - 1)\left(x^2 + x + 1\right)(x + 1)\left(x^2 - x + 1\right)$

43. $x^{-1/2} - 2x^{1/2} + x^{3/2} = x^{-1/2}\left(1 - 2x + x^2\right) = x^{-1/2}(1 - x)^2 = x^{-1/2}(x - 1)^2$

45. $4x^3 - 8x^2 + 3x - 6 = 4x^2(x - 2) + 3(x - 2) = \left(4x^2 + 3\right)(x - 2)$

47. $\left(x^2 + 2\right)^{5/2} + 2x\left(x^2 + 2\right)^{3/2} + x^2\sqrt{x^2 + 2} = \left(x^2 + 2\right)^{1/2}\left(\left(x^2 + 2\right)^2 + 2x\left(x^2 + 2\right) + x^2\right)$

    $= \sqrt{x^2 + 2}\left(x^4 + 4x^2 + 4 + 2x^3 + 4x + x^2\right) = \sqrt{x^2 + 2}\left(x^4 + 2x^3 + 5x^2 + 4x + 4\right) = \sqrt{x^2 + 2}\left(x^2 + x + 2\right)^2$

49. $(2x + 1)(3x - 2) - 5(4x - 1) = 6x^2 - 4x + 3x - 2 - 20x + 5 = 6x^2 - 21x + 3$

51. $(1 + x)(2 - x) - (3 - x)(3 + x) = 2 + x - x^2 - (9 - x^2) = 2 + x - x^2 - 9 + x^2 = -7 + x$

53. $x^2(x - 2) + x(x - 2)^2 = x^3 - 2x^2 + x(x^2 - 4x + 4) = x^3 - 2x^2 + x^3 - 4x^2 + 4x = 2x^3 - 6x^2 + 4x$

55. $\dfrac{x^2 + 2x - 3}{x^2 + 8x + 16} \cdot \dfrac{3x + 12}{x - 1} = \dfrac{(x + 3)(x - 1)}{(x + 4)(x + 4)} \cdot \dfrac{3(x + 4)}{(x - 1)} = \dfrac{3(x + 3)}{x + 4}$

57. $\dfrac{x^2 - 2x - 15}{x^2 - 6x + 5} \div \dfrac{x^2 - x - 12}{x^2 - 1} = \dfrac{(x - 5)(x + 3)}{(x - 5)(x - 1)} \cdot \dfrac{(x - 1)(x + 1)}{(x - 4)(x + 3)} = \dfrac{x + 1}{x - 4}$

59. $\dfrac{1}{x - 1} - \dfrac{2}{x^2 - 1} = \dfrac{1}{x - 1} - \dfrac{2}{(x - 1)(x + 1)} = \dfrac{x + 1}{(x - 1)(x + 1)} - \dfrac{2}{(x - 1)(x + 1)}$

$$= \dfrac{x + 1 - 2}{(x - 1)(x + 1)} = \dfrac{x - 1}{(x - 1)(x + 1)} = \dfrac{1}{x + 1}$$

61. $\dfrac{\frac{1}{x} - \frac{1}{2}}{x - 2} = \dfrac{\frac{2}{2x} - \frac{x}{2x}}{x - 2} = \dfrac{2 - x}{2x} \cdot \dfrac{1}{x - 2} = \dfrac{-1(x - 2)}{2x} \cdot \dfrac{1}{x - 2} = \dfrac{-1}{2x}$

63. $\dfrac{\sqrt{6}}{\sqrt{3} + \sqrt{2}} = \dfrac{\sqrt{6}}{\sqrt{3} + \sqrt{2}} \cdot \dfrac{\sqrt{3} - \sqrt{2}}{\sqrt{3} - \sqrt{2}} = \dfrac{3\sqrt{2} - 2\sqrt{3}}{3 - 2} = 3\sqrt{2} - 2\sqrt{3}$

65. $7x - 6 = 4x + 9 \Leftrightarrow 3x = 15 \Leftrightarrow x = 5$

67. $\dfrac{x + 1}{x - 1} = \dfrac{3x}{3x - 6} = \dfrac{3x}{3(x - 2)} = \dfrac{x}{x - 2} \Leftrightarrow (x + 1)(x - 2) = x(x - 1) \Leftrightarrow x^2 - x - 2 = x^2 - x \Leftrightarrow -2 = 0.$ Since this
last equation is never true, there is no real solution to the original equation.

69. $x^2 - 9x + 14 = 0 \Leftrightarrow (x - 7)(x - 2) = 0 \Leftrightarrow x = 7$ or $x = 2.$

71. $2x^2 + x = 1 \Leftrightarrow 2x^2 + x - 1 = 0 \Leftrightarrow (2x - 1)(x + 1) = 0.$ So either $2x - 1 = 0 \Leftrightarrow 2x = 1 \Leftrightarrow x = \frac{1}{2};$ or $x + 1 = 0 \Leftrightarrow$
$x = -1.$

73. $0 = 4x^3 - 25x = x(4x^2 - 25) = x(2x - 5)(2x + 5) = 0.$ So either $x = 0;$ or $2x - 5 = 0 \Leftrightarrow 2x = 5 \Leftrightarrow x = \frac{5}{2};$ or
$2x + 5 = 0 \Leftrightarrow 2x = -5 \Leftrightarrow x = -\frac{5}{2}.$

75. $3x^2 + 4x - 1 = 0 \Rightarrow$

$$x = \dfrac{-b \pm \sqrt{b^2 - 4ac}}{2a} = \dfrac{-(4) \pm \sqrt{(4)^2 - 4(3)(-1)}}{2(-3)} = \dfrac{-4 \pm \sqrt{16 + 12}}{-6} = \dfrac{-4 \pm \sqrt{28}}{-6} = \dfrac{-4 \pm 2\sqrt{7}}{6} = \dfrac{2(-2 \pm \sqrt{7})}{-6} = \dfrac{-2 \pm \sqrt{7}}{3}.$$

77. $\dfrac{x}{x - 2} + \dfrac{1}{x + 2} = \dfrac{8}{x^2 - 4} \Leftrightarrow x(x + 2) + (x - 2) = 8 \Leftrightarrow x^2 + 2x + x - 2 = 8 \Leftrightarrow x^2 + 3x - 10 = 0 \Leftrightarrow (x - 2)(x + 5) = 0$
$\Leftrightarrow x = 2$ or $x = -5.$ However, since $x = 2$ makes the expression undefined, we reject this solution. Hence the only
solution is $x = -5.$

79. $|x - 7| = 4 \Leftrightarrow x - 7 = \pm 4 \Leftrightarrow x = 7 \pm 4,$ so $x = 11$ or $x = 3.$

81. Let x be the number of pounds of raisins. Then the number of pounds of nuts is $50 - x.$

	Raisins	Nuts	Mixture
Pounds	x	$50 - x$	50
Rate (cost per pound)	3.20	2.40	2.72

So $3.20x + 2.40(50 - x) = 2.72(50) \Leftrightarrow 3.20x + 120 - 2.40x = 136 \Leftrightarrow 0.8x = 16 \Leftrightarrow x = 20.$ Thus the mixture uses
20 pounds of raisins and $50 - 20 = 30$ pounds of nuts.

83. Let r be the rate the woman runs in mi/h. Then she cycles at $r + 8$ mi/h.

	Rate	Time	Distance
Cycle	$r + 8$	$\dfrac{4}{r+8}$	4
Run	r	$\dfrac{2.5}{r}$	2.5

Since the total time of the workout is 1 hour, we have $\dfrac{4}{r+8} + \dfrac{2.5}{r} = 1$. Multiplying by $2r\,(r + 8)$, we

get $4\,(2r) + 2.5\,(2)\,(r + 8) = 2r\,(r + 8) \Leftrightarrow 8r + 5r + 40 = 2r^2 + 16r \Leftrightarrow 0 = 2r^2 + 3r - 40 \Rightarrow$

$r = \dfrac{-3 \pm \sqrt{(3)^2 - 4(2)(-40)}}{2(2)} = \dfrac{-3 \pm \sqrt{9 + 320}}{4} = \dfrac{-3 \pm \sqrt{329}}{4}$. Since $r \geq 0$, we reject the negative value. She runs at

$r = \dfrac{-3 + \sqrt{329}}{4} \approx 3.78$ mi/h.

85. Let t be the time it would take Abbie to paint a living room if she works alone. It would take Beth $2t$ hours to paint the

living room alone, and it would take $3t$ hours for Cathie to paint the living room. Thus Abbie does $\dfrac{1}{t}$ of the job per hour,

Beth does $\dfrac{1}{2t}$ of the job per hour, and Cathie does $\dfrac{1}{3t}$ of the job per hour. So $\dfrac{1}{t} + \dfrac{1}{2t} + \dfrac{1}{3t} = 1 \Leftrightarrow 6 + 3 + 2 = 6t \Leftrightarrow$

$6t = 11 \Leftrightarrow t = \frac{11}{6}$. So it would Abbie 1 hour 50 minutes to paint the living room alone.

87. $3x - 2 > -11 \Leftrightarrow 3x > -9 \Leftrightarrow x > -3$.
Interval: $(-3, \infty)$.

Graph:

89. $x^2 + 4x - 12 > 0 \Leftrightarrow (x - 2)\,(x + 6) > 0$. The expression on the left of the inequality changes sign where $x = 2$ and where $x = -6$. Thus we must check the intervals in the following table.

Interval	$(-\infty, -6)$	$(-6, 2)$	$(2, \infty)$
Sign of $x - 2$	$-$	$-$	$+$
Sign of $x + 6$	$-$	$+$	$+$
Sign of $(x - 2)\,(x + 6)$	$+$	$-$	$+$

Interval: $(-\infty, -6) \cup (2, \infty)$.

Graph:

91. $\dfrac{x - 4}{x^2 - 4} \leq 0 \Leftrightarrow \dfrac{x - 4}{(x - 2)\,(x + 2)} \leq 0$. The expression on the left of the inequality changes sign where $x = -2$, where $x = 2$, and where $x = 4$. Thus we must check the intervals in the following table.

Interval	$(-\infty, -2)$	$(-2, 2)$	$(2, 4)$	$(4, \infty)$
Sign of $x - 4$	$-$	$-$	$-$	$+$
Sign of $x - 2$	$-$	$-$	$+$	$+$
Sign of $x + 2$	$-$	$+$	$+$	$+$
Sign of $\dfrac{x - 4}{(x - 2)\,(x + 2)}$	$-$	$+$	$-$	$+$

Since the expression is not defined when $x = \pm 2$, we exclude these values and the solution is $(-\infty, -2) \cup (2, 4]$.

Graph:

93. $|x - 5| \leq 3 \Leftrightarrow -3 \leq x - 5 \leq 3 \Leftrightarrow 2 \leq x \leq 8$.

Interval: $[2, 8]$

Graph:

95. $x^2 - 4x = 2x + 7$. We graph the equations $y_1 = x^2 - 4x$ and $y_2 = 2x + 7$ in the viewing rectangle rectangle $[-10, 10]$ by $[-5, 25]$. Using a zoom or trace function, we get the solutions $x = -1$ and $x = 7$.

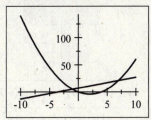

97. $4x - 3 \geq x^2$. We graph the equations $y_1 = 4x - 3$ and $y_2 = x^2$ in the viewing rectangle $[-5, 5]$ by $[0, 15]$. Using a zoom or trace function, we find the points of intersection are at $x = 1$ and $x = 3$. Since we want $4x - 3 \geq x^2$, the solution is the interval $[1, 3]$.

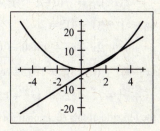

99. (a)

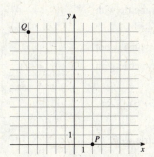

(b) The distance from P to Q is

$$d(P, Q) = \sqrt{(-5 - 2)^2 + (12 - 0)^2}$$
$$= \sqrt{49 + 144} = \sqrt{193}$$

(c) The midpoint is $\left(\dfrac{-5 + 2}{2}, \dfrac{12 + 0}{2} \right) = \left(-\dfrac{3}{2}, 6 \right)$.

(d) The line has slope $m = \dfrac{12 - 0}{-5 - 2} = -\dfrac{12}{7}$, and has equation $y - 0 = -\dfrac{12}{7}(x - 2) \Leftrightarrow$ $y = -\dfrac{12}{7}x + \dfrac{24}{7} \Leftrightarrow 12x + 7y - 24 = 0$.

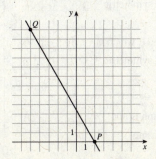

(e) The radius of this circle was found in part (b). It is $r = d(P, Q) = \sqrt{193}$. So an equation is

$$(x - 2)^2 + (y - 0)^2 = \left(\sqrt{193} \right)^2 \Leftrightarrow (x - 2)^2 + y^2 = 193.$$

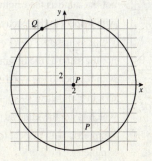

101.

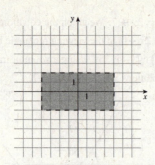

103. $d(A, C) = \sqrt{(4 - (-1))^2 + (4 - (-3))^2} = \sqrt{(4 + 1)^2 + (4 + 3)^2} = \sqrt{74}$ and

$d(B, C) = \sqrt{(5 - (-1))^2 + (3 - (-3))^2} = \sqrt{(5 + 1)^2 + (3 + 3)^2} = \sqrt{72}$. Therefore, B is closer to C.

105. The center is $C = (-5, -1)$, and the point $P = (0, 0)$ is on the circle. The radius of the circle is

$r = d(P, C) = \sqrt{(0 - (-5))^2 + (0 - (-1))^2} = \sqrt{(0 + 5)^2 + (0 + 1)^2} = \sqrt{26}$. Thus, the equation of the circle is

$(x + 5)^2 + (y + 1)^2 = 26$.

107. $x^2 + y^2 + 2x - 6y + 9 = 0 \Leftrightarrow \left(x^2 + 2x\right) + \left(y^2 - 6y\right) = -9 \Leftrightarrow \left(x^2 + 2x + 1\right) + \left(y^2 - 6y + 9\right) = -9 + 1 + 9 \Leftrightarrow$

$(x + 1)^2 + (y - 3)^2 = 1$. This equation represents a circle with center at $(-1, 3)$ and radius 1.

109. $x^2 + y^2 + 72 = 12x \Leftrightarrow \left(x^2 - 12x\right) + y^2 = -72 \Leftrightarrow \left(x^2 - 12x + 36\right) + y^2 = -72 + 36 \Leftrightarrow (x - 6)^2 + y^2 = -36$. Since

the left side of this equation must be greater than or equal to zero, this equation has no graph.

111. $y = 2 - 3x$

x	y
-2	8
0	2
$\frac{2}{3}$	0

x-axis symmetry: $(-y) = 2 - 3x \quad \Leftrightarrow \quad y = -2 + 3x$, which is not the same as the original equation, so the graph is not symmetric with respect to the x-axis.

y-axis symmetry: $y = 2 - 3(-x) \quad \Leftrightarrow \quad y = 2 + 3x$, which is not the same as the original equation, so the graph is not symmetric with respect to the y-axis.

Origin symmetry: $(-y) = 2 - 3(-x) \quad \Leftrightarrow \quad -y = 2 + 3x \quad \Leftrightarrow$

$y = -2 - 3x$, which is not the

same as the original equation, so the graph is not symmetric with respect to the origin.

Hence the graph has no symmetry.

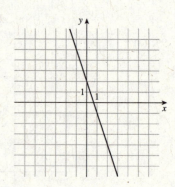

113. $x + 3y = 21 \Leftrightarrow y = -\frac{1}{3}x + 7$

x	y
-3	8
0	7
21	0

x-axis symmetry: $x + 3(-y) = 21 \quad \Leftrightarrow \quad x - 3y = 21$, which is not the same as the original equation, so the graph is not symmetric with respect to the x-axis.

y-axis symmetry: $(-x) + 3y = 21 \quad \Leftrightarrow \quad x - 3y = -21$, which is not the same as the original equation, so the graph is not symmetric with respect to the y-axis.

Origin symmetry: $(-x) + 3(-y) = 21 \quad \Leftrightarrow \quad x + 3y = -21$,

which is not the same as the original equation, so the graph is not symmetric with respect to the origin. Hence the graph has no symmetry.

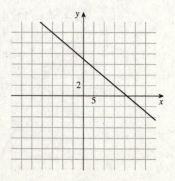

115. $y = 16 - x^2$

x	y
-3	7
-1	15
0	16
1	15
3	7

x-axis symmetry: $(-y) = 16 - x^2 \Leftrightarrow y = -16 + x^2$, which is not the same as the original equation, so the graph is not symmetric with respect to the x-axis.

y-axis symmetry: $y = 16 - (-x)^2 \Leftrightarrow y = 16 - x^2$, which is the same as the original equation, so its is symmetric with respect to the y-axis.

Origin symmetry: $(-y) = 16 - (-x)^2 \Leftrightarrow y = -16 + x^2$, which is not the same as the original equation, so the graph is not symmetric with respect to the origin.

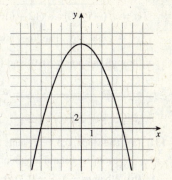

117. $x = \sqrt{y}$

x	y
0	0
1	1
2	4
3	9

x-axis symmetry: $x = \sqrt{-y}$, which is not the same as the original equation, so the graph is not symmetric with respect to the x-axis.

y-axis symmetry: $(-x) = \sqrt{y} \Leftrightarrow x = -\sqrt{y}$, which is not the same as the original equation, so the graph is not symmetric with respect to the y-axis.

Origin symmetry: $(-x) = \sqrt{-y}$, which is not the same as the original equation, so the graph is not symmetric with respect to the origin. Hence, the graph has no symmetry.

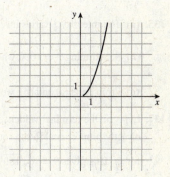

119. $y = x^2 - 6x$. Viewing rectangle $[-10, 10]$ by $[-10, 10]$.

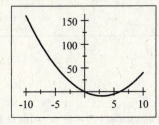

121. $y = x^3 - 4x^2 - 5$. Viewing rectangle $[-4, 10]$ by $[-30, 20]$.

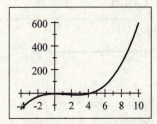

123. The line has slope $m = \dfrac{-4+6}{2+1} = \dfrac{2}{3}$, and so, by the point-slope formula, the equation is $y + 4 = \frac{2}{3}(x-2) \Leftrightarrow y = \frac{2}{3}x - \frac{16}{3}$ $\Leftrightarrow 2x - 3y - 16 = 0$.

125. The x-intercept is 4, and the y-intercept is 12, so the slope is $m = \dfrac{12-0}{0-4} = -3$. Therefore, by the slope-intercept formula, the equation of the line is $y = -3x + 12 \Leftrightarrow 3x + y - 12 = 0$.

127. We first find the slope of the line $3x + 15y = 22$. This gives $3x + 15y = 22 \Leftrightarrow 15y = -3x + 22 \Leftrightarrow y = -\frac{1}{5}x + \frac{22}{15}$. So this line has slope $m = -\frac{1}{5}$, as does any line parallel to it. Then the parallel line passing through the origin has equation $y - 0 = -\frac{1}{5}(x - 0) \Leftrightarrow x + 5y = 0$.

129. Here the center is at $(0, 0)$, and the circle passes through the point $(-5, 12)$, so the radius is

$r = \sqrt{(-5 - 0)^2 + (12 - 0)^2} = \sqrt{25 + 144} = \sqrt{169} = 13$. The equation of the circle is $x^2 + y^2 = 13^2 \Leftrightarrow$

$x^2 + y^2 = 169$. The line shown is the tangent that passes through the point $(-5, 12)$, so it is perpendicular to the line

through the points $(0, 0)$ and $(-5, 12)$. This line has slope $m_1 = \dfrac{12 - 0}{-5 - 0} = -\dfrac{12}{5}$. The slope of the line we seek is

$m_2 = -\dfrac{1}{m_1} = -\dfrac{1}{-12/5} = \dfrac{5}{12}$. Thus, an equation of the tangent line is $y - 12 = \frac{5}{12}(x + 5) \Leftrightarrow y - 12 = \frac{5}{12}x + \frac{25}{12} \Leftrightarrow$

$y = \frac{5}{12}x + \frac{169}{12} \Leftrightarrow 5x - 12y + 169 = 0$.

131. (a) The slope, 0.3, represents the increase in length of the spring for each unit increase in weight w. The S-intercept is the
resting or natural length of the spring.

(b) When $w = 5$, $S = 0.3(5) + 2.5 = 1.5 + 2.5 = 4.0$ inches.

133. Since M varies directly as z we have $M = kz$. Substituting $M = 120$ when $z = 15$, we find $120 = k(15) \Leftrightarrow k = 8$.
Therefore, $M = 8z$.

135. (a) The intensity I varies inversely as the square of the distance d, so $I = \dfrac{k}{d^2}$.

(b) Substituting $I = 1000$ when $d = 8$, we get $1000 = \dfrac{k}{(8)^2} \Leftrightarrow k = 64{,}000$.

(c) From parts (a) and (b), we have $I = \dfrac{64{,}000}{d^2}$. Substituting $d = 20$, we get $I = \dfrac{64{,}000}{(20)^2} = 160$ candles.

137. Let v be the terminal velocity of the parachutist in mi/h and w be his weight in pounds. Since the terminal velocity is
directly proportional to the square root of the weight, we have $v = k\sqrt{w}$. Substituting $v = 9$ when $w = 160$, we solve

for k. This gives $9 = k\sqrt{160} \Leftrightarrow k = \dfrac{9}{\sqrt{160}} \approx 0.712$. Thus $v = 0.712\sqrt{w}$. When $w = 240$, the terminal velocity is

$v = 0.712\sqrt{240} \approx 11$ mi/h.

CHAPTER 1 TEST

1. (a)

$(-5, 3]$ $(-5, \infty)$

(b) $x \le 3 \Leftrightarrow x \in (-\infty, 3]$; $-1 \le x < 4 \Leftrightarrow x \in [-1, 4)$ **(c)** Distance $= |-7 - 9| = |-16| = 16$

3. (a) $186{,}000{,}000{,}000 = 1.86 \times 10^{11}$

(b) $0.0000003965 = 3.965 \times 10^{-7}$

5. $\dfrac{\sqrt{10}}{\sqrt{5} - 2} = \dfrac{\sqrt{10}}{\sqrt{5} - 2} \cdot \dfrac{\sqrt{5} + 2}{\sqrt{5} + 2} = \dfrac{\sqrt{50} + 2\sqrt{10}}{5 - 4} = \dfrac{5\sqrt{2} + 2\sqrt{10}}{1} = 5\sqrt{2} + 2\sqrt{10}$

7. (a) $4x^2 - 25 = (2x - 5)(2x + 5)$

(b) $2x^2 + 5x - 12 = (2x - 3)(x + 4)$

(c) $x^3 - 3x^2 - 4x + 12 = x^2(x - 3) - 4(x - 3) = (x - 3)\left(x^2 - 4\right) = (x - 3)(x - 2)(x + 2)$

(d) $x^4 + 27x = x\left(x^3 + 27\right) = x(x + 3)\left(x^2 - 3x + 9\right)$

(e) $3x^{3/2} - 9x^{1/2} + 6x^{-1/2} = 3x^{-1/2}\left(x^2 - 3x + 2\right) = 3x^{-1/2}(x - 2)(x - 1)$

(f) $x^3y - 4xy = xy\left(x^2 - 4\right) = xy(x - 2)(x + 2)$

9. Let t be the time (in hours) it took Mary to drive from Amity to Belleville. Then $4.4 - t$ is the time it took Mary to drive from Belleville to Amity. Since then distance from Amity to Belleville equals the distance from Belleville to Amity we have $50t = 60(4.4 - t) \Leftrightarrow 50t = 264 - 60t \Leftrightarrow 110t = 264 \Leftrightarrow t = 2.4$ hours. Thus the distance is $50(2.4) = 120$ mi.

11. (a) $-4 < 5 - 3x \le 17 \Leftrightarrow -9 < -3x \le 12 \Leftrightarrow 3 > x \ge -4$. Expressing in standard form we have: $-4 \le x < 3$.

Interval: $[-4, 3)$. Graph:

(b) $x(x-1)(x+2) > 0$. The expression on the left of the inequality changes sign when $x = 0$, $x = 1$, and $x = -2$. Thus we must check the intervals in the following table.

Interval	$(-\infty, -2)$	$(-2, 0)$	$(0, 1)$	$(1, \infty)$
Sign of x	$-$	$-$	$+$	$+$
Sign of $x - 1$	$-$	$-$	$-$	$+$
Sign of $x + 2$	$-$	$+$	$+$	$+$
Sign of $x(x-1)(x-2)$	$-$	$+$	$-$	$+$

From the table, the solution set is $\{x \mid -2 < x < 0 \text{ or } 1 < x\}$. Interval: $(-2, 0) \cup (1, \infty)$.

Graph:

(c) $|x - 4| < 3$ is equivalent to $-3 < x - 4 < 3 \Leftrightarrow 1 < x < 7$. Interval: $(1, 7)$. Graph:

(d) $\dfrac{2x - 3}{x + 1} \le 1 \Leftrightarrow \dfrac{2x - 3}{x + 1} - 1 \le 0 \Leftrightarrow \dfrac{2x - 3}{x + 1} - \dfrac{x + 1}{x + 1} \le 0 \Leftrightarrow \dfrac{x - 4}{x + 1} \le 0$. The expression on the left of the inequality changes sign where $x = -4$ and where $x = -1$. Thus we must check the intervals in the following table.

Interval	$(-\infty, -1)$	$(-1, 4)$	$(4, \infty)$
Sign of $x - 4$	$-$	$-$	$+$
Sign of $x + 1$	$-$	$+$	$+$
Sign of $\dfrac{x - 4}{x + 1}$	$+$	$-$	$+$

Since $x = -1$ makes the expression in the inequality undefined, we exclude this value. Interval: $(-1, 4]$.

Graph:

13. For $\sqrt{6x - x^2}$ to be defined as a real number $6x - x^2 \ge 0 \Leftrightarrow x(6 - x) \ge 0$. The expression on the left of the inequality changes sign when $x = 0$ and $x = 6$. Thus we must check the intervals in the following table.

Interval	$(-\infty, 0)$	$(0, 6)$	$(6, \infty)$
Sign of x	$-$	$+$	$+$
Sign of $6 - x$	$+$	$+$	$-$
Sign of $x(6 - x)$	$-$	$+$	$-$

From the table, we see that $\sqrt{6x - x^2}$ is defined when $0 \le x \le 6$.

15. (a)

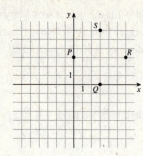

There are several ways to determine the coordinates of S. The diagonals of a square have equal length and are perpendicular. The diagonal PR is horizontal and has length is 6 units. So the diagonal QS is vertical and also has length 6. Thus, the coordinates of S are $(3, 6)$.

(b) The length of PQ is $\sqrt{(0-3)^2 + (3-0)^2} = \sqrt{18} = 3\sqrt{2}$. So the area of $PQRS$ is $\left(3\sqrt{2}\right)^2 = 18$.

17. (a)

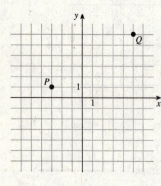

(b) The distance between P and Q is

$$d(P, Q) = \sqrt{(-3-5)^2 + (1-6)^2} = \sqrt{64 + 25} = \sqrt{89}.$$

(c) The midpoint is $\left(\dfrac{-3+5}{2}, \dfrac{1+6}{2}\right) = \left(1, \dfrac{7}{2}\right)$.

(d) The slope of the line is $\dfrac{1-6}{-3-5} = \dfrac{-5}{-8} = \dfrac{5}{8}$.

(e) The perpendicular bisector of PQ contains the midpoint, $\left(1, \dfrac{7}{2}\right)$, and it slope is the negative reciprocal of $\dfrac{5}{8}$. Thus the slope is $-\dfrac{1}{5/8} = -\dfrac{8}{5}$. Hence the equation is $y - \dfrac{7}{2} = -\dfrac{8}{5}(x - 1) \Leftrightarrow y = -\dfrac{8}{5}x + \dfrac{8}{5} + \dfrac{7}{2} = -\dfrac{8}{5}x + \dfrac{51}{10}$. That is, $y = -\dfrac{8}{5}x + \dfrac{51}{10}$.

(f) The center of the circle is the midpoint, $\left(1, \dfrac{7}{2}\right)$, and the length of the radius is $\dfrac{1}{2}\sqrt{89}$. Thus the equation of the circle whose diameter is PQ is $(x - 1)^2 + \left(y - \dfrac{7}{2}\right)^2 = \left(\dfrac{1}{2}\sqrt{89}\right)^2 \Leftrightarrow (x - 1)^2 + \left(y - \dfrac{7}{2}\right)^2 = \dfrac{89}{4}$.

19. $2x - 3y = 15 \Leftrightarrow -3y = -2x + 15 \Leftrightarrow$ $y = \dfrac{2}{3}x - 5$. The slope is $\dfrac{2}{3}$ and the y-intercept is -5.

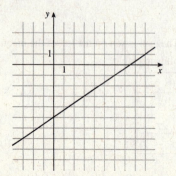

21. (a) When $x = 100$ we have $T = 0.08\,(100) - 4 = 8 - 4 = 4$, so the temperature at one meter is $4°$ C.

(c) The slope represents the raise in temperature as the depth increase. The T-intercept is the surface temperature of the soil and the x-intercept represents the depth of the "frost line", where the soil below is not frozen.

(b)

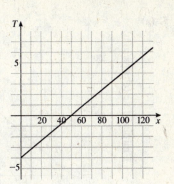

FOCUS ON MODELING   Fitting Lines to Data

1. (a)

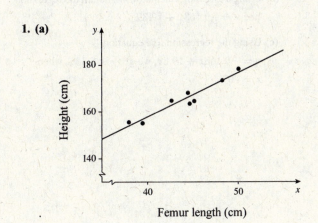

Femur length (cm)

(b) Using a graphing calculator, we obtain the regression line $y = 1.8807x + 82.65$.

(c) Using $x = 58$ in the equation $y = 1.8807x + 82.65$, we get $y = 1.8807\,(58) + 82.65 \approx 191.7$ cm.

3. (a)

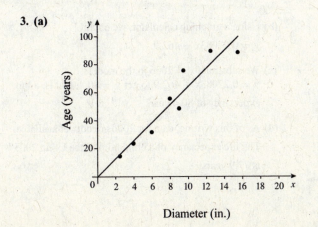

Diameter (in.)

(b) Using a graphing calculator, we obtain the regression line $y = 6.451x - 0.1523$.

(c) Using $x = 18$ in the equation $y = 6.451x - 0.1523$, we get $y = 6.451\,(18) - 0.1523 \approx 116$ years.

5. (a)

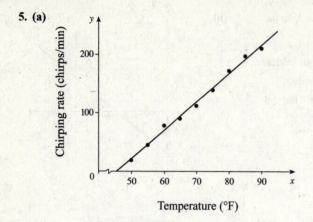

(b) Using a graphing calculator, we obtain the regression line $y = 4.857x - 220.97$.

(c) Using $x = 100°$ F in the equation $y = 4.857x - 220.97$, we get $y \approx 265$ chirps per minute.

7. (a)

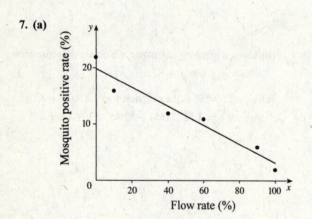

(b) Using a graphing calculator, we obtain the regression line $y = -0.168x + 19.89$.

(c) Using the regression line equation $y = -0.168x + 19.89$, we get $y \approx 8.13\%$ when $x = 70\%$.

9. (a)

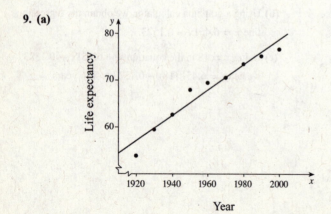

(b) Using a graphing calculator, we obtain $y = 0.27083x - 462.9$.

(c) We substitute $x = 2006$ in the model $y = 0.27083x - 462.9$ to get $y = 80.4$, that is, a life expectancy of 80.4 years.

(d) As of this writing, data for 2006 are not yet available. The life expectancy of a child born in the US in 2005 is 77.9 years.

11. (a) If we take $x = 0$ in 1900 for both men and women, then the regression equation for the men's data is $y = -0.170x + 64.61$ and the regression equation for the women's data is $y = -0.260x + 78.27$.

(b)

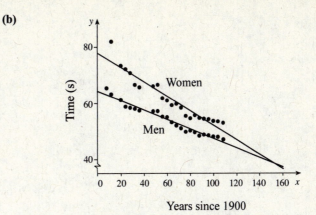

Years since 1900

These lines predict that the women will overtake the men in this event when $-0.170x + 64.61 = -0.260x + 78.27 \Leftrightarrow 0.09x = 13.66 \Leftrightarrow x = 151.78$, or in 2052. This seems unlikely, but who knows?

13. Results will depend on student surveys in each class.

2 FUNCTIONS

2.1 WHAT IS A FUNCTION?

1. If a function f is given by the formula $y = f(x)$, then $f(a)$ is the *value* of f at $x = a$.

3. (a) $f(x) = x^2 - 3x$ and $g(x) = \dfrac{x-5}{x}$ have 5 in their domain because they are defined when $x = 5$. However, $h(x) = \sqrt{x-10}$ is undefined when $x = 5$ because $\sqrt{5-10} = \sqrt{-5}$, so 5 is not in the domain of h.

(b) $f(5) = 5^2 - 3(5) = 25 - 15 = 10$ and $g(5) = \dfrac{5-5}{5} = \dfrac{0}{5} = 0$.

5. $f(x) = 2(x+3)$

7. $f(x) = (x-5)^2$

9. Square, then add 2.

11. Subtract 4, then divide by 3.

13. Machine diagram for $f(x) = \sqrt{x-1}$.

$1 \rightarrow$	subtract 1, then take square root	$\rightarrow 0$
$2 \rightarrow$	subtract 1, then take square root	$\rightarrow 1$
$5 \rightarrow$	subtract 1, then take square root	$\rightarrow 2$

15. $f(x) = 2(x-1)^2$

x	$f(x)$
-1	$2(-1-1)^2 = 8$
0	$2(-1)^2 = 2$
1	$2(1-1)^2 = 0$
2	$2(2-1)^2 = 2$
3	$2(3-1)^2 = 8$

17. $f(x) = x^2 - 6$; $f(-3) = (-3)^2 - 6 = 9 - 6 = 3$; $f(3) = 3^2 - 6 = 9 - 6 = 3$; $f(0) = 0^2 - 6 = -6$; $f\left(\frac{1}{2}\right) = \left(\frac{1}{2}\right)^2 - 6 = \frac{1}{4} - 6 = -\frac{23}{4}$; $f(10) = 10^2 - 6 = 100 - 6 = 94$.

19. $f(x) = 2x + 1$; $f(1) = 2(1) + 1 = 3$; $f(-2) = 2(-2) + 1 = -3$; $f\left(\frac{1}{2}\right) = 2\left(\frac{1}{2}\right) + 1 = 2$; $f(a) = 2(a) + 1 = 2a + 1$; $f(-a) = 2(-a) + 1 = -2a + 1$; $f(a+b) = 2(a+b) + 1 = 2a + 2b + 1$.

21. $g(x) = \dfrac{1-x}{1+x}$; $g(2) = \dfrac{1-(2)}{1+(2)} = \dfrac{-1}{3} = -\dfrac{1}{3}$; $g(-2) = \dfrac{1-(-2)}{1+(-2)} = \dfrac{3}{-1} = -3$; $g\left(\frac{1}{2}\right) = \dfrac{1-\left(\frac{1}{2}\right)}{1+\left(\frac{1}{2}\right)} = \dfrac{\frac{1}{2}}{\frac{3}{2}} = \dfrac{1}{3}$;

$g(a) = \dfrac{1-(a)}{1+(a)} = \dfrac{1-a}{1+a}$; $g(a-1) = \dfrac{1-(a-1)}{1+(a-1)} = \dfrac{1-a+1}{1+a-1} = \dfrac{2-a}{a}$; $g(-1) = \dfrac{1-(-1)}{1+(-1)} = \dfrac{2}{0}$, so $g(-1)$ is not defined.

23. $f(x) = 2x^2 + 3x - 4$; $f(0) = 2(0)^2 + 3(0) - 4 = -4$; $f(2) = 2(2)^2 + 3(2) - 4 = 8 + 6 - 4 = 10$;

$f(-2) = 2(-2)^2 + 3(-2) - 4 = 8 - 6 - 4 = -2$; $f\left(\sqrt{2}\right) = 2\left(\sqrt{2}\right)^2 + 3\left(\sqrt{2}\right) - 4 = 4 + 3\sqrt{2} - 4 = 3\sqrt{2}$;

$f(x+1) = 2(x+1)^2 + 3(x+1) - 4 = 2x^2 + 4x + 2 + 3x + 3 - 4 = 2x^2 + 7x + 1$;

$f(-x) = 2(-x)^2 + 3(-x) - 4 = 2x^2 - 3x - 4$.

25. $f(x) = 2|x-1|$; $f(-2) = 2|-2-1| = 2(3) = 6$; $f(0) = 2|0-1| = 2(1) = 2$;

$f\left(\frac{1}{2}\right) = 2\left|\frac{1}{2} - 1\right| = 2\left(\frac{1}{2}\right) = 1$; $f(2) = 2|2-1| = 2(1) = 2$; $f(x+1) = 2|(x+1) - 1| = 2|x|$;

$f\left(x^2 + 2\right) = 2\left|\left(x^2+2\right) - 1\right| = 2\left|x^2 + 1\right| = 2x^2 + 2$ (since $x^2 + 1 > 0$).

69

27. Since $-2 < 0$, we have $f(-2) = (-2)^2 = 4$. Since $-1 < 0$, we have $f(-1) = (-1)^2 = 1$. Since $0 \geq 0$, we have $f(0) = 0 + 1 = 1$. Since $1 \geq 0$, we have $f(1) = 1 + 1 = 2$. Since $2 \geq 0$, we have $f(2) = 2 + 1 = 3$.

29. Since $-4 \leq -1$, we have $f(-4) = (-4)^2 + 2(-4) = 16 - 8 = 8$. Since $-\frac{3}{2} \leq -1$, we have

$f\left(-\frac{3}{2}\right) = \left(-\frac{3}{2}\right)^2 + 2\left(-\frac{3}{2}\right) = \frac{9}{4} - 3 = -\frac{3}{4}$. Since $-1 \leq -1$, we have $f(-1) = (-1)^2 + 2(-1) = 1 - 2 = -1$. Since $-1 < 0 \leq 1$, we have $f(0) = 0$. Since $25 > 1$, we have $f(25) = -1$.

31. $f(x + 2) = (x + 2)^2 + 1 = x^2 + 4x + 4 + 1 = x^2 + 4x + 5$; $f(x) + f(2) = x^2 + 1 + (2)^2 + 1 = x^2 + 1 + 4 + 1 = x^2 + 6$.

33. $f\left(x^2\right) = x^2 + 4$; $[f(x)]^2 = [x + 4]^2 = x^2 + 8x + 16$.

35. $f(a) = 3(a) + 2 = 3a + 2$; $f(a + h) = 3(a + h) + 2 = 3a + 3h + 2$;

$\dfrac{f(a + h) - f(a)}{h} = \dfrac{(3a + 3h + 2) - (3a + 2)}{h} = \dfrac{3a + 3h + 2 - 3a - 2}{h} = \dfrac{3h}{h} = 3$.

37. $f(a) = 5$; $f(a + h) = 5$; $\dfrac{f(a + h) - f(a)}{h} = \dfrac{5 - 5}{h} = 0$.

39. $f(a) = \dfrac{a}{a + 1}$; $f(a + h) = \dfrac{a + h}{a + h + 1}$;

$\dfrac{f(a + h) - f(a)}{h} = \dfrac{\dfrac{a + h}{a + h + 1} - \dfrac{a}{a + 1}}{h} = \dfrac{\dfrac{(a + h)(a + 1)}{(a + h + 1)(a + 1)} - \dfrac{a(a + h + 1)}{(a + h + 1)(a + 1)}}{h}$

$= \dfrac{\dfrac{(a + h)(a + 1) - a(a + h + 1)}{(a + h + 1)(a + 1)}}{h} = \dfrac{a^2 + a + ah + h - \left(a^2 + ah + a\right)}{h(a + h + 1)(a + 1)}$

$= \dfrac{1}{(a + h + 1)(a + 1)}$

41. $f(a) = 3 - 5a + 4a^2$;

$f(a + h) = 3 - 5(a + h) + 4(a + h)^2 = 3 - 5a - 5h + 4\left(a^2 + 2ah + h^2\right)$

$= 3 - 5a - 5h + 4a^2 + 8ah + 4h^2$;

$\dfrac{f(a + h) - f(a)}{h} = \dfrac{\left(3 - 5a - 5h + 4a^2 + 8ah + 4h^2\right) - \left(3 - 5a + 4a^2\right)}{h}$

$= \dfrac{3 - 5a - 5h + 4a^2 + 8ah + 4h^2 - 3 + 5a - 4a^2}{h} = \dfrac{-5h + 8ah + 4h^2}{h}$

$= \dfrac{h(-5 + 8a + 4h)}{h} = -5 + 8a + 4h$.

43. $f(x) = 2x$. Since there is no restrictions, the domain is the set of real numbers, $(-\infty, \infty)$.

45. $f(x) = 2x$. The domain is restricted by the exercise to $[-1, 5]$.

47. $f(x) = \dfrac{1}{x - 3}$. Since the denominator cannot equal 0 we have $x - 3 \neq 0 \Leftrightarrow x \neq 3$. Thus the domain is $\{x \mid x \neq 3\}$. In interval notation, the domain is $(-\infty, 3) \cup (3, \infty)$.

49. $f(x) = \dfrac{x + 2}{x^2 - 1}$. Since the denominator cannot equal 0 we have $x^2 - 1 \neq 0 \Leftrightarrow x^2 \neq 1 \Rightarrow x \neq \pm 1$. Thus the domain is $\{x \mid x \neq \pm 1\}$. In interval notation, the domain is $(-\infty, -1) \cup (-1, 1) \cup (1, \infty)$.

51. $f(x) = \sqrt{x - 5}$. We require $x - 5 \geq 0 \Leftrightarrow x \geq 5$. Thus the domain is $\{x \mid x \geq 5\}$. The domain can also be expressed in interval notation as $[5, \infty)$.

53. $f(t) = \sqrt[3]{t - 1}$. Since the odd root is defined for all real numbers, the domain is the set of real numbers, $(-\infty, \infty)$.

55. $h(x) = \sqrt{2x - 5}$. Since the square root is defined as a real number only for nonnegative numbers, we require that $2x - 5 \geq 0 \Leftrightarrow 2x \geq 5 \Leftrightarrow x \geq \frac{5}{2}$. So the domain is $\{x \mid x \geq \frac{5}{2}\}$. In interval notation, the domain is $\left[\frac{5}{2}, \infty\right)$.

57. $g(x) = \dfrac{\sqrt{2+x}}{3-x}$. We require $2 + x \geq 0$, and the denominator cannot equal 0. Now $2 + x \geq 0 \Leftrightarrow x \geq -2$, and $3 - x \neq 0$

$\Leftrightarrow x \neq 3$. Thus the domain is $\{x \mid x \geq -2 \text{ and } x \neq 3\}$, which can be expressed in interval notation as $[-2, 3) \cup (3, \infty)$.

59. $g(x) = \sqrt[4]{x^2 - 6x}$. Since the input to an even root must be nonnegative, we have $x^2 - 6x \geq 0 \Leftrightarrow x(x - 6) \geq 0$. We make a table:

	$(-\infty, 0)$	$(0, 6)$	$(6, \infty)$
Sign of x	$-$	$+$	$+$
Sign of $x - 6$	$-$	$-$	$+$
Sign of $x(x - 6)$	$+$	$-$	$+$

Thus the domain is $(-\infty, 0] \cup [6, \infty)$.

61. $f(x) = \dfrac{3}{\sqrt{x-4}}$. Since the input to an even root must be nonnegative and the denominator cannot equal 0, we have

$x - 4 > 0 \Leftrightarrow x > 4$. Thus the domain is $(4, \infty)$.

63. $f(x) = \dfrac{(x+1)^2}{\sqrt{2x-1}}$. Since the input to an even root must be nonnegative and the denominator cannot equal 0, we have

$2x - 1 > 0 \Leftrightarrow x > \frac{1}{2}$. Thus the domain is $\left(\frac{1}{2}, \infty\right)$.

65. To evaluate $f(x)$, divide the input by 3 and add $\frac{2}{3}$ to the result.

(a) $f(x) = \dfrac{x}{3} + \dfrac{2}{3}$

(b)

x	$f(x)$
2	$\frac{4}{3}$
4	2
6	$\frac{8}{3}$
8	$\frac{10}{3}$

(c)

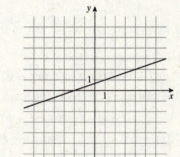

67. Let $T(x)$ be the amount of sales tax charged in Lemon County on a purchase of x dollars. To find the tax, take 8% of the purchase price.

(a) $T(x) = 0.08x$

(b)

x	$T(x)$
2	0.16
4	0.32
6	0.48
8	0.64

(c)

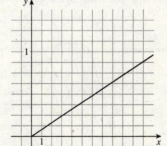

69. (a) $C(10) = 1500 + 3(10) + 0.02(10)^2 + 0.0001(10)^3 = 1500 + 30 + 2 + 0.1 = 1532.1$

$C(100) = 1500 + 3(100) + 0.02(100)^2 + 0.0001(100)^3 = 1500 + 300 + 200 + 100 = 2100$

(b) $C(10)$ represents the cost of producing 10 yards of fabric and $C(100)$ represents the cost of producing 100 yards of fabric.

(c) $C(0) = 1500 + 3(0) + 0.02(0)^2 + 0.0001(0)^3 = 1500$

71. (a) $V(0) = 50\left(1 - \frac{0}{20}\right)^2 = 50$ and $V(20) = 50\left(1 - \frac{20}{20}\right)^2 = 0$.

(b) $V(0) = 50$ represents the volume of the full tank at time $t = 0$, and $V(20) = 0$ represents the volume of the empty tank twenty minutes later.

(c)

x	$V(x)$
0	50
5	28.125
10	12.5
15	3.125
20	0

73. (a) $v(0.1) = 18500\left(0.25 - 0.1^2\right) = 4440$,

$v(0.4) = 18500\left(0.25 - 0.4^2\right) = 1665$.

(b) They tell us that the blood flows much faster (about 2.75 times faster) 0.1 cm from the center than 0.1 cm from the edge.

(c)

r	$v(r)$
0	4625
0.1	4440
0.2	3885
0.3	2960
0.4	1665
0.5	0

75. (a) $L(0.5c) = 10\sqrt{1 - \frac{(0.5c)^2}{c^2}} \approx 8.66$ m, $L(0.75c) = 10\sqrt{1 - \frac{(0.75c)^2}{c^2}} \approx 6.61$ m, and

$L(0.9c) = 10\sqrt{1 - \frac{(0.9c)^2}{c^2}} \approx 4.36$ m.

(b) It will appear to get shorter.

77. (a) $C(75) = 75 + 15 = \$90$; $C(90) = 90 + 15 = \$105$; $C(100) = \$100$; and $C(105) = \$105$.

(b) The total price of the books purchased, including shipping.

79. (a) $F(x) = \begin{cases} 15(40 - x) & \text{if } 0 < x < 40 \\ 0 & \text{if } 40 \le x \le 65 \\ 15(x - 65) & \text{if } x > 65 \end{cases}$

(b) $F(30) = 15(40 - 10) = 15 \cdot 10 = \$150$; $F(50) = \$0$; and $F(75) = 15(75 - 65) \, 15 \cdot 10 = \$150$.

(c) The fines for violating the speed limits on the freeway.

81.

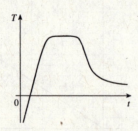

83.

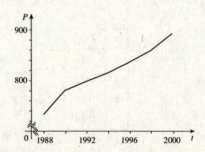

85. Answers will vary.

2.2   GRAPHS OF FUNCTIONS

1. To graph the function f we plot the points $(x, f(x))$ in a coordinate plane. To graph $f(x) = x^3 + 2$ we plot the points $\left(x, x^3 + 2\right)$. So, the point $\left(2, 2^3 + 2\right) = (2, 10)$ is on the graph of f. The height of the graph of f above the x-axis when $x = 2$ is 10.

3. If the point $(2, 3)$ is on the graph of f, then $f(2) = 3$.

5.

x	$f(x) = 2$
-9	2
-6	2
-3	2
0	2
3	2
6	2

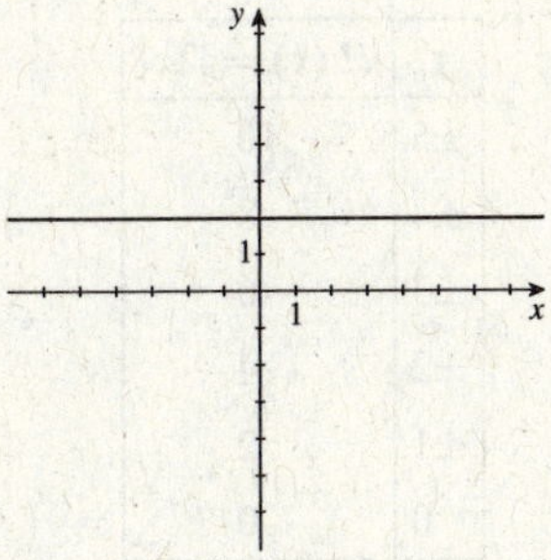

7.

x	$f(x) = 2x - 4$
-1	-6
0	-4
1	-2
2	0
3	2
4	4
5	6

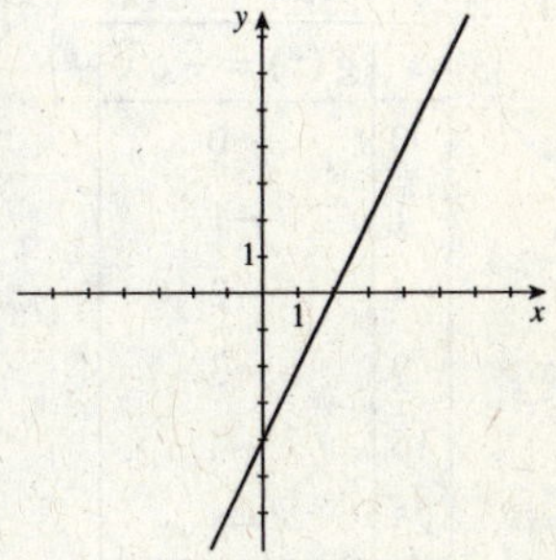

9.

x	$f(x) = -x + 3,$ $-3 \leq x \leq 3$
-3	6
-2	5
0	3
1	2
2	1
3	0

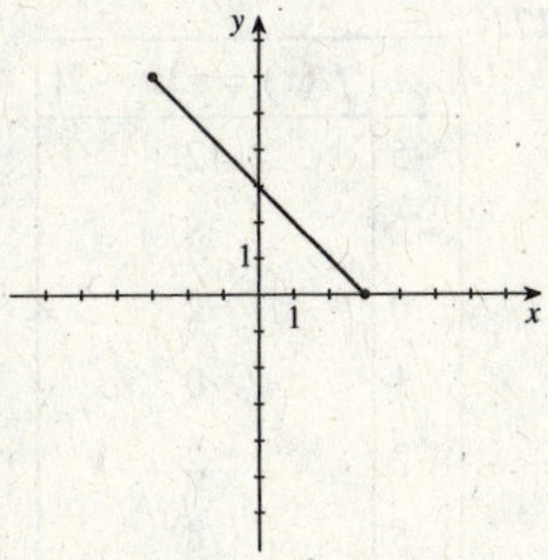

11.

x	$f(x) = -x^2$
± 4	-16
± 3	-9
± 2	-4
± 1	-1
0	0

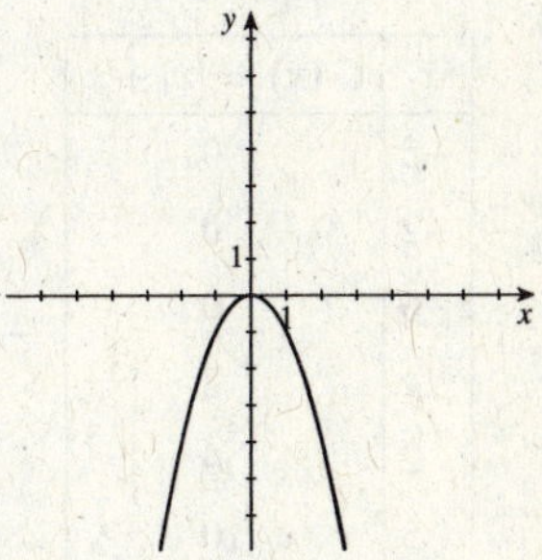

13.

x	$h(x) = 16 - x^2$
± 6	-20
± 4	0
± 2	12
0	16

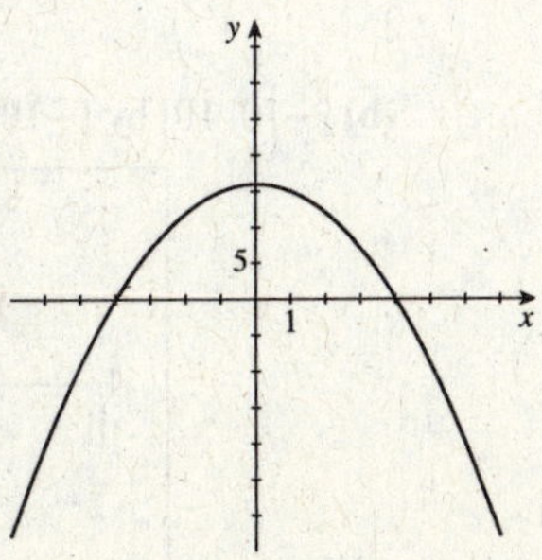

15.

x	$g(x) = x^3 - 8$
-3	-35
-2	-16
-1	-9
0	-8
1	-7
2	0
3	19

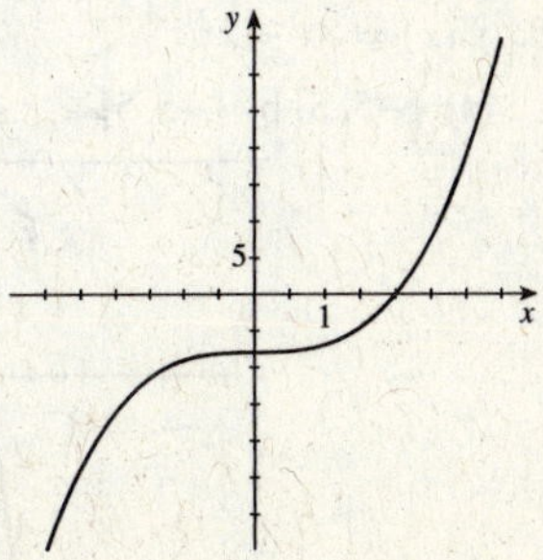

17.

x	$g(x) = x^2 - 2x$
-2	8
-1	3
0	0
1	-1
2	0
3	3
4	8

19.

x	$f(x) = 1 + \sqrt{x}$
0	1
1	2
4	3
9	4
16	5
25	6

21.

x	$g(x) = -\sqrt{x}$
0	0
1	-1
4	-2
9	-3
16	-4
25	-5

23.

x	$H(x) = |2x|$
±5	10
±4	8
±3	6
±2	4
±1	2
0	0

25.

x	$G(x) = |x| + x$
-5	0
-2	0
0	0
1	2
2	4
5	10

27.

x	$f(x) = |2x - 2|$
-5	12
-2	8
0	2
1	0
2	2
5	8

29. $f(x) = 8x - x^2$

   (a) $[-5, 5]$ by $[-5, 5]$

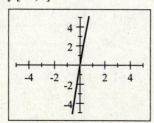

   (b) $[-10, 10]$ by $[-10, 10]$

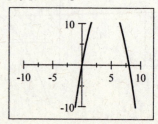

(c) $[-2, 10]$ by $[-5, 20]$

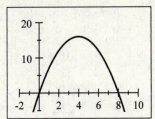

(d) $[-10, 10]$ by $[-100, 100]$

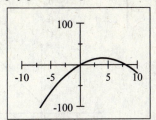

The viewing rectangle in part (c) produces the most appropriate graph of the equation.

31. $h(x) = x^3 - 5x - 4$

(a) $[-2, 2]$ by $[-2, 2]$

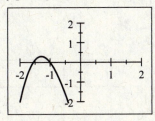

(b) $[-3, 3]$ by $[-10, 10]$

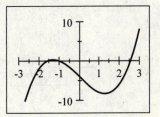

(c) $[-3, 3]$ by $[-10, 5]$

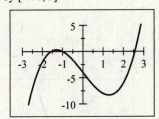

(d) $[-10, 10]$ by $[-10, 10]$

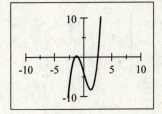

The viewing rectangle in part (c) produces the most appropriate graph of the equation.

33. $f(x) = \begin{cases} 0 & \text{if } x < 2 \\ 1 & \text{if } x \geq 2 \end{cases}$

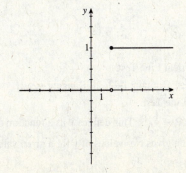

35. $f(x) = \begin{cases} 3 & \text{if } x < 2 \\ x - 1 & \text{if } x \geq 2 \end{cases}$

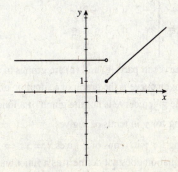

37. $f(x) = \begin{cases} x & \text{if } x \le 0 \\ x+1 & \text{if } x > 0 \end{cases}$

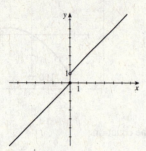

39. $f(x) = \begin{cases} -1 & \text{if } x < -1 \\ 1 & \text{if } -1 \le x \le 1 \\ -1 & \text{if } x > 1 \end{cases}$

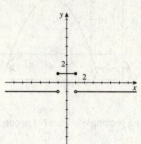

41. $f(x) = \begin{cases} 2 & \text{if } x \le -1 \\ x^2 & \text{if } x > -1 \end{cases}$

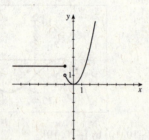

43. $f(x) = \begin{cases} 0 & \text{if } |x| \le 2 \\ 3 & \text{if } |x| > 2 \end{cases}$

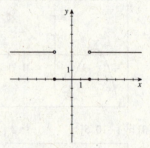

45. $f(x) = \begin{cases} 4 & \text{if } x < -2 \\ x^2 & \text{if } -2 \le x \le 2 \\ -x+6 & \text{if } x > 2 \end{cases}$

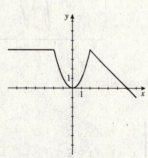

47. $f(x) = \begin{cases} x+2 & \text{if } x \le -1 \\ x^2 & \text{if } x > -1 \end{cases}$

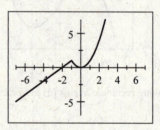

49. $f(x) = \begin{cases} -2 & \text{if } x < -2 \\ x & \text{if } -2 \le x \le 2 \\ 2 & \text{if } x > 2 \end{cases}$

51. The curves in parts (a) and (c) are graphs of a function of x, by the Vertical Line Test.

53. The given curve is the graph of a function of x, by the Vertical Line Test. Domain: $[-3, 2]$. Range: $[-2, 2]$.

55. No, the given curve is not the graph of a function of x, by the Vertical Line Test.

57. Solving for y in terms of x gives $x^2 + 2y = 4 \Leftrightarrow 2y = 4 - x^2 \Leftrightarrow y = 2 - \frac{1}{2}x^2$. This defines y as a function of x.

59. Solving for y in terms of x gives $x = y^2 \Leftrightarrow y = \pm\sqrt{x}$. The last equation gives two values of y for a given value of x. Thus, this equation does not define y as a function of x.

61. Solving for y in terms of x gives $x + y^2 = 9 \Leftrightarrow y^2 = 9 - x \Leftrightarrow y = \pm\sqrt{9 - x}$. The last equation gives two values of y for a given value of x. Thus, this equation does not define y as a function of x.

63. Solving for y in terms of x gives $x^2 y + y = 1 \Leftrightarrow y\left(x^2 + 1\right) = 1 \Leftrightarrow y = \dfrac{1}{x^2 + 1}$. This defines y as a function of x.

65. Solving for y in terms of x gives $2\,|x| + y = 0 \Leftrightarrow y = -2\,|x|$. This defines y as a function of x.

67. Solving for y in terms of x gives $x = y^3 \Leftrightarrow y = \sqrt[3]{x}$. This defines y as a function of x.

69. (a) $f(x) = x^2 + c$, for $c = 0, 2, 4$, and 6.   **(b)** $f(x) = x^2 + c$, for $c = 0, -2, -4$, and -6.

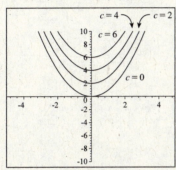   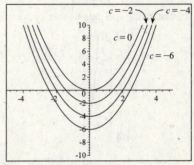

(c) The graphs in part (a) are obtained by shifting the graph of $f(x) = x^2$ upward c units, $c > 0$. The graphs in part (b) are obtained by shifting the graph of $f(x) = x^2$ downward c units.

71. (a) $f(x) = (x - c)^3$, for $c = 0, 2, 4$, and 6.   **(b)** $f(x) = (x - c)^3$, for $c = 0, -2, -4$, and -6.

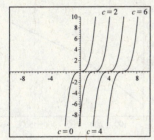   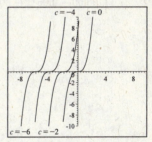

(c) The graphs in part (a) are obtained by shifting the graph of $f(x) = x^3$ to the right c units, $c > 0$. The graphs in part (b) are obtained by shifting the graph of $f(x) = x^3$ to the left $|c|$ units, $c < 0$.

73. (a) $f(x) = x^c$, for $c = \frac{1}{2}, \frac{1}{4}$, and $\frac{1}{6}$.   **(b)** $f(x) = x^c$, for $c = 1, \frac{1}{3}$, and $\frac{1}{5}$.

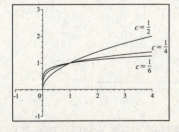   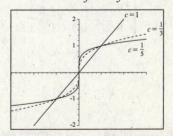

(c) Graphs of even roots are similar to $y = \sqrt{x}$, graphs of odd roots are similar to $y = \sqrt[3]{x}$. As c increases, the graph of $y = \sqrt[c]{x}$ becomes steeper near $x = 0$ and flatter when $x > 1$.

75. The slope of the line segment joining the points $(-2, 1)$ and $(4, -6)$ is $m = \dfrac{-6 - 1}{4 - (-2)} = -\dfrac{7}{6}$. Using the point-slope form,

we have $y - 1 = -\dfrac{7}{6}(x + 2) \Leftrightarrow y = -\dfrac{7}{6}x - \dfrac{7}{3} + 1 \Leftrightarrow y = -\dfrac{7}{6}x - \dfrac{4}{3}$. Thus the function is $f(x) = -\dfrac{7}{6}x - \dfrac{4}{3}$ for $-2 \le x \le 4$.

77. First solve the circle for y: $x^2 + y^2 = 9 \Leftrightarrow y^2 = 9 - x^2 \Rightarrow y = \pm\sqrt{9 - x^2}$. Since we seek the top half of the circle, we choose $y = \sqrt{9 - x^2}$. So the function is $f(x) = \sqrt{9 - x^2}$, $-3 \le x \le 3$.

79. We graph $T(r) = \dfrac{0.5}{r^2}$ for $10 \le r \le 100$. As the balloon is inflated, the skin gets thinner, as we would expect.

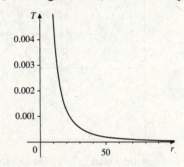

81. (a) $E(x) = \begin{cases} 6.00 + 0.10x & \text{if } 0 \le x \le 300 \\ 36.00 + 0.06(x - 300) & \text{if } 300 < x \end{cases}$

(b)

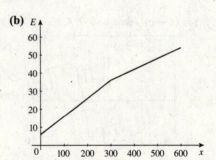

83. $P(x) = \begin{cases} 0.44 & \text{if } 0 < x \le 1 \\ 0.61 & \text{if } 1 < x \le 2 \\ 0.78 & \text{if } 2 < x \le 3 \\ 0.95 & \text{if } 3 < x \le 3.5 \end{cases}$

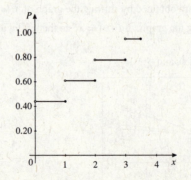

85. Answers will vary. Some examples are almost anything we purchase based on weight, volume, length, or time, for example gasoline. Although the amount delivered by the pump is continuous, the amount we pay is rounded to the penny. An example involving time would be the cost of a telephone call.

87. (a) The graphs of $f(x) = x^2 + x - 6$ and $g(x) = \left|x^2 + x - 6\right|$ are shown in the viewing rectangle $[-10, 10]$ by $[-10, 10]$.

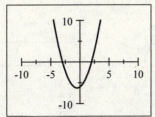   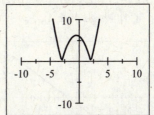

For those values of x where $f(x) \geq 0$, the graphs of f and g coincide, and for those values of x where $f(x) < 0$, the graph of g is obtained from that of f by reflecting the part below the x-axis about the x-axis.

(b) The graphs of $f(x) = x^4 - 6x^2$ and $g(x) = \left|x^4 - 6x^2\right|$ are shown in the viewing rectangle $[-5, 5]$ by $[-10, 15]$.

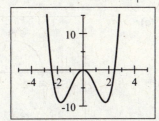   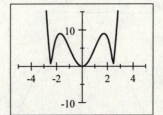

For those values of x where $f(x) \geq 0$, the graphs of f and g coincide, and for those values of x where $f(x) < 0$, the graph of g is obtained from that of f by reflecting the part below the x-axis above the x-axis.

(c) In general, if $g(x) = |f(x)|$, then for those values of x where $f(x) \geq 0$, the graphs of f and g coincide, and for those values of x where $f(x) < 0$, the graph of g is obtained from that of f by reflecting the part below the x-axis above the x-axis.

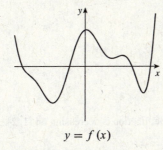   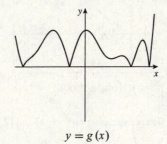

$y = f(x)$ $y = g(x)$

2.3   GETTING INFORMATION FROM THE GRAPH OF A FUNCTION

1. To find a function value $f(a)$ from the graph of f we find the height of the graph above the x-axis at $x = a$. From the graph of f we see that $f(3) = 4$.

3. (a) If f is increasing on an interval, then the y-values of the points on the graph *rise* as the x-values increase. From the graph of f we see that f is increasing on the intervals $[1, 2]$ and $[4, 5]$.

(b) If f is decreasing on an interval, then y-values of the points on the graph *fall* as the x-values increase. From the graph of f we see that f is decreasing on the intervals $[2, 4]$ and $[5, 6]$.

5. (a) $h(-2) = 1, h(0) = -1, h(2) = 3$, and $h(3) = 4$.

(b) Domain: $[-3, 4]$. Range: $[-1, 4]$.

(c) $h(-3) = 3, h(2) = 3$, and $h(4) = 3$, so $h(x) = 3$ when $x = -3, x = 2$, or $x = 4$.

(d) The graph of h lies below or on the horizontal line $y = 3$ when $-3 \le x \le 2$ or $x = 4$, so $h(x) \le 3$ for those values of x.

7. (a) $g(-4) = 3, g(-2) = 2, g(0) = -2, g(2) = 1,$ and $g(4) = 0$.

   (b) Domain: $[-4, 4]$. Range: $[-2, 3]$.

9. (a)

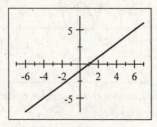

   (b) Domain: $(-\infty, \infty)$; Range: $(-\infty, \infty)$

11. (a)

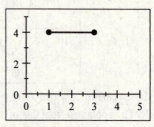

   (b) Domain: $[1, 3]$; Range: $\{4\}$

13. (a)

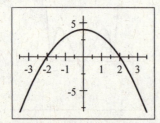

   (b) Domain: $(-\infty, \infty)$; Range: $(-\infty, 4]$

15. (a)

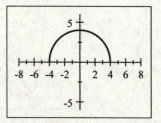

   (b) Domain: $[-4, 4]$; Range: $[0, 4]$

17. (a)

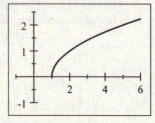

   (b) Domain: $[1, \infty)$; Range: $[0, \infty)$

19. (a) The function is increasing on $[-1, 1]$ and $[2, 4]$.   **(b)** The function is decreasing on $[1, 2]$.

21. (a) The function is increasing on $[-2, -1]$ and $[1, 2]$.   **(b)** The function is decreasing on $[-3, -2]$, $[-1, 1]$, and $[2, 3]$.

23. (a) $f(x) = x^2 - 5x$ is graphed in the viewing rectangle $[-2, 7]$ by $[-10, 10]$.

25. (a) $f(x) = 2x^3 - 3x^2 - 12x$ is graphed in the viewing rectangle $[-3, 5]$ by $[-25, 20]$.

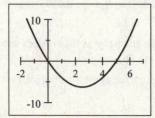

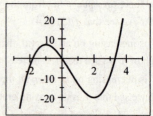

   (b) The function is increasing on $[2.5, \infty)$. It is decreasing on $(-\infty, 2.5]$.

   (b) The function is increasing on $(-\infty, -1]$ and $[2, \infty)$. It is decreasing on $[-1, 2]$.

27. (a) $f(x) = x^3 + 2x^2 - x - 2$ is graphed in the viewing rectangle $[-5, 5]$ by $[-3, 3]$.

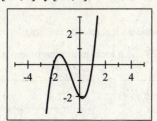

(b) The function is increasing on $(-\infty, -1.55]$ and $[0.22, \infty)$. It is decreasing on $[-1.55, 0.22]$.

29. (a) $f(x) = x^{2/5}$ is graphed in the viewing rectangle $[-10, 10]$ by $[-5, 5]$.

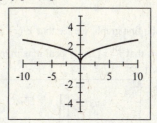

(b) The function is increasing on $[0, \infty)$. It is decreasing on $(-\infty, 0]$.

31. (a) Local maximum: 2 at $x = 0$. Local minimum: -1 at $x = -2$ and 0 at $x = 2$.

(b) The function is increasing on $[-2, 0]$ and $[2, \infty)$ and decreasing on $(-\infty, -2]$ and $[0, 2]$.

33. (a) Local maximum: 0 at $x = 0$ and 1 at $x = 3$. Local minimum: -2 at $x = -2$ and -1 at $x = 1$.

(b) The function is increasing on $[-2, 0]$ and $[1, 3]$ and decreasing on $(-\infty, -2]$, $[0, 1]$, and $[3, \infty)$.

35. (a) In the first graph, we see that $f(x) = x^3 - x$ has a local minimum and a local maximum. Smaller x- and y-ranges show that $f(x)$ has a local maximum of about 0.38 when $x \approx -0.58$ and a local minimum of about -0.38 when $x \approx 0.58$.

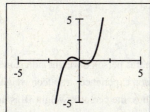 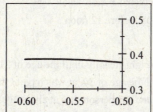 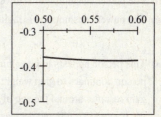

(b) The function is increasing on $(-\infty, -0.58]$ and $[0.58, \infty)$ and decreasing on $[-0.58, 0.58]$.

37. (a) In the first graph, we see that $g(x) = x^4 - 2x^3 - 11x^2$ has two local minimums and a local maximum. The local maximum is $g(x) = 0$ when $x = 0$. Smaller x- and y-ranges show that local minima are $g(x) \approx -13.61$ when $x \approx -1.71$ and $g(x) \approx -73.32$ when $x \approx 3.21$.

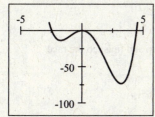 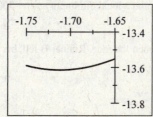 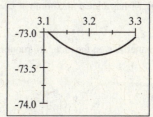

(b) The function is increasing on $[-1.71, 0]$ and $[3.21, \infty)$ and decreasing on $(-\infty, -1.71]$ and $[0, 3.21]$.

39. (a) In the first graph, we see that $U(x) = x\sqrt{6 - x}$ has only a local maximum. Smaller x- and y-ranges show that $U(x)$ has a local maximum of about 5.66 when $x \approx 4.00$.

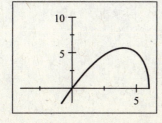 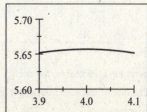

(b) The function is increasing on $(-\infty, 4.00]$ and decreasing on $[4.00, 6]$.

41. (a) In the first graph, we see that $V(x) = \dfrac{1 - x^2}{x^3}$ has a local minimum and a local maximum. Smaller x- and y-ranges show that $V(x)$ has a local maximum of about 0.38 when $x \approx -1.73$ and a local minimum of about -0.38 when $x \approx 1.73$.

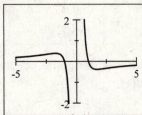 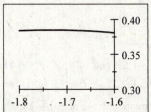 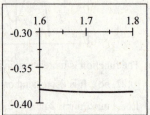

(b) The function is increasing on $(-\infty, -1.73]$ and $[1.73, \infty)$ and decreasing on $[-1.73, 0)$ and $(0, 1.73]$.

43. (a) At 6 A.M. the graph shows that the power consumption is about 500 megawatts. Since $t = 18$ represents 6 P.M., the graph shows that the power consumption at 6 P.M. is about 725 megawatts.

(b) The power consumption is lowest between 3 A.M. and 4 A.M..

(c) The power consumption is highest just before 12 noon.

45. (a) This person appears to be gaining weight steadily until the age of 21 when this person's weight gain slows down. The person continues to gain weight until the age of 30, at which point this person experiences a sudden weight loss. Weight gain resumes around the age of 32, and the person dies at about age 68. Thus, the person's weight W is increasing on $[0, 30]$ and $[32, 68]$ and decreasing on $[30, 32]$

(b) The sudden weight loss could be due to a number of reasons, among them major illness, a weight loss program, etc.

47. (a) The function W is increasing on $[0, 150]$ and $[300, \infty)$ and decreasing on $[150, 300]$.

(b) W has a local maximum at $x = 150$ and a local minimum at $x = 300$.

49. Runner A won the race. All runners finished the race. Runner B fell, but got up and finished the race.

51. (a)

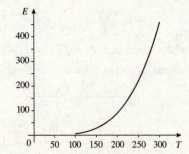

(b) As the temperature T increases, the energy E increases. The rate of increase gets larger as the temperature increases.

53. In the first graph, we see the general location of the maximum of $N(s) = \dfrac{88s}{17 + 17\left(\dfrac{s}{20}\right)^2}$. In the second graph we isolate

the maximum, and from this graph we see that at the speed of 20 mi/h the largest number of cars that can use the highway safely is 52.

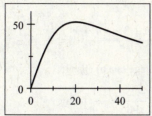
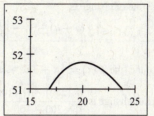

55. In the first graph, we see the general location of the maximum of $v(r) = 3.2(1 - r)r^2$ is around $r = 0.7$ cm. In the second graph, we isolate the maximum, and from this graph we see that at the maximum velocity is approximately 0.47 when $r \approx 0.67$ cm.

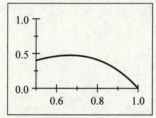
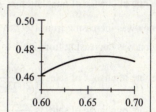

57. Numerous answers are possible.

2.4   AVERAGE RATE OF CHANGE OF A FUNCTION

1. If you travel 100 miles in two hours then your average speed for the trip is average speed $= \dfrac{100 \text{ miles}}{2 \text{ hours}} = 50$ mi/h.

3. The average rate of change of the function $f(x) = x^2$ between $x = 1$ and $x = 5$ is

average rate of change $= \dfrac{f(5) - f(1)}{5 - 1} = \dfrac{5^2 - 1^2}{4} = \dfrac{25 - 1}{4} = \dfrac{24}{4} = 6.$

5. We use the points $(1, 3)$ and $(4, 5)$, so the average rate of change is $\dfrac{5 - 3}{4 - 1} = \dfrac{2}{3}.$

7. We use the points $(0, 6)$ and $(5, 2)$, so the average rate of change is $\dfrac{2 - 6}{5 - 0} = \dfrac{-4}{5}.$

9. The average rate of change is $\dfrac{f(3) - f(2)}{3 - 2} = \dfrac{[3(3) - 2] - [3(2) - 2]}{1} = 7 - 4 = 3.$

11. The average rate of change is $\dfrac{h(4) - h(-1)}{4 - (-1)} = \dfrac{\left[4^2 + 2(4)\right] - \left[(-1)^2 + 2(-1)\right]}{5} = \dfrac{24 - (-1)}{5} = 5.$

13. The average rate of change is $\dfrac{f(10) - f(0)}{10 - 0} = \dfrac{\left[10^3 - 4\left(10^2\right)\right] - \left[0^3 - 4\left(0^2\right)\right]}{10 - 0} = \dfrac{600 - 0}{10} = 60.$

15. The average rate of change is

$\dfrac{f(2 + h) - f(2)}{(2 + h) - 2} = \dfrac{\left[3(2 + h)^2\right] - \left[3\left(2^2\right)\right]}{h} = \dfrac{12 + 12h + 3h^2 - 12}{h} = \dfrac{12h + 3h^2}{h} = \dfrac{h(12 + 3h)}{h} = 12 + 3h.$

17. The average rate of change is $\dfrac{g(1) - g(a)}{1 - a} = \dfrac{\dfrac{1}{1} - \dfrac{1}{a}}{1 - a} \cdot \dfrac{a}{a} = \dfrac{a - 1}{a(1 - a)} = \dfrac{-1(1 - a)}{a(1 - a)} = \dfrac{-1}{a}.$

19. The average rate of change is

$$\frac{f(a+h) - f(a)}{(a+h) - a} = \frac{\frac{2}{a+h} - \frac{2}{a}}{h} \cdot \frac{a(a+h)}{a(a+h)} = \frac{2a - 2(a+h)}{ah(a+h)} = \frac{-2h}{ah(a+h)} = \frac{-2}{a(a+h)}.$$

21. (a) The average rate of change is

$$\frac{f(a+h) - f(a)}{(a+h) - a} = \frac{\left[\frac{1}{2}(a+h) + 3\right] - \left[\frac{1}{2}a + 3\right]}{h} = \frac{\frac{1}{2}a + \frac{1}{2}h + 3 - \frac{1}{2}a - 3}{h} = \frac{\frac{1}{2}h}{h} = \frac{1}{2}.$$

(b) The slope of the line $f(x) = \frac{1}{2}x + 3$ is $\frac{1}{2}$, which is also the average rate of change.

23. The average rate of change is $\dfrac{W(200) - W(100)}{200 - 100} = \dfrac{50 - 75}{200 - 100} = \dfrac{-25}{100} = -\dfrac{1}{4}$ ft/day.

25. (a) The average rate of change of population is $\dfrac{1,591 - 856}{2001 - 1998} = \dfrac{735}{3} = 245$ persons/yr.

(b) The average rate of change of population is $\dfrac{826 - 1,483}{2004 - 2002} = \dfrac{-657}{2} = -328.5$ persons/yr.

(c) The population was increasing from 1997 to 2001.

(d) The population was decreasing from 2001 to 2006.

27. (a) The average rate of change of sales is $\dfrac{584 - 512}{2003 - 1993} = \dfrac{72}{10} = 7.2$ units/yr.

(b) The average rate of change of sales is $\dfrac{520 - 512}{1994 - 1993} = \dfrac{8}{1} = 8$ units/yr.

(c) The average rate of change of sales is $\dfrac{410 - 520}{1996 - 1994} = \dfrac{-110}{2} = -55$ units/yr.

(d)

Year	CD players sold	Change in sales from previous year
1993	512	—
1994	520	8
1995	413	−107
1996	410	−3
1997	468	58
1998	510	42
1999	590	80
2000	607	17
2001	732	125
2002	612	−120
2003	584	−28

Sales increased most quickly between 2000 and 2001. Sales decreased most quickly between 2001 and 2002.

29. The average rate of change of the temperature of the soup over the first 20 minutes is

$$\frac{T(20) - T(0)}{20 - 0} = \frac{119 - 200}{20 - 0} = \frac{-81}{20} = -4.05° \text{ F/min.}$$ Over the next 20 minutes, it is

$$\frac{T(40) - T(20)}{40 - 20} = \frac{89 - 119}{40 - 20} = -\frac{30}{20} = -1.5° \text{ F/min.}$$ The first 20 minutes had a higher average rate of change of temperature (in absolute value).

31. (a) For all three runners, the average rate of change is $\dfrac{d(10) - d(0)}{10 - 0} = \dfrac{100}{10} = 10$.

(b) Runner A gets a great jump out of the blocks but tires at the end of the race. Runner B runs a steady race. Runner C is slow at the beginning but accelerates down the track.

33. We first multiply both sides of the given expression by $x - a$: $c = \dfrac{f(x) - f(a)}{x - a}$ $\Leftrightarrow$ $(x - a)c = (x - a)\dfrac{f(x) - f(a)}{x - a}$

$\Leftrightarrow$ $cx - ca = f(x) - f(a)$ $\Leftrightarrow$ $f(x) = cx + (f(a) - ca)$. The expression in parentheses is a constant, so this last equation is an equation of a line in slope-intercept form. Thus, f is a linear function.

2.5   TRANSFORMATIONS OF FUNCTIONS

1. (a) The graph of $y = f(x) + 3$ is obtained from the graph of $y = f(x)$ by shifting *upward* 3 units.

   (b) The graph of $y = f(x + 3)$ is obtained from the graph of $y = f(x)$ by shifting *left* 3 units.

3. (a) The graph of $y = -f(x)$ is obtained from the graph of $y = f(x)$ by reflecting in the *x-axis*.

   (b) The graph of $y = f(-x)$ is obtained from the graph of $y = f(x)$ by reflecting in the *y-axis*.

5. (a) The graph of $y = f(x) - 5$ can be obtained by shifting the graph of $y = f(x)$ downward 5 units.

   (b) The graph of $y = f(x - 5)$ can be obtained by shifting the graph of $y = f(x)$ to the right 5 units.

7. (a) The graph of $y = -f(x)$ can be obtained by reflecting the graph of $y = f(x)$ in the x-axis.

   (b) The graph of $y = f(-x)$ can be obtained by reflecting the graph of $y = f(x)$ in the y-axis.

9. (a) The graph of $y = -f(x) + 5$ can be obtained by reflecting the graph of $y = f(x)$ in the x-axis, then shifting the resulting graph upward 5 units.

   (b) The graph of $y = 3f(x) - 5$ can be obtained by stretching the graph of $y = f(x)$ vertically by a factor of 3, then shifting the resulting graph downward 5 units.

11. (a) The graph of $y = 2f(x + 1) - 3$ can be obtained by shifting the graph of $y = f(x)$ to the left 1 unit, stretching it vertically by a factor of 2, and shifting it downward 3 units.

   (b) The graph of $y = 2f(x - 1) + 3$ can be obtained by shifting the graph of $y = f(x)$ to the right 1 unit, stretching it vertically by a factor of 2, and shifting it upward 3 units.

13. (a) The graph of $y = f(4x)$ can be obtained by shrinking the graph of $y = f(x)$ horizontally by a factor of $\frac{1}{4}$.

   (b) The graph of $y = f\left(\frac{1}{4}x\right)$ can be obtained by stretching the graph of $y = f(x)$ horizontally by a factor of 4.

15. (a) The graph of $g(x) = (x + 2)^2$ is obtained by shifting the graph of $f(x)$ to the left 2 units.

   (b) The graph of $g(x) = x^2 + 2$ is obtained by shifting the graph of $f(x)$ upward 2 units.

17. (a) The graph of $g(x) = |x + 2| - 2$ is obtained by shifting the graph of $f(x)$ to the left 2 units and downward 2 units.

   (b) The graph of $g(x) = g(x) = |x - 2| + 2$ is obtained from by shifting the graph of $f(x)$ to the right 2 units and upward 2 units.

19. (a) 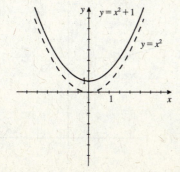   **(b)**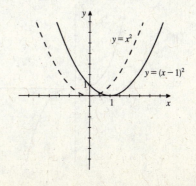

(c)

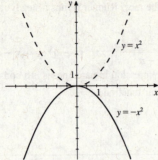

(d)

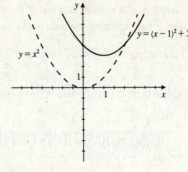

21. $f(x) = x^2 - 1$. Shift the graph of $y = x^2$ downward 1 unit.

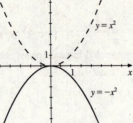

23. $f(x) = \sqrt{x} + 1$. Shift the graph of $y = \sqrt{x}$ upward 1 unit.

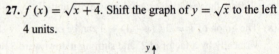

25. $f(x) = (x - 5)^2$. Shift the graph of $y = x^2$ to the right 5 units.

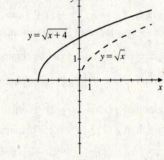

27. $f(x) = \sqrt{x + 4}$. Shift the graph of $y = \sqrt{x}$ to the left 4 units.

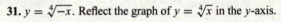

29. $f(x) = -x^3$. Reflect the graph of $y = x^3$ in the x-axis.

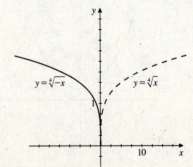

31. $y = \sqrt[4]{-x}$. Reflect the graph of $y = \sqrt[4]{x}$ in the y-axis.

33. $y = \frac{1}{4}x^2$. Shrink the graph of $y = x^2$ vertically by a factor of $\frac{1}{4}$.

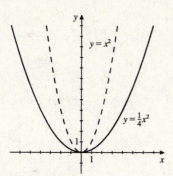

35. $y = 3|x|$. Stretch the graph of $y = |x|$ vertically by a factor of 3.

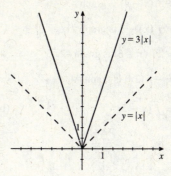

37. $y = (x - 3)^2 + 5$. Shift the graph of $y = x^2$ to the right 3 units and upward 5 units.

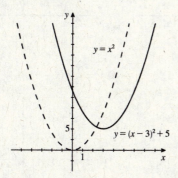

39. $y = 3 - \frac{1}{2}(x - 1)^2$. Shift the graph of $y = x^2$ to the right one unit, shrink vertically by a factor of $\frac{1}{2}$, reflect in the x-axis, then shift upward 3 units.

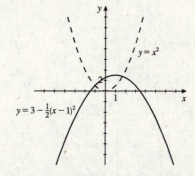

41. $y = |x + 2| + 2$. Shift the graph of $y = |x|$ to the left 2 units and upward 2 units.

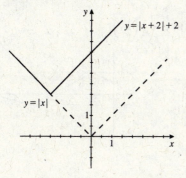

43. $y = \frac{1}{2}\sqrt{x + 4} - 3$. Shrink the graph of $y = \sqrt{x}$ vertically by a factor of $\frac{1}{2}$, then shift the result to the left 4 units and downward 3 units.

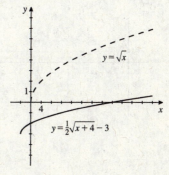

45. $y = f(x) + 3$. When $f(x) = x^2$, $y = x^2 + 3$.

47. $y = f(x + 2)$. When $f(x) = \sqrt{x}$, $y = \sqrt{x + 2}$.

49. $y = f(x - 3) + 1$. When $f(x) = |x|$, $y = |x - 3| + 1$.

51. $y = f(-x) + 1$. When $f(x) = \sqrt[4]{x}$, $y = \sqrt[4]{-x} + 1$.

53. $y = 2f(x - 3) - 2$. When $f(x) = x^2$,
$y = 2(x - 3)^2 - 2$.

55. $g(x) = f(x - 2) = (x - 2)^2 = x^2 - 4x + 4$

57. $g(x) = f(x+1) + 2 = |x+1| + 2$

59. $g(x) = -f(x+2) = -\sqrt{x+2}$

61. (a) $y = f(x-4)$ is graph #3.

    (b) $y = f(x) + 3$ is graph #1.

    (c) $y = 2f(x+6)$ is graph #2.

    (d) $y = -f(2x)$ is graph #4.

63. (a) $y = f(x-2)$ **(b)** $y = f(x) - 2$ **(c)** $y = 2f(x)$

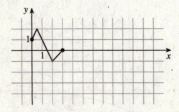
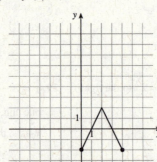
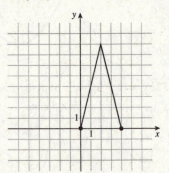

    (d) $y = -f(x) + 3$ **(e)** $y = f(-x)$ **(f)** $y = \frac{1}{2}f(x-1)$

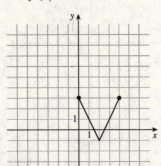
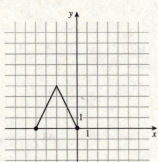
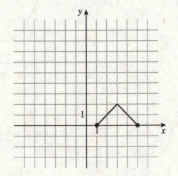

65. (a) $y = g(2x)$ **(b)** $y = g\left(\frac{1}{2}x\right)$

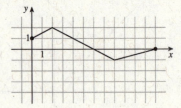

67. $y = [\![2x]\!]$

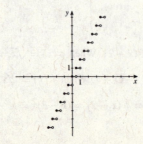

69.

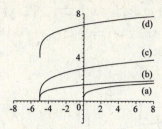

For part (b), shift the graph in (a) to the left 5 units; for part (c), shift the graph in (a) to the left 5 units, and stretch it vertically by a factor of 2; for part (d), shift the graph in (a) to the left 5 units, stretch it vertically by a factor of 2, and then shift it upward 4 units.

71.

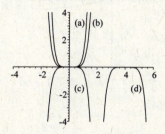

For part (b), shrink the graph in (a) vertically by a factor of $\frac{1}{3}$; for part (c), shrink the graph in (a) vertically by a factor of $\frac{1}{3}$, and reflect it in the x-axis; for part (d), shift the graph in (a) to the right 4 units, shrink vertically by a factor of $\frac{1}{3}$, and then reflect it in the x-axis.

73. (a) $y = f(x) = \sqrt{2x - x^2}$ **(b)** $y = f(2x) = \sqrt{2(2x) - (2x)^2}$ **(c)** $y = f\left(\frac{1}{2}x\right) = \sqrt{2\left(\frac{1}{2}x\right) - \left(\frac{1}{2}x\right)^2}$

$$= \sqrt{4x - 4x^2}$$ $$= \sqrt{x - \tfrac{1}{4}x^2}$$

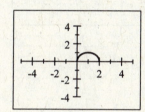   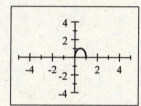   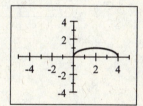

The graph in part (b) is obtained by horizontally shrinking the graph in part (a) by a factor of $\frac{1}{2}$ (so the graph is half as wide). The graph in part (c) is obtained by horizontally stretching the graph in part (a) by a factor of 2 (so the graph is twice as wide).

75. $f(x) = x^4$. $f(-x) = (-x)^4 = x^4 = f(x)$. Thus $f(x)$ is even.

77. $f(x) = x^2 + x$. $f(-x) = (-x)^2 + (-x) = x^2 - x$. Thus $f(-x) \neq f(x)$. Also, $f(-x) \neq -f(x)$, so $f(x)$ is neither odd nor even.

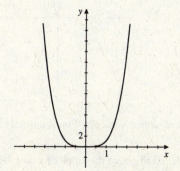   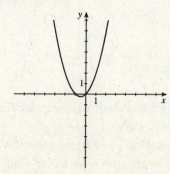

79. $f(x) = x^3 - x$.

$f(-x) = (-x)^3 - (-x) = -x^3 + x$
$= -\left(x^3 - x\right) = -f(x)$.

Thus $f(x)$ is odd.

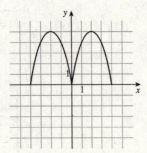

81. $f(x) = 1 - \sqrt[3]{x}$. $f(-x) = 1 - \sqrt[3]{(-x)} = 1 + \sqrt[3]{x}$. Thus $f(-x) \neq f(x)$. Also $f(-x) \neq -f(x)$, so $f(x)$ is neither odd nor even.

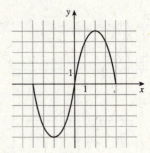

83. (a) Even

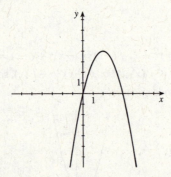

(b) Odd

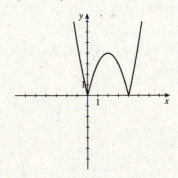

85. Since $f(x) = x^2 - 4 < 0$, for $-2 < x < 2$, the graph of $y = g(x)$ is found by sketching the graph of $y = f(x)$ for $x \leq -2$ and $x \geq 2$, then reflecting in the x-axis the part of the graph of $y = f(x)$ for $-2 < x < 2$.

87. (a) $f(x) = 4x - x^2$

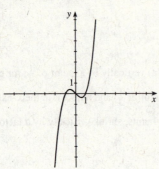

(b) $f(x) = \left|4x - x^2\right|$

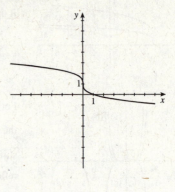

89. (a) The graph of $y = t^2$ must be shrunk vertically by a factor of 0.01 and shifted vertically 4 units upward to obtain the graph of $y = f(t)$.

(b) The graph of $y = f(t)$ must be shifted horizontally 10 units to the left to obtain the graph of $y = g(t)$. So $g(t) = f(t + 10) = 4 + 0.01(t + 10)^2 = 5 + 0.2t + 0.01t^2$.

91. f even implies $f(-x) = f(x)$; g even implies $g(-x) = g(x)$; f odd implies $f(-x) = -f(x)$; and g odd implies
$g(-x) = -g(x)$

If f and g are both even, then $(f + g)(-x) = f(-x) + g(-x) = f(x) + g(x) = (f + g)(x)$ and $f + g$ is even.

If f and g are both odd, then $(f + g)(-x) = f(-x) + g(-x) = -f(x) - g(x) = -(f + g)(x)$ and $f + g$ is odd.

If f odd and g even, then $(f + g)(-x) = f(-x) + g(-x) = -f(x) + g(x)$, which is neither odd nor even.

93. $f(x) = x^n$ is even when n is an even integer and $f(x) = x^n$ is odd when n is an odd integer.

These names were chosen because polynomials with only terms with odd powers are odd functions, and polynomials with only terms with even powers are even functions.

2.6 COMBINING FUNCTIONS

1. From the graphs of f and g in the figure, we find $(f + g)(2) = f(2) + g(2) = 3 + 5 = 8$,

$(f - g)(2) = f(2) - g(2) = 3 - 5 = -2$, $(fg)(2) = f(2)g(2) = 3 \cdot 5 = 15$, and $\left(\dfrac{f}{g}\right)(2) = \dfrac{f(2)}{g(2)} = \dfrac{3}{5}$.

3. If the rule of the function f is "add one" and the rule of the function g is "multiply by 2" then the rule of $f \circ g$ is "*multiply by 2, then add one*" and the rule of $g \circ f$ is "*add one, then multiply by 2.*"

5. $f(x) = x - 3$ has domain $(-\infty, \infty)$. $g(x) = x^2$ has domain $(-\infty, \infty)$. The intersection of the domains of f and g is $(-\infty, \infty)$.

$(f + g)(x) = (x - 3) + \left(x^2\right) = x^2 + x - 3$, and the domain is $(-\infty, \infty)$.

$(f - g)(x) = (x - 3) - \left(x^2\right) = -x^2 + x - 3$, and the domain is $(-\infty, \infty)$.

$(fg)(x) = (x - 3)\left(x^2\right) = x^3 - 3x^2$, and the domain is $(-\infty, \infty)$.

$\left(\dfrac{f}{g}\right)(x) = \dfrac{x - 3}{x^2}$, and the domain is $\{x \mid x \neq 0\}$.

7. $f(x) = \sqrt{4 - x^2}$, has domain $[-2, 2]$. $g(x) = \sqrt{1 + x}$, has domain $[-1, \infty)$. The intersection of the domains of f and g is $[-1, 2]$.

$(f + g)(x) = \sqrt{4 - x^2} + \sqrt{1 + x}$, and the domain is $[-1, 2]$.

$(f - g)(x) = \sqrt{4 - x^2} - \sqrt{1 + x}$, and the domain is $[-1, 2]$.

$(fg)(x) = \sqrt{4 - x^2}\sqrt{1 + x} = \sqrt{-x^3 - x^2 + 4x + 4}$, and the domain is $[-1, 2]$.

$\left(\dfrac{f}{g}\right)(x) = \dfrac{\sqrt{4 - x^2}}{\sqrt{1 + x}} = \sqrt{\dfrac{4 - x^2}{1 + x}}$, and the domain is $(-1, 2]$.

9. $f(x) = \dfrac{2}{x}$ has domain $x \neq 0$. $g(x) = \dfrac{4}{x + 4}$, has domain $x \neq -4$. The intersection of the domains of f and g is $\{x \mid x \neq 0, -4\}$; in interval notation, this is $(-\infty, -4) \cup (-4, 0) \cup (0, \infty)$.

$(f + g)(x) = \dfrac{2}{x} + \dfrac{4}{x + 4} = \dfrac{2}{x} + \dfrac{4}{x + 4} = \dfrac{2(3x + 4)}{x(x + 4)}$, and the domain is $(-\infty, -4) \cup (-4, 0) \cup (0, \infty)$.

$(f - g)(x) = \dfrac{2}{x} - \dfrac{4}{x + 4} = -\dfrac{2(x - 4)}{x(x + 4)}$, and the domain is $(-\infty, -4) \cup (-4, 0) \cup (0, \infty)$.

$(fg)(x) = \dfrac{2}{x} \cdot \dfrac{4}{x + 4} = \dfrac{8}{x(x + 4)}$, and the domain is $(-\infty, -4) \cup (-4, 0) \cup (0, \infty)$.

$\left(\dfrac{f}{g}\right)(x) = \dfrac{\dfrac{2}{x}}{\dfrac{4}{x + 4}} = \dfrac{x + 4}{2x}$, and the domain is $(-\infty, -4) \cup (-4, 0) \cup (0, \infty)$.

11. $f(x) = \sqrt{x} + \sqrt{1-x}$. The domain of $\sqrt{x}$ is $[0, \infty)$, and the domain of $\sqrt{1-x}$ is $(-\infty, 1]$. Thus the domain is $(-\infty, 1] \cap [0, \infty) = [0, 1]$.

13. $h(x) = (x-3)^{-1/4} = \dfrac{1}{(x-3)^{1/4}}$. Since $1/4$ is an even root and the denominator can not equal 0, $x - 3 > 0 \Leftrightarrow x > 3$. So the domain is $(3, \infty)$.

15.

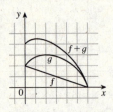

17.

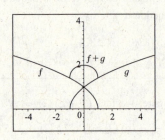

19.

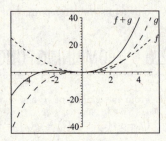

21. (a) $f(g(0)) = f\left(2 - (0)^2\right) = f(2) = 3(2) - 5 = 1$

(b) $g(f(0)) = g(3(0) - 5) = g(-5) = 2 - (-5)^2 = -23$

23. (a) $(f \circ g)(-2) = f(g(-2)) = f\left(2 - (-2)^2\right) = f(-2) = 3(-2) - 5 = -11$

(b) $(g \circ f)(-2) = g(f(-2)) = g(3(-2) - 5) = g(-11) = 2 - (-11)^2 = -119$

25. (a) $(f \circ g)(x) = f(g(x)) = f\left(2 - x^2\right) = 3\left(2 - x^2\right) - 5 = 6 - 3x^2 - 5 = 1 - 3x^2$

(b) $(g \circ f)(x) = g(f(x)) = g(3x - 5) = 2 - (3x - 5)^2 = 2 - \left(9x^2 - 30x + 25\right) = -9x^2 + 30x - 23$

27. $f(g(2)) = f(5) = 4$ **29.** $(g \circ f)(4) = g(f(4)) = g(2) = 5$

31. $(g \circ g)(-2) = g(g(-2)) = g(1) = 4$

33. $f(x) = 2x + 3$, has domain $(-\infty, \infty)$; $g(x) = 4x - 1$, has domain $(-\infty, \infty)$.

$(f \circ g)(x) = f(4x - 1) = 2(4x - 1) + 3 = 8x + 1$, and the domain is $(-\infty, \infty)$.

$(g \circ f)(x) = g(2x + 3) = 4(2x + 3) - 1 = 8x + 11$, and the domain is $(-\infty, \infty)$.

$(f \circ f)(x) = f(2x + 3) = 2(2x + 3) + 3 = 4x + 9$, and the domain is $(-\infty, \infty)$.

$(g \circ g)(x) = g(4x - 1) = 4(4x - 1) - 1 = 16x - 5$, and the domain is $(-\infty, \infty)$.

35. $f(x) = x^2$, has domain $(-\infty, \infty)$; $g(x) = x + 1$, has domain $(-\infty, \infty)$.

$(f \circ g)(x) = f(x + 1) = (x + 1)^2 = x^2 + 2x + 1$, and the domain is $(-\infty, \infty)$.

$(g \circ f)(x) = g\left(x^2\right) = \left(x^2\right) + 1 = x^2 + 1$, and the domain is $(-\infty, \infty)$.

$(f \circ f)(x) = f\left(x^2\right) = \left(x^2\right)^2 = x^4$, and the domain is $(-\infty, \infty)$.

$(g \circ g)(x) = g(x + 1) = (x + 1) + 1 = x + 2$, and the domain is $(-\infty, \infty)$.

37. $f(x) = \dfrac{1}{x}$, has domain $\{x \mid x \neq 0\}$; $g(x) = 2x + 4$, has domain $(-\infty, \infty)$.

$(f \circ g)(x) = f(2x + 4) = \dfrac{1}{2x + 4}$. $(f \circ g)(x)$ is defined for $2x + 4 \neq 0 \Leftrightarrow x \neq -2$. So the domain is
$\{x \mid x \neq -2\} = (-\infty, -2) \cup (-2, \infty)$.

$(g \circ f)(x) = g\left(\dfrac{1}{x}\right) = 2\left(\dfrac{1}{x}\right) + 4 = \dfrac{2}{x} + 4$, the domain is $\{x \mid x \neq 0\} = (-\infty, 0) \cup (0, \infty)$.

$(f \circ f)(x) = f\left(\dfrac{1}{x}\right) = \dfrac{1}{\left(\dfrac{1}{x}\right)} = x$. $(f \circ f)(x)$ is defined whenever both $f(x)$ and $f(f(x))$ are defined; that is,

whenever $\{x \mid x \neq 0\} = (-\infty, 0) \cup (0, \infty)$.

$(g \circ g)(x) = g(2x + 4) = 2(2x + 4) + 4 = 4x + 8 + 4 = 4x + 12$, and the domain is $(-\infty, \infty)$.

39. $f(x) = |x|$, has domain $(-\infty, \infty)$; $g(x) = 2x + 3$, has domain $(-\infty, \infty)$

$(f \circ g)(x) = f(2x + 4) = |2x + 3|$, and the domain is $(-\infty, \infty)$.

$(g \circ f)(x) = g(|x|) = 2|x| + 3$, and the domain is $(-\infty, \infty)$.

$(f \circ f)(x) = f(|x|) = ||x|| = |x|$, and the domain is $(-\infty, \infty)$.

$(g \circ g)(x) = g(2x + 3) = 2(2x + 3) + 3 = 4x + 6 + 3 = 4x + 9$. Domain is $(-\infty, \infty)$.

41. $f(x) = \dfrac{x}{x + 1}$, has domain $\{x \mid x \neq -1\}$; $g(x) = 2x - 1$, has domain $(-\infty, \infty)$

$(f \circ g)(x) = f(2x - 1) = \dfrac{2x - 1}{(2x - 1) + 1} = \dfrac{2x - 1}{2x}$, and the domain is $\{x \mid x \neq 0\} = (-\infty, 0) \cup (0, \infty)$.

$(g \circ f)(x) = g\left(\dfrac{x}{x + 1}\right) = 2\left(\dfrac{x}{x + 1}\right) - 1 = \dfrac{2x}{x + 1} - 1$, and the domain is $\{x \mid x \neq -1\} = (-\infty, -1) \cup (-1, \infty)$

$(f \circ f)(x) = f\left(\dfrac{x}{x + 1}\right) = \dfrac{\dfrac{x}{x + 1}}{\dfrac{x}{x + 1} + 1} \cdot \dfrac{x + 1}{x + 1} = \dfrac{x}{x + x + 1} = \dfrac{x}{2x + 1}$. $(f \circ f)(x)$ is defined whenever both $f(x)$ and

$f(f(x))$ are defined; that is, whenever $x \neq -1$ and $2x + 1 \neq 0 \Rightarrow x \neq -\frac{1}{2}$, which is $(-\infty, -1) \cup \left(-1, -\frac{1}{2}\right) \cup \left(-\frac{1}{2}, \infty\right)$.

$(g \circ g)(x) = g(2x - 1) = 2(2x - 1) - 1 = 4x - 2 - 1 = 4x - 3$, and the domain is $(-\infty, \infty)$.

43. $f(x) = \dfrac{x}{x + 1}$, has domain $\{x \mid x \neq -1\}$; $g(x) = \dfrac{1}{x}$ has domain $\{x \mid x \neq 0\}$.

$(f \circ g)(x) = f\left(\dfrac{1}{x}\right) = \dfrac{\dfrac{1}{x}}{\dfrac{1}{x} + 1} = \dfrac{1}{x\left(\dfrac{1}{x} + 1\right)} = \dfrac{1}{x + 1}$. $(f \circ g)(x)$ is defined whenever both $g(x)$ and $f(g(x))$ are

defined, so the domain is $\{x \mid x \neq -1, 0\}$.

$(g \circ f)(x) = g\left(\dfrac{x}{x + 1}\right) = \dfrac{1}{\dfrac{x}{x + 1}} = \dfrac{x + 1}{x}$. $(g \circ f)(x)$ is defined whenever both $f(x)$ and $g(f(x))$ are defined, so the

domain is $\{x \mid x \neq -1, 0\}$.

$(f \circ f)(x) = f\left(\dfrac{x}{x + 1}\right) = \dfrac{\dfrac{x}{x + 1}}{\dfrac{x}{x + 1} + 1} = \dfrac{x}{(x + 1)\left(\dfrac{x}{x + 1} + 1\right)} = \dfrac{x}{2x + 1}$. $(f \circ f)(x)$ is defined whenever both $f(x)$ and

$f(f(x))$ are defined, so the domain is $\left\{x \mid x \neq -1, -\frac{1}{2}\right\}$.

$(g \circ g)(x) = g\left(\dfrac{1}{x}\right) = \dfrac{1}{\dfrac{1}{x}} = x$. $(g \circ g)(x)$ is defined whenever both $g(x)$ and $g(g(x))$ are defined, so the domain is

$\{x \mid x \neq 0\}$.

45. $(f \circ g \circ h)(x) = f(g(h(x))) = f(g(x - 1)) = f(\sqrt{x - 1}) = \sqrt{x - 1} - 1$

47. $(f \circ g \circ h)(x) = f(g(h(x))) = f(g(\sqrt{x})) = f(\sqrt{x} - 5) = (\sqrt{x} - 5)^4 + 1$

49. $F(x) = (x - 9)^5$. Let $f(x) = x^5$ and $g(x) = x - 9$, then $F(x) = (f \circ g)(x)$.

51. $G(x) = \dfrac{x^2}{x^2 + 4}$. Let $f(x) = \dfrac{x}{x + 4}$ and $g(x) = x^2$, then $G(x) = (f \circ g)(x)$.

53. $H(x) = \left|1 - x^3\right|$. Let $f(x) = |x|$ and $g(x) = 1 - x^3$, then $H(x) = (f \circ g)(x)$.

55. $F(x) = \dfrac{1}{x^2 + 1}$. Let $f(x) = \dfrac{1}{x}$, $g(x) = x + 1$, and $h(x) = x^2$, then $F(x) = (f \circ g \circ h)(x)$.

57. $G(x) = \left(4 + \sqrt[3]{x}\right)^9$. Let $f(x) = x^9$, $g(x) = 4 + x$, and $h(x) = \sqrt[3]{x}$, then $G(x) = (f \circ g \circ h)(x)$.

59. The price per sticker is $0.15 - 0.000002x$ and the number sold is x, so the revenue is
$R(x) = (0.15 - 0.000002x)x = 0.15x - 0.000002x^2$.

61. **(a)** Because the ripple travels at a speed of 60 cm/s, the distance traveled in t seconds is the radius, so $g(t) = 60t$.

   (b) The area of a circle is πr^2, so $f(r) = \pi r^2$.

   (c) $f \circ g = \pi (g(t))^2 = \pi (60t)^2 = 3600\pi t^2$ cm^2. This function represents the area of the ripple as a function of time.

63. Let r be the radius of the spherical balloon in centimeters. Since the radius is increasing at a rate of 2 cm/s, the radius is $r = 2t$
after t seconds. Therefore, the surface area of the balloon can be written as $S = 4\pi r^2 = 4\pi (2t)^2 = 4\pi \left(4t^2\right) = 16\pi t^2$.

65. **(a)** $f(x) = 0.90x$

   (b) $g(x) = x - 100$

   (c) $f \circ g = f(x - 100) = 0.90(x - 100) = 0.90x - 90$. $f \circ g$ represents applying the $100 coupon, then the
   10% discount. $g \circ f = g(0.90x) = 0.90x - 100$. $g \circ f$ represents applying the 10% discount, then the $100 coupon.
   So applying the 10% discount, then the $100 coupon gives the lower price.

67. $A(x) = 1.05x$. $(A \circ A)(x) = A(A(x)) = A(1.05x) = 1.05(1.05x) = (1.05)^2 x$.
$(A \circ A \circ A)(x) = A(A \circ A(x)) = A\left((1.05)^2 x\right) = 1.05\left[(1.05)^2 x\right] = (1.05)^3 x$.

$(A \circ A \circ A \circ A)(x) = A(A \circ A \circ A(x)) = A\left((1.05)^3 x\right) = 1.05\left[(1.05)^3 x\right] = (1.05)^4 x$. A represents the amount in
the account after 1 year; $A \circ A$ represents the amount in the account after 2 years; $A \circ A \circ A$ represents the amount in the
account after 3 years; and $A \circ A \circ A \circ A$ represents the amount in the account after 4 years. We can see that if we compose
n copies of A, we get $(1.05)^n x$.

69. $g(x) = 2x + 1$ and $h(x) = 4x^2 + 4x + 7$.
Method 1: Notice that $(2x + 1)^2 = 4x^2 + 4x + 1$. We see that adding 6 to this quantity gives
$(2x + 1)^2 + 6 = 4x^2 + 4x + 1 + 6 = 4x^2 + 4x + 7$, which is $h(x)$. So let $f(x) = x^2 + 6$, and we have
$(f \circ g)(x) = (2x + 1)^2 + 6 = h(x)$.
Method 2: Since $g(x)$ is linear and $h(x)$ is a second degree polynomial, $f(x)$ must be a second degree polynomial,
that is, $f(x) = ax^2 + bx + c$ for some a, b, and c. Thus $f(g(x)) = f(2x + 1) = a(2x + 1)^2 + b(2x + 1) + c \Leftrightarrow$
$4ax^2 + 4ax + a + 2bx + b + c = 4ax^2 + (4a + 2b)x + (a + b + c) = 4x^2 + 4x + 7$. Comparing this with $f(g(x))$, we
have $4a = 4$ (the x^2 coefficients), $4a + 2b = 4$ (the x coefficients), and $a + b + c = 7$ (the constant terms) $\Leftrightarrow a = 1$ and
$2a + b = 2$ and $a + b + c = 7 \Leftrightarrow a = 1, b = 0, c = 6$. Thus $f(x) = x^2 + 6$.
$f(x) = 3x + 5$ and $h(x) = 3x^2 + 3x + 2$.
Note since $f(x)$ is linear and $h(x)$ is quadratic, $g(x)$ must also be quadratic. We can then use trial and error to find $g(x)$.
Another method is the following: We wish to find g so that $(f \circ g)(x) = h(x)$. Thus $f(g(x)) = 3x^2 + 3x + 2 \Leftrightarrow$
$3(g(x)) + 5 = 3x^2 + 3x + 2 \Leftrightarrow 3(g(x)) = 3x^2 + 3x - 3 \Leftrightarrow g(x) = x^2 + x - 1$.

2.7   ONE-TO-ONE FUNCTIONS AND THEIR INVERSES

1. A function f is one-to-one if different inputs produce *different* outputs. You can tell from the graph that a function is one-to-one by using the *Horizontal Line* Test.

3. **(a)** Proceeding backward through the description of f, we can describe f^{-1} as follows: "Take the third root, subtract 5, then divide by 3."

   (b) $f(x) = (3x + 5)^3$ and $f^{-1}(x) = \dfrac{\sqrt[3]{x} - 5}{3}$.

5. By the Horizontal Line Test, f is not one-to-one.   7. By the Horizontal Line Test, f is one-to-one.

9. By the Horizontal Line Test, f is not one-to-one.

11. $f(x) = -2x + 4$. If $x_1 \neq x_2$, then $-2x_1 \neq -2x_2$ and $-2x_1 + 4 \neq -2x_2 + 4$. So f is a one-to-one function.

13. $g(x) = \sqrt{x}$. If $x_1 \neq x_2$, then $\sqrt{x_1} \neq \sqrt{x_2}$ because two different numbers cannot have the same square root. Therefore, g is a one-to-one function.

15. $h(x) = x^2 - 2x$. Since $h(0) = 0$ and $h(2) = (2) - 2(2) = 0$ we have $h(0) = h(2)$. So f is not a one-to-one function.

17. $f(x) = x^4 + 5$. Every nonzero number and its negative have the same fourth power. For example, $(-1)^4 = 1 = (1)^4$, so $f(-1) = f(1)$. Thus f is not a one-to-one function.

19. $f(x) = \dfrac{1}{x^2}$. Every nonzero number and its negative have the same square. For example, $\dfrac{1}{(-1)^2} = 1 = \dfrac{1}{(1)^2}$, so $f(-1) = f(1)$. Thus f is not a one-to-one function.

21. **(a)** $f(2) = 7$. Since f is one-to-one, $f^{-1}(7) = 2$.

   (b) $f^{-1}(3) = -1$. Since f is one-to-one, $f(-1) = 3$.

23. $f(x) = 5 - 2x$. Since f is one-to-one and $f(1) = 5 - 2(1) = 3$, then $f^{-1}(3) = 1$. (Find 1 by solving the equation $5 - 2x = 3$.)

25. $f(g(x)) = f(x + 6) = (x + 6) - 6 = x$ for all x.
   $g(f(x)) = g(x - 6) = (x - 6) + 6 = x$ for all x. Thus f and g are inverses of each other.

27. $f(g(x)) = f\left(\dfrac{x + 5}{2}\right) = 2\left(\dfrac{x + 5}{2}\right) - 5 = x + 5 - 5 = x$ for all x.

   $g(f(x)) = g(2x - 5) = \dfrac{(2x - 5) + 5}{2} = x$ for all x. Thus f and g are inverses of each other.

29. $f(g(x)) = f\left(\dfrac{1}{x}\right) = \dfrac{1}{1/x} = x$ for all $x \neq 0$. Since $f(x) = g(x)$, we also have $g(f(x)) = x$ for all $x \neq 0$. Thus f and g are inverses of each other.

31. $f(g(x)) = f\left(\sqrt{x + 4}\right) = \left(\sqrt{x + 4}\right)^2 - 4 = x + 4 - 4 = x$ for all $x \geq -4$.

   $g(f(x)) = g\left(x^2 - 4\right) = \sqrt{(x^2 - 4) + 4} = \sqrt{x^2} = x$ for all $x \geq 0$. Thus f and g are inverses of each other.

33. $f(g(x)) = f\left(\dfrac{1}{x} + 1\right) = \dfrac{1}{\left(\dfrac{1}{x} + 1\right) - 1} = x$ for all $x \neq 0$.

   $g(f(x)) = g\left(\dfrac{1}{x - 1}\right) = \dfrac{1}{\left(\dfrac{1}{x - 1}\right)} + 1 = (x - 1) + 1 = x$ for all $x \neq 1$. Thus f and g are inverses of each other.

35. $f(g(x)) = f\left(\dfrac{2x+2}{x-1}\right) = \dfrac{\frac{2x+2}{x-1}+2}{\frac{2x+2}{x-1}-2} = \dfrac{2x+2+2(x-1)}{2x+2-2(x-1)} = \dfrac{4x}{4} = x$ for all $x \neq 1$.

$g(f(x)) = g\left(\dfrac{x+2}{x-2}\right) = \dfrac{2\left(\frac{x+2}{x-2}\right)+2}{\frac{x+2}{x-2}-1} = \dfrac{2(x+2)+2(x-2)}{x+2-1(x-2)} = \dfrac{4x}{4} = x$ for all $x \neq 2$. Thus f and g are inverses of each other.

37. $f(x) = 2x+1$. $y = 2x+1 \Leftrightarrow 2x = y-1 \Leftrightarrow x = \frac{1}{2}(y-1)$. So $f^{-1}(x) = \frac{1}{2}(x-1)$.

39. $f(x) = 4x+7$. $y = 4x+7 \Leftrightarrow 4x = y-7 \Leftrightarrow x = \frac{1}{4}(y-7)$. So $f^{-1}(x) = \frac{1}{4}(x-7)$.

41. $f(x) = 5-4x^3$. $y = 5-4x^3 \Leftrightarrow 4x^3 = 5-y \Leftrightarrow x^3 = \frac{1}{4}(5-y) \Leftrightarrow x = \sqrt[3]{\frac{1}{4}(5-y)}$. So $f^{-1}(x) = \sqrt[3]{\frac{1}{4}(5-x)}$.

43. $f(x) = \dfrac{1}{x+2}$. $y = \dfrac{1}{x+2} \Leftrightarrow x+2 = \dfrac{1}{y} \Leftrightarrow x = \dfrac{1}{y}-2$. So $f^{-1}(x) = \dfrac{1}{x}-2$.

45. $f(x) = \dfrac{x}{x+4}$. $y = \dfrac{x}{x+4} \Leftrightarrow y(x+4) = x \Leftrightarrow xy+4y = x \Leftrightarrow x-xy = 4y \Leftrightarrow x(1-y) = 4y \Leftrightarrow x = \dfrac{4y}{1-y}$. So $f^{-1}(x) = \dfrac{4x}{1-x}$.

47. $f(x) = \dfrac{2x+5}{x-7}$. $y = \dfrac{2x+5}{x-7} \Leftrightarrow y(x-7) = 2x+5 \Leftrightarrow xy-7y = 2x+5 \Leftrightarrow xy-2x = 7y+5 \Leftrightarrow x(y-2) = 7y+5 \Leftrightarrow x = \dfrac{7y+5}{y-2}$. So $f^{-1}(x) = \dfrac{7x+5}{x-2}$.

49. $f(x) = \dfrac{1+3x}{5-2x}$. $y = \dfrac{1+3x}{5-2x} \Leftrightarrow y(5-2x) = 1+3x \Leftrightarrow 5y-2xy = 1+3x \Leftrightarrow 3x+2xy = 5y-1 \Leftrightarrow x(3+2y) = 5y-1 \Leftrightarrow x = \dfrac{5y-1}{2y+3}$. So $f^{-1}(x) = \dfrac{5x-1}{2x+3}$.

51. $f(x) = \sqrt{2+5x}$, $x \geq -\frac{2}{5}$. $y = \sqrt{2+5x}$, $y \geq 0 \Leftrightarrow y^2 = 2+5x \Leftrightarrow 5x = y^2-2 \Leftrightarrow x = \frac{1}{5}\left(y^2-2\right)$ and $y \geq 0$. So $f^{-1}(x) = \frac{1}{5}\left(x^2-2\right)$, $x \geq 0$.

53. $f(x) = 4-x^2$, $x \geq 0$. $y = 4-x^2 \Leftrightarrow x^2 = 4-y \Leftrightarrow x = \sqrt{4-y}$. So $f^{-1}(x) = \sqrt{4-x}$, $x \leq 4$. (Note that $x \geq 0 \Rightarrow f(x) \leq 4$.)

55. $f(x) = 4+\sqrt[3]{x}$. $y = 4+\sqrt[3]{x} \Leftrightarrow \sqrt[3]{x} = y-4 \Leftrightarrow x = (y-4)^3$. So $f^{-1}(x) = (x-4)^3$.

57. $f(x) = 1+\sqrt{1+x}$. $y = 1+\sqrt{1+x}$, $y \geq 1 \Leftrightarrow \sqrt{1+x} = y-1 \Leftrightarrow 1+x = (y-1)^2 \Leftrightarrow x = (y-1)^2-1 = y^2-2y$. So $f^{-1}(x) = x^2-2x$, $x \geq 1$.

59. $f(x) = x^4$, $x \geq 0$. $y = x^4$, $y \geq 0 \Leftrightarrow x = \sqrt[4]{y}$. So $f^{-1}(x) = \sqrt[4]{x}$, $x \geq 0$.

61. (a), (b) $f(x) = 3x - 6$

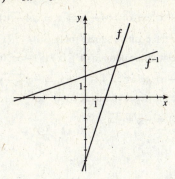

(c) $f(x) = 3x - 6$. $y = 3x - 6 \Leftrightarrow 3x = y + 6 \Leftrightarrow$ $x = \frac{1}{3}(y + 6)$. So $f^{-1}(x) = \frac{1}{3}(x + 6)$.

63. (a), (b) $f(x) = \sqrt{x + 1}$

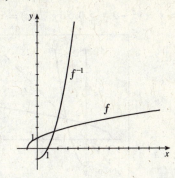

(c) $f(x) = \sqrt{x + 1}, x \geq -1$. $y = \sqrt{x + 1}, y \geq 0$ $\Leftrightarrow y^2 = x + 1 \Leftrightarrow x = y^2 - 1$ and $y \geq 0$. So $f^{-1}(x) = x^2 - 1, x \geq 0$.

65. $f(x) = x^3 - x$. Using a graphing device and the Horizontal Line Test, we see that f is not a one-to-one function. For example, $f(0) = 0 = f(-1)$.

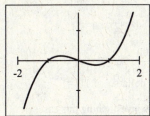

67. $f(x) = \dfrac{x + 12}{x - 6}$. Using a graphing device and the Horizontal Line Test, we see that f is a one-to-one function.

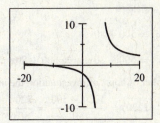

69. $f(x) = |x| - |x - 6|$. Using a graphing device and the Horizontal Line Test, we see that f is not a one-to-one function. For example $f(0) = -6 = f(-2)$.

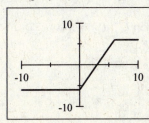

71. (a) $y = f(x) = 2 + x \Leftrightarrow x = y - 2$. So $f^{-1}(x) = x - 2$.

(b)

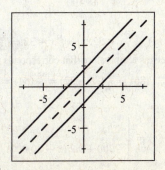

73. (a) $y = g(x) = \sqrt{x+3}$, $y \geq 0 \Leftrightarrow x + 3 = y^2$, $y \geq 0$
$\Leftrightarrow x = y^2 - 3$, $y \geq 0$. So $g^{-1}(x) = x^2 - 3$, $x \geq 0$.

(b)

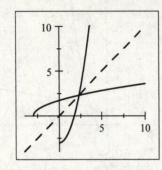

75. If we restrict the domain of $f(x)$ to $[0, \infty)$, then
$y = 4 - x^2 \Leftrightarrow x^2 = 4 - y \Rightarrow x = \sqrt{4-y}$ (since $x \geq 0$,
we take the positive square root). So $f^{-1}(x) = \sqrt{4-x}$.
If we restrict the domain of $f(x)$ to $(-\infty, 0]$, then
$y = 4 - x^2 \Leftrightarrow x^2 = 4 - y \Rightarrow x = -\sqrt{4-y}$ (since $x \leq 0$,
we take the negative square root). So
$f^{-1}(x) = -\sqrt{4-x}$.

77. If we restrict the domain of $h(x)$ to $[-2, \infty)$, then $y = (x+2)^2 \Rightarrow x + 2 = \sqrt{y}$ (since $x \geq -2$, we take the positive square
root) $\Leftrightarrow x = -2 + \sqrt{y}$. So $h^{-1}(x) = -2 + \sqrt{x}$.
If we restrict the domain of $h(x)$ to $(-\infty, -2]$, then $y = (x+2)^2 \Rightarrow x + 2 = -\sqrt{y}$ (since $x \leq -2$, we take the negative
square root) $\Leftrightarrow x = -2 - \sqrt{y}$. So $h^{-1}(x) = -2 - \sqrt{x}$.

79.

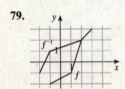

81. (a) $f(x) = 500 + 80x$.

(b) $f(x) = 500 + 80x$. $y = 500 + 80x \Leftrightarrow 80x = y - 500 \Leftrightarrow x = \dfrac{y - 500}{80}$. So $f^{-1}(x) = \dfrac{x - 500}{80}$. f^{-1} represents the
number of hours of investigation the investigate spends on a case for x dollars.

(c) $f^{-1}(1220) = \dfrac{1220 - 500}{80} = \dfrac{720}{80} = 9$. The investigator spent 9 hours investigating this case.

83. (a) $v(r) = 18{,}500\left(0.25 - r^2\right)$. $t = 18{,}500\left(0.25 - r^2\right) \Leftrightarrow t = 4625 - 18{,}500r^2 \Leftrightarrow 18500r^2 = 4625 - t \Leftrightarrow$
$r^2 = \dfrac{4625 - t}{18{,}500} \Rightarrow r = \pm\sqrt{\dfrac{4625 - t}{18{,}500}}$. Since r represents a distance, $r \geq 0$, so $v^{-1}(t) = \sqrt{\dfrac{4625 - t}{18{,}500}}$. v^{-1} represents
the radial distance from the center of the vein at which the blood has velocity v.

(b) $v^{-1}(30) = \sqrt{\dfrac{4625 - 30}{18{,}500}} \approx 0.498$ cm. The velocity is 30 at 0.498 cm from the center of the artery or vein.

85. (a) $F(x) = \frac{9}{5}x + 32$. $y = \frac{9}{5}x + 32 \Leftrightarrow \frac{9}{5}x = y - 32 \Leftrightarrow x = \frac{5}{9}(y - 32)$. So $F^{-1}(x) = \frac{5}{9}(x - 32)$. F^{-1} represents the
Celsius temperature that corresponds to the Fahrenheit temperature of F.

(b) $F^{-1}(86) = \frac{5}{9}(86 - 32) = \frac{5}{9}(54) = 30$. So $86°$ Fahrenheit is the same as $30°$ Celsius.

87. (a) $f(x) = \begin{cases} 0.1x, & \text{if } 0 \leq x \leq 20{,}000 \\ 2000 + 0.2\,(x - 20{,}000) & \text{if } x > 20{,}000 \end{cases}$

(b) We will find the inverse of each piece of the function f.

$f_1(x) = 0.1x$. $y = 0.1x \Leftrightarrow x = 10y$. So $f_1^{-1}(x) = 10x$.

$f_2(x) = 2000 + 0.2(x - 20{,}000) = 0.2x - 2000$. $y = 0.2x - 2000 \Leftrightarrow 0.2x = y + 2000 \Leftrightarrow x = 5y + 10{,}000$. So $f_2^{-1}(x) = 5x + 10{,}000$.

Since $f(0) = 0$ and $f(20{,}000) = 2000$ we have $f^{-1}(x) = \begin{cases} 10x, & \text{if } 0 \le x \le 2000 \\ 5x + 10{,}000 & \text{if } x > 2000 \end{cases}$ It represents the taxpayer's income.

(c) $f^{-1}(10{,}000) = 5(10{,}000) + 10{,}000 = 60{,}000$. The required income is €60,000.

89. $f(x) = 7 + 2x$. $y = 7 + 2x \Leftrightarrow 2x = y - 7 \Leftrightarrow x = \dfrac{y - 7}{2}$. So $f^{-1}(x) = \dfrac{x - 7}{2}$. f^{-1} is the number of toppings on a pizza that costs x dollars.

91. (a) $f(x) = \dfrac{2x + 1}{5}$ is "multiply by 2, add 1, and then divide by 5". So the reverse is "multiply by 5, subtract 1, and then divide by 2" or $f^{-1}(x) = \dfrac{5x - 1}{2}$. Check: $f \circ f^{-1}(x) = f\left(\dfrac{5x - 1}{2}\right) = \dfrac{2\left(\dfrac{5x - 1}{2}\right) + 1}{5} = \dfrac{5x - 1 + 1}{5} = \dfrac{5x}{5} = x$

and $f^{-1} \circ f(x) = f^{-1}\left(\dfrac{2x + 1}{5}\right) = \dfrac{5\left(\dfrac{2x + 1}{5}\right) - 1}{2} = \dfrac{2x + 1 - 1}{2} = \dfrac{2x}{2} = x$.

(b) $f(x) = 3 - \dfrac{1}{x} = \dfrac{-1}{x} + 3$ is "take the negative reciprocal and add 3". Since the reverse of "take the negative reciprocal" is "take the negative reciprocal", $f^{-1}(x)$ is "subtract 3 and take the negative reciprocal", that is,

$f^{-1}(x) = \dfrac{-1}{x - 3}$. Check: $f \circ f^{-1}(x) = f\left(\dfrac{-1}{x - 3}\right) = 3 - \dfrac{1}{\dfrac{-1}{x - 3}} = 3 - \left(1 \cdot \dfrac{x - 3}{-1}\right) = 3 + x - 3 = x$ and

$f^{-1} \circ f(x) = f^{-1}\left(3 - \dfrac{1}{x}\right) = \dfrac{-1}{\left(3 - \dfrac{1}{x}\right) - 3} = \dfrac{-1}{-\dfrac{1}{x}} = -1 \cdot \dfrac{x}{-1} = x$.

(c) $f(x) = \sqrt{x^3 + 2}$ is "cube, add 2, and then take the square root". So the reverse is "square, subtract 2, then take the cube root" or $f^{-1}(x) = \sqrt[3]{x^2 - 2}$. Domain for $f(x)$ is $\left[-\sqrt[3]{2}, \infty\right)$; domain for $f^{-1}(x)$ is $[0, \infty)$. Check:

$f \circ f^{-1}(x) = f\left(\sqrt[3]{x^2 - 2}\right) = \sqrt{\left(\sqrt[3]{x^2 - 2}\right)^3 + 2} = \sqrt{x^2 - 2 + 2} = \sqrt{x^2} = x$ (on the appropriate domain) and

$f^{-1} \circ f(x) = f^{-1}\left(\sqrt{x^3 + 2}\right) = \sqrt[3]{\left(\sqrt{x^3 + 2}\right)^2 - 2} = \sqrt[3]{x^3 + 2 - 2} = \sqrt[3]{x^3} = x$ (on the appropriate domain).

(d) $f(x) = (2x-5)^3$ is "double, subtract 5, and then cube". So the reverse is "take the cube root, add

5, and divide by 2" or $f^{-1}(x) = \dfrac{\sqrt[3]{x}+5}{2}$ Domain for both $f(x)$ and $f^{-1}(x)$ is $(-\infty, \infty)$. Check:

$$f \circ f^{-1}(x) = f\left(\frac{\sqrt[3]{x}+5}{2}\right) = \left[2\left(\frac{\sqrt[3]{x}+5}{2}\right)-5\right]^3 = (\sqrt[3]{x}+5-5)^3 = (\sqrt[3]{x})^3 = \sqrt[3]{x^3} = x \text{ and}$$

$$f^{-1} \circ f(x) = f^{-1}\left((2x-5)^3\right) = \frac{\sqrt[x]{(2x-5)^3}+5}{2} = \frac{(2x-5)+5}{2} = \frac{2x}{2} = x.$$

In a function like $f(x) = 3x - 2$, the variable occurs only once and it easy to see how to reverse the operations step by step. But in $f(x) = x^3 + 2x + 6$, you apply two different operations to the variable x (cubing and multiplying by 2) and then add 6, so it is not possible to reverse the operations step by step.

93. (a) We find $g^{-1}(x)$: $y = 2x + 1 \Leftrightarrow 2x = y - 1 \Leftrightarrow x = \frac{1}{2}(y-1)$. So $g^{-1}(x) = \frac{1}{2}(x-1)$. Thus

$$f(x) = h \circ g^{-1}(x) = h\left(\tfrac{1}{2}(x-1)\right) = 4\left[\tfrac{1}{2}(x-1)\right]^2 + 4\left[\tfrac{1}{2}(x-1)\right] + 7 = x^2 - 2x + 1 + 2x - 2 + 7 = x^2 + 6.$$

(b) $f \circ g = h \Leftrightarrow f^{-1} \circ f \circ g = f^{-1} \circ h \Leftrightarrow I \circ g = f^{-1} \circ h \Leftrightarrow g = f^{-1} \circ h$. Note that we compose with f^{-1} on the left on each side of the equation. We find f^{-1}: $y = 3x + 5 \Leftrightarrow 3x = y - 5 \Leftrightarrow x = \frac{1}{3}(y-5)$. So $f^{-1}(x) = \frac{1}{3}(x-5)$.

Thus $g(x) = f^{-1} \circ h(x) = f^{-1}\left(3x^2 + 3x + 2\right) = \frac{1}{3}\left[\left(3x^2 + 3x + 2\right) - 5\right] = \frac{1}{3}\left[3x^2 + 3x - 3\right] = x^2 + x - 1$.

CHAPTER 2 REVIEW

1. "Square, then subtract 5" can be represented by the function $f(x) = x^2 - 5$.

3. $f(x) = 3(x+10)$: "Add 10, then multiply by 3."

5. $g(x) = x^2 - 4x$

x	$g(x)$
-1	5
0	0
1	-3
2	-4
3	-3

7. $C(x) = 5000 + 30x - 0.001x^2$

(a) $C(1000) = 5000 + 30(1000) - 0.001(1000)^2 = \$34{,}000$ and

$C(10{,}000) = 5000 + 30(10{,}000) - 0.001(10{,}000)^2 = \$205{,}000$.

(b) From part (a), we see that the total cost of printing 1000 copies of the book is $34,000 and the total cost of printing 10,000 copies is $205,000.

(c) $C(0) = 5000 + 30(0) - 0.001(0)^2 = \$5000$. This represents the fixed costs associated with getting the print run ready.

9. $f(x) = x^2 - 4x + 6$; $f(0) = (0)^2 - 4(0) + 6 = 6$; $f(2) = (2)^2 - 4(2) + 6 = 2$;

$f(-2) = (-2)^2 - 4(-2) + 6 = 18$; $f(a) = (a)^2 - 4(a) + 6 = a^2 - 4a + 6$; $f(-a) = (-a)^2 - 4(-a) + 6 = a^2 + 4a + 6$;

$f(x+1) = (x+1)^2 - 4(x+1) + 6 = x^2 + 2x + 1 - 4x - 4 + 6 = x^2 - 2x + 3$; $f(2x) = (2x)^2 - 4(2x) + 6 = 4x^2 - 8x + 6$;

$2f(x) - 2 = 2\left(x^2 - 4x + 6\right) - 2 = 2x^2 - 8x + 12 - 2 = 2x^2 - 8x + 10$.

11. By the Vertical Line Test, figures (b) and (c) are graphs of functions. By the Horizontal Line Test, figure (c) is the graph of a one-to-one function.

13. Domain: We must have $x + 3 \geq 0 \Leftrightarrow x \geq -3$. In interval notation, the domain is $[-3, \infty)$.

Range: For x in the domain of f, we have $x \geq -3 \Leftrightarrow x + 3 \geq 0 \Leftrightarrow \sqrt{x + 3} \geq 0 \Leftrightarrow f(x) \geq 0$. So the range is $[0, \infty)$.

15. $f(x) = 7x + 15$. The domain is all real numbers, $(-\infty, \infty)$.

17. $f(x) = \sqrt{x + 4}$. We require $x + 4 \geq 0 \Leftrightarrow x \geq -4$. Thus the domain is $[-4, \infty)$.

19. $f(x) = \dfrac{1}{x} + \dfrac{1}{x + 1} + \dfrac{1}{x + 2}$. The denominators cannot equal 0, therefore the domain is $\{x \mid x \neq 0, -1, -2\}$.

21. $h(x) = \sqrt{4 - x} + \sqrt{x^2 - 1}$. We require the expression inside the radicals be nonnegative. So $4 - x \geq 0 \Leftrightarrow 4 \geq x$; also $x^2 - 1 \geq 0 \Leftrightarrow (x - 1)(x + 1) \geq 0$. We make a table:

Interval	$(-\infty, -1)$	$(-1, 1)$	$(1, \infty)$
Sign of $x - 1$	−	−	+
Sign of $x + 1$	−	+	+
Sign of $(x - 1)(x + 1)$	+	−	+

Thus the domain is $(-\infty, 4] \cap \{(-\infty, -1] \cup [1, \infty)\} = (-\infty, -1] \cup [1, 4]$.

23. $f(x) = 1 - 2x$

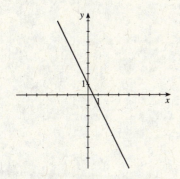

25. $f(t) = 1 - \frac{1}{2}t^2$

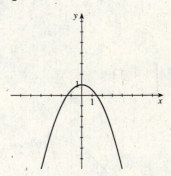

27. $f(x) = x^2 - 6x + 6$

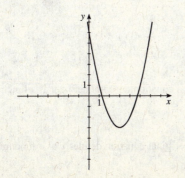

29. $g(x) = 1 - \sqrt{x}$

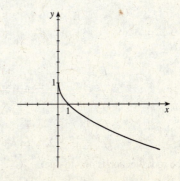

31. $h(x) = \frac{1}{2}x^3$

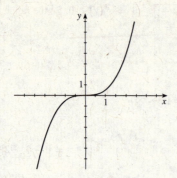

33. $h(x) = \sqrt[3]{x}$

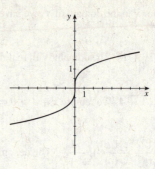

35. $g(x) = \dfrac{1}{x^2}$

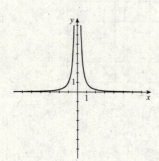

37. $f(x) = \begin{cases} 1 - x & \text{if } x < 0 \\ 1 & \text{if } x \geq 0 \end{cases}$

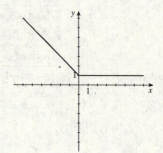

39. $f(x) = \begin{cases} x + 6 & \text{if } x < -2 \\ x^2 & \text{if } x \geq -2 \end{cases}$

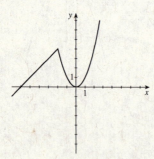

41. $x + y^2 = 14 \;\Rightarrow\; y^2 = 14 - x \;\Rightarrow\; y = \pm\sqrt{14 - x}$, so the original equation does not define y as a function of x.

43. $x^3 - y^3 = 27 \;\Leftrightarrow\; y^3 = x^3 - 27 \;\Leftrightarrow\; y = \left(x^3 - 27\right)^{1/3}$, so the original equation defines y as a function of x (since the cube root function is one-to-one).

45. $f(x) = 6x^3 - 15x^2 + 4x - 1$

(i) $[-2, 2]$ by $[-2, 2]$

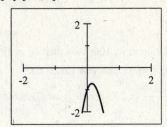

(ii) $[-8, 8]$ by $[-8, 8]$

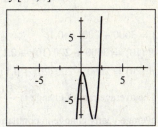

(iii) $[-4, 4]$ by $[-12, 12]$

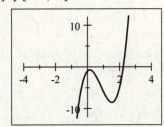

(iv) $[-100, 100]$ by $[-100, 100]$

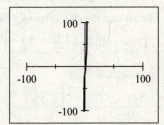

From the graphs, we see that the viewing rectangle in (iii) produces the most appropriate graph.

47. $f(x) = x^2 + 25x + 173$

$\quad = \left(x^2 + 25x + \frac{625}{4} \right) + 173 - \frac{625}{4}$

$\quad = \left(x + \frac{25}{2} \right)^2 + \frac{67}{4}$

We use the viewing rectangle $[-30, 5]$ by $[-20, 250]$.

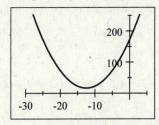

49. $f(x) = \dfrac{x}{\sqrt{x^2 + 16}}$. Since $\sqrt{x^2 + 16} \geq \sqrt{x^2} = |x|$, it

follows that y should behave like $\dfrac{x}{|x|}$. Thus we use the

viewing rectangle $[-20, 20]$ by $[-2, 2]$.

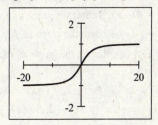

51. $f(x) = \sqrt{x^3 - 4x + 1}$. The domain consists of all x
where $x^3 - 4x + 1 \geq 0$. Using a graphing device, we see
that the domain is approximately $[-2.1, 0.2] \cup [1.9, \infty)$.

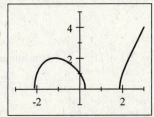

53. $f(x) = x^3 - 4x^2$ is graphed in the viewing rectangle
$[-5, 5]$ by $[-20, 10]$. $f(x)$ is increasing on $(-\infty, 0]$ and
$[2.67, \infty)$. It is decreasing on $[0, 2.67]$.

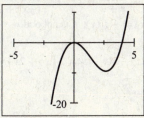

55. The average rate of change is $\dfrac{f(2) - f(0)}{2 - 0} = \dfrac{\left[(2)^2 + 3(2) \right] - \left[0^2 + 3(0) \right]}{2} = \dfrac{4 + 6 - 0}{2} = 5.$

57. The average rate of change is $\dfrac{f(3+h)-f(3)}{(3+h)-3} = \dfrac{\frac{1}{3+h}-\frac{1}{3}}{h} \cdot \dfrac{3(3+h)}{3(3+h)} = \dfrac{3-(3+h)}{3h(3+h)} = \dfrac{-h}{3h(3+h)} = -\dfrac{1}{3(3+h)}$.

59. $P(t) = 3000 + 200t + 0.1t^2$

    (a) $P(10) = 3000 + 200(10) + 0.1(10)^2 = 5010$ represents the population in its 10th year (that is, in 1995), and

        $P(20) = 3000 + 200(20) + 0.1(20)^2 = 7040$ represents its population in its 20th year (in 2005).

    (b) The average rate of change is $\dfrac{P(20)-P(10)}{20-10} = \dfrac{7040-5010}{10} = \dfrac{2030}{10} = 203$ people/year. This represents the

    average yearly change in population between 1995 and 2005.

61. $f(x) = \frac{1}{2}x - 6$

    (a) The average rate of change of f between $x = 0$ and $x = 2$ is

$$\frac{f(2)-f(0)}{2-0} = \frac{\left[\frac{1}{2}(2)-6\right] - \left[\frac{1}{2}(0)-6\right]}{2} = \frac{-5-(-6)}{2} = \frac{1}{2}, \text{ and the average rate of change of } f \text{ between } x = 15$$

    and $x = 50$ is

$$\frac{f(50)-f(15)}{50-15} = \frac{\left[\frac{1}{2}(50)-6\right] - \left[\frac{1}{2}(15)-6\right]}{35} = \frac{19-\frac{3}{2}}{35} = \frac{1}{2}.$$

    (b) The rates of change are the same because f is a linear function.

63. **(a)** $y = f(x) + 8$. Shift the graph of $f(x)$ upward 8 units.

    (b) $y = f(x+8)$. Shift the graph of $f(x)$ to the left 8 units.

    (c) $y = 1 + 2f(x)$. Stretch the graph of $f(x)$ vertically by a factor of 2, then shift it upward 1 unit.

    (d) $y = f(x-2) - 2$. Shift the graph of $f(x)$ to the right 2 units, then downward 2 units.

    (e) $y = f(-x)$. Reflect the graph of $f(x)$ about the y-axis.

    (f) $y = -f(-x)$. Reflect the graph of $f(x)$ first about the y-axis, then reflect about the x-axis.

    (g) $y = -f(x)$. Reflect the graph of $f(x)$ about the x-axis.

    (h) $y = f^{-1}(x)$. Reflect the graph of $f(x)$ about the line $y = x$.

65. **(a)** $f(x) = 2x^5 - 3x^2 + 2$. $f(-x) = 2(-x)^5 - 3(-x)^2 + 2 = -2x^5 - 3x^2 + 2$. Since $f(x) \neq f(-x)$, f is not even.

    $-f(x) = -2x^5 + 3x^2 - 2$. Since $-f(x) \neq f(-x)$, f is not odd.

    (b) $f(x) = x^3 - x^7$. $f(-x) = (-x)^3 - (-x)^7 = -\left(x^3 - x^7\right) = -f(x)$, hence f is odd.

    (c) $f(x) = \dfrac{1-x^2}{1+x^2}$. $f(-x) = \dfrac{1-(-x)^2}{1+(-x)^2} = \dfrac{1-x^2}{1+x^2} = f(x)$. Since $f(x) = f(-x)$, f is even.

    (d) $f(x) = \dfrac{1}{x+2}$. $f(-x) = \dfrac{1}{(-x)+2} = \dfrac{1}{2-x}$. $-f(x) = -\dfrac{1}{x+2}$. Since $f(x) \neq f(-x)$, f is not even, and since

    $f(-x) \neq -f(x)$, f is not odd.

67. $g(x) = 2x^2 + 4x - 5 = 2\left(x^2 + 2x\right) - 5 = 2\left(x^2 + 2x + 1\right) - 5 - 2 = 2(x+1)^2 - 7$. So the minimum value is

    $g(-1) = -7$.

69. $h(t) = -16t^2 + 48t + 32 = -16\left(t^2 - 3t\right) + 32 = -16\left(t^2 - 3t + \frac{9}{4}\right) + 32 + 36$

      $= -16\left(t^2 - 3t + \frac{9}{4}\right) + 68 = -16\left(t - \frac{3}{2}\right)^2 + 68$

    The stone reaches a maximum height of 68 feet.

71. $f(x) = 3.3 + 1.6x - 2.5x^3$. In the first viewing rectangle, $[-2, 2]$ by $[-4, 8]$, we see that $f(x)$ has a local maximum and a local minimum. In the next viewing rectangle, $[0.4, 0.5]$ by $[3.78, 3.80]$, we isolate the local maximum value as approximately 3.79 when $x \approx 0.46$. In the last viewing rectangle, $[-0.5, -0.4]$ by $[2.80, 2.82]$, we isolate the local minimum value as 2.81 when $x \approx -0.46$.

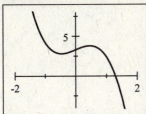 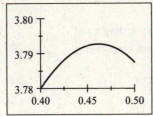 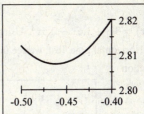

73. $f(x) = x + 2$, $g(x) = x^2$

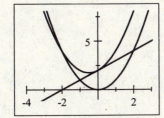

75. $f(x) = x^2 - 3x + 2$ and $g(x) = 4 - 3x$.

   (a) $(f + g)(x) = \left(x^2 - 3x + 2\right) + (4 - 3x) = x^2 - 6x + 6$

   (b) $(f - g)(x) = \left(x^2 - 3x + 2\right) - (4 - 3x) = x^2 - 2$

   (c) $(fg)(x) = \left(x^2 - 3x + 2\right)(4 - 3x) = 4x^2 - 12x + 8 - 3x^3 + 9x^2 - 6x = -3x^3 + 13x^2 - 18x + 8$

   (d) $\left(\dfrac{f}{g}\right)(x) = \dfrac{x^2 - 3x + 2}{4 - 3x}$, $x \neq \frac{4}{3}$

   (e) $(f \circ g)(x) = f(4 - 3x) = (4 - 3x)^2 - 3(4 - 3x) + 2 = 16 - 24x + 9x^2 - 12 + 9x + 2 = 9x^2 - 15x + 6$

   (f) $(g \circ f)(x) = g\left(x^2 - 3x + 2\right) = 4 - 3\left(x^2 - 3x + 2\right) = -3x^2 + 9x - 2$

77. $f(x) = 3x - 1$ and $g(x) = 2x - x^2$.

   $(f \circ g)(x) = f\left(2x - x^2\right) = 3\left(2x - x^2\right) - 1 = -3x^2 + 6x - 1$, and the domain is $(-\infty, \infty)$.

   $(g \circ f)(x) = g(3x - 1) = 2(3x - 1) - (3x - 1)^2 = 6x - 2 - 9x^2 + 6x - 1 = -9x^2 + 12x - 3$, and the domain is $(-\infty, \infty)$

   $(f \circ f)(x) = f(3x - 1) = 3(3x - 1) - 1 = 9x - 4$, and the domain is $(-\infty, \infty)$.

   $(g \circ g)(x) = g\left(2x - x^2\right) = 2\left(2x - x^2\right) - \left(2x - x^2\right)^2 = 4x - 2x^2 - 4x^2 + 4x^3 - x^4 = -x^4 + 4x^3 - 6x^2 + 4x$, and domain is $(-\infty, \infty)$.

79. $f(x) = \sqrt{1 - x}$, $g(x) = 1 - x^2$ and $h(x) = 1 + \sqrt{x}$.

   $(f \circ g \circ h)(x) = f(g(h(x))) = f\left(g\left(1 + \sqrt{x}\right)\right) = f\left(1 - \left(1 + \sqrt{x}\right)^2\right) = f\left(1 - \left(1 + 2\sqrt{x} + x\right)\right)$

       $= f\left(-x - 2\sqrt{x}\right) = \sqrt{1 - \left(-x - 2\sqrt{x}\right)} = \sqrt{1 + 2\sqrt{x} + x} = \sqrt{\left(1 + \sqrt{x}\right)^2} = 1 + \sqrt{x}$

81. $f(x) = 3 + x^3$. If $x_1 \neq x_2$, then $x_1^3 \neq x_2^3$ (unequal numbers have unequal cubes), and therefore $3 + x_1^3 \neq 3 + x_2^3$. Thus f is a one-to-one function.

83. $h(x) = \dfrac{1}{x^4}$. Since the fourth powers of a number and its negative are equal, h is not one-to-one. For example,

$h(-1) = \dfrac{1}{(-1)^4} = 1$ and $h(1) = \dfrac{1}{(1)^4} = 1$, so $h(-1) = h(1)$.

85. $p(x) = 3.3 + 1.6x - 2.5x^3$. Using a graphing device and the Horizontal Line Test, we see that p is not a one-to-one function.

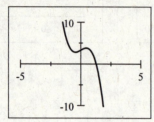

87. $f(x) = 3x - 2 \Leftrightarrow y = 3x - 2 \Leftrightarrow 3x = y + 2 \Leftrightarrow x = \frac{1}{3}(y + 2)$. So $f^{-1}(x) = \frac{1}{3}(x + 2)$.

89. $f(x) = (x + 1)^3 \Leftrightarrow y = (x + 1)^3 \Leftrightarrow x + 1 = \sqrt[3]{y} \Leftrightarrow x = \sqrt[3]{y} - 1$. So $f^{-1}(x) = \sqrt[3]{x} - 1$.

91. (a), (b) $f(x) = x^2 - 4$, $x \geq 0$

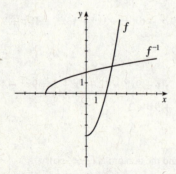

(c) $f(x) = x^2 - 4$, $x \geq 0 \Leftrightarrow y = x^2 - 4$, $y \geq -4$

$\Leftrightarrow x^2 = y + 4 \Leftrightarrow x = \sqrt{y + 4}$. So

$f^{-1}(x) = \sqrt{x + 4}$, $x \geq -4$.

CHAPTER 2 TEST

1. By the Vertical Line Test, figures (a) and (b) are graphs of functions. By the Horizontal Line Test, only figure (a) is the graph of a one-to-one function.

3. (a) "Subtract 2, then cube the result" can be expressed algebraically as $f(x) = (x - 2)^3$.

(c)

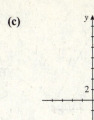

(b)

x	$f(x)$
-1	-27
0	-8
1	-1
2	0
3	1
4	8

(d) We know that f has an inverse because it passes the Horizontal Line Test. A verbal description for f^{-1} is, "Take the cube root, then add 2."

(e) $y = (x - 2)^3$ $\Leftrightarrow$ $\sqrt[3]{y} = x - 2$ $\Leftrightarrow$ $x = \sqrt[3]{y} + 2$. Thus, a formula for f^{-1} is $f^{-1}(x) = \sqrt[3]{x} + 2$.

5. The average rate of change is $\dfrac{f(2) - f(5)}{2 - 5} = \dfrac{\left[2^2 - 2(2)\right] - \left[5^2 - 2(5)\right]}{-3} = \dfrac{4 - 4 - (25 - 10)}{-3} = \dfrac{-15}{-3} = 5$.

7. (a) $y = f(x - 3) + 2$. Shift the graph of $f(x)$ to the right 3 units, then shift the graph upward 2 units.

(b) $y = f(-x)$. Reflect the graph of $f(x)$ about the y-axis.

9. $f(x) = x^2 + 1$; $g(x) = x - 3$.

(a) $(f \circ g)(x) = f(g(x)) = f(x - 3) = (x - 3)^2 + 1 = x^2 - 6x + 9 + 1 = x^2 - 6x + 10$

(b) $(g \circ f)(x) = g(f(x)) = g\left(x^2 + 1\right) = \left(x^2 + 1\right) - 3 = x^2 - 2$

(c) $f(g(2)) = f(-1) = (-1)^2 + 1 = 2$. (We have used the fact that $g(2) = (2) - 3 = -1$.)

(d) $g(f(2)) = g(5) = 5 - 3 = 2$. (We have used the fact that $f(2) = 2^2 + 1 = 5$.)

(e) $(g \circ g \circ g)(x) = g(g(g(x))) = g(g(x - 3)) = g(x - 6) = (x - 6) - 3 = x - 9$. (We have used the fact that $g(x - 3) = (x - 3) - 3 = x - 6$.)

11. (a) The domain of f is $[0, 6]$, and the range of f is $[1, 7]$.

(c) The average rate of change is $\dfrac{f(6) - f(2)}{6 - 2} = \dfrac{7 - 2}{4} = \dfrac{5}{4}$.

(b)

FOCUS ON MODELING   Modeling with Functions

1. Let w be the width of the building lot. Then the length of the lot is $3w$. So the area of the building lot is $A(w) = 3w^2$, $w > 0$.

3. Let w be the width of the base of the rectangle. Then the height of the rectangle is $\frac{1}{2}w$. Thus the volume of the box is given by the function $V(w) = \frac{1}{2}w^3$, $w > 0$.

5. Let P be the perimeter of the rectangle and y be the length of the other side. Since $P = 2x + 2y$ and the perimeter is 20, we have $2x + 2y = 20 \Leftrightarrow x + y = 10 \Leftrightarrow y = 10 - x$. Since area is $A = xy$, substituting gives $A(x) = x(10 - x) = 10x - x^2$, and since A must be positive, the domain is $0 < x < 10$.

7.

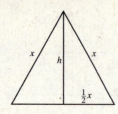

Let h be the height of an altitude of the equilateral triangle whose side has length x, as shown in the diagram. Thus the area is given by $A = \frac{1}{2}xh$. By the Pythagorean Theorem, $h^2 + \left(\frac{1}{2}x\right)^2 = x^2 \Leftrightarrow h^2 + \frac{1}{4}x^2 = x^2 \Leftrightarrow h^2 = \frac{3}{4}x^2 \Leftrightarrow h = \frac{\sqrt{3}}{2}x$. Substituting into the area of a triangle, we get

$$A(x) = \frac{1}{2}xh = \frac{1}{2}x\left(\frac{\sqrt{3}}{2}x\right) = \frac{\sqrt{3}}{4}x^2, \, x > 0.$$

9. We solve for r in the formula for the area of a circle. This gives $A = \pi r^2 \Leftrightarrow r^2 = \frac{A}{\pi} \Rightarrow r = \sqrt{\frac{A}{\pi}}$, so the model is

$$r(A) = \sqrt{\frac{A}{\pi}}, \, A > 0.$$

11. Let h be the height of the box in feet. The volume of the box is $V = 60$. Then $x^2h = 60 \Leftrightarrow h = \frac{60}{x^2}$. The surface area, S, of the box is the sum of the area of the 4 sides and the area of the base and top. Thus $S = 4xh + 2x^2 = 4x\left(\frac{60}{x^2}\right) + 2x^2 = \frac{240}{x} + 2x^2$, so the model is $S(x) = \frac{240}{x} + 2x^2, \, x > 0$.

13.

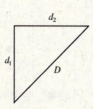

Let d_1 be the distance traveled south by the first ship and d_2 be the distance traveled east by the second ship. The first ship travels south for t hours at 5 mi/h, so $d_1 = 15t$ and, similarly, $d_2 = 20t$. Since the ships are traveling at right angles to each other, we can apply the Pythagorean Theorem to get

$$D(t) = \sqrt{d_1^2 + d_2^2} = \sqrt{(15t)^2 + (20t)^2} = \sqrt{225t^2 + 400t^2} = 25t.$$

15.

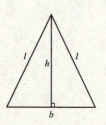

Let b be the length of the base, l be the length of the equal sides, and h be the height in centimeters. Since the perimeter is 8, $2l + b = 8 \Leftrightarrow 2l = 8 - b \Leftrightarrow l = \frac{1}{2}(8 - b)$. By the Pythagorean Theorem, $h^2 + \left(\frac{1}{2}b\right)^2 = l^2 \Leftrightarrow h = \sqrt{l^2 - \frac{1}{4}b^2}$. Therefore the area of the triangle is

$$A = \frac{1}{2} \cdot b \cdot h = \frac{1}{2} \cdot b\sqrt{l^2 - \frac{1}{4}b^2} = \frac{b}{2}\sqrt{\frac{1}{4}(8 - b)^2 - \frac{1}{4}b^2}$$
$$= \frac{b}{4}\sqrt{64 - 16b + b^2 - b^2} = \frac{b}{4}\sqrt{64 - 16b} = \frac{b}{4} \cdot 4\sqrt{4 - b} = b\sqrt{4 - b}$$

so the model is $A(b) = b\sqrt{4 - b}, \, 0 < b < 4$.

17. Let w be the length of the rectangle. By the Pythagorean Theorem, $\left(\frac{1}{2}w\right)^2 + h^2 = 10^2 \Leftrightarrow \frac{w^2}{4} + h^2 = 10^2 \Leftrightarrow w^2 = 4\left(100 - h^2\right) \Leftrightarrow w = 2\sqrt{100 - h^2}$ (since $w > 0$). Therefore, the area of the rectangle is $A = wh = 2h\sqrt{100 - h^2}$, so the model is $A(h) = 2h\sqrt{100 - h^2}, \, 0 < h < 10$.

19. (a) We complete the table.

First number	Second number	Product
1	18	18
2	17	34
3	16	48
4	15	60
5	14	70
6	13	78
7	12	84
8	11	88
9	10	90
10	9	90
11	8	88

From the table we conclude that the numbers is still increasing, the numbers whose product is a maximum should both be 9.5.

(b) Let x be one number: then $19 - x$ is the other number, and so the product, p, is
$$p\,(x) = x\,(19 - x) = 19x - x^2.$$

(c) $p\,(x) = 19x - x^2 = -\left(x^2 - 19x\right)$
$$= -\left[x^2 - 19x + \left(\tfrac{19}{2}\right)^2\right] + \left(\tfrac{19}{2}\right)^2$$
$$= -\,(x - 9.5)^2 + 90.25$$

So the product is maximized when the numbers are both 9.5.

21. (a) Let x be the width of the field (in feet) and l be the length of the field (in feet). Since the farmer has 2400 ft of fencing we must have $2x + l = 2400$.

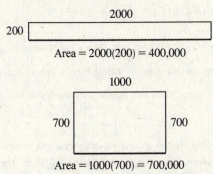

Area = 2000(200) = 400,000

Area = 1000(700) = 700,000

Area = 400(1000) = 400,000

Width	Length	Area
200	2000	400,000
300	1800	540,000
400	1600	640,000
500	1400	700,000
600	1200	720,000
700	1000	700,000
800	800	640,000

It appears that the field of largest area is about 600 ft × 1200 ft.

(b) Let x be the width of the field (in feet) and l be the length of the field (in feet). Since the farmer has 2400 ft of fencing we must have $2x + l = 2400 \Leftrightarrow l = 2400 - 2x$. The area of the fenced-in field is given by
$$A\,(x) = l \cdot x = (2400 - 2x)\,x = -2x^2 + 2400x = -2\left(x^2 - 1200x\right).$$

(c) The area is $A\,(x) = -2\left(x^2 - 1200x + 600^2\right) + 2\left(600^2\right) = -2\,(x - 600)^2 + 720{,}000$. So the maximum area occurs when $x = 600$ feet and $l = 2400 - 2\,(600) = 1200$ feet.

23. (a) Let x be the length of the fence along the road. If the area is 1200, we have $1200 = x \cdot$ width, so the width of the garden is $\dfrac{1200}{x}$. Then the cost of the fence is given by the function $C\,(x) = 5\,(x) + 3\left[x + 2 \cdot \dfrac{1200}{x}\right] = 8x + \dfrac{7200}{x}$.

(b) We graph the function $y = C(x)$ in the viewing rectangle $[0, 75] \times [0, 800]$. From this we get the cost is minimized when $x = 30$ ft. Then the width is $\frac{1200}{30} = 40$ ft. So the length is 30 ft and the width is 40 ft.

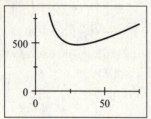

(c) We graph the function $y = C(x)$ and $y = 600$ in the viewing rectangle $[10, 65] \times [450, 650]$. From this we get that the cost is at most $600 when $15 \leq x \leq 60$. So the range of lengths he can fence along the road is 15 feet to 60 feet.

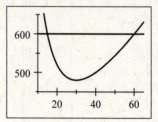

25. (a) Let h be the height in feet of the straight portion of the window. The circumference of the semicircle is $C = \frac{1}{2}\pi x$. Since the perimeter of the window is 30 feet, we have $x + 2h + \frac{1}{2}\pi x = 30$. Solving for h, we get $2h = 30 - x - \frac{1}{2}\pi x \Leftrightarrow h = 15 - \frac{1}{2}x - \frac{1}{4}\pi x$. The area of the window is
$$A(x) = xh + \frac{1}{2}\pi\left(\frac{1}{2}x\right)^2 = x\left(15 - \frac{1}{2}x - \frac{1}{4}\pi x\right) + \frac{1}{8}\pi x^2 = 15x - \frac{1}{2}x^2 - \frac{1}{8}\pi x^2.$$

(b) $A(x) = 15x - \frac{1}{8}(\pi+4)x^2 = -\frac{1}{8}(\pi+4)\left[x^2 - \frac{120}{\pi+4}x\right]$
$$= -\frac{1}{8}(\pi+4)\left[x^2 - \frac{120}{\pi+4}x + \left(\frac{60}{\pi+4}\right)^2\right] + \frac{450}{\pi+4} = -\frac{1}{8}(\pi+4)\left(x - \frac{60}{\pi+4}\right)^2 + \frac{450}{\pi+4}$$

The area is maximized when $x = \frac{60}{\pi+4} \approx 8.40$, and hence $h \approx 15 - \frac{1}{2}(8.40) - \frac{1}{4}\pi(8.40) \approx 4.20$.

27. (a) Let x be the length of one side of the base and let h be the height of the box in feet. Since the volume of the box is $V = x^2h = 12$, we have $x^2h = 12 \Leftrightarrow h = \frac{12}{x^2}$. The surface area, A, of the box is sum of the area of the four sides and the area of the base. Thus the surface area of the box is given by the formula
$$A(x) = 4xh + x^2 = 4x\left(\frac{12}{x^2}\right) + x^2 = \frac{48}{x} + x^2, x > 0.$$

(b) The function $y = A(x)$ is shown in the first viewing rectangle below. In the second viewing rectangle, we isolate the minimum, and we see that the amount of material is minimized when x (the length and width) is 2.88 ft. Then the height is $h = \frac{12}{x^2} \approx 1.44$ ft.

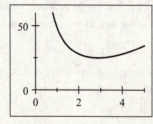

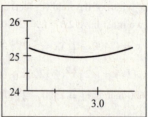

29. (a) Let w be the width of the pen and l be the length in meters. We use the area to establish a relationship between w and l. Since the area is 100 m^2, we have $l \cdot w = 100 \Leftrightarrow l = \frac{100}{w}$. So the amount of fencing used is
$$F = 2l + 2w = 2\left(\frac{100}{w}\right) + 2w = \frac{200 + 2w^2}{w}.$$

(b) Using a graphing device, we first graph F in the viewing rectangle $[0, 40]$ by $[0, 100]$, and locate the approximate location of the minimum value. In the second viewing rectangle, $[8, 12]$ by $[39, 41]$, we see that the minimum value of F occurs when $w = 10$. Therefore the pen should be a square with side 10 m.

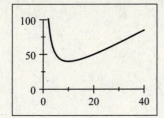
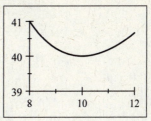

31. (a) Let x be the distance from point B to C, in miles. Then the distance from A to C is $\sqrt{x^2 + 25}$, and the energy used in flying from A to C then C to D is $f(x) = 14\sqrt{x^2 + 25} + 10(12 - x)$.

(b) By using a graphing device, the energy expenditure is minimized when the distance from B to C is about 5.1 miles.

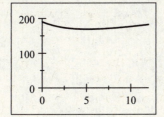
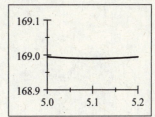

3 POLYNOMIAL AND RATIONAL FUNCTIONS

3.1 QUADRATIC FUNCTIONS AND MODELS

1. To put the quadratic function $f(x) = ax^2 + bx + c$ in standard form we complete the *square*.

3. The graph of $f(x) = 2(x-3)^2 + 5$ is a parabola that opens *upward*, with its vertex at $(3, 5)$, and $f(3) = 5$ is the *minimum* value of f.

5. (a) Vertex: $(3, 4)$

    (b) Maximum value of f: 4

    (c) Domain $(-\infty, \infty)$, range: $(-\infty, 4]$

7. (a) Vertex: $(1, -3)$

    (b) Minimum value of f: -3

    (c) Domain: $(-\infty, \infty)$, range: $[-3, \infty)$

9. (a) $f(x) = x^2 - 6x = (x-3)^2 - 9$

    (b) Vertex: $y = x^2 - 6x = x^2 - 6x + 9 - 9 = (x-3)^2 - 9$. So the vertex is at $(3, -9)$.

    x-intercepts: $y = 0 \Rightarrow 0 = x^2 - 6x = x(x-6)$. So $x = 0$ or $x = 6$. The x-intercepts are $x = 0$ and $x = 6$.

    y-intercept: $x = 0 \Rightarrow y = 0$. The y-intercept is $y = 0$.

(c)

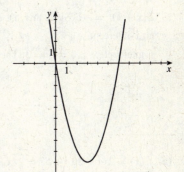

11. (a) $f(x) = 2x^2 + 6x = 2\left(x + \frac{3}{2}\right)^2 - \frac{9}{2}$

    (b) The vertex is $\left(-\frac{3}{2}, -\frac{9}{2}\right)$.

    x-intercepts: $y = 0 \Rightarrow 0 = 2x^2 + 6x = 2x(x+3) \Rightarrow x = 0$ or $x = -3$. The x-intercepts are $x = 0$ and $x = -3$.

    y-intercept: $x = 0 \Rightarrow y = 0$. The y-intercept is $y = 0$.

(c)

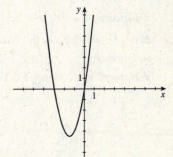

13. (a) $f(x) = x^2 + 4x + 3 = (x+2)^2 - 1$

    (b) The vertex is $(-2, -1)$.

    x-intercepts: $y = 0 \Rightarrow 0 = x^2 + 4x + 3 = (x+1)(x+3)$. So $x = -1$ or $x = -3$. The x-intercepts are $x = -1$ and $x = -3$.

    y-intercept: $x = 0 \Rightarrow y = 3$. The y-intercept is $y = 3$.

(c)

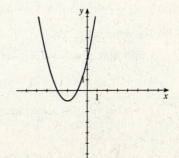

15. (a) $f(x) = -x^2 + 6x + 4 = -(x-3)^2 + 13$

(c)

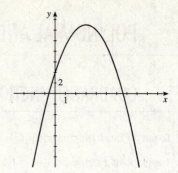

(b) The vertex is $(3, 13)$.

x-intercepts: $y = 0 \Rightarrow 0 = -(x-3)^2 + 13 \Leftrightarrow (x-3)^2 = 13 \Rightarrow$

$x - 3 = \pm\sqrt{13} \Leftrightarrow x = 3 \pm \sqrt{13}$. The x-intercepts are $x = 3 - \sqrt{13}$

and $x = 3 + \sqrt{13}$.

y-intercept: $x = 0 \Rightarrow y = 4$. The y-intercept is $y = 4$.

17. (a) $f(x) = 2x^2 + 4x + 3 = 2(x+1)^2 + 1$

(c)

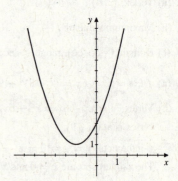

(b) The vertex is $(-1, 1)$.

x-intercepts: $y = 0 \Rightarrow 0 = 2x^2 + 4x + 3 = 2(x+1)^2 + 1 \Leftrightarrow$

$2(x+1)^2 = -1$. Since this last equation has no real solution, there is

no x-intercept.

y-intercept: $x = 0 \Rightarrow y = 3$. The y-intercept is $y = 3$.

19. (a) $f(x) = 2x^2 - 20x + 57 = 2(x-5)^2 + 7$

(c)

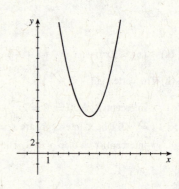

(b) The vertex is $(5, 7)$.

x-intercepts: $y = 0 \Rightarrow 0 = 2x^2 - 20x + 57 = 2(x-5)^2 + 7 \Leftrightarrow$

$2(x-5)^2 = -7$. Since this last equation has no real solution, there is

no x-intercept.

y-intercept: $x = 0 \Rightarrow y = 57$. The y-intercept is $y = 57$.

21. (a) $f(x) = -4x^2 - 16x + 3 = -4(x+2)^2 + 19$

(c)

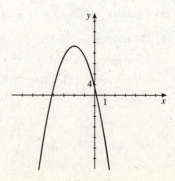

(b) The vertex is $(-2, 19)$.

x-intercepts: $y = 0 \Rightarrow 0 = -4x^2 - 16x + 3 = -4(x+2)^2 + 19 \Leftrightarrow$

$4(x+2)^2 = 19 \Leftrightarrow (x+2)^2 = \frac{19}{4} \Rightarrow x + 2 = \pm\sqrt{\frac{19}{4}} = \pm\frac{\sqrt{19}}{2} \Leftrightarrow$

$x = -2 \pm \frac{\sqrt{19}}{2}$. The x-intercepts are $x = -2 - \frac{\sqrt{19}}{2}$ and

$x = -2 + \frac{\sqrt{19}}{2}$.

y-intercept: $x = 0 \Rightarrow y = 3$. The y-intercept is $y = 3$.

23. (a) $f(x) = x^2 + 2x - 1 = \left(x^2 + 2x\right) - 1$

$\qquad = \left(x^2 + 2x + 1\right) - 1 - 1 = (x + 1)^2 - 2$

(b)

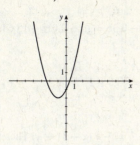

(c) The minimum value is $f(-1) = -2$.

25. (a) $f(x) = 3x^2 - 6x + 1 = 3\left(x^2 - 2x\right) + 1$

$\qquad = 3\left(x^2 - 2x + 1\right) + 1 - 3$

$\qquad = 3(x - 1)^2 - 2$

(b)

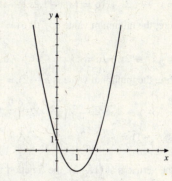

(c) The minimum value is $f(1) = -2$.

27. (a) $f(x) = -x^2 - 3x + 3 = -\left(x^2 + 3x\right) + 3$

$\qquad = -\left(x^2 + 3x + \frac{9}{4}\right) + 3 + \frac{9}{4}$

$\qquad = -\left(x + \frac{3}{2}\right)^2 + \frac{21}{4}$

(b)

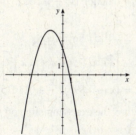

(c) The maximum value is $f\left(-\frac{3}{2}\right) = \frac{21}{4}$.

29. (a) $g(x) = 3x^2 - 12x + 13 = 3\left(x^2 - 4x\right) + 13$

$\qquad = 3\left(x^2 - 4x + 4\right) + 13 - 12$

$\qquad = 3(x - 2)^2 + 1$

(b)

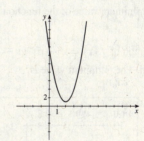

(c) The minimum value is $g(2) = 1$.

31. (a) $h(x) = 1 - x - x^2 = -\left(x^2 + x\right) + 1$

$\qquad = -\left(x^2 + x + \frac{1}{4}\right) + 1 + \frac{1}{4}$

$\qquad = -\left(x + \frac{1}{2}\right)^2 + \frac{5}{4}$

(b)

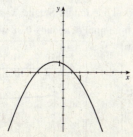

(c) The maximum value is $h\left(-\frac{1}{2}\right) = \frac{5}{4}$.

33. $f(x) = x^2 + x + 1 = \left(x^2 + x\right) + 1$

$\qquad = \left(x^2 + x + \frac{1}{4}\right) + 1 + \frac{1}{4} = \left(x + \frac{1}{2}\right)^2 + \frac{3}{4}$

Therefore, the minimum value is $f\left(-\frac{1}{2}\right) = \frac{3}{4}$.

35. $f(t) = 100 - 49t - 7t^2 = -7\left(t^2 + 7t\right) + 100 = -7\left(t^2 + 7t + \frac{49}{4}\right) + 100 + \frac{343}{4} = -7\left(t + \frac{7}{2}\right)^2 + \frac{743}{4}$.

Therefore, the maximum value is $f\left(-\frac{7}{2}\right) = \frac{743}{4} = 185.75$.

37. $f(s) = s^2 - 1.2s + 16 = \left(s^2 - 1.2s\right) + 16 = \left(s^2 - 1.2s + 0.36\right) + 16 - 0.36 = (s - 0.6)^2 + 15.64$.

Therefore, the minimum value is $f(0.6) = 15.64$.

39. $h(x) = \frac{1}{2}x^2 + 2x - 6 = \frac{1}{2}\left(x^2 + 4x\right) - 6 = \frac{1}{2}\left(x^2 + 4x + 4\right) - 6 - 2 = \frac{1}{2}(x + 2)^2 - 8$.

Therefore, the minimum value is $h(-2) = -8$.

41. $f(x) = 3 - x - \frac{1}{2}x^2 = -\frac{1}{2}\left(x^2 + 2x\right) + 3 = -\frac{1}{2}\left(x^2 + 2x + 1\right) + 3 + \frac{1}{2} = -\frac{1}{2}(x + 1) + \frac{7}{2}$. Therefore, the maximum

value is $f(-1) = \frac{7}{2}$.

43. Since the vertex is at $(1, -2)$, the function is of the form $f(x) = a(x - 1)^2 - 2$. Substituting the point $(4, 16)$, we get
$16 = a(4 - 1)^2 - 2 \Leftrightarrow 16 = 9a - 2 \Leftrightarrow 9a = 18 \Leftrightarrow a = 2$. So the function is $f(x) = 2(x - 1)^2 - 2 = 2x^2 - 4x$.

45. $f(x) = -x^2 + 4x - 3 = -\left(x^2 - 4x\right) - 3 = -\left(x^2 - 4x + 4\right) - 3 + 4 = -(x - 2)^2 + 1$. So the domain of $f(x)$ is
$(-\infty, \infty)$. Since $f(x)$ has a maximum value of 1, the range is $(-\infty, 1]$.

47. $f(x) = 2x^2 + 6x - 7 = 2\left(x + \frac{3}{2}\right)^2 - 7 - \frac{9}{2} = 2\left(x + \frac{3}{2}\right)^2 - \frac{23}{2}$. The domain of the function is all real numbers, and

since the minimum value of the function is $f\left(-\frac{3}{2}\right) = -\frac{23}{2}$, the range of the function is $\left[-\frac{23}{2}, \infty\right)$.

49. (a) The graph of $f(x) = x^2 + 1.79x - 3.21$ is
shown. The minimum value is
$f(x) \approx -4.01$.

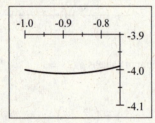

(b) $f(x) = x^2 + 1.79x - 3.21$
$= \left[x^2 + 1.79x + \left(\frac{1.79}{2}\right)^2\right] - 3.21 - \left(\frac{1.79}{2}\right)^2$
$= (x + 0.895)^2 - 4.011025$

Therefore, the exact minimum of $f(x)$ is -4.011025.

51. Local maximum: 2 at $x = 0$. Local minimum: -1 at $x = -2$ and 0 at $x = 2$.

53. Local maximum: 0 at $x = 0$ and 1 at $x = 3$. Local minimum: -2 at $x = -2$ and -1 at $x = 1$.

55. In the first graph, we see that $f(x) = x^3 - x$ has a local minimum and a local maximum. Smaller x- and y-ranges show
that $f(x)$ has a local maximum of about 0.38 when $x \approx -0.58$ and a local minimum of about -0.38 when $x \approx 0.58$.

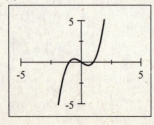
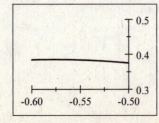
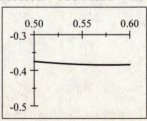

57. In the first graph, we see that $g(x) = x^4 - 2x^3 - 11x^2$ has two local minimums and a local maximum. The local maximum is $g(x) = 0$ when $x = 0$. Smaller x- and y-ranges show that local minima are $g(x) \approx -13.61$ when $x \approx -1.71$ and $g(x) \approx -73.32$ when $x \approx 3.21$.

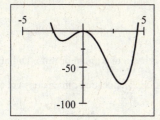
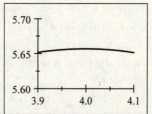
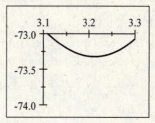

59. In the first graph, we see that $U(x) = x\sqrt{6 - x}$ only has a local maximum. Smaller x- and y-ranges show that $U(x)$ has a local maximum of about 5.66 when $x \approx 4.00$.

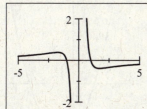
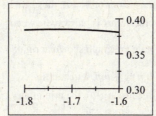

61. In the first graph, we see that $V(x) = \dfrac{1 - x^2}{x^3}$ has a local minimum and a local maximum. Smaller x- and y-ranges show that $V(x)$ has a local maximum of about 0.38 when $x \approx -1.73$ and a local minimum of about -0.38 when $x \approx 1.73$.

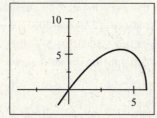
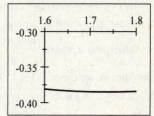

63. $y = f(t) = 40t - 16t^2 = -16\left(t^2 - \frac{5}{2}\right) = -16\left[t^2 - \frac{5}{2}t + \left(\frac{5}{4}\right)^2\right] + 16\left(\frac{5}{4}\right)^2 = -16\left(t - \frac{5}{4}\right)^2 + 25$. Thus the maximum

height attained by the ball is $f\left(\frac{5}{4}\right) = 25$ feet.

65. $R(x) = 80x - 0.4x^2 = -0.4\left(x^2 - 200x\right) = -0.4\left(x^2 - 200x + 10{,}000\right) + 4{,}000 = -0.4(x - 100)^2 + 4{,}000$. So revenue is maximized at $\$4{,}000$ when 100 units are sold.

67. $E(n) = \frac{2}{3}n - \frac{1}{90}n^2 = -\frac{1}{90}\left(n^2 - 60n\right) = -\frac{1}{90}\left(n^2 - 60n + 900\right) + 10 = -\frac{1}{90}(n - 30)^2 + 10$. Since the maximum of the function occurs when $n = 30$, the viewer should watch the commercial 30 times for maximum effectiveness.

69. $A(n) = n(900 - 9n) = -9n^2 + 900n$ is a quadratic function with $a = -9$ and $b = 900$, so by the formula, the maximum or minimum value occurs at $n = -\dfrac{b}{2a} = -\dfrac{900}{2(-9)} = 50$ trees, and because $a < 0$, this gives a maximum value.

71. The area of the fenced-in field is given by $A(x) = (2400 - 2x)x = -2x^2 + 2400x$. Thus, by the formula in this section, the maximum or minimum value occurs at $x = -\dfrac{b}{2a} = -\dfrac{2400}{2(-2)} = 600$. The maximum area occurs when $x = 600$ feet and $l = 2400 - 2(600) = 1200$ feet.

73. $A(x) = 15x - \frac{1}{8}(\pi + 4)x^2$, so by the formula, the maximum area occurs when $x = -\dfrac{b}{2a} = -\dfrac{-15}{2\left[\frac{1}{8}(\pi+4)\right]} \approx 8.4$ ft

and $h \approx 15 - \frac{1}{2}(8.40) - \frac{1}{4}\pi(8.40) \approx 4.2$ ft.

75. (a) The area of the corral is $A(x) = x(1200 - x) = 1200x - x^2 = -x^2 + 1200x$.

(b) A is a quadratic function with $a = -1$ and $b = 1200$, so by the formula, it has a maximum or minimum at

$$x = -\frac{b}{2a} = -\frac{1200}{2(-1)} = 600,$$ and because $a < 0$, this gives a maximum value. The desired dimensions are 600 ft by 600 ft.

77. (a) To model the revenue, we need to find the total attendance. Let x be the ticket price. Then the amount by which the ticket price is lowered is $10 - x$, and we are given that for every dollar it is lowered, the attendance increases by 3000; that is, the increase in attendance is $3000(10 - x)$. Thus, the attendance is $27{,}000 + 3000(10 - x)$, and since each spectator pays $x, the revenue is $R(x) = x[27{,}000 + 3000(10 - x)] = -3000x^2 + 57{,}000x$.

(b) Since R is a quadratic function with $a = -3000$ and $b = 57{,}000$, the maximum occurs at

$$x = -\frac{b}{2a} = -\frac{57{,}000}{2(-3000)} = 9.5;$$ that is, when admission is \$9.50.

(c) We solve $R(x) = 0$ for x: $-3000x^2 + 57{,}000x = 0 \Leftrightarrow -3000x(x - 19) = 0 \Leftrightarrow x = 0$ or $x = 19$. Thus, if admission is \$19, nobody will attend and no revenue will be generated.

79. Because $f(x) = (x - m)(x - n) = 0$ when $x = m$ or $x = n$, those are its x-intercepts. By symmetry, we expect that the vertex is halfway between these values; that is, at $x = \dfrac{m+n}{2}$. We obtain the graph shown at right.

Expanding, we see that $f(x) = x^2 - (m + n)x + mn$, a quadratic function with $a = 1$ and $b = -(m + n)$. Because $a > 0$, the minimum value occurs at $x = -\dfrac{b}{2a} = \dfrac{m+n}{2}$, the x-value of the vertex, as expected.

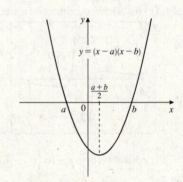

3.2 POLYNOMIAL FUNCTIONS AND THEIR GRAPHS

1. Graph I cannot be that of a polynomial because it is not smooth (it has a cusp.) Graph II could be that of a polynomial function, because it is smooth and continuous. Graph III could not be that of a polynomial function because it has a break.

3. (a) By definition, if c is a zero of P, then $P(c) = 0$.

(b) This is false in general. For example, take $P(x) = x - 1$. Then $c = 1$ is a zero of P, but $P(0) = 0 - 1 = -1 \neq c$.

(c) This is true. If c is a zero of P, then $x - c$ is a factor of $P(x)$

(d) This is false in general. For example, take $P(x) = x - 1$. Then $c = 1$ is a zero, but the y-intercept of P is -1. (It is true that if c is a root of P, it is also an x-intercept of the graph of P.)

5. (a) $P(x) = x^2 - 4$

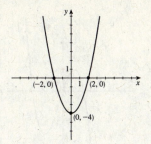

(b) $Q(x) = (x-4)^2$

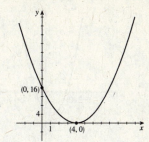

(c) $R(x) = 2x^2 - 2$

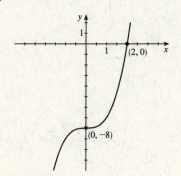

(d) $S(x) = 2(x-2)^2$

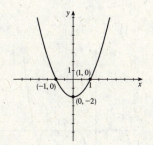

7. (a) $P(x) = x^3 - 8$

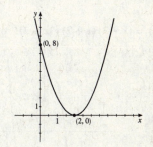

(b) $Q(x) = -x^3 + 27$

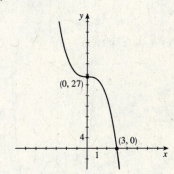

(c) $R(x) = -(x+2)^3$

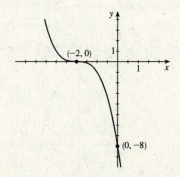

(d) $S(x) = \frac{1}{2}(x-1)^3 + 4$

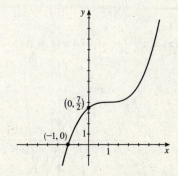

9. III   **11.** V   **13.** VI

15. $P(x) = (x - 1)(x + 2)$

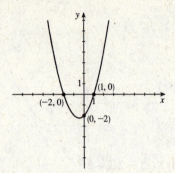

17. $P(x) = x(x - 3)(x + 2)$

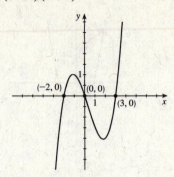

19. $P(x) = (x - 3)(x + 2)(3x - 2)$

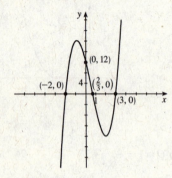

21. $P(x) = (x - 1)^2(x - 3)$

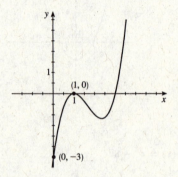

23. $P(x) = \frac{1}{12}(x + 2)^2(x - 3)^2$

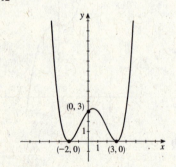

25. $P(x) = x^3(x + 2)(x - 3)^2$

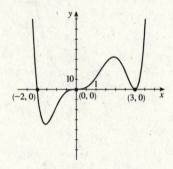

27. $P(x) = x^3 - x^2 - 6x = x(x + 2)(x - 3)$

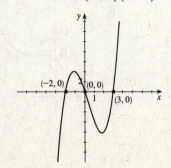

29. $P(x) = -x^3 + x^2 + 12x = -x(x + 3)(x - 4)$

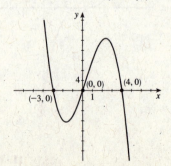

31. $P(x) = x^4 - 3x^3 + 2x^2 = x^2(x-1)(x-2)$

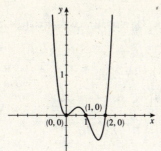

33. $P(x) = x^3 + x^2 - x - 1 = (x-1)(x+1)^2$

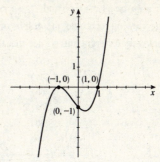

35. $P(x) = 2x^3 - x^2 - 18x + 9$

$= (x-3)(2x-1)(x+3)$

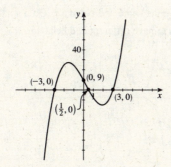

37. $P(x) = x^4 - 2x^3 - 8x + 16$

$= (x-2)^2(x^2 + 2x + 4)$

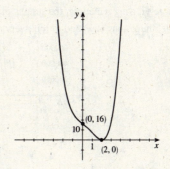

39. $P(x) = x^4 - 3x^2 - 4 = (x-2)(x+2)(x^2+1)$

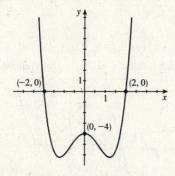

41. $P(x) = 3x^3 - x^2 + 5x + 1$; $Q(x) = 3x^3$. Since P has odd degree and positive leading coefficient, it has the following end behavior: $y \to \infty$ as $x \to \infty$ and $y \to -\infty$ as $x \to -\infty$.

On a large viewing rectangle, the graphs of P and Q look almost the same. On a small viewing rectangle, we see that the graphs of P and Q have different intercepts.

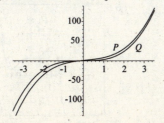
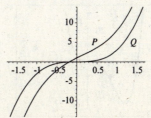

43. $P(x) = x^4 - 7x^2 + 5x + 5$; $Q(x) = x^4$. Since P has even degree and positive leading coefficient, it has the following end behavior: $y \to \infty$ as $x \to \infty$ and $y \to \infty$ as $x \to -\infty$.

On a large viewing rectangle, the graphs of P and Q look almost the same. On a small viewing rectangle, the graphs of P and Q look very different and we see that they have different intercepts.

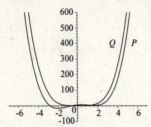 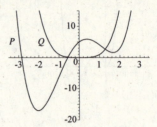

45. $P(x) = x^{11} - 9x^9$; $Q(x) = x^{11}$. Since P has odd degree and positive leading coefficient, it has the following end behavior: $y \to \infty$ as $x \to \infty$ and $y \to -\infty$ as $x \to -\infty$.

On a large viewing rectangle, the graphs of P and Q look like they have the same end behavior. On a small viewing rectangle, the graphs of P and Q look very different and seem (wrongly) to have different end behavior.

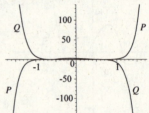 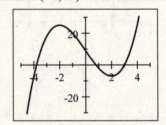

47. (a) x-intercepts at 0 and 4, y-intercept at 0.

(b) Local maximum at $(2, 4)$, no local minimum.

49. (a) x-intercepts at -2 and 1, y-intercept at -1.

(b) Local maximum at $(1, 0)$, local minimum at $(-1, -2)$.

51. $y = -x^2 + 8x$, $[-4, 12]$ by $[-50, 30]$

No local minimum. Local maximum at $(4, 16)$.

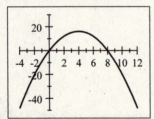

53. $y = x^3 - 12x + 9$, $[-5, 5]$ by $[-30, 30]$

Local maximum at $(-2, 25)$. Local minimum at $(2, -7)$.

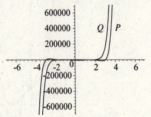

55. $y = x^4 + 4x^3$, $[-5, 5]$ by $[-30, 30]$

Local minimum at $(-3, -27)$. No local maximum.

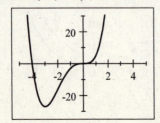

57. $y = 3x^5 - 5x^3 + 3$, $[-3, 3]$ by $[-5, 10]$

Local maximum at $(-1, 5)$. Local minimum at $(1, 1)$.

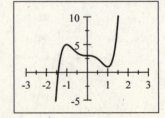

59. $y = -2x^2 + 3x + 5$ has one local maximum at $(0.75, 6.13)$.

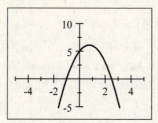

61. $y = x^3 - x^2 - x$ has one local maximum at $(-0.33, 0.19)$ and one local minimum at $(1.00, -1.00)$.

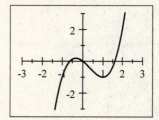

63. $y = x^4 - 5x^2 + 4$ has one local maximum at $(0, 4)$ and two local minima at $(-1.58, -2.25)$ and $(1.58, -2.25)$.

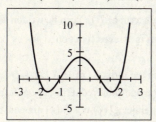

65. $y = (x - 2)^5 + 32$ has no maximum or minimum.

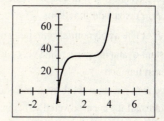

67. $y = x^8 - 3x^4 + x$ has one local maximum at $(0.44, 0.33)$ and two local minima at $(1.09, -1.15)$ and $(-1.12, -3.36)$.

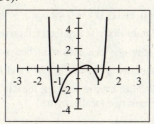

69. $y = cx^3$; $c = 1, 2, 5, \frac{1}{2}$. Increasing the value of c stretches the graph vertically.

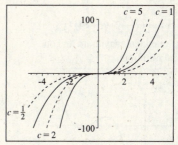

71. $P(x) = x^4 + c$; $c = -1, 0, 1,$ and 2. Increasing the value of c moves the graph up.

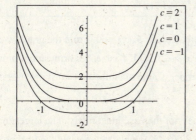

73. $P(x) = x^4 - cx$; $c = 0, 1, 8,$ and 27. Increasing the value of c causes a deeper dip in the graph, in the fourth quadrant, and moves the positive x-intercept to the right.

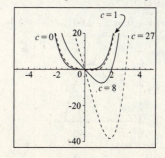

75. (a)

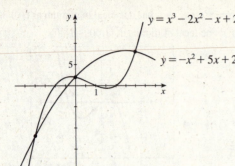

$y = x^3 - 2x^2 - x + 2$

$y = -x^2 + 5x + 2$

(b) The two graphs appear to intersect at 3 points.

(c) $x^3 - 2x^2 - x + 2 = -x^2 + 5x + 2 \Leftrightarrow x^3 - x^2 - 6x = 0 \Leftrightarrow$ $x\left(x^2 - x - 6\right) = 0 \Leftrightarrow x(x-3)(x+2) = 0$. Then either $x = 0$, $x = 3$, or $x = -2$. If $x = 0$, then $y = 2$; if $x = 3$ then $y = 8$; if $x = -2$, then $y = -12$. Hence the points where the two graphs intersect are $(0, 2)$, $(3, 8)$, and $(-2, -12)$.

77. (a) Let $P(x)$ be a polynomial containing only odd powers of x. Then each term of $P(x)$ can be written as Cx^{2n+1}, for some constant C and integer n. Since $C(-x)^{2n+1} = -Cx^{2n+1}$, each term of $P(x)$ is an odd function. Thus by part (a), $P(x)$ is an odd function.

(b) Let $P(x)$ be a polynomial containing only even powers of x. Then each term of $P(x)$ can be written as Cx^{2n}, for some constant C and integer n. Since $C(-x)^{2n} = Cx^{2n}$, each term of $P(x)$ is an even function. Thus by part (b), $P(x)$ is an even function.

(c) Since $P(x)$ contains both even and odd powers of x, we can write it in the form $P(x) = R(x) + Q(x)$, where $R(x)$ contains all the even-powered terms in $P(x)$ and $Q(x)$ contains all the odd-powered terms. By part (d), $Q(x)$ is an odd function, and by part (e), $R(x)$ is an even function. Thus, since neither $Q(x)$ nor $R(x)$ are constantly 0 (by assumption), by part (c), $P(x) = R(x) + Q(x)$ is neither even nor odd.

(d) $P(x) = x^5 + 6x^3 - x^2 - 2x + 5 = \left(x^5 + 6x^3 - 2x\right) + \left(-x^2 + 5\right) = P_O(x) + P_E(x)$ where $P_O(x) = x^5 + 6x^3 - 2x$ and $P_E(x) = -x^2 + 5$. Since $P_O(x)$ contains only odd powers of x, it is an odd function, and since $P_E(x)$ contains only even powers of x, it is an even function.

79. (a) $P(x) = (x-2)(x-4)(x-5)$ has one local maximum and one local minimum.

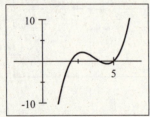

(b) Since $P(a) = P(b) = 0$, and $P(x) > 0$ for $a < x < b$ (see the table below), the graph of P must first rise and then fall on the interval (a, b), and so P must have at least one local maximum between a and b. Using similar reasoning, the fact that $P(b) = P(c) = 0$ and $P(x) < 0$ for $b < x < c$ shows that P must have at least one local minimum between b and c. Thus P has at least two local extrema.

Interval	$(-\infty, a)$	(a, b)	(b, c)	(c, ∞)
Sign of $x - a$	$-$	$+$	$+$	$+$
Sign of $x - b$	$-$	$-$	$+$	$+$
Sign of $x - c$	$-$	$-$	$-$	$+$
Sign of $(x-a)(x-b)(x-c)$	$-$	$+$	$-$	$+$

81. $P(x) = 8x + 0.3x^2 - 0.0013x^3 - 372$

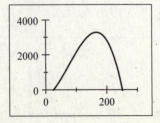

(a) For the firm to break even, $P(x) = 0$. From the graph, we see that $P(x) = 0$ when $x \approx 25.2$. Of course, the firm cannot produce fractions of a blender, so the manufacturer must produce at least 26 blenders a year.

(b) No, the profit does not increase indefinitely. The largest profit is approximately $\$3276.22$, which occurs when the firm produces 166 blenders per year.

83. (a) The length of the bottom is $40 - 2x$, the width of the bottom is $20 - 2x$, and the height is x, so the volume of the box is

$$V = x\,(20 - 2x)\,(40 - 2x) = 4x^3 - 120x^2 + 800x.$$

(b) Since the height and width must be positive, we must have $x > 0$ and $20 - 2x > 0$, and so the domain of V is $0 < x < 10$.

(c) Using the domain from part (b), we graph V in the viewing rectangle $[0, 10]$ by $[0, 1600]$. The maximum volume is $V \approx 1539.6$ when $x = 4.23$.

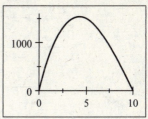

85.

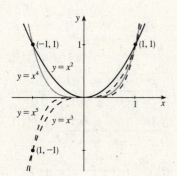

The graph of $y = x^{100}$ is close to the x-axis for $|x| < 1$, but passes through the points $(1, 1)$ and $(-1, 1)$. The graph of $y = x^{101}$ behaves similarly except that the y-values are negative for negative values of x, and it passes through $(-1, -1)$ instead of $(-1, 1)$.

87. No, it is impossible. The end behavior of a third degree polynomial is the same as that of $y = kx^3$, and for this function, the values of y go off in opposite directions as $x \to \infty$ and $x \to -\infty$. But for a function with just one extremum, the values of y would head off in the same direction (either both up or both down) on either side of the extremum. An nth-degree polynomial can have $n - 1$ extrema or $n - 3$ extrema or $n - 5$ extrema, and so on (decreasing by 2). A polynomial that has six local extrema must be of degree 7 or higher. For example, $P\,(x) = (x - 1)\,(x - 2)\,(x - 3)\,(x - 4)\,(x - 5)\,(x - 6)\,(x - 7)$ has six local extrema.

3.3 DIVIDING POLYNOMIALS

1. If we divide the polynomial P by the factor $x - c$, and we obtain the equation $P\,(x) = (x - c)\,Q\,(x) + R\,(x)$, then we say that $x - c$ is the *divisor*, $Q\,(x)$ is the *quotient*, and $R\,(x)$ is the *remainder*.

3.

$$
\begin{array}{r}
3x-\ 4 \\
x+3 \enclose{longdiv}{3x^2+5x-4} \\
\underline{3x^2+9x} \\
-4x-4 \\
\underline{-4x-12} \\
8
\end{array}
$$

Thus the quotient is $3x - 4$ and the remainder is 8, so $P\,(x) = 3x^2 + 5x - 4 = (x + 3) \cdot (3x - 4) + 8.$

5.

$$
\begin{array}{r}
x^2\qquad -\ 1 \\
2x-3 \enclose{longdiv}{2x^3-3x^2-2x} \\
\underline{2x^3-3x^2} \\
-2x \\
\underline{-2x+3} \\
-3
\end{array}
$$

Thus the quotient is $x^2 - 1$ and the remainder is -3, and $P\,(x) = 2x^3 - 3x^2 - 2x = \left(x^2 - 1\right)(2x - 3) - 3.$

7.

$$\begin{array}{r}
x^2 - x - 3 \\
x^2+3 \ \overline{\smash{\big)}\ x^4 - x^3 + 0x^2 + 4x + 2} \\
\underline{x^4 \qquad\quad + 3x^2} \\
-x^3 - 3x^2 + 4x \\
\underline{-x^2 \qquad\quad - 3x} \\
-3x^2 + 7x + 2 \\
\underline{-3x^2 \qquad\quad - 9} \\
7x + 11
\end{array}$$

Thus the quotient is $x^2 - x - 3$ and the remainder is $7x + 11$, and

$$P(x) = x^3 + 4x^2 - 6x + 1$$
$$= \left(x^2 + 3\right) \cdot \left(x^2 - x - 3\right) + (7x + 11)$$

9.

$$\begin{array}{r|rrr}
-3 & 1 & 4 & -8 \\
   &   & -3 & -3 \\
\hline
   & 1 & 1 & -11
\end{array}$$

Thus the quotient is $x + 1$ and the remainder is -11, and

$$\frac{P(x)}{D(x)} = \frac{x^2 + 4x - 8}{x + 3} = (x + 1) + \frac{-11}{x + 3}.$$

11.

$$\begin{array}{r}
2x - \frac{1}{2} \\
2x - 1 \ \overline{\smash{\big)}\ 4x^2 - 3x - 7} \\
\underline{4x^2 - 2x} \\
-x - 7 \\
\underline{-x + \frac{1}{2}} \\
-\frac{15}{2}
\end{array}$$

Thus the quotient is $2x - \frac{1}{2}$ and the remainder is $-\frac{15}{2}$, and

$$\frac{P(x)}{D(x)} = \frac{4x^2 - 3x - 7}{2x - 1} = \left(2x - \frac{1}{2}\right) + \frac{-\frac{15}{2}}{2x - 1}.$$

13.

$$\begin{array}{r}
2x^2 - x + 1 \\
x^2 + 4 \ \overline{\smash{\big)}\ 2x^4 - x^3 + 9x^2} \\
\underline{2x^4 \qquad\quad + 8x^2} \\
-x^3 + x^2 \\
\underline{-x^3 \qquad\quad - 4x} \\
x^2 + 4x \\
\underline{x^2 \qquad\quad + 4} \\
4x - 4
\end{array}$$

Thus the quotient is $2x^2 - x + 1$ and the remainder is $4x - 4$, and

$$\frac{P(x)}{D(x)} = \frac{2x^4 - x^3 + 9x^2}{x^2 + 4} = \left(2x^2 - x + 1\right) + \frac{4x - 4}{x^2 + 4}.$$

15.

$$\begin{array}{r}
x - 2 \\
x - 4 \ \overline{\smash{\big)}\ x^2 - 6x - 8} \\
\underline{x^2 - 4x} \\
-2x - 8 \\
\underline{-2x + 8} \\
-16
\end{array}$$

Thus the quotient is $x - 2$ and the remainder is -16.

17.

$$\begin{array}{r}
2x^2 \qquad\quad - 1 \\
2x + 1 \ \overline{\smash{\big)}\ 4x^3 + 2x^2 - 2x - 3} \\
\underline{4x^3 + 2x^2} \\
-2x - 3 \\
\underline{-2x - 1} \\
-2
\end{array}$$

Thus the quotient is $2x^2 - 1$ and the remainder is -2.

19.

$$\begin{array}{r}
x + 2 \\
x^2 - 2x + 2 \ \overline{\smash{\big)}\ x^3 + 0x^2 + 6x + 3} \\
\underline{x^3 - 2x^2 + 2x} \\
2x^2 + 4x + 3 \\
\underline{2x^2 - 4x + 4} \\
8x - 1
\end{array}$$

Thus the quotient is $x + 2$, and the remainder is $8x - 1$.

21.

$$\begin{array}{r}
3x + 1 \\
2x^2 + 0x + 5 \ \overline{\smash{\big)}\ 6x^3 + 2x^2 + 22x + 0} \\
\underline{6x^3 \qquad\quad + 15x} \\
2x^2 + 7x + 0 \\
\underline{2x^2 \qquad\quad + 5} \\
7x - 5
\end{array}$$

Thus the quotient is $3x + 1$, and the remainder is $7x - 5$.

23.

$$
\begin{array}{r}
x^4 \qquad\qquad\qquad + 1 \\
x^2 + 1 \;\overline{\big)\; x^6 + 0x^5 + x^4 + 0x^3 + x^2 + 0x + 1} \\
\underline{x^6 \qquad\quad + x^4} \\
0 \qquad\quad + x^2 \qquad + 1 \\
\underline{x^2 \qquad\quad + 1} \\
0
\end{array}
$$

Thus the quotient is $x^4 + 1$, and the remainder is 0.

25. The synthetic division table for this problem takes the following form.

$$
\begin{array}{r|rrr}
3 & 1 & -5 & 4 \\
  &   & 3 & -6 \\
\hline
  & 1 & -2 & -2
\end{array}
$$

Thus the quotient is $x - 2$, and the remainder is -2.

27. The synthetic division table for this problem takes the following form.

$$
\begin{array}{r|rrr}
6 & 3 & 5 & 0 \\
  &   & 18 & 138 \\
\hline
  & 3 & 23 & 138
\end{array}
$$

Thus the quotient is $3x + 23$, and the remainder is 138.

29. Since $x + 2 = x - (-2)$, the synthetic division table for this problem takes the following form.

$$
\begin{array}{r|rrrr}
-2 & 1 & 2 & 2 & 1 \\
   &   & -2 & 0 & -4 \\
\hline
   & 1 & 0 & 2 & -3
\end{array}
$$

Thus the quotient is $x^2 + 2$, and the remainder is -3.

31. Since $x + 3 = x - (-3)$ and $x^3 - 8x + 2 = x^3 + 0x^2 - 8x + 2$, the synthetic division table for this problem takes the following form.

$$
\begin{array}{r|rrrr}
-3 & 1 & 0 & -8 & 2 \\
   &   & -3 & 9 & -3 \\
\hline
   & 1 & -3 & 1 & -1
\end{array}
$$

Thus the quotient is $x^2 - 3x + 1$, and the remainder is -1.

33. Since $x^5 + 3x^3 - 6 = x^5 + 0x^4 + 3x^3 + 0x^2 + 0x - 6$, the synthetic division table for this problem takes the following form.

$$
\begin{array}{r|rrrrrr}
1 & 1 & 0 & 3 & 0 & 0 & -6 \\
  &   & 1 & 1 & 4 & 4 & 4 \\
\hline
  & 1 & 1 & 4 & 4 & 4 & -2
\end{array}
$$

Thus the quotient is $x^4 + x^3 + 4x^2 + 4x + 4$, and the remainder is -2.

35. The synthetic division table for this problem takes the following form.

$$
\begin{array}{r|rrrr}
\frac{1}{2} & 2 & 3 & -2 & 1 \\
            &   & 1 & 2 & 0 \\
\hline
            & 2 & 4 & 0 & 1
\end{array}
$$

Thus the quotient is $2x^2 + 4x$, and the remainder is 1.

37. Since $x^3 - 27 = x^3 + 0x^2 + 0x - 27$, the synthetic division table for this problem takes the following form.

$$
\begin{array}{r|rrrr}
3 & 1 & 0 & 0 & -27 \\
  &   & 3 & 9 & 27 \\
\hline
  & 1 & 3 & 9 & 0
\end{array}
$$

Thus the quotient is $x^2 + 3x + 9$, and the remainder is 0.

39. $P(x) = 4x^2 + 12x + 5, c = -1$

$$
\begin{array}{r|rrr}
-1 & 4 & 12 & 5 \\
   &   & -4 & -8 \\
\hline
   & 4 & 8 & -3
\end{array}
$$

Therefore, by the Remainder Theorem, $P(-1) = -3$.

41. $P(x) = x^3 + 3x^2 - 7x + 6, c = 2$

$$
\begin{array}{r|rrrr}
2 & 1 & 3 & -7 & 6 \\
  &   & 2 & 10 & 6 \\
\hline
  & 1 & 5 & 3 & 12
\end{array}
$$

Therefore, by the Remainder Theorem, $P(2) = 12$.

43. $P(x) = x^3 + 2x^2 - 7, c = -2$

$$
\begin{array}{r|rrrr}
-2 & 1 & 2 & 0 & -7 \\
   &   & -2 & 0 & 0 \\
\hline
   & 1 & 0 & 0 & -7
\end{array}
$$

Therefore, by the Remainder Theorem, $P(-2) = -7$.

45. $P(x) = 5x^4 + 30x^3 - 40x^2 + 36x + 14, c = -7$

$$
\begin{array}{r|rrrrr}
-7 & 5 & 30 & -40 & 36 & 14 \\
   &   & -35 & 35 & 35 & -497 \\
\hline
   & 5 & -5 & -5 & 71 & -483
\end{array}
$$

Therefore, by the Remainder Theorem, $P(-7) = -483$.

47. $P(x) = x^7 - 3x^2 - 1$

$$= x^7 + 0x^6 + 0x^5 + 0x^4 + 0x^3 - 3x^2 + 0x - 1$$

$c = 3$

$$
\begin{array}{r|rrrrrrr}
3 & 1 & 0 & 0 & 0 & 0 & -3 & 0 & -1 \\
& & 3 & 9 & 27 & 81 & 243 & 720 & 2160 \\
\hline
& 1 & 3 & 9 & 27 & 81 & 240 & 720 & 2159
\end{array}
$$

Therefore by the Remainder Theorem, $P(3) = 2159$.

49. $P(x) = 3x^3 + 4x^2 - 2x + 1, c = \frac{2}{3}$

$$
\begin{array}{r|rrrr}
\frac{2}{3} & 3 & 4 & -2 & 1 \\
& & 2 & 4 & \frac{4}{3} \\
\hline
& 3 & 6 & 2 & \frac{7}{3}
\end{array}
$$

Therefore, by the Remainder Theorem, $P\left(\frac{2}{3}\right) = \frac{7}{3}$.

51. $P(x) = x^3 + 2x^2 - 3x - 8, c = 0.1$

$$
\begin{array}{r|rrrr}
0.1 & 1 & 2 & -3 & -8 \\
& & 0.1 & 0.21 & -0.279 \\
\hline
& 1 & 2.1 & -2.79 & -8.279
\end{array}
$$

Therefore, by the Remainder Theorem, $P(0.1) = -8.279$.

53. $P(x) = x^3 - 3x^2 + 3x - 1, c = 1$

$$
\begin{array}{r|rrrr}
1 & 1 & -3 & 3 & -1 \\
& & 1 & -2 & 1 \\
\hline
& 1 & -2 & 1 & 0
\end{array}
$$

Since the remainder is 0, $x - 1$ is a factor.

55. $P(x) = 2x^3 + 7x^2 + 6x - 5, c = \frac{1}{2}$

$$
\begin{array}{r|rrrr}
\frac{1}{2} & 2 & 7 & 6 & -5 \\
& & 1 & 4 & 5 \\
\hline
& 2 & 8 & 10 & 0
\end{array}
$$

Since the remainder is 0, $x - \frac{1}{2}$ is a factor.

57. $P(x) = x^3 - x^2 - 11x + 15, c = 3$

$$
\begin{array}{r|rrrr}
3 & 1 & -1 & -11 & 15 \\
& & 3 & 6 & -15 \\
\hline
& 1 & 2 & -5 & 0
\end{array}
$$

Since the remainder is 0, we know that 3 is a zero

$$x^3 - x^2 - 11x + 15 = (x - 3)\left(x^2 + 2x - 5\right). \text{ Now}$$

$x^2 + 2x - 5 = 0$ when $x = \frac{-2 \pm \sqrt{2^2 + 4(1)(5)}}{2} = -1 \pm \sqrt{6}$.

Hence, the zeros are $-1 - \sqrt{6}, -1 + \sqrt{6}$, and 3.

59. Since the zeros are $x = -1, x = 1$, and $x = 3$, the factors are $x + 1, x - 1$, and $x - 3$.

Thus

$$P(x) = (x + 1)(x - 1)(x - 3) = x^3 - 3x^2 - x + 3.$$

61. Since the zeros are $x = -1, x = 1, x = 3$, and $x = 5$, the factors are $x + 1, x - 1, x - 3$, and $x - 5$.

Thus $P(x) = (x + 1)(x - 1)(x - 3)(x - 5) = x^4 - 8x^3 + 14x^2 + 8x - 15$.

63. Since the zeros of the polynomial are 1, -2, and 3, it follows that

$P(x) = C(x - 1)(x + 2)(x - 3) = C\left(x^3 - 2x^2 - 5x + 6\right) = Cx^3 - 2Cx^2 - 5Cx + 6C$. Since the coefficient of x^2 is

to be 3, $-2C = 3$ so $C = -\frac{3}{2}$. Therefore, $P(x) = -\frac{3}{2}\left(x^3 - 2x^2 - 5x + 6\right) = -\frac{3}{2}x^3 + 3x^2 + \frac{15}{2}x - 9$ is the polynomial.

65. The y-intercept is 2 and the zeros of the polynomial are $-1, 1$, and 2.

It follows that $P(x) = C(x + 1)(x - 1)(x - 2) = C\left(x^3 - 2x^2 - x + 2\right)$. Since $P(0) = 2$ we have

$2 = C\left[(0)^3 - 2(0)^2 - (0) + 2\right] \Leftrightarrow 2 = 2C \Leftrightarrow C = 1$ and $P(x) = (x + 1)(x - 1)(x - 2) = x^3 - 2x^2 - x + 2$.

67. The y-intercept is 4 and the zeros of the polynomial are -2 and 1 both being degree two.

It follows that $P(x) = C(x+2)^2(x-1)^2 = C\left(x^4 + 2x^3 - 3x^2 - 4x + 4\right)$. Since $P(0) = 4$ we have

$4 = C\left[(0)^4 + 2(0)^3 - 3(0)^2 - 4(0) + 4\right] \Leftrightarrow 4 = 4C \Leftrightarrow C = 1$.

Thus $P(x) = (x+2)^2(x-1)^2 = x^4 + 2x^3 - 3x^2 - 4x + 4$.

69. A. By the Remainder Theorem, the remainder when $P(x) = 6x^{1000} - 17x^{562} + 12x + 26$ is divided by $x + 1$ is

$P(-1) = 6(-1)^{1000} - 17(-1)^{562} + 12(-1) + 26 = 6 - 17 - 12 + 26 = 3$.

B. If $x - 1$ is a factor of $Q(x) = x^{567} - 3x^{400} + x^9 + 2$, then $Q(1)$ must equal 0.

$Q(1) = (1)^{567} - 3(1)^{400} + (1)^9 + 2 = 1 - 3 + 1 + 2 = 1 \neq 0$, so $x - 1$ is not a factor.

3.4   REAL ZEROS OF POLYNOMIALS

1. If the polynomial function $P(x) = a_n x^n + a_{n-1}x^{n-1} + \cdots a_1 x + a_0$ has integer coefficients, then the only numbers that

could possibly be rational zeros of P are all of the form $\dfrac{p}{q}$, where p is a factor of *the constant coefficient* a_0 and q is a factor

of *the leading coefficient* a_n. The possible rational zeros of $P(x) = 6x^3 + 5x^2 - 19x - 10$ are ± 1, $\pm\frac{1}{2}$, $\pm\frac{1}{3}$, $\pm\frac{1}{6}$, ± 2, $\pm\frac{2}{3}$,

± 5, $\pm\frac{5}{2}$, $\pm\frac{5}{3}$, $\pm\frac{5}{6}$, ± 10, and $\pm\frac{10}{3}$.

3. This is true. If c is a real zero of the polynomial P, then $P(x) = (x - c)Q(x)$, and any other zero of $P(x)$ is also a zero of $Q(x) = P(x)/(x - c)$.

5. $P(x) = x^3 - 4x^2 + 3$ has possible rational zeros ± 1 and ± 3.

7. $R(x) = 2x^5 + 3x^3 + 4x^2 - 8$ has possible rational zeros ± 1, ± 2, ± 4, ± 8, $\pm\frac{1}{2}$.

9. $T(x) = 4x^4 - 2x^2 - 7$ has possible rational zeros ± 1, ± 7, $\pm\frac{1}{2}$, $\pm\frac{7}{2}$, $\pm\frac{1}{4}$, $\pm\frac{7}{4}$.

11. (a) $P(x) = 5x^3 - x^2 - 5x + 1$ has possible rational zeros ± 1, $\pm\frac{1}{5}$.

(b) From the graph, the actual zeroes are -1, $\frac{1}{5}$, and 1.

13. (a) $P(x) = 2x^4 - 9x^3 + 9x^2 + x - 3$ has possible rational zeros ± 1, ± 3, $\pm\frac{1}{2}$, $\pm\frac{3}{2}$.

(b) From the graph, the actual zeroes are $-\frac{1}{2}$, 1, and 3.

15. $P(x) = x^3 + 3x^2 - 4$. The possible rational zeros are ± 1, ± 2, ± 4. $P(x)$ has 1 variation in sign and hence 1 positive real zero. $P(-x) = -x^3 + 3x^2 - 4$ has 2 variations in sign and hence 0 or 2 negative real zeros.

$$
\begin{array}{r|rrrr}
1 & 1 & 3 & 0 & -4 \\
  &   & 1 & 4 & 4 \\
\hline
  & 1 & 4 & 4 & 0 \\
\end{array}
\Rightarrow x = 1 \text{ is a zero.}
$$

$P(x) = x^3 + 3x^2 - 4 = (x - 1)\left(x^2 + 4x + 4\right)$

$\qquad = (x - 1)(x + 2)^2$

Therefore, the zeros are $x = -2$ and 1.

17. $P(x) = x^3 - 3x - 2$. The possible rational zeros are ± 1, ± 2. $P(x)$ has 1 variation in sign and hence 1 positive real zero. $P(-x) = -x^3 + 3x - 2$ has 2 variations in sign and hence 0 or 2 negative real zeros.

$$
\begin{array}{r|rrrr}
1 & 1 & 0 & -3 & -2 \\
  &   & 1 & 1 & -2 \\
\hline
  & 1 & 1 & -2 & -4 \\
\end{array}
\Rightarrow x = 1 \text{ is not a zero.}
\qquad
\begin{array}{r|rrrr}
2 & 1 & 0 & -3 & -2 \\
  &   & 2 & 4 & 2 \\
\hline
  & 1 & 2 & 1 & 0 \\
\end{array}
\Rightarrow x = 2 \text{ is a zero.}
$$

$P(x) = x^3 - 3x - 2 = (x - 2)\left(x^2 + 2x + 1\right) = (x - 2)(x + 1)^2$. Therefore, the zeros are $x = 2$ and -1.

19. $P(x) = x^3 - 6x^2 + 12x - 8$. The possible rational zeros are $\pm 1, \pm 2, \pm 4, \pm 8$. $P(x)$ has 3 variations in sign and hence 1 or 3 positive real zeros. $P(-x) = -x^3 - 6x^2 - 12x - 8$ has no variations in sign and hence 0 negative real zeros.

$$
\begin{array}{r|rrrr}
1 & 1 & -6 & 12 & -8 \\
  &   & 1 & -5 & 7 \\
\hline
  & 1 & -5 & 7 & -1
\end{array}
\Rightarrow x = 1 \text{ is not a zero.}
\qquad
\begin{array}{r|rrrr}
2 & 1 & -6 & 12 & -8 \\
  &   & 2 & -8 & 8 \\
\hline
  & 1 & -4 & 4 & 0
\end{array}
\Rightarrow x = 2 \text{ is a zero.}
$$

$P(x) = x^3 - 6x^2 + 12x - 8 = (x - 2)\left(x^2 - 4x + 4\right) = (x - 2)^3$. Therefore, the zero is $x = 2$.

21. $P(x) = x^3 - 4x^2 + x + 6$. The possible rational zeros are $\pm 1, \pm 2, \pm 3, \pm 6$. $P(x)$ has 2 variations in sign and hence 0 or 2 positive real zeros. $P(-x) = -x^3 - 4x^2 - x + 6$ has 1 variation in sign and hence 1 negative real zeros.

$$
\begin{array}{r|rrrr}
-1 & 1 & -4 & 1 & 6 \\
   &   & -1 & 5 & -6 \\
\hline
   & 1 & -5 & 6 & 0
\end{array}
\Rightarrow x + 1 \text{ is a factor.}
$$

So

$$P(x) = x^3 - 4x^2 + x + 6 = (x + 1)\left(x^2 - 5x + 6\right)$$
$$= (x + 1)(x - 3)(x - 2)$$

Therefore, the zeros are $x = -1, 2, 3$.

23. $P(x) = x^3 + 3x^2 - x - 3$. The possible rational zeros are $\pm 1, \pm 3$. $P(x)$ has 1 variation in sign and hence 1 positive real zero. $P(-x) = -x^3 + 3x^2 + x - 4$ has 2 variations in sign and hence 0 or 2 negative real zeros.

$$
\begin{array}{r|rrrr}
-1 & 1 & 3 & -1 & -3 \\
   &   & -1 & -2 & 3 \\
\hline
   & 1 & 2 & -3 & 0
\end{array}
\Rightarrow x + 1 \text{ is a factor.}
$$

So $P(x) = x^3 + 3x^2 - x - 3 = (x + 1)\left(x^2 + 2x - 3\right) = (x + 1)(x + 3)(x - 1)$. Therefore, the zeros are $-1, -3$, and 1.

25. *Method 1:* $P(x) = x^4 - 5x^2 + 4$. The possible rational zeros are $\pm 1, \pm 2, \pm 4$. $P(x)$ has 1 variation in sign and hence 1 positive real zero. $P(-x) = x^4 - 5x^2 + 4$ has 2 variations in sign and hence 0 or 2 negative real zeros.

$$
\begin{array}{r|rrrrr}
1 & 1 & 0 & -5 & 0 & 4 \\
  &   & 1 & 1 & -4 & -4 \\
\hline
  & 1 & 1 & -4 & -4 & 0
\end{array}
\Rightarrow x = 1 \text{ is a zero.}
$$

Thus $P(x) = x^4 - 5x^2 + 4 = (x - 1)\left(x^3 + x^2 - 4x - 4\right)$. Continuing with the quotient we have:

$$
\begin{array}{r|rrrr}
-1 & 1 & 1 & -4 & -4 \\
   &   & -1 & 0 & 4 \\
\hline
   & 1 & 0 & -4 & 0
\end{array}
\Rightarrow x = -1 \text{ is a zero.}
$$

$P(x) = x^4 - 5x^2 + 4 = (x - 1)(x + 1)\left(x^2 - 4\right) = (x - 1)(x + 1)(x - 2)(x + 2)$. Therefore, the zeros are $x = \pm 1$, ± 2.

Method 2: Substituting $u = x^2$, the polynomial becomes $P(u) = u^2 - 5u + 4$, which factors:

$u^2 - 5u + 4 = (u - 1)(u - 4) = \left(x^2 - 1\right)\left(x^2 - 4\right)$, so either $x^2 = 1$ or $x^2 = 4$. If $x^2 = 1$, then $x = \pm 1$; if $x^2 = 4$, then $x = \pm 2$. Therefore, the zeros are $x = \pm 1$ and ± 2.

27. $P(x) = x^4 + 6x^3 + 7x^2 - 6x - 8$. The possible rational zeros are $\pm 1, \pm 2, \pm 4, \pm 8$. $P(x)$ has 1 variation in sign and hence 1 positive real zero. $P(-x) = x^4 - 6x^3 + 7x^2 + 6x - 8$ has 3 variations in sign and hence 1 or 3 negative real zeros.

$$
\begin{array}{r|rrrrr}
1 & 1 & 6 & 7 & -6 & -8 \\
  &   & 1 & 7 & 14 & 8 \\
\hline
  & 1 & 7 & 14 & 8 & 0 \\
\end{array}
\quad \Rightarrow x = 1 \text{ is a zero}
$$

and there are no other positive zeros. Thus $P(x) = x^4 + 6x^3 + 7x^2 - 6x - 8 = (x - 1)\left(x^3 + 7x^2 + 14x + 8\right)$. Continuing by factoring the quotient, we have:

$$
\begin{array}{r|rrrr}
-1 & 1 & 7 & 14 & 8 \\
   &   & -1 & -6 & -8 \\
\hline
   & 1 & 6 & 8 & 0 \\
\end{array}
\quad \Rightarrow x = -1 \text{ is a zero.}
$$

So $P(x) = x^4 + 6x^3 + 7x^2 - 6x - 8 = (x - 1)(x + 1)\left(x^2 + 6x + 8\right) = (x - 1)(x + 1)(x + 2)(x + 4)$. Therefore, the zeros are $x = -4, -2$, and ± 1.

29. $P(x) = 4x^4 - 25x^2 + 36$ has possible rational zeros $\pm 1, \pm 2, \pm 3, \pm 4, \pm 6, \pm 9, \pm 12, \pm 18, \pm 36, \pm \frac{1}{2}, \pm \frac{1}{4}, \pm \frac{3}{2}, \pm \frac{3}{4}, \pm \frac{9}{2}, \pm \frac{9}{4}$. Since $P(x)$ has 2 variations in sign, there are 0 or 2 positive real zeros. Since $P(-x) = 4x^4 - 25x^2 + 36$ has 2 variations in sign, there are 0 or 2 negative real zeros.

$$
\begin{array}{r|rrrrr}
1 & 4 & 0 & -25 & 0 & 36 \\
  &   & 4 & 4 & -21 & -21 \\
\hline
  & 4 & 4 & -21 & -21 & 15 \\
\end{array}
\qquad
\begin{array}{r|rrrrr}
2 & 4 & 0 & -25 & 0 & 36 \\
  &   & 8 & 16 & -18 & -36 \\
\hline
  & 4 & 8 & -9 & -18 & 0 \\
\end{array}
\quad \Rightarrow x = 2 \text{ is a zero.}
$$

$$
\begin{array}{r|rrrr}
2 & 4 & 8 & -9 & -18 \\
  &   & 8 & 32 & 46 \\
\hline
  & 4 & 16 & 23 & 28 \\
\end{array}
\quad \Rightarrow \text{all positive, } x = 2 \text{ is an upper bound.}
\qquad
\begin{array}{r|rrrr}
\frac{1}{2} & 4 & 8 & -9 & -18 \\
  &   & 2 & 5 & -2 \\
\hline
  & 4 & 10 & -4 & -20 \\
\end{array}
$$

$$
\begin{array}{r|rrrr}
\frac{1}{4} & 4 & 8 & -9 & -18 \\
  &   & 1 & \frac{9}{4} & -\frac{27}{16} \\
\hline
  & 4 & 9 & -\frac{27}{4} & -\frac{315}{16} \\
\end{array}
\qquad
\begin{array}{r|rrrr}
\frac{3}{2} & 4 & 8 & -9 & -18 \\
  &   & 6 & 21 & 18 \\
\hline
  & 4 & 14 & 12 & 0 \\
\end{array}
\quad \Rightarrow x = \frac{3}{2} \text{ is a zero.}
$$

$P(x) = (x - 2)(2x - 3)\left(2x^2 + 7x + 6\right) = (x - 2)(2x - 3)(2x + 3)(x + 2)$. Therefore, the zeros are $x = \pm 2$ and $\pm \frac{3}{2}$.

Note: Since $P(x)$ has only even terms, factoring by substitution also works. Let $x^2 = u$; then $P(u) = 4u^2 - 25u + 36 = (u - 4)(4u - 9) = \left(x^2 - 4\right)\left(4x^2 - 9\right)$, which gives the same results.

31. $P(x) = 3x^4 - 10x^3 - 9x^2 + 40x - 12$. The possible rational zeros are $\pm 1, \pm 2, \pm 3, \pm 4, \pm 6, \pm 12$. $P(x)$ has 3 variations in sign and hence 1, 3, or 5 positive real zeros. $P(-x) = 3x^4 + 10x^3 - 9x^2 - 40x - 12$ has 1 variation in sign and hence 1 negative real zero.

$$
\begin{array}{r|rrrrr}
1 & 3 & -10 & -9 & 40 & -12 \\
  &   & 3 & -7 & -16 & 24 \\
\hline
  & 3 & -7 & -16 & 24 & 24
\end{array}
\Rightarrow x = 1 \text{ is not a zero.}
\qquad
\begin{array}{r|rrrrr}
2 & 3 & -10 & -9 & 40 & -12 \\
  &   & 6 & -8 & -34 & 12 \\
\hline
  & 3 & -4 & -17 & 6 & 0
\end{array}
\Rightarrow x = 2 \text{ is a zero.}
$$

Thus $P(x) = 3x^4 - 10x^3 - 9x^2 + 40x - 12 = (x-2)\left(3x^3 - 4x^2 - 17x + 6\right)$. Continuing by factoring the quotient, we have

$$
\begin{array}{r|rrrr}
3 & 3 & -4 & -17 & 6 \\
  &   & 9 & 15 & -6 \\
\hline
  & 3 & 5 & -2 & 0
\end{array}
\Rightarrow x = 3 \text{ is a zero.}
$$

Thus $P(x) = (x-3)(x-2)\left(3x^2 + 5x - 2\right) = (x-3)(x-2)(3x-1)(x+2)$. Therefore, the zeros are $-2, \frac{1}{3}, 2$, and 3.

33. Factoring by grouping can be applied to this exercise. $4x^3 + 4x^2 - x - 1 = 4x^2(x+1) - (x+1) = (x+1)\left(4x^2 - 1\right) = (x+1)(2x+1)(2x-1)$. Therefore, the zeros are $x = -1$ and $\pm\frac{1}{2}$.

35. $P(x) = 4x^3 - 7x + 3$. The possible rational zeros are $\pm 1, \pm 3, \pm\frac{1}{2}, \pm\frac{3}{2}, \pm\frac{1}{4}, \pm\frac{3}{4}$. Since $P(x)$ has 2 variations in sign, there are 0 or 2 positive zeros. Since $P(-x) = -4x^3 + 7x + 3$ has 1 variation in sign, there is 1 negative zero.

$$
\begin{array}{r|rrrr}
\frac{1}{2} & 4 & 0 & -7 & 3 \\
  &   & 2 & 1 & -3 \\
\hline
  & 4 & 2 & -6 & 0
\end{array}
\Rightarrow x = \frac{1}{2} \text{ is a zero.}
$$

$P(x) = \left(x - \frac{1}{2}\right)\left(4x^2 + 2x - 6\right) = (2x-1)\left(2x^2 + x - 3\right) = (2x-1)(x-1)(2x+3) = 0$. Thus, the zeros are $x = -\frac{3}{2}, \frac{1}{2}$, and 1.

37. $P(x) = 4x^3 + 8x^2 - 11x - 15$. The possible rational zeros are ± 1, ± 3, ± 5, $\pm \frac{1}{2}$, $\pm \frac{1}{4}$, $\pm \frac{3}{2}$, $\pm \frac{3}{4}$, $\pm \frac{5}{2}$, $\pm \frac{5}{4}$. $P(x)$ has 1 variation in sign and hence 1 positive real zero. $P(-x) = -4x^3 + 8x^2 + 11x - 15$ has 2 variations in sign, so P has 0 or 2 negative real zeros.

1	4	8	-11	-15	
		4	12	1	
	4	12	1	-14	$\Rightarrow x = 1$ is not a zero.

3	4	8	-11	-15	
		12	60	147	
	4	20	49	132	$\Rightarrow x = 3$ is not a zero.

5	4	8	-11	-15	
		20	140	645	
	4	28	129	630	$\Rightarrow x = 5$ is not a zero.

3	4	8	-11	-15	
		12	60	147	
	4	20	49	132	$\Rightarrow x = 3$ is not a zero.

$\frac{1}{2}$	4	8	-11	-15	
		2	5	-3	
	4	10	-6	-18	$\Rightarrow x = \frac{1}{2}$ is not a zero.

$\frac{1}{4}$	4	8	-11	-15	
		1	$\frac{9}{4}$	$-\frac{35}{16}$	
	4	9	$-\frac{35}{4}$	$-\frac{275}{16}$	$\Rightarrow x = \frac{1}{4}$ is not a zero.

$\frac{3}{2}$	4	8	-11	-15	
		6	21	15	
	4	14	10	0	$\Rightarrow x = \frac{3}{2}$ is a zero.

Thus $P(x) = 4x^3 + 8x^2 - 11x - 15 = \left(x - \frac{3}{2}\right)\left(4x^2 + 14x + 10\right)$. Continuing by factoring the quotient, whose possible rational zeros are -1, -5, $-\frac{1}{2}$, $-\frac{1}{4}$, $-\frac{5}{2}$, and $-\frac{5}{4}$, we have

-1	4	14	10	
		-4	-10	
	4	10	0	$\Rightarrow x = -1$ is a zero.

Thus $P(x) = (2x - 3)(x + 1)(2x + 5)$ has zeros $\frac{3}{2}$, -1, and $-\frac{5}{2}$.

39. $P(x) = 20x^3 - 8x^2 - 5x + 2$. The possible rational zeros are ± 1, ± 2, $\pm\frac{1}{2}$, $\pm\frac{1}{4}$, $\pm\frac{1}{5}$, $\pm\frac{1}{10}$, $\pm\frac{1}{20}$, $\pm\frac{2}{5}$. $P(x)$ has 2 variations in sign and hence 0 or 2 positive real zeros. $P(-x) = -20x^3 - 8x^2 + 5x + 2$ has 1 variations in sign and hence 1 negative real zero.

$$
\begin{array}{r|rrrr}
1 & 20 & -8 & -5 & 2 \\
  &   & 20 & 12 & 7 \\
\hline
  & 20 & 12 & 7 & 9
\end{array}
\Rightarrow x = 1 \text{ is not a zero.}
\qquad
\begin{array}{r|rrrr}
2 & 20 & -8 & -5 & 2 \\
  &   & 40 & 64 & 118 \\
\hline
  & 20 & 32 & 59 & 120
\end{array}
\Rightarrow x = 2 \text{ is not a zero.}
$$

$$
\begin{array}{r|rrrr}
\frac{1}{2} & 20 & -8 & -5 & 2 \\
 &   & 10 & 1 & -2 \\
\hline
 & 20 & 2 & -4 & 0
\end{array}
\Rightarrow x = \tfrac{1}{2} \text{ is a zero.}
$$

Thus, $P(x) = 20x^3 - 8x^2 - 5x + 2 = \left(x - \frac{1}{2}\right)\left(20x^2 + 2x - 4\right)$. Continuing:

$$
\begin{array}{r|rrr}
\frac{1}{4} & 20 & 2 & -4 \\
 &   & 5 & \frac{7}{4} \\
\hline
 & 20 & 7 & -\frac{9}{4}
\end{array}
\Rightarrow x = \tfrac{1}{4} \text{ is not a zero.}
\qquad
\begin{array}{r|rrr}
\frac{1}{5} & 20 & 2 & -4 \\
 &   & 4 & \frac{6}{5} \\
\hline
 & 20 & 6 & -\frac{14}{5}
\end{array}
\Rightarrow x = \tfrac{1}{5} \text{ is not a zero.}
$$

$$
\begin{array}{r|rrr}
\frac{1}{10} & 20 & 2 & -4 \\
 &   & 2 & \frac{2}{5} \\
\hline
 & 20 & 4 & -\frac{18}{5}
\end{array}
\Rightarrow x = \tfrac{1}{10} \text{ is not a zero.}
\qquad
\begin{array}{r|rrr}
\frac{1}{20} & 20 & 2 & -4 \\
 &   & 1 & \frac{3}{20} \\
\hline
 & 20 & 3 & -\frac{77}{20}
\end{array}
\Rightarrow x = \tfrac{1}{20} \text{ is not a zero.}
$$

$$
\begin{array}{r|rrr}
\frac{2}{5} & 20 & 2 & -4 \\
 &   & 8 & 4 \\
\hline
 & 20 & 10 & 0
\end{array}
\Rightarrow x = \tfrac{2}{5} \text{ is a zero.}
$$

Thus, $P(x) = (2x - 1)(5x - 2)(2x + 1)$ has zeros $\frac{1}{2}$, $\frac{2}{5}$, and $-\frac{1}{2}$.

41. $P(x) = 2x^4 - 7x^3 + 3x^2 + 8x - 4$. The possible rational zeros are ± 1, ± 2, ± 4, $\pm\frac{1}{2}$. $P(x)$ has 3 variations in sign and hence 1 or 3 positive real zeros. $P(-x) = 2x^4 + 7x^3 + 3x^2 - 8x - 4$ has 1 variation in sign and hence 1 negative real zero.

$$
\begin{array}{r|rrrrr}
1 & 2 & -7 & 3 & 8 & -4 \\
  &   & 2 & -5 & -2 & 6 \\
\hline
  & 2 & -5 & -2 & 6 & 2
\end{array}
\Rightarrow x = 1 \text{ is not a zero.}
\qquad
\begin{array}{r|rrrrr}
\frac{1}{2} & 2 & -7 & 3 & 8 & -4 \\
  &   & 1 & -3 & 0 & 4 \\
\hline
  & 2 & -6 & 0 & 8 & 0
\end{array}
\Rightarrow x = \tfrac{1}{2} \text{ is a zero.}
$$

Thus $P(x) = 2x^4 - 7x^3 + 3x^2 + 8x - 4 = \left(x - \frac{1}{2}\right)\left(2x^3 - 6x^2 + 8\right)$. Continuing by factoring the quotient, we have:

$$
\begin{array}{r|rrrr}
2 & 2 & -6 & 0 & 8 \\
  &   & 4 & -4 & -8 \\
\hline
  & 2 & -2 & -4 & 0
\end{array}
\Rightarrow x = 2 \text{ is a zero.}
$$

$P(x) = \left(x - \frac{1}{2}\right)(x - 2)\left(2x^2 - 2x - 4\right) = 2\left(x - \frac{1}{2}\right)(x - 2)\left(x^2 - x - 2\right) = 2\left(x - \frac{1}{2}\right)(x - 2)^2(x + 1)$. Thus, the zeros are $x = \frac{1}{2}$, 2, and -1.

43. $P(x) = x^5 + 3x^4 - 9x^3 - 31x^2 + 36$. The possible rational zeros are ±1, ±2, ±3, ±4, ±6, ±8, ±9, ±12, ±18. $P(x)$ has 2 variations in sign and hence 0 or 2 positive real zeros. $P(-x) = -x^5 + 3x^4 + 9x^3 - 31x^2 + 36$ has 3 variations in sign and hence 1 or 3 negative real zeros.

$$
\begin{array}{r|rrrrrr}
1 & 1 & 3 & -9 & -31 & 0 & 36 \\
  &   & 1 & 4 & -5 & -36 & -36 \\
\hline
  & 1 & 4 & -5 & -36 & -36 & 0 \\
\end{array}
\Rightarrow x = 1 \text{ is a zero.}
$$

So $P(x) = x^5 + 3x^4 - 9x^3 - 31x^2 + 36 = (x-1)\left(x^4 + 4x^3 - 5x^2 - 36x - 36\right)$. Continuing by factoring the quotient, we have:

$$
\begin{array}{r|rrrrr}
1 & 1 & 4 & -5 & -36 & -36 \\
  &   & 1 & 5 & 0 & -36 \\
\hline
  & 1 & 1 & 0 & -36 & -72 \\
\end{array}
\qquad
\begin{array}{r|rrrrr}
2 & 1 & 4 & -5 & -36 & -36 \\
  &   & 2 & 12 & 14 & -44 \\
\hline
  & 1 & 6 & 7 & -22 & -80 \\
\end{array}
$$

$$
\begin{array}{r|rrrrr}
3 & 1 & 4 & -5 & -36 & -36 \\
  &   & 3 & 21 & 48 & 36 \\
\hline
  & 1 & 7 & 16 & 12 & 0 \\
\end{array}
\Rightarrow x = 3 \text{ is a zero.}
$$

So $P(x) = (x-1)(x-3)\left(x^3 + 7x^2 + 16x + 12\right)$. Since we have 2 positive zeros, there are no more positive zeros, so we continue by factoring the quotient with possible negative zeros.

$$
\begin{array}{r|rrrr}
-1 & 1 & 7 & 16 & 12 \\
   &   & -1 & -6 & -10 \\
\hline
   & 1 & 6 & 10 & 2 \\
\end{array}
\qquad
\begin{array}{r|rrrr}
-2 & 1 & 7 & 16 & 12 \\
   &   & -2 & -10 & -12 \\
\hline
   & 1 & 5 & 6 & 0 \\
\end{array}
\Rightarrow x = -2 \text{ is a zero.}
$$

Then $P(x) = (x-1)(x-3)(x+2)\left(x^2 + 5x + 6\right) = (x-1)(x-3)(x+2)^2(x+3)$. Thus, the zeros are $x = 1, 3, -2,$ and -3.

45. $P(x) = 3x^5 - 14x^4 - 14x^3 + 36x^2 + 43x + 10$ has possible rational zeros ±1, ±2, ±5, ±10, $\pm\frac{1}{3}$, $\pm\frac{2}{3}$, $\pm\frac{5}{3}$, $\pm\frac{10}{3}$. Since $P(x)$ has 2 variations in sign, there are 0 or 2 positive real zeros. Since $P(-x) = -3x^5 - 14x^4 + 14x^3 + 36x^2 - 43x + 10$ has 3 variations in sign, there are 1 or 3 negative real zeros.

$$
\begin{array}{r|rrrrrr}
1 & 3 & -14 & -14 & 36 & 43 & 10 \\
  &   & 3 & -11 & -25 & 11 & 54 \\
\hline
  & 3 & -11 & -25 & 11 & 54 & 64 \\
\end{array}
\qquad
\begin{array}{r|rrrrrr}
2 & 3 & -14 & -14 & 36 & 43 & 10 \\
  &   & 6 & -16 & -60 & -48 & -10 \\
\hline
  & 3 & -8 & -30 & -24 & -5 & 0 \\
\end{array}
\Rightarrow x = 2 \text{ is a zero.}
$$

$P(x) = (x-2)\left(3x^4 - 8x^3 - 30x^2 - 24x - 5\right)$

$$
\begin{array}{r|rrrrr}
2 & 3 & -8 & -30 & -24 & -5 \\
  &   & 6 & -4 & -68 & -184 \\
\hline
  & 3 & -2 & -34 & -92 & -189 \\
\end{array}
\qquad
\begin{array}{r|rrrrr}
5 & 3 & -8 & -30 & -24 & -5 \\
  &   & 15 & 35 & 25 & 5 \\
\hline
  & 3 & 7 & 5 & 1 & 0 \\
\end{array}
\Rightarrow x = 5 \text{ is a zero.}
$$

$P(x) = (x-2)(x-5)\left(3x^3 + 7x^2 + 5x + 1\right)$. Since $3x^3 + 7x^2 + 5x + 1$ has no variation in sign, there are no more positive zeros.

$$
\begin{array}{r|rrrr}
-1 & 3 & 7 & 5 & 1 \\
   &   & -3 & -4 & -1 \\
\hline
   & 3 & 4 & 1 & 0 \\
\end{array}
\Rightarrow x = -1 \text{ is a zero.}
$$

$P(x) = (x-2)(x-5)(x+1)\left(3x^2 + 4x + 1\right) = (x-2)(x-5)(x+1)(x+1)(3x+1)$. Therefore, the zeros are $x = -1, -\frac{1}{3}, 2,$ and 5.

47. $P(x) = x^3 + 4x^2 + 3x - 2$. The possible rational zeros are $\pm 1, \pm 2$. $P(x)$ has 1 variation in sign and hence 1 positive real zero. $P(-x) = -x^3 + 4x^2 - 3x - 2$ has 2 variations in sign and hence 0 or 2 negative real zeros.

$$
\begin{array}{r|rrrr}
1 & 1 & 4 & 3 & -2 \\
  &   & 1 & 5 & 8 \\
\hline
  & 1 & 5 & 8 & 6
\end{array}
\quad \Rightarrow x = 1 \text{ is an upper bound.}
\qquad
\begin{array}{r|rrrr}
-1 & 1 & 4 & 3 & -2 \\
   &   & -1 & -3 & 0 \\
\hline
   & 1 & 3 & 0 & -2
\end{array}
$$

$$
\begin{array}{r|rrrr}
-2 & 1 & 4 & 3 & -2 \\
   &   & -2 & -4 & 2 \\
\hline
   & 1 & 2 & -1 & 0
\end{array}
\quad \Rightarrow x = -2 \text{ is a zero.}
$$

So $P(x) = (x+2)\left(x^2 + 2x - 1\right)$. Using the quadratic formula on the second factor, we have:

$x = \dfrac{-2 \pm \sqrt{2^2 - 4(1)(-1)}}{2(1)} = \dfrac{-2 \pm \sqrt{8}}{2} = \dfrac{-2 \pm 2\sqrt{2}}{2} = -1 \pm \sqrt{2}$. Therefore, the zeros are $x = -2, -1 + \sqrt{2}$, and $-1 - \sqrt{2}$.

49. $P(x) = x^4 - 6x^3 + 4x^2 + 15x + 4$. The possible rational zeros are $\pm 1, \pm 2, \pm 4$. $P(x)$ has 2 variations in sign and hence 0 or 2 positive real zeros. $P(-x) = x^4 + 6x^3 + 4x^2 - 15x + 4$ has 2 variations in sign and hence 0 or 2 negative real zeros.

$$
\begin{array}{r|rrrrr}
1 & 1 & -6 & 4 & 15 & 4 \\
  &   & 1 & -5 & -1 & 14 \\
\hline
  & 1 & -5 & -1 & 14 & 18
\end{array}
\qquad
\begin{array}{r|rrrrr}
2 & 1 & -6 & 4 & 15 & 4 \\
  &   & 2 & -8 & -8 & 14 \\
\hline
  & 1 & -4 & -4 & 7 & 18
\end{array}
$$

$$
\begin{array}{r|rrrrr}
4 & 1 & -6 & 4 & 15 & 4 \\
  &   & 4 & -8 & -16 & -4 \\
\hline
  & 1 & -2 & -4 & -1 & 0
\end{array}
\quad \Rightarrow x = 4 \text{ is a zero.}
$$

So $P(x) = (x-4)\left(x^3 - 2x^2 - 4x - 1\right)$. Continuing by factoring the quotient, we have:

$$
\begin{array}{r|rrrr}
4 & 1 & -2 & -4 & -1 \\
  &   & 4 & 8 & 16 \\
\hline
  & 1 & 2 & 4 & 15
\end{array}
\quad \Rightarrow x = 4 \text{ is an upper bound.}
\qquad
\begin{array}{r|rrrr}
-1 & 1 & -2 & -4 & -1 \\
   &   & -1 & 3 & 1 \\
\hline
   & 1 & -3 & -1 & 0
\end{array}
\quad \Rightarrow x = -1 \text{ is a zero.}
$$

So $P(x) = (x-4)(x+1)\left(x^2 - 3x - 1\right)$. Using the quadratic formula on the third factor, we have:

$x = \dfrac{-(-3) \pm \sqrt{(-3)^2 - 4(1)(-1)}}{2(1)} = \dfrac{3 \pm \sqrt{13}}{2}$. Therefore, the zeros are $x = 4, -1$, and $\dfrac{3 \pm \sqrt{13}}{2}$.

51. $P(x) = x^4 - 7x^3 + 14x^2 - 3x - 9$. The possible rational zeros are $\pm 1, \pm 3, \pm 9$. $P(x)$ has 3 variations in sign and hence 1 or 3 positive real zeros. $P(-x) = x^4 + 7x^3 + 14x^2 + 3x - 4$ has 1 variation in sign and hence 1 negative real zero.

$$
\begin{array}{r|rrrrr}
1 & 1 & -7 & 14 & -3 & -9 \\
  &   & 1 & -6 & 8 & 5 \\
\hline
  & 1 & -6 & 8 & 5 & 4
\end{array}
\qquad
\begin{array}{r|rrrrr}
3 & 1 & -7 & 14 & -3 & -9 \\
  &   & 3 & -12 & 6 & 9 \\
\hline
  & 1 & -4 & 2 & 3 & 0
\end{array}
\quad \Rightarrow x = 3 \text{ is a zero.}
$$

So $P(x) = (x-3)\left(x^3 - 4x^2 + 2x + 3\right)$. Since the constant term of the second term is 3, ± 9 are no longer possible zeros.

Continuing by factoring the quotient, we have:
$$
\begin{array}{r|rrrr}
3 & 1 & -4 & 2 & 3 \\
  &   & 3 & -3 & -3 \\
\hline
  & 1 & -1 & -1 & 0
\end{array}
\quad \Rightarrow x = 3 \text{ is a zero again.}
$$

So $P(x) = (x-3)^2\left(x^2 - x - 1\right)$. Using the quadratic formula on the second factor, we have:

$x = \dfrac{-(-1) \pm \sqrt{(-1)^2 - 4(1)(-1)}}{2(1)} = \dfrac{1 \pm \sqrt{5}}{2}$. Therefore, the zeros are $x = 3$ and $\dfrac{1 \pm \sqrt{5}}{2}$.

53. $P(x) = 4x^3 - 6x^2 + 1$. The possible rational zeros are ± 1, $\pm \frac{1}{2}$, $\pm \frac{1}{4}$. $P(x)$ has 2 variations in sign and hence 0 or 2 positive real zeros. $P(-x) = -4x^3 - 6x^2 + 1$ has 1 variation in sign and hence 1 negative real zero.

$$
\begin{array}{r|rrrr}
1 & 4 & -6 & 0 & 1 \\
  &   & 4 & -2 & -2 \\
\hline
  & 4 & -2 & -2 & -1
\end{array}
\qquad
\begin{array}{r|rrrr}
\frac{1}{2} & 4 & -6 & 0 & 1 \\
  &   & 2 & -2 & -1 \\
\hline
  & 4 & -4 & -2 & 0
\end{array}
\;\Rightarrow x = \tfrac{1}{2} \text{ is a zero.}
$$

So $P(x) = \left(x - \frac{1}{2}\right)\left(4x^2 - 4x - 2\right)$. Using the quadratic formula on the second factor, we have:

$x = \frac{-(-4)\pm\sqrt{(-4)^2-4(4)(-2)}}{2(4)} = \frac{4\pm\sqrt{48}}{8} = \frac{4\pm4\sqrt{3}}{8} = \frac{1\pm\sqrt{3}}{2}$. Therefore, the zeros are $x = \frac{1}{2}$ and $\frac{1\pm\sqrt{3}}{2}$.

55. $P(x) = 2x^4 + 15x^3 + 17x^2 + 3x - 1$. The possible rational zeros are ± 1, $\pm \frac{1}{2}$. $P(x)$ has 1 variation in sign and hence 1 positive real zero. $P(-x) = 2x^4 - 15x^3 + 17x^2 - 3x - 1$ has 3 variations in sign and hence 1 or 3 negative real zeros.

$$
\begin{array}{r|rrrrr}
\frac{1}{2} & 2 & 15 & 17 & 3 & -1 \\
  &   & 1 & 8 & \frac{25}{2} & \frac{31}{4} \\
\hline
  & 2 & 16 & 25 & \frac{31}{2} & \frac{27}{4}
\end{array}
\;\Rightarrow x = \tfrac{1}{2} \text{ is an upper bound.}
$$

$$
\begin{array}{r|rrrrr}
-\frac{1}{2} & 2 & 15 & 17 & 3 & -1 \\
  &   & -1 & -7 & -5 & 1 \\
\hline
  & 2 & 14 & 10 & -2 & 0
\end{array}
\;\Rightarrow x = -\tfrac{1}{2} \text{ is a zero.}
$$

So $P(x) = \left(x + \frac{1}{2}\right)\left(2x^3 + 14x^2 + 10x - 2\right) = 2\left(x + \frac{1}{2}\right)\left(x^3 + 7x^2 + 5x - 1\right)$.

$$
\begin{array}{r|rrrr}
-1 & 1 & 7 & 5 & -1 \\
  &   & -1 & -6 & 1 \\
\hline
  & 1 & 6 & -1 & 0
\end{array}
\;\Rightarrow x = -1 \text{ is a zero.}
$$

So $P(x) = \left(x + \frac{1}{2}\right)\left(2x^3 + 14x^2 + 10x - 2\right) = 2\left(x + \frac{1}{2}\right)(x + 1)\left(x^2 + 6x - 1\right)$ Using the quadratic formula on the

third factor, we have $x = \frac{-(6)\pm\sqrt{(6)^2-4(1)(-1)}}{2(1)} = \frac{-6\pm\sqrt{40}}{2} = \frac{-6\pm2\sqrt{10}}{2} = -3 \pm \sqrt{10}$. Therefore, the zeros are $x = -1$, $-\frac{1}{2}$, and $-3 \pm \sqrt{10}$.

57. (a) $P(x) = x^3 - 3x^2 - 4x + 12$ has possible rational zeros ± 1, ± 2, ± 3, ± 4, ± 6, ± 12.

(b)

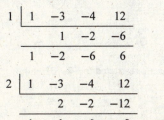

$$
\begin{array}{r|rrrr}
1 & 1 & -3 & -4 & 12 \\
  &   & 1 & -2 & -6 \\
\hline
  & 1 & -2 & -6 & 6
\end{array}
$$

$$
\begin{array}{r|rrrr}
2 & 1 & -3 & -4 & 12 \\
  &   & 2 & -2 & -12 \\
\hline
  & 1 & -1 & -6 & 0
\end{array}
\;\Rightarrow x = 2 \text{ is a zero.}
$$

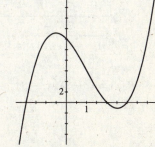

So $P(x) = (x - 2)\left(x^2 - x - 6\right) = (x - 2)(x + 2)(x - 3)$. The real zeros of P are -2, 2, and 3.

59. (a) $P(x) = 2x^3 - 7x^2 + 4x + 4$ has possible rational zeros ± 1, ± 2, ± 4, **(b)**
$\pm \frac{1}{2}$.

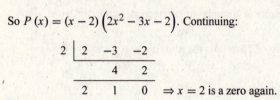

$$\begin{array}{r|rrrr} 1 & 2 & -7 & 4 & 4 \\ & & 2 & -5 & -1 \\ \hline & 2 & -5 & -1 & -3 \end{array} \qquad \begin{array}{r|rrrr} 2 & 2 & -7 & 4 & 4 \\ & & 4 & -6 & -4 \\ \hline & 2 & -3 & -2 & 0 \end{array} \Rightarrow x = 2 \text{ is a zero.}$$

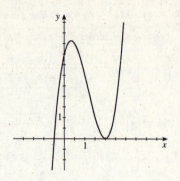

So $P(x) = (x - 2)\left(2x^2 - 3x - 2\right)$. Continuing:

$$\begin{array}{r|rrr} 2 & 2 & -3 & -2 \\ & & 4 & 2 \\ \hline & 2 & 1 & 0 \end{array} \Rightarrow x = 2 \text{ is a zero again.}$$

Thus $P(x) = (x - 2)^2 (2x + 1)$. The real zeros of P are 2 and $-\frac{1}{2}$.

61. (a) $P(x) = x^4 - 5x^3 + 6x^2 + 4x - 8$ has possible rational zeros ± 1, **(b)**
± 2, ± 4, ± 8.

$$\begin{array}{r|rrrrr} 1 & 1 & -5 & 6 & 4 & -8 \\ & & 1 & -4 & 2 & 6 \\ \hline & 1 & -4 & 2 & 6 & -2 \end{array}$$

$$\begin{array}{r|rrrrr} 2 & 1 & -5 & 6 & 4 & -8 \\ & & 2 & -6 & 0 & 8 \\ \hline & 1 & -3 & 0 & 4 & 0 \end{array} \Rightarrow x = 2 \text{ is a zero.}$$

So $P(x) = (x - 2)\left(x^3 - 3x^2 + 4\right)$ and the possible rational zeros are restricted to -1, ± 2, ± 4.

$$\begin{array}{r|rrrr} 2 & 1 & -3 & 0 & 4 \\ & & 2 & -2 & -4 \\ \hline & 1 & -1 & -2 & 0 \end{array} \Rightarrow x = 2 \text{ is a zero again.}$$

$P(x) = (x - 2)^2 \left(x^2 - x - 2\right) = (x - 2)^2 (x - 2)(x + 1) = (x - 2)^3 (x + 1)$. So the real zeros of P are -1 and 2.

63. (a) $P(x) = x^5 - x^4 - 5x^3 + x^2 + 8x + 4$ has possible rational zeros $\pm 1, \pm 2, \pm 4$.

(b)

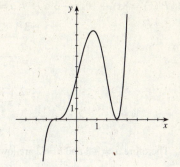

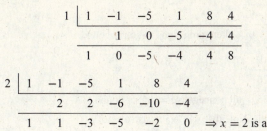

$$
\begin{array}{r|rrrrrr}
1 & 1 & -1 & -5 & 1 & 8 & 4 \\
  &   & 1 & 0 & -5 & -4 & 4 \\
\hline
  & 1 & 0 & -5 & -4 & 4 & 8
\end{array}
$$

$$
\begin{array}{r|rrrrrr}
2 & 1 & -1 & -5 & 1 & 8 & 4 \\
  &   & 2 & 2 & -6 & -10 & -4 \\
\hline
  & 1 & 1 & -3 & -5 & -2 & 0
\end{array} \Rightarrow x = 2 \text{ is a zero.}
$$

So $P(x) = (x-2)\left(x^4 + x^3 - 3x^2 - 5x - 2\right)$, and the possible

rational zeros are restricted to $-1, \pm 2$.

$$
\begin{array}{r|rrrrr}
2 & 1 & 1 & -3 & -5 & -2 \\
  &   & 2 & 6 & 6 & 2 \\
\hline
  & 1 & 3 & 3 & 1 & 0
\end{array} \Rightarrow x = 2 \text{ is a zero again.}
$$

So $P(x) = (x-2)^2\left(x^3 + 3x^2 + 3x + 1\right)$, and the possible rational zeros are restricted to -1.

$$
\begin{array}{r|rrrr}
-1 & 1 & 3 & 3 & 1 \\
   &   & -1 & -2 & -1 \\
\hline
   & 1 & 2 & 1 & 0
\end{array} \Rightarrow x = -1 \text{ is a zero.}
$$

So $P(x) = (x-2)^2(x+1)\left(x^2 + 2x + 1\right) = (x-2)^2(x+1)^3$., and the real zeros of P are -1 and 2.

65. $P(x) = x^3 - x^2 - x - 3$. Since $P(x)$ has 1 variation in sign, P has 1 positive real zero. Since $P(-x) = -x^3 - x^2 + x - 3$ has 2 variations in sign, P has 2 or 0 negative real zeros. Thus, P has 1 or 3 real zeros.

67. $P(x) = 2x^6 + 5x^4 - x^3 - 5x - 1$. Since $P(x)$ has 1 variation in sign, P has 1 positive real zero. Since $P(-x) = 2x^6 + 5x^4 + x^3 + 5x - 1$ has 1 variation in sign, P has 1 negative real zero. Therefore, P has 2 real zeros.

69. $P(x) = x^5 + 4x^3 - x^2 + 6x$. Since $P(x)$ has 2 variations in sign, P has 2 or 0 positive real zeros. Since $P(-x) = -x^5 - 4x^3 - x^2 - 6x$ has no variation in sign, P has no negative real zero. Therefore, P has a total of 1 or 3 real zeros (since $x = 0$ is a zero, but is neither positive nor negative).

71. $P(x) = 2x^3 + 5x^2 + x - 2; a = -3, b = 1$

$$
\begin{array}{r|rrrr}
-3 & 2 & 5 & 1 & -2 \\
   &   & -6 & 3 & -12 \\
\hline
   & 2 & -1 & 4 & -14
\end{array} \text{ alternating signs} \Rightarrow \text{lower bound.}
$$

$$
\begin{array}{r|rrrr}
1 & 2 & 5 & 1 & -2 \\
  &   & 2 & 7 & 8 \\
\hline
  & 2 & 7 & 8 & 6
\end{array} \text{ all nonnegative} \Rightarrow \text{upper bound.}
$$

Therefore $a = -3$ and $b = 1$ are lower and upper bounds.

73. $P(x) = 8x^3 + 10x^2 - 39x + 9$; $a = -3, b = 2$

$$
\begin{array}{r|rrrr}
-3 & 8 & 10 & -39 & 9 \\
   &   & -24 & 42 & -9 \\
\hline
   & 8 & -14 & 3 & 0
\end{array}
$$ alternating signs $\Rightarrow$ lower bound.

$$
\begin{array}{r|rrrr}
2 & 8 & 10 & -39 & 9 \\
  &   & 16 & 52 & 26 \\
\hline
  & 8 & 26 & 13 & 35
\end{array}
$$ all nonnegative $\Rightarrow$ upper bound.

Therefore $a = -3$ and $b = 2$ are lower and upper bounds. Note that $x = -3$ is also a zero.

75. $P(x) = x^3 - 3x^2 + 4$ and use the Upper and Lower Bounds Theorem:

$$
\begin{array}{r|rrrr}
-1 & 1 & -3 & 0 & 4 \\
   &   & -1 & 4 & -4 \\
\hline
   & 1 & -4 & 4 & 0
\end{array}
$$ alternating signs $\Rightarrow$ lower bound.

$$
\begin{array}{r|rrrr}
3 & 1 & -3 & 0 & 4 \\
  &   & 3 & 0 & 0 \\
\hline
  & 1 & 0 & 0 & 4
\end{array}
$$ all nonnegative $\Rightarrow$ upper bound.

Therefore -1 is a lower bound (and a zero) and 3 is an upper bound. (There are many possible solutions.)

77. $P(x) = x^4 - 2x^3 + x^2 - 9x + 2$.

$$
\begin{array}{r|rrrrr}
1 & 1 & -2 & 1 & -9 & 2 \\
  &   & 1 & -1 & 0 & -9 \\
\hline
  & 1 & -1 & 0 & -9 & -7
\end{array}
\qquad
\begin{array}{r|rrrrr}
2 & 1 & -2 & 1 & -9 & 2 \\
  &   & 2 & 0 & 2 & -14 \\
\hline
  & 1 & 0 & 1 & -7 & -12
\end{array}
$$

$$
\begin{array}{r|rrrrr}
3 & 1 & -2 & 1 & -9 & 2 \\
  &   & 3 & 3 & 12 & 9 \\
\hline
  & 1 & 1 & 4 & 3 & 11
\end{array}
$$ all positive $\Rightarrow$ upper bound.

$$
\begin{array}{r|rrrrr}
-1 & 1 & -2 & 1 & -9 & 2 \\
   &   & -1 & 3 & -4 & 13 \\
\hline
   & 1 & -3 & 4 & -13 & 15
\end{array}
$$ alternating signs $\Rightarrow$ lower bound.

Therefore -1 is a lower bound and 3 is an upper bound. (There are many possible solutions.)

79. $P(x) = 2x^4 + 3x^3 - 4x^2 - 3x + 2$.

$$
\begin{array}{r|rrrrr}
1 & 2 & 3 & -4 & -3 & 2 \\
  &   & 2 & 5 & 1 & -2 \\
\hline
  & 2 & 5 & 1 & -2 & 0
\end{array}
$$ $\Rightarrow x = 1$ is a zero.

$P(x) = (x - 1)\left(2x^3 + 5x^2 + x - 2\right)$

$$
\begin{array}{r|rrrr}
-1 & 2 & 5 & 1 & -2 \\
   &   & -2 & -3 & 2 \\
\hline
   & 2 & 3 & -2 & 0
\end{array}
$$ $\Rightarrow x = -1$ is a zero.

$P(x) = (x - 1)(x + 1)\left(2x^2 + 3x - 2\right) = (x - 1)(x + 1)(2x - 1)(x + 2)$. Therefore, the zeros are $x = -2, \frac{1}{2}, \pm 1$.

81. *Method 1:* $P(x) = 4x^4 - 21x^2 + 5$ has 2 variations in sign, so by Descartes' rule of signs there are either 2 or 0 positive zeros. If we replace x with $(-x)$, the function does not change, so there are either 2 or 0 negative zeros. Possible rational zeros are $\pm 1, \pm\frac{1}{2}, \pm\frac{1}{4}, \pm 5, \pm\frac{5}{2}, \pm\frac{5}{4}$. By inspection, ± 1 and ± 5 are not zeros, so we must look for non-integer solutions:

$$
\begin{array}{r|rrrrr}
\frac{1}{2} & 4 & 0 & -21 & 0 & 5 \\
& & 2 & 1 & -10 & -5 \\
\hline
& 4 & 2 & -20 & -10 & 0
\end{array}
\Rightarrow x = \tfrac{1}{2} \text{ is a zero.}
$$

$P(x) = \left(x - \frac{1}{2}\right)\left(4x^3 + 2x^2 - 20x - 10\right)$, continuing with the quotient, we have:

$$
\begin{array}{r|rrrr}
-\frac{1}{2} & 4 & 2 & -20 & -10 \\
& & -2 & 0 & 10 \\
\hline
& 4 & 0 & -20 & 0
\end{array}
\Rightarrow x = -\tfrac{1}{2} \text{ is a zero.}
$$

$P(x) = \left(x - \frac{1}{2}\right)\left(x + \frac{1}{2}\right)\left(4x^2 - 20\right) = 0$. If $4x^2 - 20 = 0$, then $x = \pm\sqrt{5}$. Thus the zeros are $x = \pm\frac{1}{2}, \pm\sqrt{5}$.

Method 2: Substituting $u = x^2$, the equation becomes $4u^2 - 21u + 5 = 0$, which factors: $4u^2 - 21u + 5 = (4u - 1)(u - 5) = \left(4x^2 - 1\right)\left(x^2 - 5\right)$. Then either we have $x^2 = 5$, so that $x = \pm\sqrt{5}$, or we have $x^2 = \frac{1}{4}$, so that $x = \pm\sqrt{\frac{1}{4}} = \pm\frac{1}{2}$. Thus the zeros are $x = \pm\frac{1}{2}, \pm\sqrt{5}$.

83. $P(x) = x^5 - 7x^4 + 9x^3 + 23x^2 - 50x + 24$. The possible rational zeros are $\pm 1, \pm 2, \pm 3, \pm 4, \pm 6, \pm 8, \pm 12, \pm 24$. $P(x)$ has 4 variations in sign and hence 0, 2, or 4 positive real zeros. $P(-x) = -x^5 - 7x^4 - 9x^3 + 23x^2 + 50x + 24$ has 1 variation in sign, and hence 1 negative real zero.

$$
\begin{array}{r|rrrrrr}
1 & 1 & -7 & 9 & 23 & -50 & 24 \\
& & 1 & -6 & 3 & 26 & -24 \\
\hline
& 1 & -6 & 3 & 26 & -24 & 0
\end{array}
\Rightarrow x = 1 \text{ is a zero.}
$$

$P(x) = (x - 1)\left(x^4 - 6x^3 + 3x^2 + 26x - 24\right)$; continuing with the quotient, we try 1 again.

$$
\begin{array}{r|rrrrr}
1 & 1 & -6 & 3 & 26 & -24 \\
& & 1 & -5 & -2 & 24 \\
\hline
& 1 & -5 & -2 & 24 & 0
\end{array}
\Rightarrow x = 1 \text{ is a zero again.}
$$

$P(x) = (x - 1)^2\left(x^3 - 5x^2 - 2x + 24\right)$; continuing with the quotient, we start by trying 1 again.

$$
\begin{array}{r|rrrr}
1 & 1 & -5 & -2 & 24 \\
& & 1 & -4 & -6 \\
\hline
& 1 & -4 & -6 & 18
\end{array}
\qquad
\begin{array}{r|rrrr}
2 & 1 & -5 & -2 & 24 \\
& & 2 & -6 & -16 \\
\hline
& 1 & -3 & -8 & 8
\end{array}
\qquad
\begin{array}{r|rrrr}
3 & 1 & -5 & -2 & 24 \\
& & 3 & -6 & -24 \\
\hline
& 1 & -2 & -8 & 0
\end{array}
\Rightarrow x = 3 \text{ is a zero.}
$$

$P(x) = (x - 1)^2(x - 3)\left(x^2 - 2x - 8\right) = (x - 1)^2(x - 3)(x - 4)(x + 2)$. Therefore, the zeros are $x = -2, 1, 3, 4$.

85. $P(x) = x^3 - x - 2$. The only possible rational zeros of $P(x)$ are ± 1 and ± 2.

$$
\begin{array}{r|rrrr}
1 & 1 & 0 & -1 & -2 \\
& & 1 & 1 & 0 \\
\hline
& 1 & 1 & 0 & -2
\end{array}
\qquad
\begin{array}{r|rrrr}
2 & 1 & 0 & -1 & -2 \\
& & 2 & 4 & 6 \\
\hline
& 1 & 2 & 3 & 4
\end{array}
\qquad
\begin{array}{r|rrrr}
-1 & 1 & 0 & -1 & -2 \\
& & -1 & 1 & 0 \\
\hline
& 1 & -1 & 0 & -2
\end{array}
$$

Since the row that contains -1 alternates between nonnegative and nonpositive, -1 is a lower bound and there is no need to try -2. Therefore, $P(x)$ does not have any rational zeros.

87. $P(x) = 3x^3 - x^2 - 6x + 12$ has possible rational zeros $\pm 1, \pm 2, \pm 3, \pm 4, \pm 6, \pm 12, \pm\frac{1}{3}, \pm\frac{2}{3}, \pm\frac{4}{3}$.

	3	-1	-6	12
1	3	2	-4	8
2	3	5	4	20
-1	3	-4	-2	14
-2	3	-7	8	-4

2 | all positive $\Rightarrow x = 2$ is an upper bound

-2 | alternating signs $\Rightarrow x = -2$ is a lower bound

	3	-1	-6	12
$\frac{1}{3}$	3	0	-6	10
$\frac{2}{3}$	3	1	$-\frac{16}{3}$	$\frac{76}{9}$
$\frac{4}{3}$	3	3	-2	$\frac{28}{3}$
$-\frac{1}{3}$	3	-2	$-\frac{16}{3}$	$\frac{124}{9}$
$-\frac{2}{3}$	3	-3	-4	$\frac{44}{3}$
$-\frac{4}{3}$	3	-5	$\frac{2}{3}$	$\frac{100}{9}$

Therefore, there is no rational zero.

89. $P(x) = x^3 - 3x^2 - 4x + 12$, $[-4, 4]$ by $[-15, 15]$. The possible rational zeros are $\pm 1, \pm 2, \pm 3, \pm 4, \pm 6, \pm 12$. By observing the graph of P, the rational zeros are $x = -2, 2, 3$.

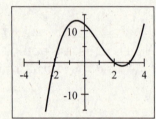

91. $P(x) = 2x^4 - 5x^3 - 14x^2 + 5x + 12$, $[-2, 5]$ by $[-40, 40]$. The possible rational zeros are $\pm 1, \pm 2, \pm 3, \pm 4, \pm 6, \pm 12, \pm\frac{1}{2}, \pm\frac{3}{2}$. By observing the graph of P, the zeros are $x = -\frac{3}{2}, -1, 1, 4$.

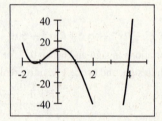

93. $x^4 - x - 4 = 0$. Possible rational solutions are $\pm 1, \pm 2, \pm 4$.

1	1	0	0	-1	-4
		1	1	1	0
	1	1	1	0	-4

-1	1	0	0	-1	-4
		-1	1	-1	2
	1	-1	1	-2	-2

2	1	0	0	-1	-4
		2	4	8	14
	1	2	4	7	10

$\Rightarrow x = 2$ is an upper bound.

-2	1	0	0	-1	-4
		-2	4	-8	18
	1	-2	4	-9	14

$\Rightarrow x = -2$ is a lower bound.

Therefore, we graph the function $P(x) = x^4 - x - 4$ in the viewing rectangle $[-2, 2]$ by $[-5, 20]$ and see there are two solutions. In the viewing rectangle $[-1.3, -1.25]$ by $[-0.1, 0.1]$, we find the solution $x \approx -1.28$. In the viewing rectangle $[1.5, 1.6]$ by $[-0.1, 0.1]$, we find the solution $x \approx 1.53$. Thus the solutions are $x \approx -1.28, 1.53$.

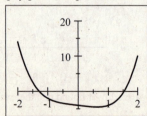
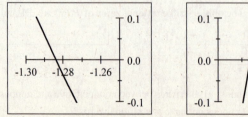
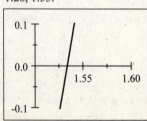

95. $4.00x^4 + 4.00x^3 - 10.96x^2 - 5.88x + 9.09 = 0.$

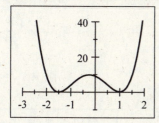

$$\begin{array}{r|rrrrr} 1 & 4 & 4 & -10.96 & -5.88 & 9.09 \\ & & 4 & 8 & -2.96 & -8.84 \\ \hline & 4 & 8 & -2.96 & -8.84 & 0.25 \end{array}$$

$$\begin{array}{r|rrrrr} 2 & 4 & 4 & -10.96 & -5.88 & 9.09 \\ & & 8 & 24 & 26.08 & 40.40 \\ \hline & 4 & 12 & 13.04 & 20.2 & 49.49 \end{array} \Rightarrow x = 2 \text{ is an upper bound.}$$

$$\begin{array}{r|rrrrr} -2 & 4 & 4 & -10.96 & -5.88 & 9.09 \\ & & -8 & 8 & 5.92 & -0.08 \\ \hline & 4 & -4 & -2.96 & 0.04 & 9.01 \end{array}$$

$$\begin{array}{r|rrrrr} -3 & 4 & 4 & -10.96 & -5.88 & 9.09 \\ & & -12 & 24 & -39.12 & 135 \\ \hline & 4 & -8 & 13.04 & -45 & 144.09 \end{array} \Rightarrow x = -3 \text{ is a lower bound.}$$

Therefore, we graph the function $P(x) = 4.00x^4 + 4.00x^3 - 10.96x^2 - 5.88x + 9.09$ in the viewing rectangle $[-3, 2]$ by $[-10, 40]$. There appear to be two solutions. In the viewing rectangle $[-1.6, -1.4]$ by $[-0.1, 0.1]$, we find the solution $x \approx -1.50$. In the viewing rectangle $[0.8, 1.2]$ by $[0, 1]$, we see that the graph comes close but does not go through the x-axis. Thus there is no solution here. Therefore, the only solution is $x \approx -1.50$.

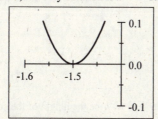
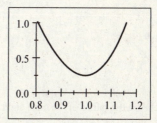

97. (a) Since $z > b$, we have $z - b > 0$. Since all the coefficients of $Q(x)$ are nonnegative, and since $z > 0$, we have $Q(z) > 0$ (being a sum of positive terms). Thus, $P(z) = (z - b) \cdot Q(z) + r > 0$, since the sum of a positive number and a nonnegative number.

(b) In part (a), we showed that if b satisfies the conditions of the first part of the Upper and Lower Bounds Theorem and $z > b$, then $P(z) > 0$. This means that no real zero of P can be larger than b, so b is an upper bound for the real zeros.

(c) Suppose $-b$ is a negative lower bound for the real zeros of $P(x)$. Then clearly b is an upper bound for $P_1(x) = P(-x)$. Thus, as in Part (a), we can write $P_1(x) = (x - b) \cdot Q(x) + r$, where $r > 0$ and the coefficients of Q are all nonnegative, and $P(x) = P_1(-x) = (-x - b) \cdot Q(-x) + r = (x + b) \cdot [-Q(-x)] + r$. Since the coefficients of $Q(x)$ are all nonnegative, the coefficients of $-Q(-x)$ will be alternately nonpositive and nonnegative, which proves the second part of the Upper and Lower Bounds Theorem.

99. Let r be the radius of the silo. The volume of the hemispherical roof is $\frac{1}{2}\left(\frac{4}{3}\pi r^3\right) = \frac{2}{3}\pi r^3$. The volume of the cylindrical section is $\pi\left(r^2\right)(30) = 30\pi r^2$. Because the total volume of the silo is $15{,}000$ ft^3, we get the following equation:

$\frac{2}{3}\pi r^3 + 30\pi r^2 = 15000 \Leftrightarrow \frac{2}{3}\pi r^3 + 30\pi r^2 - 15000 = 0 \Leftrightarrow \pi r^3 + 45\pi r^2 - 22500 = 0.$ Using a graphing device, we first graph the polynomial in the viewing rectangle $[0, 15]$ by $[-10000, 10000]$. The solution, $r \approx 11.28$ ft., is shown in the viewing rectangle $[11.2, 11.4]$ by $[-1, 1]$.

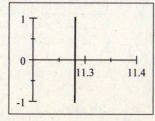
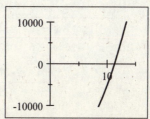

101. $h(t) = 11.60t - 12.41t^2 + 6.20t^3$
$$- 1.58t^4 + 0.20t^5 - 0.01t^6$$
is shown in the viewing rectangle
[0, 10] by [0, 6].

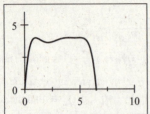

(a) It started to snow again.

(b) No, $h(t) \leq 4$.

(c) The function $h(t)$ is shown in the viewing rectangle [6, 6.5] by [0, 0.5]. The x-intercept of the function is a little less than 6.5, which means that the snow melted just before midnight on Saturday night.

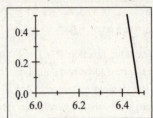

103. Let r be the radius of the cone and cylinder and let h be the height of the cone. Since the height and diameter are equal, we get $h = 2r$. So the volume of the cylinder is $V_1 = \pi r^2 \cdot$ (cylinder height) $= 20\pi r^2$, and the volume of the cone is $V_2 = \frac{1}{3}\pi r^2 h = \frac{1}{3}\pi r^2 (2r) = \frac{2}{3}\pi r^3$. Since the total volume is $\frac{500\pi}{3}$, it follows that $\frac{2}{3}\pi r^3 + 20\pi r^2 = \frac{500\pi}{3} \Leftrightarrow r^3 + 30r^2 - 250 = 0$. By Descartes' Rule of Signs, there is 1 positive zero. Since r is between 2.76 and 2.765 (see the table), the radius should be 2.76 m (correct to two decimals).

r	$r^3 + 30r^2 - 250$
1	−219
2	−122
3	47
2.7	−11.62
2.76	−2.33
2.77	1.44
2.765	1.44
2.8	7.15

105. Let b be the width of the base, and let l be the length of the box. Then the length plus girth is $l + 4b = 108$, and the volume is $V = lb^2 = 2200$. Solving the first equation for l and substituting this value into the second equation yields $l = 108 - 4b$ $\Rightarrow V = (108 - 4b)b^2 = 2200 \Leftrightarrow 4b^3 - 108b^2 + 2200 = 0 \Leftrightarrow 4(b^3 - 27b^2 + 550) = 0$. Now $P(b) = b^3 - 27b^2 + 550$ has two variations in sign, so there are 0 or 2 positive real zeros. We also observe that since $l > 0$, $b < 27$, so $b = 27$ is an upper bound. Thus the possible positive rational real zeros are 1, 2, 3, 10, 11, 22, 25.

```
1 | 1   -27    0    550
  |       1  -26   -26
  ------------------------
    1   -26  -26    524
```

```
2 | 1   -27    0    550
  |       2  -50  -100
  ------------------------
    1   -25  -50    450
```

```
5 | 1   -27    0    550
  |       5 -110  -550
  ------------------------
    1   -22 -110      0   ⇒ b = 5 is a zero.
```

$P(b) = (b - 5)\left(b^2 - 22b - 110\right)$. The other zeros are $b = \frac{22 \pm \sqrt{484 - 4(1)(-110)}}{2} = \frac{22 \pm \sqrt{924}}{2} = \frac{22 \pm 30.397}{2}$. The positive answer from this factor is $b \approx 26.20$. Thus we have two possible solutions, $b = 5$ or $b \approx 26.20$. If $b = 5$, then $l = 108 - 4(5) = 88$; if $b \approx 26.20$, then $l = 108 - 4(26.20) = 3.20$. Thus the length of the box is either 88 in. or 3.20 in.

107. (a) Substituting $X - \frac{a}{3}$ for x we have

$$
\begin{aligned}
x^3 + ax^2 + bx + c &= \left(X - \frac{a}{3}\right)^3 + a\left(X - \frac{a}{3}\right)^2 + b\left(X - \frac{a}{3}\right) + c \\
&= X^3 - aX^2 + \frac{a^2}{3}X + \frac{a^3}{27} + a\left(X^2 - \frac{2a}{3}X + \frac{a^2}{9}\right) + bX - \frac{ab}{3} + c \\
&= X^3 - aX^2 + \frac{a^2}{3}X + \frac{a^3}{27} + aX^2 - \frac{2a^2}{3}X + \frac{a^3}{9} + bX - \frac{ab}{3} + c \\
&= X^3 + (-a + a)X^2 + \left(-\frac{a^2}{3} - \frac{2a^2}{3} + b\right)X + \left(\frac{a^3}{27} + \frac{a^3}{9} - \frac{ab}{3} + c\right) \\
&= X^3 + \left(b - a^2\right)X + \left(\frac{4a^3}{27} - \frac{ab}{3} + c\right)
\end{aligned}
$$

(b) $x^3 + 6x^2 + 9x + 4 = 0$. Setting $a = 6$, $b = 9$, and $c = 4$, we have: $X^3 + \left(9 - 6^2\right)X + (32 - 18 + 4) = X^3 - 27X + 18$.

3.5 COMPLEX NUMBERS

1. The imaginary number i has the property that $i^2 = -1$.

3. (a) The complex conjugate of $3 + 4i$ is $\overline{3 + 4i} = 3 - 4i$.

   (b) $(3 + 4i)\left(\overline{3 + 4i}\right) = 3^2 + 4^2 = 25$

5. $5 - 7i$: real part 5, imaginary part -7.

7. $\frac{-2 - 5i}{3} = -\frac{2}{3} - \frac{5}{3}i$: real part $-\frac{2}{3}$, imaginary part $-\frac{5}{3}$.

9. 3: real part 3, imaginary part 0.

11. $-\frac{2}{3}i$: real part 0, imaginary part $-\frac{2}{3}$.

13. $\sqrt{3} + \sqrt{-4} = \sqrt{3} + 2i$: real part $\sqrt{3}$, imaginary part 2.

15. $(2 - 5i) + (3 + 4i) = (2 + 3) + (-5 + 4)i = 5 - i$

17. $(-6 + 6i) + (9 - i) = (-6 + 9) + (6 - 1)i = 3 + 5i$

19. $\left(7 - \frac{1}{2}i\right) - \left(5 + \frac{3}{2}i\right) = (7 - 5) + \left(-\frac{1}{2} - \frac{3}{2}\right)i = 2 - 2i$

21. $(-12 + 8i) - (7 + 4i) = -12 + 8i - 7 - 4i = (-12 - 7) + (8 - 4)i = -19 + 4i$

23. $4(-1 + 2i) = -4 + 8i$

25. $(7 - i)(4 + 2i) = 28 + 14i - 4i - 2i^2 = (28 + 2) + (14 - 4)i = 30 + 10i$

27. $(3 - 4i)(5 - 12i) = 15 - 36i - 20i + 48i^2 = (15 - 48) + (-36 - 20)i = -33 - 56i$

29. $(6 + 5i)(2 - 3i) = 12 - 18i + 10i - 15i^2 = (12 + 15) + (-18 + 10)i = 27 - 8i$

31. $i^3 = i^2 \cdot i = -1 \cdot i = -i$

33. $i^{100} = \left(i^4\right)^{25} = (1)^{25} = 1$

35. $\frac{1}{i} = \frac{1}{i} \cdot \frac{i}{i} = \frac{i}{i^2} = \frac{i}{-1} = -i$

37. $\frac{2 - 3i}{1 - 2i} = \frac{2 - 3i}{1 - 2i} \cdot \frac{1 + 2i}{1 + 2i} = \frac{2 + 4i - 3i - 6i^2}{1 - 4i^2} = \frac{(2 + 6) + (4 - 3)i}{1 + 4} = \frac{8 + i}{5}$ or $\frac{8}{5} + \frac{1}{5}i$

39. $\dfrac{26+39i}{2-3i} = \dfrac{26+39i}{2-3i} \cdot \dfrac{2+3i}{2+3i} = \dfrac{52+78i+78i+117i^2}{4-9i^2} = \dfrac{(52-117)+(78+78)i}{4+9} = \dfrac{-65+156i}{13}$

$\qquad = \dfrac{13(-5+12i)}{13} = -5+12i$

41. $\dfrac{10i}{1-2i} = \dfrac{10i}{1-2i} \cdot \dfrac{1+2i}{1+2i} = \dfrac{10i+20i^2}{1-4i^2} = \dfrac{-20+10i}{1+4} = \dfrac{5(-4+2i)}{5} = -4+2i$

43. $\dfrac{4+6i}{3i} = \dfrac{4+6i}{3i} \cdot \dfrac{3i}{3i} = \dfrac{12i+18i^2}{9i^2} = \dfrac{-18+12i}{-9} = \dfrac{-18}{-9} + \dfrac{12}{-9}i = 2 - \dfrac{4}{3}i$

45. $\dfrac{1}{1+i} - \dfrac{1}{1-i} = \dfrac{1}{1+i} \cdot \dfrac{1-i}{1-i} - \dfrac{1}{1-i} \cdot \dfrac{1+i}{1+i} = \dfrac{1-i}{1-i^2} - \dfrac{1+i}{1-i^2} = \dfrac{1-i}{2} + \dfrac{-1-i}{2} = -i$

47. $\sqrt{-25} = 5i$

49. $\sqrt{-3}\sqrt{-12} = i\sqrt{3} \cdot 2i\sqrt{3} = 6i^2 = -6$

51. $(3-\sqrt{-5})(1+\sqrt{-1}) = (3-i\sqrt{5})(1+i) = 3+3i-i\sqrt{5}-i^2\sqrt{5} = (3+\sqrt{5}) + (3-\sqrt{5})i$

53. $\dfrac{2+\sqrt{-8}}{1+\sqrt{-2}} = \dfrac{2+2i\sqrt{2}}{1+i\sqrt{2}} = \dfrac{2\left(1+i\sqrt{2}\right)}{1+i\sqrt{2}} = 2$

55. $\dfrac{\sqrt{-36}}{\sqrt{-2}\sqrt{-9}} = \dfrac{6i}{i\sqrt{2}\cdot 3i} = \dfrac{2}{i\sqrt{2}} \cdot \dfrac{i\sqrt{2}}{i\sqrt{2}} = \dfrac{2i\sqrt{2}}{2i^2} = \dfrac{i\sqrt{2}}{-1} = -i\sqrt{2}$

57. $x^2+49=0 \iff x^2=-49 \implies x = \pm 7i$

59. $x^2-4x+5=0 \implies x = \dfrac{-(-4)\pm\sqrt{(-4)^2-4(1)(5)}}{2(1)} = \dfrac{4\pm\sqrt{16-20}}{2} = \dfrac{4\pm\sqrt{-4}}{2} = \dfrac{4\pm 2i}{2} = 2\pm i$

61. $x^2+2x+5=0 \implies x = \dfrac{-(2)\pm\sqrt{(2)^2-4(1)(5)}}{2(1)} = \dfrac{-2\pm\sqrt{4-20}}{2} = \dfrac{-2\pm\sqrt{-16}}{2} = \dfrac{-2\pm 4i}{2} = -1\pm 2i$

63. $x^2+x+1=0 \implies x = \dfrac{-(1)\pm\sqrt{(1)^2-4(1)(1)}}{2(1)} = \dfrac{-1\pm\sqrt{1-4}}{2} = \dfrac{-1\pm\sqrt{-3}}{2} = \dfrac{-1\pm i\sqrt{3}}{2} = -\dfrac{1}{2} \pm \dfrac{\sqrt{3}}{2}i$

65. $2x^2-2x+1=0 \implies x = \dfrac{-(-2)\pm\sqrt{(-2)^2-4(2)(1)}}{2(2)} = \dfrac{2\pm\sqrt{4-8}}{4} = \dfrac{2\pm\sqrt{-4}}{4} = \dfrac{2\pm 2i}{4} = \dfrac{1}{2} \pm \dfrac{1}{2}i$

67. $t+3+\dfrac{3}{t}=0 \iff t^2+3t+3=0 \implies t = \dfrac{-(3)\pm\sqrt{(3)^2-4(1)(3)}}{2(1)} = \dfrac{-3\pm\sqrt{9-12}}{2} = \dfrac{-3\pm\sqrt{-3}}{2} = \dfrac{-3\pm i\sqrt{3}}{2} = -\dfrac{3}{2} \pm \dfrac{\sqrt{3}}{2}i$

69. $6x^2+12x+7=0 \implies$

$\qquad x = \dfrac{-(12)\pm\sqrt{(12)^2-4(6)(7)}}{2(6)} = \dfrac{-12\pm\sqrt{144-168}}{12} = \dfrac{-12\pm\sqrt{-24}}{12} = \dfrac{-12\pm 2i\sqrt{6}}{12} = \dfrac{-12}{12} \pm \dfrac{2i\sqrt{6}}{12} = -1 \pm \dfrac{\sqrt{6}}{6}i$

71. $\dfrac{1}{2}x^2-x+5=0 \implies x = \dfrac{-(-1)\pm\sqrt{(-1)^2-4\left(\frac{1}{2}\right)(5)}}{2\left(\frac{1}{2}\right)} = \dfrac{1\pm\sqrt{1-10}}{1} = 1\pm\sqrt{-9} = 1\pm 3i$

73. LHS $= \overline{z} + \overline{w} = \overline{(a+bi)} + \overline{(c+di)} = a-bi+c-di = (a+c)+(-b-d)i = (a+c)-(b+d)i$.

      RHS $= \overline{z+w} = \overline{(a+bi)+(c+di)} = \overline{(a+c)+(b+d)i} = (a+c)-(b+d)i$.

      Since LHS $=$ RHS, this proves the statement.

75. LHS $= (\overline{z})^2 = \left(\overline{(a+bi)}\right)^2 = (a-bi)^2 = a^2-2abi+b^2i^2 = \left(a^2-b^2\right)-2abi$.

      RHS $= \overline{z^2} = \overline{(a+bi)^2} = \overline{a^2+2abi+b^2i^2} = \overline{(a^2-b^2)+2abi} = \left(a^2-b^2\right)-2abi$.

      Since LHS $=$ RHS, this proves the statement.

77. $z + \overline{z} = (a+bi) + \overline{(a+bi)} = a+bi+a-bi = 2a$, which is a real number.

79. $z \cdot \overline{z} = (a + bi) \cdot \overline{(a + bi)} = (a + bi) \cdot (a - bi) = a^2 - b^2 i^2 = a^2 + b^2$, which is a real number.

81. Using the quadratic formula, the solutions to the equation are $x = \dfrac{-b \pm \sqrt{b^2 - 4ac}}{2a}$. Since both solutions are imaginary,

we have $b^2 - 4ac < 0 \Leftrightarrow 4ac - b^2 > 0$, so the solutions are $x = \dfrac{-b}{2a} \pm \dfrac{\sqrt{4ac - b^2}}{2a} i$, where $\sqrt{4ac - b^2}$ is a real number.

Thus the solutions are complex conjugates of each other.

3.6 COMPLEX ZEROS AND THE FUNDAMENTAL THEOREM OF ALGEBRA

1. The polynomial $P(x) = 3(x - 5)^3 (x - 3)(x + 2)$ has degree 5. It has zeros 5, 3, and -2. The zero 5 has multiplicity 3 and the zero 3 has multiplicity 1.

3. A polynomial of degree $n \geq 1$ has exactly n zeros, if a zero of multiplicity m is counted m times.

5. (a) $x^4 + 4x^2 = 0 \Leftrightarrow x^2 (x^2 + 4) = 0$. So $x = 0$ or $x^2 + 4 = 0$. If $x^2 + 4 = 0$ then $x^2 = -4 \Leftrightarrow x = \pm 2i$. Therefore, the solutions are $x = 0$ and $\pm 2i$.

(b) To get the complete factorization, we factor the remaining quadratic factor $P(x) = x^2 (x + 4) = x^2 (x - 2i)(x + 2i)$.

7. (a) $x^3 - 2x^2 + 2x = 0 \Leftrightarrow x (x^2 - 2x + 2) = 0$. So $x = 0$ or $x^2 - 2x + 2 = 0$. If $x^2 - 2x + 2 = 0$ then

$x = \dfrac{-(-2) \pm \sqrt{(-2)^2 - 4(1)(2)}}{2} = \dfrac{2 \pm \sqrt{-4}}{2} = \dfrac{2 \pm 2i}{2} = 1 \pm i$. Therefore, the solutions are $x = 0, 1 \pm i$.

(b) Since $1 - i$ and $1 + i$ are zeros, $x - (1 - i) = x - 1 + i$ and $x - (1 + i) = x - 1 - i$ are the factors of $x^2 - 2x + 2$. Thus the complete factorization is $P(x) = x (x^2 - 2x + 2) = x (x - 1 + i)(x - 1 - i)$.

9. (a) $x^4 + 2x^2 + 1 = 0 \Leftrightarrow (x^2 + 1)^2 = 0 \Leftrightarrow x^2 + 1 = 0 \Leftrightarrow x^2 = -1 \Leftrightarrow x = \pm i$. Therefore the zeros of P are $x = \pm i$.

(b) Since $-i$ and i are zeros, $x + i$ and $x - i$ are the factors of $x^2 + 1$. Thus the complete factorization is

$P(x) = (x^2 + 1)^2 = [(x + i)(x - i)]^2 = (x + i)^2 (x - i)^2$.

11. (a) $x^4 - 16 = 0 \Leftrightarrow 0 = (x^2 - 4)(x^2 + 4) = (x - 2)(x + 2)(x^2 + 4)$. So $x = \pm 2$ or $x^2 + 4 = 0$. If $x^2 + 4 = 0$ then $x^2 = -4 \Rightarrow x = \pm 2i$. Therefore the zeros of P are $x = \pm 2, \pm 2i$.

(b) Since $-i$ and i are zeros, $x + i$ and $x - i$ are the factors of $x^2 + 1$. Thus the complete factorization is

$P(x) = (x - 2)(x + 2)(x^2 + 4) = (x - 2)(x + 2)(x - 2i)(x + 2i)$.

13. (a) $x^3 + 8 = 0 \Leftrightarrow (x + 2)(x^2 - 2x + 4) = 0$. So $x = -2$ or $x^2 - 2x + 4 = 0$. If $x^2 - 2x + 4 = 0$ then

$x = \dfrac{-(-2) \pm \sqrt{(-2)^2 - 4(1)(4)}}{2} = \dfrac{2 \pm \sqrt{-12}}{2} = \dfrac{2 \pm 2i\sqrt{3}}{2} = 1 \pm i\sqrt{3}$. Therefore, the zeros of P are $x = -2, 1 \pm i\sqrt{3}$.

(b) Since $1 - i\sqrt{3}$ and $1 + i\sqrt{3}$ are the zeros from the $x^2 - 2x + 4 = 0$, $x - (1 - i\sqrt{3})$ and $x - (1 + i\sqrt{3})$ are the factors of $x^2 - 2x + 4$. Thus the complete factorization is

$P(x) = (x + 2)(x^2 - 2x + 4) = (x + 2)\left[x - (1 - i\sqrt{3})\right]\left[x - (1 + i\sqrt{3})\right]$

$= (x + 2)(x - 1 + i\sqrt{3})(x - 1 - i\sqrt{3})$

15. (a) $x^6 - 1 = 0 \Leftrightarrow 0 = (x^3 - 1)(x^3 + 1) = (x - 1)(x^2 + x + 1)(x + 1)(x^2 - x + 1)$. Clearly, $x = \pm 1$ are solutions.

If $x^2 + x + 1 = 0$, then $x = \dfrac{-1 \pm \sqrt{1 - 4(1)(1)}}{2} = \dfrac{-1 \pm \sqrt{-3}}{2} = -\dfrac{1}{2} \pm \dfrac{\sqrt{-3}}{2}$ so $x = -\dfrac{1}{2} \pm i\dfrac{\sqrt{3}}{2}$. And if $x^2 - x + 1 = 0$, then

$x = \dfrac{1 \pm \sqrt{1 - 4(1)(1)}}{2} = \dfrac{1 \pm \sqrt{-3}}{2} = \dfrac{1}{2} \pm \dfrac{\sqrt{-3}}{2} = \dfrac{1}{2} \pm i\dfrac{\sqrt{3}}{2}$. Therefore, the zeros of P are $x = \pm 1, -\dfrac{1}{2} \pm i\dfrac{\sqrt{3}}{2}, \dfrac{1}{2} \pm i\dfrac{\sqrt{3}}{2}$.

(b) The zeros of $x^2 + x + 1 = 0$ are $-\frac{1}{2} - i\frac{\sqrt{3}}{2}$ and $-\frac{1}{2} + i\frac{\sqrt{3}}{2}$, so $x^2 + x + 1$ factors as

$$\left[x - \left(-\frac{1}{2} - i\frac{\sqrt{3}}{2}\right)\right]\left[x - \left(-\frac{1}{2} + i\frac{\sqrt{3}}{2}\right)\right] = \left(x + \frac{1}{2} + i\frac{\sqrt{3}}{2}\right)\left(x + \frac{1}{2} - i\frac{\sqrt{3}}{2}\right). \quad \text{Similarly, since}$$

the zeros of $x^2 - x + 1 = 0$ are $\frac{1}{2} - i\frac{\sqrt{3}}{2}$ and $\frac{1}{2} + i\frac{\sqrt{3}}{2}$, so $x^2 - x + 1$ factors as

$$\left[x - \left(\frac{1}{2} - i\frac{\sqrt{3}}{2}\right)\right]\left[x - \left(\frac{1}{2} + i\frac{\sqrt{3}}{2}\right)\right] = \left(x - \frac{1}{2} + i\frac{\sqrt{3}}{2}\right)\left(x - \frac{1}{2} - i\frac{\sqrt{3}}{2}\right). \quad \text{Thus the complete}$$

factorization is

$$P(x) = (x - 1)\left(x^2 + x + 1\right)(x + 1)\left(x^2 - x + 1\right)$$

$$= (x - 1)(x + 1)\left(x + \frac{1}{2} + i\frac{\sqrt{3}}{2}\right)\left(x + \frac{1}{2} - i\frac{\sqrt{3}}{2}\right)\left(x - \frac{1}{2} + i\frac{\sqrt{3}}{2}\right)\left(x - \frac{1}{2} - i\frac{\sqrt{3}}{2}\right)$$

17. $P(x) = x^2 + 25 = (x - 5i)(x + 5i)$. The zeros of P are $5i$ and $-5i$, both multiplicity 1.

19. $Q(x) = x^2 + 2x + 2$. Using the quadratic formula $x = \frac{-(2) \pm \sqrt{(2)^2 - 4(1)(2)}}{2(1)} = \frac{-2 \pm \sqrt{-4}}{2} = \frac{-2 \pm 2i}{2} = -1 \pm i$. So $Q(x) = (x + 1 - i)(x + 1 + i)$. The zeros of Q are $-1 - i$ (multiplicity 1) and $-1 + i$ (multiplicity 1).

21. $P(x) = x^3 + 4x = x\left(x^2 + 4\right) = x(x - 2i)(x + 2i)$. The zeros of P are $0, 2i$, and $-2i$ (all multiplicity 1).

23. $Q(x) = x^4 - 1 = \left(x^2 - 1\right)\left(x^2 + 1\right) = (x - 1)(x + 1)\left(x^2 + 1\right) = (x - 1)(x + 1)(x - i)(x + i)$. The zeros of Q are $1, -1, i$, and $-i$ (all of multiplicity 1).

25. $P(x) = 16x^4 - 81 = \left(4x^2 - 9\right)\left(4x^2 + 9\right) = (2x - 3)(2x + 3)(2x - 3i)(2x + 3i)$. The zeros of P are $\frac{3}{2}, -\frac{3}{2}, \frac{3}{2}i$, and $-\frac{3}{2}i$ (all of multiplicity 1).

27. $P(x) = x^3 + x^2 + 9x + 9 = x^2(x + 1) + 9(x + 1) = (x + 1)\left(x^2 + 9\right) = (x + 1)(x - 3i)(x + 3i)$. The zeros of P are $-1, 3i$, and $-3i$ (all of multiplicity 1).

29. $Q(x) = x^4 + 2x^2 + 1 = \left(x^2 + 1\right)^2 = (x - i)^2(x + i)^2$. The zeros of Q are i and $-i$ (both of multiplicity 2).

31. $P(x) = x^4 + 3x^2 - 4 = \left(x^2 - 1\right)\left(x^2 + 4\right) = (x - 1)(x + 1)(x - 2i)(x + 2i)$. The zeros of P are $1, -1, 2i$, and $-2i$ (all of multiplicity 1).

33. $P(x) = x^5 + 6x^3 + 9x = x\left(x^4 + 6x^2 + 9\right) = x\left(x^2 + 3\right)^2 = x\left(x - i\sqrt{3}\right)^2\left(x + i\sqrt{3}\right)^2$. The zeros of P are 0 (multiplicity 1), $i\sqrt{3}$ (multiplicity 2), and $-i\sqrt{3}$ (multiplicity 2).

35. Since $1 + i$ and $1 - i$ are conjugates, the factorization of the polynomial must be

$$P(x) = a(x - [1 + i])(x - [1 - i]) = a\left(x^2 - 2x + 2\right). \text{ If we let } a = 1, \text{ we get } P(x) = x^2 - 2x + 2.$$

37. Since $2i$ and $-2i$ are conjugates, the factorization of the polynomial must be

$$Q(x) = b(x - 3)(x - 2i)(x + 2i) = b(x - 3)\left(x^2 + 4\right) = b\left(x^3 - 3x^2 + 4x - 12\right). \text{ If we let } b = 1, \text{ we get}$$

$$Q(x) = x^3 - 3x^2 + 4x - 12.$$

39. Since i is a zero, by the Conjugate Roots Theorem, $-i$ is also a zero. So the factorization of the polynomial must be

$$P(x) = a(x - 2)(x - i)(x + i) = a\left(x^3 - 2x^2 + x - 2\right). \text{ If we let } a = 1, \text{ we get } P(x) = x^3 - 2x^2 + x - 2.$$

41. Since the zeros are $1 - 2i$ and 1 (with multiplicity 2), by the Conjugate Roots Theorem, the other zero is $1 + 2i$. So a factorization is

$$R(x) = c(x - [1 - 2i])(x - [1 + 2i])(x - 1)^2 = c([x - 1] + 2i)([x - 1] - 2i)(x - 1)^2$$

$$= c\left([x - 1]^2 - [2i]^2\right)\left(x^2 - 2x + 1\right) = c\left(x^2 - 2x + 1 + 4\right)\left(x^2 - 2x + 1\right) = c\left(x^2 - 2x + 5\right)\left(x^2 - 2x + 1\right)$$

$$= c\left(x^4 - 2x^3 + x^2 - 2x^3 + 4x^2 - 2x + 5x^2 - 10x + 5\right) = c\left(x^4 - 4x^3 + 10x^2 - 12x + 5\right)$$

If we let $c = 1$ we get $R(x) = x^4 - 4x^3 + 10x^2 - 12x + 5$.

43. Since the zeros are i and $1 + i$, by the Conjugate Roots Theorem, the other zeros are $-i$ and $1 - i$. So a factorization is

$$
\begin{aligned}
T(x) &= C(x - i)(x + i)(x - [1 + i])(x - [1 - i]) \\
&= C\left(x^2 - i^2\right)([x - 1] - i)([x - 1] + i) = C\left(x^2 + 1\right)\left(x^2 - 2x + 1 - i^2\right) = C\left(x^2 + 1\right)\left(x^2 - 2x + 2\right) \\
&= C\left(x^4 - 2x^3 + 2x^2 + x^2 - 2x + 2\right) = C\left(x^4 - 2x^3 + 3x^2 - 2x + 2\right) = Cx^4 - 2Cx^3 + 3Cx^2 - 2Cx + 2C
\end{aligned}
$$

Since the constant coefficient is 12, it follows that $2C = 12 \Leftrightarrow C = 6$, and so
$T(x) = 6\left(x^4 - 2x^3 + 3x^2 - 2x + 2\right) = 6x^4 - 12x^3 + 18x^2 - 12x + 12$.

45. $P(x) = x^3 + 2x^2 + 4x + 8 = x^2(x + 2) + 4(x + 2) = (x + 2)\left(x^2 + 4\right) = (x + 2)(x - 2i)(x + 2i)$. Thus the zeros are -2 and $\pm 2i$.

47. $P(x) = x^3 - 2x^2 + 2x - 1$. By inspection, $P(1) = 1 - 2 + 2 - 1 = 0$, and hence $x = 1$ is a zero.

$$
\begin{array}{r|rrrr}
1 & 1 & -2 & 2 & -1 \\
  &   & 1 & -1 & 1 \\
\hline
  & 1 & -1 & 1 & 0
\end{array}
$$

Thus $P(x) = (x - 1)\left(x^2 - x + 1\right)$. So $x = 1$ or $x^2 - x + 1 = 0$.

Using the quadratic formula, we have $x = \frac{1 \pm \sqrt{1 - 4(1)(1)}}{2} = \frac{1 \pm i\sqrt{3}}{2}$. Hence, the zeros are 1 and $\frac{1 \pm i\sqrt{3}}{2}$.

49. $P(x) = x^3 - 3x^2 + 3x - 2$.

$$
\begin{array}{r|rrrr}
2 & 1 & -3 & 3 & -2 \\
  &   & 2 & -2 & 2 \\
\hline
  & 1 & -1 & 1 & 0
\end{array}
$$

Thus $P(x) = (x - 2)\left(x^2 - x + 1\right)$. So $x = 2$ or $x^2 - x + 1 = 0$

Using the quadratic formula we have $x = \frac{1 \pm \sqrt{1 - 4(1)(1)}}{2} = \frac{1 \pm i\sqrt{3}}{2}$. Hence, the zeros are 2, and $\frac{1 \pm i\sqrt{3}}{2}$.

51. $P(x) = 2x^3 + 7x^2 + 12x + 9$ has possible rational zeros ± 1, ± 3, ± 9, $\pm \frac{1}{2}$, $\pm \frac{3}{2}$, $\pm \frac{9}{2}$. Since all coefficients are positive, there are no positive real zeros.

$$
\begin{array}{r|rrrr}
-1 & 2 & 7 & 12 & 9 \\
   &   & -2 & -5 & -7 \\
\hline
   & 2 & 5 & 7 & 2
\end{array}
\qquad
\begin{array}{r|rrrr}
-2 & 2 & 7 & 12 & 9 \\
   &   & -4 & -6 & -12 \\
\hline
   & 2 & 3 & 6 & -3
\end{array}
$$

There is a zero between -1 and -2.

$$
\begin{array}{r|rrrr}
-\frac{3}{2} & 2 & 7 & 12 & 9 \\
             &   & -3 & -6 & -9 \\
\hline
             & 2 & 4 & 6 & 0
\end{array}
\quad \Rightarrow x = -\frac{3}{2} \text{ is a zero.}
$$

$P(x) = \left(x + \frac{3}{2}\right)\left(2x^2 + 4x + 6\right) = 2\left(x + \frac{3}{2}\right)\left(x^2 + 2x + 3\right)$. Now $x^2 + 2x + 3$ has zeros

$x = \frac{-2 \pm \sqrt{4 - 4(3)(1)}}{2} = \frac{-2 \pm 2\sqrt{-2}}{2} = -1 \pm i\sqrt{2}$. Hence, the zeros are $-\frac{3}{2}$ and $-1 \pm i\sqrt{2}$.

53. $P(x) = x^4 + x^3 + 7x^2 + 9x - 18$. Since $P(x)$ has one change in sign, we are guaranteed a positive zero, and since $P(-x) = x^4 - x^3 + 7x^2 - 9x - 18$, there are 1 or 3 negative zeros.

$$\begin{array}{r|rrrrr} 1 & 1 & 1 & 7 & 9 & -18 \\ & & 1 & 2 & 9 & 18 \\ \hline & 1 & 2 & 9 & 18 & 0 \end{array}$$

Therefore, $P(x) = (x - 1)\left(x^3 + 2x^2 + 9x + 18\right)$. Continuing with the quotient, we try negative zeros.

$$\begin{array}{r|rrrr} -1 & 1 & 2 & 9 & 18 \\ & & -1 & -1 & -8 \\ \hline & 1 & 1 & 8 & 10 \end{array} \qquad \begin{array}{r|rrrr} -2 & 1 & 2 & 9 & 18 \\ & & -2 & 0 & -18 \\ \hline & 1 & 0 & 9 & 0 \end{array}$$

$P(x) = (x - 1)(x + 2)\left(x^2 + 9\right) = (x - 1)(x + 2)(x - 3i)(x + 3i)$. Therefore, the zeros are 1, -2, and $\pm 3i$.

55. We see a pattern and use it to factor by grouping. This gives

$$P(x) = x^5 - x^4 + 7x^3 - 7x^2 + 12x - 12 = x^4(x - 1) + 7x^2(x - 1) + 12(x - 1) = (x - 1)\left(x^4 + 7x^2 + 12\right)$$

$$= (x - 1)\left(x^2 + 3\right)\left(x^2 + 4\right) = (x - 1)\left(x - i\sqrt{3}\right)\left(x + i\sqrt{3}\right)(x - 2i)(x + 2i)$$

Therefore, the zeros are 1, $\pm i\sqrt{3}$, and $\pm 2i$.

57. $P(x) = x^4 - 6x^3 + 13x^2 - 24x + 36$ has possible rational zeros ± 1, ± 2, ± 3, ± 4, ± 6, ± 9, ± 12, ± 18. $P(x)$ has 4 variations in sign and $P(-x)$ has no variation in sign.

$$\begin{array}{r|rrrrr} 1 & 1 & -6 & 13 & -24 & 36 \\ & & 1 & -5 & 8 & -16 \\ \hline & 1 & -5 & 8 & -16 & 20 \end{array} \quad \begin{array}{r|rrrrr} 2 & 1 & -6 & 13 & -24 & 36 \\ & & 2 & -8 & 10 & -28 \\ \hline & 1 & -4 & 5 & -14 & 8 \end{array} \quad \begin{array}{r|rrrrr} 3 & 1 & -6 & 13 & -24 & 36 \\ & & 3 & -9 & 12 & -36 \\ \hline & 1 & -3 & 4 & -12 & 0 \end{array} \Rightarrow x = 3 \text{ is a zero.}$$

Continuing:

$$\begin{array}{r|rrrr} 3 & 1 & -3 & 4 & -12 \\ & & 3 & 0 & 12 \\ \hline & 1 & 0 & 4 & 0 \end{array} \Rightarrow x = 3 \text{ is a zero.}$$

$P(x) = (x - 3)^2\left(x^2 + 4\right) = (x - 3)^2(x - 2i)(x + 2i)$. Therefore, the zeros are 3 (multiplicity 2) and $\pm 2i$.

59. $P(x) = 4x^4 + 4x^3 + 5x^2 + 4x + 1$ has possible rational zeros ± 1, $\pm \frac{1}{2}$, $\pm \frac{1}{4}$. Since there is no variation in sign, all real zeros (if there are any) are negative.

$$\begin{array}{r|rrrrr} -1 & 4 & 4 & 5 & 4 & 1 \\ & & -4 & 0 & -5 & 1 \\ \hline & 4 & 0 & 5 & -1 & 2 \end{array} \qquad \begin{array}{r|rrrrr} -\frac{1}{2} & 4 & 4 & 5 & 4 & 1 \\ & & -2 & -1 & -2 & -1 \\ \hline & 4 & 2 & 4 & 2 & 0 \end{array} \Rightarrow x = -\frac{1}{2} \text{ is a zero.}$$

$P(x) = \left(x + \frac{1}{2}\right)\left(4x^3 + 2x^2 + 4x + 2\right)$. Continuing:

$$\begin{array}{r|rrrr} -\frac{1}{2} & 4 & 2 & 4 & 2 \\ & & -2 & 0 & -2 \\ \hline & 4 & 0 & 4 & 0 \end{array} \Rightarrow x = -\frac{1}{2} \text{ is a zero again.}$$

$P(x) = \left(x + \frac{1}{2}\right)^2\left(4x^2 + 4\right)$. Thus, the zeros of $P(x)$ are $-\frac{1}{2}$ (multiplicity 2) and $\pm i$.

61. $P(x) = x^5 - 3x^4 + 12x^3 - 28x^2 + 27x - 9$ has possible rational zeros $\pm 1, \pm 3, \pm 9$. $P(x)$ has 4 variations in sign and $P(-x)$ has 1 variation in sign.

$$
\begin{array}{r|rrrrrr}
1 & 1 & -3 & 12 & -28 & 27 & -9 \\
  &   & 1 & -2 & 10 & -18 & 9 \\
\hline
  & 1 & -2 & 10 & -18 & 9 & 0
\end{array}
\Rightarrow x = 1 \text{ is a zero.}
$$

$$
\begin{array}{r|rrrrr}
1 & 1 & -2 & 10 & -18 & 9 \\
  &   & 1 & -1 & 9   & -9 \\
\hline
  & 1 & -1 & 9 & -9 & 0
\end{array}
\Rightarrow x = 1 \text{ is a zero.}
\qquad
\begin{array}{r|rrrr}
1 & 1 & -1 & 9 & -9 \\
  &   & 1 & 0 & 9 \\
\hline
  & 1 & 0 & 9 & 0
\end{array}
\Rightarrow x = 1 \text{ is a zero.}
$$

$P(x) = (x-1)^3 \left(x^2 + 9\right) = (x-1)^3 (x-3i)(x+3i)$. Therefore, the zeros are 1 (multiplicity 3) and $\pm 3i$.

63. **(a)** $P(x) = x^3 - 5x^2 + 4x - 20 = x^2(x-5) + 4(x-5) = (x-5)\left(x^2 + 4\right)$

(b) $P(x) = (x-5)(x-2i)(x+2i)$

65. **(a)** $P(x) = x^4 + 8x^2 - 9 = \left(x^2 - 1\right)\left(x^2 + 9\right) = (x-1)(x+1)\left(x^2 + 9\right)$

(b) $P(x) = (x-1)(x+1)(x-3i)(x+3i)$

67. **(a)** $P(x) = x^6 - 64 = \left(x^3 - 8\right)\left(x^3 + 8\right) = (x-2)\left(x^2 + 2x + 4\right)(x+2)\left(x^2 - 2x + 4\right)$

(b) $P(x) = (x-2)(x+2)\left(x+1-i\sqrt{3}\right)\left(x+1+i\sqrt{3}\right)\left(x-1-i\sqrt{3}\right)\left(x-1+i\sqrt{3}\right)$

69. **(a)** $x^4 - 2x^3 - 11x^2 + 12x = x\left(x^3 - 2x^2 - 11x + 12\right) = 0$. We first find the bounds for our viewing rectangle.

$$
\begin{array}{r|rrrr}
  & 1 & -2 & -11 & 12 \\
5 & 1 & 3 & 4   & 32 \\
\end{array}
\Rightarrow x = 5 \text{ is an upper bound.}
$$
$$
\begin{array}{r|rrrr}
-4 & 1 & -6 & 13 & -50 \\
\end{array}
\Rightarrow x = -4 \text{ is a lower bound.}
$$

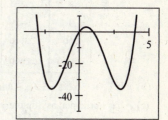

We graph $P(x) = x^4 - 2x^3 - 11x^2 + 12x$ in the viewing rectangle $[-4, 5]$ by $[-50, 10]$ and see that it has 4 real solutions. Since this matches the degree of $P(x)$, $P(x)$ has no imaginary solution.

(b) $x^4 - 2x^3 - 11x^2 + 12x - 5 = 0$. We use the same bounds for our viewing rectangle, $[-4, 5]$ by $[-50, 10]$, and see that $R(x) = x^4 - 2x^3 - 11x^2 + 12x - 5$ has 2 real solutions. Since the degree of $R(x)$ is 4, $R(x)$ must have 2 imaginary solutions.

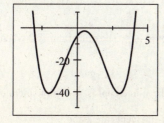

(c) $x^4 - 2x^3 - 11x^2 + 12x + 40 = 0$. We graph $T(x) = x^4 - 2x^3 - 11x^2 + 12x + 40$ in the viewing rectangle $[-4, 5]$ by $[-10, 50]$, and see that T has no real solution. Since the degree of T is 4, T must have 4 imaginary solutions.

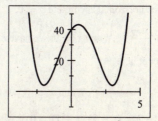

71. (a) $P(x) = x^2 - (1+i)x + (2+2i)$. So $P(2i) = (2i)^2 - (1+i)(2i) + 2 + 2i = -4 - 2i + 2 + 2 + 2i = 0$,

and $P(1-i) = (1-i)^2 - (1+i)(1-i) + (2+2i) = 1 - 2i - 1 - 1 - 1 + 2 + 2i = 0$.

Therefore, $2i$ and $1-i$ are solutions of the equation $x^2 - (1+i)x + (2+2i) = 0$. However,

$P(-2i) = (-2i)^2 - (1+i)(-2i) + 2 + 2i = -4 + 2i - 2 + 2 + 2i = -4 + 4i$, and

$P(1+i) = (1+i)^2 - (1+i)(1+i) + 2 + 2i = 2 + 2i$. Since, $P(-2i) \neq 0$ and $P(1+i) \neq 0$, $-2i$ and $1+i$ are

not solutions.

(b) This does not violate the Conjugate Roots Theorem because the coefficients of the polynomial $P(x)$ are not all real.

73. Because P has real coefficients, the imaginary zeros come in pairs: $a \pm bi$ (by the Conjugate Roots Theorem), where $b \neq 0$.

Thus there must be an even number of imaginary zeros. Since P is of odd degree, it has an odd number of zeros (counting

multiplicity). It follows that P has at least one real zero.

3.7   RATIONAL FUNCTIONS

1. If the rational function $y = r(x)$ has the vertical asymptote $x = 2$, then as $x \to 2^+$, either $y \to \infty$ or $y \to -\infty$.

3. The function $r(x) = \dfrac{(x+1)(x-2)}{(x+2)(x-3)}$ has x-intercepts -1 and 2.

5. The function r has vertical asymptotes $x = -2$ and $x = 3$.

7. $r(x) = \dfrac{x}{x-2}$

(a)

x	$r(x)$
1.5	-3
1.9	-19
1.99	-199
1.999	-1999

x	$r(x)$
2.5	5
2.1	21
2.01	201
2.001	2001

x	$r(x)$
10	1.25
50	1.042
100	1.020
1000	1.002

x	$r(x)$
-10	0.833
-50	0.962
-100	0.980
-1000	0.998

(b) $r(x) \to -\infty$ as $x \to 2^-$ and $r(x) \to \infty$ as $x \to 2^+$.

(c) r has horizontal asymptote $y = 1$.

9. $r(x) = \dfrac{3x - 10}{(x-2)^2}$

(a)

x	$r(x)$
1.5	-22
1.9	-430
1.99	$-40{,}300$
1.999	$-4{,}003{,}000$

x	$r(x)$
2.5	-10
2.1	-370
2.01	$-39{,}700$
2.001	$-3{,}997{,}000$

x	$r(x)$
10	0.3125
50	0.0608
100	0.0302
1000	0.0030

x	$r(x)$
-10	-0.2778
-50	-0.0592
-100	-0.0298
-1000	-0.0030

(b) $r(x) \to -\infty$ as $x \to 2$.

(c) r has horizontal asymptote $y = 0$.

11. $r(x) = \dfrac{x-1}{x+4}$. When $x = 0$, we have $r(0) = -\frac{1}{4}$, so the y-intercept is $-\frac{1}{4}$. The numerator is 0 when $x = 1$, so the

x-intercept is 1.

13. $t(x) = \dfrac{x^2 - x - 2}{x - 6}$. When $x = 0$, we have $t(0) = \dfrac{-2}{-6} = \frac{1}{3}$, so the y-intercept is $\frac{1}{3}$. The numerator is 0 when

$x^2 - x - 2 = (x-2)(x+1) = 0$ or when $x = 2$ or $x = -1$, so the x-intercepts are 2 and -1.

15. $r(x) = \dfrac{x^2 - 9}{x^2}$. Since 0 is not in the domain of $r(x)$, there is no y-intercept. The numerator is 0 when

$x^2 - 9 = (x - 3)(x + 3) = 0$ or when $x = \pm 3$, so the x-intercepts are ± 3.

17. From the graph, the x-intercept is 3, the y-intercept is 3, the vertical asymptote is $x = 2$, and the horizontal asymptote is $y = 2$.

19. From the graph, the x-intercepts are -1 and 1, the y-intercept is about $\frac{1}{4}$, the vertical asymptotes are $x = -2$ and $x = 2$, and the horizontal asymptote is $y = 1$.

21. $r(x) = \dfrac{5}{x - 2}$ has a vertical asymptote where $x - 2 = 0 \iff x = 2$, and $y = 0$ is a horizontal asymptote because the degree of the denominator is greater than that of the numerator.

23. $r(x) = \dfrac{6x}{x^2 + 2}$ has no vertical asymptote because $x^2 + 2 > 0$ for all x. $y = 0$ is a horizontal asymptote because the degree of the denominator is greater than that of the numerator.

25. $s(x) = \dfrac{6x^2 + 1}{2x^2 + x - 1}$ has vertical asymptotes where $2x^2 + x - 1 = 0 \iff (x + 1)(2x - 1) = 0 \iff x = -1$ or $x = \frac{1}{2}$, and horizontal asymptote $y = \frac{6}{2} = 3$.

27. $s(x) = \dfrac{(5x - 1)(x + 1)}{(3x - 1)(x + 2)}$ has vertical asymptotes $x = \frac{1}{3}$ and $x = -2$, and horizontal asymptote $y = \frac{5}{3}$.

29. $r(x) = \dfrac{6x^3 - 2}{2x^3 + 5x^2 + 6x} = \dfrac{6x^3 - 2}{x\left(2x^2 + 5x + 6\right)}$ has vertical asymptote $x = 0$ and horizontal asymptote $y = \frac{6}{2} = 3$.

31. $y = \dfrac{x^2 + 2}{x - 1}$. A vertical asymptote occurs when $x - 1 = 0 \iff x = 1$. There are no horizontal asymptotes because the degree of the numerator is greater than the degree of the denominator.

33. $r(x) = \dfrac{1}{x - 1} = f(x - 1)$. From this form we see that the graph of r is obtained

from the graph of f by shifting 1 unit to the right. Thus r has vertical asymptote
$x = 1$ and horizontal asymptote $y = 0$.

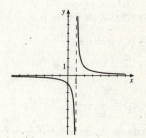

35. $s(x) = \dfrac{3}{x + 1} = 3\left(\dfrac{1}{x + 1}\right) = 3f(x + 1)$. From this form we see that the graph

of s is obtained from the graph of f by shifting 1 unit to the left and stretching
vertically by a factor of 3. Thus s has vertical asymptote $x = -1$ and horizontal
asymptote $y = 0$.

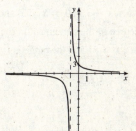

37. $t(x) = \dfrac{2x - 3}{x - 2} = 2 + \dfrac{1}{x - 2} = f(x - 2) + 2$ (see long

division below). From this form we see that the graph of t is
obtained from the graph of f by shifting 2 units to the right
and 2 units vertically. Thus t has vertical asymptote $x = 2$
and horizontal asymptote $y = 2$.

$$
\begin{array}{r}
2 \\
x - 2 \overline{\smash{\big)}\ 2x - 3} \\
\underline{2x - 2} \\
1
\end{array}
$$

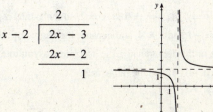

39. $r(x) = \dfrac{x+2}{x+3} = 1 - \dfrac{1}{x+3} = -f(x+3) + 1$ (see long

division below). From this form we see that the graph of r is
obtained from the graph of f by shifting 3 units to the left,
reflect about the x-axis, and then shifting vertically 1 unit.
Thus r has vertical asymptote $x = -3$ and horizontal
asymptote $y = 1$.

$$
\begin{array}{r}
1 \\
x+3 \overline{)\, x+2} \\
x+3 \\
\hline
-1
\end{array}
$$

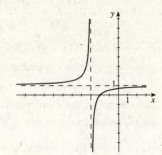

41. $y = \dfrac{4x-4}{x+2}$. When $x = 0$, $y = -2$, so the y-intercept is -2. When $y = 0$,

$4x - 4 = 0 \Leftrightarrow x = 1$, so the x-intercept is 1. Since the degree of the numerator and
denominator are the same the horizontal asymptote is $y = \frac{4}{1} = 4$. A vertical

asymptote occurs when $x = -2$. As $x \to -2^+$, $y = \dfrac{4x-4}{x+2} \to -\infty$, and as

$x \to -2^-$, $y = \dfrac{4x-4}{x+2} \to \infty$. The domain is $\{x \mid x \neq -2\}$ and the range is

$\{y \mid y \neq 4\}$.

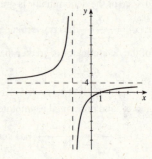

43. $s(x) = \dfrac{4-3x}{x+7}$. When $x = 0$, $y = \frac{4}{7}$, so the y-intercept is $\frac{4}{7}$. The x-intercepts

occur when $y = 0 \Leftrightarrow 4 - 3x = 0 \Leftrightarrow x = \frac{4}{3}$. A vertical asymptote occurs when
$x = -7$. Since the degree of the numerator and denominator are the same the

horizontal asymptote is $y = \dfrac{-3}{1} = -3$. The domain is $\{x \mid x \neq -7\}$ and the range

is $\{y \mid y \neq -3\}$.

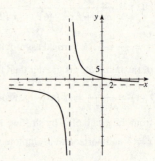

45. $r(x) = \dfrac{18}{(x-3)^2}$. When $x = 0$, $y = \frac{18}{9} = 2$, and so the y-intercept is 2. Since the

numerator can never be zero, there is no x-intercept. There is a vertical asymptote
when $x - 3 = 0 \Leftrightarrow x = 3$, and because the degree of the numerator is less than the
degree of the denominator, the horizontal asymptote is $y = 0$. The domain is
$\{x \mid x \neq 3\}$ and the range is $\{y \mid y > 0\}$.

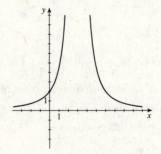

47. $s(x) = \dfrac{4x-8}{(x-4)(x+1)}$. When $x = 0$, $y = \dfrac{-8}{(-4)(1)} = 2$, so the y-intercept is 2.

When $y = 0$, $4x - 8 = 0 \Leftrightarrow x = 2$, so the x-intercept is 2. The vertical asymptotes
are $x = -1$ and $x = 4$, and because the degree of the numerator is less than the
degree of the denominator, the horizontal asymptote is $y = 0$. The domain is
$\{x \mid x \neq -1, 4\}$ and the range is $\mathbb{R}$.

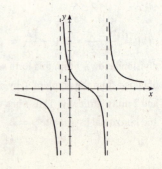

49. $s(x) = \dfrac{6}{x^2 - 5x - 6}$. When $x = 0$, $y = \dfrac{6}{-6} = -1$, so the y-intercept is -1.

Since the numerator is never zero, there is no x-intercept. The vertical asymptotes occur when $x^2 - 5x - 6 = (x + 1)(x - 6) \Leftrightarrow x = -1$ and $x = 6$, and because the degree of the numerator is less less than the degree of the denominator, the horizontal asymptote is $y = 0$. The domain is $\{x \mid x \neq -1, 6\}$ and the range is $\{y \mid y \leq -0.5 \text{ or } y > 0\}$.

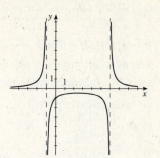

51. $t(x) = \dfrac{3x + 6}{x^2 + 2x - 8}$. When $x = 0$, $y = \dfrac{6}{-8} = -\dfrac{3}{4}$, so the y-intercept is $-\dfrac{3}{4}$.

When $y = 0$, $3x + 6 = 0 \Leftrightarrow x = -2$, so the x-intercept is -2. The vertical asymptotes occur when $x^2 + 2x - 8 = (x - 2)(x + 4) = 0 \Leftrightarrow x = 2$ and $x = -4$. Since the degree of the numerator is less than the degree of the denominator, the horizontal asymptote is $y = 0$. The domain is $\{x \mid x \neq -4, 2\}$ and the range is $\mathbb{R}$.

53. $r(x) = \dfrac{(x - 1)(x + 2)}{(x + 1)(x - 3)}$. When $x = 0$, $y = \dfrac{2}{3}$, so the y-intercept is $\dfrac{2}{3}$. When

$y = 0$, $(x - 1)(x + 2) = 0 \Rightarrow x = -2, 1$, so, the x-intercepts are -2 and 1. The vertical asymptotes are $x = -1$ and $x = 3$, and because the degree of the numerator and denominator are the same the horizontal asymptote is $y = \dfrac{1}{1} = 1$. The domain is $\{x \mid x \neq -1, 3\}$ and the range is $\mathbb{R}$.

55. $r(x) = \dfrac{x^2 - 2x + 1}{x^2 + 2x + 1} = \dfrac{(x - 1)^2}{(x + 1)^2} = \left(\dfrac{x - 1}{x + 1}\right)^2$. When $x = 0$, $y = 1$, so the

y-intercept is 1. When $y = 0$, $x = 1$, so the x-intercept is 1. A vertical asymptote occurs at $x + 1 = 0 \Leftrightarrow x = -1$. Because the degree of the numerator and denominator are the same the horizontal asymptote is $y = \dfrac{1}{1} = 1$. The domain is $\{x \mid x \neq -1\}$ and the range is $\{y \mid y \geq 0\}$.

intercept can be plus or minus!

57. $r(x) = \dfrac{2x^2 + 10x - 12}{x^2 + x - 6} = \dfrac{2(x - 1)(x + 6)}{(x - 2)(x + 3)}$. When $x = 0$, $y = \dfrac{2(-1)(6)}{(-2)(3)} = 2$,

so the y-intercept is 2. When $y = 0$, $2(x - 1)(x + 6) = 0 \Rightarrow x = -6, 1$, so the x-intercepts are -6 and 1. Vertical asymptotes occur when $(x - 2)(x + 3) = 0 \Leftrightarrow x = -3$ or $x = 2$. Because the degree of the numerator and denominator are the same the horizontal asymptote is $y = \dfrac{2}{1} = 2$. The domain is $\{x \mid x \neq -3, 2\}$ and the range is $\mathbb{R}$.

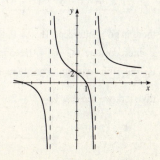

59. $y = \dfrac{x^2 - x - 6}{x^2 + 3x} = \dfrac{(x-3)(x+2)}{x(x+3)}$. The x-intercept occurs when $y = 0 \Leftrightarrow$

$(x-3)(x+2) = 0 \Rightarrow x = -2, 3$, so the x-intercepts are -2 and 3. There is no

y-intercept because y is undefined when $x = 0$. The vertical asymptotes are $x = 0$

and $x = -3$. Because the degree of the numerator and denominator are the same,

the horizontal asymptotes is $y = \frac{1}{1} = 1$. The domain is $\{x \mid x \neq -3, 0\}$ and the

range is $\mathbb{R}$.

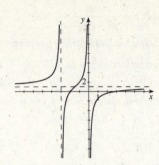

61. $r(x) = \dfrac{3x^2 + 6}{x^2 - 2x - 3} = \dfrac{3\left(x^2 + 2\right)}{(x-3)(x+1)}$. When $x = 0$, $y = -2$, so the y-intercept

is -2. Since the numerator can never equal zero, there is no x-intercept. Vertical

asymptotes occur when $x = -1, 3$. Because the degree of the numerator and

denominator are the same, the horizontal asymptote is. $y = \frac{3}{1} = 3$. The domain is

$\{x \mid x \neq -1, 3\}$ and the range is $\{y \mid y \leq -1.5 \text{ or } y \geq 2.4\}$.

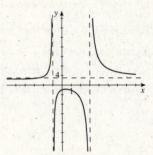

63. $s(x) = \dfrac{x^2 - 2x + 1}{x^3 - 3x^2} = \dfrac{(x-1)^2}{x^2(x-3)}$. Since $x = 0$ is not in the domain of $s(x)$,

there is no y-intercept. The x-intercept occurs when $y = 0 \Leftrightarrow$

$x^2 - 2x + 1 = (x-1)^2 = 0 \Rightarrow x = 1$, so the x-intercept is 1. Vertical asymptotes

occur when $x = 0, 3$. Since the degree of the numerator is less than the degree of

the denominator, the horizontal asymptote is $y = 0$. The domain is $\{x \mid x \neq 0, 3\}$

and the range is $\mathbb{R}$.

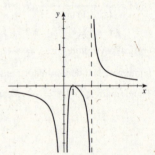

65. $r(x) = \dfrac{x^2}{x - 2}$. When $x = 0$, $y = 0$, so the graph passes through the origin. There

is a vertical asymptote when $x - 2 = 0 \Leftrightarrow x = 2$, with $y \to \infty$ as $x \to 2^+$, and

$y \to -\infty$ as $x \to 2^-$. Because the degree of the numerator is greater than the

degree of the denominator, there is no horizontal asymptotes. By using long

division, we see that $y = x + 2 + \dfrac{4}{x - 2}$, so $y = x + 2$ is a slant asymptote.

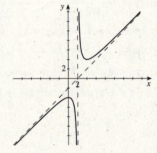

67. $r(x) = \dfrac{x^2 - 2x - 8}{x} = \dfrac{(x-4)(x+2)}{x}$. The vertical asymptote is $x = 0$, thus,

there is no y-intercept. If $y = 0$, then $(x-4)(x+2) = 0 \Rightarrow x = -2, 4$, so the

x-intercepts are -2 and 4. Because the degree of the numerator is greater than the

degree of the denominator, there are no horizontal asymptotes. By using long

division, we see that $y = x - 2 - \dfrac{8}{x}$, so $y = x - 2$ is a slant asymptote.

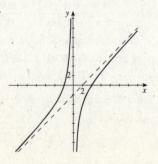

69. $r(x) = \dfrac{x^2 + 5x + 4}{x - 3} = \dfrac{(x+4)(x+1)}{x-3}$. When $x = 0$, $y = -\frac{4}{3}$, so the y-intercept

is $-\frac{4}{3}$. When $y = 0$, $(x+4)(x+1) = 0 \Leftrightarrow x = -4, -1$, so the two x-intercepts

are -4 and -1. A vertical asymptote occurs when $x = 3$, with $y \to \infty$ as

$x \to 3^+$, and $y \to -\infty$ as $x \to 3^-$. Using long division, we see that

$y = x + 8 + \dfrac{28}{x - 3}$, so $y = x + 8$ is a slant asymptote.

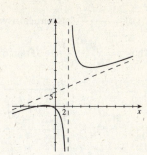

71. $r(x) = \dfrac{x^3 + x^2}{x^2 - 4} = \dfrac{x^2(x+1)}{(x-2)(x+2)}$. When $x = 0$, $y = 0$, so the graph passes

through the origin. Moreover, when $y = 0$, we have $x^2(x+1) = 0 \Rightarrow x = 0, -1$,

so the x-intercepts are 0 and -1. Vertical asymptotes occur when $x = \pm 2$; as

$x \to \pm 2^-$, $y = -\infty$ and as $x \to \pm 2^+$, $y \to \infty$. Because the degree of the

numerator is greater than the degree of the denominator, there is no horizontal

asymptote. Using long division, we see that $y = x + 1 + \dfrac{4x + 4}{x^2 - 4}$, so $y = x + 1$ is a

slant asymptote.

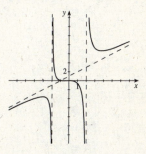

73. $f(x) = \dfrac{2x^2 + 6x + 6}{x + 3}$, $g(x) = 2x$. f has vertical asymptote $x = -3$.

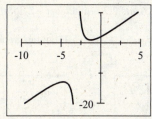
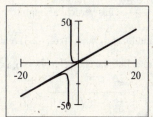

75. $f(x) = \dfrac{x^3 - 2x^2 + 16}{x - 2}$, $g(x) = x^2$. f has vertical asymptote $x = 2$.

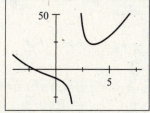
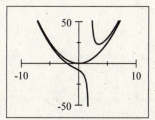

77. $f(x) = \dfrac{2x^2 - 5x}{2x + 3}$ has vertical asymptote $x = -1.5$, x-intercepts 0 and 2.5, y-intercept 0, local maximum $(-3.9, -10.4)$,

and local minimum $(0.9, -0.6)$. Using long division, we get $f(x) = x - 4 + \dfrac{12}{2x + 3}$. From the graph, we see that the end

behavior of $f(x)$ is like the end behavior of $g(x) = x - 4$.

$$
\begin{array}{r}
x - 4 \\
2x + 3 \overline{\smash{\big)}\ 2x^2 - 5x } \\
\underline{2x^2 + 3x } \\
-8x \\
\underline{-8x - 12} \\
12
\end{array}
$$

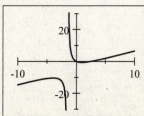
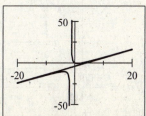

79. $f(x) = \dfrac{x^5}{x^3 - 1}$ has vertical asymptote $x = 1$, x-intercept 0, y-intercept 0, and local minimum $(1.4, 3.1)$.

Thus $y = x^2 + \dfrac{x^2}{x^3 - 1}$. From the graph we see that the end behavior of $f(x)$ is like the end behavior of $g(x) = x^2$.

$$\begin{array}{r|l} & x^2 \\ \hline x^3 - 1 & x^5 \\ & x^5 - x^2 \\ \hline & \phantom{x^5 -{}} x^2 \end{array}$$

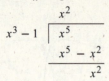

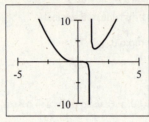

Graph of f

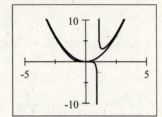

Graph of f and g

81. $f(x) = \dfrac{x^4 - 3x^3 + 6}{x - 3}$ has vertical asymptote $x = 3$, x-intercepts 1.6 and 2.7, y-intercept -2, local maxima $(-0.4, -1.8)$ and $(2.4, 3.8)$, and local minima $(0.6, -2.3)$ and $(3.4, 54.3)$. Thus $y = x^3 + \dfrac{6}{x - 3}$. From the graphs, we see that the end behavior of $f(x)$ is like the end behavior of $g(x) = x^3$.

$$\begin{array}{r|l} & x^3 \\ \hline x - 3 & x^4 - 3x^3 + 6 \\ & x^4 - 3x^3 \\ \hline & \phantom{x^4 - 3x^3 +{}} 6 \end{array}$$

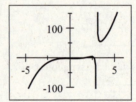

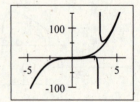

83. (a)

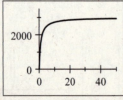

(b) $p(t) = \dfrac{3000t}{t + 1} = 3000 - \dfrac{3000}{t + 1}$. So as $t \to \infty$, we have $p(t) \to 3000$.

85. $c(t) = \dfrac{5t}{t^2 + 1}$

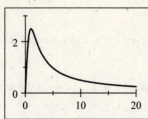

(a) The highest concentration of drug is 2.50 mg/L, and it is reached 1 hour after the drug is administered.

(b) The concentration of the drug in the bloodstream goes to 0.

(c) From the first viewing rectangle, we see that an approximate solution is near $t = 15$. Thus we graph $y = \dfrac{5t}{t^2 + 1}$ and $y = 0.3$ in the viewing rectangle [14, 18] by [0, 0.5]. So it takes about 16.61 hours for the concentration to drop below 0.3 mg/L.

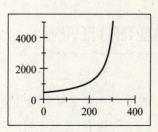

87. $P(v) = P_0 \left(\dfrac{s_0}{s_0 - v} \right) \Rightarrow P(v) = 440 \left(\dfrac{332}{332 - v} \right)$

If the speed of the train approaches the speed of sound, the pitch of the whistle becomes very loud. This would be experienced as a "sonic boom"— an effect seldom heard with trains.

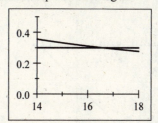

89. Vertical asymptote $x = 3$: $p(x) = \dfrac{1}{x - 3}$. Vertical asymptote $x = 3$ and horizontal asymptote $y = 2$: $r(x) = \dfrac{2x}{x - 3}$.

Vertical asymptotes $x = 1$ and $x = -1$, horizontal asymptote 0, and x-intercept 4: $q(x) = \dfrac{x - 4}{(x - 1)(x + 1)}$. Of course, other answers are possible.

91. (a) $r(x) = \dfrac{3x^2 - 3x - 6}{x - 2} = \dfrac{3(x - 2)(x + 1)}{x - 2} = 3(x + 1)$, for $x \neq 2$. Therefore,

$r(x) = 3x + 3$, $x \neq 2$. Since $3(2) + 3 = 9$, the graph is the line $y = 3x + 3$ with the point $(2, 9)$ removed.

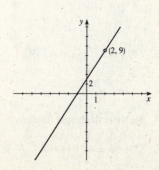

(b) $s(x) = \dfrac{x^2 + x - 20}{x + 5} = \dfrac{(x - 4)(x + 5)}{x + 5} = x - 4$, for $x \neq -5$. Therefore, $s(x) = x - 4$, $x \neq -5$. Since $(-5) - 4 = -9$, the graph is the line $y = x - 4$ with the point $(-5, -9)$ removed.

$t(x) = \dfrac{2x^2 - x - 1}{x - 1} = \dfrac{(2x + 1)(x - 1)}{x - 1} = 2x + 1$, for $x \neq 1$. Therefore, $t(x) = 2x + 1$, $x \neq 1$. Since $2(1) + 1 = 3$, the graph is the line $y = 2x + 1$ with the point $(1, 3)$ removed.

$u(x) = \dfrac{x - 2}{x^2 - 2x} = \dfrac{x - 2}{x(x - 2)} = \dfrac{1}{x}$, for $x \neq 2$. Therefore, $u(x) = \dfrac{1}{x}$, $x \neq 2$. When $x = 2$, $\dfrac{1}{x} = \dfrac{1}{2}$, so the graph is the curve $y = \dfrac{1}{x}$ with the point $\left(2, \frac{1}{2}\right)$ removed.

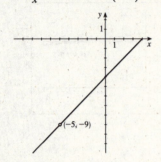
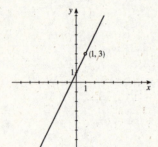
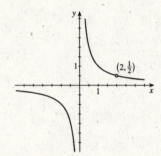

CHAPTER 3 REVIEW

1. (a) $f(x) = x^2 + 4x + 1 = \left(x^2 + 4x\right) + 1$
$= \left(x^2 + 4x + 4\right) + 1 - 4$
$= (x + 2)^2 - 3$

(b)

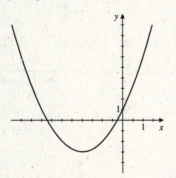

3. (a) $f(x) = 1 + 8x - x^2 = -\left(x^2 - 8x\right) + 1$
$= -\left(x^2 - 8x + 16\right) + 1 + 16$
$= -(x - 4)^2 + 17$

(b)

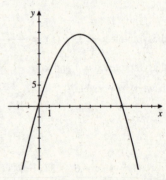

5. $f(x) = 2x^2 + 4x - 5 = 2\left(x^2 + 2x\right) - 5 = 2\left(x^2 + 2x + 1\right) - 5 - 2 = 2(x + 1)^2 - 7$ has the minimum value -7 when $x = -1$.

7. We write the height function in standard form: $h(t) = -16t^2 + 48t + 32 = -16\left(t^2 - 3t\right) + 32 = -16\left(t^2 - 3t + \frac{9}{4}\right) + 32 + 36 = -16\left(t - \frac{3}{2}\right)^2 + 68$. The stone reaches a maximum height of 68 ft.

9. $P(x) = -x^3 + 64$

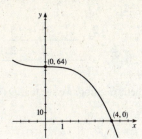

11. $P(x) = 2(x+1)^4 - 32$

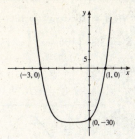

13. $P(x) = 32 + (x-1)^5$

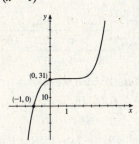

15. (a) $P(x) = x^3(x-2)^2$. The zeros of P are 0 and 2, with multiplicities 3 and 2, respectively.

(b) We sketch the graph using the guidelines on page 283.

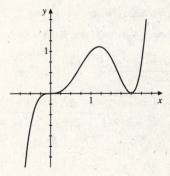

17. $P(x) = x^3 - 4x + 1$. x-intercepts: -2.1, 0.3, and 1.9. y-intercept: 1. Local maximum is $(-1.2, 4.1)$. Local minimum is $(1.2, -2.1)$. $y \to \infty$ as $x \to \infty$; $y \to -\infty$ as $x \to -\infty$.

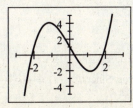

19. $P(x) = 3x^4 - 4x^3 - 10x - 1$. x-intercepts; -0.1 and 2.1. y-intercept: -1. Local maximum is $(1.4, -14.5)$. There is no local maximum. $y \to \infty$ as $x \to \pm\infty$.

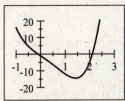

21. (a) Use the Pythagorean Theorem and solving for y^2 we have, $x^2 + y^2 = 10^2 \Leftrightarrow y^2 = 100 - x^2$. Substituting we get $S = 13.8x \left(100 - x^2\right) = 1380x - 13.8x^3$.

(b) Domain is $[0, 10]$.

(c)

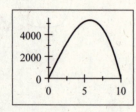

(d) The strongest beam has width 5.8 inches.

25. $\dfrac{x^3 - x^2 + 11x + 2}{x - 4}$

$$
\begin{array}{r|rrrr}
4 & 1 & -1 & 11 & 2 \\
  &   & 4 & 12 & 92 \\
\hline
  & 1 & 3 & 23 & 94
\end{array}
$$

Using synthetic division, we see that $Q(x) = x^2 + 3x + 23$ and $R(x) = 94$.

29. $\dfrac{2x^3 + x^2 - 8x + 15}{x^2 + 2x - 1}$

$$
\begin{array}{r}
2x - 3 \\
x^2 + 2x - 1 \overline{\smash{\big)}\ 2x^3 + x^2 - 8x + 15} \\
\underline{2x^3 + 4x^2 - 2x} \\
-3x^2 - 6x + 15 \\
\underline{-3x^2 - 6x + 3} \\
12
\end{array}
$$

Therefore, $Q(x) = 2x - 3$, and $R(x) = 12$.

33. $\frac{1}{2}$ is a zero of $P(x) = 2x^4 + x^3 - 5x^2 + 10x - 4$ if $P\left(\frac{1}{2}\right) = 0$.

$$
\begin{array}{r|rrrrr}
\frac{1}{2} & 2 & 1 & -5 & 10 & -4 \\
            &   & 1 & 1 & -2 & 4 \\
\hline
            & 2 & 2 & -4 & 8 & 0
\end{array}
$$

Since $P\left(\frac{1}{2}\right) = 0$, $\frac{1}{2}$ is a zero of the polynomial.

23. $\dfrac{x^2 - 3x + 5}{x - 2}$

$$
\begin{array}{r|rrr}
2 & 1 & -3 & 5 \\
  &   & 2 & -2 \\
\hline
  & 1 & -1 & 3
\end{array}
$$

Using synthetic division, we see that $Q(x) = x - 1$ and $R(x) = 3$.

27. $\dfrac{x^4 - 8x^2 + 2x + 7}{x + 5}$

$$
\begin{array}{r|rrrrr}
-5 & 1 & 0 & -8 & 2   & 7 \\
   &   & -5 & 25 & -85 & 415 \\
\hline
   & 1 & -5 & 17 & -83 & 422
\end{array}
$$

Using synthetic division, we see that $Q(x) = x^3 - 5x^2 + 17x - 83$ and $R(x) = 422$.

31. $P(x) = 2x^3 - 9x^2 - 7x + 13$; find $P(5)$.

$$
\begin{array}{r|rrrr}
5 & 2 & -9 & -7 & 13 \\
  &   & 10 & 5 & -10 \\
\hline
  & 2 & 1 & -2 & 3
\end{array}
$$

Therefore, $P(5) = 3$.

35. $P(x) = x^{500} + 6x^{201} - x^2 - 2x + 4$. The remainder from dividing $P(x)$ by $x - 1$ is

$$P(1) = (1)^{500} + 6(1)^{201} - (1)^2 - 2(1) + 4 = 8.$$

37. (a) $P(x) = x^5 - 6x^3 - x^2 + 2x + 18$ has possible rational zeros $\pm 1, \pm 2, \pm 3, \pm 6, \pm 9, \pm 18$.

(b) Since $P(x)$ has 2 variations in sign, there are either 0 or 2 positive real zeros. Since $P(-x) = -x^5 + 6x^3 - x^2 - 2x + 18$ has 3 variations in sign, there are 1 or 3 negative real zeros.

39. (a) $P(x) = x^3 - 16x = x\left(x^2 - 16\right)$

$= x(x-4)(x+4)$

has zeros $-4, 0, 4$ (all of multiplicity 1).

(b)

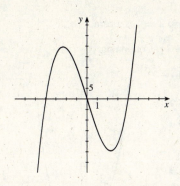

41. (a) $P(x) = x^4 + x^3 - 2x^2 = x^2\left(x^2 + x - 2\right)$

$= x^2(x+2)(x-1)$

The zeros are 0 (multiplicity 2), -2 (multiplicity 1), and 1 (multiplicity 1)

(b)

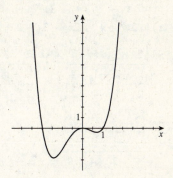

43. (a) $P(x) = x^4 - 2x^3 - 7x^2 + 8x + 12$. The possible rational zeros are $\pm1, \pm2, \pm3, \pm4, \pm6, \pm12$. P has 2 variations in sign, so it has either 2 or 0 positive real zeros.

$$
\begin{array}{r|rrrrr}
1 & 1 & -2 & -7 & 8 & 12 \\
  &   & 1 & -1 & -8 & 0 \\
\hline
  & 1 & -1 & -8 & 0 & 12
\end{array}
\qquad
\begin{array}{r|rrrrr}
2 & 1 & -2 & -7 & 8 & 12 \\
  &   & 2 & 0 & -14 & -12 \\
\hline
  & 2 & 0 & -7 & -6 & 0
\end{array}
\Rightarrow x = 2 \text{ is a root.}
$$

$P(x) = x^4 - 2x^3 - 7x^2 + 8x + 12 = (x-2)\left(x^3 - 7x - 6\right)$. Continuing:

$$
\begin{array}{r|rrrr}
2 & 1 & 0 & -6 & -6 \\
  &   & 2 & 4 & -4 \\
\hline
  & 1 & 2 & -2 & -10
\end{array}
\qquad
\begin{array}{r|rrrr}
3 & 1 & 0 & -7 & -6 \\
  &   & 3 & 9 & 6 \\
\hline
  & 1 & 3 & 2 & 0
\end{array}
$$

(b)

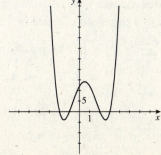

so $x = 3$ is a root and

$P(x) = (x-2)(x-3)\left(x^2 + 3x + 2\right)$

$= (x-2)(x-3)(x+1)(x+2)$

Therefore the real roots are $-2, -1, 2,$ and 3 (all of multiplicity 1).

45. **(a)** $P(x) = 2x^4 + x^3 + 2x^2 - 3x - 2$. The possible rational roots are ± 1, ± 2, $\pm \frac{1}{2}$. P has one variation in sign, and hence 1 positive real root. $P(-x)$ has 3 variations in sign and hence either 3 or 1 negative real roots.

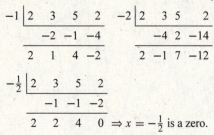

$$
\begin{array}{r|rrrrr}
1 & 2 & 1 & 2 & -3 & -2 \\
  &   & 2 & 3 & 5 & 2 \\
\hline
  & 2 & 3 & 5 & 2 & 0
\end{array} \quad \Rightarrow x = 1 \text{ is a zero.}
$$

$P(x) = 2x^4 + x^3 + 2x^2 - 3x - 2 = (x - 1)\left(2x^3 + 3x^2 + 5x + 2\right)$. Continuing:

$$
\begin{array}{r|rrrr}
-1 & 2 & 3 & 5 & 2 \\
   &   & -2 & -1 & -4 \\
\hline
   & 2 & 1 & 4 & -2
\end{array}
\qquad
\begin{array}{r|rrrr}
-2 & 2 & 3 & 5 & 2 \\
   &   & -4 & 2 & -14 \\
\hline
   & 2 & -1 & 7 & -12
\end{array}
$$

(b)

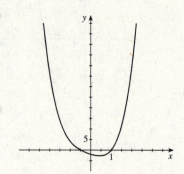

$$
\begin{array}{r|rrrr}
-\frac{1}{2} & 2 & 3 & 5 & 2 \\
   &   & -1 & -1 & -2 \\
\hline
   & 2 & 2 & 4 & 0
\end{array} \quad \Rightarrow x = -\frac{1}{2} \text{ is a zero.}
$$

$P(x) = (x - 1)\left(x + \frac{1}{2}\right)\left(2x^2 + 2x + 4\right)$. The quadratic is irreducible, so the real zeros are 1 and $-\frac{1}{2}$ (each of multiplicity 1).

47. $(2 - 3i) + (1 + 4i) = (2 + 1) + (-3 + 4)i = 3 + i$

49. $(2 + i)(3 - 2i) = 6 - 4i + 3i - 2i^2 = 6 - i + 2 = 8 - i$

51. $\dfrac{4 + 2i}{2 - i} = \dfrac{4 + 2i}{2 - i} \cdot \dfrac{2 + i}{2 + i} = \dfrac{8 + 8i + 2i^2}{4 - i^2} = \dfrac{8 + 8i - 2}{4 + 1} = \dfrac{6 + 8i}{5} = \frac{6}{5} + \frac{8}{5}i$

53. $i^{25} = i^{24}i = \left(i^4\right)^6 i = (1)^6 i = i$

55. $\left(1 - \sqrt{-1}\right)\left(1 + \sqrt{-1}\right) = (1 - i)(1 + i) = 1 + i - i - i^2 = 1 + 1 = 2$

57. Since the zeros are $-\frac{1}{2}$, 2, and 3, a factorization is

$$
\begin{aligned}
P(x) &= C\left(x + \tfrac{1}{2}\right)(x - 2)(x - 3) = \tfrac{1}{2}C(2x + 1)\left(x^2 - 5x + 6\right) \\
&= \tfrac{1}{2}C\left(2x^3 - 10x^2 + 12x + x^2 - 5x + 6\right) = \tfrac{1}{2}C\left(2x^3 - 9x^2 + 7x + 6\right)
\end{aligned}
$$

Since the constant coefficient is 12, $\frac{1}{2}C(6) = 12 \Leftrightarrow C = 4$, and so the polynomial is $P(x) = 4x^3 - 18x^2 + 14x + 12$.

59. No, there is no polynomial of degree 4 with integer coefficients that has zeros i, $2i$, $3i$ and $4i$. Since the imaginary zeros of polynomial equations with real coefficients come in complex conjugate pairs, there would have to be 8 zeros, which is impossible for a polynomial of degree 4.

61. $P(x) = x^3 - 3x^2 - 13x + 15$ has possible rational zeros ± 1, ± 3, ± 5, ± 15.

$$
\begin{array}{r|rrrr}
1 & 1 & -3 & -13 & 15 \\
  &   & 1 & -2 & -15 \\
\hline
  & 1 & -2 & -15 & 0
\end{array} \quad \Rightarrow x = 1 \text{ is a zero.}
$$

So $P(x) = x^3 - 3x^2 - 13x + 15 = (x - 1)\left(x^2 - 2x - 15\right) = (x - 1)(x - 5)(x + 3)$. Therefore, the zeros are -3, 1, and 5.

63. $P(x) = x^4 + 6x^3 + 17x^2 + 28x + 20$ has possible rational zeros $\pm 1, \pm 2, \pm 4, \pm 5, \pm 10, \pm 20$. Since all of the coefficients are positive, there are no positive real zeros.

$$
\begin{array}{r|rrrrr}
-1 & 1 & 6 & 17 & 28 & 20 \\
   &   & -1 & -5 & -12 & -16 \\
\hline
   & 1 & 5 & 12 & 16 & 4
\end{array}
\qquad
\begin{array}{r|rrrrr}
-2 & 1 & 6 & 17 & 28 & 20 \\
   &   & -2 & -8 & -18 & -20 \\
\hline
   & 1 & 4 & 9 & 10 & 0
\end{array}
\Rightarrow x = -2 \text{ is a zero.}
$$

$P(x) = x^4 + 6x^3 + 17x^2 + 28x + 20 = (x+2)\left(x^3 + 4x^2 + 9x + 10\right)$. Continuing with the quotient, we have

$$
\begin{array}{r|rrrr}
-2 & 1 & 4 & 9 & 10 \\
   &   & -2 & -4 & -10 \\
\hline
   & 1 & 2 & 5 & 0
\end{array}
\Rightarrow x = -2 \text{ is a zero.}
$$

Thus $P(x) = x^4 + 6x^3 + 17x^2 + 28x + 20 = (x+2)^2\left(x^2 + 2x + 5\right)$. Now $x^2 + 2x + 5 = 0$ when $x = \frac{-2 \pm \sqrt{4-4(5)(1)}}{2} = \frac{-2 \pm 4i}{2} = -1 \pm 2i$. Thus, the zeros are -2 (multiplicity 2) and $-1 \pm 2i$.

65. $P(x) = x^5 - 3x^4 - x^3 + 11x^2 - 12x + 4$ has possible rational zeros $\pm 1, \pm 2, \pm 4$.

$$
\begin{array}{r|rrrrrr}
1 & 1 & -3 & -1 & 11 & -12 & 4 \\
  &   & 1 & -2 & -3 & 8 & -4 \\
\hline
  & 1 & -2 & -3 & 8 & -4 & 0
\end{array}
\Rightarrow x = 1 \text{ is a zero.}
$$

$P(x) = x^5 - 3x^4 - x^3 + 11x^2 - 12x + 4 = (x-1)\left(x^4 - 2x^3 - 3x^2 + 8x - 4\right)$. Continuing with the quotient, we have

$$
\begin{array}{r|rrrrr}
1 & 1 & -2 & -3 & 8 & -4 \\
  &   & 1 & -1 & -4 & 4 \\
\hline
  & 1 & -1 & -4 & 4 & 0
\end{array}
\Rightarrow x = 1 \text{ is a zero.}
$$

$$
\begin{aligned}
x^5 - 3x^4 - x^3 + 11x^2 - 12x + 4 &= (x-1)^2\left(x^3 - x^2 - 4x + 4\right) = (x-1)^3\left(x^2 - 4\right) \\
&= (x-1)^3(x-2)(x+2)
\end{aligned}
$$

Therefore, the zeros are 1 (multiplicity 3), -2, and 2.

67. $P(x) = x^6 - 64 = \left(x^3 - 8\right)\left(x^3 + 8\right) = (x-2)\left(x^2 + 2x + 4\right)(x+2)\left(x^2 - 2x + 4\right)$. Now using the quadratic formula to find the zeros of $x^2 + 2x + 4$, we have $x = \frac{-2 \pm \sqrt{4-4(4)(1)}}{2} = \frac{-2 \pm 2i\sqrt{3}}{2} = -1 \pm i\sqrt{3}$, and using the quadratic formula to find the zeros of $x^2 - 2x + 4$, we have $x = \frac{2 \pm \sqrt{4-4(4)(1)}}{2} = \frac{2 \pm 2i\sqrt{3}}{2} = 1 \pm i\sqrt{3}$. Therefore, the zeros are $2, -2, 1 \pm i\sqrt{3}$, and $-1 \pm i\sqrt{3}$.

69. $P(x) = 6x^4 - 18x^3 + 6x^2 - 30x + 36 = 6\left(x^4 - 3x^3 + x^2 - 5x + 6\right)$ has possible rational zeros $\pm 1, \pm 2, \pm 3, \pm 6$.

$$
\begin{array}{r|rrrrr}
1 & 6 & -18 & 6 & -30 & 36 \\
  &   & 6 & -12 & -6 & -36 \\
\hline
  & 6 & -12 & -6 & -36 & 0 \\
\end{array}
\Rightarrow x = 1 \text{ is a zero.}
$$

So $P(x) = 6x^4 - 18x^3 + 6x^2 - 30x + 36 = (x-1)\left(6x^3 - 12x^2 - 6x - 36\right) = 6(x-1)\left(x^3 - 2x^2 - x - 6\right)$.

Continuing with the quotient we have

$$
\begin{array}{r|rrrr}
1 & 1 & -2 & -1 & -6 \\
  &   & 1 & -1 & -2 \\
\hline
  & 1 & -1 & -2 & -8 \\
\end{array}
\qquad
\begin{array}{r|rrrr}
2 & 1 & -2 & -1 & -6 \\
  &   & 2 & 0 & -2 \\
\hline
  & 1 & 0 & -1 & -8 \\
\end{array}
\qquad
\begin{array}{r|rrrr}
3 & 1 & -2 & -1 & -6 \\
  &   & 3 & 3 & 6 \\
\hline
  & 1 & 1 & 2 & 0 \\
\end{array}
\Rightarrow x = 3 \text{ is a zero.}
$$

So $P(x) = 6x^4 - 18x^3 + 6x^2 - 30x + 36 = 6(x-1)(x-3)\left(x^2 + x + 2\right)$. Now $x^2 + x + 2 = 0$ when

$x = \dfrac{-1 \pm \sqrt{1 - 4(1)(2)}}{2} = \dfrac{-1 \pm i\sqrt{7}}{2}$, and so the zeros are 1, 3, and $\dfrac{-1 \pm i\sqrt{7}}{2}$.

71. $2x^2 = 5x + 3 \Leftrightarrow 2x^2 - 5x - 3 = 0$. The solutions are $x = -0.5, 3$.

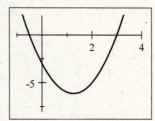

73. $x^4 - 3x^3 - 3x^2 - 9x - 2 = 0$ has solutions $x \approx -0.24$, 4.24.

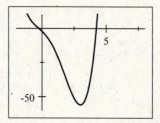

75. $P(x) = x^3 - 2x - 4$

$$
\begin{array}{r|rrrr}
1 & 1 & 0 & -2 & -4 \\
  &   & 1 & 1 & -1 \\
\hline
  & 1 & 1 & -1 & -5 \\
\end{array}
\qquad
\begin{array}{r|rrrr}
2 & 1 & 0 & -2 & -4 \\
  &   & 2 & 4 & 4 \\
\hline
  & 1 & 2 & 2 & 0 \\
\end{array}
$$

$P(x) = x^3 - 2x - 4 = (x-2)\left(x^2 + 2x + 2\right)$. Since $x^2 + 2x + 2 = 0$ has no real solution, the only real zero of P is $x = 2$.

77. $r(x) = \dfrac{3x - 12}{x + 1}$. When $x = 0$, we have $r(0) = \dfrac{-12}{1} = -12$, so the y-intercept is

-12. Since $y = 0$, when $3x - 12 = 0 \Leftrightarrow x = 4$, the x-intercept is 4. The vertical

asymptote is $x = -1$. Because the degree of the denominator and numerator are

the same, the horizontal asymptote is $y = \dfrac{3}{1} = 3$.

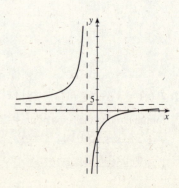

79. $r(x) = \dfrac{x-2}{x^2 - 2x - 8} = \dfrac{x-2}{(x+2)(x-4)}$. When $x = 0$, we have $r(0) = \dfrac{-2}{-8} = \dfrac{1}{4}$,

so the y-intercept is $\dfrac{1}{4}$. When $y = 0$, we have $x - 2 = 0 \Leftrightarrow x = 2$, so the

x-intercept is 2. There are vertical asymptotes at $x = -2$ and $x = 4$.

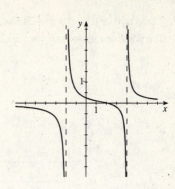

81. $r(x) = \dfrac{x^2 - 9}{2x^2 + 1} = \dfrac{(x+3)(x-3)}{2x^2 + 1}$. When $x = 0$, we have $r(0) = \dfrac{-9}{1}$, so the

y-intercept is -9. When $y = 0$, we have $x^2 - 9 = 0 \Leftrightarrow x = \pm 3$ so the x-intercepts

are -3 and 3. Since $2x^2 + 1 > 0$, the denominator is never zero so there are no

vertical asymptotes. The horizontal asymptote is at $y = \dfrac{1}{2}$ because the degree of

the denominator and numerator are the same.

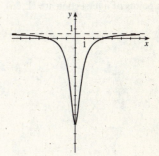

83. $r(x) = \dfrac{x-3}{2x+6}$. From the graph we see that the

x-intercept is 3, the y-intercept is -0.5, there is a vertical

asymptote at $x = -3$ and a horizontal asymptote at

$y = 0.5$, and there is no local extremum.

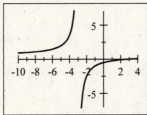

85. $r(x) = \dfrac{x^3 + 8}{x^2 - x - 2}$. From the graph we see that the x-intercept is -2, the

y-intercept is -4, there are vertical asymptotes at $x = -1$ and $x = 2$, there is no

horizontal asymptote, the local maximum is $(0.425, -3.599)$, and the local

minimum is $(4.216, 7.175)$. By using long division, we see that

$f(x) = x + 1 + \dfrac{10 - x}{x^2 - x - 2}$, so f has a slant asymptote of $y = x + 1$.

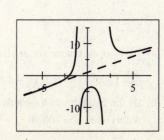

87. The graphs of $y = x^4 + x^2 + 24x$ and $y = 6x^3 + 20$ intersect when $x^4 + x^2 + 24x = 6x^3 + 20 \Leftrightarrow x^4 - 6x^3 + x^2 + 24x - 20 = 0$. The possible rational zeros are $\pm 1, \pm 2, \pm 4, \pm 5, \pm 10, \pm 20$.

$$
\begin{array}{r|rrrrr}
1 & 1 & -6 & 1 & 24 & -20 \\
  &   & 1 & -5 & -4 & 20 \\
\hline
  & 1 & -5 & -4 & 20 & 0 \\
\end{array} \Rightarrow x = 1 \text{ is a zero.}
$$

So $x^4 - 6x^3 + x^2 + 24x - 20 = (x - 1)\left(x^3 - 5x^2 - 4x + 20\right) = 0$. Continuing with the quotient:

$$
\begin{array}{r|rrrr}
1 & 1 & -5 & -4 & 20 \\
  &   & 1 & -4 & -8 \\
\hline
  & 1 & -4 & -8 & 12 \\
\end{array}
\qquad
\begin{array}{r|rrrr}
2 & 1 & -5 & -4 & 20 \\
  &   & 2 & -6 & -20 \\
\hline
  & 1 & -3 & -10 & 0 \\
\end{array} \Rightarrow x = 2 \text{ is a zero.}
$$

So $x^4 - 6x^3 + x^2 + 24x - 20 = (x - 1)(x - 2)\left(x^2 - 3x - 10\right) = (x - 1)(x - 2)(x - 5)(x + 2) = 0$. Hence, the points of intersection are $(1, 26)$, $(2, 68)$, $(5, 770)$, and $(-2, -28)$.

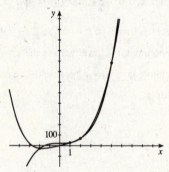

CHAPTER 3 TEST

1. $f(x) = x^2 - x - 6$

$\qquad = \left(x^2 - x\right) - 6$

$\qquad = \left(x^2 - x + \frac{1}{4}\right) - 6 - \frac{1}{4}$

$\qquad = \left(x - \frac{1}{2}\right)^2 - \frac{25}{4}$

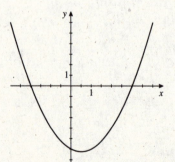

3. (a) We write the function in standard form: $h(x) = 10x - 0.01x^2 = -0.01\left(x^2 - 1000x\right) = -0.01\left(x^2 - 1000x + 500^2\right) +$

$\qquad 0.01\left(500^2\right) = -0.01(x - 500)^2 + 2500$. Thus, the maximum height reached by the cannonball is 2500 feet.

(b) By the symmetry of the parabola, we see that the cannonball's height will be 0 again (and thus it will splash into the water) when $x = 1000$ ft.

5. (a)

$$\begin{array}{r|rrrrr} 2 & 1 & 0 & -4 & 2 & 5 \\ & & 2 & 4 & 0 & 4 \\ \hline & 1 & 2 & 0 & 2 & 9 \end{array}$$

Therefore, the quotient is

$Q(x) = x^3 + 2x^2 + 2$, and the remainder is

$R(x) = 9$.

(b)

$$\begin{array}{r} x^3 + 2x^2 \qquad\qquad + \tfrac{1}{2} \\ 2x^2 - 1 \overline{\smash{\big)}\, 2x^5 + 4x^4 - x^3 - x^2 + 0x + 7} \\ \underline{2x^5 \qquad\quad - x^3} \\ 4x^4 \qquad\quad - x^2 \\ \underline{4x^4 \qquad\quad - 2x^2} \\ x^2 \qquad + 7 \\ \underline{x^2 \qquad - \tfrac{1}{2}} \\ \tfrac{15}{2} \end{array}$$

Therefore, the quotient is $Q(x) = x^3 + 2x^2 + \tfrac{1}{2}$ and the

remainder is $R(x) = \tfrac{15}{2}$.

7. (a) $(3 - 2i) + (4 + 3i) = (3 + 4) + (-2 + 3)i = 7 + i$

(b) $(3 - 2i) - (4 + 3i) = (3 - 4) + (-2 - 3)i = -1 - 5i$

(c) $(3 - 2i)(4 + 3i) = 12 + 9i - 8i - 6i^2 = 12 + i - 6(-1) = 18 + i$

(d) $\dfrac{3 - 2i}{4 + 3i} = \dfrac{(3 - 2i)(4 - 3i)}{(4 + 3i)(4 - 3i)} = \dfrac{12 - 17i - 6i^2}{16 - 9i^2} = \dfrac{6}{25} - \dfrac{17}{25}i$

(e) $i^{48} = \left(i^2\right)^{24} = (-1)^{24} = 1$

(f) $\left(\sqrt{2} - \sqrt{-2}\right)\left(\sqrt{8} + \sqrt{-2}\right) = \left(\sqrt{2} - i\sqrt{2}\right)\left(\sqrt{8} + i\sqrt{2}\right) = \sqrt{2}\sqrt{8} + i\left(\sqrt{2}\right)^2 - i\sqrt{2}\sqrt{8} - i^2\left(\sqrt{2}\right)^2$

$= \sqrt{16} + 2i - i\sqrt{16} + 2 = 6 - 2i$

9. $P(x) = x^4 - 2x^3 + 5x^2 - 8x + 4$. The possible rational zeros of P are: $\pm 1, \pm 2,$ and ± 4. Since there are four changes in sign, P has 4, 2, or 0 positive real zeros.

$$\begin{array}{r|rrrrr} 1 & 1 & -2 & 5 & -8 & 4 \\ & & 1 & -1 & 4 & -4 \\ \hline & 1 & -1 & 4 & -4 & 0 \end{array}$$

So $P(x) = (x - 1)\left(x^3 - x^2 + 4x - 4\right)$. Factoring the second factor by grouping, we have

$P(x) = (x - 1)\left[x^2(x - 1) + 4(x - 1)\right] = (x - 1)\left(x^2 + 4\right)(x - 1) = (x - 1)^2(x - 2i)(x + 2i)$.

11. $P(x) = 2x^4 - 7x^3 + x^2 - 18x + 3$.

(a) Since $P(x)$ has 4 variations in sign, $P(x)$ can have 4, 2, or 0 positive real zeros. Since $P(-x) = 2x^4 + 7x^3 + x^2 + 18x + 3$ has no variations in sign, there are no negative real zeros.

(b)

$$\begin{array}{r|rrrrr} 4 & 2 & -7 & 1 & -18 & 3 \\ & & 8 & 4 & 20 & 8 \\ \hline & 2 & 1 & 5 & 2 & 11 \end{array}$$

Since the last row contains no negative entry, 4 is an upper bound for the real zeros of $P(x)$.

$$\begin{array}{r|rrrrr} -1 & 2 & -7 & 1 & -18 & 3 \\ & & -1 & 9 & -10 & 28 \\ \hline & 2 & -9 & 10 & -28 & 31 \end{array}$$

Since the last row alternates in sign, -1 is a lower bound for the real zeros of $P(x)$.

(c) Using the upper and lower limit from part (b), we graph $P(x)$ in the viewing rectangle $[-1, 4]$ by $[-1, 1]$. The two real zeros are 0.17 and 3.93.

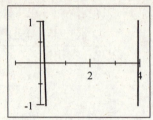

(d) Local minimum $(2.8, -70.3)$.

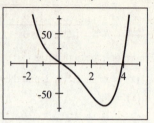

FOCUS ON MODELING   Fitting Polynomial Curves to Data

1. (a) Using a graphing calculator, we obtain the quadratic polynomial

$$y = -0.275428x^2 + 19.7485x - 273.5523 \text{ (where}$$

miles are measured in thousands).

(b)

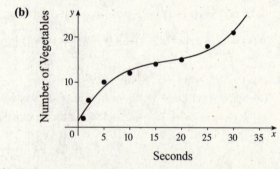

(c) Moving the cursor along the path of the polynomial, we find that 35.85 lb/in² gives the longest tire life.

3. (a) Using a graphing calculator, we obtain the cubic polynomial

$$y = 0.00203709x^3 - 0.104522x^2$$
$$+ 1.966206x + 1.45576.$$

(b)

(c) Moving the cursor along the path of the polynomial, we find that the subjects could name about 43 vegetables in 40 seconds.

(d) Moving the cursor along the path of the polynomial, we find that the subjects could name 5 vegetables in about 2.0 seconds.

5. (a)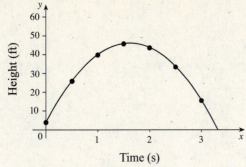

Time (s)

A quadratic model seems appropriate.

(b) Using a graphing calculator, we obtain the quadratic polynomial $y = -16.0x^2 + 51.8429x + 4.20714$.

(c) Moving the cursor along the path of the polynomial, we find that the ball is 20 ft. above the ground 0.3 seconds and 2.9 seconds after it is thrown upward.

(d) Again, moving the cursor along the path of the polynomial, we find that the maximum height is 46.2 ft.

4 EXPONENTIAL AND LOGARITHMIC FUNCTIONS

4.1 EXPONENTIAL FUNCTIONS

1. The function $f(x) = 5^x$ is an exponential function with base 5; $f(-2) = 5^{-2} = \frac{1}{25}$, $f(0) = 5^0 = 1$, $f(2) = 5^2 = 25$, and $f(6) = 5^6 = 15{,}625$.

3. (a) To obtain the graph of $g(x) = 2^x - 1$ we start with the graph of $f(x) = 2^x$ and shift it *downward* 1 unit.

(b) To obtain the graph of $h(x) = 2^{x-1}$ we start with the graph of $f(x) = 2^x$ and shift it to the *right* 1 unit.

5. $f(x) = 4^x$; $f(0.5) = 2$, $f\left(\sqrt{2}\right) \approx 7.103$, $f(\pi) \approx 77.880$, $f\left(\frac{1}{3}\right) \approx 1.587$

7. $g(x) = \left(\frac{2}{3}\right)^{x-1}$; $g(1.3) \approx 0.885$, $g\left(\sqrt{5}\right) \approx 0.606$, $g(2\pi) \approx 0.117$, $g\left(-\frac{1}{2}\right) \approx 1.837$

9. $f(x) = 2^x$

x	y
-4	$\frac{1}{16}$
-2	$\frac{1}{4}$
0	1
2	4
4	16

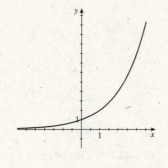

11. $f(x) = \left(\frac{1}{3}\right)^x$

x	y
-2	9
-1	3
0	1
1	$\frac{1}{3}$
2	$\frac{1}{9}$

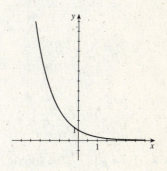

13. $g(x) = 3(1.3)^x$

x	y
-2	1.775
-1	2.308
0	3.0
1	3.9
2	5.07
3	6.591
4	8.568

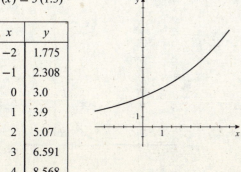

15. $f(x) = 2^x$ and $g(x) = 2^{-x}$

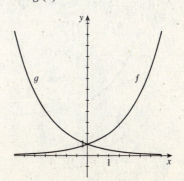

173

17. $f(x) = 4^x$ and $g(x) = 7^x$.

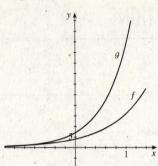

19. From the graph, $f(2) = a^2 = 9$, so $a = 3$. Thus
$f(x) = 3^x$.

21. From the graph, $f(2) = a^2 = \frac{1}{16}$, so $a = \frac{1}{4}$. Thus $f(x) = \left(\frac{1}{4}\right)^x$.

23. The graph of $f(x) = 5^{x+1}$ is obtained from that of $y = 5^x$ by shifting 1 unit to the left, so it has graph II.

25. The graph of $f(x) = -3^x$ is obtained by reflecting the
graph of $y = 3^x$ about the x-axis. Domain: $(-\infty, \infty)$.
Range: $(-\infty, 0)$. Asymptote: $y = 0$.

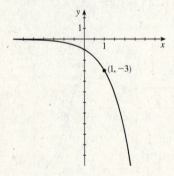

27. $g(x) = 2^x - 3$. The graph of g is obtained by shifting the
graph of $y = 2^x$ downward 3 units. Domain: $(-\infty, \infty)$.
Range: $(-3, \infty)$. Asymptote: $y = -3$.

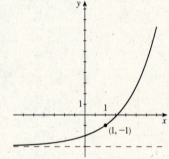

29. $h(x) = 4 + \left(\frac{1}{2}\right)^x$. The graph of h is obtained by shifting
the graph of $y = \left(\frac{1}{2}\right)^x$ upward 4 units. Domain:
$(-\infty, \infty)$. Range: $(4, \infty)$. Asymptote: $y = 4$.

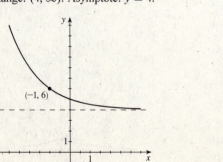

31. $f(x) = 10^{x+3}$. The graph of f is obtained by shifting the
graph of $y = 10^x$ to the left 3 units. Domain: $(-\infty, \infty)$.
Range: $(0, \infty)$. Asymptote: $y = 0$.

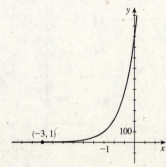

33. $y = 5^{-x} + 1$. The graph of y is obtained by reflecting the graph of $y = 5^x$ about the x-axis and then shifting upward 1 unit. Domain: $(-\infty, \infty)$. Range: $(1, \infty)$. Asymptote: $y = 1$.

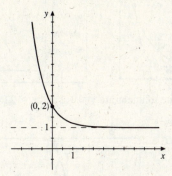

35. $y = 3 - 10^{x-1} = -10^{x-1} + 3$. The graph of y is obtained by reflecting the graph of $y = 10^x$ about the y-axis, then shifting to the right 1 unit and upward 3 units. Domain: $(-\infty, \infty)$. Range: $(-\infty, 3)$. Asymptote: $y = 3$.

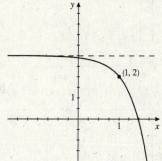

37. (a)

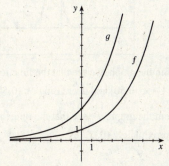

(b) Since $g(x) = 3(2^x) = 3f(x)$ and $f(x) > 0$, the height of the graph of $g(x)$ is always three times the height of the graph of $f(x) = 2^x$, so the graph of g is steeper than the graph of f.

39.

x	$f(x) = x^3$	$g(x) = 3^x$
0	0	1
1	1	3
2	8	9
3	27	27
4	64	81
5	125	243
6	216	729
7	343	2187
8	512	6561
9	729	19,683
10	1000	59,049
15	3375	14,348,907
20	8000	3,486,784,401

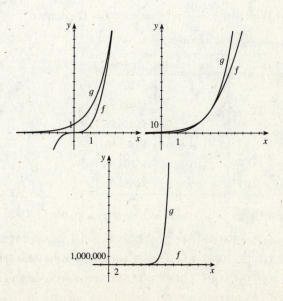

41. (a) From the graphs below, we see that the graph of f ultimately increases much more quickly than the graph of g.

(i) $[0, 5]$ by $[0, 20]$

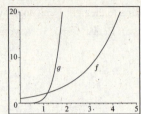

(ii) $[0, 25]$ by $[0, 10^7]$

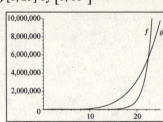

(iii) $[0, 50]$ by $\left[0, 10^8\right]$

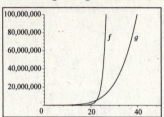

(b) From the graphs in parts (a)(i) and (a)(ii), we see that the approximate solutions are $x \approx 1.2$ and $x \approx 22.4$.

43.

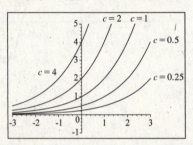

The larger the value of c, the more rapidly the graph of $f(x) = c2^x$ increases. Also notice that the graphs are just shifted horizontally 1 unit. This is because of our choice of c; each c in this exercise is of the form 2^k. So $f(x) = 2^k \cdot 2^x = 2^{x+k}$.

45. $y = 10^{x-x^2}$

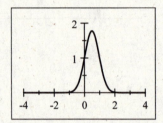

(a) From the graph, we see that the function is increasing on $(-\infty, 0.50]$ and decreasing on $[0.50, \infty)$.

(b) From the graph, we see that the range is approximately $(0, 1.78]$.

47. (a) After 1 hour, there are $1500 \cdot 2 = 3000$ bacteria. After 2 hours, there are $(1500 \cdot 2) \cdot 2 = 6000$ bacteria. After 3 hours, there are $(1500 \cdot 2 \cdot 2) \cdot 2 = 12,000$ bacteria. We see that after t hours, there are $N(t) = 1500 \cdot 2^t$ bacteria.

(b) After 24 hours, there are $N(24) = 1500 \cdot 2^{24} = 25,165,824,000$ bacteria.

49. Using the formula $A(t) = P(1 + i)^k$ with $P = 5000$,

$i = 4\%$ per year $= \dfrac{0.04}{12}$ per month, and $k = 12 \cdot$ number of years, we fill in the table:

Time (years)	Amount
1	$5203.71
2	$5415.71
3	$5636.36
4	$5865.99
5	$6104.98
6	$6353.71

51. $P = 10,000$, $r = 0.03$, and $n = 2$. So $A(t) = 10,000\left(1 + \frac{0.03}{2}\right)^{2t} = 10,000 \cdot 1.015^{2t}$.

(a) $A(5) = 10000 \cdot 1.015^{10} \approx 11,605.41$, and so the value of the investment is $11,605.41.

(b) $A(10) = 10000 \cdot 1.015^{20} \approx 13,468.55$, and so the value of the investment is $13,468.55.

(c) $A(15) = 10000 \cdot 1.015^{30} \approx 15,630.80$, and so the value of the investment is $15,630.80.

53. $P = 500$, $r = 0.0375$, and $n = 4$. So $A(t) = 500\left(1 + \frac{0.0375}{4}\right)^{4t}$.

  (a) $A(1) = 500\left(1 + \frac{0.0375}{4}\right)^{4} \approx 519.02$, and so the value of the investment is $519.02.

  (b) $A(2) = 500\left(1 + \frac{0.0375}{4}\right)^{8} \approx 538.75$, and so the value of the investment is $538.75.

  (c) $A(10) = 500\left(1 + \frac{0.0375}{4}\right)^{40} \approx 726.23$, and so the value of the investment is $726.23.

55. We must solve for P in the equation $10000 = P\left(1 + \frac{0.09}{2}\right)^{2(3)} = P(1.045)^{6} \Leftrightarrow 10000 = 1.3023P \Leftrightarrow P = 7678.96$.
Thus, the present value is $7,678.96.

57. $r_{\text{APY}} = \left(1 + \frac{r}{n}\right)^{n} - 1$. Here $r = 0.08$ and $n = 12$, so $r_{\text{APY}} = \left(1 + \frac{0.08}{12}\right)^{12} - 1 \approx (1.0066667)^{12} - 1 \approx 0.083000$.
Thus, the annual percentage yield is about 8.3%.

59. (a) In this case the payment is $1 million.

  (b) In this case the total pay is $2 + 2^2 + 2^3 + \cdots + 2^{30} > 2^{30}$ cents $= \$10,737,418.24$. Since this is much more than method (a), method (b) is more profitable.

4.2   THE NATURAL EXPONENTIAL FUNCTION

1. The function $f(x) = e^x$ is called the *natural* exponential function. The number e is approximately equal to 2.71828.

3. $h(x) = e^x$; $h(3) = 20.086$, $h(0.23) = 1.259$, $h(1) = 2.718$, $h(-2) = 0.135$

5. $f(x) = 3e^x$

x	y
-2	0.41
-1	1.10
-0.5	1.82
0	3
0.5	4.95
1	8.15
2	22.17

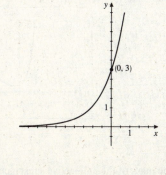

7. $y = -e^x$. The graph of $y = -e^x$ is obtained from the graph of $y = e^x$ by reflecting it about the x-axis. Domain: $(-\infty, \infty)$. Range: $(-\infty, 0)$. Asymptote: $y = 0$.

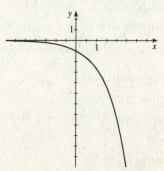

9. $y = e^{-x} - 1$. The graph of $y = e^{-x} - 1$ is obtained from the graph of $y = e^x$ by reflecting it about the y-axis then shifting downward 1 unit. Domain: $(-\infty, \infty)$. Range: $(-1, \infty)$. Asymptote: $y = -1$.

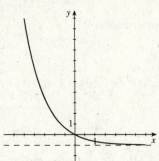

11. $y = e^{x-2}$. The graph of $y = e^{x-2}$ is obtained from the graph of $y = e^x$ by shifting it to the right 2 units. Domain: $(-\infty, \infty)$. Range: $(0, \infty)$. Asymptote: $y = 0$.

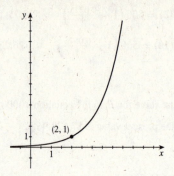

13. $h(x) = e^{x+1} - 3$. The graph of h is obtained from the graph of $y = e^x$ by shifting it to the left 1 unit and downward 3 units. Domain: $(-\infty, \infty)$. Range: $(-3, \infty)$. Asymptote: $y = -3$.

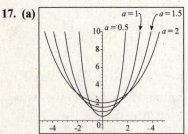

15. (a)

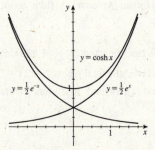

(b) $\cosh(-x) = \dfrac{e^{-x} + e^{-(-x)}}{2} = \dfrac{e^{-x} + e^x}{2}$

$\qquad = \dfrac{e^x + e^{-x}}{2} = \cosh x$

17. (a)

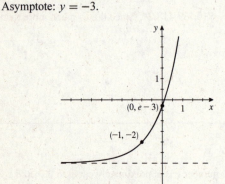

(b) As a increases the curve $y = \dfrac{a}{2}\left(e^{x/a} + e^{-x/a}\right)$

flattens out and the y intercept increases.

19. $g(x) = e^x + e^{-3x}$. The graph of $g(x)$ is shown in the viewing rectangle $[-4, 4]$ by $[0, 20]$. From the graph, we see that there is a local minimum of approximately 1.75 when $x \approx 0.27$.

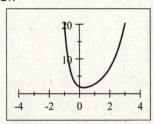

21. $m(t) = 13e^{-0.015t}$

    (a) $m(0) = 13$ kg.

    (b) $m(45) = 13e^{-0.015(45)} = 13e^{-0.675} = 6.619$ kg. Thus the mass of the radioactive substance after 45 days is about 6.6 kg.

23. $v(t) = 80\left(1 - e^{-0.2t}\right)$

(a) $v(0) = 80\left(1 - e^0\right) = 80(1-1) = 0.$

(b) $v(5) = 80\left(1 - e^{-0.2(5)}\right) \approx 80(0.632) = 50.57$ ft/s. So the velocity

after 5 s is about 50.6 ft/s.

$v(10) = 80\left(1 - e^{-0.2(10)}\right) \approx 80(0.865) = 69.2$ ft/s. So the

velocity after 10 s is about 69.2 ft/s.

(d) The terminal velocity is 80 ft/s.

(c)

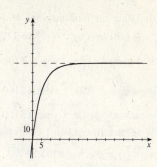

25. $P(t) = \dfrac{1200}{1 + 11e^{-0.2t}}$

(a) $P(0) = \dfrac{1200}{1 + 11e^{-0.2(0)}} = \dfrac{1200}{1 + 11} = 100.$

(b) $P(10) = \dfrac{1200}{1 + 11e^{-0.2(10)}} \approx 482.$ $P(20) = \dfrac{1200}{1 + 11e^{-0.2(20)}} \approx 999.$ $P(30) = \dfrac{1200}{1 + 11e^{-0.2(30)}} \approx 1168.$

(c) As $t \to \infty$ we have $e^{-0.2t} \to 0$, so $P(t) \to \dfrac{1200}{1+0} = 1200$. The graph shown confirms this.

27. $P(t) = \dfrac{73.2}{6.1 + 5.9e^{-0.02t}}$

(a) In the year 2200, $t = 2200 - 2000 = 200$, and the population is

predicted to be $P(200) = \dfrac{73.2}{6.1 + 5.9e^{-0.02(200)}} \approx 11.79$ billion. In

2300, $t = 300$, and $P(300) = \dfrac{73.2}{6.1 + 5.9e^{-0.02(300)}} \approx 11.97$ billion.

(c) As t increases, the denominator approaches 6.1, so according to this

model, the world population approaches $\frac{73.2}{6.1} = 12$ billion people.

(b)

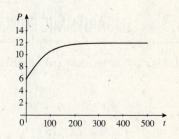

29. Using the formula $A(t) = Pe^{rt}$ with $P = 7000$ and

$r = 3\% = 0.03$, we fill in the table:

Time (years)	Amount
1	$7213.18
2	$7432.86
3	$7659.22
4	$7892.48
5	$8132.84
6	$8380.52

31. We use the formula $A(t) = Pe^{rt}$ with $P = 2000$ and $r = 3.5\% = 0.035$.

(a) $A(2) = 2000e^{0.035 \cdot 2} \approx \$2145.02$ **(b)** $A(42) = 2000e^{0.035 \cdot 4} \approx \$2300.55$ **(c)** $A(12) = 2000e^{0.035 \cdot 12} \approx \$3043.92$

33. (a) Using the formula $A(t) = P(1 + i)^k$ with $P = 600$, $i = 2.5\%$ per year $= 0.025$, and $k = 10$, we calculate
   $A(10) = 600(1.025)^{10} \approx \$768.05$.

(b) Here $i = \dfrac{0.025}{2}$ semiannually and $k = 10 \cdot 2 = 20$, so $A(10) = 600\left(1 + \frac{0.025}{2}\right)^{20} \approx \$769.22$.

(c) Here $i = 2.5\%$ per year $= \dfrac{0.025}{4}$ quarterly and $k = 10 \cdot 4 = 40$, so $A(10) = 600\left(1 + \frac{0.025}{4}\right)^{40} \approx \$769.82$.

(d) Using the formula $A(t) = Pe^{rt}$ with $P = 600$, $r = 2.5\% = 0.025$, and $t = 10$, we have
$$A(10) = 600e^{0.025 \cdot 10} \approx \$770.42.$$

35. *Investment 1:* After 1 year, a $100 investment grows to $A(1) = 100\left(1 + \frac{0.025}{2}\right)^2 \approx 102.52$.

Investment 2: After 1 year, a $100 investment grows to $A(1) = 100\left(1 + \frac{0.0225}{4}\right)^4 = 102.27$.

Investment 3: After 1 year, a $100 investment grows to $A(1) = 100e^{0.02} \approx 102.02$.

We see that Investment 1 yields the highest return.

37. (a) $A(t) = Pe^{rt} = 5000e^{0.09t}$

(b)

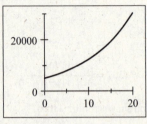

(c) $A(t) = 25{,}000$ when $t \approx 17.88$ years.

4.3   LOGARITHMIC FUNCTIONS

1. $\log x$ is the exponent to which the base 10 must be raised in order to get x.

x	10^3	10^2	10^1	10^0	10^{-1}	10^{-2}	10^{-3}	$10^{1/2}$
$\log x$	3	2	1	0	-1	-2	-3	$1/2$

3. (a) $5^3 = 125$, so $\log_5 125 = 3$.   **(b)** $\log_5 25 = 2$, so $5^2 = 25$.

5.

Logarithmic form	Exponential form
$\log_8 8 = 1$	$8^1 = 8$
$\log_8 64 = 2$	$8^2 = 64$
$\log_8 4 = \frac{2}{3}$	$8^{2/3} = 4$
$\log_8 512 = 3$	$8^3 = 512$
$\log_8 \frac{1}{8} = -1$	$8^{-1} = \frac{1}{8}$
$\log_8 \frac{1}{64} = -2$	$8^{-2} = \frac{1}{64}$

7. (a) $5^2 = 25$

(b) $5^0 = 1$

9. (a) $8^{1/3} = 2$

(b) $2^{-3} = \frac{1}{8}$

11. (a) $e^x = 5$

(b) $e^5 = y$

13. (a) $\log_5 125 = 3$

(b) $\log_{10} 0.0001 = -4$

15. (a) $\log_8 \frac{1}{8} = -1$

(b) $\log_2 \left(\frac{1}{8}\right) = -3$

17. (a) $\ln 2 = x$

(b) $\ln y = 3$

19. (a) $\log_3 3 = 1$

(b) $\log_3 1 = \log_3 3^0 = 0$

(c) $\log_3 3^2 = 2$

21. (a) $\log_6 36 = \log_6 6^2 = 2$

(b) $\log_9 81 = \log_9 9^2 = 2$

(c) $\log_7 7^{10} = 10$

23. (a) $\log_3 \left(\frac{1}{27}\right) = \log_3 3^{-3} = -3$

(b) $\log_{10} \sqrt{10} = \log_{10} 10^{1/2} = \frac{1}{2}$

(c) $\log_5 0.2 = \log_5 \left(\frac{1}{5}\right) = \log_5 5^{-1} = -1$

25. (a) $2^{\log_2 37} = 37$

(b) $3^{\log_3 8} = 8$

(c) $e^{\ln \sqrt{5}} = \sqrt{5}$

27. (a) $\log_8 0.25 = \log_8 8^{-2/3} = -\frac{2}{3}$

(b) $\ln e^4 = 4$

(c) $\ln\left(\dfrac{1}{e}\right) = \ln e^{-1} = -1$

29. (a) $\log_2 x = 5 \Leftrightarrow x = 2^5 = 32$

(b) $x = \log_2 16 = \log_2 2^4 = 4$

31. (a) $x = \log_3 243 = \log_3 3^5 = 5$

(b) $\log_3 x = 3 \Leftrightarrow x = 3^3 = 27$

33. (a) $\log_{10} x = 2 \Leftrightarrow x = 10^2 = 100$

(b) $\log_5 x = 2 \Leftrightarrow x = 5^2 = 25$

35. (a) $\log_x 16 = 4 \Leftrightarrow x^4 = 16 \Leftrightarrow x = 2$

(b) $\log_x 8 = \frac{3}{2} \Leftrightarrow x^{3/2} = 8 \Leftrightarrow x = 8^{2/3} = 4$

37. (a) $\log 2 \approx 0.3010$

(b) $\log 35.2 \approx 1.5465$

(c) $\log\left(\frac{2}{3}\right) \approx -0.1761$

39. (a) $\ln 5 \approx 1.6094$

(b) $\ln 25.3 \approx 3.2308$

(c) $\ln\left(1 + \sqrt{3}\right) \approx 1.0051$

41.

x	$f(x)$
$\dfrac{1}{3^3}$	-3
$\dfrac{1}{3^2}$	-2
$\dfrac{1}{3}$	-1
1	0
3	1
3^2	2

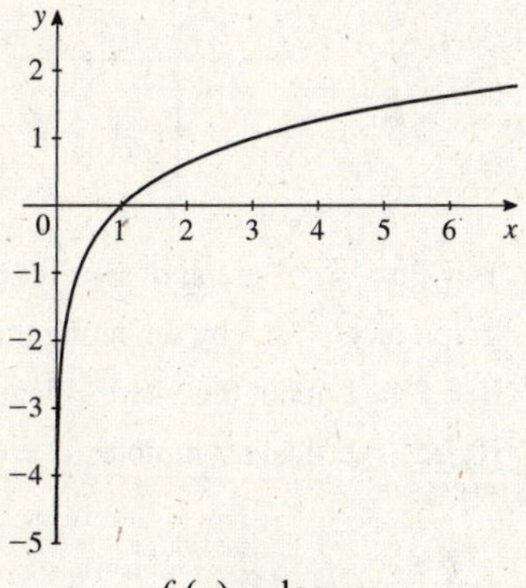

$f(x) = \log_3 x$

43.

x	$f(x)$
$\dfrac{1}{10^3}$	-6
$\dfrac{1}{10^2}$	-4
$\dfrac{1}{10}$	-2
1	0
10	2
10^2	4

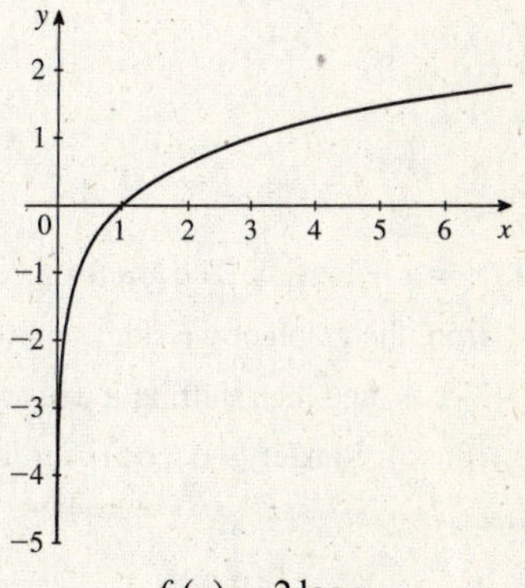

$f(x) = 2\log x$

45. Since the point $(5, 1)$ is on the graph, we have $1 = \log_a 5 \Leftrightarrow a^1 = 5$. Thus the function is $y = \log_5 x$.

47. Since the point $\left(3, \frac{1}{2}\right)$ is on the graph, we have $\frac{1}{2} = \log_a 3 \Leftrightarrow a^{1/2} = 3 \Leftrightarrow a = 9$. Thus the function is $y = \log_9 x$.

49. I

51. The graph of $y = \log_4 x$ is obtained from the graph of $y = 4^x$ by reflecting it about the line $y = x$.

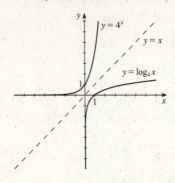

53. $f(x) = \log_2(x - 4)$. The graph of f is obtained from the graph of $y = \log_2 x$ by shifting it to the right 4 units. Domain: $(4, \infty)$. Range: $(-\infty, \infty)$. Vertical asymptote: $x = 4$.

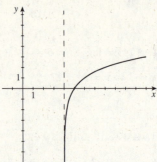

55. $g(x) = \log_5(-x)$. The graph of g is obtained from the graph of $y = \log_5 x$ by reflecting it about the y-axis. Domain: $(-\infty, 0)$. Range: $(-\infty, \infty)$. Vertical asymptote: $x = 0$.

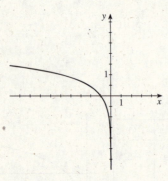

57. $y = 2 + \log_3 x$. The graph of $y = 2 + \log_3 x$ is obtained from the graph of $y = \log_3 x$ by shifting it upward 2 units. Domain: $(0, \infty)$. Range: $(-\infty, \infty)$. Vertical asymptote: $x = 0$.

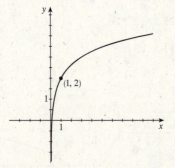

59. $y = 1 - \log_{10} x$. The graph of $y = 1 - \log_{10} x$ is obtained from the graph of $y = \log_{10} x$ by reflecting it about the x-axis, and then shifting it upward 1 unit. Domain: $(0, \infty)$. Range: $(-\infty, \infty)$. Vertical asymptote: $x = 0$.

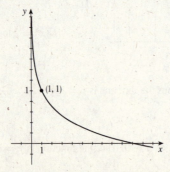

61. $y = |\ln x|$. The graph of $y = |\ln x|$ is obtained from the graph of $y = \ln x$ by reflecting the part of the graph for $0 < x < 1$ about the x-axis. Domain: $(0, \infty)$. Range: $[0, \infty)$. Vertical asymptote: $x = 0$.

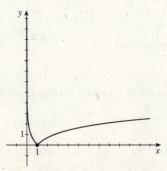

63. $f(x) = \log_{10}(x + 3)$. We require that $x + 3 > 0 \Leftrightarrow x > -3$, so the domain is $(-3, \infty)$.

65. $g(x) = \log_3\left(x^2 - 1\right)$. We require that $x^2 - 1 > 0 \Leftrightarrow x^2 > 1 \Rightarrow x < -1$ or $x > 1$, so the domain is $(-\infty, -1) \cup (1, \infty)$.

67. $h(x) = \ln x + \ln(2 - x)$. We require that $x > 0$ and $2 - x > 0 \Leftrightarrow x > 0$ and $x < 2 \Leftrightarrow 0 < x < 2$, so the domain is $(0, 2)$.

69. $y = \log_{10}\left(1 - x^2\right)$ has domain $(-1, 1)$, vertical
asymptotes $x = -1$ and $x = 1$, and local maximum $y = 0$
at $x = 0$.

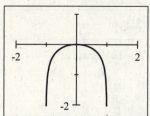

71. $y = x + \ln x$ has domain $(0, \infty)$, vertical asymptote
$x = 0$, and no local maximum or minimum.

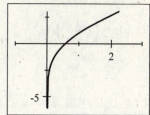

73. $y = \dfrac{\ln x}{x}$ has domain $(0, \infty)$, vertical asymptote $x = 0$,
horizontal asymptote $y = 0$, and local maximum $y \approx 0.37$
at $x \approx 2.72$.

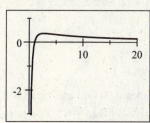

75. $f(x) = 2^x$ and $g(x) = x + 1$ both have domain
$(-\infty, \infty)$, so $(f \circ g)(x) = f(g(x)) = 2^{g(x)} = 2^{x+1}$
with domain $(-\infty, \infty)$ and
$(g \circ f)(x) = g(f(x)) = 2^x + 1$ with domain $(-\infty, \infty)$.

77. $f(x) = \log_2 x$ has domain $(0, \infty)$ and $g(x) = x - 2$ has domain $(-\infty, \infty)$, so $(f \circ g)(x) = f(g(x)) = \log_2(x - 2)$
with domain $(2, \infty)$ and $(g \circ f)(x) = g(f(x)) = \log_2 x - 2$ with domain $(0, \infty)$.

79. The graph of $g(x) = \sqrt{x}$ grows faster than the graph of
$f(x) = \ln x$.

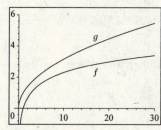

81. (a)

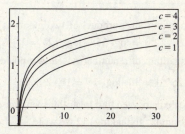

(b) Notice that $f(x) = \log(cx) = \log c + \log x$, so
as c increases, the graph of $f(x) = \log(cx)$ is
shifted upward $\log c$ units.

83. (a) $f(x) = \log_2(\log_{10} x)$. Since the domain of $\log_2 x$ is the positive real numbers, we have: $\log_{10} x > 0 \Leftrightarrow x > 10^0 = 1$.
Thus the domain of $f(x)$ is $(1, \infty)$.

(b) $y = \log_2(\log_{10} x) \Leftrightarrow 2^y = \log_{10} x \Leftrightarrow 10^{2^y} = x$. Thus $f^{-1}(x) = 10^{2^x}$.

85. (a) $f(x) = \dfrac{2^x}{1+2^x}$. $y = \dfrac{2^x}{1+2^x} \Leftrightarrow y + y2^x = 2^x$

$\Leftrightarrow y = 2^x - y2^x = 2^x(1-y) \Leftrightarrow 2^x = \dfrac{y}{1-y}$

$\Leftrightarrow x = \log_2\left(\dfrac{y}{1-y}\right)$. Thus

$f^{-1}(x) = \log_2\left(\dfrac{x}{1-x}\right)$.

(b) $\dfrac{x}{1-x} > 0$. Solving this using the methods from Chapter 1, we start with the endpoints, 0 and 1.

Interval	$(-\infty, 0)$	$(0, 1)$	$(1, \infty)$
Sign of x	$-$	$+$	$+$
Sign of $1-x$	$+$	$+$	$-$
Sign of $\dfrac{x}{1-x}$	$-$	$+$	$-$

Thus the domain of $f^{-1}(x)$ is $(0, 1)$.

87. Using $D = 0.73D_0$ we have $A = -8267 \ln\left(\dfrac{D}{D_0}\right) = -8267 \ln 0.73 \approx 2601$ years.

89. When $r = 6\%$ we have $t = \dfrac{\ln 2}{0.06} \approx 11.6$ years. When $r = 7\%$ we have $t = \dfrac{\ln 2}{0.07} \approx 9.9$ years. And when $r = 8\%$ we have

$t = \dfrac{\ln 2}{0.08} \approx 8.7$ years.

91. Using $A = 100$ and $W = 5$ we find the ID to be $\dfrac{\log(2A/W)}{\log 2} = \dfrac{\log(2 \cdot 100/5)}{\log 2} = \dfrac{\log 40}{\log 2} \approx 5.32$. Using $A = 100$ and

$W = 10$ we find the ID to be $\dfrac{\log(2A/W)}{\log 2} = \dfrac{\log(2 \cdot 100/10)}{\log 2} = \dfrac{\log 20}{\log 2} \approx 4.32$. So the smaller icon is $\dfrac{5.23}{4.32} \approx 1.23$ times

harder.

93. $\log\left(\log 10^{100}\right) = \log 100 = 2$

$\log\left(\log\left(\log 10^{\text{googol}}\right)\right) = \log(\log(\text{googol})) = \log\left(\log 10^{100}\right) = \log(100) = 2$

95. The numbers between 1000 and 9999 (inclusive) each have 4 digits, while $\log 1000 = 3$ and $\log 10,000 = 4$. Since $[\![\log x]\!] = 3$ for all integers x where $1000 \le x < 10,000$, the number of digits is $[\![\log x]\!] + 1$. Likewise, if x is an integer where $10^{n-1} \le x < 10^n$, then x has n digits and $[\![\log x]\!] = n - 1$. Since $[\![\log x]\!] = n - 1 \Leftrightarrow n = [\![\log x]\!] + 1$, the number of digits in x is $[\![\log x]\!] + 1$.

4.4   LAWS OF LOGARITHMS

1. The logarithm of a product of two numbers is the same as the *sum* of the logarithms of these numbers. So $\log_5(25 \cdot 125) = \log_5 25 + \log_5 125 = 2 + 3 = 5$.

3. The logarithm of a number raised to a power is the same as the power *times* the logarithm of the number. So $\log_5\left(25^{10}\right) = 10 \cdot \log_5 25 = 10 \cdot 2 = 20$.

5. Most calculators can find logarithms with base 10 and base e. To find logarithms with different bases we use the *change of base* formula. To find $\log_7 12$ we write

$$\log_7 12 = \frac{\log 12}{\log 7} \approx \frac{1.079}{0.845} \approx 1.277$$

7. $\log_3 \sqrt{27} = \log_3 3^{3/2} = \frac{3}{2}$

9. $\log 4 + \log 25 = \log(4 \cdot 25) = \log 100 = 2$

11. $\log_4 192 - \log_4 3 = \log_4 \frac{192}{3} = \log_4 64 = \log_4 4^3 = 3$

13. $\log_2 6 - \log_2 15 + \log_2 20 = \log_2 \frac{6}{15} + \log_2 20 = \log_2\left(\frac{2}{5} \cdot 20\right) = \log_2 8 = \log_2 2^3 = 3$

15. $\log_4 16^{100} = \log_4 \left(4^2\right)^{100} = \log_4 4^{200} = 200$

17. $\log\left(\log 10^{10,000}\right) = \log\left(10,000 \log 10\right) = \log\left(10,000 \cdot 1\right) = \log\left(10,000\right) = \log 10^4 = 4 \log 10 = 4$

19. $\log_2 2x = \log_2 2 + \log_2 x = 1 + \log_2 x$ **21.** $\log_2\left[x\left(x-1\right)\right] = \log_2 x + \log_2\left(x-1\right)$

23. $\log 6^{10} = 10 \log 6$ **25.** $\log_2\left(AB^2\right) = \log_2 A + \log_2 B^2 = \log_2 A + 2 \log_2 B$

27. $\log_3\left(x\sqrt{y}\right) = \log_3 x + \log_3\sqrt{y} = \log_3 x + \frac{1}{2}\log_3 y$ **29.** $\log_5\sqrt[3]{x^2+1} = \frac{1}{3}\log_5\left(x^2+1\right)$

31. $\ln\sqrt{ab} = \frac{1}{2}\ln ab = \frac{1}{2}\left(\ln a + \ln b\right)$

33. $\log\left(\dfrac{x^3 y^4}{z^6}\right) = \log\left(x^3 y^4\right) - \log z^6 = 3 \log x + 4 \log y - 6 \log z$

35. $\log_2\left(\dfrac{x\left(x^2+1\right)}{\sqrt{x^2-1}}\right) = \log_2 x + \log_2\left(x^2+1\right) - \frac{1}{2}\log_2\left(x^2-1\right)$

37. $\ln\left(x\sqrt{\dfrac{y}{z}}\right) = \ln x + \frac{1}{2}\ln\left(\dfrac{y}{z}\right) = \ln x + \frac{1}{2}\left(\ln y - \ln z\right)$

39. $\log\sqrt[4]{x^2+y^2} = \frac{1}{4}\log\left(x^2+y^2\right)$

41. $\log\sqrt{\dfrac{x^2+4}{\left(x^2+1\right)\left(x^3-7\right)^2}} = \frac{1}{2}\log\dfrac{x^2+4}{\left(x^2+1\right)\left(x^3-7\right)^2} = \frac{1}{2}\left[\log\left(x^2+4\right) - \log\left(x^2+1\right)\left(x^3-7\right)^2\right]$

$$= \frac{1}{2}\left[\log\left(x^2+4\right) - \log\left(x^2+1\right) - 2\log\left(x^3-7\right)\right]$$

43. $\ln\dfrac{x^3\sqrt{x-1}}{3x+4} = \ln\left(x^3\sqrt{x-1}\right) - \ln\left(3x+4\right) = 3\ln x + \frac{1}{2}\ln\left(x-1\right) - \ln\left(3x+4\right)$

45. $\log_3 5 + 5\log_3 2 = \log_3 5 + \log_3 2^5 = \log_3\left(5 \cdot 2^5\right) = \log_3 160$

47. $\log_2 A + \log_2 B - 2\log_2 C = \log_2\left(AB\right) - \log_2\left(C^2\right) = \log_2\left(\dfrac{AB}{C^2}\right)$

49. $4\log x - \frac{1}{3}\log\left(x^2+1\right) + 2\log\left(x-1\right) = \log x^4 - \log\sqrt[3]{x^2+1} + \log\left(x-1\right)^2$

$$= \log\left(\dfrac{x^4}{\sqrt[3]{x^2+1}}\right) + \log\left(x-1\right)^2 = \log\left(\dfrac{x^4\left(x-1\right)^2}{\sqrt[3]{x^2+1}}\right)$$

51. $\ln 5 + 2\ln x + 3\ln\left(x^2+5\right) = \ln\left(5x^2\right) + \ln\left(x^2+5\right)^3 = \ln\left[5x^2\left(x^2+5\right)^3\right]$

53. $\frac{1}{3}\log\left(x+2\right)^3 + \frac{1}{2}\left[\log x^4 - \log\left(x^2-x-6\right)^2\right] = 3 \cdot \frac{1}{3}\log\left(x+2\right) + \frac{1}{2}\log\dfrac{x^4}{\left(x^2-x-6\right)^2}$

$$= \log\left(x+2\right) + \log\left(\dfrac{x^4}{\left[\left(x-3\right)\left(x+2\right)\right]^2}\right)^{1/2} = \log\left(x+2\right) + \log\dfrac{x^2}{\left(x-3\right)\left(x+2\right)} = \log\dfrac{x^2\left(x+2\right)}{\left(x-3\right)\left(x+2\right)} = \log\dfrac{x^2}{x-3}$$

55. $\log_2 5 = \dfrac{\log 5}{\log 2} \approx 2.321928$ **57.** $\log_3 16 = \dfrac{\log 16}{\log 3} \approx 2.523719$

59. $\log_7 2.61 = \dfrac{\log 2.61}{\log 7} \approx 0.493008$ **61.** $\log_4 125 = \dfrac{\log 125}{\log 4} \approx 3.482892$

63. $\log_3 x = \dfrac{\log_e x}{\log_e 3} = \dfrac{\ln x}{\ln 3} = \dfrac{1}{\ln 3}\ln x$. The graph of $y = \dfrac{1}{\ln 3}\ln x$ is

shown in the viewing rectangle $[-1, 4]$ by $[-3, 2]$.

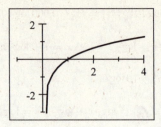

65. $\log e = \dfrac{\ln e}{\ln 10} = \dfrac{1}{\ln 10}$

67. $-\ln\left(x - \sqrt{x^2-1}\right) = \ln\left(\dfrac{1}{x-\sqrt{x^2-1}}\right) = \ln\left(\dfrac{1}{x-\sqrt{x^2-1}} \cdot \dfrac{x+\sqrt{x^2-1}}{x+\sqrt{x^2-1}}\right) = \ln\left(\dfrac{x+\sqrt{x^2-1}}{x^2-(x^2-1)}\right)$

$= \ln\left(x + \sqrt{x^2-1}\right)$

69. (a) $\log P = \log c - k\log W \Leftrightarrow \log P = \log c - \log W^k \Leftrightarrow \log P = \log\left(\dfrac{c}{W^k}\right) \Leftrightarrow P = \dfrac{c}{W^k}$.

(b) Using $k = 2.1$ and $c = 8000$, when $W = 2$ we have $P = \dfrac{8000}{2^{2.1}} \approx 1866$ and when $W = 10$ we have $P = \dfrac{8000}{10^{2.1}} \approx 64$.

71. (a) $M = -2.5\log(B/B_0) = -2.5\log B + 2.5\log B_0$.

(b) Suppose B_1 and B_2 are the brightness of two stars such that $B_1 < B_2$ and let M_1 and M_2 be their respective magnitudes. Since log is an increasing function, we have $\log B_1 < \log B_2$. Then $\log B_1 < \log B_2 \Leftrightarrow$ $\log B_1 - \log B_0 < \log B_2 - \log B_0 \Leftrightarrow \log(B_1/B_0) < \log(B_2/B_0) \Leftrightarrow -2.5\log(B_1/B_0) > -2.5\log(B_2/B_0) \Leftrightarrow$ $M_1 > M_2$. Thus the brighter star has less magnitudes.

(c) Let B_1 be the brightness of the star Albiero. Then $100B_1$ is the brightness of Betelgeuse, and its magnitude is

$M = -2.5\log(100B_1/B_0) = -2.5\left[\log 100 + \log(B_1/B_0)\right] = -2.5\left[2 + \log(B_1/B_0)\right] = -5 - 2.5\log(B_1/B_0)$

$= -5 + $ magnitude of Albiero

73. The error is on the first line: $\log 0.1 < 0$, so $2\log 0.1 < \log 0.1$.

4.5   EXPONENTIAL AND LOGARITHMIC EQUATIONS

1. (a) First we isolate e^x to get the equivalent equation $e^x = 25$.

(b) Next, we take the natural logarithm of each side to get the equivalent equation $x = \ln 25$.

(c) Now we use a calculator to find $x \approx 3.219$.

3. $10^x = 25 \Leftrightarrow \log 10^x = \log 25 \Leftrightarrow x\log 10 = \log 25 \Leftrightarrow x \approx 1.398$

5. $e^{-2x} = 7 \Leftrightarrow \ln e^{-2x} = \ln 7 \Leftrightarrow -2x\ln e = \ln 7 \Leftrightarrow -2x = \ln 7 \Leftrightarrow x = -\tfrac{1}{2}\ln 7 \approx -0.9730$

7. $2^{1-x} = 3 \Leftrightarrow \log 2^{1-x} = \log 3 \Leftrightarrow (1-x)\log 2 = \log 3 \Leftrightarrow 1 - x = \dfrac{\log 3}{\log 2} \Leftrightarrow x = 1 - \dfrac{\log 3}{\log 2} \approx -0.5850$

9. $3e^x = 10 \Leftrightarrow e^x = \tfrac{10}{3} \Leftrightarrow x = \ln\left(\tfrac{10}{3}\right) \approx 1.2040$

11. $e^{1-4x} = 2 \Leftrightarrow 1 - 4x = \ln 2 \Leftrightarrow -4x = -1 + \ln 2 \Leftrightarrow x = \dfrac{1-\ln 2}{4} = 0.0767$

13. $4 + 3^{5x} = 8 \Leftrightarrow 3^{5x} = 4 \Leftrightarrow \log 3^{5x} = \log 4 \Leftrightarrow 5x\log 3 = \log 4 \Leftrightarrow 5x = \dfrac{\log 4}{\log 3} \Leftrightarrow x = \dfrac{\log 4}{5\log 3} \approx 0.2524$

15. $8^{0.4x} = 5 \Leftrightarrow \log 8^{0.4x} = \log 5 \Leftrightarrow 0.4x\log 8 = \log 5 \Leftrightarrow 0.4x = \dfrac{\log 5}{\log 8} \Leftrightarrow x = \dfrac{\log 5}{0.4\log 8} \approx 1.9349$

17. $5^{-x/100} = 2 \Leftrightarrow \log 5^{-x/100} = \log 2 \Leftrightarrow -\dfrac{x}{100}\log 5 = \log 2 \Leftrightarrow x = -\dfrac{100\log 2}{\log 5} \approx -43.0677$

19. $e^{2x+1} = 200 \Leftrightarrow 2x + 1 = \ln 200 \Leftrightarrow 2x = -1 + \ln 200 \Leftrightarrow x = \dfrac{-1 + \ln 200}{2} \approx 2.1492$

21. $5^x = 4^{x+1} \Leftrightarrow \log 5^x = \log 4^{x+1} \Leftrightarrow x \log 5 = (x+1) \log 4 = x \log 4 + \log 4 \Leftrightarrow x \log 5 - x \log 4 = \log 4 \Leftrightarrow$

$x (\log 5 - \log 4) = \log 4 \Leftrightarrow x = \dfrac{\log 4}{\log 5 - \log 4} \approx 6.2126$

23. $2^{3x+1} = 3^{x-2} \Leftrightarrow \log 2^{3x+1} = \log 3^{x-2} \Leftrightarrow (3x+1) \log 2 = (x-2) \log 3 \Leftrightarrow 3x \log 2 + \log 2 = x \log 3 - 2 \log 3 \Leftrightarrow$

$3x \log 2 - x \log 3 = -\log 2 - 2 \log 3 \Leftrightarrow x (3 \log 2 - \log 3) = -(\log 2 + 2 \log 3) \Leftrightarrow s = -\dfrac{\log 2 + 2 \log 3}{3 \log 2 - \log 3} \approx -2.9469$

25. $\dfrac{50}{1 + e^{-x}} = 4 \Leftrightarrow 50 = 4 + 4e^{-x} \Leftrightarrow 46 = 4e^{-x} \Leftrightarrow 11.5 = e^{-x} \Leftrightarrow \ln 11.5 = -x \Leftrightarrow x = -\ln 11.5 \approx -2.4423$

27. $100 (1.04)^{2t} = 300 \Leftrightarrow 1.04^{2t} = 3 \Leftrightarrow \log 1.04^{2t} = \log 3 \Leftrightarrow 2t \log 1.04 = \log 3 \Leftrightarrow t = \dfrac{\log 3}{2 \log 1.04} \approx 14.0055$

29. $e^{2x} - 3e^x + 2 = 0 \Leftrightarrow (e^x - 1)(e^x - 2) = 0 \Rightarrow e^x - 1 = 0$ or $e^x - 2 = 0$. If $e^x - 1 = 0$, then $e^x = 1 \Leftrightarrow x = \ln 1 = 0$. If

$e^x - 2 = 0$, then $e^x = 2 \Leftrightarrow x = \ln 2 \approx 0.6931$. So the solutions are $x = 0$ and $x \approx 0.6931$.

31. $e^{4x} + 4e^{2x} - 21 = 0 \Leftrightarrow \left(e^{2x} + 7\right)\left(e^{2x} - 3\right) = 0 \Rightarrow e^{2x} = -7$ or $e^{2x} = 3$. Now $e^{2x} = -7$ has no solution, since $e^{2x} > 0$

for all x. But we can solve $e^{2x} = 3 \Leftrightarrow 2x = \ln 3 \Leftrightarrow x = \frac{1}{2} \ln 3 \approx 0.5493$. So the only solution is $x \approx 0.5493$.

33. $x^2 2^x - 2^x = 0 \Leftrightarrow 2^x \left(x^2 - 1\right) = 0 \Rightarrow 2^x = 0$ (never) or $x^2 - 1 = 0$. If $x^2 - 1 = 0$, then $x^2 = 1 \Rightarrow x = \pm 1$. So the only

solutions are $x = \pm 1$.

35. $4x^3 e^{-3x} - 3x^4 e^{-3x} = 0 \Leftrightarrow x^3 e^{-3x} (4 - 3x) = 0 \Rightarrow x = 0$ or $e^{-3x} = 0$ (never) or $4 - 3x = 0$. If $4 - 3x = 0$, then

$3x = 4 \Leftrightarrow x = \frac{4}{3}$. So the solutions are $x = 0$ and $x = \frac{4}{3}$.

37. $\ln x = 10 \Leftrightarrow x = e^{10} \approx 22{,}026$

39. $\log x = -2 \Leftrightarrow x = 10^{-2} = 0.01$

41. $\log (3x + 5) = 2 \Leftrightarrow 3x + 5 = 10^2 = 100 \Leftrightarrow 3x = 95 \Leftrightarrow x = \frac{95}{3} \approx 31.6667$

43. $4 - \log (3 - x) = 3 \Leftrightarrow \log (3 - x) = 1 \Leftrightarrow 3 - x = 10 \Leftrightarrow x = -7$

45. $\log_2 3 + \log_2 x = \log_2 5 + \log_2 (x - 2) \Leftrightarrow \log_2 (3x) = \log_2 (5x - 10) \Leftrightarrow 3x = 5x - 10 \Leftrightarrow 2x = 10 \Leftrightarrow x = 5$

47. $\log x + \log (x - 1) = \log (4x) \Leftrightarrow \log [x (x - 1)] = \log (4x) \Leftrightarrow x^2 - x = 4x \Leftrightarrow x^2 - 5x = 0 \Leftrightarrow x (x - 5) = 0 \Rightarrow x = 0$

or $x = 5$. So the possible solutions are $x = 0$ and $x = 5$. However, when $x = 0$, $\log x$ is undefined. Thus the only solution

is $x = 5$.

49. $\log_5 (x + 1) - \log_5 (x - 1) = 2 \Leftrightarrow \log_5 \left(\dfrac{x+1}{x-1}\right) = 2 \Leftrightarrow \dfrac{x+1}{x-1} = 5^2 \Leftrightarrow x + 1 = 25x - 25 \Leftrightarrow 24x = 26 \Leftrightarrow x = \frac{13}{12}$

51. $\log_2 x + \log_2 (x - 3) = 2 \Leftrightarrow \log_2 [x (x - 3)] = 2 \Leftrightarrow x^2 - 3x = 2^2 \Leftrightarrow x^2 - 3x - 4 = 0 \Leftrightarrow (x - 4)(x + 1) \Leftrightarrow x = -1$ or

$x = 4$. Since $\log (-1 - 3) = \log (-4)$ is undefined, the only solution is $x = 4$.

53. $\log_9 (x - 5) + \log_9 (x + 3) = 1 \Leftrightarrow \log_9 [(x - 5)(x + 3)] = 1 \Leftrightarrow (x - 5)(x + 3) = 9^1 \Leftrightarrow x^2 - 2x - 24 = 0 \Leftrightarrow$

$(x - 6)(x + 4) = 0 \Rightarrow x = 6$ or -4. However, $x = -4$ is inadmissible, so $x = 6$ is the only solution.

55. $\log (x + 3) = \log x + \log 3 \Leftrightarrow \log (x + 3) = \log (3x) \Leftrightarrow x + 3 = 3x \Leftrightarrow 2x = 3 \Leftrightarrow x = \frac{3}{2}$

57. $2^{2/\log_5 x} = \frac{1}{16} \Leftrightarrow \log_2 2^{2/\log_5 x} = \log_2 \left(\frac{1}{16}\right) \Leftrightarrow \dfrac{2}{\log_5 x} = -4 \Leftrightarrow \log_5 x = -\frac{1}{2} \Leftrightarrow x = 5^{-1/2} = \frac{1}{\sqrt{5}} \approx 0.4472$

59. $\ln x = 3 - x \Leftrightarrow \ln x + x - 3 = 0$. Let
$f(x) = \ln x + x - 3$. We need to solve the equation
$f(x) = 0$. From the graph of f, we get $x \approx 2.21$.

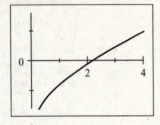

61. $x^3 - x = \log_{10}(x + 1) \Leftrightarrow x^3 - x - \log_{10}(x + 1) = 0$.
Let $f(x) = x^3 - x - \log_{10}(x + 1)$. We need to solve the
equation $f(x) = 0$. From the graph of f, we get $x = 0$ or
$x \approx 1.14$.

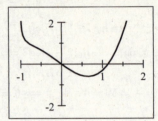

63. $e^x = -x \Leftrightarrow e^x + x = 0$. Let $f(x) = e^x + x$. We need to
solve the equation $f(x) = 0$. From the graph of f, we get
$x \approx -0.57$.

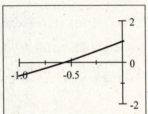

65. $4^{-x} = \sqrt{x} \Leftrightarrow 4^{-x} - \sqrt{x} = 0$. Let $f(x) = 4^{-x} - \sqrt{x}$.
We need to solve the equation $f(x) = 0$. From the graph
of f, we get $x \approx 0.36$.

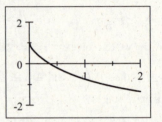

67. $\log(x - 2) + \log(9 - x) < 1 \Leftrightarrow \log[(x - 2)(9 - x)] < 1 \Leftrightarrow \log\left(-x^2 + 11x - 18\right) < 1 \Rightarrow -x^2 + 11x - 18 < 10^1$
$\Leftrightarrow 0 < x^2 - 11x + 28 \Leftrightarrow 0 < (x - 7)(x - 4)$. Also, since the domain of a logarithm is positive we must have
$0 < -x^2 + 11x - 18 \Leftrightarrow 0 < (x - 2)(9 - x)$. Using the methods from Chapter 1 with the endpoints 2, 4, 7, 9 for the
intervals, we make the following table:

Interval	$(-\infty, 2)$	$(2, 4)$	$(4, 7)$	$(7, 9)$	$(9, \infty)$
Sign of $x - 7$	$-$	$-$	$-$	$+$	$+$
Sign of $x - 4$	$-$	$-$	$+$	$+$	$+$
Sign of $x - 2$	$-$	$+$	$+$	$+$	$+$
Sign of $9 - x$	$+$	$+$	$+$	$+$	$-$
Sign of $(x - 7)(x - 4)$	$+$	$+$	$-$	$+$	$+$
Sign of $(x - 2)(9 - x)$	$-$	$+$	$+$	$+$	$-$

Thus the solution is $(2, 4) \cup (7, 9)$.

69. $2 < 10^x < 5 \Leftrightarrow \log 2 < x < \log 5 \Leftrightarrow 0.3010 < x < 0.6990$. Hence the solution to the inequality is approximately the
interval $(0.3010, 0.6990)$.

71. To find the inverse of $f(x) = 2^{2x}$, we set $y = f(x)$ and solve for x. $y = 2^{2x} \Leftrightarrow \ln y = \ln\left(2^{2x}\right) = 2x \ln 2 \Leftrightarrow x = \dfrac{\ln y}{2 \ln 2}$.
Interchange x and y: $y = \dfrac{\ln x}{2 \ln 2}$. Thus, $f^{-1}(x) = \dfrac{\ln x}{2 \ln 2}$.

73. To find the inverse of $f(x) = \log_2(x - 1)$, we set $y = f(x)$ and solve for x. $y = \log_2(x - 1) \Leftrightarrow 2^y = 2^{\log_2(x-1)} = x - 1$
$\Leftrightarrow x = 2^y + 1$. Interchange x and y: $y = 2^x + 1$. Thus, $f^{-1}(x) = 2^x + 1$.

75. (a) $A(3) = 5000 \left(1 + \dfrac{0.085}{4}\right)^{4(3)} = 5000 \left(1.02125^{12}\right) = 6435.09$. Thus the amount after 3 years is \$6,435.09.

(b) $10000 = 5000 \left(1 + \dfrac{0.085}{4}\right)^{4t} = 5000 \left(1.02125^{4t}\right) \Leftrightarrow 2 = 1.02125^{4t} \Leftrightarrow \log 2 = 4t \log 1.02125 \Leftrightarrow$

$t = \dfrac{\log 2}{4 \log 1.02125} \approx 8.24$ years. Thus the investment will double in about 8.24 years.

77. $8000 = 5000 \left(1 + \dfrac{0.075}{4}\right)^{4t} = 5000 \left(1.01875^{4t}\right) \Leftrightarrow 1.6 = 1.01875^{4t} \Leftrightarrow \log 1.6 = 4t \log 1.01875 \Leftrightarrow$

$t = \dfrac{\log 1.6}{4 \log 1.01875} \approx 6.33$ years. The investment will increase to \$8000 in approximately 6 years and 4 months.

79. $2 = e^{0.085t} \Leftrightarrow \ln 2 = 0.085t \Leftrightarrow t = \dfrac{\ln 2}{0.085} \approx 8.15$ years. Thus the investment will double in about 8.15 years.

81. $15e^{-0.087t} = 5 \Leftrightarrow e^{-0.087t} = \frac{1}{3} \Leftrightarrow -0.087t = \ln\left(\frac{1}{3}\right) = -\ln 3 \Leftrightarrow t = \dfrac{\ln 3}{0.087} \approx 12.6277$. So only 5 grams remain after

approximately 13 days.

83. (a) $P(3) = \dfrac{10}{1 + 4e^{-0.8(3)}} = 7.337$, so there are approximately 7337 fish after 3 years.

(b) We solve for t. $\dfrac{10}{1 + 4e^{-0.8t}} = 5 \Leftrightarrow 1 + 4e^{-0.8t} = \frac{10}{5} = 2 \Leftrightarrow 4e^{-0.8t} = 1 \Leftrightarrow e^{-0.8t} = 0.25 \Leftrightarrow -0.8t = \ln 0.25 \Leftrightarrow$

$t = \dfrac{\ln 0.25}{-0.8} = 1.73$. So the population will reach 5000 fish in about 1 year and 9 months.

85. (a) $\ln\left(\dfrac{P}{P_0}\right) = -\dfrac{h}{k} \Leftrightarrow \dfrac{P}{P_0} = e^{-h/k} \Leftrightarrow P = P_0 e^{-h/k}$. Substituting $k = 7$ and $P_0 = 100$ we get $P = 100e^{-h/7}$.

(b) When $h = 4$ we have $P = 100e^{-4/7} \approx 56.47$ kPa.

87. (a) $I = \frac{60}{13}\left(1 - e^{-13t/5}\right) \Leftrightarrow \frac{13}{60}I = 1 - e^{-13t/5} \Leftrightarrow e^{-13t/5} = 1 - \frac{13}{60}I \Leftrightarrow -\frac{13}{5}t = \ln\left(1 - \frac{13}{60}I\right) \Leftrightarrow$

$t = -\frac{5}{13}\ln\left(1 - \frac{13}{60}I\right)$.

(b) Substituting $I = 2$, we have $t = -\frac{5}{13}\ln\left[1 - \frac{13}{60}(2)\right] \approx 0.218$ seconds.

89. Since $9^1 = 9$, $9^2 = 81$, and $9^3 = 729$, the solution of $9^x = 20$ must be between 1 and 2 (because 20 is between 9 and 81), whereas the solution to $9^x = 100$ must be between 2 and 3 (because 100 is between 81 and 729).

91. (a) $(x-1)^{\log(x-1)} = 100(x-1) \Leftrightarrow \log\left((x-1)^{\log(x-1)}\right) = \log(100(x-1)) \Leftrightarrow$

$[\log(x-1)]\log(x-1) = \log 100 + \log(x-1) \Leftrightarrow [\log(x-1)]^2 - \log(x-1) - 2 = 0 \Leftrightarrow$

$[\log(x-1) - 2][\log(x-1) + 1] = 0$. Thus either $\log(x-1) = 2 \Leftrightarrow x = 101$ or $\log(x-1) = -1 \Leftrightarrow x = \frac{11}{10}$.

(b) $\log_2 x + \log_4 x + \log_8 x = 11 \Leftrightarrow \log_2 x + \log_2 \sqrt{x} + \log_2 \sqrt[3]{x} = 11 \Leftrightarrow \log_2\left(x\sqrt{x}\sqrt[3]{x}\right) = 11 \Leftrightarrow \log_2\left(x^{11/6}\right) = 11$

$\Leftrightarrow \frac{11}{6}\log_2 x = 11 \Leftrightarrow \log_2 x = 6 \Leftrightarrow x = 2^6 = 64$

(c) $4^x - 2^{x+1} = 3 \Leftrightarrow (2^x)^2 - 2(2^x) - 3 = 0 \Leftrightarrow (2^x - 3)(2^x + 1) = 0 \Leftrightarrow$ either $2^x = 3 \Leftrightarrow x = \dfrac{\ln 3}{\ln 2}$ or $2^x = -1$, which

has no real solution. So $x = \dfrac{\ln 3}{\ln 2}$ is the only real solution.

4.6   MODELING WITH EXPONENTIAL AND LOGARITHMIC FUNCTIONS

1. (a) Here $n_0 = 10$ and $a = 1.5$ hours, so $n(t) = 10 \cdot 2^{t/1.5} = 10 \cdot 2^{2t/3}$.

(b) After 35 hours, there will be $n(35) = 10 \cdot 2^{2(35)/3} \approx 1.06 \times 10^8$ bacteria.

(c) $n(t) = 10 \cdot 2^{2t/3} = 10,000 \Leftrightarrow 2^{2t/3} = 1000 \Leftrightarrow \ln\left(2^{2t/3}\right) = \ln 1000 \Leftrightarrow \dfrac{2t}{3}\ln 2 = \ln 1000 \Leftrightarrow t = \dfrac{3}{2}\dfrac{\ln 1000}{\ln 2} \approx 14.9$,

so the bacteria count will reach 10,000 in about 14.9 hours.

3. (a) A model for the squirrel population is $n(t) = n_0 \cdot 2^{t/6}$. We are given

that $n(30) = 100,000$, so $n_0 \cdot 2^{30/6} = 100,000 \Leftrightarrow$

$n_0 = \dfrac{100,000}{2^5} = 3125$. Initially, there were approximately

3125 squirrels.

(b) In 10 years, we will have $t = 40$, so the population will be

$n(40) = 3125 \cdot 2^{40/6} \approx 317,480$ squirrels.

(c)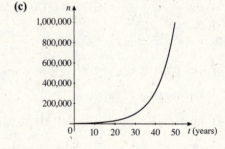

5. (a) $r = 0.08$ and $n(0) = 18,000$. Thus the population is given by the

formula $n(t) = 18,000e^{0.08t}$.

(b) $t = 2013 - 2005 = 8$. Then we have

$n(8) = 18,000e^{0.08(8)} = 18000e^{0.64} \approx 34,137$. Thus there should be

34,137 foxes in the region by the year 2013.

(c)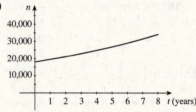

7. $n(t) = n_0 e^{rt}$; $n_0 = 110$ million, $t = 2020 - 1995 = 25$.

(a) $r = 0.03$; $n(25) = 110,000,000e^{0.03(25)} = 110,000,000e^{0.75} \approx 232,870,000$. Thus at a 3% growth rate, the projected population will be approximately 233 million people by the year 2020.

(b) $r = 0.02$; $n(25) = 110,000,000e^{0.02(25)} = 110,000,000e^{0.50} \approx 181,359,340$. Thus at a 2% growth rate, the projected population will be approximately 181 million people by the year 2020.

9. (a) The doubling time is 18 years and the initial population is 112,000, so

a model is $n(t) = 112,000 \cdot 2^{t/18}$.

(b) We need to find the relative growth rate r. Since the population is

$2 \cdot 112,000 = 224,000$ when $t = 18$, we have $224,000 = 112,000e^{18r}$

$\Leftrightarrow 2 = e^{18r} \Leftrightarrow \ln 2 = 18r \Leftrightarrow r = \dfrac{\ln 2}{18} \approx 0.0385$. Thus, a model is

$n(t) = 112,000e^{0.0385t}$.

(c)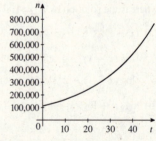

(d) Using the model in part (a), we solve the equation $n(t) = 112,000 \cdot 2^{t/18} = 500,000 \Leftrightarrow 2^{t/18} = \dfrac{125}{28} \Leftrightarrow$

$\ln 2^{t/18} = \ln \dfrac{125}{28} \Leftrightarrow \dfrac{t}{18}\ln 2 = \ln \dfrac{125}{28} \Leftrightarrow t = \dfrac{18\ln\frac{125}{28}}{\ln 2} \approx 38.85$. Therefore, the population should reach 500,000 in the

year 2045.

11. (a) The deer population in 2003 was 20,000.

(b) Using the model $n(t) = 20,000e^{rt}$ and the point $(4, 31000)$, we have $31,000 = 20,000e^{4r} \Leftrightarrow 1.55 = e^{4r} \Leftrightarrow$

$4r = \ln 1.55 \Leftrightarrow r = \frac{1}{4}\ln 1.55 \approx 0.1096$. Thus $n(t) = 20,000e^{0.1096t}$

(c) $n(8) = 20,000e^{0.1096(8)} \approx 48,218$, so the projected deer population in 2011 is about 48,000.

(d) $100{,}000 = 20{,}000e^{0.1096t} \Leftrightarrow 5 = e^{0.1096t} \Leftrightarrow 0.1096t = \ln 5 \Leftrightarrow t = \dfrac{\ln 5}{0.1096} \approx 14.63.$ Since $2003 + 14.63 = 2017.63$, the deer population will reach 100,000 during the year 2017.

13. (a) Using the formula $n(t) = n_0 e^{rt}$ with $n_0 = 8600$ and $n(1) = 10000$, we solve for r, giving $10000 = n(1) = 8600e^r$
$\Leftrightarrow \frac{50}{43} = e^r \Leftrightarrow r = \ln\left(\frac{50}{43}\right) \approx 0.1508.$ Thus $n(t) = 8600e^{0.1508t}$.

(b) $n(2) = 8600e^{0.1508(2)} \approx 11627.$ Thus the number of bacteria after two hours is about 11,600.

(c) $17200 = 8600e^{0.1508t} \Leftrightarrow 2 = e^{0.1508t} \Leftrightarrow 0.1508t = \ln 2 \Leftrightarrow t = \dfrac{\ln 2}{0.1508} \approx 4.596.$ Thus the number of bacteria will double in about 4.6 hours.

15. (a) Calculating dates relative to 1990 gives $n_0 = 29.76$ and $n(10) = 33.87.$ Then $n(10) = 29.76e^{10r} = 33.87 \Leftrightarrow$
$e^{10r} = \frac{33.87}{29.76} \approx 1.1381 \Leftrightarrow 10r = \ln 1.1381 \Leftrightarrow r = \frac{1}{10}\ln 1.1381 \approx 0.012936.$ Thus $n(t) = 29.76e^{0.012936t}$ million people.

(b) $2(29.76) = 29.76e^{0.012936t} \Leftrightarrow 2 = e^{0.012936t} \Leftrightarrow \ln 2 = 0.012936t \Leftrightarrow t = \dfrac{\ln 2}{0.012936} \approx 53.58,$ so the population doubles in about 54 years.

(c) $t = 2010 - 1990 = 20,$ so our model gives the 2010 population as $n(20) \approx 29.76e^{0.012936(20)} \approx 38.55$ million. The actual population was estimated at 36.96 million in 2009.

17. (a) Because the half-life is 1600 years and the sample weighs 22 mg initially, a suitable model is $m(t) = 22 \cdot 2^{-t/1600}$.

(b) From the formula for radioactive decay, we have $m(t) = m_0 e^{-rt}$, where $m_0 = 22$ and $r = \dfrac{\ln 2}{h} = \dfrac{\ln 2}{1600} \approx 0.000433.$
Thus, the amount after t years is given by $m(t) = 22e^{-0.000433t}$.

(c) $m(4000) = 22e^{-0.000433(4000)} \approx 3.89,$ so the amount after 4000 years is about 4 mg.

(d) We have to solve for t in the equation $18 = 22\,e^{-0.000433t}$. This gives $18 = 22e^{-0.000433t} \Leftrightarrow \frac{9}{11} = e^{-0.000433t} \Leftrightarrow$

$-0.000433t = \ln\left(\frac{9}{11}\right) \Leftrightarrow t = \dfrac{\ln\left(\frac{9}{11}\right)}{-0.000433} \approx 463.4,$ so it takes about 463 years.

19. By the formula in the text, $m(t) = m_0 e^{-rt}$ where $r = \dfrac{\ln 2}{h},$ so $m(t) = 50e^{-[(\ln 2)/28]t}.$ We need to solve for t in the

equation $32 = 50e^{-[(\ln 2)/28]t}$. This gives $e^{-[(\ln 2)/28]t} = \frac{32}{50} \Leftrightarrow -\dfrac{\ln 2}{28}t = \ln\left(\frac{32}{50}\right) \Leftrightarrow t = -\dfrac{28}{\ln 2}\cdot\ln\left(\frac{32}{50}\right) \approx 18.03,$ so it takes about 18 years.

21. By the formula for radioactive decay, we have $m(t) = m_0 e^{-rt}$, where $r = \dfrac{\ln 2}{h},$ in other words $m(t) = m_0 e^{-[(\ln 2)/h]t}.$ In

this exercise we have to solve for h in the equation $200 = 250e^{-[(\ln 2)/h]\cdot 48} \Leftrightarrow 0.8 = e^{-[(\ln 2)/h]\cdot 48} \Leftrightarrow \ln(0.8) = -\dfrac{\ln 2}{h}\cdot 48$

$\Leftrightarrow h = -\dfrac{\ln 2}{\ln 0.8}\cdot 48 \approx 149.1$ hours. So the half-life is approximately 149 hours.

23. By the formula in the text, $m(t) = m_0 e^{-[(\ln 2)/h]\cdot t},$ so we have $0.65 = 1\cdot e^{-[(\ln 2)/5730]\cdot t} \Leftrightarrow \ln(0.65) = -\dfrac{\ln 2}{5730}t \Leftrightarrow$

$t = -\dfrac{5730\ln 0.65}{\ln 2} \approx 3561.$ Thus the artifact is about 3560 years old.

25. (a) $T(0) = 65 + 145e^{-0.05(0)} = 65 + 145 = 210°$ F.

(b) $T(10) = 65 + 145e^{-0.05(10)} \approx 152.9.$ Thus the temperature after 10 minutes is about 153° F.

(c) $100 = 65 + 145e^{-0.05t} \Leftrightarrow 35 = 145e^{-0.05t} \Leftrightarrow 0.2414 = e^{-0.05t} \Leftrightarrow \ln 0.2414 = -0.05t \Leftrightarrow t = -\dfrac{\ln 0.2414}{0.05} \approx 28.4.$
Thus the temperature will be 100° F in about 28 minutes.

27. Using Newton's Law of Cooling, $T(t) = T_s + D_0 e^{-kt}$ with $T_s = 75$ and $D_0 = 185 - 75 = 110$. So $T(t) = 75 + 110 e^{-kt}$.

   (a) Since $T(30) = 150$, we have $T(30) = 75 + 110 e^{-30k} = 150 \Leftrightarrow 110 e^{-30k} = 75 \Leftrightarrow e^{-30k} = \frac{15}{22} \Leftrightarrow -30k = \ln\left(\frac{15}{22}\right)$

   $\Leftrightarrow k = -\frac{1}{30}\ln\left(\frac{15}{22}\right)$. Thus we have $T(45) = 75 + 110 e^{(45/30)\ln(15/22)} \approx 136.9$, and so the temperature of the turkey after 45 minutes is about $137°$ F.

   (b) The temperature will be $100°$F when $75 + 110 e^{(t/30)\ln(15/22)} = 100 \Leftrightarrow e^{(t/30)\ln(15/22)} = \frac{25}{110} = \frac{5}{22} \Leftrightarrow$

   $\left(\frac{t}{30}\right)\ln\left(\frac{15}{22}\right) = \ln\left(\frac{5}{22}\right) \Leftrightarrow t = 30\dfrac{\ln\left(\frac{5}{22}\right)}{\ln\left(\frac{15}{22}\right)} \approx 116.1$. So the temperature will be $100°$ F after 116 minutes.

29. (a) $\text{pH} = -\log[\text{H}^+] = -\log\left(5.0 \times 10^{-3}\right) \approx 2.3$

   (b) $\text{pH} = -\log[\text{H}^+] = -\log\left(3.2 \times 10^{-4}\right) \approx 3.5$

   (c) $\text{pH} = -\log[\text{H}^+] = -\log\left(5.0 \times 10^{-9}\right) \approx 8.3$

31. (a) $\text{pH} = -\log[\text{H}^+] = 3.0 \Leftrightarrow [\text{H}^+] = 10^{-3}$ M

   (b) $\text{pH} = -\log[\text{H}^+] = 6.5 \Leftrightarrow [\text{H}^+] = 10^{-6.5} \approx 3.2 \times 10^{-7}$ M

33. $4.0 \times 10^{-7} \le [\text{H}^+] \le 1.6 \times 10^{-5} \Leftrightarrow \log\left(4.0 \times 10^{-7}\right) \le \log[\text{H}^+] \le \log\left(1.6 \times 10^{-5}\right) \Leftrightarrow$

   $-\log\left(4.0 \times 10^{-7}\right) \ge \text{pH} \ge -\log\left(1.6 \times 10^{-5}\right) \Leftrightarrow 6.4 \ge \text{pH} \ge 4.8$. Therefore the range of pH readings for cheese is approximately 4.8 to 6.4.

35. Let I_0 be the intensity of the smaller earthquake and I_1 the intensity of the larger earthquake. Then $I_1 = 20I_0$.

   Notice that $M_0 = \log\left(\dfrac{I_0}{S}\right) = \log I_0 - \log S$ and $M_1 = \log\left(\dfrac{I_1}{S}\right) = \log\left(\dfrac{20I_0}{S}\right) = \log 20 + \log I_0 - \log S$. Then

   $M_1 - M_0 = \log 20 + \log I_0 - \log S - \log I_0 + \log S = \log 20 \approx 1.3$. Therefore the magnitude is 1.3 times larger.

37. Let the subscript A represent the Alaska earthquake and S represent the San Francisco earthquake. Then

   $M_A = \log\left(\dfrac{I_A}{S}\right) = 8.6 \Leftrightarrow I_A = S \cdot 10^{8.6}$; also, $M_S = \log\left(\dfrac{I_S}{S}\right) = 8.3 \Leftrightarrow I_S = S \cdot 10^{8.6}$. So

   $\dfrac{I_A}{I_S} = \dfrac{S \cdot 10^{8.6}}{S \cdot 10^{8.3}} = 10^{0.3} \approx 1.995$, and hence the Alaskan earthquake was roughly twice as intense as the San Francisco earthquake.

39. Let the subscript M represent the Mexico City earthquake, and T represent the Tangshan earthquake. We have $\dfrac{I_T}{I_M} = 1.26$

   $\Leftrightarrow \log 1.26 = \log\dfrac{I_T}{I_M} = \log\dfrac{I_T/S}{I_M/S} = \log\dfrac{I_T}{S} - \log\dfrac{I_M}{S} = M_T - M_M$. Therefore $M_T = M_M + \log 1.26 \approx 8.1 + 0.1 = 8.2$. Thus the magnitude of the Tangshan earthquake was roughly 8.2.

41. $\beta = 10\log\left(\dfrac{I}{I_0}\right) = 10\log\left(\dfrac{2.0 \times 10^{-5}}{1.0 \times 10^{-12}}\right) = 10\log\left(2 \times 10^7\right) = 10\left(\log 2 + \log 10^7\right) = 10\left(\log 2 + 7\right) \approx 73$. Therefore the intensity level was 73 dB.

43. (a) $\beta_1 = 10\log\left(\dfrac{I_1}{I_0}\right)$ and $I_1 = \dfrac{k}{d_1^2} \Leftrightarrow \beta_1 = 10\log\left(\dfrac{k}{d_1^2 I_0}\right) = 10\left[\log\left(\dfrac{k}{I_0}\right) - 2\log d_1\right] = 10\log\left(\dfrac{k}{I_0}\right) - 20\log d_1$.

   Similarly, $\beta_2 = 10\log\left(\dfrac{k}{I_0}\right) - 20\log d_2$. Substituting the expression for β_1 gives

   $\beta_2 = 10\log\left(\dfrac{k}{I_0}\right) - 20\log d_1 + 20\log d_1 - 20\log d_2 = \beta_1 + 20\log d_1 - 20\log d_2 = \beta_1 + 20\log\left(\dfrac{d_1}{d_2}\right)$.

(b) $\beta_1 = 120$, $d_1 = 2$, and $d_2 = 10$. Then $\beta_2 = \beta_1 + 20\log\left(\dfrac{d_1}{d_2}\right) = 120 + 20\log\left(\frac{2}{10}\right) = 120 + 20\log 0.2 \approx 106$, and

so the intensity level at 10 m is approximately 106 dB.

CHAPTER 4 REVIEW

1. $f(x) = 5^x$; $f(-1.5) \approx 0.0894$, $f\left(\sqrt{2}\right) \approx 9.739$, $f(2.5) \approx 55.902$

3. $g(x) = 4 \cdot \left(\frac{2}{3}\right)^{x-2}$; $g(-0.7) \approx 11.954$, $g(e) \approx 2.989$, $g(\pi) \approx 2.518$

5. $f(x) = 2^{-x+1}$. Domain $(-\infty, \infty)$, range $(0, \infty)$, asymptote $y = 0$.

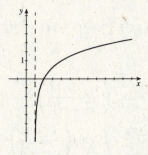

7. $g(x) = 3 + 2^x$. Domain $(-\infty, \infty)$, range $(3, \infty)$, asymptote $y = 3$.

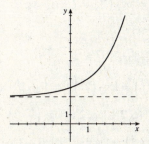

9. $f(x) = \log_3(x - 1)$. Domain $(1, \infty)$, range $(-\infty, \infty)$, asymptote $x = 1$.

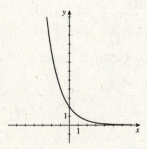

11. $f(x) = 2 - \log_2 x$. Domain $(0, \infty)$, range $(-\infty, \infty)$, asymptote $x = 0$.

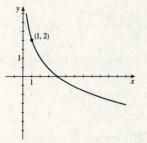
$(1, 2)$

13. $F(x) = e^x - 1$. Domain $(-\infty, \infty)$, range $(-1, \infty)$, asymptote $y = -1$.

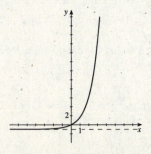

15. $g(x) = 2\ln x$. Domain $(0, \infty)$, range $(-\infty, \infty)$, asymptote $x = 0$.

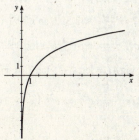

17. $f(x) = 10^{x^2} + \log(1 - 2x)$. Since $\log u$ is defined only for $u > 0$, we require $1 - 2x > 0 \Leftrightarrow -2x > -1 \Leftrightarrow x < \frac{1}{2}$, and so the domain is $\left(-\infty, \frac{1}{2}\right)$.

19. $h(x) = \ln\left(x^2 - 4\right)$. We must have $x^2 - 4 > 0$ (since $\ln y$ is defined only for $y > 0$) $\Leftrightarrow x^2 - 4 > 0 \Leftrightarrow (x - 2)(x + 2) > 0$. The endpoints of the intervals are -2 and 2.

Interval	$(-\infty, -2)$	$(-2, 2)$	$(2, \infty)$
Sign of $x - 2$	$-$	$-$	$+$
Sign of $x + 2$	$-$	$+$	$+$
Sign of $(x - 2)(x + 2)$	$+$	$-$	$+$

Thus the domain is $(-\infty, -2) \cup (2, \infty)$.

21. $\log_2 1024 = 10 \Leftrightarrow 2^{10} = 1024$

23. $\log x = y \Leftrightarrow 10^y = x$

25. $2^6 = 64 \Leftrightarrow \log_2 64 = 6$

27. $10^x = 74 \Leftrightarrow \log_{10} 74 = x \Leftrightarrow \log 74 = x$

29. $\log_2 128 = \log_2 \left(2^7\right) = 7$

31. $10^{\log 45} = 45$

33. $\ln\left(e^6\right) = 6$

35. $\log_3 \frac{1}{27} = \log_3 3^{-3} = -3$

37. $\log_5 \sqrt{5} = \log_5 5^{1/2} = \frac{1}{2}$

39. $\log 25 + \log 4 = \log(25 \cdot 4) = \log 10^2 = 2$

41. $\log_2 \left(16^{23}\right) = \log_2 \left(2^4\right)^{23} = \log_2 2^{92} = 92$

43. $\log_8 6 - \log_8 3 + \log_8 2 = \log_8 \left(\frac{6}{3} \cdot 2\right) = \log_8 4 = \log_8 8^{2/3} = \frac{2}{3}$

45. $\log\left(AB^2 C^3\right) = \log A + 2 \log B + 3 \log C$

47. $\ln \sqrt{\dfrac{x^2 - 1}{x^2 + 1}} = \frac{1}{2} \ln\left(\dfrac{x^2 - 1}{x^2 + 1}\right) = \frac{1}{2}\left[\ln\left(x^2 - 1\right) - \ln\left(x^2 + 1\right)\right]$

49. $\log_5 \left(\dfrac{x^2 (1 - 5x)^{3/2}}{\sqrt{x^3 - x}}\right) = \log_5 x^2 (1 - 5x)^{3/2} - \log_5 \sqrt{x(x^2 - 1)} = 2 \log_5 x + \frac{3}{2} \log_5 (1 - 5x) - \frac{1}{2} \log_5 \left(x^3 - x\right)$

51. $\log 6 + 4 \log 2 = \log 6 + \log 2^4 = \log\left(6 \cdot 2^4\right) = \log 96$

53. $\frac{3}{2} \log_2 (x - y) - 2 \log_2 \left(x^2 + y^2\right) = \log_2 (x - y)^{3/2} - \log_2 \left(x^2 + y^2\right)^2 = \log_2 \left(\dfrac{(x - y)^{3/2}}{(x^2 + y^2)^2}\right)$

55. $\log(x - 2) + \log(x + 2) - \frac{1}{2} \log\left(x^2 + 4\right) = \log[(x - 2)(x + 2)] - \log \sqrt{x^2 + 4} = \log\left(\dfrac{x^2 - 4}{\sqrt{x^2 + 4}}\right)$

57. $3^{2x-7} = 27 \Leftrightarrow 3^{2x-7} = 3^3 \Leftrightarrow 2x - 7 = 3 \Leftrightarrow 2x = 10 \Leftrightarrow x = 5$

59. $2^{3x-5} = 7 \Leftrightarrow \log_2 \left(2^{3x-5}\right) = \log_2 7 \Leftrightarrow 3x - 5 = \log_2 7 \Leftrightarrow x = \frac{1}{3}\left(\log_2 7 + 5\right)$. Using the Change of Base Formula, we have $\log_2 7 = \frac{\log 7}{\log 2} \approx 2.807$, so $x \approx \frac{1}{3}(2.807 + 5) \approx 2.602$.

61. $4^{1-x} = 3^{2x+5} \Leftrightarrow \log 4^{1-x} = \log 3^{2x+5} \Leftrightarrow (1 - x) \log 4 = (2x + 5) \log 3 \Leftrightarrow \log 4 - 5 \log 3 = 2x \log 3 + x \log 4 \Leftrightarrow$
$x(\log 3 + \log 4) = \log 4 - 5 \log 3 \Leftrightarrow x = \dfrac{\log 4 - 5 \log 3}{2 \log 3 + \log 4} \approx -1.146$

63. $x^2 e^{2x} + 2x e^{2x} = 8e^{2x} \Leftrightarrow e^{2x}\left(x^2 + 2x - 8\right) = 0 \Leftrightarrow x^2 + 2x - 8 = 0$ (since $e^{2x} \neq 0$) $\Leftrightarrow (x + 4)(x - 2) = 0 \Leftrightarrow x = -4$
or $x = 2$

65. $\log_2 (1 - x) = 4 \Leftrightarrow 1 - x = 2^4 \Leftrightarrow x = 1 - 16 = -15$

67. $\log_8 (x + 5) - \log_8 (x - 2) = 1 \Leftrightarrow \log_8 \dfrac{x + 5}{x - 2} = 1 \Leftrightarrow \dfrac{x + 5}{x - 2} = 8 \Leftrightarrow x + 5 = 8(x - 2) \Leftrightarrow x + 5 = 8x - 16 \Leftrightarrow 7x = 21$
$\Leftrightarrow x = 3$

69. $5^{-2x/3} = 0.63 \Leftrightarrow \dfrac{-2x}{3} \log 5 = \log 0.63 \Leftrightarrow x = -\dfrac{3 \log 0.63}{2 \log 5} \approx 0.430618$

71. $5^{2x+1} = 3^{4x-1} \Leftrightarrow (2x+1) \log 5 = (4x-1) \log 3 \Leftrightarrow 2x \log 5 + \log 5 = 4x \log 3 - \log 3 \Leftrightarrow$

$x(2 \log 5 - 4 \log 3) = -\log 3 - \log 5 \Leftrightarrow x = \dfrac{\log 3 + \log 5}{4 \log 3 - 2 \log 5} \approx 2.303600$

73. $y = e^{x/(x+2)}$. Vertical asymptote $x = -2$, horizontal asymptote $y = 2.72$, no maximum or minimum.

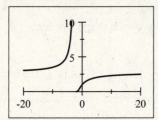

75. $y = \log\left(x^3 - x\right)$. Vertical asymptotes $x = -1$, $x = 0$, $x = 1$, no horizontal asymptote, local maximum of about -0.41 when $x \approx -0.58$.

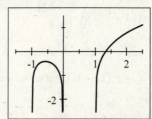

77. $3 \log x = 6 - 2x$. We graph $y = 3 \log x$ and $y = 6 - 2x$ in the same viewing rectangle. The solution occurs where the two graphs intersect. From the graphs, we see that the solution is $x \approx 2.42$.

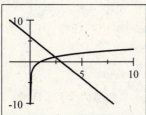

79. $\ln x > x - 2$. We graph the function $f(x) = \ln x - x + 2$, and we see that the graph lies above the x-axis for $0.16 < x < 3.15$. So the approximate solution of the given inequality is $0.16 < x < 3.15$.

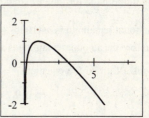

81. $f(x) = e^x - 3e^{-x} - 4x$. We graph the function $f(x)$, and we see that the function is increasing on $(-\infty, 0]$ and $[1.10, \infty)$ and that it is decreasing on $[0, 1.10]$.

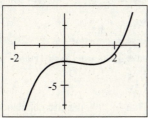

83. $\log_4 15 = \dfrac{\log 15}{\log 4} = 1.953445$

85. $\log_9 0.28 = \dfrac{\log 0.28}{\log 9} \approx -0.579352$

87. Notice that $\log_4 258 > \log_4 256 = \log_4 4^4 = 4$ and so $\log_4 258 > 4$. Also $\log_5 620 < \log_5 625 = \log_5 5^4 = 4$ and so $\log_5 620 < 4$. Then $\log_4 258 > 4 > \log_5 620$ and so $\log_4 258$ is larger.

89. $P = 12{,}000$, $r = 0.10$, and $t = 3$. Then $A = P\left(1 + \dfrac{r}{n}\right)^{nt}$.

(a) For $n = 2$, $A = 12{,}000\left(1 + \dfrac{0.10}{2}\right)^{2(3)} = 12{,}000\left(1.05^6\right) \approx \$16{,}081.15$.

(b) For $n = 12$, $A = 12{,}000 \left(1 + \frac{0.10}{12}\right)^{12(3)} \approx \$16{,}178.18$.

(c) For $n = 365$, $A = 12{,}000 \left(1 + \frac{0.10}{365}\right)^{365(3)} \approx \$16{,}197.64$.

(d) For $n = \infty$, $A = Pe^{rt} = 12{,}000 e^{0.10(3)} \approx \$16{,}198.31$.

91. We use the formula $A = P \left(1 + \frac{r}{n}\right)^{nt}$ with $P = 100{,}000$, $r = 0.052$, $n = 365$, and $A = 100{,}000 + 10{,}000 = 110{,}000$,

and solve for t: $110{,}000 = 100{,}000 \left(1 + \frac{0.052}{365}\right)^{365t} \Leftrightarrow 1.1 = \left(1 + \frac{0.052}{365}\right)^{365t} \Leftrightarrow \log 1.1 = 365t \log \left(1 + \frac{0.052}{365}\right) \Leftrightarrow$

$t = \dfrac{\log 1.1}{365 \log \left(1 + \frac{0.052}{365}\right)} \approx 1.833$. The account will accumulate \$10,000 in interest in approximately 1.8 years.

93. After one year, a principal P will grow to the amount $A = P \left(1 + \dfrac{0.0425}{365}\right)^{365} = P(1.04341)$. The formula for simple

interest is $A = P(1 + r)$. Comparing, we see that $1 + r = 1.04341$, so $r = 0.04341$. Thus the annual percentage yield is 4.341%.

95. (a) Using the model $n(t) = n_0 e^{rt}$, with $n_0 = 30$ and $r = 0.15$, we have the formula $n(t) = 30 e^{0.15t}$.

(b) $n(4) = 30 e^{0.15(4)} \approx 55$.

(c) $500 = 30 e^{0.15t} \Leftrightarrow \frac{50}{3} = e^{0.15t} \Leftrightarrow 0.15t = \ln\left(\frac{50}{3}\right) \Leftrightarrow t = \frac{1}{0.15} \ln\left(\frac{50}{3}\right) \approx 18.76$. So the stray cat population will reach 500 in about 19 years.

97. (a) From the formula for radioactive decay, we have $m(t) = 10 e^{-rt}$, where $r = -\dfrac{\ln 2}{2.7 \times 10^5}$. So after 1000 years

the amount remaining is $m(1000) = 10 \cdot e^{[-\ln 2 /(2.7 \times 10^5)] \cdot 1000} = 10 e^{-(\ln 2)/(2.7 \times 10^2)} = 10 e^{-(\ln 2)/270} \approx 9.97$. Therefore the amount remaining is about 9.97 mg.

(b) We solve for t in the equation $7 = 10 e^{-[\ln 2/(2.7 \times 10^5)] \cdot t}$. We have $7 = 10 e^{-[\ln 2/(2.7 \times 10^5)] \cdot t} \Leftrightarrow$

$0.7 = e^{-[\ln 2/(2.7 \times 10^5)] \cdot t} \Leftrightarrow \ln 0.7 = -\dfrac{\ln 2}{2.7 \times 10^5} \cdot t \Leftrightarrow t = -\dfrac{\ln 0.7}{\ln 2} \cdot 2.7 \times 10^5 \approx 138{,}934.75$. Thus it takes about 139,000 years.

99. (a) From the formula for radioactive decay, $r = \dfrac{\ln 2}{1590} \approx 0.0004359$ and $n(t) = 150 \cdot e^{-0.0004359t}$.

(b) $n(1000) = 150 \cdot e^{-0.0004359 \cdot 1000} \approx 97.00$, and so the amount remaining is about 97.00 mg.

(c) Find t so that $50 = 150 \cdot e^{-0.0004359t}$. We have $50 = 150 \cdot e^{-0.0004359t} \Leftrightarrow \frac{1}{3} = e^{-0.0004359t} \Leftrightarrow$

$t = -\dfrac{1}{0.0004359} \ln\left(\frac{1}{3}\right) \approx 2520$. Thus only 50 mg remain after about 2520 years.

101. (a) Using $n_0 = 1500$ and $n(5) = 3200$ in the formula $n(t) = n_0 e^{rt}$, we have $3200 = n(5) = 1500 e^{5r} \Leftrightarrow e^{5r} = \frac{32}{15} \Leftrightarrow$

$5r = \ln\left(\frac{32}{15}\right) \Leftrightarrow r = \frac{1}{5} \ln\left(\frac{32}{15}\right) \approx 0.1515$. Thus $n(t) = 1500 \cdot e^{0.1515t}$.

(b) We have $t = 1999 - 1988 = 11$ so $n(11) = 1500 e^{0.1515 \cdot 11} \approx 7940$. Thus in 1999 the bird population should be about 7940.

103. $[H^+] = 1.3 \times 10^{-8}$ M. Then $pH = -\log[H^+] = -\log\left(1.3 \times 10^{-8}\right) \approx 7.9$, and so fresh egg whites are basic.

105. Let I_0 be the intensity of the smaller earthquake and I_1 be the intensity of the larger earthquake. Then $I_1 = 35 I_0$. Since

$M = \log\left(\dfrac{I}{S}\right)$, we have $M_0 = \log\left(\dfrac{I_0}{S}\right) = 6.5$ and $M_1 = \log\left(\dfrac{I_1}{S}\right) = \log\left(\dfrac{35 I_0}{S}\right) = \log 35 + \log\left(\dfrac{I_0}{S}\right) = \log 35 +$

$M_0 = \log 35 + 6.5 \approx 8.04$. So the magnitude on the Richter scale of the larger earthquake is approximately 8.0.

CHAPTER 4 TEST

1. (a)

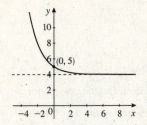

$f(x) = 2^{-x} + 4$ has domain $(-\infty, \infty)$, range $(4, \infty)$, and horizontal asymptote $y = 4$.

(b)

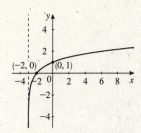

$g(x) = \log_3(x + 3)$ has domain $(-3, \infty)$, range $(-\infty, \infty)$, and vertical asymptote $x = -3$.

3. Find the exact value of each expression.

(a) $10^{\log 36} = 36$

(b) $\ln e^3 = 3$

(c) $\log_3 \sqrt{27} = \log_3 \left(3^3\right)^{1/2} = \log_3 3^{3/2} = \frac{3}{2}$

(d) $\log_2 80 - \log_2 10 = \log_2 \left(\frac{80}{10}\right) = \log_2 8 = \log_2 2^3 = 3$

(e) $\log_8 4 = \log_8 8^{2/3} = \frac{2}{3}$

(f) $\log_6 4 + \log_6 9 = \log_6 (4 \cdot 9) = \log_6 6^2 = 2$

5. $\ln x - 2\ln\left(x^2 + 1\right) + \frac{1}{2}\ln\left(3 - x^4\right) = \ln\left(x\sqrt{3 - x^4}\right) - \ln\left(x^2 + 1\right)^2 = \ln\left(\dfrac{x\sqrt{3 - x^4}}{\left(x^2 + 1\right)^2}\right)$

7. (a) From the formula for population growth, we have $8000 = 1000e^{r \cdot 1}$
$\Leftrightarrow 8 = e^r \Leftrightarrow r = \ln 8 \approx 2.07944$. Thus $n(t) = 1000e^{2.07944t}$.

(b) $n(1.5) = 1000e^{2.07944(1.5)} \approx 22{,}627$

(c) $15000 = 1000e^{2.07944t} \Leftrightarrow 15 = e^{2.07944t} \Leftrightarrow \ln 15 = 2.07944t \Leftrightarrow$
$t = \dfrac{\ln 15}{2.07944} \approx 1.3$. Thus the population will reach 15,000 after approximately 1.3 hours.

(d)

9. (a) The initial mass is $m_0 = 3$ and the half-life is $h = 10$, so using the radioactive decay model with $m_0 = 3$ and $r = \frac{\ln 2}{h} = \frac{\ln 2}{10}$, we have $A(t) = 3e^{-[(\ln 2)/10]t} \approx 3e^{-0.069t}$.

(b) After 1 minute $= 60$ seconds, the amount remaining is $A(60) = 3e^{-0.069(60)} \approx 0.048$ g.

(c) We solve $3e^{-0.069t} = 10^{-6} \Leftrightarrow e^{-0.069t} = \dfrac{10^{-6}}{3} \Leftrightarrow \ln e^{-0.069t} = \ln\left(\dfrac{10^{-6}}{3}\right) \Leftrightarrow -0.069t = \ln\left(\dfrac{10^{-6}}{3}\right) \Leftrightarrow$

$t = -\dfrac{1}{0.069}\ln\left(\dfrac{10^{-6}}{3}\right) \approx 216$, so there is 1 μg of ^{91}Kr remaining after about 216 seconds, or 3.6 minutes.

FOCUS ON MODELING   Fitting Exponential and Power Curves to Data

1. (a)

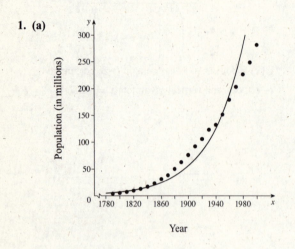

Year

(b) Using a graphing calculator, we obtain the model
$y = ab^t$, where $a = 1.1806094 \times 10^{-15}$ and
$b = 1.0204139$, and y is the population (in millions) in
the year t.

(c) Substituting $t = 2010$ into the model of part (b), we get
$y = ab^{2010} \approx 515.9$ million.

(d) According to the model, the population in 1965 should
have been about $y = ab^{1965} \approx 207.8$ million.

(e) The values given by the model are clearly much too
large. This means that an exponential model is *not*
appropriate for these data.

3. (a) Yes.

(b)

Year t	Health Expenditures E ($bn)	ln E
1970	74.3	4.30811
1980	251.1	5.52585
1985	434.5	6.07420
1987	506.2	6.22693
1990	696.6	6.54621
1992	820.3	6.70967
1994	937.2	6.84290
1996	1039.4	6.94640
1998	1150.0	7.04752
2000	1310.0	7.17778
2001	1424.5	7.26158

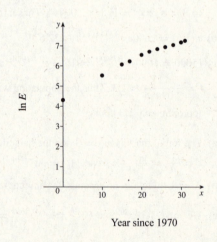

Year since 1970

Yes, the scatter plot appears to be roughly linear.

(c) Let t be the number of years elapsed since 1970 . Then $\ln E = 4.551437 + 0.09238268t$, where E is expenditure in
billions of dollars.

(d) $E = e^{4.551437+0.09238268t} = 94.76849e^{0.09238268t}$

(e) In 2009 we have $t = 2009 - 1970 = 39$, so the estimated 2009 health-care expenditures are
$94.76849e^{0.09238268(39)} \approx 3478.5$ billion dollars.

5. (a) Using a graphing calculator, we find that
$I_0 = 22.7586444$ and $k = 0.1062398$.

(c) We solve $0.15 = 22.7586444e^{-0.1062398x}$ for x:

$0.15 = 22.7586444e^{-0.1062398x} \Leftrightarrow$

$0.006590902 = e^{-0.1062398x} \Leftrightarrow$
$-5.022065 = -0.1062398x \Leftrightarrow x \approx 47.27$. So light

intensity drops below 0.15 lumens below around

47.27 feet.

(b)

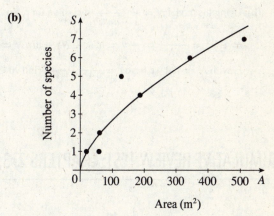

Depth (ft)

7. (a) Let A be the area of the cave and S the number of
species of bat. Using a graphing calculator, we obtain

the power function model $S = 0.14A^{0.64}$.

(c) According to the model, there are

$S = 0.14(205)^{0.64} \approx 4$ species of bat living in the El

Sapo cave.

(b)

Area (m²)

The model fits the data reasonably well.

9. (a)

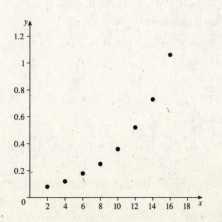

(b)

x	y	$\ln x$	$\ln y$
2	0.08	0.69315	-2.52573
4	0.12	1.38629	-2.12026
6	0.18	1.79176	-1.71480
8	0.25	2.07944	-1.38629
10	0.36	2.30259	-1.02165
12	0.52	2.48491	-0.65393
14	0.73	2.63906	-0.31471
16	1.06	2.77259	0.05827

(b) (cont'd)

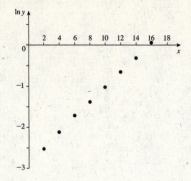

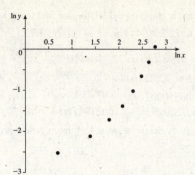

(c) The exponential function.

(d) $y = a \cdot b^x$ where $a = 0.057697$ and $b = 1.200236$.

11. (a) Using the `Logistic` command on a TI-83 we get $y = \dfrac{c}{1 + ae^{-bx}}$ where $a = 49.10976596$, $b = 0.4981144989$, and

$c = 500.855793$.

(b) Using the model $N = \dfrac{c}{1 + ae^{-bt}}$ we solve for t. So $N = \dfrac{c}{1 + ae^{-bt}} \Leftrightarrow 1 + ae^{-bt} = \dfrac{c}{N} \Leftrightarrow ae^{-bt} = \left(\dfrac{c}{N}\right) - 1 = \dfrac{c - N}{N}$

$\Leftrightarrow e^{-bt} = \dfrac{c - N}{aN} \Leftrightarrow -bt = \ln(c - N) - \ln aN \Leftrightarrow t = \dfrac{1}{b}[\ln aN - \ln(c - N)]$. Substituting the values for a, b, and

c, with $N = 400$ we have $t = \dfrac{1}{0.4981144989}(\ln 19643.90638 - \ln 100.855793) \approx 10.58$ days.

CUMULATIVE REVIEW TEST: CHAPTERS 2, 3, and 4

1. $f(x) = x^2 - 4x$, $g(x) = \sqrt{x + 4}$

(a) The domain of f is $(-\infty, \infty)$.

(b) The domain of g is the set of all x for which $x + 4 \geq 0 \Leftrightarrow x \geq -4$, that is, $[-4, \infty)$.

(c) $f(-2) = (-2)^2 - 4(-2) = 12$, $f(0) = 0^2 - 4(0) = 0$, $f(4) = 4^2 - 4(4) = 0$, $g(0) = \sqrt{0 + 4} = 2$,

$g(8) = \sqrt{8 + 4} = 2\sqrt{3}$, $g(-6) = \sqrt{-6 + 4} = \sqrt{-2}$, which is undefined.

(d) $f(x + 2) = (x + 2)^2 - 4(x + 2) = x^2 + 4x + 4 - 4x - 8 = x^2 - 4$, $g(x + 2) = \sqrt{(x + 2) + 4} = \sqrt{x + 6}$,

$f(2 + h) = (2 + h)^2 - 4(2 + h) = 4 + 4h + h^2 - 8 - 4h = h^2 - 4$

(e) $\dfrac{g(21) - g(5)}{21 - 5} = \dfrac{\sqrt{21 + 4} - \sqrt{5 + 4}}{16} = \dfrac{5 - 3}{16} = \dfrac{1}{8}$

(f) $f \circ g(x) = f(g(x)) = (\sqrt{x + 4})^2 - 4(\sqrt{x + 4}) = x + 4 - 4\sqrt{x + 4}$,

$g \circ f(x) = g(f(x)) = \sqrt{x^2 - 4x + 4} = \sqrt{(x - 2)^2} = |x - 2|$, $f(g(12)) = 12 + 4 - 4\sqrt{12 + 4} = 16 - 4\sqrt{16} = 0$,

$g(f(12)) = 12 - 2 = 10$

(g) $y = \sqrt{x + 4} \Rightarrow y^2 = x + 4 \Leftrightarrow x = y^2 - 4$. Reverse x and y: $y = x^2 - 4$. Thus, the inverse of g is $g^{-1}(x) = x^2 - 4$,

$x \geq 0$.

3. $f(x) = -2x^2 + 8x + 5$

(a) $f(x) = -2\left(x^2 - 4x\right) + 5 = -2\left(x^2 - 4x + 4\right) + 8 +$

  $5 = -2(x - 2)^2 + 13$

(b) Because $a = -2 < 0$, f has a maximum value of 13 at $x = 2$.

(d) f is increasing on $(-\infty, 2]$ and decreasing on $[2, \infty)$.

(e) $g(x) = -2x^2 + 8x + 10 = f(x) + 5$, so its graph is obtained by shifting that of f upward 5 units.

(f) $h(x) = -2(x + 3)^2 + 8(x + 3) + 5 = f(x + 3)$, so its graph is obtained by shifting that of f to the left 3 units.

(c)

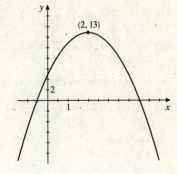

5. (a) $P(x) = 2x^3 - 11x^2 + 10x + 8$ has possible rational zeros $\pm\frac{1}{2}, \pm 1,$

  $\pm 2, \pm 4, \pm 8$.

(b) P has 2 variations in sign and hence 0 or 2 positive real zeros.

  $P(-x) = -x^3 - x^2 + 8x + 12$ has 1 variation in sign and hence

  1 negative real zero.

$$
\begin{array}{r|rrrr}
2 & 2 & -11 & 10 & 8 \\
  &   & 4 & -14 & -8 \\
\hline
  & 2 & -7 & -4 & 0
\end{array}
\quad \Rightarrow x = 2 \text{ is a zero.}
$$

  $P(x) = 2x^3 - 11x^2 + 10x + 8 = (x - 2)\left(2x^2 - 7x - 4\right)$

    $= (x - 2)(2x + 1)(x - 4)$

  Therefore, the zeros are $-\frac{1}{2}$, 2, and 4.

(c) See part (b).

(d)

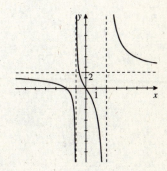

7. $r(x) = \dfrac{3x^2 + 6x}{x^2 - x - 2} = \dfrac{3x(x + 2)}{(x + 1)(x - 2)}$ has x-intercepts 0 and -2, y-intercept 0, horizontal asymptote $y = 3$, and vertical asymptotes -1 and 2.

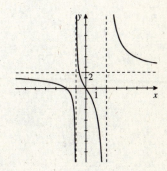

9. (a) $\log_3 16 - 2\log_3 36 = \log_3 16 - \log_3 36^2 = \log_3 \dfrac{16}{36^2} = \log_3 \dfrac{2^4}{\left(2^2 \cdot 3^2\right)^2} = \log_3 \left(3^{-4}\right) = -4$

(b) $\log\left(\dfrac{x^5 \sqrt{x - 1}}{2x - 3}\right) = \log\left(\dfrac{x^5 (x - 1)^{1/2}}{2x - 3}\right) = \log x^5 + \log(x - 1)^{1/2} - \log(2x - 3)$

  $= 5\log x + \frac{1}{2}\log(x - 1) - \log(2x - 3)$

11. $P = 25{,}000$, $r = 0.054$, and $n = 365$. So $A(t) = 25{,}000 \left(1 + \frac{0.054}{365}\right)^{365t}$.

(a) $A(3) = 25{,}000 \left(1 + \frac{0.054}{365}\right)^{365(3)} \approx 29{,}396.15$, so after 3 years the account contains \$29,396.15.

(b) $A(t) = 35{,}000 \Leftrightarrow 25\left(1 + \frac{0.054}{365}\right)^{365t} = 35 \Leftrightarrow \ln 25 + 365t \ln\left(1 + \frac{0.054}{365}\right) = \ln 35 \Leftrightarrow t = \dfrac{\ln 35 - \ln 25}{365 \ln\left(1 + \frac{0.054}{365}\right)} \approx 6.23$,

so the account will have grown to \$35,000 after about 6.23 years.

(c) $A(t) = 50{,}000 \Leftrightarrow \left(1 + \frac{0.054}{365}\right)^{365t} = 2 \Leftrightarrow t = \dfrac{\ln 2}{365 \ln\left(1 + \frac{0.054}{365}\right)} \approx 12.84$, so the initial deposit will double in about

12.84 years.

5 TRIGONOMETRIC FUNCTIONS: UNIT CIRCLE APPROACH

5.1 THE UNIT CIRCLE

1. (a) The unit circle is the circle centered at $(0, 0)$ with radius 1.

(b) The equation of the unit circle is $x^2 + y^2 = 1$.

(c) (i) Since $1^2 + 0^2 = 1$, the point is $P(1, 0)$.　　**(ii)** $P(0, 1)$　　**(iii)** $P(-1, 0)$　　**(iv)** $P(0, -1)$

3. Since $\left(\frac{4}{5}\right)^2 + \left(-\frac{3}{5}\right)^2 = \frac{16}{25} + \frac{9}{25} = 1$, $P\left(\frac{4}{5}, -\frac{3}{5}\right)$ lies on the unit circle.

5. Since $\left(\frac{7}{25}\right)^2 + \left(\frac{24}{25}\right)^2 = \frac{49}{625} + \frac{576}{625} = 1$, $P\left(\frac{7}{25}, \frac{24}{25}\right)$ lies on the unit circle.

7. Since $\left(-\frac{\sqrt{5}}{3}\right)^2 + \left(\frac{2}{3}\right)^2 = \frac{5}{9} + \frac{4}{9} = 1$, $P\left(-\frac{\sqrt{5}}{3}, \frac{2}{3}\right)$ lies on the unit circle.

9. $\left(-\frac{3}{5}\right)^2 + y^2 = 1 \Leftrightarrow y^2 = 1 - \frac{9}{25} \Leftrightarrow y^2 = \frac{16}{25} \Leftrightarrow y = \pm\frac{4}{5}$. Since $P(x, y)$ is in quadrant III, y is negative, so the point is

$P\left(-\frac{3}{5}, -\frac{4}{5}\right)$.

11. $x^2 + \left(\frac{1}{3}\right)^2 = 1 \Leftrightarrow x^2 = 1 - \frac{1}{9} \Leftrightarrow x^2 = \frac{8}{9} \Leftrightarrow x = \pm\frac{2\sqrt{2}}{3}$. Since P is in quadrant II, x is negative, so the point is

$P\left(-\frac{2\sqrt{2}}{3}, \frac{1}{3}\right)$.

13. $x^2 + \left(-\frac{2}{7}\right)^2 = 1 \Leftrightarrow x^2 = 1 - \frac{4}{49} \Leftrightarrow x^2 = \frac{45}{49} \Leftrightarrow x = \pm\frac{3\sqrt{5}}{7}$. Since $P(x, y)$ is in quadrant IV, x is positive, so the point is

$P\left(\frac{3\sqrt{5}}{7}, -\frac{2}{7}\right)$.

15. $\left(\frac{4}{5}\right)^2 + y^2 = 1 \Leftrightarrow y^2 = 1 - \frac{16}{25} \Leftrightarrow y^2 = \frac{9}{25} \Leftrightarrow y = \pm\frac{3}{5}$. Since its y-coordinate is positive, the point is $P\left(\frac{4}{5}, \frac{3}{5}\right)$.

17. $x^2 + \left(\frac{2}{3}\right)^2 = 1 \Leftrightarrow x^2 = 1 - \frac{4}{9} \Leftrightarrow x^2 = \frac{5}{9} \Leftrightarrow x = \pm\frac{\sqrt{5}}{3}$. Since its x-coordinate is negative, the point is $P\left(-\frac{\sqrt{5}}{3}, \frac{2}{3}\right)$.

19. $\left(-\frac{\sqrt{2}}{3}\right)^2 + y^2 = 1 \Leftrightarrow y^2 = 1 - \frac{2}{9} \Leftrightarrow y^2 = \frac{7}{9} \Leftrightarrow y = \pm\frac{\sqrt{7}}{3}$. Since P lies below the x-axis, its y-coordinate is negative, so

the point is $P\left(-\frac{\sqrt{2}}{3}, -\frac{\sqrt{7}}{3}\right)$.

21.

t	Terminal Point
0	$(1, 0)$
$\frac{\pi}{4}$	$\left(\frac{\sqrt{2}}{2}, \frac{\sqrt{2}}{2}\right)$
$\frac{\pi}{2}$	$(0, 1)$
$\frac{3\pi}{4}$	$\left(-\frac{\sqrt{2}}{2}, \frac{\sqrt{2}}{2}\right)$
π	$(-1, 0)$

t	Terminal Point
π	$(-1, 0)$
$\frac{5\pi}{4}$	$\left(-\frac{\sqrt{2}}{2}, -\frac{\sqrt{2}}{2}\right)$
$\frac{3\pi}{2}$	$(0, -1)$
$\frac{7\pi}{4}$	$\left(\frac{\sqrt{2}}{2}, -\frac{\sqrt{2}}{2}\right)$
2π	$(1, 0)$

23. $P(x, y) = (0, 1)$

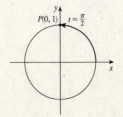

203

25. $P(x, y) = \left(-\frac{\sqrt{3}}{2}, \frac{1}{2}\right)$

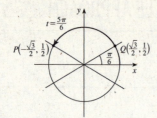

27. $P(x, y) = \left(\frac{1}{2}, -\frac{\sqrt{3}}{2}\right)$

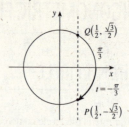

29. $P(x, y) = \left(-\frac{1}{2}, \frac{\sqrt{3}}{2}\right)$

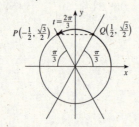

31. $P(x, y) = \left(-\frac{\sqrt{2}}{2}, -\frac{\sqrt{2}}{2}\right)$

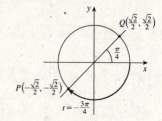

33. Let $Q(x, y) = \left(\frac{3}{5}, \frac{4}{5}\right)$ be the terminal point determined by t.

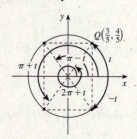

(a) $\pi - t$ determines the point $P(-x, y) = \left(-\frac{3}{5}, \frac{4}{5}\right)$.

(b) $-t$ determines the point $P(x, -y) = \left(\frac{3}{5}, -\frac{4}{5}\right)$.

(c) $\pi + t$ determines the point $P(-x, -y) = \left(-\frac{3}{5}, -\frac{4}{5}\right)$.

(d) $2\pi + t$ determines the point $P(x, y) = \left(\frac{3}{5}, \frac{4}{5}\right)$.

35. (a) $\bar{t} = \frac{5\pi}{4} - \pi = \frac{\pi}{4}$

    (b) $\bar{t} = \frac{7\pi}{3} - 2\pi = \frac{\pi}{3}$

    (c) $\bar{t} = \frac{4\pi}{3} - \pi = \frac{\pi}{3}$

    (d) $\bar{t} = \frac{\pi}{6}$

37. (a) $\bar{t} = \pi - \frac{5\pi}{7} = \frac{2\pi}{7}$

    (b) $\bar{t} = \pi - \frac{7\pi}{9} = \frac{2\pi}{9}$

    (c) $\bar{t} = \pi - 3 \approx 0.142$

    (d) $\bar{t} = 2\pi - 5 \approx 1.283$

39. (a) $\bar{t} = \pi - \frac{2\pi}{3} = \frac{\pi}{3}$

    (b) $P\left(-\frac{1}{2}, \frac{\sqrt{3}}{2}\right)$

41. (a) $\bar{t} = \pi - \frac{3\pi}{4} = \frac{\pi}{4}$

    (b) $P\left(-\frac{\sqrt{2}}{2}, \frac{\sqrt{2}}{2}\right)$

43. (a) $\bar{t} = \pi - \frac{2\pi}{3} = \frac{\pi}{3}$

    (b) $P\left(-\frac{1}{2}, -\frac{\sqrt{3}}{2}\right)$

45. (a) $\bar{t} = \frac{13\pi}{4} - 3\pi = \frac{\pi}{4}$

    (b) $P\left(-\frac{\sqrt{2}}{2}, -\frac{\sqrt{2}}{2}\right)$

47. (a) $\bar{t} = \frac{7\pi}{6} - \pi = \frac{\pi}{6}$

    (b) $P\left(-\frac{\sqrt{3}}{2}, -\frac{1}{2}\right)$

49. (a) $\bar{t} = 4\pi - \frac{11\pi}{3} = \frac{\pi}{3}$

    (b) $P\left(\frac{1}{2}, \frac{\sqrt{3}}{2}\right)$

51. (a) $\bar{t} = \frac{16\pi}{3} - 5\pi = \frac{\pi}{3}$

    (b) $P\left(-\frac{1}{2}, -\frac{\sqrt{3}}{2}\right)$

53. $t = 1 \implies (0.5, 0.8)$

55. $t = -1.1 \implies (0.5, -0.9)$

57. The distances PQ and PR are equal because they both subtend arcs of length $\frac{\pi}{3}$. Since $P(x, y)$ is

a point on the unit circle, $x^2 + y^2 = 1$. Now $d(P, Q) = \sqrt{(x-x)^2 + (y-(-y))^2} = 2y$ and

$d(R, S) = \sqrt{(x-0)^2 + (y-1)^2} = \sqrt{x^2 + y^2 - 2y + 1} = \sqrt{2 - 2y}$ (using the fact that $x^2 + y^2 = 1$). Setting these equal

gives $2y = \sqrt{2 - 2y} \implies 4y^2 = 2 - 2y \Leftrightarrow 4y^2 + 2y - 2 = 0 \Leftrightarrow 2(2y-1)(y+1) = 0$. So $y = -1$ or $y = \frac{1}{2}$. Since

P is in quadrant I, $y = \frac{1}{2}$ is the only viable solution. Again using $x^2 + y^2 = 1$ we have $x^2 + \left(\frac{1}{2}\right)^2 = 1 \Leftrightarrow x^2 = \frac{3}{4} \implies$

$x = \pm\frac{\sqrt{3}}{2}$. Again, since P is in quadrant I the coordinates must be $\left(\frac{\sqrt{3}}{2}, \frac{1}{2}\right)$.

5.2 TRIGONOMETRIC FUNCTIONS OF REAL NUMBERS

1. If $P(x, y)$ is the terminal point on the unit circle determined by t, then $\sin t = y$, $\cos t = x$, and $\tan t = y/x$.

3.

t	$\sin t$	$\cos t$
0	0	1
$\frac{\pi}{4}$	$\frac{\sqrt{2}}{2}$	$\frac{\sqrt{2}}{2}$
$\frac{\pi}{2}$	1	0
$\frac{3\pi}{4}$	$\frac{\sqrt{2}}{2}$	$-\frac{\sqrt{2}}{2}$
π	0	-1
$\frac{5\pi}{4}$	$-\frac{\sqrt{2}}{2}$	$-\frac{\sqrt{2}}{2}$
$\frac{3\pi}{2}$	-1	0
$\frac{7\pi}{4}$	$-\frac{\sqrt{2}}{2}$	$\frac{\sqrt{2}}{2}$
2π	0	1

5. (a) $\sin \frac{2\pi}{3} = \frac{\sqrt{3}}{2}$

(b) $\cos \frac{2\pi}{3} = -\frac{1}{2}$

(c) $\tan \frac{2\pi}{3} = -\sqrt{3}$

7. (a) $\sin \frac{7\pi}{6} = -\frac{1}{2}$

(b) $\sin(-\frac{\pi}{6}) = -\frac{1}{2}$

(c) $\sin \frac{11\pi}{6} = -\frac{1}{2}$

9. (a) $\cos \frac{3\pi}{4} = -\frac{\sqrt{2}}{2}$

(b) $\cos \frac{5\pi}{4} = -\frac{\sqrt{2}}{2}$

(c) $\cos \frac{7\pi}{4} = \frac{\sqrt{2}}{2}$

11. (a) $\sin \frac{7\pi}{3} = \frac{\sqrt{3}}{2}$

(b) $\csc \frac{7\pi}{3} = \frac{2\sqrt{3}}{3}$

(c) $\cot \frac{7\pi}{3} = \frac{\sqrt{3}}{3}$

13. (a) $\sin(-\frac{\pi}{2}) = -1$

(b) $\cos(-\frac{\pi}{2}) = 0$

(c) $\cot(-\frac{\pi}{2}) = 0$

15. (a) $\sec \frac{11\pi}{3} = 2$

(b) $\csc \frac{11\pi}{3} = -\frac{2\sqrt{3}}{3}$

(c) $\sec(-\frac{\pi}{3}) = 2$

17. (a) $\tan \frac{5\pi}{6} = -\frac{\sqrt{3}}{3}$

(b) $\tan \frac{7\pi}{6} = \frac{\sqrt{3}}{3}$

(c) $\tan \frac{11\pi}{6} = -\frac{\sqrt{3}}{3}$

19. (a) $\cos\left(-\frac{\pi}{4}\right) = \frac{\sqrt{2}}{2}$

(b) $\csc\left(-\frac{\pi}{4}\right) = -\sqrt{2}$

(c) $\cot\left(-\frac{\pi}{4}\right) = -1$

21. (a) $\csc\left(-\frac{\pi}{2}\right) = -1$

(b) $\csc \frac{\pi}{2} = 1$

(c) $\csc \frac{3\pi}{2} = -1$

23. (a) $\sin 13\pi = 0$

(b) $\cos 14\pi = 1$

(c) $\tan 15\pi = 0$

25. $t = 0 \implies \sin t = 0$, $\cos t = 1$, $\tan t = 0$, $\sec t = 1$, $\csc t$ and $\cot t$ are undefined.

27. $t = \pi \implies \sin t = 0$, $\cos t = -1$, $\tan t = 0$, $\sec t = -1$, $\csc t$ and $\cot t$ are undefined.

29. $\left(\frac{3}{5}\right)^2 + \left(\frac{4}{5}\right)^2 = \frac{9}{25} + \frac{1}{2} = 1$. So $\sin t = \frac{4}{5}$, $\cos t = \frac{3}{5}$, and $\tan t = \frac{\frac{4}{5}}{\frac{3}{5}} = \frac{4}{3}$.

31. $\left(\frac{\sqrt{5}}{4}\right)^2 + \left(-\frac{\sqrt{11}}{4}\right)^2 = \frac{5}{16} + \frac{11}{16} = 1$. So $\sin t = -\frac{\sqrt{11}}{4}$, $\cos t = \frac{\sqrt{5}}{4}$, and $\tan t = \frac{-\frac{\sqrt{11}}{4}}{\frac{\sqrt{5}}{4}} = -\frac{\sqrt{11}}{\sqrt{5}} = -\frac{\sqrt{55}}{5}$.

33. $\left(-\frac{6}{7}\right)^2 + \left(\frac{\sqrt{13}}{7}\right)^2 = \frac{36}{49} + \frac{13}{49} = 1$. So $\sin t = \frac{\sqrt{13}}{7}$, $\cos t = -\frac{6}{7}$, and $\tan t = \frac{\frac{\sqrt{13}}{7}}{-\frac{6}{7}} = -\frac{\sqrt{13}}{6}$.

35. $\left(-\frac{5}{13}\right)^2 + \left(-\frac{12}{13}\right)^2 = \frac{25}{169} + \frac{144}{169} = 1$. So $\sin t = -\frac{12}{13}$, $\cos t = -\frac{5}{13}$, and $\tan t = \frac{-\frac{12}{13}}{-\frac{5}{13}} = \frac{12}{5}$.

37. $\left(-\frac{20}{29}\right)^2 + \left(\frac{21}{29}\right)^2 = \frac{400}{841} + \frac{441}{841} = 1$. So $\sin t = \frac{21}{29}$, $\cos t = -\frac{20}{29}$, and $\tan t = \frac{\frac{21}{29}}{-\frac{20}{29}} = -\frac{21}{20}$.

39. (a) 0.8   **41. (a)** 0.9   **43. (a)** 1.0   **45. (a)** -0.6

    (b) 0.84147   **(b)** 0.93204   **(b)** 1.02964   **(b)** -0.57482

47. $\sin t \cdot \cos t$. Since $\sin t$ is positive in quadrant II and $\cos t$ is negative in quadrant II, their product is negative.

49. $\dfrac{\tan t \cdot \sin t}{\cot t} = \tan t \cdot \dfrac{1}{\cot t} \cdot \sin t = \tan t \cdot \tan t \cdot \sin t = \tan^2 t \cdot \sin t$. Since $\tan^2 t$ is always positive and $\sin t$ is negative in quadrant III, the expression is negative in quadrant III.

51. Quadrant II   **53.** Quadrant II

55. $\sin t = \sqrt{1 - \cos^2 t}$ **57.** $\tan t = \dfrac{\sin t}{\cos t} = \dfrac{\sin t}{\sqrt{1 - \sin^2 t}}$

59. $\sec t = -\sqrt{1 + \tan^2 t}$ **61.** $\tan t = \sqrt{\sec^2 t - 1}$

63. $\tan^2 t = \dfrac{\sin^2 t}{\cos^2 t} = \dfrac{\sin^2 t}{1 - \sin^2 t}$

65. $\sin t = \frac{3}{5}$ and t is in quadrant II, so the terminal point determined by t is $P\left(x, \frac{3}{5}\right)$. Since P is on the unit circle $x^2 + \left(\frac{3}{5}\right)^2 = 1$. Solving for x gives $x = \pm\sqrt{1 - \frac{9}{25}} = \pm\sqrt{\frac{1}{2}} = \pm\frac{4}{5}$. Since t is in quadrant III, $x = -\frac{4}{5}$. Thus the terminal point is $P\left(-\frac{4}{5}, \frac{3}{5}\right)$. Thus, $\cos t = -\frac{4}{5}$, $\tan t = -\frac{3}{4}$, $\csc t = \frac{5}{3}$, $\sec t = -\frac{5}{4}$, $\cot t = -\frac{4}{3}$.

67. $\sec t = 3$ and t lies in quadrant IV. Thus, $\cos t = \frac{1}{3}$ and the terminal point determined by t is $P\left(\frac{1}{3}, y\right)$. Since P is on the unit circle $\left(\frac{1}{3}\right)^2 + y^2 = 1$. Solving for y gives $y = \pm\sqrt{1 - \frac{1}{9}} = \pm\sqrt{\frac{8}{9}} = \pm\frac{2\sqrt{2}}{3}$. Since t is in quadrant IV, $y = -\frac{2\sqrt{2}}{3}$. Thus the terminal point is $P\left(\frac{1}{3}, -\frac{2\sqrt{2}}{3}\right)$. Therefore, $\sin t = -\frac{2\sqrt{2}}{3}$, $\cos t = \frac{1}{3}$, $\tan t = -2\sqrt{2}$, $\csc t = -\frac{3}{2\sqrt{2}} = -\frac{3\sqrt{2}}{4}$, $\cot t = -\frac{1}{2\sqrt{2}} = -\frac{\sqrt{2}}{4}$.

69. $\tan t = -\frac{3}{4}$ and $\cos t > 0$, so t is in quadrant IV. Since $\sec^2 t = \tan^2 t + 1$ we have $\sec^2 t = \left(-\frac{3}{4}\right)^2 + 1 = \frac{9}{16} + 1 = \frac{25}{16}$. Thus $\sec t = \pm\sqrt{\frac{25}{16}} = \pm\frac{5}{4}$. Since $\cos t > 0$, we have $\cos t = \dfrac{1}{\sec t} = \dfrac{1}{\frac{5}{4}} = \frac{4}{5}$. Let $P\left(\frac{4}{5}, y\right)$. Since $\tan t \cdot \cos t = \sin t$ we have $\sin t = \left(-\frac{3}{4}\right)\left(\frac{4}{5}\right) = -\frac{3}{5}$. Thus, the terminal point determined by t is $P\left(\frac{4}{5}, -\frac{3}{5}\right)$, and so $\sin t = -\frac{3}{5}$, $\cos t = \frac{4}{5}$, $\csc t = -\frac{5}{3}$, $\sec t = \frac{5}{4}$, $\cot t = -\frac{4}{3}$.

71. $\sin t = -\frac{1}{4}$, $\sec t < 0$, so t is in quadrant III. So the terminal point determined by t is $P\left(x, -\frac{1}{4}\right)$. Since P is on the unit circle $x^2 + \left(-\frac{1}{4}\right)^2 = 1$. Solving for x gives $x = \pm\sqrt{1 - \frac{1}{16}} = \pm\sqrt{\frac{15}{16}} = \pm\frac{\sqrt{15}}{4}$. Since t is in quadrant III, $x = -\frac{\sqrt{15}}{4}$. Thus, the terminal point determined by t is $P\left(-\frac{\sqrt{15}}{4}, -\frac{1}{4}\right)$, and so $\cos t = -\frac{\sqrt{15}}{4}$, $\tan t = \frac{1}{\sqrt{15}} = \frac{\sqrt{15}}{15}$, $\csc t = -4$, $\sec t = -\frac{4}{\sqrt{15}} = -\frac{4\sqrt{15}}{15}$, $\cot t = \sqrt{15}$.

73. $f(-x) = (-x)^2 \sin(-x) = -x^2 \sin x = -f(x)$, so f is odd.

75. $f(-x) = \sin(-x)\cos(-x) = -\sin x \cos x = -f(x)$, so f is odd.

77. $f(-x) = |-x|\cos(-x) = |x|\cos x = f(x)$, so f is even.

79. $f(-x) = (-x)^3 + \cos(-x) = -x^3 + \cos x$ which is neither $f(x)$ nor $-f(x)$, so f is neither even nor odd.

81.

t	0	0.25	0.50	0.75	1.00	1.25
$y(t)$	4	-2.83	0	2.83	-4	2.83

83. (a) $I(0.1) = 0.8e^{-0.3} \sin 1 \approx 0.499$ A

(b) $I(0.5) = 0.8e^{-1.5} \sin 5 \approx -0.171$ A

85. Notice that if $P(t) = (x, y)$, then $P(t + \pi) = (-x, -y)$. Thus,

(a) $\sin(t + \pi) = -y$ and $\sin t = y$. Therefore, $\sin(t + \pi) = -\sin t$.

(b) $\cos(t + \pi) = -x$ and $\cos t = x$. Therefore, $\cos(t + \pi) = -\cos t$.

(c) $\tan(t + \pi) = \dfrac{\sin(t + \pi)}{\cos(t + \pi)} = \dfrac{-y}{-x} = \dfrac{y}{x} = \dfrac{\sin t}{\cos t} = \tan t$.

5.3   TRIGONOMETRIC GRAPHS

1. The trigonometric functions $y = \sin x$ and $y = \cos x$ have amplitude 1 and period 2π.

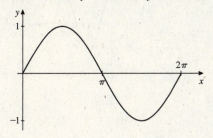

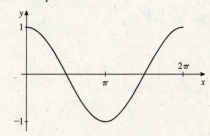

3. $f(x) = 1 + \cos x$

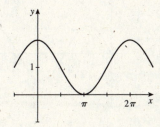

5. $f(x) = -\sin x$

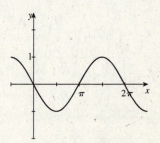

7. $f(x) = -2 + \sin x$

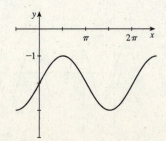

9. $g(x) = 3\cos x$

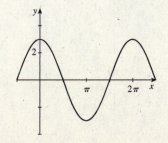

11. $g(x) = -\frac{1}{2}\sin x$

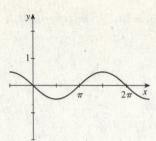

13. $g(x) = 3 + 3\cos x$

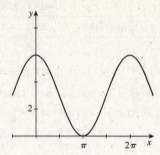

15. $h(x) = |\cos x|$

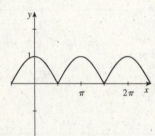

17. $y = \cos 2x$ has amplitude 1 and period π.

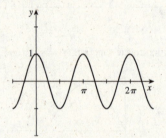

19. $y = -3\sin 3x$ has amplitude 3 and period $\frac{2\pi}{3}$.

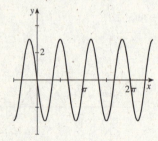

21. $y = 10\sin\frac{1}{2}x$ has amplitude 10 and period 4π.

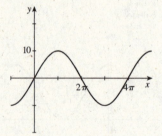

23. $y = -\frac{1}{3}\cos\frac{1}{3}x$ has amplitude $\frac{1}{3}$ and period 6π.

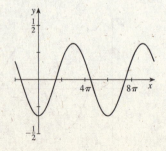

25. $y = -2\sin 2\pi x$ has amplitude 2 and period 1.

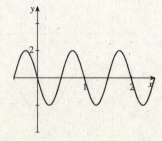

27. $y = 1 + \frac{1}{2} \cos \pi x$ has amplitude $\frac{1}{2}$ and period 2.

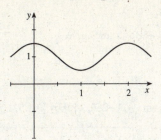

29. $y = \cos\left(x - \frac{\pi}{2}\right)$ has amplitude 1, period 2π, and phase shift $\frac{\pi}{2}$.

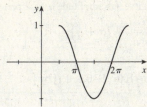

31. $y = -2 \sin\left(x - \frac{\pi}{6}\right)$ has amplitude 2, period 2π, and phase shift $\frac{\pi}{6}$.

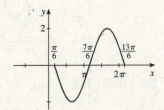

33. $y = -4 \sin 2\left(x + \frac{\pi}{2}\right)$ has amplitude 4, period π, and phase shift $-\frac{\pi}{2}$.

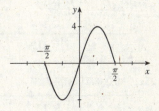

35. $y = 5 \cos\left(3x - \frac{\pi}{4}\right) = 5 \cos 3\left(x - \frac{\pi}{12}\right)$ has amplitude 5, period $\frac{2\pi}{3}$, and phase shift $\frac{\pi}{12}$.

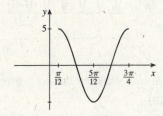

37. $y = \frac{1}{2} - \frac{1}{2} \cos\left(2x - \frac{\pi}{3}\right) = \frac{1}{2} - \frac{1}{2} \cos 2\left(x - \frac{\pi}{6}\right)$ has amplitude $\frac{1}{2}$, period π, and phase shift $\frac{\pi}{6}$.

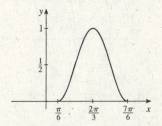

39. $y = 3 \cos \pi \left(x + \frac{1}{2}\right)$ has amplitude 3, period 2, and phase shift $-\frac{1}{2}$.

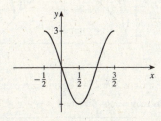

41. $y = \sin(3x + \pi) = \sin 3\left(x + \frac{\pi}{3}\right)$ has amplitude 1, period $\frac{2\pi}{3}$, and phase shift $-\frac{\pi}{3}$

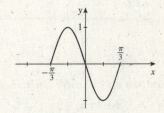

43. (a) This function has amplitude $a = 4$, period $\frac{2\pi}{k} = 2\pi$, and phase shift $b = 0$ as a sine curve.

(b) $y = a \sin k (x - b) = 4 \sin x$

45. (a) This curve has amplitude $a = \frac{3}{2}$, period $\frac{2\pi}{k} = \frac{2\pi}{3}$, and phase shift $b = 0$ as a cosine curve.

(b) $y = a \cos k (x - b) = \frac{3}{2} \cos 3x$

47. (a) This curve has amplitude $a = \frac{1}{2}$, period $\frac{2\pi}{k} = \pi$, and phase shift $b = -\frac{\pi}{3}$ as a cosine curve.

(b) $y = -\frac{1}{2} \cos 2 \left(x + \frac{\pi}{3} \right)$

49. (a) This curve has amplitude $a = 4$, period $\frac{2\pi}{k} = \frac{3}{2}$, and phase shift $b = -\frac{1}{2}$ as a sine curve.

(b) $y = 4 \sin \frac{4\pi}{3} \left(x + \frac{1}{2} \right)$

51. $f(x) = \cos 100x$, $[-0.1, 0.1]$ by $[-1.5, 1.5]$

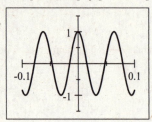

53. $f(x) = \sin \frac{x}{40}$, $[-250, 250]$ by $[-1.5, 1.5]$

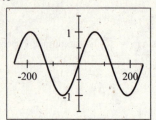

55. $y = \tan 25x$, $[-0.2, 0.2]$ by $[-3, 3]$

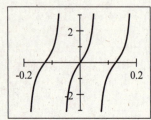

57. $y = \sin^2 20x$, $[-0.5, 0.5]$ by $[-0.2, 1.2]$

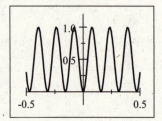

59. $f(x) = x$, $g(x) = \sin x$

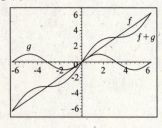

61. $y = x^2 \sin x$ is a sine curve that lies between the graphs of $y = x^2$ and $y = -x^2$.

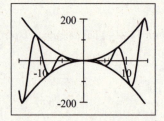

63. $y = \sqrt{x} \sin 5\pi x$ is a sine curve that lies between the graphs of $y = \sqrt{x}$ and $y = -\sqrt{x}$.

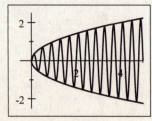

65. $y = \cos 3\pi x \cos 21\pi x$ is a cosine curve that lies between the graphs of $y = \cos 3\pi x$ and $y = -\cos 3\pi x$.

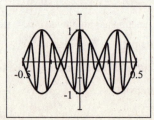

67. $y = \sin x + \sin 2x$. The period is 2π, so we graph the function over one period, $(-\pi, \pi)$. Maximum value 1.76 when $x \approx 0.94 + 2n\pi$, minimum value -1.76 when $x \approx -0.94 + 2n\pi$, n any integer.

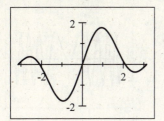

69. $y = 2\sin x + \sin^2 x$. The period is 2π, so we graph the function over one period, $(-\pi, \pi)$. Maximum value 3.00 when $x \approx 1.57 + 2n\pi$, minimum value -1.00 when $x \approx -1.57 + 2n\pi$, n any integer.

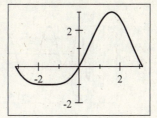

71. $\cos x = 0.4$, $x \in [0, \pi]$. The solution is $x \approx 1.16$.

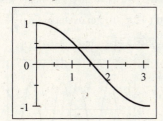

73. $\csc x = 3$, $x \in [0, \pi]$. The solutions are $x \approx 0.34, 2.80$.

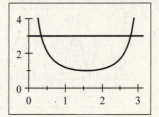

75. $f(x) = \dfrac{1 - \cos x}{x}$

   (a) Since $f(-x) = \dfrac{1 - \cos(-x)}{-x} = \dfrac{1 - \cos x}{-x} = -f(x)$, the function is odd.

   (b) The x-intercepts occur when $1 - \cos x = 0 \iff \cos x = 1 \iff x = 0, \pm 2\pi, \pm 4\pi, \pm 6\pi, \ldots$

   (d) As $x \to \pm\infty$, $f(x) \to 0$.

   (e) As $x \to 0$, $f(x) \to 0$.

   (c)

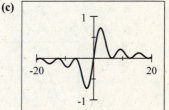

77. **(a)** The period of the wave is $\dfrac{2\pi}{\pi/10} = 20$ seconds.

   (b) Since $h(0) = 3$ and $h(10) = -3$, the wave height is $3 - (-3) = 6$ feet.

79. **(a)** The period of p is $\dfrac{2\pi}{160\pi} = \dfrac{1}{80}$ minute.

   (b) Since each period represents a heart beat, there are 80 heart beats per minute.

   (d) The maximum (or systolic) is $115 + 25 = 140$ and the minimum (or diastolic) is $115 - 25 = 90$. The read would be $140/90$ which higher than normal.

   (c)

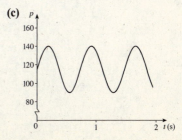

81. (a) $y = \sin\left(\sqrt{x}\right)$. This graph looks like a sine function which has been stretched horizontally (stretched more for larger values of x). It is defined only for $x \geq 0$, so it is neither even nor odd.

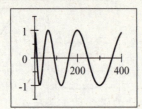

(b) $y = \sin\left(x^2\right)$. This graph looks like a graph of $\sin|x|$ which has been shrunk for $|x| > 1$ (shrunk more for larger values of x) and stretched for $|x| < 1$. It is an even function, whereas $\sin x$ is odd.

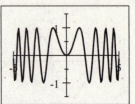

83. (a) The graph of $y = |\sin x|$ is shown in the viewing rectangle $[-6.28, 6.28]$ by $[-0.5, 1.5]$. This function is periodic with period π.

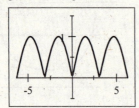

(b) The graph of $y = \sin|x|$ is shown in the viewing rectangle $[-10, 10]$ by $[-1.5, 1.5]$. The function is not periodic. Note that while $\sin|x + 2\pi| = \sin|x|$ for many values of x, it is false for $x \in (-2\pi, 0)$. For example $\sin\left|-\frac{\pi}{2}\right| = \sin\frac{\pi}{2} = 1$ while $\sin\left|-\frac{\pi}{2} + 2\pi\right| = \sin\frac{3\pi}{2} = -1$.

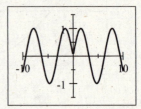

(c) The graph of $y = 2^{\cos x}$ is shown in the viewing rectangle $[-10, 10]$ by $[-1, 3]$. This function is periodic with period $= 2\pi$.

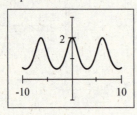

(d) The graph of $y = x - [\![x]\!]$ is shown in the viewing rectangle $[-7.5, 7.5]$ by $[-0.5, 1.5]$. This function is periodic with period 1. Be sure to turn off "connected" mode when graphing functions with gaps in their graph.

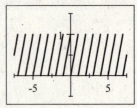

5.4   MORE TRIGONOMETRIC GRAPHS

1. The trigonometric function $y = \tan x$ has period π and asymptotes $x = \frac{\pi}{2} + n\pi$, n an integer.

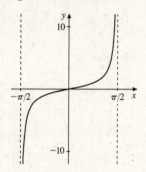

3. $f(x) = \tan\left(x + \frac{\pi}{4}\right)$ corresponds to Graph II. f is undefined at $x = \frac{\pi}{4}$ and $x = \frac{3\pi}{4}$, and Graph II has the shape of a graph of a tangent function.

5. $f(x) = \cot 2x$ corresponds to Graph VI.

7. $f(x) = 2\sec x$ corresponds to Graph IV.

9. $y = 4\tan x$ has period π.

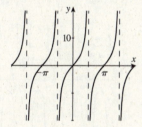

11. $y = -\frac{1}{2}\tan x$ has period π.

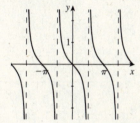

13. $y = -\cot x$ has period π.

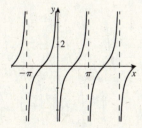

15. $y = 2\csc x$ has period 2π.

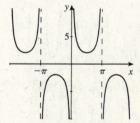

17. $y = 3\sec x$ has period 2π.

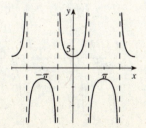

19. $y = \tan\left(x + \frac{\pi}{2}\right)$ has period π.

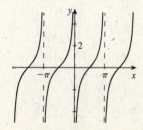

21. $y = \csc\left(x - \frac{\pi}{2}\right)$ has period 2π.

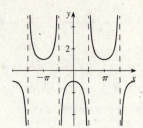

23. $y = \cot\left(x + \frac{\pi}{4}\right)$ has period π.

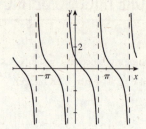

25. $y = \frac{1}{2}\sec\left(x - \frac{\pi}{6}\right)$ has period 2π.

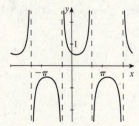

27. $y = \tan 4x$ has period $\frac{\pi}{4}$.

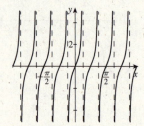

29. $y = \tan\left(\frac{\pi}{4}x\right)$ has period $\frac{\pi}{\frac{\pi}{4}} = 4$.

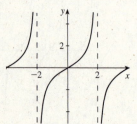

31. $y = \sec 2x$ has period $\frac{2\pi}{2} = \pi$.

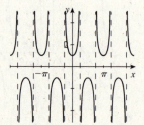

33. $y = \csc 4x$ has period $\frac{2\pi}{4} = \frac{\pi}{2}$.

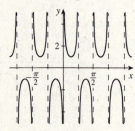

35. $y = 2\tan 3\pi x$ has period $\frac{1}{3}$.

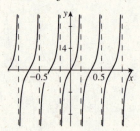

37. $y = 5\csc\frac{3\pi}{2}x$ has period $\frac{2\pi}{\frac{3\pi}{2}} = \frac{4}{3}$.

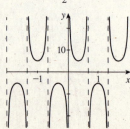

39. $y = \tan 2\left(x + \frac{\pi}{2}\right)$ has period $\frac{\pi}{2}$.

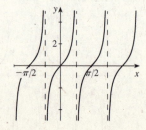

41. $y = \tan 2 (x - \pi) = \tan 2x$ has period $\frac{\pi}{2}$.

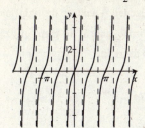

43. $y = \cot \left(2x - \frac{\pi}{2}\right) = \cot 2 \left(x - \frac{\pi}{4}\right)$ has period $\frac{\pi}{2}$.

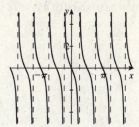

45. $y = 2 \csc \left(\pi x - \frac{\pi}{3}\right) = 2 \csc \pi \left(x - \frac{1}{3}\right)$ has period $\frac{2\pi}{\pi} = 2$.

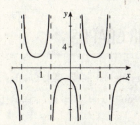

47. $y = 5 \sec \left(3x - \frac{\pi}{2}\right) = 5 \sec 3 \left(x - \frac{\pi}{6}\right)$ has period $\frac{2\pi}{3}$.

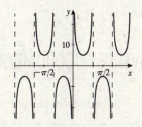

49. $y = \tan \left(\frac{2}{3}x - \frac{\pi}{6}\right) = \tan \frac{2}{3} \left(x - \frac{\pi}{4}\right)$ has period $\pi / \left(\frac{2}{3}\right) = \frac{3\pi}{2}$.

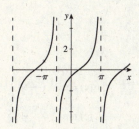

51. $y = 3 \sec \pi \left(x + \frac{1}{2}\right)$ has period $\frac{2\pi}{\pi} = 2$.

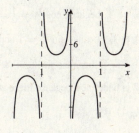

53. $y = -2 \tan \left(2x - \frac{\pi}{3}\right) = -2 \tan 2 \left(x - \frac{\pi}{6}\right)$ has period $\frac{\pi}{2}$.

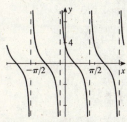

55. (a) If f is periodic with period p, then by the definition of a period, $f(x + p) = f(x)$ for all x in the domain of f. Therefore, $\dfrac{1}{f(x + p)} = \dfrac{1}{f(x)}$ for all $f(x) \neq 0$. Thus, $\dfrac{1}{f}$ is also periodic with period p.

(b) Since $\sin x$ has period 2π, it follows from part (a) that $\csc x = \dfrac{1}{\sin x}$ also has period 2π. Similarly, since $\cos x$ has period 2π, we conclude $\sec x = \dfrac{1}{\cos x}$ also has period 2π.

57. (a) $d(t) = 3 \tan \pi t$, so $d(0.15) \approx 1.53$,

$d(0.25) \approx 3.00$, and $d(0.45) \approx 18.94$.

(b)

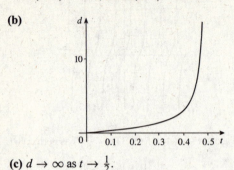

(c) $d \to \infty$ as $t \to \frac{1}{2}$.

59. The graph of $y = -\cot x$ is the same as the graph of $y = \tan x$ shifted $\frac{\pi}{2}$ units to the right, and the graph of $y = \csc x$ is the same as the graph of $y = \sec x$ shifted $\frac{\pi}{2}$ units to the right.

5.5   INVERSE TRIGONOMETRIC FUNCTIONS AND THEIR GRAPHS

1. (a) To define the inverse sine function we restrict the domain of sine to the interval $\left[-\frac{\pi}{2}, \frac{\pi}{2}\right]$. On this interval the sine function is one-to-one and its inverse function $\sin^{-1}$ is defined by $\sin^{-1} x = y \Leftrightarrow \sin y = x$. For example, $\sin^{-1} \frac{1}{2} = \frac{\pi}{6}$ because $\sin \frac{\pi}{6} = \frac{1}{2}$.

(b) To define the inverse cosine function we restrict the domain of cosine to the interval $[0, \pi]$. On this interval the cosine function is one-to-one and its inverse function $\cos^{-1}$ is defined by $\cos^{-1} x = y \Leftrightarrow \cos y = x$. For example, $\cos^{-1} \frac{1}{2} = \frac{\pi}{3}$ because $\cos \frac{\pi}{3} = \frac{1}{2}$.

3. (a) $\sin^{-1} 1 = \frac{\pi}{2}$ because $\sin \frac{\pi}{2} = 1$ and $\frac{\pi}{2}$ lies in $\left[-\frac{\pi}{2}, \frac{\pi}{2}\right]$.

(b) $\sin^{-1} \frac{\sqrt{3}}{2} = \frac{\pi}{3}$ because $\sin \frac{\pi}{3} = \frac{\sqrt{3}}{2}$ and $\frac{\pi}{3}$ lies in $\left[-\frac{\pi}{2}, \frac{\pi}{2}\right]$.

(c) $\sin^{-1} 2$ is undefined because there is no real number x such that $\sin x = 2$.

5. (a) $\cos^{-1}(-1) = \pi$ **(b)** $\cos^{-1} \frac{1}{2} = \frac{\pi}{3}$ **(c)** $\cos^{-1}\left(-\frac{\sqrt{3}}{2}\right) = \frac{5\pi}{6}$

7. (a) $\tan^{-1}(-1) = -\frac{\pi}{4}$ **(b)** $\tan^{-1} \sqrt{3} = \frac{\pi}{3}$ **(c)** $\tan^{-1} \frac{\sqrt{3}}{3} = \frac{\pi}{6}$

9. (a) $\cos^{-1}\left(-\frac{1}{2}\right) = \frac{2\pi}{3}$ **(b)** $\sin^{-1}\left(-\frac{\sqrt{2}}{2}\right) = -\frac{\pi}{4}$ **(c)** $\tan^{-1} 1 = \frac{\pi}{4}$

11. $\sin^{-1} \frac{2}{3} = 0.72973$ **13.** $\cos^{-1}\left(-\frac{3}{7}\right) = 2.01371$

15. $\cos^{-1}(-0.92761) = 2.75876$ **17.** $\tan^{-1} 10 = 1.47113$

19. $\tan^{-1}(1.23456) = 0.88998$ **21.** $\sin^{-1}(-0.25713) = -0.26005$

23. $\sin\left(\sin^{-1} \frac{1}{4}\right) = \frac{1}{4}$ **25.** $\tan\left(\tan^{-1} 5\right) = 5$

27. $\sin\left(\sin^{-1}\left(\frac{3}{2}\right)\right)$ is undefined because $\frac{3}{2} > 1$.   **29.** $\cos^{-1}\left(\cos \frac{5\pi}{6}\right) = \frac{5\pi}{6}$ because $\frac{5\pi}{6}$ lies in $[0, 2\pi]$.

31. $\sin^{-1}\left(\sin\left(-\frac{\pi}{6}\right)\right) = -\frac{\pi}{6}$ because $-\frac{\pi}{6}$ lies in $[-\pi, \pi]$.

33. $\sin^{-1}\left(\sin\left(\frac{5\pi}{6}\right)\right) = \frac{\pi}{6}$ because $\sin \frac{\pi}{6} = \sin \frac{5\pi}{6}$ and $\frac{\pi}{6}$ lies in $\left[-\frac{\pi}{2}, \frac{\pi}{2}\right]$.

35. $\cos^{-1}\left(\cos\left(\frac{17\pi}{6}\right)\right) = \frac{\pi}{6}$ because $\cos \frac{\pi}{6} = \cos \frac{17\pi}{6}$ and $\frac{\pi}{6}$ lies in $[0, 2\pi]$.

37. $\tan^{-1}\left(\tan \frac{2\pi}{3}\right) = -\frac{\pi}{3}$ because $\tan\left(-\frac{\pi}{3}\right) = \tan \frac{2\pi}{3}$ and $-\frac{\pi}{3}$ lies in $\left(-\frac{\pi}{2}, \frac{\pi}{2}\right)$.

39. $\tan\left(\sin^{-1}\frac{1}{2}\right) = \tan\frac{\pi}{6} = \frac{\sqrt{3}}{3}$

41. $\cos\left(\sin^{-1}\frac{\sqrt{3}}{2}\right) = \cos\frac{\pi}{3} = \frac{1}{2}$

43. $\sin\left(\tan^{-1}(-1)\right) = \sin\left(-\frac{\pi}{4}\right) = -\frac{\sqrt{2}}{2}$

45. The domain of $f(x) = \sin\left(\sin^{-1}x\right)$ is the same as that of $\sin^{-1}x$, $[-1, 1]$, and the graph of f is the same as that of $y = x$ on $[1, 1]$.

The domain of $g(x) = \sin^{-1}(\sin x)$ is the same as that of $\sin x$, $(-\infty, \infty)$, because for all x, the value of $\sin x$ lies within the domain of $\sin^{-1}x$. $g(x) = \sin^{-1}(\sin x) = x$ for $-\frac{\pi}{2} \le x \le \frac{\pi}{2}$. Because the graph of $y = \sin x$ is symmetric about the line $x = \frac{\pi}{2}$, we can obtain the part of the graph of g for $\frac{\pi}{2} \le x \le \frac{3\pi}{2}$ by reflecting the graph of $y = x$ about this vertical line. The graph of g is periodic with period 2π.

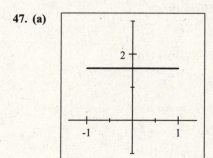

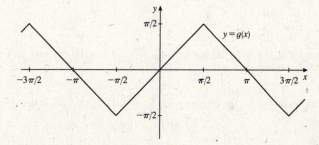

47. (a)

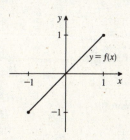

From the graph of $y = \sin^{-1}x + \cos^{-1}x$, it appears that $y \approx 1.57$. We suspect that the actual value is $\frac{\pi}{2}$.

(b) To show that $\sin^{-1}x + \cos^{-1}x = \frac{\pi}{2}$, start with the identity

$\sin\left(a - \frac{\pi}{2}\right) = -\cos a$ and take arcsin of both sides to obtain

$a - \frac{\pi}{2} = \sin^{-1}(-\cos a)$. Now let $a = \cos^{-1}x$. Then

$\cos^{-1}x - \frac{\pi}{2} = \sin^{-1}\left(-\cos\left(\cos^{-1}x\right)\right) = \sin^{-1}(-x) = -\sin^{-1}x$, so

$\sin^{-1}x + \cos^{-1}x = \frac{\pi}{2}$.

5.6 MODELING HARMONIC MOTION

1. (a) Because $y = 0$ at time $t = 0$, $y = a\sin\omega t$ is an appropriate model.

(b) Because $y = a$ at time $t = 0$, $y = a\cos\omega t$ is an appropriate model.

3. $y = 2\sin 3t$

(a) Amplitude 2, period $\frac{2\pi}{3}$, frequency $\frac{1}{\text{period}} = \frac{3}{2\pi}$.

(b)

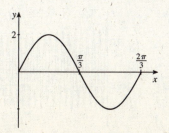

5. $y = -\cos 0.3t$

(a) Amplitude 1, period $\frac{2\pi}{0.3} = \frac{20\pi}{3}$, frequency $\frac{3}{20\pi}$.

(b)

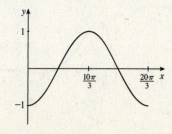

7. $y = -0.25 \cos\left(1.5t - \frac{\pi}{3}\right) = -0.25 \cos\left(\frac{3}{2}t - \frac{\pi}{3}\right)$

$= -0.25 \cos \frac{3}{2}\left(t - \frac{2\pi}{9}\right)$

(a) Amplitude 0.25, period $\frac{2\pi}{3/2} = \frac{4\pi}{3}$, frequency $\frac{3}{4\pi}$.

(b)

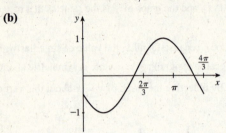

9. $y = 5 \cos\left(\frac{2}{3}t + \frac{3}{4}\right) = 5 \cos \frac{2}{3}\left(t + \frac{9}{8}\right)$

(a) Amplitude 5, period $\frac{2\pi}{2/3} = 3\pi$, frequency $\frac{1}{3\pi}$.

(b)

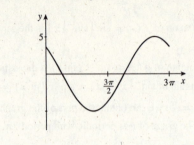

11. The amplitude is $a = 10$ cm, the period is $\frac{2\pi}{k} = 3$ s, and $f(0) = 0$, so $f(x) = 10 \sin \frac{2\pi}{3}t$.

13. The amplitude is 6 in., the frequency is $\frac{k}{2\pi} = \frac{5}{\pi}$ Hz, and $f(0) = 0$, so $f(x) = 6 \sin 10t$.

15. The amplitude is 60 ft, the period is $\frac{2\pi}{k} = 0.5$ min, and $f(0) = 60$, so $f(x) = 60 \cos 4\pi t$.

17. The amplitude is 2.4 m, the frequency is $\frac{k}{2\pi} = 750$ Hz, and $f(0) = 2.4$, so $f(x) = 2.4 \cos 1500\pi t$.

19. (a) $k = 2$, $c = 1.5$, and $f = 3 \Rightarrow \omega = 6\pi$, so we have

$y = 2e^{-1.5t} \cos 6\pi t$.

(b)

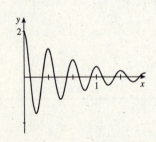

21. (a) $k = 100$, $c = 0.05$, and $p = 4 \Rightarrow \omega = \frac{\pi}{2}$, so we have $y = 100e^{-0.05t} \cos \frac{\pi}{2}t$.

(b)

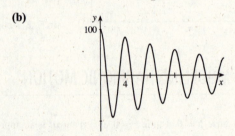

23. (a) $k = 7$, $c = 10$, and $p = \frac{\pi}{6} \Rightarrow \omega = 12$, so we have

$y = 7e^{-10t} \sin 12t$.

(b)

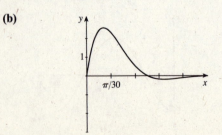

25. (a) $k = 0.3$, $c = 0.2$, and $f = 20 \Rightarrow \omega = 40\pi$, so we have $y = 0.3e^{-0.2t} \sin 40\pi t$.

(b)

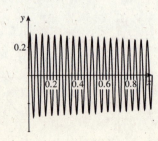

27. $y = 0.2 \cos 20\pi t + 8$

(a) The frequency is $\frac{20\pi}{2\pi} = 10$ cycles/min.

(b)

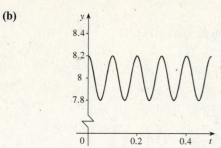

(c) Since $y = 0.2 \cos 20\pi t + 8 \le 0.2(1) + 8 = 8.2$ and when $t = 0$, $y = 8.2$, the maximum displacement is 8.2 m.

29. $p(t) = 115 + 25 \sin(160\pi t)$

(a) Amplitude 25, period $\frac{2\pi}{160\pi} = \frac{1}{80} = 0.0125$, frequency $\frac{1}{\text{period}} = 80$.

(b)

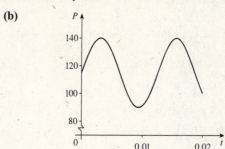

(c) The period decreases and the frequency increases.

31. The graph resembles a sine wave with an amplitude of 5, a period of $\frac{2}{5}$, and no phase shift. Therefore, $a = 5$, $\frac{2\pi}{\omega} = \frac{2}{5}$ $\Leftrightarrow$ $\omega = 5\pi$, and a formula is $d(t) = 5 \sin 5\pi t$.

33. $a = 21$, $f = \frac{1}{12}$ cycle/hour $\Rightarrow$ $\frac{\omega}{2\pi} = \frac{1}{12}$ $\Leftrightarrow$ $\omega = \frac{\pi}{6}$. So, $y = 21 \sin\left(\frac{\pi}{6}t\right)$ (assuming the tide is at mean level and rising when $t = 0$).

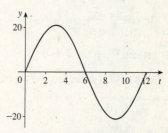

35. Since the mass travels from its highest point (compressed spring) to its lowest point in $\frac{1}{2}$ s, it completes half a period in $\frac{1}{2}$ s. So, $\frac{1}{2}$ (one period) $= \frac{1}{2}$ s $\Rightarrow$ $\frac{1}{2} \cdot \frac{2\pi}{\omega} = \frac{1}{2}$ $\Leftrightarrow$ $\omega = 2\pi$. Also, $a = 5$. So $y = 5 \cos 2\pi t$.

37. Since the Ferris wheel has a radius of 10 m and the bottom of the wheel is 1 m above the ground, the minimum height is 1 m and the maximum height is 21 m. Then $a = 10$ and $\frac{2\pi}{\omega} = 20$ s $\Leftrightarrow$ $\omega = \frac{\pi}{10}$, and so $y = 11 + 10 \sin\left(\frac{\pi}{10}t\right)$, where t is in seconds.

39. $a = 0.2$, $\frac{2\pi}{\omega} = 10$ $\Leftrightarrow$ $\omega = \frac{\pi}{5}$. Then $y = 3.8 + 0.2 \sin\left(\frac{\pi}{5}t\right)$.

41. The amplitude is $\frac{1}{2}(100 - 80) = 10$ mmHG, the period is 24 hours, and the phase shift is 8 hours, so $f(t) = 10 \sin\left(\frac{\pi}{12}(t - 8)\right) + 90$.

43. (a) The maximum voltage is the amplitude, that is, $V_{\max} = a = 45$ V.

(b) From the graph we see that 4 cycles are completed every 0.1 seconds, or equivalently, 40 cycles are completed every second, so $f = 40$.

(c) The number of revolutions per second of the armature is the frequency, that is, $\frac{\omega}{2\pi} = f = 40$.

(d) $a = 45$, $f = \frac{\omega}{2\pi} = 40$ $\Leftrightarrow$ $\omega = 80\pi$. Then $V(t) = 45 \cos 80\pi t$.

45. $k = 1$, $c = 0.9$, and $\frac{\omega}{2\pi} = \frac{1}{2} \Leftrightarrow \omega = \pi$. Since $f(0) = 0$, $f(t) = e^{-0.9t} \sin \pi t$.

47. $\frac{ke^{-ct}}{ke^{-c(t+3)}} = 4 \Leftrightarrow e^{-ct+c(t+3)} = 4$ $\Leftrightarrow$ $e^{3c} = 4$ $\Leftrightarrow$ $3c = \ln 4$ $\Leftrightarrow$ $c = \frac{1}{3}\ln 4 \approx 0.46$.

CHAPTER 5 REVIEW

1. **(a)** Since $\left(-\frac{\sqrt{3}}{2}\right)^2 + \left(\frac{1}{2}\right)^2 = \frac{3}{4} + \frac{1}{4} = 1$, the point $P\left(-\frac{\sqrt{3}}{2}, \frac{1}{2}\right)$ lies on the unit circle.

   (b) $\sin t = \frac{1}{2}$, $\cos t = -\frac{\sqrt{3}}{2}$, $\tan t = \dfrac{\frac{1}{2}}{-\frac{\sqrt{3}}{2}} = -\frac{\sqrt{3}}{3}$.

3. $t = \frac{2\pi}{3}$

   (a) $\bar{t} = \pi - \frac{2\pi}{3} = \frac{\pi}{3}$

   (b) $P\left(-\frac{1}{2}, \frac{\sqrt{3}}{2}\right)$

   (c) $\sin t = \frac{\sqrt{3}}{2}$, $\cos t = -\frac{1}{2}$, $\tan t = -\sqrt{3}$, $\csc t = \frac{2\sqrt{3}}{3}$, $\sec t = -2$, and $\cot t = -\frac{\sqrt{3}}{3}$.

5. $t = -\frac{11\pi}{4}$

   (a) $\bar{t} = 3\pi + \left(-\frac{11\pi}{4}\right) = \frac{\pi}{4}$

   (b) $P\left(-\frac{\sqrt{2}}{2}, -\frac{\sqrt{2}}{2}\right)$

   (c) $\sin t = -\frac{\sqrt{2}}{2}$, $\cos t = -\frac{\sqrt{2}}{2}$, $\tan t = 1$, $\csc t = -\sqrt{2}$, $\sec t = -\sqrt{2}$, and $\cot t = 1$.

7. **(a)** $\sin \frac{3\pi}{4} = \sin \frac{\pi}{4} = \frac{\sqrt{2}}{2}$

   (b) $\cos \frac{3\pi}{4} = -\cos \frac{\pi}{4} = -\frac{\sqrt{2}}{2}$

9. **(a)** $\sin 1.1 \approx 0.89121$

   (b) $\cos 1.1 \approx 0.45360$

11. **(a)** $\cos \frac{9\pi}{2} = \cos \frac{\pi}{2} = 0$

    (b) $\sec \frac{9\pi}{2}$ is undefined

13. **(a)** $\tan \frac{5\pi}{2}$ is undefined

    (b) $\cot \frac{5\pi}{2} = \cot \frac{\pi}{2} = 0$

15. **(a)** $\tan \frac{5\pi}{6} = -\frac{\sqrt{3}}{3}$

    (b) $\cot \frac{5\pi}{6} = -\sqrt{3}$

17. $\dfrac{\tan t}{\cos t} = \dfrac{\frac{\sin t}{\cos t}}{\cos t} = \dfrac{\sin t}{\cos^2 t} = \dfrac{\sin t}{1 - \sin^2 t}$

19. $\tan t = \dfrac{\sin t}{\cos t} = \dfrac{\sin t}{\pm\sqrt{1 - \sin^2 t}} = \dfrac{\sin t}{\sqrt{1 - \sin^2 t}}$ (because t is in quadrant IV, $\cos t$ is positive).

21. $\sin t = \frac{5}{13}$, $\cos t = -\frac{12}{13}$. Then $\tan t = \dfrac{\frac{5}{13}}{-\frac{12}{13}} = -\frac{5}{12}$, $\csc t = \frac{13}{5}$, $\sec t = -\frac{13}{12}$, and $\cot t = -\frac{12}{5}$.

23. $\cot t = -\frac{1}{2}$, $\csc t = \frac{\sqrt{5}}{2}$. Since $\csc t = \dfrac{1}{\sin t}$, we know $\sin t = \dfrac{2}{\sqrt{5}} = \dfrac{2\sqrt{5}}{5}$. Now $\cot t = \dfrac{\cos t}{\sin t}$, so $\cos t = \sin t \cdot \cot t = \dfrac{2\sqrt{5}}{5} \cdot \left(-\frac{1}{2}\right) = -\frac{\sqrt{5}}{5}$, and $\tan t = \dfrac{1}{\left(-\frac{1}{2}\right)} = -2$ while $\sec t = \dfrac{1}{\cos t} = \dfrac{1}{\left(-\frac{\sqrt{5}}{5}\right)} = -\frac{5}{\sqrt{5}} = -\sqrt{5}$.

25. $\tan t = \frac{1}{4}$, t is in quadrant III $\Rightarrow$

    $\sec t + \cot t = -\sqrt{\tan^2 t + 1} + \dfrac{1}{\tan t} = -\sqrt{\left(\frac{1}{4}\right)^2 + 1} + 4 = -\sqrt{\frac{17}{16}} + 4 = 4 - \dfrac{\sqrt{17}}{4} = \dfrac{16 - \sqrt{17}}{4}$

27. $\cos t = \frac{3}{5}$, t is in quadrant I $\Rightarrow$

    $\tan t + \sec t = \dfrac{\sin t}{\cos t} + \dfrac{1}{\cos t} = \dfrac{\sqrt{1 - \cos^2 t}}{\cos t} + \dfrac{1}{\cos t} = \dfrac{\sqrt{1 - \left(\frac{3}{5}\right)^2}}{\frac{3}{5}} + \dfrac{5}{3} = \dfrac{\sqrt{\frac{1}{2}}}{\frac{3}{5}} + \dfrac{5}{3} = \dfrac{4}{5} \cdot \dfrac{5}{3} + \dfrac{5}{3} = \dfrac{4}{3} + \dfrac{5}{3} = \dfrac{9}{3} = 3$

29. $y = 10 \cos \frac{1}{2} x$

(a) This function has amplitude 10, period $\frac{2\pi}{\frac{1}{2}} = 4\pi$, and phase shift 0.

(b)

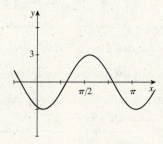

31. $y = -\sin \frac{1}{2} x$

(a) This function has amplitude 1, period $\frac{2\pi}{1/2} = 4\pi$, and phase shift 0.

(b)

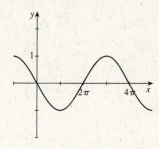

33. $y = 3 \sin (2x - 2) = 3 \sin 2 (x - 1)$

(a) This function has amplitude 3, period $\frac{2\pi}{2} = \pi$, and phase shift 1.

(b)

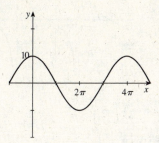

35. $y = -\cos \left(\frac{\pi}{2} x + \frac{\pi}{6} \right) = -\cos \frac{\pi}{2} \left(x + \frac{1}{3} \right)$

(a) This function has amplitude 1, period $\frac{2\pi}{\pi/2} = 4$, and phase shift $-\frac{1}{3}$.

(b)

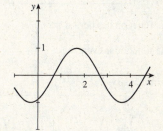

37. From the graph we see that the amplitude is 5, the period is $\frac{\pi}{2}$, and there is no phase shift. Therefore, the function is $y = 5 \sin 4x$.

39. From the graph we see that the amplitude is $\frac{1}{2}$, the period is 1, and there is a phase shift of $-\frac{1}{3}$. Therefore, the function is $y = \frac{1}{2} \sin 2\pi \left(x + \frac{1}{3} \right)$.

41. $y = 3 \tan x$ has period π.

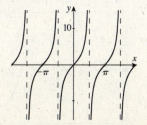

43. $y = 2 \cot \left(x - \frac{\pi}{2} \right)$ has period π.

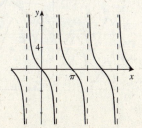

45. $y = 4\csc(2x + \pi) = 4\csc 2\left(x + \frac{\pi}{2}\right)$ has period $\frac{2\pi}{2} = \pi$.

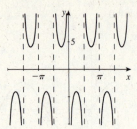

47. $y = \tan\left(\frac{1}{2}x - \frac{\pi}{8}\right) = \tan\frac{1}{2}\left(x - \frac{\pi}{4}\right)$ has period $\frac{\pi}{\frac{1}{2}} = 2\pi$.

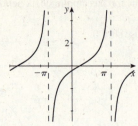

49. $\sin^{-1} 1 = \frac{\pi}{2}$

51. $\sin^{-1}\left(\sin\frac{13\pi}{6}\right) = \frac{\pi}{6}$

53. (a) $y = |\cos x|$

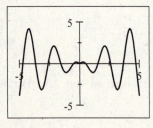

(b) This function has period π.

(c) This function is even.

55. (a) $y = \cos\left(2^{0.1x}\right)$

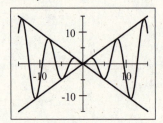

(b) This function is not periodic.

(c) This function is neither even nor odd.

57. (a) $y = |x|\cos 3x$

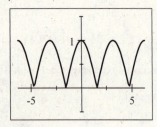

(b) This function is not periodic.

(c) This function is even.

59. $y = x\sin x$ is a sine function whose graph lies between those of $y = x$ and $y = -x$.

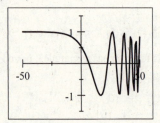

61. $y = x + \sin 4x$ is the sum of the two functions $y = x$ and $y = \sin 4x$.

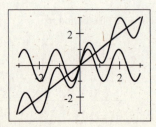

63. $y = \cos x + \sin 2x$. Since the period is 2π, we graph over the interval $[-\pi, \pi]$. The maximum value is 1.76 when $x \approx 0.63 \pm 2n\pi$, the minimum value is -1.76 when $x \approx 2.51 \pm 2n\pi$, n an integer.

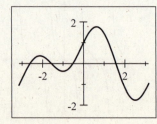

65. We want to find solutions to $\sin x = 0.3$ in the interval $[0, 2\pi]$, so we plot the functions $y = \sin x$ and $y = 0.3$ and look for their intersection. We see that $x \approx 0.305$ or $x \approx 2.837$.

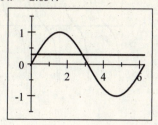

67. $f(x) = \dfrac{\sin^2 x}{x}$

(a) The function is odd.

(b) The graph intersects the x-axis at $x = 0, \pm\pi, \pm 2\pi, \pm 3\pi, \ldots$

(d) As $x \to \pm\infty$, $f(x) \to 0$.

(e) As $x \to 0$, $f(x) \to 0$.

(c)

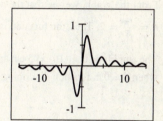

69. The amplitude is $a = 50$ cm. The frequency is 8 Hz, so $\omega = 8(2\pi) = 16\pi$. Since the mass is at its maximum displacement when $t = 0$, the motion follows a cosine curve. So a function describing the motion of P is $f(t) = 50\cos 16\pi t$.

71. From the graph, we see that the amplitude is 4 ft, the period is 12 hours, and there is no phase shift. Thus, the variation in water level is described by $y = 4\cos\frac{\pi}{6}t$.

CHAPTER 5 TEST

1. Since $P(x, y)$ lies on the unit circle, $x^2 + y^2 = 1 \Rightarrow y = \pm\sqrt{1 - \left(\frac{\sqrt{11}}{6}\right)^2} = \pm\sqrt{\frac{25}{36}} = \pm\frac{5}{6}$. But $P(x, y)$ lies in the fourth quadrant. Therefore y is negative $\Rightarrow y = -\frac{5}{6}$.

3. (a) $\sin\frac{7\pi}{6} = -0.5$

(b) $\cos\dfrac{13\pi}{4} = -\dfrac{\sqrt{2}}{2}$

(c) $\tan\left(-\frac{5\pi}{3}\right) = \sqrt{3}$

(d) $\csc\left(\frac{3\pi}{2}\right) = -1$

5. $\cos t = -\frac{8}{17}$, t in quadrant III $\Rightarrow \tan t \cdot \cot t + \csc t = 1 + \dfrac{1}{-\sqrt{1 - \cos^2 t}}$ (since t is in quadrant III)$= 1 - \dfrac{1}{-\sqrt{1 - \frac{64}{289}}} = 1 - \dfrac{1}{\frac{15}{17}} = -\dfrac{2}{15}$.

7. $y = 2\sin\left(\frac{1}{2}x - \frac{\pi}{6}\right) = \sin\frac{1}{2}\left(x - \frac{\pi}{3}\right)$

(a) This function has amplitude 2, period $\dfrac{2\pi}{\frac{1}{2}} = 4\pi$, and

phase shift $\frac{\pi}{3}$.

(b)

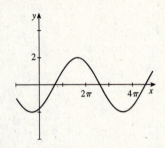

9. $y = \tan 2\left(x - \frac{\pi}{4}\right)$ has period $\frac{\pi}{2}$.

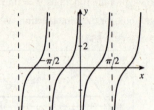

11. From the graph, we see that the amplitude is 2 and the phase shift is $-\frac{\pi}{3}$. Also, the period is π, so $\frac{2\pi}{k} = \pi \implies$

   $k = \frac{2\pi}{\pi} = 2$. Thus, the function is $y = 2\sin 2\left(x + \frac{\pi}{3}\right)$.

13. The amplitude is $\frac{1}{2}(10) = 5$ cm and the frequency is 2 Hz. Assuming that the mass is at its rest position and moving upward

   when $t = 0$, a function describing the distance of the mass from its rest position is $f(t) = 5\sin 4\pi t$.

FOCUS ON MODELING   Fitting Sinusoidal Curves to Data

1. (a) See the graph in part (c).

   (b) Using the method of Example 1, we find the vertical shift

   $b = \frac{1}{2}$ (maximum value + minimum value) $= \frac{1}{2}(2.1 - 2.1) = 0$, the amplitude

   $a = \frac{1}{2}$ (maximum value − minimum value) $= \frac{1}{2}(2.1 - (-2.1)) = 2.1$, the period $\dfrac{2\pi}{\omega} = 2(6 - 0) = 12$ (so

   $\omega \approx 0.5236$), and the phase shift $c = 0$. Thus, our model is $y = 2.1\cos\frac{\pi}{6}t$.

(c)

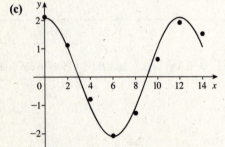

The curve fits the data quite well.

(d) Using the SinReg command on the TI-83, we find

   $y = 2.048714222\sin(0.5030795477t + 1.551856108)$

   $- 0.0089616507.$

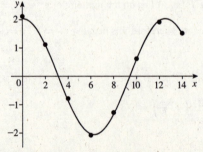

(e) Our model from part (d) is equivalent to $y = 2.05\cos\left(0.50t + 1.55 - \frac{\pi}{2}\right) - 0.01 \approx 2.05\cos(0.50t - 0.02) - 0.01$.

   This is the same as the function in part (b), correct to one decimal place.

3. (a) See the graph in part (c).

(b) Using the method of Example 1, we find the vertical shift

$b = \frac{1}{2}$ (maximum value + minimum value) $= \frac{1}{2}$ (25.1 + 1.0) $= 13.05$, the amplitude

$a = \frac{1}{2}$ (maximum value − minimum value) $= \frac{1}{2}$ (25.1 − 1.0) $= 12.05$, the period $\dfrac{2\pi}{\omega} = 2\,(1.5 - 0.9) = 1.2$ (so

$\omega \approx 5.236$), and the phase shift $c = 0.3$. Thus, our model is $y = 12.05 \cos\,(5.236\,(t - 0.3)) + 13.05$.

(c)

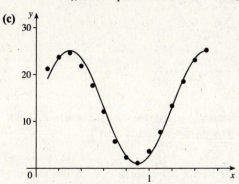

The curve fits the data fairly well.

(d) Using the `SinReg` command on the TI-83, we find

$$y = 11.71905062 \sin\,(5.048853286t + 0.2388957877)$$
$$+ 12.96070536.$$

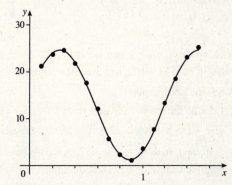

(e) Our model from part (d) is equivalent to $y = 11.72 \cos\left(5.05t + 0.24 - \frac{\pi}{2}\right) + 12.96 \approx 11.72 \cos\,(5.05t - 1.33) + 12.96$. This is close but not identical to the function in part (b).

5. (a) See the graph in part (c).

(b) Let t be the time (in months) from January. We find a function of the form $y = a \cos \omega\,(t - c) + b$, where y is the temperature in °F.

$a = \frac{1}{2}\,(85.8 - 40) = 22.9$. The period is $2\,(\text{Jul} - \text{Jan}) = 2\,(6 - 0) = 12$, and so $\omega = \frac{2\pi}{12} \approx 0.52$.

$b = \frac{1}{2}\,(85.8 + 40) = 62.9$. Because the maximum value occurs in July, $c = 6$. Thus the function is $y = 22.9 \cos 0.52\,(t - 6) + 62.9$.

(c)

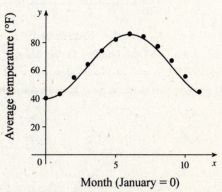

Month (January = 0)

(d) Using the `SinReg` command on the TI-83 we find that for the function $y = a \sin\,(bt + c) + d$, where $a = 23.4$, $b = 0.48$, $c = -1.36$, and $d = 62.2$. Thus we get the model $y = 23.4 \sin\,(0.48t - 1.36) + 62.2$.

7. (a) See the graph in part (c).

(b) Let t be the time years. We find a function of the form $y = a \sin \omega\,(t - c) + b$, where y is the owl population.

$a = \frac{1}{2}\,(80 - 20) = 30$. The period is $2\,(9 - 3) = 12$ and so $\omega = \frac{2\pi}{12} \approx 0.52$. $b = \frac{1}{2}\,(80 + 20) = 50$. Because the values start at the middle we have $c = 0$. Thus the function is $y = 30 \sin 0.52t + 50$.

(c)

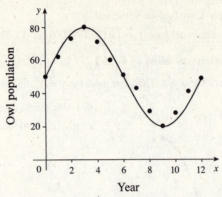

(d) Using the `SinReg` command on the TI–83 we find that for the function $y = a \sin(bt + c) + d$, where $a = 25.8$, $b = 0.52$, $c = -0.02$, and $d = 50.6$. Thus we get the model $y = 25.8 \sin(0.52t - 0.02) + 50.6$.

9. (a) See the graph in part (c).

(b) Let t be the number of years since 1975. We find a function of the form $y = a \cos \omega (t - c) + b$. We have $a = \frac{1}{2}(85.8 - 40) = 22.9$. The number of sunspots varies over a cycle of approximately 11 years (as the problem indicates), so $\omega = \frac{2\pi}{11} \approx 0.57$. Also, we have $b = \frac{1}{2}(158 + 9) = 83.5$. The first maximum value occurs at approximately $c = 4.5$. Thus, our model is $y = 74.5 \cos(0.57(t - 4.5)) + 83.5$, where y is the average daily sunspot count and t is the number of years since 1975.

(c)

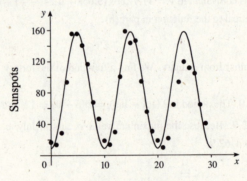

Year since 1975

(d) Using the `SinReg` command on the TI-83, we find that for the function $y = a \sin(bt + c) + d$, where $a = 67.65094323$, $b = 0.6205550572$, $c = -1.654463632$, and $d = 74.50460325$. Thus we get the model $y = 67.65 \sin(0.62t - 1.65) + 74.5$. This model is more accurate for more recent years than the one we found in part (b).

6 TRIGONOMETRIC FUNCTIONS: RIGHT TRIANGLE APPROACH

6.1   ANGLE MEASURE

1. **(a)** The radian measure of an angle θ is the length of the *arc* that subtends the angle in a circle of radius 1.

   (b) To convert degrees to radians we multiply by $\frac{\pi}{180}$.

   (c) To convert radians to degrees we multiply by $\frac{180}{\pi}$.

3. $72° = 72° \cdot \frac{\pi}{180°}$ rad $= \frac{2\pi}{5}$ rad ≈ 1.257 rad

5. $-45° = -45° \cdot \frac{\pi}{180°}$ rad $= -\frac{\pi}{4}$ rad ≈ -0.785 rad

7. $-75° = -75° \cdot \frac{\pi}{180°}$ rad $= -\frac{5\pi}{12}$ rad ≈ -1.309 rad

9. $1080° = 1080° \cdot \frac{\pi}{180°}$ rad $= 6\pi$ rad ≈ 18.850 rad

11. $96° = 96° \cdot \frac{\pi}{180°}$ rad $= \frac{8\pi}{15}$ rad ≈ 1.676 rad

13. $7.5° = 7.5° \cdot \frac{\pi}{180°}$ rad $= \frac{\pi}{24}$ rad ≈ 0.131 rad

15. $\frac{7\pi}{6} = \frac{7\pi}{6} \cdot \frac{180°}{\pi} = 210°$

15. $-\frac{5\pi}{4} = -\frac{5\pi}{4} \cdot \frac{180°}{\pi} = -225°$

19. $3 = 3 \cdot \frac{180°}{\pi} = \frac{540°}{\pi} \approx 171.9°$

21. $-1.2 = -1.2 \cdot \frac{180°}{\pi} = -\frac{216°}{\pi} = -68.8°$

23. $\frac{\pi}{10} = \frac{\pi}{10} \cdot \frac{180°}{\pi} = 18°$

25. $-\frac{2\pi}{15} = -\frac{2\pi}{15} \cdot \frac{180°}{\pi} = -24°$

27. $50°$ is coterminal with $50° + 360° = 410°$, $50° + 720° = 770°$, $50° - 360° = -310°$, and $50° - 720° = -670°$. (Other answers are possible.)

29. $\frac{3\pi}{4}$ is coterminal with $\frac{3\pi}{4} + 2\pi = \frac{11\pi}{4}$, $\frac{3\pi}{4} + 4\pi = \frac{19\pi}{4}$, $\frac{3\pi}{4} - 2\pi = -\frac{5\pi}{4}$, and $\frac{3\pi}{4} - 4\pi = -\frac{13\pi}{4}$. (Other answers are possible.)

31. $-\frac{\pi}{4}$ is coterminal with $-\frac{\pi}{4} + 2\pi = \frac{7\pi}{4}$, $-\frac{\pi}{4} + 4\pi = \frac{15\pi}{4}$, $-\frac{\pi}{4} - 2\pi = -\frac{9\pi}{4}$, and $-\frac{\pi}{4} - 4\pi = -\frac{17\pi}{4}$. (Other answers are possible.)

33. Since $430° - 70° = 360°$, the angles are coterminal.

35. Since $\frac{17\pi}{6} - \frac{5\pi}{6} = \frac{12\pi}{6} = 2\pi$; the angles are coterminal.

37. Since $875° - 155° = 720° = 2 \cdot 360°$, the angles are coterminal.

39. Since $733° - 2 \cdot 360° = 13°$, the angles $733°$ and $13°$ are coterminal.

41. Since $1110° - 3 \cdot 360° = 30°$, the angles $1110°$ and $30°$ are coterminal.

43. Since $-800° + 3 \cdot 360° = 280°$, the angles $-800°$ and $280°$ are coterminal.

45. Since $\frac{17\pi}{6} - 2\pi = \frac{5\pi}{6}$, the angles $\frac{17\pi}{6}$ and $\frac{5\pi}{6}$ are coterminal.

47. Since $87\pi - 43 \cdot 2\pi = \pi$, the angles 87π and π are coterminal.

49. Since $\frac{17\pi}{4} - 2 \cdot 2\pi = \frac{\pi}{4}$, the angles $\frac{17\pi}{4}$ and $\frac{\pi}{4}$ are coterminal.

51. Using the formula $s = \theta r$, the length of the arc is $s = \left(220° \cdot \frac{\pi}{180°}\right) \cdot 5 = \frac{55\pi}{9} \approx 19.2$.

53. Solving for r we have $r = \frac{s}{\theta}$, so the radius of the circle is $r = \frac{8}{2} = 4$.

55. Using the formula $s = \theta r$, the length of the arc is $s = 2 \cdot 2 = 4$ mi.

57. Solving for θ, we have $\theta = \frac{s}{r}$, so the measure of the central angle is $\theta = \frac{100}{50} = 2$ rad. Converting to degrees we have
$\theta = 2 \cdot \frac{180°}{\pi} \approx 114.6°$

59. Solving for r, we have $r = \frac{s}{\theta}$, so the radius of the circle is $r = \frac{6}{\pi/6} = \frac{36}{\pi} \approx 11.46$ m.

61. **(a)** $A = \frac{1}{2}r^2\theta = \frac{1}{2} \cdot 8^2 \cdot 80° \cdot \frac{\pi}{180°} = 32 \cdot \frac{4\pi}{9} = \frac{128\pi}{9} \approx 44.68$

   (b) $A = \frac{1}{2}r^2\theta = \frac{1}{2} \cdot 10^2 \cdot 0.5 = 25$

63. $A = \frac{1}{2}r^2\theta = \frac{1}{2} \cdot 10^2 \cdot 1 = 50 \text{ m}^2$

65. $\theta = 2$ rad, $A = 16 \text{ m}^2$. Since $A = \frac{1}{2}r^2\theta$, we have $r = \sqrt{2A/\theta} = \sqrt{2 \cdot 16/2} = \sqrt{16} = 4$ m.

67. Since the area of the circle is 72 cm^2, the radius of the circle is $r = \sqrt{A/\pi} = \sqrt{72/\pi}$. Then the area of the sector is
$A = \frac{1}{2}r^2\theta = \frac{1}{2} \cdot \frac{72}{\pi} \cdot \frac{\pi}{6} = 6 \text{ cm}^2$.

69. The circumference of each wheel is $\pi d = 28\pi$ in. If the wheels revolve 10,000 times, the distance traveled is
$10{,}000 \cdot 28\pi \text{ in.} \cdot \dfrac{1 \text{ ft}}{12 \text{ in.}} \cdot \dfrac{1 \text{ mi}}{5280 \text{ ft}} \approx 13.88$ mi.

71. We find the measure of the angle in degrees and then convert to radians. $\theta = 40.5° - 25.5° = 15°$ and $15 \cdot \frac{\pi}{180°}$ rad $= \frac{\pi}{12}$ rad.
Then using the formula $s = \theta r$, we have $s = \frac{\pi}{12} \cdot 3960 = 330\pi \approx 1036.725$ and so the distance between the two cities is
roughly 1037 mi.

73. In one day, the earth travels $\frac{1}{365}$ of its orbit which is $\frac{2\pi}{365}$ rad. Then $s = \theta r = \frac{2\pi}{365} \cdot 93{,}000{,}000 \approx 1{,}600{,}911.3$, so the
distance traveled is approximately 1.6 million miles.

75. The central angle is 1 minute $= \left(\frac{1}{60}\right)° = \frac{1}{60} \cdot \frac{\pi}{180°}$ rad $= \frac{\pi}{10{,}800}$ rad. Then $s = \theta r = \frac{\pi}{10{,}800} \cdot 3960 \approx 1.152$, and so a
nautical mile is approximately 1.152 mi.

77. The area is equal to the area of the large sector (with radius 34 in.) minus the area of the small sector (with radius 14 in.)
Thus, $A = \frac{1}{2}r_1^2\theta - \frac{1}{2}r_2^2\theta = \frac{1}{2}\left(34^2 - 14^2\right)\left(135° \cdot \frac{\pi}{180°}\right) \approx 1131 \text{ in.}^2$.

79. (a) The angular speed is $\omega = \dfrac{45 \cdot 2\pi \text{ rad}}{1 \text{ min}} = 90\pi$ rad/min.

   (b) The linear speed is $v = \dfrac{45 \cdot 2\pi \cdot 16}{1} = 1440\pi$ in./min ≈ 4523.9 in./min.

81. $v = \dfrac{8 \cdot 2\pi \cdot 2}{15} = \dfrac{32\pi}{15} \approx 6.702$ ft/s.

83. 23 h 56 min 4 s = 23.9344 hr. So the linear speed is $\dfrac{1 \cdot 2\pi \cdot (3960)}{1 \text{ day}} \cdot \dfrac{1 \text{ day}}{23.9344 \text{ hr}} \approx 1039.57$ mi/h.

85. $v = \dfrac{100 \cdot 2\pi \cdot 0.20 \text{ m}}{60 \text{ s}} = \dfrac{2\pi}{3} \approx 2.09$ m/s.

87. (a) The circumference of the opening is the length of the arc subtended by the angle θ on the flat piece of paper, that is,
$C = s = r\theta = 6 \cdot \frac{5\pi}{3} = 10\pi \approx 31.4$ cm.

   (b) Solving for r, we find $r = \dfrac{C}{2\pi} = \dfrac{10\pi}{2\pi} = 5$ cm.

   (c) By the Pythagorean Theorem, $h^2 = 6^2 - 5^2 = 11$, so $h = \sqrt{11} \approx 3.3$ cm.

   (d) The volume of a cone is $V = \frac{1}{3}\pi r^2 h$. In this case $V = \frac{1}{3}\pi \cdot 5^2 \cdot \sqrt{11} \approx 86.8 \text{ cm}^3$.

89. Answers will vary, although of course everyone prefers radians.

6.2 TRIGONOMETRY OF RIGHT TRIANGLES

1. (a)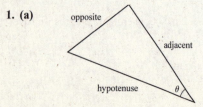

(b) $\sin\theta = \dfrac{\text{opposite}}{\text{hypotenuse}}$, $\cos\theta = \dfrac{\text{adjacent}}{\text{hypotenuse}}$, and $\tan\theta = \dfrac{\text{opposite}}{\text{adjacent}}$.

(c) The trigonometric ratios do not depend on the size of the triangle because all right triangles with angle θ are *similar*.

3. $\sin\theta = \frac{4}{5}$, $\cos\theta = \frac{3}{5}$, $\tan\theta = \frac{4}{3}$, $\csc\theta = \frac{5}{4}$, $\sec\theta = \frac{5}{3}$, $\cot\theta = \frac{3}{4}$

5. The remaining side is obtained by the Pythagorean Theorem: $\sqrt{41^2 - 40^2} = \sqrt{81} = 9$. Then $\sin\theta = \frac{40}{41}$, $\cos\theta = \frac{9}{41}$, $\tan\theta = \frac{40}{9}$, $\csc\theta = \frac{41}{40}$, $\sec\theta = \frac{41}{9}$, $\cot\theta = \frac{9}{40}$

7. The remaining side is obtained by the Pythagorean Theorem: $\sqrt{3^2 + 2^2} = \sqrt{13}$. Then $\sin\theta = \frac{2}{\sqrt{13}} = \frac{2\sqrt{13}}{13}$, $\cos\theta = \frac{3}{\sqrt{13}} = \frac{3\sqrt{13}}{13}$, $\tan\theta = \frac{2}{3}$, $\csc\theta = \frac{\sqrt{13}}{2}$, $\sec\theta = \frac{\sqrt{13}}{3}$, $\cot\theta = \frac{3}{2}$

9. $c = \sqrt{5^2 + 3^2} = \sqrt{34}$

   (a) $\sin\alpha = \cos\beta = \frac{3}{\sqrt{34}} = \frac{3\sqrt{34}}{34}$ (b) $\tan\alpha = \cot\beta = \frac{3}{5}$ (c) $\sec\alpha = \csc\beta = \frac{\sqrt{34}}{5}$

11. Since $\sin 30° = \frac{x}{25}$, we have $x = 25\sin 30° = 25 \cdot \frac{1}{2} = \frac{25}{2}$.

13. Since $\sin 60° = \frac{x}{13}$, we have $x = 13\sin 60° = 13 \cdot \frac{\sqrt{3}}{2} = \frac{13\sqrt{3}}{2}$.

15. Since $\tan 36° = \frac{12}{x}$, we have $x = \frac{12}{\tan 36°} \approx 16.51658$.

17. $\frac{x}{28} = \cos\theta \Leftrightarrow x = 28\cos\theta$, and $\frac{y}{28} = \sin\theta \Leftrightarrow y = 28\sin\theta$.

19. $\sin\theta = \frac{3}{5}$. Then the third side is $x = \sqrt{5^2 - 3^2} = 4$. The other five ratios are $\cos\theta = \frac{4}{5}$, $\tan\theta = \frac{3}{4}$, $\csc\theta = \frac{5}{3}$, $\sec\theta = \frac{5}{4}$, and $\cot\theta = \frac{4}{3}$.

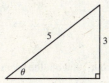

21. $\cot\theta = 1$. Then the third side is $r = \sqrt{1^2 + 1^2} = \sqrt{2}$. The other five ratios are $\sin\theta = \frac{1}{\sqrt{2}} = \frac{\sqrt{2}}{2}$, $\cos\theta = \frac{1}{\sqrt{2}} = \frac{\sqrt{2}}{2}$, $\tan\theta = 1$, $\csc\theta = \sqrt{2}$, and $\sec\theta = \sqrt{2}$.

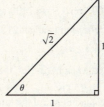

23. $\sec\theta = 7$. The third side is $y = \sqrt{7^2 - 2^2} = \sqrt{45} = 3\sqrt{5}$. The other five ratios are $\sin\theta = \frac{3\sqrt{5}}{7}$, $\cos\theta = \frac{2}{7}$, $\tan\theta = \frac{3\sqrt{5}}{2}$, $\csc\theta = \frac{7}{3\sqrt{5}} = \frac{7\sqrt{5}}{15}$, and $\cot\theta = \frac{2}{3\sqrt{5}} = \frac{2\sqrt{5}}{15}$.

25. $\sin\frac{\pi}{6} + \cos\frac{\pi}{6} = \frac{1}{2} + \frac{\sqrt{3}}{2} = \frac{1+\sqrt{3}}{2}$

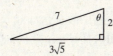

27. $\sin 30°\cos 60° + \sin 60°\cos 30° = \frac{1}{2} \cdot \frac{1}{2} + \frac{\sqrt{3}}{2} \cdot \frac{\sqrt{3}}{2} = \frac{1}{4} + \frac{3}{4} = 1$

29. $(\cos 30°)^2 - (\sin 30°)^2 = \left(\frac{\sqrt{3}}{2}\right)^2 - \left(\frac{1}{2}\right)^2 = \frac{3}{4} - \frac{1}{4} = \frac{1}{2}$

31. This is an isosceles right triangle, so the other leg has length $16\tan 45° = 16$, the hypotenuse has length $\frac{16}{\sin 45°} = 16\sqrt{2} \approx 22.63$, and the other angle is $90° - 45° = 45°$.

33. The other leg has length $35 \tan 52° \approx 44.79$, the hypotenuse has length $\dfrac{35}{\cos 52°} \approx 56.85$, and the other angle is $90° - 52° = 38°$.

35. The adjacent leg has length $33.5 \cos \dfrac{\pi}{8} \approx 30.95$, the opposite leg has length $33.5 \sin \dfrac{\pi}{8} \approx 12.82$, and the other angle is $\dfrac{\pi}{2} - \dfrac{\pi}{8} = \dfrac{3\pi}{8}$.

37. The adjacent leg has length $\dfrac{106}{\tan \frac{\pi}{5}} \approx 145.90$, the hypotenuse has length $\dfrac{106}{\sin \frac{\pi}{5}} \approx 180.34$, and the other angle is $\dfrac{\pi}{2} - \dfrac{\pi}{5} = \dfrac{3\pi}{10}$.

39. $\sin \theta \approx \dfrac{1}{2.24} \approx 0.45.$ $\cos \theta \approx \dfrac{2}{2.24} \approx 0.89$, $\tan \theta = \dfrac{1}{2}$, $\csc \theta \approx 2.24$, $\sec \theta \approx \dfrac{2.24}{2} \approx 1.12$, $\cot \theta \approx 2.00$.

41. $x = \dfrac{100}{\tan 60°} + \dfrac{100}{\tan 30°} \approx 230.9$

43. Let h be the length of the shared side. Then $\sin 60° = \dfrac{50}{h} \Leftrightarrow h = \dfrac{50}{\sin 60°} \approx 57.735 \Leftrightarrow \sin 65° = \dfrac{x}{h} \Leftrightarrow x = \dfrac{h}{\sin 65°} \approx 63.7$

45.

From the diagram, $\sin \theta = \dfrac{x}{y}$ and $\tan \theta = \dfrac{y}{10}$, so $x = y \sin \theta = 10 \sin \theta \tan \theta$.

47. Let h be the height, in feet, of the Empire State Building. Then $\tan 11° = \dfrac{h}{5280} \Leftrightarrow h = 5280 \cdot \tan 11° \approx 1026$ ft.

49. (a) Let h be the distance, in miles, that the beam has diverged. Then $\tan 0.5° = \dfrac{h}{240{,}000} \Leftrightarrow$ $h = 240{,}000 \cdot \tan 0.5° \approx 2100$ mi.

(b) Since the deflection is about 2100 mi whereas the radius of the moon is about 1000 mi, the beam will not strike the moon.

51. Let h represent the height, in feet, that the ladder reaches on the building. Then $\sin 72° = \dfrac{h}{20} \Leftrightarrow h = 20 \sin 72° \approx 19$ ft.

53. Let h be the height, in feet, of the kite above the ground. Then $\sin 50° = \dfrac{h}{450} \Leftrightarrow h = 450 \sin 50° \approx 345$ ft.

55. Let h_1 be the height of the window in feet and h_2 be the height from the window to the top of the tower. Then $\tan 25° = \dfrac{h_1}{325}$ $\Leftrightarrow h_1 = 325 \cdot \tan 25° \approx 152$ ft. Also, $\tan 39° = \dfrac{h_2}{325} \Leftrightarrow h_2 = 325 \cdot \tan 39° \approx 263$ ft. Therefore, the height of the window is approximately 152 ft and the height of the tower is approximately $152 + 263 = 415$ ft.

57. Let d_1 be the distance, in feet, between a point directly below the plane and one car, and d_2 be the distance, in feet, between the same point and the other car. Then $\tan 52° = \dfrac{d_1}{5150} \Leftrightarrow d_1 = 5150 \cdot \tan 52° \approx 6591.7$ ft. Also, $\tan 38° = \dfrac{d_2}{5150} \Leftrightarrow$ $d_2 = 5150 \cdot \tan 38° \approx 4023.6$ ft. So in this case, the distance between the two cars is about 2570 ft.

59. Let x be the horizontal distance, in feet, between a point on the ground directly below the top of the mountain and the point on the plain closest to the mountain. Let h be the height, in feet, of the mountain. Then $\tan 35° = \dfrac{h}{x}$ and $\tan 32° = \dfrac{h}{x + 1000}$. So $h = x \tan 35° = (x + 1000) \tan 32° \Leftrightarrow x = \dfrac{1000 \cdot \tan 32°}{\tan 35° - \tan 32°} \approx 8294.2$. Thus $h \approx 8294.2 \cdot \tan 35° \approx 5808$ ft.

61. Let d be the distance, in miles, from the earth to the sun. Then $\sec 89.85° = \dfrac{d}{240{,}000} \Leftrightarrow$ $d = 240{,}000 \cdot \sec 89.85° \approx 91.7$ million miles.

63. Let r represent the radius, in miles, of the earth. Then $\sin 60.276° = \frac{r}{r+600} \Leftrightarrow (r+600)\sin 60.276° = r \Leftrightarrow$

$600\sin 60.276° = r(1-\sin 60.276°) \Leftrightarrow r = \frac{600\sin 60.276°}{1-\sin 60.276°} \approx 3960.099$. So the earth's radius is about 3960 mi.

65. Let d be the distance, in AU, between Venus and the sun. Then $\sin 46.3° = \frac{d}{1} = d$, so $d = \sin 46.3° \approx 0.723$ AU.

6.3   TRIGONOMETRIC FUNCTIONS OF ANGLES

1. If the angle θ is in standard position and $P(x, y)$ is a point on the terminal side of θ, and r is the distance from the origin to P, then $\sin\theta = \dfrac{y}{r}$, $\cos\theta = \dfrac{x}{r}$, and $\tan\theta = \dfrac{y}{x}$.

3. (a) The reference angle for $150°$ is $180° - 150° = 30°$.

   (b) The reference angle for $330°$ is $360° - 330° = 30°$.

   (c) The reference angle for $-30°$ is $-(-30°) = 30°$.

5. (a) The reference angle for $225°$ is $225° - 180° = 45°$.

   (b) The reference angle for $810°$ is $810° - 720° = 90°$.

   (c) The reference angle for $-105°$ is
$$180° - 105° = 75°.$$

7. (a) The reference angle for $\frac{11\pi}{4}$ is $3\pi - \frac{11\pi}{4} = \frac{\pi}{4}$.

   (b) The reference angle for $-\frac{11\pi}{6}$ is $2\pi - \frac{11\pi}{6} = \frac{\pi}{6}$.

   (c) The reference angle for $\frac{11\pi}{3}$ is $4\pi - \frac{11\pi}{3} = \frac{\pi}{3}$.

9. (a) The reference angle for $\frac{5\pi}{7}$ is $\pi - \frac{5\pi}{7} = \frac{2\pi}{7}$.

   (b) The reference angle for -1.4π is $1.4\pi - \pi = 0.4\pi$.

   (c) The reference angle for 1.4 is 1.4 because $1.4 < \frac{\pi}{2}$.

11. $\sin 150° = \sin 30° = \frac{1}{2}$

13. $\cos 210° = -\cos 30° = -\frac{\sqrt{3}}{2}$

15. $\tan(-60°) = -\tan 60° = -\sqrt{3}$

17. $\csc(-630°) = \csc 90° = \frac{1}{\sin 90°} = 1$

19. $\cos 570° = -\cos 30° = -\frac{\sqrt{3}}{2}$

21. $\tan 750° = \tan 30° = \frac{1}{\sqrt{3}} = \frac{\sqrt{3}}{3}$

23. $\sin \frac{2\pi}{3} = \sin \frac{\pi}{3} = \frac{\sqrt{3}}{2}$

25. $\sin \frac{3\pi}{2} = -\sin \frac{\pi}{2} = -1$

27. $\cos\left(-\frac{7\pi}{3}\right) = \cos \frac{\pi}{3} = \frac{1}{2}$

29. $\sec \frac{17\pi}{3} = \sec \frac{\pi}{3} = \frac{1}{\cos \frac{\pi}{3}} = 2$

31. $\cot\left(-\frac{\pi}{4}\right) = -\cot \frac{\pi}{4} = \frac{-1}{\tan \frac{\pi}{4}} = -1$

33. $\tan \frac{5\pi}{2} = \tan \frac{\pi}{2}$ which is undefined.

35. Since $\sin\theta < 0$ and $\cos\theta < 0$, θ is in quadrant III.

37. $\sec\theta > 0 \Rightarrow \cos\theta > 0$. Also $\tan\theta < 0 \Rightarrow \dfrac{\sin\theta}{\cos\theta} < 0 \Leftrightarrow \sin\theta < 0$ (since $\cos\theta > 0$). Since $\sin\theta < 0$ and $\cos\theta > 0$, θ is in quadrant IV.

39. $\sec^2\theta = 1 + \tan^2\theta \Leftrightarrow \tan^2\theta = \dfrac{1}{\cos^2\theta} - 1 \Leftrightarrow \tan\theta = \sqrt{\dfrac{1}{\cos^2\theta} - 1} = \sqrt{\dfrac{1-\cos^2\theta}{\cos^2\theta}} = \dfrac{\sqrt{1-\cos^2\theta}}{|\cos\theta|} = \dfrac{\sqrt{1-\cos^2\theta}}{-\cos\theta}$

(since $\cos\theta < 0$ in quadrant III, $|\cos\theta| = -\cos\theta$). Thus $\tan\theta = -\dfrac{\sqrt{1-\cos^2\theta}}{\cos\theta}$.

41. $\cos^2\theta + \sin^2\theta = 1 \Leftrightarrow \cos\theta = \sqrt{1-\sin^2\theta}$ because $\cos\theta > 0$ in quadrant IV.

43. $\sec^2\theta = 1 + \tan^2\theta \Leftrightarrow \sec\theta = -\sqrt{1+\tan^2\theta}$ because $\sec\theta < 0$ in quadrant II.

45. $\sin\theta = \frac{3}{5}$. Then $x = -\sqrt{5^2 - 3^2} = -\sqrt{16} = -4$, since θ is in quadrant II. Thus, $\cos\theta = -\frac{4}{5}$, $\tan\theta = -\frac{3}{4}$, $\csc\theta = \frac{5}{3}$, $\sec\theta = -\frac{5}{4}$, and $\cot\theta = -\frac{4}{3}$.

47. $\tan\theta = -\frac{3}{4}$. Then $r = \sqrt{3^2 + 4^2} = 5$, and so $\sin\theta = -\frac{3}{5}$, $\cos\theta = \frac{4}{5}$, $\csc\theta = -\frac{5}{3}$, $\sec\theta = \frac{5}{4}$, and $\cot\theta = -\frac{4}{3}$.

49. $\csc \theta = 2$. Then $\sin \theta = \frac{1}{2}$ and $x = \sqrt{2^2 - 1^2} = \sqrt{3}$. So $\sin \theta = \frac{1}{2}$, $\cos \theta = \frac{\sqrt{3}}{2}$, $\tan \theta = \frac{1}{\sqrt{3}} = \frac{\sqrt{3}}{3}$, $\sec \theta = \frac{2}{\sqrt{3}} = \frac{2\sqrt{3}}{3}$, and $\cot \theta = \sqrt{3}$.

51. $\cos \theta = -\frac{2}{7}$. Then $y = \sqrt{7^2 - 2^2} = \sqrt{45} = 3\sqrt{5}$, and so $\sin \theta = \frac{3\sqrt{5}}{7}$, $\tan \theta = -\frac{3\sqrt{5}}{2}$, $\csc \theta = \frac{7}{3\sqrt{5}} = \frac{7\sqrt{5}}{15}$, $\sec \theta = -\frac{7}{2}$, and $\cot \theta = -\frac{2}{3\sqrt{5}} = -\frac{2\sqrt{5}}{15}$.

53. (a) $\sin 2\theta = \sin \left(2 \cdot \frac{\pi}{3}\right) = \sin \frac{2\pi}{3} = \sin \frac{\pi}{3} = \frac{\sqrt{3}}{2}$, while $2 \sin \theta = 2 \sin \frac{\pi}{3} = 2 \cdot \frac{\sqrt{3}}{2} = \sqrt{3}$.

(b) $\sin \frac{1}{2}\theta = \sin \left(\frac{1}{2} \cdot \frac{\pi}{3}\right) = \sin \frac{\pi}{6} = \frac{1}{2}$, while $\frac{1}{2} \sin \theta = \frac{1}{2} \sin \frac{\pi}{3} = \frac{1}{2} \cdot \frac{\sqrt{3}}{2} = \frac{\sqrt{3}}{4}$.

(c) $\sin^2 \theta = \left(\sin \frac{\pi}{3}\right)^2 = \left(\frac{\sqrt{3}}{2}\right)^2 = \frac{3}{4}$, while $\sin \left(\theta^2\right) = \sin \left(\frac{\pi}{3}\right)^2 = \sin \frac{\pi^2}{9} \approx 0.88967$.

55. $a = 10$, $b = 22$, and $\theta = 10°$. Thus, the area of the triangle is $A = \frac{1}{2}(10)(22) \sin 10° = 110 \sin 10° \approx 19.1$.

57. $A = 16$, $a = 5$, and $b = 7$. So $\sin \theta = \frac{2A}{ab} = \frac{2 \cdot 16}{5 \cdot 7} = \frac{32}{35} \Leftrightarrow \theta = \sin^{-1} \frac{32}{35} \approx 66.1°$.

59. For the sector defined by the two sides, $A_1 = \frac{1}{2}r^2\theta = \frac{1}{2} \cdot 2^2 \cdot 120° \cdot \frac{\pi}{180°} = \frac{4\pi}{3}$. For the triangle defined by the two sides, $A_2 = \frac{1}{2}ab \sin \theta = \frac{1}{2} \cdot 2 \cdot 2 \cdot \sin 120° = 2 \sin 60° = \sqrt{3}$. Thus the area of the region is $A_1 - A_2 = \frac{4\pi}{3} - \sqrt{3} \approx 2.46$.

61. $\sin^2 \theta + \cos^2 \theta = 1 \Leftrightarrow \left(\sin^2 \theta + \cos^2 \theta\right) \cdot \frac{1}{\cos^2 \theta} = 1 \cdot \frac{1}{\cos^2 \theta} \Leftrightarrow \tan^2 \theta + 1 = \sec^2 \theta$

63. (a) $\tan \theta = \dfrac{h}{1 \text{ mile}}$, so $h = \tan \theta \cdot 1 \text{ mile} \cdot \dfrac{5280 \text{ ft}}{1 \text{ mile}} = 5280 \tan \theta$ ft.

(b)

θ	20°	60°	80°	85°
h	1922	9145	29,944	60,351

65. (a) From the figure in the text, we express depth and width in terms of θ.

Since $\sin \theta = \dfrac{\text{depth}}{20}$ and $\cos \theta = \dfrac{\text{width}}{20}$, we have depth $= 20 \sin \theta$ and width $= 20 \cos \theta$. Thus, the cross-section area of the beam is $A(\theta) = (\text{depth})(\text{width}) = (20 \cos \theta)(20 \sin \theta) = 400 \cos \theta \sin \theta$.

(b)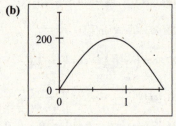

(c) The beam with the largest cross-sectional area is the square beam, $10\sqrt{2}$ by $10\sqrt{2}$ (about 14.14 by 14.14).

67. (a) On Earth, the range is $R = \dfrac{v_0^2 \sin (2\theta)}{g} = \dfrac{12^2 \sin \frac{\pi}{3}}{32} = \dfrac{9\sqrt{3}}{4} \approx 3.897$ ft and the height is

$H = \dfrac{v_0^2 \sin^2 \theta}{2g} = \dfrac{12^2 \sin^2 \frac{\pi}{6}}{2 \cdot 32} = \dfrac{9}{16} = 0.5625$ ft.

(b) On the moon, $R = \dfrac{12^2 \sin \frac{\pi}{3}}{5.2} \approx 23.982$ ft and $H = \dfrac{12^2 \sin^2 \frac{\pi}{6}}{2 \cdot 5.2} \approx 3.462$ ft

69. (a) $W = 3.02 - 0.38 \cot \theta + 0.65 \csc \theta$

(b) From the graph, it appears that W has its minimum value at about $\theta = 0.946 \approx 54.2°$.

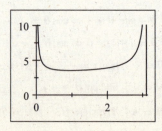

71. We have $\sin\alpha = k\sin\beta$, where $\alpha = 59.4°$ and $k = 1.33$. Substituting, $\sin 59.4° = 1.33\sin\beta$

$\Rightarrow \sin\beta = \dfrac{\sin 59.4°}{1.33} \approx 0.6472$. Using a calculator, we find that $\beta \approx \sin^{-1} 0.6472 \approx 40.3°$, so

$\theta = 4\beta - 2\alpha \approx 4\,(40.3°) - 2\,(59.4°) = 42.4°$.

73. $\cos\theta = \dfrac{\text{adj}}{\text{hyp}} = \dfrac{|OP|}{|OR|} = \dfrac{|OP|}{1} = |OP|$. Since QS is tangent to the circle at R, $\triangle ORQ$ is a right triangle. Then

$\tan\theta = \dfrac{\text{opp}}{\text{adj}} = \dfrac{|RQ|}{|OR|} = |RQ|$ and $\sec\theta = \dfrac{\text{hyp}}{\text{adj}} = \dfrac{|OQ|}{|OR|} = |OQ|$. Since $\angle SOQ$ is a right angle $\triangle SOQ$ is a right

triangle and $\angle OSR = \theta$. Then $\csc\theta = \dfrac{\text{hyp}}{\text{opp}} = \dfrac{|OS|}{|OR|} = |OS|$ and $\cot\theta = \dfrac{\text{adj}}{\text{opp}} = \dfrac{|SR|}{|OR|} = |SR|$. Summarizing, we have

$\sin\theta = |PR|$, $\cos\theta = |OP|$, $\tan\theta = |RQ|$, $\sec\theta = |OQ|$, $\csc\theta = |OS|$, and $\cot\theta = |SR|$.

6.4   INVERSE TRIGONOMETRIC FUNCTIONS AND TRIANGLES

1. (a) The function $\sin^{-1}$ has domain $[-1, 1]$ and range $\left[-\frac{\pi}{2}, \frac{\pi}{2}\right]$.

   (b) The function $\cos^{-1}$ has domain $[-1, 1]$ and range $[0, \pi]$.

   (c) The function $\tan^{-1}$ has domain $\mathbb{R}$ and range $\left(-\frac{\pi}{2}, \frac{\pi}{2}\right)$.

3. (a) $\sin^{-1}\frac{1}{2} = \frac{\pi}{6}$ because $\sin\frac{\pi}{6} = \frac{1}{2}$ and $\frac{\pi}{6}$ lies in the range of $\sin^{-1}$, $\left[-\frac{\pi}{2}, \frac{\pi}{2}\right]$.

   (b) $\cos^{-1}\left(-\frac{\sqrt{3}}{2}\right) = \frac{5\pi}{6}$ because $\cos\frac{5\pi}{6} = -\frac{\sqrt{3}}{2}$ and $\frac{5\pi}{6}$ lies in the range of $\cos^{-1}$, $[0, \pi]$.

   (c) $\tan^{-1}(-1) = -\frac{\pi}{4}$ because $\tan\left(-\frac{\pi}{4}\right) = -1$ and $-\frac{\pi}{4}$ lies in the range of $\tan^{-1}$, $\left(-\frac{\pi}{2}, \frac{\pi}{2}\right)$.

5. (a) $\sin^{-1}\left(-\frac{1}{2}\right) = -\frac{\pi}{6}$ **(b)** $\cos^{-1}\frac{1}{2} = \frac{\pi}{3}$ **(c)** $\tan^{-1}\left(\frac{\sqrt{3}}{3}\right) = \frac{\pi}{6}$

7. $\sin^{-1}(0.45) \approx 0.46677$ **9.** $\cos^{-1}\left(-\frac{1}{4}\right) \approx 1.82348$ **11.** $\tan^{-1} 3 \approx 1.24905$ **13.** $\cos^{-1} 3$ is undefined.

15. $\sin\theta = \frac{6}{10} = \frac{3}{5}$, so $\theta = \sin^{-1}\frac{3}{5} \approx 36.9°$.   **17.** $\tan\theta = \frac{9}{13}$, so $\theta = \tan^{-1}\frac{9}{13} \approx 34.7°$.

19. $\sin\theta = \frac{4}{7}$, so $\theta = \sin^{-1}\frac{4}{7} \approx 34.8°$.

21. We use $\sin^{-1}$ to find one solution in the interval $\left[-\frac{\pi}{2}, \frac{\pi}{2}\right]$. $\sin\theta = \frac{1}{2} \Rightarrow \theta = \sin^{-1}\frac{1}{2} = 30°$. Another solution with
   θ between $0°$ and $180°$ is obtained by taking the supplement of the angle: $180° - 30° = 150°$. So the solutions of the
   equation with θ between $0°$ and $180°$ are $\theta = 30°$ and $\theta = 150°$.

23. $\sin\theta = 0.7$, so the solutions with $0° < \theta < 180°$ are $\theta = \sin^{-1} 0.7 \approx 44.4°$ and $\theta = 180° - \sin^{-1} 0.7 \approx 135.6°$.

25. The cosine function is one-to-one on the interval $[0, \pi]$, so there is only one solution of the equation with θ between $0°$ and
   $180°$. We find that solution by taking $\cos^{-1}$ of each side: $\cos\theta = 0.7 \Leftrightarrow \theta = \cos^{-1} 0.7 \approx 45.6°$.

27. To find $\sin\left(\cos^{-1}\frac{3}{5}\right)$, first let $\theta = \cos^{-1}\frac{3}{5}$. Then θ is the number in the interval

$\left[-\frac{\pi}{2}, \frac{\pi}{2}\right]$ whose cosine is $\frac{3}{5}$. We draw a right triangle with θ as one of its acute

angles, with adjacent side 3 and hypotenuse 5. The remaining leg of the triangle is

found by the Pythagorean Theorem to be 4. From the figure we get

$\sin\left(\cos^{-1}\frac{3}{5}\right) = \sin\theta = \frac{4}{5}$.

Another method: By the cancellation properties of inverse functions, $\cos\left(\cos^{-1}\frac{3}{5}\right)$ is exactly $\frac{3}{5}$. To find $\sin\left(\cos^{-1}\frac{3}{5}\right)$, we

first write the sine function in terms of the cosine function. Let $u = \cos^{-1}\frac{3}{5}$. Since $0 \le u \le \pi$, $\sin u$ is positive, and since

$\cos^2 u + \sin^2 u = 1$, we can write $\sin u = \sqrt{1 - \cos^2 u} = \sqrt{1 - \cos^2\left(\cos^{-1}\frac{3}{5}\right)} = \sqrt{1 - \left(\frac{3}{5}\right)^2} = \sqrt{1 - \frac{9}{25}} = \sqrt{\frac{16}{25}} = \frac{4}{5}$.

Therefore, $\sin\left(\cos^{-1}\frac{3}{5}\right) = \frac{4}{5}$.

29. To find $\sec\left(\sin^{-1}\frac{12}{13}\right)$, we draw a right triangle with

angle θ, opposite side 12, and hypotenuse 13. From the

figure we see that $\sec\left(\sin^{-1}\frac{12}{13}\right) = \sec\theta = \frac{13}{5}$.

31. To find $\tan\left(\sin^{-1}\frac{12}{13}\right)$, we draw a right triangle with angle

θ, opposite side 12, and hypotenuse 13. From the figure

we see that $\tan\left(\sin^{-1}\frac{12}{13}\right) = \tan\theta = \frac{12}{5}$.

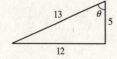

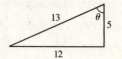

33. We want to find $\cos\left(\sin^{-1}x\right)$. Let $\theta = \sin^{-1}x$, so $\sin\theta = x$. We sketch a right

triangle with an acute angle θ, opposite side x, and hypotenuse 1. By the

Pythagorean Theorem, the remaining leg is $\sqrt{1 - x^2}$. From the figure we have

$\cos\left(\sin^{-1}x\right) = \cos\theta = \sqrt{1 - x^2}$.

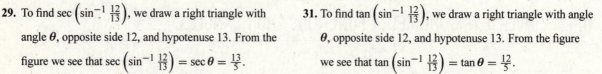

Another method: Let $u = \sin^{-1}x$. We need to find $\cos u$ in terms of x. To do so, we write cosine in terms of sine. Note

that $-\frac{\pi}{2} \le u \le \frac{\pi}{2}$ because $u = \sin^{-1}x$. Now $\cos u = \sqrt{1 - \sin^2 u}$ is positive because u lies in the interval $\left[-\frac{\pi}{2}, \frac{\pi}{2}\right]$.

Substituting $u = \sin^{-1}x$ and using the cancellation property $\sin(\sin^{-1}x) = x$ gives $\cos(\sin^{-1}x) = \sqrt{1 - x^2}$.

35. We want to find $\tan\left(\sin^{-1}x\right)$. Let $\theta = \sin^{-1}x$, so $\sin\theta = x$. We sketch a right

triangle with an acute angle θ, opposite side x, and hypotenuse 1. By the

Pythagorean Theorem, the remaining leg is $\sqrt{1 - x^2}$. From the figure we have

$\tan\left(\sin^{-1}x\right) = \tan\theta = \dfrac{x}{\sqrt{1 - x^2}}$.

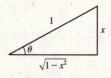

37. Let θ represent the angle of elevation of the ladder. Let h represent the height, in feet, that the ladder reaches on the

building. Then $\cos\theta = \frac{6}{20} = 0.3 \Leftrightarrow \theta = \cos^{-1} 0.3 \approx 1.266$ rad $\approx 72.5°$. By the Pythagorean Theorem, $h^2 + 6^2 = 20^2 \Leftrightarrow$

$h = \sqrt{400 - 36} = \sqrt{364} \approx 19$ ft.

39. (a) Solving $\tan\theta = h/2$ for h, we have $h = 2\tan\theta$.

(b) Solving $\tan\theta = h/2$ for θ we have $\theta = \tan^{-1}(h/2)$.

41. (a) Solving $\sin\theta = h/680$ for θ we have $\theta = \sin^{-1}(h/680)$.

(b) Set $h = 500$ to get $\theta = \sin^{-1}\left(\frac{500}{680}\right) \approx 0.826$ rad.

43. (a) $\theta = \sin^{-1}\left(\dfrac{1}{(2 \cdot 3 + 1)\tan 10°}\right) = \sin^{-1}\left(\dfrac{1}{7\tan 10°}\right) \approx \sin^{-1} 0.8102 \approx 54.1°$

(b) For $n = 2$, $\theta = \sin^{-1}\left(\dfrac{1}{5\tan 15°}\right) \approx 48.3°$. For $n = 3$, $\theta = \sin^{-1}\left(\dfrac{1}{7\tan 15°}\right) \approx 32.2°$. For $n = 4$,

$\theta = \sin^{-1}\left(\dfrac{1}{9\tan 15°}\right) \approx 24.5°$. $n = 0$ and $n = 1$ are outside of the domain for $\beta = 15°$, because $\dfrac{1}{\tan 15°} \approx 3.732$

and $\dfrac{1}{3\tan 15°} \approx 1.244$, neither of which is in the domain of $\sin^{-1}$.

6.5   THE LAW OF SINES

1. In triangle ABC with sides a, b, and c the Law of Sines states that $\dfrac{\sin A}{a} = \dfrac{\sin B}{b} = \dfrac{\sin C}{c}$.

3. $\angle C = 180° - 98.4° - 24.6° = 57°$. $x = \dfrac{376 \sin 57°}{\sin 98.4°} \approx 318.75$.

5. $\angle C = 180° - 52° - 70° = 58°$. $x = \dfrac{26.7 \sin 52°}{\sin 58°} \approx 24.8$.

7. $\sin C = \dfrac{36 \sin 120°}{45} \approx 0.693 \Leftrightarrow \angle C \approx \sin^{-1} 0.693 \approx 44°$.

9. $\angle C = 180° - 46° - 20° = 114°$. Then $a = \dfrac{65 \sin 46°}{\sin 114°} \approx 51$ and $b = \dfrac{65 \sin 20°}{\sin 114°} \approx 24$.

11. $\angle B = 68°$, so $\angle A = 180° - 68° - 68° = 44°$ and $a = \dfrac{12 \sin 44°}{\sin 68°} \approx 8.99$.

13. $\angle C = 180° - 50° - 68° = 62°$. Then

$a = \dfrac{230 \sin 50°}{\sin 62°} \approx 200$ and $b = \dfrac{230 \sin 68°}{\sin 62°} \approx 242$.

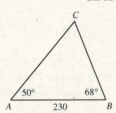

15. $\angle B = 180° - 30° - 65° = 85°$. Then

$a = \dfrac{10 \sin 30°}{\sin 85°} \approx 5.0$ and $c = \dfrac{10 \sin 65°}{\sin 85°} \approx 9$.

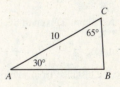

17. $\angle A = 180° - 51° - 29° = 100°$. Then

$a = \dfrac{44 \sin 100°}{\sin 29°} \approx 89$ and $c = \dfrac{44 \sin 51°}{\sin 29°} \approx 71$.

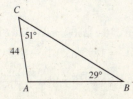

19. Since $\angle A > 90°$ there is only one triangle.

$\sin B = \dfrac{15 \sin 110°}{28} \approx 0.503 \Leftrightarrow$

$\angle B \approx \sin^{-1} 0.503 \approx 30°$. Then

$\angle C \approx 180° - 110° - 30° = 40°$, and so

$c = \dfrac{28 \sin 40°}{\sin 110°} \approx 19$. Thus $\angle B \approx 30°$, $\angle C \approx 40°$, and

$c \approx 19$.

21. $\angle A = 125°$ is the largest angle, but since side a is not the longest side, there can be no such triangle.

23. $\sin C = \dfrac{30\sin 25°}{25} \approx 0.507 \Leftrightarrow \angle C_1 \approx \sin^{-1} 0.507 \approx 30.47°$ or $\angle C_2 \approx 180° - 39.47° = 149.53°$.

If $\angle C_1 = 30.47°$, then $\angle A_1 \approx 180° - 25° - 30.47° = 124.53°$ and $a_1 = \dfrac{25\sin 124.53°}{\sin 25°} \approx 48.73$.

If $\angle C_2 = 149.53°$, then $\angle A_2 \approx 180° - 25° - 149.53° = 5.47°$ and $a_2 = \dfrac{25\sin 5.47°}{\sin 25°} \approx 5.64$.

Thus, one triangle has $\angle A_1 \approx 125°$, $\angle C_1 \approx 30°$, and $a_1 \approx 49$; the other has $\angle A_2 \approx 5°$, $\angle C_2 \approx 150°$, and $a_2 \approx 5.6$.

25. $\sin B = \dfrac{100\sin 50°}{50} \approx 1.532$. Since $|\sin\theta| \le 1$ for all θ, there can be no such angle B, and thus no such triangle.

27. $\sin A = \dfrac{26\sin 29°}{15} \approx 0.840 \Leftrightarrow \angle A_1 \approx \sin^{-1} 0.840 \approx 57.2°$ or $\angle A_2 \approx 180° - 57.2° = 122.8°$.

If $\angle A_1 \approx 57.2°$, then $\angle B_1 = 180° - 29° - 57.2° = 93.8°$ and $b_1 \approx \dfrac{15\sin 93.8°}{\sin 29°} \approx 30.9$.

If $\angle A_2 \approx 122.8°$, then $\angle B_2 = 180 - 29° - 122.8° = 28.2°$ and $b_2 \approx \dfrac{15\sin 28.1°}{\sin 29°} \approx 14.6$.

Thus, one triangle has $\angle A_1 \approx 57.2°$, $\angle B_1 \approx 93.8°$, and $b_1 \approx 30.9$; the other has $\angle A_2 \approx 122.8°$, $\angle B_2 \approx 28.2°$, and $b_2 \approx 14.6$.

29. **(a)** From $\triangle ABC$ and the Law of Sines we get $\dfrac{\sin 30°}{20} = \dfrac{\sin B}{28} \Leftrightarrow \sin B = \dfrac{28\sin 30°}{20} = 0.7$, so

$\angle B \approx \sin^{-1} 0.7 \approx 44.427°$. Since $\triangle BCD$ is isosceles, $\angle B = \angle BDC \approx 44.427°$. Thus, $\angle BCD = 180° - 2\angle B \approx 91.146° \approx 91.1°$.

(b) From $\triangle ABC$ we get $\angle BCA = 180° - \angle A - \angle B \approx 180° - 30° - 44.427° = 105.573°$. Hence $\angle DCA = \angle BCA - \angle BCD \approx 105.573° - 91.146° = 14.4°$.

31. **(a)** $\sin B = \dfrac{20\sin 40°}{15} \approx 0.857 \Leftrightarrow \angle B_1 \approx \sin^{-1} 0.857 \approx 58.99°$ or $\angle B_2 \approx 180° - 58.99° \approx 121.01°$.

If $\angle B_1 = 58.99°$, then $\angle C_1 \approx 180° - 40° - 58.99° \approx 81.01°$ and $c_1 = \dfrac{15\sin 81.01°}{\sin 40°} \approx 23.05$.

If $\angle B_2 = 121.01°$, then $\angle C_2 \approx 180° - 40° - 121.01° = 18.99°$ and $c_2 = \dfrac{15\sin 18.99°}{\sin 40°} \approx 7.59$. Thus there are two triangles.

(b) By the area formula given in Section 6.3, $\dfrac{\text{Area of } \triangle ABC}{\text{Area of } \triangle A'B'C'} = \dfrac{\frac{1}{2}ab\sin C}{\frac{1}{2}ab\sin C'} = \dfrac{\sin C}{\sin C'}$, because a and b are the same in both triangles.

33. **(a)** Let a be the distance from satellite to the tracking station A in miles. Then the subtended angle at the satellite is

$\angle C = 180° - 93° - 84.2° = 2.8°$, and so $a = \dfrac{50\sin 84.2°}{\sin 2.8°} \approx 1018$ mi.

(b) Let d be the distance above the ground in miles. Then $d = 1018.3\sin 87° \approx 1017$ mi.

35. $\angle C = 180° - 82° - 52° = 46°$, so by the Law of Sines, $\dfrac{|AC|}{\sin 52°} = \dfrac{|AB|}{\sin 46°} \Leftrightarrow |AC| = \dfrac{|AB|\sin 52°}{\sin 46°}$, so substituting we have $|AC| = \dfrac{200\sin 52°}{\sin 46°} \approx 219$ ft.

37.

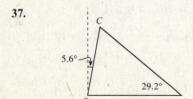

We draw a diagram. A is the position of the tourist and C is the top of the tower. $\angle B = 90° - 5.6° = 84.4°$ and so $\angle C = 180° - 29.2° - 84.4° = 66.4°$. Thus, by the Law of Sines, the length of the tower is $|BC| = \dfrac{105\sin 29.2°}{\sin 66.4°} \approx 55.9$ m.

39. The angle subtended by the top of the tree and the sun's rays is $\angle A = 180° - 90° - 52° = 38°$. Thus the height of the tree is $h = \dfrac{215 \sin 30°}{\sin 38°} \approx 175$ ft.

41. Call the balloon's position R. Then in $\triangle PQR$, we see that $\angle P = 62° - 32° = 30°$, and $\angle Q = 180° - 71° + 32° = 141°$. Therefore, $\angle R = 180° - 30° - 141° = 9°$. So by the Law of Sines, $\dfrac{|QR|}{\sin 30°} = \dfrac{|PQ|}{\sin 9°} \Leftrightarrow |QR| = 60 \cdot \dfrac{\sin 30°}{\sin 9°} \approx 192$ m.

43. Let d be the distance from the earth to Venus, and let β be the angle formed by sun, Venus, and earth. By the Law of Sines, $\dfrac{\sin \beta}{1} = \dfrac{\sin 39.4°}{0.723} \approx 0.878$, so either $\beta \approx \sin^{-1} 0.878 \approx 61.4°$ or $\beta \approx 180° - \sin^{-1} 0.878 \approx 118.6°$.

In the first case, $\dfrac{d}{\sin(180° - 39.4° - 61.4°)} = \dfrac{0.723}{\sin 39.4°} \Leftrightarrow d \approx 1.119$ AU; in the second case,

$\dfrac{d}{\sin(180° - 39.4° - 118.6°)} = \dfrac{0.723}{\sin 39.4°} \Leftrightarrow d \approx 0.427$ AU.

45.

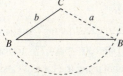 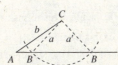 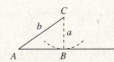 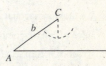

$a \geq b$: One solution $b > a > b \sin A$: Two solutions $a = b \sin A$: One solution $a < b \sin A$: No solution

$\angle A = 30°$, $b = 100$, $\sin A = \frac{1}{2}$. If $a \geq b = 100$ then there is one triangle. If $100 > a > 100 \sin 30° = 50$, then there are two possible triangles. If $a = 50$, then there is one (right) triangle. And if $a < 50$, then no triangle is possible.

6.6 THE LAW OF COSINES

1. For triangle ABC with sides a, b, and c the Law of Cosines states $c^2 = a^2 + b^2 - 2ab \cos C$.

3. $x^2 = 21^2 + 42^2 - 2 \cdot 21 \cdot 42 \cdot \cos 39° = 441 + 1764 - 1764 \cos 39° \approx 834.115$ and so $x \approx \sqrt{834.115} \approx 28.9$.

5. $x^2 = 25^2 + 25^2 - 2 \cdot 25 \cdot 25 \cdot \cos 140° = 625 + 625 - 1250 \cos 140° \approx 2207.556$ and so $x \approx \sqrt{2207.556} \approx 47$.

7. $37.83^2 = 68.01^2 + 42.15^2 - 2 \cdot 68.01 \cdot 42.15 \cdot \cos \theta$. Then $\cos \theta = \dfrac{37.83^2 - 68.01^2 - 42.15^2}{-2 \cdot 68.01 \cdot 42.15} \approx 0.867 \Leftrightarrow \theta \approx \cos^{-1} 0.867 \approx 29.89°$.

9. $x^2 = 24^2 + 30^2 - 2 \cdot 24 \cdot 30 \cdot \cos 30° = 576 + 900 - 1440 \cos 30° \approx 228.923$ and so $x \approx \sqrt{228.923} \approx 15$.

11. $c^2 = 10^2 + 18^2 - 2 \cdot 10 \cdot 18 \cdot \cos 120° = 100 + 324 - 360 \cos 120° = 604$ and so $c \approx \sqrt{604} \approx 24.576$. Then $\sin A \approx \dfrac{18 \sin 120°}{24.576} \approx 0.634295 \Leftrightarrow \angle A \approx \sin^{-1} 0.634295 \approx 39.4°$, and $\angle B \approx 180° - 120° - 39.4° = 20.6°$.

13. $c^2 = 3^2 + 4^2 - 2 \cdot 3 \cdot 4 \cdot \cos 53° = 9 + 16 - 24 \cos 53° \approx 10.556 \Leftrightarrow c \approx \sqrt{10.556} \approx 3.2$. Then $\sin B = \dfrac{4 \sin 53°}{3.25} \approx 0.983$ $\Leftrightarrow \angle B \approx \sin^{-1} 0.983 \approx 79°$ and $\angle A \approx 180° - 53° - 79° = 48°$.

15. $20^2 = 25^2 + 22^2 - 2 \cdot 25 \cdot 22 \cdot \cos A \Leftrightarrow \cos A = \dfrac{20^2 - 25^2 - 22^2}{-2 \cdot 25 \cdot 22} \approx 0.644 \Leftrightarrow \angle A \approx \cos^{-1} 0.644 \approx 50°$. Then $\sin B \approx \dfrac{25 \sin 49.9°}{20} \approx 0.956 \Leftrightarrow \angle B \approx \sin^{-1} 0.956 \approx 73°$, and so $\angle C \approx 180° - 50° - 73° = 57°$.

17. $\sin C = \dfrac{162 \sin 40°}{125} \approx 0.833 \Leftrightarrow \angle C_1 \approx \sin^{-1} 0.833 \approx 56.4°$ or $\angle C_2 \approx 180° - 56.4° \approx 123.6°$.

If $\angle C_1 \approx 56.4°$, then $\angle A_1 \approx 180° - 40° - 56.4° = 83.6°$ and $a_1 = \dfrac{125 \sin 83.6°}{\sin 40°} \approx 193$.

If $\angle C_2 \approx 123.6°$, then $\angle A_2 \approx 180° - 40° - 123.6° = 16.4°$ and $a_2 = \dfrac{125 \sin 16.4°}{\sin 40°} \approx 54.9$.

Thus, one triangle has $\angle A \approx 83.6°$, $\angle C \approx 56.4°$, and $a \approx 193$; the other has $\angle A \approx 16.4°$, $\angle C \approx 123.6°$, and $a \approx 54.9$.

19. $\sin B = \dfrac{65 \sin 55°}{50} \approx 1.065$. Since $|\sin \theta| \leq 1$ for all θ, there is no such $\angle B$, and hence there is no such triangle.

21. $\angle B = 180° - 35° - 85° = 60°$. Then $x = \frac{3 \sin 35°}{\sin 60°} \approx 2$.

23. $x = \frac{50 \sin 30°}{\sin 100°} \approx 25.4$

25. $b^2 = 110^2 + 138^2 - 2\,(110)\,(138) \cdot \cos 38° = 12{,}100 + 19{,}044 - 30{,}360 \cos 38° \approx 7220.0$ and so $b \approx 85.0$. Therefore, using the Law of Cosines again, we have $\cos \theta = \frac{110^2 + 85^2 - 128^2}{2(110)(138)} \iff \theta \approx 89.15°$.

27. $x^2 = 38^2 + 48^2 - 2 \cdot 38 \cdot 48 \cdot \cos 30° = 1444 + 2304 - 3648 \cos 30° \approx 588.739$ and so $x \approx 24.3$.

29. The semiperimeter is $s = \frac{9+12+15}{2} = 18$, so by Heron's Formula the area is
$A = \sqrt{18\,(18-9)\,(18-12)\,(18-15)} = \sqrt{2916} = 54$.

31. The semiperimeter is $s = \frac{7+8+9}{2} = 12$, so by Heron's Formula the area is
$A = \sqrt{12\,(12-7)\,(12-8)\,(12-9)} = \sqrt{720} = 12\sqrt{5} \approx 26.8$.

33. The semiperimeter is $s = \frac{3+4+6}{2} = \frac{13}{2}$, so by Heron's Formula the area is
$A = \sqrt{\frac{13}{2}\left(\frac{13}{2}-3\right)\left(\frac{13}{2}-4\right)\left(\frac{13}{2}-6\right)} = \sqrt{\frac{455}{16}} = \frac{\sqrt{455}}{4} \approx 5.33$.

35. We draw a diagonal connecting the vertices adjacent to the $100°$ angle. This forms two triangles. Consider the triangle with sides of length 5 and 6 containing the $100°$ angle. The area of this triangle is $A_1 = \frac{1}{2}\,(5)\,(6) \sin 100° \approx 14.77$. To use Heron's Formula to find the area of the second triangle, we need to find the length of the diagonal using the Law of Cosines: $c^2 = a^2 + b^2 - 2ab \cos C = 5^2 + 6^2 - 2 \cdot 5 \cdot 6 \cos 100° \approx 71.419 \Rightarrow c \approx 8.45$. Thus the second triangle has semiperimeter $s = \frac{8+7+8.45}{2} \approx 11.7255$ and area $A_2 = \sqrt{11.7255\,(11.7255 - 8)\,(11.7255 - 7)\,(11.7255 - 8.45)} \approx 26.00$. The area of the quadrilateral is the sum of the areas of the two triangles: $A = A_1 + A_2 \approx 14.77 + 26.00 = 40.77$.

37.

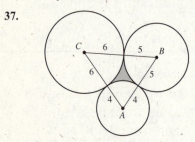

Label the centers of the circles A, B, and C, as in the figure. By the Law of Cosines, $\cos A = \frac{AB^2 + AC^2 - BC^2}{2\,(AB)\,(AC)} = \frac{9^2 + 10^2 - 11^2}{2\,(9)\,(10)} = \frac{1}{3} \Rightarrow \angle A \approx 70.53°$.

Now, by the Law of Sines, $\frac{\sin 70.53°}{11} = \frac{\sin B}{AC} = \frac{\sin C}{AB}$. So

$\sin B = \frac{10}{11} \sin 70.53° \approx 0.85710 \Rightarrow B \approx \sin^{-1} 0.85710 \approx 58.99°$ and

$\sin C = \frac{9}{11} \sin 70.53° \approx 0.77139 \Rightarrow C \approx \sin^{-1} 0.77139 \approx 50.48°$. The area of

$\triangle ABC$ is $\frac{1}{2}\,(AB)\,(AC) \sin A = \frac{1}{2}\,(9)\,(10)\,(\sin 70.53°) \approx 42.426$.

The area of sector A is given by $S_A = \pi R^2 \cdot \frac{\theta}{360°} = \pi\,(4)^2 \cdot \frac{70.53°}{360°} \approx 9.848$. Similarly, the areas of sectors B and C are $S_B \approx 12.870$ and $S_C \approx 15.859$. Thus, the area enclosed between the circles is $A = \triangle ABC - S_A - S_B - S_C \Rightarrow A \approx 42.426 - 9.848 - 12.870 - 15.859 \approx 3.85 \text{ cm}^2$.

39. Let c be the distance across the lake, in miles. Then $c^2 = 2.82^2 + 3.56^2 - 2\,(2.82)\,(3.56) \cdot \cos 40.3° \approx 5.313 \iff c \approx 2.30$ mi.

41. In half an hour, the faster car travels 25 miles while the slower car travels 15 miles. The distance between them is given by the Law of Cosines: $d^2 = 25^2 + 15^2 - 2\,(25)\,(15) \cdot \cos 65° \Rightarrow$
$d = \sqrt{25^2 + 15^2 - 2\,(25)\,(15) \cdot \cos 65°} = 5\sqrt{25 + 9 - 30 \cdot \cos 65°} \approx 23.1$ mi.

43. The pilot travels a distance of $625 \cdot 1.5 = 937.5$ miles in her original direction and $625 \cdot 2 = 1250$ miles in the new direction. Since she makes a course correction of $10°$ to the right, the included angle is $180° - 10° = 170°$. From the figure, we use the Law of Cosines to get the expression $d^2 = 937.5^2 + 1250^2 - 2\,(937.5)\,(1250) \cdot \cos 170° \approx 4{,}749{,}549.42$, so $d \approx 2179$ miles. Thus, the pilot's distance from her original position is approximately 2179 miles.

45. (a) The angle subtended at Egg Island is $100°$. Thus using the Law of Cosines, the distance from Forrest Island to the fisherman's home port is

$$x^2 = 30^2 + 50^2 - 2 \cdot 30 \cdot 50 \cdot \cos 100°$$
$$= 900 + 2500 - 3000 \cos 100° \approx 3920.945$$

and so $x \approx \sqrt{3920.945} \approx 62.62$ miles.

(b) Let θ be the angle shown in the figure. Using the Law of Sines,

$$\sin\theta = \frac{50 \sin 100°}{62.62} \approx 0.7863 \Leftrightarrow \theta \approx \sin^{-1} 0.7863 \approx 51.8°. \text{ Then}$$

$\gamma = 90° - 20° - 51.8° = 18.2°$. Thus the bearing to his home port is S $18.2°$ E.

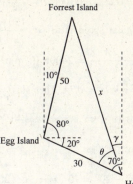

47. The largest angle is the one opposite the longest side; call this angle θ. Then by the Law of Cosines,

$$44^2 = 36^2 + 22^2 - 2(36)(22) \cdot \cos\theta \Leftrightarrow \cos\theta = \frac{36^2 + 22^2 - 44^2}{2(36)(22)} = -0.09848 \Rightarrow \theta \approx \cos^{-1}(-0.09848) \approx 96°.$$

49. Let d be the distance between the kites. Then $d^2 \approx 380^2 + 420^2 - 2(380)(420) \cdot \cos 30° \Rightarrow$
$d \approx \sqrt{380^2 + 420^2 - 2(380)(420) \cdot \cos 30°} \approx 211$ ft.

51. *Solution 1:* From the figure, we see that $\gamma = 106°$ and $\sin 74° = \dfrac{3400}{b} \Leftrightarrow$

$b = \dfrac{3400}{\sin 74°} \approx 3537$. Thus, $x^2 = 800^2 + 3537^2 - 2(800)(3537)\cos 106° \Rightarrow$

$x = \sqrt{800^2 + 3537^2 - 2(800)(3537)\cos 106°} \Rightarrow x \approx 3835$ ft.

Solution 2: Notice that $\tan 74° = \dfrac{3400}{a} \Leftrightarrow a = \dfrac{3400}{\tan 74°} \approx 974.9$. By the

Pythagorean Theorem, $x^2 = (a + 800)^2 + 3400^2$. So

$x = \sqrt{(974.9 + 800)^2 + 3400^2} \approx 3835$ ft.

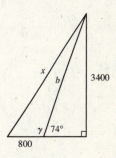

53. By Heron's formula, $A = \sqrt{s(s-a)(s-b)(s-c)}$, where $s = \dfrac{a+b+c}{2} = \dfrac{112+148+190}{2} = 225$. Thus,
$A = \sqrt{225(225-112)(225-148)(225-190)} \approx 8277.7$ ft^2. Since the land value is \$20 per square foot, the value of the lot is approximately $8277.7 \cdot 20 = \$165,554$.

CHAPTER 6 REVIEW

1. (a) $60° = 60 \cdot \dfrac{\pi}{180} = \dfrac{\pi}{3} \approx 1.05$ rad

(b) $330° = 330 \cdot \dfrac{\pi}{180} = \dfrac{11\pi}{6} \approx 5.76$ rad

(c) $-135° = -135 \cdot \dfrac{\pi}{180} = -\dfrac{3\pi}{4} \approx -3.36$ rad

(d) $-90° = -90 \cdot \dfrac{\pi}{180} = -\dfrac{\pi}{2} \approx -1.57$ rad

3. (a) $\dfrac{5\pi}{2}$ rad $= \dfrac{5\pi}{2} \cdot \dfrac{180}{\pi} = 450°$

(b) $-\dfrac{\pi}{6}$ rad $= -\dfrac{\pi}{6} \cdot \dfrac{180}{\pi} = -30°$

(c) $\dfrac{9\pi}{4}$ rad $= \dfrac{9\pi}{4} \cdot \dfrac{180}{\pi} = 405°$

(d) 3.1 rad $= 3.1 \cdot \dfrac{180}{\pi} = \dfrac{558}{\pi} \approx 177.6°$

5. $r = 8$ m, $\theta = 1$ rad. Then $s = r\theta = 8 \cdot 1 = 8$ m.

7. $s = 100$ ft, $\theta = 70° = 70 \cdot \dfrac{\pi}{180} = \dfrac{7\pi}{18}$ rad. Then $r = \dfrac{s}{\theta} = 100 \cdot \dfrac{18}{7\pi} = \dfrac{1800}{7\pi} \approx 82$ ft.

9. $r = 3960$ miles, $s = 2450$ miles. Then $\theta = \dfrac{s}{r} = \dfrac{2450}{3960} \approx 0.619$ rad $= 0.619 \cdot \dfrac{\pi}{180} \approx 35.448°$ and so the angle is approximately $35.4°$.

11. $A = \frac{1}{2}r^2\theta = \frac{1}{2}(200)^2\left(52° \cdot \frac{\pi}{180°}\right) \approx 18{,}151 \text{ ft}^2$

13. The angular speed is $\omega = \dfrac{150 \cdot 2\pi \text{ rad}}{1 \text{ min}} = 300\pi$ rad/min ≈ 942.5 rad/min. The linear speed is

$v = \dfrac{150 \cdot 2\pi \cdot 8}{1} = 2400\pi$ in./min ≈ 7539.8 in./min.

15. $r = \sqrt{5^2 + 7^2} = \sqrt{74}$. Then $\sin\theta = \frac{5}{\sqrt{74}}$, $\cos\theta = \frac{7}{\sqrt{74}}$, $\tan\theta = \frac{5}{7}$, $\csc\theta = \frac{\sqrt{74}}{5}$, $\sec\theta = \frac{\sqrt{74}}{7}$, and $\cot\theta = \frac{7}{5}$.

17. $\frac{x}{5} = \cos 40° \;\Leftrightarrow\; x = 5\cos 40° \approx 3.83$, and $\frac{y}{5} = \sin 40° \;\Leftrightarrow\; y = 5\sin 40° \approx 3.21$.

19. $\frac{1}{x} = \sin 20° \;\Leftrightarrow\; x = \frac{1}{\sin 20°} \approx 2.92$, and $\frac{x}{y} = \cos 20° \;\Leftrightarrow\; y = \frac{x}{\cos 20°} \approx \frac{2.924}{0.9397} \approx 3.11$.

21. $A = 90° - 20° = 70°$, $a = 3\cos 20° \approx 2.819$, and
$b = 3\sin 20° \approx 1.026$.

23. $c = \sqrt{25^2 - 7^2} = 24$, $A = \sin^{-1}\frac{7}{24} \approx 0.2960 \approx 17.0°$,
and $C = \sin^{-1}\frac{24}{25} \approx 1.2870 \approx 73.7°$.

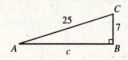

25. $\tan\theta = \frac{1}{a} \Leftrightarrow a = \frac{1}{\tan\theta} = \cot\theta$, $\sin\theta = \frac{1}{b} \Leftrightarrow b = \frac{1}{\sin\theta} = \csc\theta$

27.

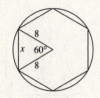

One side of the hexagon together with radial line segments through its endpoints forms a triangle with two sides of length 8 m and subtended angle 60°. Let x be the length of one such side (in meters). By the Law of Cosines,

$x^2 = 8^2 + 8^2 - 2 \cdot 8 \cdot 8 \cdot \cos 60° = 64 \;\Leftrightarrow\; x = 8$. Thus the perimeter of the hexagon is $6x = 6 \cdot 8 = 48$ m.

29. Let r represent the radius, in miles, of the moon. Then $\tan\dfrac{\theta}{2} = \dfrac{r}{r + |AB|}$, $\theta = 0.518° \;\Leftrightarrow\; r = (r + 236{,}900)\cdot\tan 0.259°$

$\Leftrightarrow\; r\left(1 - \tan 0.259°\right) = 236{,}900 \cdot \tan 0.259° \Leftrightarrow r = \dfrac{236{,}900 \cdot \tan 0.259°}{1 - \tan 0.259°} \approx 1076$ and so the radius of the moon is roughly 1076 miles.

31. $\sin 315° = -\sin 45° = -\frac{1}{\sqrt{2}} = -\frac{\sqrt{2}}{2}$

33. $\tan(-135°) = \tan 45° = 1$

35. $\cot\left(-\frac{22\pi}{3}\right) = \cot\frac{2\pi}{3} = \cot\frac{\pi}{3} = -\frac{1}{\sqrt{3}} = -\frac{\sqrt{3}}{3}$

37. $\cos 585° = \cos 225° = -\cos 45° = -\frac{1}{\sqrt{2}} = -\frac{\sqrt{2}}{2}$

39. $\csc\frac{8\pi}{3} = \csc\frac{2\pi}{3} = \csc\frac{\pi}{3} = \frac{2}{\sqrt{3}} = \frac{2\sqrt{3}}{3}$

41. $\cot(-390°) = \cot(-30°) = -\cot 30° = -\sqrt{3}$

43. $r = \sqrt{(-5)^2 + 12^2} = \sqrt{169} = 13$. Then $\sin\theta = \frac{12}{13}$, $\cos\theta = -\frac{5}{13}$, $\tan\theta = -\frac{12}{5}$, $\csc\theta = \frac{13}{12}$, $\sec\theta = -\frac{13}{5}$, and
$\cot\theta = -\frac{5}{12}$.

45. $y - \sqrt{3}x + 1 = 0 \Leftrightarrow y = \sqrt{3}x - 1$, so the slope of the line is $m = \sqrt{3}$. Then $\tan\theta = m = \sqrt{3} \Leftrightarrow \theta = 60°$.

47. $\sec^2\theta = 1 + \tan^2\theta \Leftrightarrow \tan^2\theta = \frac{1}{\cos^2\theta} - 1 \Leftrightarrow \tan\theta = \sqrt{\frac{1}{\cos^2\theta} - 1} = \sqrt{\frac{1 - \cos^2\theta}{\cos^2\theta}} = \frac{\sqrt{1 - \cos^2\theta}}{|\cos\theta|} = \frac{\sqrt{1 - \cos^2\theta}}{-\cos\theta}$

(because $\cos\theta < 0$ in quadrant III, $|\cos\theta| = -\cos\theta$). Thus $\tan\theta = -\dfrac{\sqrt{1 - \cos^2\theta}}{\cos\theta}$.

49. $\tan^2\theta = \dfrac{\sin^2\theta}{\cos^2\theta} = \dfrac{\sin^2\theta}{1 - \sin^2\theta}$

51. $\tan \theta = \frac{\sqrt{7}}{3}$, $\sec \theta = \frac{4}{3}$. Then $\cos \theta = \frac{3}{4}$ and $\sin \theta = \tan \theta \cdot \cos \theta = \frac{\sqrt{7}}{4}$, $\csc \theta = \frac{4}{\sqrt{7}} = \frac{4\sqrt{7}}{7}$, and $\cot \theta = \frac{3}{\sqrt{7}} = \frac{3\sqrt{7}}{7}$.

53. $\sin \theta = \frac{3}{5}$. Since $\cos \theta < 0$, θ is in quadrant II. Thus, $x = -\sqrt{5^2 - 3^2} = -\sqrt{16} = -4$ and so $\cos \theta = -\frac{4}{5}$, $\tan \theta = -\frac{3}{4}$, $\csc \theta = \frac{5}{3}$, $\sec \theta = -\frac{5}{4}$, $\cot \theta = -\frac{4}{3}$.

55. $\tan \theta = -\frac{1}{2}$. $\sec^2 \theta = 1 + \tan^2 \theta = 1 + \frac{1}{4} = \frac{5}{4} \Leftrightarrow \cos^2 \theta = \frac{4}{5} \Rightarrow \cos \theta = -\sqrt{\frac{4}{5}} = -\frac{2}{\sqrt{5}}$ since

$\cos \theta < 0$ in quadrant II. But $\tan \theta = \dfrac{\sin \theta}{\cos \theta} = -\frac{1}{2} \Leftrightarrow \sin \theta = -\frac{1}{2}\cos \theta = -\frac{1}{2}\left(-\frac{2}{\sqrt{5}}\right) = \frac{1}{\sqrt{5}}$. Therefore,

$\sin \theta + \cos \theta = \frac{1}{\sqrt{5}} + \left(-\frac{2}{\sqrt{5}}\right) = -\frac{1}{\sqrt{5}} = -\frac{\sqrt{5}}{5}$.

57. By the Pythagorean Theorem, $\sin^2 \theta + \cos^2 \theta = 1$ for any angle θ.

59. $\sin^{-1} \frac{\sqrt{3}}{2} = \frac{\pi}{3}$

61. Let $u = \sin^{-1}\frac{2}{5}$ and so $\sin u = \frac{2}{5}$. Then from the triangle,

$\tan\left(\sin^{-1}\frac{2}{5}\right) = \tan u = \frac{2}{\sqrt{21}}$.

63. Let $\theta = \tan^{-1} x \quad \Leftrightarrow \quad \tan \theta = x$. Then from the triangle, we have $\sin\left(\tan^{-1} x\right) = \sin \theta = \dfrac{x}{\sqrt{1+x^2}}$.

65. $\cos \theta = \frac{x}{3} \Rightarrow \theta = \cos^{-1}\left(\frac{x}{3}\right)$

67. $\angle B = 180° - 30° - 80° = 70°$, and so by the Law of Sines, $x = \dfrac{10 \sin 30°}{\sin 70°} \approx 5.32$.

69. $x^2 = 100^2 + 210^2 - 2 \cdot 100 \cdot 210 \cdot \cos 40° \approx 21{,}926.133 \Leftrightarrow x \approx 148.07$

71. $x^2 = 2^2 + 8^2 - 2(2)(8)\cos 120° = 84 \Leftrightarrow x \approx \sqrt{84} \approx 9.17$

73. By the Law of Sines, $\dfrac{\sin \theta}{23} = \dfrac{\sin 25°}{12} \Rightarrow \sin \theta = \dfrac{23 \sin 25°}{12} \Rightarrow \theta = \sin^{-1}\left(\dfrac{23 \sin 25°}{12}\right) \approx 54.1°$ or

$\theta \approx 180° - 54.1° = 125.9°$.

75. By the Law of Cosines, $120^2 = 100^2 + 85^2 - 2(100)(85)\cos \theta$, so $\cos \theta = \dfrac{120^2 - 100^2 - 85^2}{-2(100)(85)} \approx 0.16618$. Thus,

$\theta \approx \cos^{-1}(0.16618) \approx 80.4°$.

77. After 2 hours the ships have traveled distances $d_1 = 40$ mi and $d_2 = 56$ mi. The subtended angle is $180° - 32° - 42° = 106°$. Let d be the distance between the two ships in miles. Then by the Law of Cosines, $d^2 = 40^2 + 56^2 - 2(40)(56)\cos 106° \approx 5970.855 \Leftrightarrow d \approx 77.3$ miles.

79. Let d be the distance, in miles, between the points A and B. Then by the Law of Cosines, $d^2 = 3.2^2 + 5.6^2 - 2(3.2)(5.6)\cos 42° \approx 14.966 \Leftrightarrow d \approx 3.9$ mi.

81. $A = \frac{1}{2}ab \sin \theta = \frac{1}{2}(8)(14)\sin 35° \approx 32.12$

CHAPTER 6 TEST

1. $330° = 330 \cdot \frac{\pi}{180} = \frac{11\pi}{6}$ rad. $-135° = -135 \cdot \frac{\pi}{180} = -\frac{3\pi}{4}$ rad.

3. (a) The angular speed is

$$\omega = \frac{120 \cdot 2\pi \text{ rad}}{1 \text{ min}} = 240\pi \text{ rad/min} \approx 753.98 \text{ rad/min}.$$

(b) The linear speed is

$$v = \frac{120 \cdot 2\pi \cdot 16}{1} = 3840\pi \text{ ft/min}$$
$$\approx 12{,}063.7 \text{ ft/min}$$
$$\approx 137 \text{ mi/h}$$

5. $r = \sqrt{3^2 + 2^2} = \sqrt{13}$. Then

$$\tan\theta + \sin\theta = \frac{2}{3} + \frac{2}{\sqrt{13}} = \frac{2\left(\sqrt{13}+3\right)}{3\sqrt{13}} = \frac{26 + 6\sqrt{13}}{39}.$$

7. $\cos\theta = -\frac{1}{3}$ and θ is in quadrant III, so $r = 3$, $x = -1$, and $y = -\sqrt{3^2 - 1^2} = -2\sqrt{2}$. Then

$$\tan\theta\cot\theta + \csc\theta = \tan\theta \cdot \frac{1}{\tan\theta} + \csc\theta = 1 - \frac{3}{2\sqrt{2}} = \frac{2\sqrt{2}-3}{2\sqrt{2}} = \frac{4-3\sqrt{2}}{4}.$$

9. $\sec^2\theta = 1 + \tan^2\theta \Leftrightarrow \tan\theta = \pm\sqrt{\sec^2\theta - 1}$. Thus, $\tan\theta = -\sqrt{\sec^2\theta - 1}$ since $\tan\theta < 0$ in quadrant II.

11. (a) $\tan\theta = \frac{x}{4} \Rightarrow \theta = \tan^{-1}\left(\frac{x}{4}\right)$

(b) $\cos\theta = \frac{3}{x} \Rightarrow \theta = \cos^{-1}\left(\frac{3}{x}\right)$

13. By the Law of Cosines, $x^2 = 10^2 + 12^2 - 2(10)(12) \cdot \cos 48° \approx 8.409 \Leftrightarrow x \approx 9.1$.

15. Let h be the height of the shorter altitude. Then $\tan 20° = \frac{h}{50} \Leftrightarrow h = 50\tan 20°$ and $\tan 28° = \frac{x+h}{50} \Leftrightarrow x+h = 50\tan 28°$
$\Leftrightarrow \quad x = 50\tan 28° - h = 50\tan 28° - 50\tan 20° \approx 8.4$.

17. By the Law of Cosines, $9^2 = 8^2 + 6^2 - 2(8)(6)\cos\theta \Rightarrow \cos\theta = \frac{8^2 + 6^2 - 9^2}{2(8)(6)} \approx 0.1979$, so $\theta \approx \cos^{-1}(0.1979) \approx 78.6°$.

19. (a) $A\text{ (sector)} = \frac{1}{2}r^2\theta = \frac{1}{2} \cdot 10^2 \cdot 72 \cdot \frac{\pi}{180} = 50 \cdot \frac{72\pi}{180}$. $A\text{ (triangle)} = \frac{1}{2}r \cdot r\sin\theta = \frac{1}{2} \cdot 10^2 \sin 72°$. Thus, the area of the
shaded region is $A\text{ (shaded)} = A\text{ (sector)} - A\text{ (triangle)} = 50\left(\frac{72\pi}{180} - \sin 72°\right) \approx 15.3 \text{ m}^2$.

(b) The shaded region is bounded by two pieces: one piece is part of the triangle, the other is part of the circle.
The first part has length $l = \sqrt{10^2 + 10^2 - 2(10)(10) \cdot \cos 72°} = 10\sqrt{2 - 2 \cdot \cos 72°}$. The second has length
$s = 10 \cdot 72 \cdot \frac{\pi}{180} = 4\pi$. Thus, the perimeter of the shaded region is $p = l + s = 10\sqrt{2 - 2\cos 72°} + 4\pi \approx 24.3 \text{ m}$.

21. Label the figure as shown. Now $\angle\beta = 85° - 75° = 10°$, so by the Law of Sines,

$$\frac{x}{\sin 75°} = \frac{100}{\sin 10°} \quad \Leftrightarrow \quad x = 100 \cdot \frac{\sin 75°}{\sin 10°}. \text{ Now } \sin 85° = \frac{h}{x} \quad \Leftrightarrow$$

$$h = x\sin 85° = 100 \cdot \frac{\sin 75°}{\sin 10°}\sin 85° \approx 554.$$

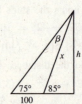

FOCUS ON MODELING   Surveying

1. Let x be the distance between the church and City Hall. To apply the Law of Sines to the triangle with vertices at City Hall, the church, and the first bridge, we first need the measure of the angle at the first bridge, which is $180° - 25° - 30° = 125°$.

Then $\dfrac{x}{\sin 125°} = \dfrac{0.86}{\sin 30°} \Leftrightarrow x = \dfrac{0.86 \sin 125°}{\sin 30°} \approx 1.4089$. So the distance between the church and City Hall is about 1.41 miles.

3. First notice that $\angle DBC = 180° - 20° - 95° = 65°$ and $\angle DAC = 180° - 60° - 45° = 75°$.

From $\triangle ACD$ we get $\dfrac{|AC|}{\sin 45°} = \dfrac{20}{\sin 75°} \Leftrightarrow |AC| = \dfrac{20 \sin 45°}{\sin 75°} \approx 14.6°$. From $\triangle BCD$ we get

$\dfrac{|BC|}{\sin 95°} = \dfrac{20}{\sin 65°} \Leftrightarrow |BC| = \dfrac{20 \sin 95°}{\sin 65°} \approx 22.0$. By applying the Law of Cosines to $\triangle ABC$ we get

$|AB|^2 = |AC|^2 + |BC|^2 - 2|AC||BC| \cos 40° \approx 14.6^2 + 22.0^2 - 2 \cdot 14.6 \cdot 22.0 \cdot \cos 40° \approx 205$, so $|AB| \approx \sqrt{205} \approx 14.3$ m. Therefore, the distance between A and B is approximately 14.3 m.

5. (a) In $\triangle ABC$, $\angle B = 180° - \beta$, so $\angle C = 180° - \alpha - (180° - \beta) = \beta - \alpha$. By the Law of Sines, $\dfrac{|BC|}{\sin \alpha} = \dfrac{|AB|}{\sin (\beta - \alpha)}$

$\Rightarrow |BC| = |AB| \dfrac{\sin \alpha}{\sin (\beta - \alpha)} = \dfrac{d \sin \alpha}{\sin (\beta - \alpha)}$.

(b) From part (a) we know that $|BC| = \dfrac{d \sin \alpha}{\sin (\beta - \alpha)}$. But $\sin \beta = \dfrac{h}{|BC|} \Leftrightarrow |BC| = \dfrac{h}{\sin \beta}$. Therefore,

$|BC| = \dfrac{d \sin \alpha}{\sin (\beta - \alpha)} = \dfrac{h}{\sin \beta} \Rightarrow h = \dfrac{d \sin \alpha \sin \beta}{\sin (\beta - \alpha)}$.

(c) $h = \dfrac{d \sin \alpha \sin \beta}{\sin (\beta - \alpha)} = \dfrac{800 \sin 25° \sin 29°}{\sin 4°} \approx 2350$ ft

7. We start by labeling the edges and calculating the remaining angles, as shown in the first figure. Using the Law of Sines, we find the following: $\dfrac{a}{\sin 29°} = \dfrac{150}{\sin 60°} \Leftrightarrow a = \dfrac{150 \sin 29°}{\sin 60°} \approx 83.97$, $\dfrac{b}{\sin 91°} = \dfrac{150}{\sin 60°} \Leftrightarrow b = \dfrac{150 \sin 91°}{\sin 60°} \approx 173.18$,

$\dfrac{c}{\sin 32°} = \dfrac{173.18}{\sin 87°} \Leftrightarrow c = \dfrac{173.18 \sin 32°}{\sin 87°} \approx 91.90$, $\dfrac{d}{\sin 61°} = \dfrac{173.18}{\sin 87°} \Leftrightarrow e = \dfrac{173.18 \sin 61°}{\sin 87°} \approx 151.67$,

$\dfrac{e}{\sin 41°} = \dfrac{151.67}{\sin 51°} \Leftrightarrow e = \dfrac{151.67 \sin 41°}{\sin 51°} \approx 128.04$, $\dfrac{f}{\sin 88°} = \dfrac{151.67}{\sin 51°} \Leftrightarrow f = \dfrac{151.67 \sin 88°}{\sin 51°} \approx 195.04$,

$\dfrac{g}{\sin 50°} = \dfrac{195.04}{\sin 92°} \Leftrightarrow g = \dfrac{195.04 \sin 50°}{\sin 92°} \approx 149.50$, and $\dfrac{h}{\sin 38°} = \dfrac{195.04}{\sin 92°} \Leftrightarrow h = \dfrac{195.04 \sin 38°}{\sin 92°} \approx 120.15$. Note that we used two decimal places throughout our calculations. Our results are shown (to one decimal place) in the second figure.

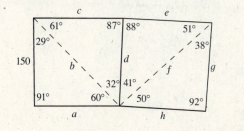
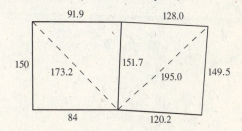

7 ANALYTIC TRIGONOMETRY

7.1 TRIGONOMETRIC IDENTITIES

1. An equation is called an identity if it is valid for *all* values of the variable. The equation $2x = x + x$ is an algebraic identity and the equation $\sin^2 x + \cos^2 x = 1$ is a trigonometric identity.

3. $\cos t \tan t = \cos t \cdot \dfrac{\sin t}{\cos t} = \sin t$

5. $\sin \theta \sec \theta = \sin \theta \cdot \dfrac{1}{\cos \theta} = \tan \theta$

7. $\tan^2 x - \sec^2 x = \dfrac{\sin^2 x}{\cos^2 x} - \dfrac{1}{\cos^2 x} = \dfrac{\sin^2 x - 1}{\cos^2 x} = \dfrac{-\cos^2 x}{\cos^2 x} = -1$

9. $\sin u + \cot u \cos u = \sin u + \dfrac{\cos u}{\sin u} \cdot \cos u = \dfrac{\sin^2 u + \cos^2 u}{\sin u} = \dfrac{1}{\sin u} = \csc u$

11. $\dfrac{\sec \theta - \cos \theta}{\sin \theta} = \dfrac{\dfrac{1}{\cos \theta} - \cos \theta}{\sin \theta} = \dfrac{1 - \cos^2 \theta}{\sin \theta \cos \theta} = \dfrac{\sin^2 \theta}{\sin \theta \cos \theta} = \dfrac{\sin \theta}{\cos \theta} = \tan \theta$

13. $\dfrac{\sin x \sec x}{\tan x} = \dfrac{\sin x \cdot \dfrac{1}{\cos x}}{\dfrac{\cos x}{\sin x}} = 1$

15. $\dfrac{1 + \cos y}{1 + \sec y} = \dfrac{1 + \cos y}{1 + \dfrac{1}{\cos y}} = \dfrac{1 + \cos y}{\dfrac{\cos y + 1}{\cos y}} = \dfrac{1 + \cos y}{1} \cdot \dfrac{\cos y}{\cos y + 1} = \cos y$

17. $\dfrac{\sec^2 x - 1}{\sec^2 x} = \dfrac{\tan^2 x}{\sec^2 x} = \dfrac{\sin^2 x}{\cos^2 x} \cdot \cos^2 x = \sin^2 x$. *Another method:* $\dfrac{\sec^2 x - 1}{\sec^2 x} = 1 - \dfrac{1}{\sec^2 x} = 1 - \cos^2 x = \sin^2 x$

19. $\dfrac{1 + \csc x}{\cos x + \cot x} = \dfrac{1 + \dfrac{1}{\sin x}}{\cos x + \dfrac{\cos x}{\sin x}} = \dfrac{1 + \dfrac{1}{\sin x}}{\cos x + \dfrac{\cos x}{\sin x}} \cdot \dfrac{\sin x}{\sin x} = \dfrac{\sin x + 1}{\cos x (\sin x + 1)} = \dfrac{1}{\cos x} = \sec x$

21. $\dfrac{1 + \sin u}{\cos u} + \dfrac{\cos u}{1 + \sin u} = \dfrac{(1 + \sin u)^2 + \cos^2 u}{\cos u (1 + \sin u)} = \dfrac{1 + 2\sin u + \sin^2 u + \cos^2 u}{\cos u (1 + \sin u)} = \dfrac{1 + 2\sin u + 1}{\cos u (1 + \sin u)}$

$= \dfrac{2 + 2\sin u}{\cos u (1 + \sin u)} = \dfrac{2(1 + \sin u)}{\cos u (1 + \sin u)} = \dfrac{2}{\cos u} = 2 \sec u$

23. $\dfrac{2 + \tan^2 x}{\sec^2 x} - 1 = \dfrac{1 + 1 + \tan^2 x}{\sec^2 x} - 1 = \dfrac{1}{\sec^2 x} + \dfrac{1 + \tan^2 x}{\sec^2 x} - 1 = \dfrac{1}{\sec^2 x} + \dfrac{\sec^2 x}{\sec^2 x} - 1$

$= \dfrac{1}{\sec^2 x} + 1 - 1 = \dfrac{1}{\sec^2 x} = \cos^2 x$

25. $\tan \theta + \cos(-\theta) + \tan(-\theta) = \tan \theta + \cos \theta - \tan \theta = \cos \theta$

27. (a) $\dfrac{\cos x}{\sec x \sin x} = \dfrac{\cos x}{\dfrac{1}{\cos x} \cdot \sin x} = \dfrac{\cos^2 x}{\sin x} = \dfrac{1 - \sin^2 x}{\sin x}$

$\qquad\qquad = \dfrac{1}{\sin x} - \dfrac{\sin^2 x}{\sin x} = \csc x - \sin x$

(b) We graph each side of the equation and see that the

graphs of $y = \dfrac{\cos x}{\sec x \sin x}$ and $y = \csc x - \sin x$ are

identical, confirming that the equation is an identity.

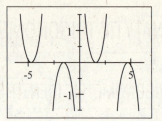

29. $\dfrac{\sin \theta}{\tan \theta} = \dfrac{\sin \theta}{\dfrac{\sin \theta}{\cos \theta}} = \sin \theta \cdot \dfrac{\cos \theta}{\sin \theta} = \cos \theta$
    31. $\dfrac{\cos u \sec u}{\tan u} = \cos u \dfrac{1}{\cos u} \cot u = \cot u$

33. $\sin B + \cos B \cot B = \sin B + \cos B \dfrac{\cos B}{\sin B} = \dfrac{\sin^2 B + \cos^2 B}{\sin B} = \dfrac{1}{\sin B} = \csc B$

35. $\cot (-\alpha) \cos (-\alpha) + \sin (-\alpha) = -\dfrac{\cos \alpha}{\sin \alpha} \cos \alpha - \sin \alpha = \dfrac{-\cos^2 \alpha - \sin^2 \alpha}{\sin \alpha} = \dfrac{-1}{\sin \alpha} = -\csc \alpha$

37. $\tan \theta + \cot \theta = \dfrac{\sin \theta}{\cos \theta} + \dfrac{\cos \theta}{\sin \theta} = \dfrac{\sin^2 \theta + \cos^2 \theta}{\cos \theta \sin \theta} = \dfrac{1}{\cos \theta \sin \theta} = \sec \theta \csc \theta$

39. $(1 - \cos \beta)(1 + \cos \beta) = 1 - \cos^2 \beta = \sin^2 \beta = \dfrac{1}{\csc^2 \beta}$

41. $\dfrac{(\sin x + \cos x)^2}{\sin^2 x - \cos^2 x} = \dfrac{(\sin x + \cos x)^2}{(\sin x + \cos x)(\sin x - \cos x)} = \dfrac{\sin x + \cos x}{\sin x - \cos x} = \dfrac{(\sin x + \cos x)(\sin x - \cos x)}{(\sin x - \cos x)(\sin x - \cos x)} = \dfrac{\sin^2 x - \cos^2 x}{(\sin x - \cos x)^2}$

43. $\dfrac{\sec t - \cos t}{\sec t} = \dfrac{\dfrac{1}{\cos t} - \cos t}{\dfrac{1}{\cos t}} = \dfrac{\dfrac{1}{\cos t} - \cos t}{\dfrac{1}{\cos t}} \cdot \dfrac{\cos t}{\cos t} = \dfrac{1 - \cos^2 t}{1} = \sin^2 t$

45. $\dfrac{1}{1 - \sin^2 y} = \dfrac{1}{\cos^2 y} = \sec^2 y = 1 + \tan^2 y$

47. $(\cot x - \csc x)(\cos x + 1) = \cot x \cos x + \cot x - \csc x \cos x - \csc x = \dfrac{\cos^2 x}{\sin x} + \dfrac{\cos x}{\sin x} - \dfrac{\cos x}{\sin x} - \dfrac{1}{\sin x}$

$\qquad\qquad = \dfrac{\cos^2 x - 1}{\sin x} = \dfrac{-\sin^2 x}{\sin x} = -\sin x$

49. $\left(1 - \cos^2 x\right)\left(1 + \cot^2 x\right) = \sin^2 x \left(1 + \dfrac{\cos^2 x}{\sin^2 x}\right) = \sin^2 x + \cos^2 x = 1$

51. $2 \cos^2 x - 1 = 2\left(1 - \sin^2 x\right) - 1 = 2 - 2 \sin^2 x - 1 = 1 - 2 \sin^2 x$

53. $\dfrac{1 - \cos \alpha}{\sin \alpha} = \dfrac{1 - \cos \alpha}{\sin \alpha} \cdot \dfrac{1 + \cos \alpha}{1 + \cos \alpha} = \dfrac{1 - \cos^2 \alpha}{\sin \alpha (1 + \cos \alpha)} = \dfrac{\sin^2 \alpha}{\sin \alpha (1 + \cos \alpha)} = \dfrac{\sin \alpha}{1 + \cos \alpha}$

55. $\tan^2 \theta \sin^2 \theta = \tan^2 \theta \left(1 - \cos^2 \theta\right) = \tan^2 \theta - \dfrac{\sin^2 \theta}{\cos^2 \theta} \cos^2 \theta = \tan^2 \theta - \sin^2 \theta$

57. $\dfrac{\sin x - 1}{\sin x + 1} = \dfrac{\sin x - 1}{\sin x + 1} \cdot \dfrac{\sin x + 1}{\sin x + 1} = \dfrac{\sin^2 x - 1}{(\sin x + 1)^2} = \dfrac{-\cos^2 x}{(\sin x + 1)^2}$

59. $\dfrac{(\sin t + \cos t)^2}{\sin t \cos t} = \dfrac{\sin^2 t + 2 \sin t \cos t + \cos^2 t}{\sin t \cos t} = \dfrac{\sin^2 t + \cos^2 t}{\sin t \cos t} + \dfrac{2 \sin t \cos t}{\sin t \cos t} = \dfrac{1}{\sin t \cos t} + 2 = 2 + \sec t \cos t$

61. $\dfrac{1 + \tan^2 u}{1 - \tan^2 u} = \dfrac{1 + \dfrac{\sin^2 u}{\cos^2 u}}{1 - \dfrac{\sin^2 u}{\cos^2 u}} = \dfrac{1 + \dfrac{\sin^2 u}{\cos^2 u}}{1 - \dfrac{\sin^2 u}{\cos^2 u}} \cdot \dfrac{\cos^2 u}{\cos^2 u} = \dfrac{\cos^2 u + \sin^2 u}{\cos^2 u - \sin^2 u} = \dfrac{1}{\cos^2 u - \sin^2 u}$

63. $\dfrac{\sec x}{\sec x - \tan x} = \dfrac{\sec x}{\sec x - \tan x} \cdot \dfrac{\sec x + \tan x}{\sec x + \tan x} = \dfrac{\sec x\,(\sec x + \tan x)}{\sec^2 x - \tan^2 x} = \dfrac{\sec x\,(\sec x + \tan x)}{1} = \sec x\,(\sec x + \tan x)$

65. $\sec v - \tan v = (\sec v - \tan v) \cdot \dfrac{\sec v + \tan v}{\sec v + \tan v} = \dfrac{\sec^2 v - \tan^2 v}{\sec v + \tan v} = \dfrac{1}{\sec v + \tan v}$

67. $\dfrac{\sin x + \cos x}{\sec x + \csc x} = \dfrac{\sin x + \cos x}{\dfrac{1}{\cos x} + \dfrac{1}{\sin x}} = \dfrac{\sin x + \cos x}{\dfrac{\sin x + \cos x}{\cos x \sin x}} = (\sin x + \cos x)\dfrac{\cos x \sin x}{\sin x + \cos x} = \cos x \sin x$

69. $\dfrac{\csc x - \cot x}{\sec x - 1} = \dfrac{\dfrac{1}{\sin x} - \dfrac{\cos x}{\sin x}}{\dfrac{1}{\cos x} - 1} = \dfrac{\dfrac{1}{\sin x} - \dfrac{\cos x}{\sin x}}{\dfrac{1}{\cos x} - 1} \cdot \dfrac{\sin x \cos x}{\sin x \cos x} = \dfrac{\cos x\,(1 - \cos x)}{\sin x\,(1 - \cos x)} = \dfrac{\cos x}{\sin x} = \cot x$

71. $\tan^2 u - \sin^2 u = \dfrac{\sin^2 u}{\cos^2 u} - \dfrac{\sin^2 u \cos^2 u}{\cos^2 u} = \dfrac{\sin^2 u}{\cos^2 u}\left(1 - \cos^2 u\right) = \tan^2 u \sin^2 u$

73. $\sec^4 x - \tan^4 x = \left(\sec^2 x - \tan^2 x\right)\left(\sec^2 x + \tan^2 x\right) = 1\left(\sec^2 x + \tan^2 x\right) = \sec^2 x + \tan^2 x$

75. $\dfrac{\sin \theta - \csc \theta}{\cos \theta - \cot \theta} = \dfrac{\sin \theta - \dfrac{1}{\sin \theta}}{\cos \theta - \dfrac{\cos \theta}{\sin \theta}} = \dfrac{\dfrac{\sin^2 \theta - 1}{\sin \theta}}{\dfrac{\cos \theta \sin \theta - \cos \theta}{\sin \theta}} = \dfrac{-\cos^2 \theta}{\cos \theta\,(\sin \theta - 1)} = \dfrac{\cos \theta}{\sin \theta - 1}$

77. $\dfrac{\cos^2 t + \tan^2 t - 1}{\sin^2 t} = \dfrac{-\sin^2 t + \tan^2 t}{\sin^2 t} = -1 + \dfrac{\sin^2 t}{\cos^2 t} \cdot \dfrac{1}{\sin^2 t} = -1 + \sec^2 t = \tan^2 t$

79. $\dfrac{1}{\sec x + \tan x} + \dfrac{1}{\sec x - \tan x} = \dfrac{\sec x - \tan x + \sec x + \tan x}{(\sec x + \tan x)\,(\sec x - \tan x)} = \dfrac{2 \sec x}{\sec^2 x - \tan^2 x} = \dfrac{2 \sec x}{1} = 2 \sec x$

81. $(\tan x + \cot x)^2 = \tan^2 x + 2 \tan x \cot x + \cot^2 x = \tan^2 x + 2 + \cot^2 x = \left(\tan^2 x + 1\right) + \left(\cot^2 x + 1\right) = \sec^2 x + \csc^2 x$

83. $\dfrac{\sec u - 1}{\sec u + 1} = \dfrac{\dfrac{1}{\cos u} - 1}{\dfrac{1}{\cos u} + 1} \cdot \dfrac{\cos u}{\cos u} = \dfrac{1 - \cos u}{1 + \cos u}$

85. $\dfrac{\sin^3 x + \cos^3 x}{\sin x + \cos x} = \dfrac{(\sin x + \cos x)\left(\sin^2 x - \sin x \cos x + \cos^2 x\right)}{\sin x + \cos x} = \sin^2 - \sin x \cos x + \cos^2 x = 1 - \sin x \cos x$

87. $\dfrac{1 + \sin x}{1 - \sin x} = \dfrac{1 + \sin x}{1 - \sin x} \cdot \dfrac{1 + \sin x}{1 + \sin x} = \dfrac{(1 + \sin x)^2}{1 - \sin^2 x} = \dfrac{(1 + \sin x)^2}{\cos^2 x} = \left(\dfrac{1 + \sin x}{\cos x}\right)^2 = (\tan x + \sec x)^2$

89. $(\tan x + \cot x)^4 = \left(\dfrac{\sin x}{\cos x} + \dfrac{\cos x}{\sin x}\right)^4 = \left(\dfrac{\sin^2 x + \cos^2 x}{\sin x \cos x}\right)^4 = \left(\dfrac{1}{\sin x \cos x}\right)^4 = \sec^4 x \csc^4 x$

91. $x = \sin \theta$; then $\dfrac{x}{\sqrt{1 - x^2}} = \dfrac{\sin \theta}{\sqrt{1 - \sin^2 \theta}} = \dfrac{\sin \theta}{\sqrt{\cos^2 \theta}} = \dfrac{\sin \theta}{\cos \theta} = \tan \theta$ (since $\cos \theta \geq 0$ for $0 \leq \theta \leq \frac{\pi}{2}$).

93. $x = \sec \theta$; then $\sqrt{x^2 - 1} = \sqrt{\sec^2 \theta - 1} = \sqrt{(\tan^2 \theta + 1) - 1} = \sqrt{\tan^2 \theta} = \tan \theta$ (since $\tan \theta \geq 0$ for $0 \leq \theta < \frac{\pi}{2}$)

95. $x = 3 \sin \theta$; then $\sqrt{9 - x^2} = \sqrt{9 - (3 \sin \theta)^2} = \sqrt{9 - 9 \sin^2 \theta} = \sqrt{9\left(1 - \sin^2 \theta\right)} = 3\sqrt{\cos^2 \theta} = 3 \cos \theta$ (since

$\cos \theta \geq 0$ for $0 \leq \theta < \frac{\pi}{2}$).

97. $f(x) = \cos^2 x - \sin^2 x$, $g(x) = 1 - 2 \sin^2 x$. From the graph, $f(x) = g(x)$ this

appears to be an identity. *Proof:*

$f(x) = \cos^2 x - \sin^2 x = \cos^2 x + \sin^2 x - 2 \sin^2 x = 1 - 2 \sin^2 x = g(x)$. Since

$f(x) = g(x)$ for all x, this is an identity.

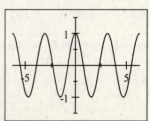

99. $f(x) = (\sin x + \cos x)^2$, $g(x) = 1$. From the graph, $f(x) = g(x)$ does not

appear to be an identity. In order to show this, we can set $x = \frac{\pi}{4}$. Then we have

$$f\left(\frac{\pi}{4}\right) = \left(\frac{1}{\sqrt{2}} + \frac{1}{\sqrt{2}}\right)^2 = \left(\frac{2}{\sqrt{2}}\right)^2 = \left(\sqrt{2}\right)^2 = 2 \neq 1 = g\left(\frac{\pi}{4}\right). \text{ Since}$$

$f\left(\frac{\pi}{4}\right) \neq g\left(\frac{\pi}{4}\right)$, this is not an identity.

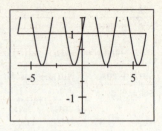

101. (a) Choose $x = \frac{\pi}{2}$. Then $\sin 2x = \sin \pi = 0$ whereas $2 \sin x = 2 \sin \frac{\pi}{2} = 2$.

(b) Choose $x = \frac{\pi}{4}$ and $y = \frac{\pi}{4}$. Then $\sin(x+y) = \sin \frac{\pi}{2} = 1$ whereas $\sin x + \sin y = \sin \frac{\pi}{4} + \sin \frac{\pi}{4} = \frac{1}{\sqrt{2}} + \frac{1}{\sqrt{2}} = \frac{2}{\sqrt{2}}$.
Since these are not equal, the equation is not an identity.

(c) Choose $\theta = \frac{\pi}{4}$. Then $\sec^2\theta + \csc^2\theta = \left(\sqrt{2}\right)^2 + \left(\sqrt{2}\right)^2 = 4 \neq 1$.

(d) Choose $x = \frac{\pi}{4}$. Then $\dfrac{1}{\sin x + \cos x} = \dfrac{1}{\sin \frac{\pi}{4} + \cos \frac{\pi}{4}} = \dfrac{1}{\frac{1}{\sqrt{2}} + \frac{1}{\sqrt{2}}} = \dfrac{1}{\sqrt{2}}$ whereas

$\csc x + \sec x = \csc \frac{\pi}{4} + \sec \frac{\pi}{4} = \sqrt{2} + \sqrt{2}$. Since these are not equal, the equation is not an identity.

103. No. All this proves is that $f(x) = g(x)$ for x in the range of the viewing rectangle. It does not prove that these functions

are equal for all values of x. For example, let $f(x) = 1 - \dfrac{x^2}{2} + \dfrac{x^4}{24} - \dfrac{x^6}{720}$ and $g(x) = \cos x$. In the first viewing rectangle

the graphs of these two functions appear identical. However, when the domain is expanded in the second viewing rectangle,

you can see that these two functions are not identical.

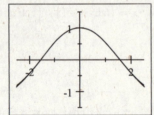
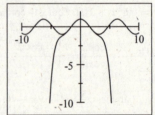

7.2 · ADDITION AND SUBTRACTION FORMULAS

1. If we know the values of the sine and cosine of x and y we can find the value of $\sin(x+y)$ using the *addition* formula for
sine, $\sin(x+y) = \sin x \cos y + \cos x \sin y$.

3. $\sin 75° = \sin(45° + 30°) = \sin 45° \cos 30° + \cos 45° \sin 30° = \frac{\sqrt{2}}{2} \cdot \frac{\sqrt{3}}{2} + \frac{\sqrt{2}}{2} \cdot \frac{1}{2} = \frac{\sqrt{6} + \sqrt{2}}{4}$

5. $\cos 105° = \cos(60° + 45°) = \cos 60° \cos 45° - \sin 60° \sin 45° = \frac{1}{2} \cdot \frac{\sqrt{2}}{2} - \frac{\sqrt{3}}{2} \cdot \frac{\sqrt{2}}{2} = \frac{\sqrt{2} - \sqrt{6}}{4}$

7. $\tan 15° = \tan(45° - 30°) = \dfrac{\tan 45° - \tan 30°}{1 + \tan 45° \tan 30°} = \dfrac{1 - \frac{\sqrt{3}}{3}}{1 + 1 \cdot \frac{\sqrt{3}}{3}} = \dfrac{3 - \sqrt{3}}{3 + \sqrt{3}} = 2 - \sqrt{3}$

9. $\sin \frac{19\pi}{12} = -\sin \frac{7\pi}{12} = -\sin\left(\frac{\pi}{4} + \frac{\pi}{3}\right) = -\sin \frac{\pi}{4} \cos \frac{\pi}{3} - \cos \frac{\pi}{4} \sin \frac{\pi}{3} = -\frac{\sqrt{2}}{2} \cdot \frac{1}{2} - \frac{\sqrt{2}}{2} \cdot \frac{\sqrt{3}}{2} = -\frac{\sqrt{6} + \sqrt{2}}{4}$

11. $\tan\left(-\frac{\pi}{12}\right) = -\tan \frac{\pi}{12} = -\tan\left(\frac{\pi}{3} - \frac{\pi}{4}\right) = -\dfrac{\tan \frac{\pi}{3} - \tan \frac{\pi}{4}}{1 + \tan \frac{\pi}{3} \tan \frac{\pi}{4}} = \dfrac{1 - \sqrt{3}}{1 + \sqrt{3}} = \sqrt{3} - 2$

13. $\cos \frac{11\pi}{12} = -\cos \frac{\pi}{12} = -\cos\left(\frac{\pi}{3} - \frac{\pi}{4}\right) = -\cos \frac{\pi}{3} \cos \frac{\pi}{4} - \sin \frac{\pi}{3} \sin \frac{\pi}{4} = -\frac{\sqrt{3}}{2} \cdot \frac{\sqrt{2}}{2} - \frac{1}{2} \cdot \frac{\sqrt{2}}{2} = -\frac{\sqrt{6} + \sqrt{2}}{4}$

15. $\sin 18° \cos 27° + \cos 18° \sin 27° = \sin(18° + 27°) = \sin 45° = \frac{1}{\sqrt{2}} = \frac{\sqrt{2}}{2}$

17. $\cos \frac{3\pi}{7} \cos \frac{2\pi}{21} + \sin \frac{3\pi}{7} \sin \frac{2\pi}{21} = \cos\left(\frac{3\pi}{7} - \frac{2\pi}{21}\right) = \cos \frac{7\pi}{21} = \cos \frac{\pi}{3} = \frac{1}{2}$

19. $\dfrac{\tan 73^\circ - \tan 13^\circ}{1 + \tan 73^\circ \tan 13^\circ} = \tan(73^\circ - 13^\circ) = \tan 60^\circ = \sqrt{3}$

21. $\tan\left(\frac{\pi}{2} - u\right) = \dfrac{\sin\left(\frac{\pi}{2} - u\right)}{\cos\left(\frac{\pi}{2} - u\right)} = \dfrac{\sin \frac{\pi}{2} \cos u - \cos \frac{\pi}{2} \sin u}{\cos \frac{\pi}{2} \cos u + \sin \frac{\pi}{2} \sin u} = \dfrac{1 \cdot \cos u - 0 \cdot \sin u}{0 \cdot \cos u + 1 \cdot \sin u} = \dfrac{\cos u}{\sin u} = \cot u$

23. $\sec\left(\frac{\pi}{2} - u\right) = \dfrac{1}{\cos\left(\frac{\pi}{2} - u\right)} = \dfrac{1}{\cos \frac{\pi}{2} \cos u + \sin \frac{\pi}{2} \sin u} = \dfrac{1}{0 \cdot \cos u + 1 \cdot \sin u} = \dfrac{1}{\sin u} = \csc u$

25. $\sin\left(x - \frac{\pi}{2}\right) = \sin x \cos \frac{\pi}{2} - \cos x \sin \frac{\pi}{2} = 0 \cdot \sin x - 1 \cdot \cos x = -\cos x$

27. $\sin(x - \pi) = \sin x \cos \pi - \cos x \sin \pi = -1 \cdot \sin x - 0 \cdot \cos x = -\sin x$

29. $\tan(x - \pi) = \dfrac{\tan x - \tan \pi}{1 + \tan x \tan \pi} = \dfrac{\tan x - 0}{1 + \tan x \cdot 0} = \tan x$

31. $\cos\left(x + \frac{\pi}{6}\right) + \sin\left(x - \frac{\pi}{3}\right) = \cos x \cos \frac{\pi}{6} - \sin x \sin \frac{\pi}{6} + \sin x \cos \frac{\pi}{3} - \cos x \sin \frac{\pi}{3}$

33. $\sin(x + y) - \sin(x - y) = \sin x \cos y + \cos x \sin y - (\sin x \cos y - \cos x \sin y) = 2 \cos x \sin y$

35. $\cot(x - y) = \dfrac{1}{\tan(x - y)} = \dfrac{1 + \tan x \tan y}{\tan x - \tan y} = \dfrac{1 + \dfrac{1}{\cot x}\dfrac{1}{\cot y}}{\dfrac{1}{\cot x} - \dfrac{1}{\cot y}} \cdot \dfrac{\cot x \cot y}{\cot x \cot y} = \dfrac{\cot x \cot y + 1}{\cot y - \cot x}$

37. $\tan x - \tan y = \dfrac{\sin x}{\cos x} - \dfrac{\sin y}{\cos y} = \dfrac{\sin x \cos y - \cos x \sin y}{\cos x \cos y} = \dfrac{\sin(x - y)}{\cos x \cos y}$

39. $\dfrac{\sin(x + y) - \sin(x - y)}{\cos(x + y) + \cos(x - y)} = \dfrac{\sin x \cos y + \cos x \sin y - (\sin x \cos y - \cos x \sin y)}{\cos x \cos y - \sin x \sin y + \cos x \cos y + \sin x \sin y} = \dfrac{2 \cos x \sin y}{2 \cos x \cos y} = \tan y$

41. $\sin(x + y + z) = \sin((x + y) + z) = \sin(x + y) \cos z + \cos(x + y) \sin z$

$= \cos z (\sin x \cos y + \cos x \sin y) + \sin z (\cos x \cos y - \sin x \sin y)$

$= \sin x \cos y \cos z + \cos x \sin y \cos z + \cos x \cos y \sin z - \sin x \sin y \sin z$

43. We want to write $\cos(\sin^{-1} x - \tan^{-1} y)$ in terms of x and y only. We let

$\theta = \sin^{-1} x$ and $\phi = \tan^{-1} y$ and sketch triangles with angles θ and ϕ such

that $\sin \theta = x$ and $\tan \phi = y$. From the triangles, we have $\cos \theta = \sqrt{1 - x^2}$,

$\cos \phi = \dfrac{1}{\sqrt{1 + y^2}}$, and $\sin \phi = \dfrac{y}{\sqrt{1 + y^2}}$.

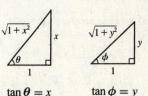

$\sin \theta = x \qquad \tan \phi = y$

From the subtraction formula for cosine we have

$\cos\left(\sin^{-1} x - \tan^{-1} y\right) = \cos(\theta - \phi) = \cos \theta \cos \phi + \sin \theta \sin \phi = \sqrt{1 - x^2} \cdot \dfrac{1}{\sqrt{1 + y^2}} + x \cdot \dfrac{y}{\sqrt{1 + y^2}} = \dfrac{\sqrt{1 - x^2} + xy}{\sqrt{1 + y^2}}$

45. Let $\theta = \tan^{-1} x$ and $\phi = \tan^{-1} y$. From the triangles, $\cos \theta = \dfrac{1}{\sqrt{1 + x^2}}$,

$\sin \theta = \dfrac{x}{\sqrt{1 + x^2}}$, $\cos \phi = \dfrac{1}{\sqrt{1 + y^2}}$, and $\sin \phi = \dfrac{y}{\sqrt{1 + y^2}}$, so using the

$\tan \theta = x \qquad \tan \phi = y$

subtraction formula for sine, we have

$\sin\left(\tan^{-1} x - \tan^{-1} y\right) = \sin(\theta - \phi) = \sin \theta \cos \phi - \cos \theta \sin \phi$

$= \dfrac{x}{\sqrt{1 + x^2}} \cdot \dfrac{1}{\sqrt{1 + y^2}} - \dfrac{1}{\sqrt{1 + x^2}} \cdot \dfrac{y}{\sqrt{1 + y^2}} = \dfrac{x - y}{\sqrt{1 + x^2}\sqrt{1 + y^2}}$

47. We know that $\cos^{-1} \frac{1}{2} = \frac{\pi}{3}$ and $\tan^{-1} 1 = \frac{\pi}{4}$, so the addition formula for sine gives

$\sin\left(\cos^{-1} \frac{1}{2} + \tan^{-1} 1\right) = \sin\left(\frac{\pi}{3} + \frac{\pi}{4}\right) = \sin \frac{\pi}{3} \cos \frac{\pi}{4} + \cos \frac{\pi}{3} \sin \frac{\pi}{4} = \dfrac{\sqrt{3}}{2} \cdot \dfrac{\sqrt{2}}{2} + \dfrac{1}{2} \cdot \dfrac{\sqrt{2}}{2} = \dfrac{\sqrt{6} + \sqrt{2}}{4}$.

49. We sketch triangles such that $\theta = \sin^{-1} \frac{3}{4}$ and $\phi = \cos^{-1} \frac{1}{3}$. From the triangles, we have $\tan\theta = \frac{3}{\sqrt{7}}$ and $\tan\phi = 2\sqrt{2}$, so the subtraction formula for tangent gives

$$\tan\left(\sin^{-1}\frac{3}{4} - \cos^{-1}\frac{1}{3}\right) = \frac{\tan\theta - \tan\phi}{1 + \tan\theta\tan\phi} = \frac{\frac{3}{\sqrt{7}} - 2\sqrt{2}}{1 + \frac{3}{\sqrt{7}} \cdot 2\sqrt{2}}$$

$$= \frac{3 - 2\sqrt{14}}{\sqrt{7} + 6\sqrt{2}}$$

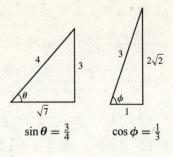

$\sin\theta = \frac{3}{4}$ $\cos\phi = \frac{1}{3}$

51. As in Example 7, we sketch the angles θ and ϕ in standard position with terminal sides in the appropriate quadrants and find the remaining sides using the Pythagorean Theorem. To find $\cos(\theta - \phi)$, we use the addition formula for sine and the triangles we have sketched:

$$\cos(\theta - \phi) = \cos\theta\cos\phi + \sin\theta\sin\phi$$

$$= \frac{3}{5}\left(-\frac{1}{2}\right) + \left(-\frac{4}{5}\right)\frac{\sqrt{3}}{2} = -\frac{3 + 4\sqrt{3}}{10}$$

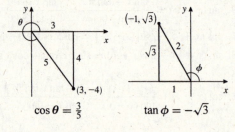

$\cos\theta = \frac{3}{5}$ $\tan\phi = -\sqrt{3}$

53. Using the addition formula for sine and the triangles shown, we have

$$\sin(\theta + \phi) = \sin\theta\cos\phi + \cos\theta\sin\phi$$

$$= \frac{5}{13}\left(-\frac{2\sqrt{5}}{5}\right) + \frac{12}{13}\left(\frac{\sqrt{5}}{5}\right) = \frac{2\sqrt{5}}{65}$$

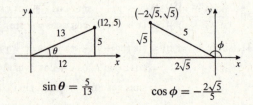

$\sin\theta = \frac{5}{13}$ $\cos\phi = -\frac{2\sqrt{5}}{5}$

55. $k = \sqrt{A^2 + B^2} = \sqrt{\left(-\sqrt{3}\right)^2 + 1^2} = \sqrt{4} = 2$. Thus, $\sin\phi = \frac{1}{2}$ and $\cos\phi = \frac{-\sqrt{3}}{2} \Rightarrow \phi = \frac{5\pi}{6}$, so

$$-\sqrt{3}\sin x + \cos x = k\sin(x + \phi) = 2\sin\left(x + \frac{5\pi}{6}\right).$$

57. $k = \sqrt{A^2 + B^2} = \sqrt{5^2 + (-5)^2} = \sqrt{50} = 5\sqrt{2}$. Thus, $\sin\phi = -\frac{5}{5\sqrt{2}} = -\frac{1}{\sqrt{2}}$ and $\cos\phi = \frac{5}{5\sqrt{2}} = \frac{1}{\sqrt{2}} \Rightarrow \phi = \frac{7\pi}{4}$, so

$$5(\sin 2x - \cos 2x) = k\sin(2x + \phi) = 5\sqrt{2}\sin\left(2x + \frac{7\pi}{4}\right).$$

59. (a) $g(x) = \cos 2x + \sqrt{3}\sin 2x \Rightarrow$

$k = \sqrt{1^2 + \left(\sqrt{3}\right)^2} = \sqrt{4} = 2$, and ϕ satisfies

$\sin\phi = \frac{1}{2}$, $\cos\phi = \frac{\sqrt{3}}{2} \Rightarrow \phi = \frac{\pi}{6}$. Thus, we can write

$g(x) = k\sin(2x + \phi) = 2\sin\left(2x + \frac{\pi}{6}\right) = 2\sin 2\left(x + \frac{\pi}{12}\right).$

(b) This is a sine curve with amplitude 2, period π, and phase shift $-\frac{\pi}{12}$.

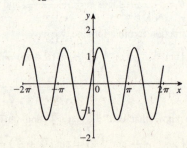

61. $g(x) = \cos x$. Now

$$\frac{g(x+h) - g(x)}{h} = \frac{\cos(x+h) - \cos x}{h} = \frac{\cos x \cos h - \sin x \sin h - \cos x}{h}$$

$$= \frac{-\cos x(1 - \cos h) - \sin h(\sin x)}{h} = -\cos x\left(\frac{1 - \cos h}{h}\right) - \left(\frac{\sin h}{h}\right)\sin x$$

63. Let $\angle A$ and $\angle B$ be the two angles shown in the diagram. Then
$180° = \gamma + A + B$, $90° = \alpha + A$, and $90° = \beta + B$. Subtracting
the second and third equation from the first, we get
$180° - 90° - 90° = \gamma + A + B - (\alpha + A) - (\beta + B) \Leftrightarrow$
$\alpha + \beta = \gamma$. Then

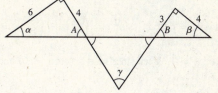

$$\tan \gamma = \tan(\alpha + \beta) = \frac{\tan \alpha + \tan \beta}{1 - \tan \alpha \tan \beta} = \frac{\frac{4}{6} + \frac{3}{4}}{1 - \frac{4}{6} \cdot \frac{3}{4}} = \frac{\frac{8}{12} + \frac{9}{12}}{1 - \frac{1}{2}} = 2 \cdot \frac{17}{12} = \frac{17}{6}.$$

65. (a) $y = \sin^2\left(x + \frac{\pi}{4}\right) + \sin^2\left(x - \frac{\pi}{4}\right)$. From the graph we see that the value of y
seems to always be equal to 1.

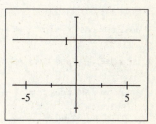

(b) $y = \sin^2\left(x + \frac{\pi}{4}\right) + \sin^2\left(x - \frac{\pi}{4}\right) = \left(\sin x \cos \frac{\pi}{4} + \cos x \sin \frac{\pi}{4}\right)^2 + \left(\sin x \cos \frac{\pi}{4} - \cos x \sin \frac{\pi}{4}\right)^2$

$$= \left[\frac{1}{\sqrt{2}}(\sin x + \cos x)\right]^2 + \left[\frac{1}{\sqrt{2}}(\sin x - \cos x)\right]^2 = \frac{1}{2}\left[(\sin x + \cos x)^2 + (\sin x - \cos x)^2\right]$$

$$= \frac{1}{2}\left[\left(\sin^2 x + 2\sin x \cos x + \cos^2 x\right) + \left(\sin^2 x - 2\sin x \cos x + \cos^2 x\right)\right]$$

$$= \frac{1}{2}\left[(1 + 2\sin x \cos x) + (1 - 2\sin x \cos x)\right] = \frac{1}{2} \cdot 2 = 1$$

67. Clearly $C = \frac{\pi}{4}$. Now $\tan(A + B) = \frac{\tan A + \tan B}{1 - \tan A \tan B} = \frac{\frac{1}{3} + \frac{1}{2}}{1 - \frac{1}{3} \cdot \frac{1}{2}} = 1$. Thus $A + B = \frac{\pi}{4}$, so $A + B + C = \frac{\pi}{4} + \frac{\pi}{4} = \frac{\pi}{2}$.

69. (a) $f(t) = C\sin\omega t + C\sin(\omega t + \alpha) = C\sin\omega t + C(\sin\omega t \cos\alpha + \cos\omega t \sin\alpha)$

$$= C(1 + \cos\alpha)\sin\omega t + C\sin\alpha \cos\omega t = A\sin\omega t + B\cos\omega t$$

where $A = C(1 + \cos\alpha)$ and $B = C\sin\alpha$.

(b) In this case, $f(t) = 10\left(1 + \cos\frac{\pi}{3}\right)\sin\omega t + 10\sin\frac{\pi}{3}\cos\omega t = 15\sin\omega t + 5\sqrt{3}\cos\omega t$. Thus

$$k = \sqrt{15^2 + \left(5\sqrt{3}\right)^2} = 10\sqrt{3} \text{ and } \phi \text{ has } \cos\phi = \frac{15}{10\sqrt{3}} = \frac{\sqrt{3}}{2} \text{ and } \sin\phi = \frac{5\sqrt{3}}{10\sqrt{3}} = \frac{1}{2}, \text{ so } \phi = \frac{\pi}{6}. \text{ Therefore,}$$
$$f(t) = 10\sqrt{3}\sin\left(\omega t + \frac{\pi}{6}\right).$$

71. $\tan(s + t) = \frac{\sin(s + t)}{\cos(s + t)} = \frac{\sin s \cos t + \cos s \sin t}{\cos s \cos t + \sin s \sin t}$

$$= \frac{\sin s \cos t + \cos s \sin t}{\cos s \cos t - \sin s \sin t} \cdot \frac{\dfrac{1}{\cos s \cos t}}{\dfrac{1}{\cos s \cos t}} = \frac{\dfrac{\sin s}{\cos s} + \dfrac{\sin t}{\cos t}}{1 - \dfrac{\sin s}{\cos s} \cdot \dfrac{\sin t}{\cos t}} = \frac{\tan s + \tan t}{1 - \tan s \tan t}$$

7.3   DOUBLE-ANGLE, HALF-ANGLE, AND PRODUCT-SUM FORMULAS

1. If we know the values of $\sin x$ and $\cos x$, we can find the value of $\sin 2x$ using the *double-angle* formula for sine, $\sin 2x = 2 \sin x \cos x$.

3. $\sin x = \frac{5}{13}$, x in quadrant I $\Rightarrow \cos x = \frac{12}{13}$ and $\tan x = \frac{5}{12}$. Thus, $\sin 2x = 2 \sin x \cos x = 2\left(\frac{5}{13}\right)\left(\frac{12}{13}\right) = \frac{120}{169}$,

$\cos 2x = \cos^2 x - \sin^2 x = \left(\frac{12}{13}\right)^2 - \left(\frac{5}{13}\right)^2 = \frac{144-25}{169} = \frac{119}{169}$, and $\tan 2x = \frac{\sin 2x}{\cos 2x} = \frac{\frac{120}{169}}{\frac{119}{169}} = \frac{120}{169} \cdot \frac{169}{119} = \frac{120}{119}$.

5. $\cos x = \frac{4}{5}$. Then $\sin x = -\frac{3}{5}$ ($\csc x < 0$) and $\tan x = -\frac{3}{4}$. Thus, $\sin 2x = 2 \sin x \cos x = 2\left(-\frac{3}{5}\right) \cdot \frac{4}{5} = -\frac{24}{25}$,

$\cos 2x = \cos^2 x - \sin^2 x = \left(\frac{4}{5}\right)^2 - \left(-\frac{3}{5}\right)^2 = \frac{16-9}{25} = \frac{7}{25}$, and $\tan 2x = \frac{\sin 2x}{\cos 2x} = \frac{-\frac{24}{25}}{\frac{7}{25}} = -\frac{24}{25} \cdot \frac{25}{7} = -\frac{24}{7}$.

7. $\sin x = -\frac{3}{5}$. Then, $\cos x = -\frac{4}{5}$ and $\tan x = \frac{3}{4}$ (x is in quadrant III). Thus, $\sin 2x = 2 \sin x \cos x = 2\left(-\frac{3}{5}\right)\left(-\frac{4}{5}\right) = \frac{24}{25}$,

$\cos 2x = \cos^2 x - \sin^2 x = \left(-\frac{4}{5}\right)^2 - \left(-\frac{3}{5}\right)^2 = \frac{16-9}{25} = \frac{7}{25}$, and $\tan 2x = \frac{\sin 2x}{\cos 2x} = \frac{\frac{24}{25}}{\frac{7}{25}} = \frac{24}{25} \cdot \frac{25}{7} = \frac{24}{7}$.

9. $\tan x = -\frac{1}{3}$ and $\cos x > 0$, so $\sin x < 0$. Thus, $\sin x = -\frac{1}{\sqrt{10}}$ and $\cos x = \frac{3}{\sqrt{10}}$. Thus,

$\sin 2x = 2 \sin x \cos x = 2\left(-\frac{1}{\sqrt{10}}\right)\left(\frac{3}{\sqrt{10}}\right) = -\frac{6}{10} = -\frac{3}{5}$, $\cos 2x = \cos^2 x - \sin^2 x = \left(\frac{3}{\sqrt{10}}\right)^2 - \left(-\frac{1}{\sqrt{10}}\right)^2 = \frac{8}{10} = \frac{4}{5}$,

and $\tan 2x = \frac{\sin 2x}{\cos 2x} = \frac{-\frac{3}{5}}{\frac{4}{5}} = -\frac{3}{5} \cdot \frac{5}{4} = -\frac{3}{4}$.

11. $\sin^4 x = \left(\sin^2 x\right)^2 = \left(\frac{1 - \cos 2x}{2}\right)^2 = \frac{1}{4} - \frac{1}{2}\cos 2x + \frac{1}{4}\cos^2 2x$

$= \frac{1}{4} - \frac{1}{2}\cos 2x + \frac{1}{4} \cdot \frac{1 + \cos 4x}{2} = \frac{1}{4} - \frac{1}{2}\cos 2x + \frac{1}{8} + \frac{1}{8}\cos 4x = \frac{3}{8} - \frac{1}{2}\cos 2x + \frac{1}{8}\cos 4x$

$= \frac{1}{2}\left(\frac{3}{4} - \cos 2x + \frac{1}{4}\cos 4x\right)$

13. We use the result of Example 4 to get

$\cos^2 x \sin^4 x = \left(\sin^2 x \cos^2 x\right)\sin^2 x = \left(\frac{1}{8} - \frac{1}{8}\cos 4x\right) \cdot \left(\frac{1}{2} - \frac{1}{2}\cos 2x\right) = \frac{1}{16}\left(1 - \cos 2x - \cos 4x + \cos 2x \cos 4x\right)$.

15. Since $\sin^4 x \cos^4 x = \left(\sin^2 x \cos^2 x\right)^2$ we can use the result of Example 4 to get

$\sin^4 x \cos^4 x = \left(\frac{1}{8} - \frac{1}{8}\cos 4x\right)^2 = \frac{1}{64} - \frac{1}{32}\cos 4x + \frac{1}{64}\cos^2 4x$

$= \frac{1}{64} - \frac{1}{32}\cos 4x + \frac{1}{64} \cdot \frac{1}{2}(1 + \cos 8x) = \frac{1}{64} - \frac{1}{32}\cos 4x + \frac{1}{128} + \frac{1}{128}\cos 8x$

$= \frac{3}{128} - \frac{1}{32}\cos 4x + \frac{1}{128}\cos 8x = \frac{1}{32}\left(\frac{3}{4} - \cos 4x + \frac{1}{4}\cos 8x\right)$

17. $\sin 15° = \sqrt{\frac{1}{2}(1 - \cos 30°)} = \sqrt{\frac{1}{2}\left(1 - \frac{\sqrt{3}}{2}\right)} = \sqrt{\frac{1}{4}\left(2 - \sqrt{3}\right)} = \frac{1}{2}\sqrt{2 - \sqrt{3}}$

19. $\tan 22.5° = \frac{1 - \cos 45°}{\sin 45°} = \frac{1 - \frac{\sqrt{2}}{2}}{\frac{\sqrt{2}}{2}} = \sqrt{2} - 1$

21. $\cos 165° = -\sqrt{\frac{1}{2}(1 + \cos 330°)} = -\sqrt{\frac{1}{2}(1 + \cos 30°)} = -\sqrt{\frac{1}{2}\left(1 + \frac{\sqrt{3}}{2}\right)} = -\frac{1}{2}\sqrt{2 + \sqrt{3}}$

23. $\tan \frac{\pi}{8} = \frac{1 - \cos \frac{\pi}{4}}{\sin \frac{\pi}{4}} = \frac{1 - \frac{\sqrt{2}}{2}}{\frac{\sqrt{2}}{2}} = \sqrt{2} - 1$

25. $\cos \frac{\pi}{12} = \sqrt{\frac{1}{2}\left(1 + \cos \frac{\pi}{6}\right)} = \sqrt{\frac{1}{2}\left(1 + \frac{\sqrt{3}}{2}\right)} = \frac{1}{2}\sqrt{2 + \sqrt{3}}$

27. $\sin \frac{9\pi}{8} = -\sqrt{\frac{1}{2}\left(1 - \cos \frac{9\pi}{4}\right)} = -\sqrt{\frac{1}{2}\left(1 - \frac{\sqrt{2}}{2}\right)} = -\frac{1}{2}\sqrt{2 - \sqrt{2}}$. We have chosen the negative root because $\frac{9\pi}{8}$ is in

quadrant III, so $\sin \frac{9\pi}{8} < 0$.

29. (a) $2\sin 18° \cos 18° = \sin 36°$

   (b) $2\sin 3\theta \cos 3\theta = \sin 6\theta$

31. (a) $\cos^2 34° - \sin^2 34° = \cos 68°$

   (b) $\cos^2 5\theta - \sin^2 5\theta = \cos 10\theta$

33. (a) $\dfrac{\sin 8°}{1 + \cos 8°} = \tan \dfrac{8°}{2} = \tan 4°$

   (b) $\dfrac{1 - \cos 4\theta}{\sin 4\theta} = \tan \dfrac{4\theta}{2} = \tan 2\theta$

35. $\sin(x + x) = \sin x \cos x + \cos x \sin x = 2\sin x \cos x$

37. $\sin x = \frac{3}{5}$.   Since x is in quadrant I, $\cos x = \frac{4}{5}$ and $\frac{x}{2}$ is also in quadrant I. Thus,

$\sin \frac{x}{2} = \sqrt{\frac{1}{2}(1 - \cos x)} = \sqrt{\frac{1}{2}\left(1 - \frac{4}{5}\right)} = \frac{1}{\sqrt{10}} = \frac{\sqrt{10}}{10}$, $\cos \frac{x}{2} = \sqrt{\frac{1}{2}(1 + \cos x)} = \sqrt{\frac{1}{2}\left(1 + \frac{4}{5}\right)} = \frac{3}{\sqrt{10}} = \frac{3\sqrt{10}}{10}$, and

$\tan \frac{x}{2} = \dfrac{\sin \frac{x}{2}}{\cos \frac{x}{2}} = \frac{1}{\sqrt{10}} \cdot \frac{\sqrt{10}}{3} = \frac{1}{3}$.

39. $\csc x = 3$. Then, $\sin x = \frac{1}{3}$ and since x is in quadrant II, $\cos x = -\dfrac{2\sqrt{2}}{3}$. Since $90° \le x \le 180°$, we have

$45° \le \frac{x}{2} \le 90°$ and so $\frac{x}{2}$ is in quadrant I. Thus, $\sin \frac{x}{2} = \sqrt{\frac{1}{2}(1 - \cos x)} = \sqrt{\frac{1}{2}\left(1 + \frac{2\sqrt{2}}{3}\right)} = \sqrt{\frac{1}{6}\left(3 + 2\sqrt{2}\right)}$,

$\cos \frac{x}{2} = \sqrt{\frac{1}{2}(1 + \cos x)} = \sqrt{\frac{1}{2}\left(1 - \frac{2\sqrt{2}}{3}\right)} = \sqrt{\frac{1}{6}\left(3 - 2\sqrt{2}\right)}$, and $\tan \frac{x}{2} = \dfrac{\sin \frac{x}{2}}{\cos \frac{x}{2}} = \sqrt{\dfrac{3 + 2\sqrt{2}}{3 - 2\sqrt{2}}} = 3 + 2\sqrt{2}$.

41. $\sec x = \frac{3}{2}$. Then $\cos x = \frac{2}{3}$ and since x is in quadrant IV, $\sin x = -\dfrac{\sqrt{5}}{3}$. Since $270° \le x \le 360°$, we have

$135° \le \frac{x}{2} \le 180°$ and so $\frac{x}{2}$ is in quadrant II. Thus, $\sin \frac{x}{2} = \sqrt{\frac{1}{2}(1 - \cos x)} = \sqrt{\frac{1}{2}\left(1 - \frac{2}{3}\right)} = \frac{1}{\sqrt{6}} = \frac{\sqrt{6}}{6}$,

$\cos \frac{x}{2} = -\sqrt{\frac{1}{2}(1 + \cos x)} = -\sqrt{\frac{1}{2}\left(1 + \frac{2}{3}\right)} = -\frac{\sqrt{5}}{\sqrt{6}} = -\dfrac{\sqrt{30}}{6}$, and $\tan \frac{x}{2} = \dfrac{\sin \frac{x}{2}}{\cos \frac{x}{2}} = \frac{1}{\sqrt{6}} \cdot \frac{\sqrt{6}}{-\sqrt{5}} = -\frac{1}{\sqrt{5}} = -\dfrac{\sqrt{5}}{5}$.

43. To write $\sin\left(2\tan^{-1} x\right)$ as an algebraic expression in x, we let $\theta = \tan^{-1} x$

and sketch a suitable triangle. We see that $\sin \theta = \dfrac{x}{\sqrt{1 + x^2}}$ and

$\cos \theta = \dfrac{1}{\sqrt{1 + x^2}}$, so using the double-angle formula for sine, we have

$\sin\left(2\tan^{-1} x\right) = \sin 2\theta = 2\sin\theta\cos\theta = 2 \cdot \dfrac{x}{\sqrt{1 + x^2}} \cdot \dfrac{1}{\sqrt{1 + x^2}} = \dfrac{2x}{1 + x^2}$.

45. Using the half-angle formula for sine, we have $\sin\left(\frac{1}{2}\cos^{-1} x\right) = \pm\sqrt{\dfrac{1 - \cos\left(\cos^{-1} x\right)}{2}} = \pm\sqrt{\dfrac{1 - x}{2}}$. Because $\cos^{-1}$

has range $[0, \pi]$, $\frac{1}{2}\cos^{-1} x$ lies in $\left[0, \frac{\pi}{2}\right]$ and so $\sin\left(\frac{1}{2}\cos^{-1} x\right)$ is positive. Thus, $\sin\left(\frac{1}{2}\cos^{-1} x\right) = \sqrt{\dfrac{1 - x}{2}}$.

47. We sketch a triangle with $\theta = \cos^{-1}\frac{7}{25}$ and find that $\sin\theta = \frac{24}{25}$. Thus, using

the double-angle formula for sine,

$\sin\left(2\cos^{-1}\frac{7}{25}\right) = \sin 2\theta = 2\sin\theta\cos\theta = 2 \cdot \frac{24}{25} \cdot \frac{7}{25} = \frac{336}{625}$.

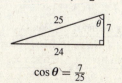

$\cos\theta = \frac{7}{25}$

49. Rewriting the given expression and using a double-angle formula for cosine, we have

$$\sec\left(2\sin^{-1}\tfrac{1}{4}\right) = \frac{1}{\cos\left(2\sin^{-1}\tfrac{1}{4}\right)} = \frac{1}{1 - 2\sin^2\left(\sin^{-1}\tfrac{1}{4}\right)} = \frac{1}{1 - 2\left(\tfrac{1}{4}\right)^2} = \frac{8}{7}.$$

51. Using a double-angle formula for cosine, we have $\cos 2\theta = 1 - 2\sin^2\theta = 1 - 2\left(-\tfrac{3}{5}\right)^2 = \tfrac{7}{25}$.

53. To evaluate $\sin 2\theta$, we first sketch the angle θ in standard position with terminal side in quadrant II and find the remaining side using the Pythagorean Theorem. Using the double-angle formula for sine, we have

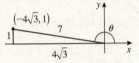

$$\sin 2\theta = 2\sin\theta\cos\theta = 2\left(\tfrac{1}{7}\right)\left(-\tfrac{4\sqrt{3}}{7}\right) = -\tfrac{8\sqrt{3}}{49}.$$

55. $\sin 2x \cos 3x = \tfrac{1}{2}\left[\sin(2x + 3x) + \sin(2x - 3x)\right] = \tfrac{1}{2}\left(\sin 5x - \sin x\right)$

57. $\cos x \sin 4x = \tfrac{1}{2}\left[\sin(4x + x) + \sin(4x - x)\right] = \tfrac{1}{2}\left(\sin 5x + \sin 3x\right)$

59. $3\cos 4x \cos 7x = 3 \cdot \tfrac{1}{2}\left[\cos(4x + 7x) + \cos(4x - 7x)\right] = \tfrac{3}{2}\left(\cos 11x + \cos 3x\right)$

61. $\sin 5x + \sin 3x = 2\sin\left(\dfrac{5x + 3x}{2}\right)\cos\left(\dfrac{5x - 3x}{2}\right) = 2\sin 4x \cos x$

63. $\cos 4x - \cos 6x = -2\sin\left(\dfrac{4x + 6x}{2}\right)\sin\left(\dfrac{4x - 6x}{2}\right) = -2\sin 5x \sin(-x) = 2\sin 5x \sin x$

65. $\sin 2x - \sin 7x = 2\cos\left(\dfrac{2x + 7x}{2}\right)\sin\left(\dfrac{2x - 7x}{2}\right) = 2\cos\dfrac{9x}{2}\sin\left(-\dfrac{5x}{2}\right) = -2\cos\dfrac{9x}{2}\sin\dfrac{5x}{2}$

67. $2\sin 52.5° \sin 97.5° = 2 \cdot \tfrac{1}{2}\left[\cos(52.5° - 97.5°) - \cos(52.5° + 97.5°)\right] = \cos(-45°) - \cos 150°$
$$= \cos 45° - \cos 150° = \tfrac{\sqrt{2}}{2} + \tfrac{\sqrt{3}}{2} = \tfrac{1}{2}\left(\sqrt{2} + \sqrt{3}\right)$$

69. $\cos 37.5° \sin 7.5° = \tfrac{1}{2}\left(\sin 45° - \sin 30°\right) = \tfrac{1}{2}\left(\tfrac{\sqrt{2}}{2} - \tfrac{1}{2}\right) = \tfrac{1}{4}\left(\sqrt{2} - 1\right)$

71. $\cos 255° - \cos 195° = -2\sin\left(\dfrac{255° + 195°}{2}\right)\sin\left(\dfrac{255° - 195°}{2}\right) = -2\sin 225° \sin 30° = -2\left(-\tfrac{\sqrt{2}}{2}\right)\tfrac{1}{2} = \tfrac{\sqrt{2}}{2}$

73. $\cos^2 5x - \sin^2 5x = \cos(2 \cdot 5x) = \cos 10x$

75. $(\sin x + \cos x)^2 = \sin^2 x + 2\sin x \cos x + \cos^2 x = 1 + 2\sin x \cos x = 1 + \sin 2x$

77. $\dfrac{\sin 4x}{\sin x} = \dfrac{2\sin 2x \cos 2x}{\sin x} = \dfrac{2(2\sin x \cos x)(\cos 2x)}{\sin x} = 4\cos x \cos 2x$

79. $\dfrac{2(\tan x - \cot x)}{\tan^2 x - \cot^2 x} = \dfrac{2(\tan x - \cot x)}{(\tan x + \cot x)(\tan x - \cot x)} = \dfrac{2}{\tan x + \cot x} = \dfrac{2}{\dfrac{\sin x}{\cos x} + \dfrac{\cos x}{\sin x}}$

$$= \dfrac{2}{\dfrac{\sin x}{\cos x} + \dfrac{\cos x}{\sin x}} \cdot \dfrac{\sin x \cos x}{\sin x \cos x} = \dfrac{2\sin x \cos x}{\sin^2 x + \cos^2 x} = 2\sin x \cos x = \sin 2x$$

81. $\tan 3x = \tan(2x + x) = \dfrac{\tan 2x + \tan x}{1 - \tan 2x \tan x} = \dfrac{\dfrac{2\tan x}{1 - \tan^2 x} + \tan x}{1 - \dfrac{2\tan x}{1 - \tan^2 x}\tan x} = \dfrac{2\tan x + \tan x\left(1 - \tan^2 x\right)}{1 - \tan^2 x - 2\tan x \tan x}$

$$= \dfrac{3\tan x - \tan^3 x}{1 - 3\tan^2 x}$$

83. $\cos^4 x - \sin^4 x = \left(\cos^2 x + \sin^2 x\right)\left(\cos^2 x - \sin^2 x\right) = \cos^2 x - \sin^2 x = \cos 2x$

85. $\dfrac{\sin x + \sin 5x}{\cos x + \cos 5x} = \dfrac{2\sin 3x \cos 2x}{2\cos 3x \cos 2x} = \dfrac{\sin 3x}{\cos 3x} = \tan 3x$

87. $\dfrac{\sin 10x}{\sin 9x + \sin x} = \dfrac{2\sin 5x \cos 5x}{2\sin 5x \cos 4x} = \dfrac{\cos 5x}{\cos 4x}$

89. $\dfrac{\sin x + \sin y}{\cos x + \cos y} = \dfrac{2 \sin \left(\dfrac{x+y}{2}\right) \cos \left(\dfrac{x-y}{2}\right)}{2 \cos \left(\dfrac{x+y}{2}\right) \cos \left(\dfrac{x-y}{2}\right)} = \dfrac{\sin \left(\dfrac{x+y}{2}\right)}{\cos \left(\dfrac{x+y}{2}\right)} = \tan \left(\dfrac{x+y}{2}\right)$

91. $\sin 130° - \sin 110° = 2 \cos \dfrac{130° + 110°}{2} \sin \dfrac{130° - 110°}{2} = 2 \cos 120° \sin 10° = 2 \left(-\tfrac{1}{2}\right) \sin 10° = -\sin 10°$

93. $\sin 45° + \sin 15° = 2 \sin \left(\dfrac{45° + 15°}{2}\right) \cos \left(\dfrac{45° - 15°}{2}\right) = 2 \sin 30° \cos 15° = 2 \cdot \tfrac{1}{2} \cdot \cos 15°$

$\qquad = \cos 15° = \sin (90° - 15°) = \sin 75°$ (applying the cofunction identity)

95. $\dfrac{\sin x + \sin 2x + \sin 3x + \sin 4x + \sin 5x}{\cos x + \cos 2x + \cos 3x + \cos 4x + \cos 5x} = \dfrac{(\sin x + \sin 5x) + (\sin 2x + \sin 4x) + \sin 3x}{(\cos x + \cos 5x) + (\cos 2x + \cos 4x) + \cos 3x}$

$\qquad = \dfrac{2 \sin 3x \cos 2x + 2 \sin 3x \cos x + \sin 3x}{2 \cos 3x \cos 2x + 2 \cos 3x \cos x + \cos 3x} = \dfrac{\sin 3x \, (2 \cos 2x + 2 \cos x + 1)}{\cos 3x \, (2 \cos 2x + 2 \cos x + 1)} = \tan 3x$

97. (a) $f(x) = \dfrac{\sin 3x}{\sin x} - \dfrac{\cos 3x}{\cos x}$

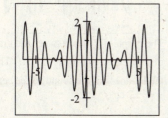

The function appears to have a constant value of 2
wherever it is defined.

(b) $f(x) = \dfrac{\sin 3x}{\sin x} - \dfrac{\cos 3x}{\cos x}$

$\qquad = \dfrac{\sin 3x \cos x - \cos 3x \sin x}{\sin x \cos x} = \dfrac{\sin (3x - x)}{\sin x \cos x}$

$\qquad = \dfrac{\sin 2x}{\sin x \cos x} = \dfrac{2 \sin x \cos x}{\sin x \cos x} = 2$

for all x for which the function is defined.

99. (a) $y = \sin 6x + \sin 7x$

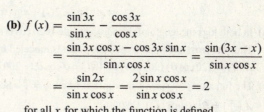

(b) By a sum-to-product formula,

$\qquad y = \sin 6x + \sin 7x$

$\qquad = 2 \sin \left(\dfrac{6x + 7x}{2}\right) \cos \left(\dfrac{6x - 7x}{2}\right)$

$\qquad = 2 \sin \left(\tfrac{13}{2}x\right) \cos \left(-\tfrac{1}{2}x\right)$

$\qquad = 2 \sin \tfrac{13}{2}x \cos \tfrac{1}{2}x$

(c) We graph $y = \sin 6x + \sin 7x$,

$y = 2 \cos \left(\tfrac{1}{2}x\right)$, and

$y = -2 \cos \left(\tfrac{1}{2}x\right)$.

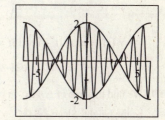

The graph of $y = f(x)$ lies
between the other two graphs.

101. (a) $\cos 4x = \cos (2x + 2x) = 2 \cos^2 2x - 1 = 2 \left(2 \cos^2 x - 1\right)^2 - 1 = 8 \cos^4 x - 8 \cos^2 x + 1$. Thus the desired

polynomial is $P(t) = 8t^4 - 8t^2 + 1$.

(b) $\cos 5x = \cos(4x + x) = \cos 4x \cos x - \sin 4x \sin x = \cos x \left(8\cos^4 x - 8\cos^2 x + 1\right) - 2\sin 2x \cos 2x \sin x$

$\qquad = 8\cos^5 x - 8\cos^3 x + \cos x - 4\sin x \cos x \left(2\cos^2 x - 1\right) \sin x \quad$ [from part (a)]

$\qquad = 8\cos^5 x - 8\cos^3 x + \cos x - 4\cos x \left(2\cos^2 x - 1\right) \sin^2 x$

$\qquad = 8\cos^5 x - 8\cos^3 x + \cos x - 4\cos x \left(2\cos^2 x - 1\right)\left(1 - \cos^2 x\right)$

$\qquad = 8\cos^5 x - 8\cos^3 x + \cos x + 8\cos^5 x - 12\cos^3 x + 4\cos x = 16\cos^5 x - 20\cos^3 x + 5\cos x$

Thus, the desired polynomial is $P(t) = 16t^5 - 20t^3 + 5t$.

103. Using a product-to-sum formula,

RHS $= 4\sin A \sin B \sin C = 4\sin A \left\{\frac{1}{2}\left[\cos(B - C) - \cos(B + C)\right]\right\} = 2\sin A \cos(B - C) - 2\sin A \cos(B + C)$.

Using another product-to-sum formula, this is equal to

$2\left\{\frac{1}{2}\left[\sin(A + B - C) + \sin(A - B + C)\right]\right\} - 2\left\{\frac{1}{2}\left[\sin(A + B + C) + \sin(A - B - C)\right]\right\}$

$\qquad = \sin(A + B - C) + \sin(A - B + C) - \sin(A + B + C) - \sin(A - B - C)$

Now $A + B + C = \pi$, so $A + B - C = \pi - 2C$, $A - B + C = \pi - 2B$, and $A - B - C = 2A - \pi$.

Thus our expression simplifies to

$\sin(A + B - C) + \sin(A - B + C) - \sin(A + B + C) - \sin(A - B - C)$

$\qquad = \sin(\pi - 2C) + \sin(\pi - 2B) + 0 - \sin(2A - \pi) = \sin 2C + \sin 2B + \sin 2A =$ LHS

105. (a) In both logs the length of the adjacent side is $20\cos\theta$ and the length of the opposite side is $20\sin\theta$.

Thus the cross-sectional area of the beam is modeled by

$A(\theta) = (20\cos\theta)(20\sin\theta) = 400\sin\theta\cos\theta = 200(2\sin\theta\cos\theta) = 200\sin 2\theta$.

(b) The function $y = \sin u$ is maximized when $u = \frac{\pi}{2}$. So $2\theta = \frac{\pi}{2} \Leftrightarrow \theta = \frac{\pi}{4}$. Thus the maximum cross-sectional area is

$A\left(\frac{\pi}{4}\right) = 200\sin 2\left(\frac{\pi}{4}\right) = 200$.

107. (a) $y = f_1(t) + f_2(t) = \cos 11t + \cos 13t$

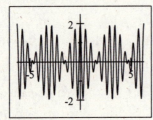

(b) Using the identity

$\cos\alpha + \cos y = 2 \cdot \cos\left(\dfrac{\alpha + y}{2}\right)\cos\left(\dfrac{\alpha - y}{2}\right)$, we have

$f(t) = \cos 11t + \cos 13t = 2 \cdot \cos\left(\dfrac{11t + 13t}{2}\right)\cos\left(\dfrac{11t - 13t}{2}\right)$

$\qquad = 2 \cdot \cos 12t \cdot \cos(-t) = 2\cos 12t \cos t$

(c) We graph $y = \cos 11t + \cos 13t$,

$y = 2\cos t$, and $y = -2\cos t$.

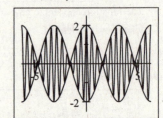

The graph of f lies between the graphs of $y = 2\cos t$ and $y = -2\cos t$. Thus, the loudness of the sound varies between $y = \pm 2\cos t$.

109. We find the area of $\triangle ABC$ in two different ways. First, let AB be the base and CD be the height. Since $\angle BOC = 2\theta$ we see that $CD = \sin 2\theta$. So the area is

$\frac{1}{2}$ (base) (height) $= \frac{1}{2} \cdot 2 \cdot \sin 2\theta = \sin 2\theta$. On the other hand, in $\triangle ABC$ we see that $\angle C$ is a right angle. So $BC = 2\sin\theta$ and $AC = 2\cos\theta$, and the area is

$\frac{1}{2}$ (base) (height) $= \frac{1}{2} \cdot (2\sin\theta)(2\cos\theta) = 2\sin\theta\cos\theta$. Equating the two expressions for the area of $\triangle ABC$, we get $\sin 2\theta = 2\sin\theta\cos\theta$.

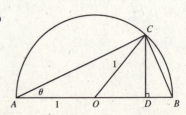

7.4   BASIC TRIGONOMETRIC EQUATIONS

1. Because the trigonometric functions are periodic, if a basic trigonometric equation has one solution, it has *infinitely many* solutions.

3. We can find some of the solutions of $\sin x = 0.3$ graphically by graphing $y = \sin x$ and $y = 0.3$. The solutions shown are $x \approx -9.729$, $x \approx -5.978$, $x \approx -3.446$, $x \approx 0.3047$, $x \approx 2.837$, $x \approx 6.588$, and $x \approx 9.120$.

5. Because sine has period 2π, we first find the solutions in the interval

$[0, 2\pi)$. From the unit circle shown, we see that $\sin \theta = \frac{\sqrt{3}}{2}$ in quadrants I

and II, so the solutions are $\theta = \frac{\pi}{3}$ and $\theta = \frac{2\pi}{3}$. We get all solutions of the

equation by adding integer multiples of 2π to these solutions:

$\theta = \frac{\pi}{3} + 2k\pi$ and $\theta = \frac{2\pi}{3} + 2k\pi$ for any integer k.

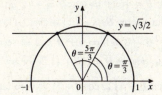

7. The cosine function is negative in quadrants II and III, so the solution of $\cos \theta = -1$ on the interval $[0, 2\pi)$ is $\theta = \pi$. Adding integer multiples of 2π to this solution gives all solutions: $\theta = \pi + 2k\pi = (2k + 1)\pi$ for any integer k.

9. The cosine function is positive in quadrants I and IV, so the solutions of $\cos \theta = \frac{1}{4}$ on the interval $[0, 2\pi)$ are $\theta = \cos^{-1} \frac{1}{4} \approx 1.32$ and $\theta = 2\pi - \cos^{-1} \frac{1}{4} \approx 4.97$. Adding integer multiples of 2π to these solutions gives all solutions: $\theta \approx 1.32 + 2k\pi, 4.97 + 2k\pi$ for any integer k.

11. The sine function is negative in quadrants III and IV, so the solutions of $\sin \theta = -0.45$ on the interval $[0, 2\pi)$ are $\theta = \pi + \sin^{-1}(0.45) \approx 3.61$ and $\theta = 2\pi - \sin^{-1}(0.45) \approx 5.82$. Adding integer multiples of 2π to these solutions gives all solutions, $\theta \approx 3.61 + 2k\pi, 5.82 + 2k\pi$ for any integer k.

13. We first find one solution by taking $\tan^{-1}$ of each side of the equation: $\theta = \tan^{-1}\left(-\sqrt{3}\right) = -\frac{\pi}{3}$. By definition, this is the only solution in the interval $\left(-\frac{\pi}{2}, \frac{\pi}{2}\right)$. Since tangent has period π, we get all solutions of the equation by adding integer multiples of π: $\theta = -\frac{\pi}{3} + k\pi$ for any integer k.

15. One solution of $\tan \theta = 5$ is $\theta = \tan^{-1} 5 \approx 1.37$. Adding integer multiples of π to this solution gives all solutions: $\theta \approx 1.37 + k\pi$ for any integer k.

17. One solution of $\cos \theta = -\frac{\sqrt{3}}{2}$ is $\theta = \cos^{-1}\left(-\frac{\sqrt{3}}{2}\right) = \frac{5\pi}{6}$ and another is $\theta = 2\pi - \frac{5\pi}{6} = \frac{7\pi}{6}$. All solutions are $\theta = \frac{5\pi}{6} + 2k\pi, \frac{7\pi}{6} + 2k\pi$ for any integer k. Specific solutions include $\theta = \frac{5\pi}{6} - 2\pi = -\frac{7\pi}{6}, \theta = \frac{7\pi}{6} - 2\pi = -\frac{5\pi}{6}, \theta = \frac{5\pi}{6}, \theta = \frac{7\pi}{6}, \theta = \frac{5\pi}{6} + 2\pi = \frac{17\pi}{6}$, and $\theta = \frac{7\pi}{6} + 2\pi = \frac{19\pi}{6}$.

19. One solution of $\sin \theta = \frac{\sqrt{2}}{2}$ is $\theta = \sin^{-1} \frac{\sqrt{2}}{2} = \frac{\pi}{4}$ and another is $\theta = \pi - \frac{\pi}{4} = \frac{3\pi}{4}$. All solutions are $\theta = \frac{\pi}{4} + 2k\pi$ and $\theta = \frac{3\pi}{4} + 2k\pi$ for any integer k. Specific solutions include $\theta = -\frac{7\pi}{4}, -\frac{5\pi}{4}, \frac{\pi}{4}, \frac{3\pi}{4}, \frac{9\pi}{4}$, and $\frac{11\pi}{4}$.

21. One solution of $\cos \theta = 0.28$ is $\theta = \cos^{-1} 0.28 \approx 1.29$ and another is $\theta = 2\pi - \cos^{-1} 0.28 \approx 5.00$. All solutions are $\theta \approx 1.29 + 2k\pi$ and $\theta \approx 5.00 + 2k\pi$ for any integer k. Specific solutions include $\theta \approx -5.00, -1.28, 1.29, 5.00, 7.57$, and 11.28.

23. One solution of $\tan \theta = -10$ is $\theta = \tan^{-1}(-10) \approx -1.47$. All solutions are $\theta \approx -1.47 + k\pi$ for any integer k. Specific solutions include $\theta \approx -7.75, -4.61, -1.47, 1.67, 4.81$, and 7.95.

25. $\cos \theta + 1 = 0 \Leftrightarrow \cos \theta = -1$. In the interval $[0, 2\pi)$ the only solution is $\theta = \pi$. Thus the solutions are $\theta = (2k + 1)\pi$ for any integer k.

27. $\sqrt{2} \sin \theta + 1 = 0 \Rightarrow \sqrt{2} \sin \theta = -1 \Leftrightarrow \sin \theta = -\frac{1}{\sqrt{2}}$ The solutions in the interval $[0, 2\pi)$ are $\theta = \frac{5\pi}{4}, \frac{7\pi}{4}$. Thus the solutions are $\theta = \frac{5\pi}{4} + 2k\pi, \frac{7\pi}{4} + 2k\pi$ for any integer k.

29. $5 \sin \theta - 1 = 0 \Leftrightarrow \sin \theta = \frac{1}{5}$. The solutions in the interval $[0, 2\pi)$ are $\theta = \sin^{-1} \frac{1}{5} \approx 0.20$ and $\theta = \pi - \sin^{-1} \frac{1}{5} \approx 2.94$. Thus the solutions are $\theta \approx 0.20 + 2k\pi$, $2.94 + 2k\pi$ for any integer k.

31. $3 \tan^2 \theta - 1 = 0 \Leftrightarrow \tan^2 \theta = \frac{1}{3} \Leftrightarrow \tan \theta = \pm \frac{\sqrt{3}}{3}$. The solutions in the interval $\left(-\frac{\pi}{2}, \frac{\pi}{2}\right)$ are $\theta = \pm \frac{\pi}{6}$, so all solutions are $\theta = -\frac{\pi}{6} + k\pi$, $\frac{\pi}{6} + k\pi$ for any integer k.

33. $2 \cos^2 \theta - 1 = 0 \Leftrightarrow \cos^2 \theta = \frac{1}{2} \Leftrightarrow \cos \theta = \pm \frac{1}{\sqrt{2}} \Leftrightarrow \theta = \frac{\pi}{4}, \frac{3\pi}{4}, \frac{5\pi}{4}, \frac{7\pi}{4}$ in $[0, 2\pi)$. Thus, the solutions are $\theta = \frac{\pi}{4} + k\pi$, $\frac{3\pi}{4} + k\pi$ for any integer k.

35. $\tan^2 \theta - 4 = 0 \Leftrightarrow \tan^2 \theta = 4 \Leftrightarrow \tan \theta = \pm 2 \Leftrightarrow \theta = \tan^{-1}(-2) \approx -1.11$ or $\theta = \tan^{-1} 2 \approx 1.11$ in $\left(-\frac{\pi}{2}, \frac{\pi}{2}\right)$. Thus, the solutions are $\theta \approx -1.11 + k\pi$, $1.11 + k\pi$ for any integer k.

37. $\sec^2 \theta - 2 = 0 \Leftrightarrow \sec^2 \theta = 2 \Leftrightarrow \sec \theta = \pm \sqrt{2}$. In the interval $[0, 2\pi)$ the solutions are $\theta = \frac{\pi}{4}, \frac{3\pi}{4}, \frac{5\pi}{4}, \frac{7\pi}{4}$. Thus, the solutions are $\theta = (2k + 1) \frac{\pi}{4}$ for any integer k.

39. $\left(\tan^2 \theta - 4\right)(2 \cos \theta + 1) = 0 \Leftrightarrow \tan^2 \theta = 4$ or $2 \cos \theta = -1$. From Exercise 35, we know that the first equation has solutions $\theta \approx -1.11 + k\pi$, $1.11 + k\pi$ for any integer k. $2 \cos \theta = -1 \Leftrightarrow \cos \theta = -\frac{1}{2}$ has solutions $\cos^{-1}\left(-\frac{1}{2}\right) = \frac{2\pi}{3}$ and $\frac{4\pi}{3}$ on $[0, 2\pi)$, so all solutions are $\theta \approx \frac{2\pi}{3} + 2k\pi$ and $\frac{4\pi}{3} + 2k\pi$ for any integer k. Thus, the original equation has solutions $\theta \approx -1.11 + k\pi$, $1.11 + k\pi$, $\frac{2\pi}{3} + 2k\pi$, and $\frac{4\pi}{3} + 2k\pi$ for any integer k.

41. $4 \cos^2 \theta - 4 \cos \theta + 1 = 0 \Leftrightarrow (2 \cos \theta - 1)^2 = 0 \Leftrightarrow 2 \cos \theta - 1 = 0 \Leftrightarrow \cos \theta = \frac{1}{2} \Leftrightarrow \theta = \frac{\pi}{3} + 2k\pi$, $\frac{5\pi}{3} + 2k\pi$ for any integer k.

43. $3 \sin^2 \theta - 7 \sin \theta + 2 = 0 \Rightarrow (3 \sin \theta - 1)(\sin \theta - 2) = 0 \Rightarrow 3 \sin \theta - 1 = 0$ or $\sin \theta - 2 = 0$. Since $|\sin \theta| \le 1$, $\sin \theta - 2 = 0$ has no solution. Thus $3 \sin \theta - 1 = 0 \Rightarrow \sin \theta = \frac{1}{3} \Rightarrow \theta \approx 0.33984$ and $\theta \approx \pi - 0.33984 \approx 2.80176$ are the solutions in $[0, 2\pi)$, and all solutions are $\theta \approx 0.33984 + 2k\pi$, $2.80176 + 2k\pi$ for any integer k.

45. $2 \cos^2 \theta - 7 \cos \theta + 3 = 0 \Leftrightarrow (2 \cos \theta - 1)(\cos \theta - 3) = 0 \Leftrightarrow \cos \theta = \frac{1}{2}$ or $\cos \theta = 3$ (which is inadmissible) $\Leftrightarrow \theta = \frac{\pi}{3}$, $\frac{5\pi}{3}$. Therefore, the solutions are $\theta = \frac{\pi}{3} + 2k\pi$, $\frac{5\pi}{3} + 2k\pi$ for any integer k.

47. $\cos^2 \theta - \cos \theta - 6 = 0 \Leftrightarrow (\cos \theta + 2)(\cos x - 3) = 0 \Leftrightarrow \cos x = -2$ or $\cos x = 3$, neither of which has a solution. Thus, the original equation has no solution.

49. $\sin^2 \theta = 2 \sin \theta + 3 \Leftrightarrow \sin^2 \theta - 2 \sin \theta - 3 = 0 \Leftrightarrow (\sin \theta - 3)(\sin \theta + 1) = 0 \Leftrightarrow \sin \theta - 3 = 0$ or $\sin \theta + 1 = 0$. Since $|\sin \theta| \le 1$ for all θ, there is no solution for $\sin \theta - 3 = 0$. Hence $\sin \theta + 1 = 0 \Leftrightarrow \sin \theta = -1 \Leftrightarrow \theta = \frac{3\pi}{2} + 2k\pi$ for any integer k.

51. $\cos \theta (2 \sin \theta + 1) = 0 \Leftrightarrow \cos \theta = 0$ or $\sin \theta = -\frac{1}{2} \Leftrightarrow \theta = \frac{\pi}{2} + k\pi$, $\frac{7\pi}{6} + 2k\pi$, $\frac{11\pi}{6} + 2k\pi$ for any integer k.

53. $\cos \theta \sin \theta - 2 \cos \theta = 0 \Leftrightarrow \cos \theta (\sin \theta - 2) = 0 \Leftrightarrow \cos \theta = 0$ or $\sin \theta - 2 = 0$. Since $|\sin \theta| \le 1$ for all θ, there is no solution for $\sin \theta - 2 = 0$. Hence, $\cos \theta = 0 \Leftrightarrow \theta = \frac{\pi}{2} + 2k\pi$, $\frac{3\pi}{2} + 2k\pi \Leftrightarrow \theta = \frac{\pi}{2} + k\pi$ for any integer k.

55. $3 \tan \theta \sin \theta - 2 \tan \theta = 0 \Leftrightarrow \tan \theta (3 \sin \theta - 2) = 0 \Leftrightarrow \tan \theta = 0$ or $\sin \theta = \frac{2}{3}$. $\tan \theta = 0$ has solution $\theta = 0$ on $\left(-\frac{\pi}{2}, \frac{\pi}{2}\right)$ and $\sin \theta = \frac{2}{3}$ has solutions $\theta = \sin^{-1} \frac{2}{3} \approx 0.73$ and $\theta = \pi - \sin^{-1} \frac{2}{3} \approx 2.41$ on $[0, 2\pi)$, so the original equation has solutions $\theta = k\pi$, $\theta \approx 0.73 + 2k\pi$, $2.41 + 2k\pi$ for any integer k.

57. We substitute $\theta_1 = 70°$ and $\frac{v_1}{v_2} = 1.33$ into Snell's Law to get $\frac{\sin 70°}{\sin \theta_2} = 1.33 \Leftrightarrow \sin \theta_2 = \frac{\sin 70°}{1.33} = 0.7065 \Rightarrow \theta_2 \approx 44.95°$.

59. (a) $F = \frac{1}{2}(1 - \cos \theta) = 0 \Rightarrow \cos \theta = 1 \Rightarrow \theta = 0$

(b) $F = \frac{1}{2}(1 - \cos \theta) = 0.25 \Rightarrow 1 - \cos \theta = 0.5 \Rightarrow \cos \theta = 0.5 \Rightarrow \theta = 60°$ or $120°$

(c) $F = \frac{1}{2}(1 - \cos \theta) = 0.5 \Rightarrow 1 - \cos \theta = 1 \Rightarrow \cos \theta = 0 \Rightarrow \theta = 90°$ or $270°$

(d) $F = \frac{1}{2}(1 - \cos \theta) = 1 \Rightarrow 1 - \cos \theta = 2 \Rightarrow \cos \theta = -1 \Rightarrow \theta = 180°$

7.5   MORE TRIGONOMETRIC EQUATIONS

1. Using a Pythagorean identity, we calculate $\sin x + \sin^2 x + \cos^2 x = 1 \Leftrightarrow \sin x + 1 = 1 \Leftrightarrow \sin x = 0$, whose solutions are $x = k\pi$ for any integer k.

3. $2\cos^2\theta + \sin\theta = 1 \Leftrightarrow 2\left(1 - \sin^2\theta\right) + \sin\theta - 1 = 0 \Leftrightarrow -2\sin^2\theta + \sin\theta + 1 = 0 \Leftrightarrow 2\sin^2\theta - \sin\theta - 1 = 0$. From Exercise 7.4.42, the solutions are $\theta = \frac{7\pi}{6} + 2k\pi, \frac{11\pi}{6} + 2k\pi, \frac{\pi}{2} + 2k\pi$ for any integer k.

5. $\tan^2\theta - 2\sec\theta = 2 \Leftrightarrow \sec^2\theta - 1 - 2\sec\theta = 2 \Leftrightarrow \sec^2\theta - 2\sec\theta - 3 = 0 \Leftrightarrow (\sec\theta - 3)(\sec\theta + 1) = 0 \Leftrightarrow \sec\theta = 3$ or $\sec\theta = -1$. If $\sec\theta = 3$, then $\cos\theta = \frac{1}{3}$, which has solutions $\theta = \cos^{-1}\frac{1}{3} \approx 1.23$ and $\theta = 2\pi - \cos^{-1}\frac{1}{3} \approx 5.05$ on $[0, 2\pi)$. If $\sec\theta = -1$, then $\cos\theta = -1$, which has solution $\theta = \pi$ on $[0, 2\pi)$. Thus, solutions are $\theta = (2k+1)\pi$, $\theta \approx 1.23 + 2k\pi, 5.05 + 2k\pi$ for any integer k.

7. $2\sin 2\theta - 3\sin\theta = 0 \Leftrightarrow 2(2\sin\theta\cos\theta) - 3\sin\theta = 0 \Leftrightarrow \sin\theta(4\cos\theta - 3) = 0 \Leftrightarrow \sin\theta = 0$ or $\cos\theta = \frac{3}{4}$. The first equation has solutions $\theta = 0, \pi$ on $[0, 2\pi)$, and the second has solutions $\theta = \cos^{-1}\frac{3}{4} \approx 0.72$ and $\theta = 2\pi - \cos^{-1}\frac{3}{4} \approx 5.56$ on $[0, 2\pi)$. Thus, solutions are $\theta = k\pi, \theta \approx 0.72 + 2k\pi, 5.56 + 2k\pi$ for any integer k.

9. $\cos 2\theta = 3\sin\theta - 1 \Leftrightarrow 1 - 2\sin^2\theta = 3\sin\theta - 1 \Leftrightarrow 2\sin^2\theta + 3\sin\theta - 2 = 0 \Leftrightarrow (\sin\theta + 2)(2\sin\theta - 1) = 0 \Leftrightarrow \sin\theta = -2$ or $\sin\theta = \frac{1}{2}$. The first equation has no solution and the second has solutions $\theta = \sin^{-1}\frac{1}{2} = \frac{\pi}{6}$ and $\theta = \pi - \sin^{-1}\frac{1}{2} = \frac{5\pi}{6}$ on $[0, 2\pi)$, so the original equation has solutions $\theta = \frac{\pi}{6} + 2k\pi, \frac{5\pi}{6} + 2k\pi$ for any integer k.

11. $2\sin^2\theta - \cos\theta = 1 \Leftrightarrow 2\left(1 - \cos^2\theta\right) - \cos\theta - 1 = 0 \Leftrightarrow -2\cos^2\theta - \cos\theta + 1 = 0 \Leftrightarrow (2\cos\theta - 1)(\cos\theta + 1) = 0 \Leftrightarrow 2\cos\theta - 1 = 0$ or $\cos\theta + 1 = 0 \Leftrightarrow \cos\theta = \frac{1}{2}$ or $\cos\theta = -1 \Leftrightarrow \theta = \frac{\pi}{3} + 2k\pi, \frac{5\pi}{3} + 2k\pi, (2k+1)\pi$ for any integer k.

13. $\sin\theta - 1 = \cos\theta \Leftrightarrow \sin\theta + \cos\theta = 1$. Squaring both sides, we have $\sin^2\theta + \cos^2\theta + 2\sin\theta\cos\theta = 1 \Leftrightarrow \sin 2\theta = 0$, which has solutions $\theta = 0, \frac{\pi}{2}, \pi, \frac{3\pi}{2}$ in $[0, 2\pi)$. Checking in the original equation, we see that only $\theta = \frac{\pi}{2}$ and $\theta = \pi$ are valid. (The extraneous solutions were introduced by squaring both sides.) Thus, the solutions are $\theta = (2k+1)\pi, \frac{\pi}{2} + 2k\pi$ for any integer k.

15. $\tan\theta + 1 = \sec\theta \Leftrightarrow \frac{\sin\theta}{\cos\theta} + 1 = \frac{1}{\cos\theta} \Leftrightarrow \sin\theta + \cos\theta = 1$. Squaring both sides, we have $\sin^2\theta + \cos^2\theta + 2\sin\theta\cos\theta = 1 \Leftrightarrow \sin 2\theta = 0$, which has solutions $\theta = 0, \frac{\pi}{2}, \pi, \frac{3\pi}{2}$ on $[0, 2\pi)$. Checking in the original equation, we see that only $\theta = 0$ is valid. Thus, the solutions are $\theta = 2k\pi$ for any integer k.

17. (a) $2\cos 3\theta = 1 \Leftrightarrow \cos 3\theta = \frac{1}{2} \Rightarrow 3\theta = \frac{\pi}{3}, \frac{5\pi}{3}$ for 3θ in $[0, 2\pi)$. Thus, solutions are $\frac{\pi}{9} + \frac{2}{3}k\pi, \frac{5\pi}{9} + \frac{2}{3}k\pi$ for any integer k.

   (b) We take $k = 0, 1, 2$ in the expressions in part (a) to obtain the solutions $\theta = \frac{\pi}{9}, \frac{5\pi}{9}, \frac{7\pi}{9}, \frac{11\pi}{9}, \frac{13\pi}{9}, \frac{17\pi}{9}$ in $[0, 2\pi)$.

19. (a) $2\cos 2\theta + 1 = 0 \Leftrightarrow \cos 2\theta = -\frac{1}{2} \Leftrightarrow 2\theta = \frac{2\pi}{3} + 2k\pi, \frac{4\pi}{3} + 2k\pi \Leftrightarrow \theta = \frac{\pi}{3} + k\pi, \frac{2\pi}{3} + k\pi$ for any integer k.

   (b) The solutions in $[0, 2\pi)$ are $\frac{\pi}{3}, \frac{2\pi}{3}, \frac{4\pi}{3}, \frac{5\pi}{3}$.

21. (a) $\sqrt{3}\tan 3\theta + 1 = 0 \Leftrightarrow \tan 3\theta = -\frac{1}{\sqrt{3}} \Leftrightarrow 3\theta = \frac{5\pi}{6} + k\pi \Leftrightarrow \theta = \frac{5\pi}{18} + \frac{1}{3}k\pi$ for any integer k.

   (b) The solutions in $[0, 2\pi)$ are $\frac{5\pi}{18}, \frac{11\pi}{18}, \frac{17\pi}{18}, \frac{23\pi}{18}, \frac{29\pi}{18}, \frac{35\pi}{18}$.

23. (a) $\cos\frac{\theta}{2} - 1 = 0 \Leftrightarrow \cos\frac{\theta}{2} = 1 \Leftrightarrow \frac{\theta}{2} = 2k\pi \Leftrightarrow \theta = 4k\pi$ for any integer k.

   (b) The only solution in $[0, 2\pi)$ is $\theta = 0$.

25. (a) $2\sin\frac{\theta}{3} + \sqrt{3} = 0 \Leftrightarrow 2\sin\frac{\theta}{3} = -\sqrt{3} \Leftrightarrow \sin\frac{\theta}{3} = -\frac{\sqrt{3}}{2} \Leftrightarrow \frac{\theta}{3} = \frac{4\pi}{3} + 2k\pi, \frac{5\pi}{3} + 2k\pi \Leftrightarrow \theta = 4\pi + 6k\pi, 5\pi + 6k\pi$ for any integer k.

   (b) There is no solution in $[0, 2\pi)$.

27. (a) $\sin 2\theta = 3\cos 2\theta \Leftrightarrow \tan 2\theta = 3 \Leftrightarrow \theta = \frac{1}{2}\tan^{-1} 3 \approx 0.62$ on $\left(-\frac{\pi}{4}, \frac{\pi}{4}\right)$. Thus, solutions are $\theta \approx 0.62 + \frac{1}{2}k\pi$ for any integer k.

(b) The solutions in $[0, 2\pi)$ are $\theta \approx 0.62, 2.19, 3.76, 5.33$.

29. (a) $\sec\theta - \tan\theta = \cos\theta \Leftrightarrow \cos\theta (\sec\theta - \tan\theta) = \cos\theta (\cos\theta) \Leftrightarrow 1 - \sin\theta = \cos^2\theta \Leftrightarrow 1 - \sin\theta = 1 - \sin^2\theta \Leftrightarrow$
$\sin\theta = \sin^2\theta \Leftrightarrow \sin^2\theta - \sin\theta = 0 \Leftrightarrow \sin\theta(\sin\theta - 1) = 0 \Leftrightarrow \sin\theta = 0$ or $\sin\theta = 1 \Leftrightarrow \theta = 0, \pi$ or $\theta = \frac{\pi}{2}$ in
$[0, 2\pi)$. However, since the equation is undefined when $\theta = \frac{\pi}{2}$, the solutions are $\theta = k\pi$ for any integer k.

(b) The solutions in $[0, 2\pi)$ are $\theta = 0, \pi$.

31. (a) $3\tan^3\theta - 3\tan^2\theta - \tan\theta + 1 = 0 \Leftrightarrow (\tan\theta - 1)\left(3\tan^2\theta - 1\right) = 0 \Leftrightarrow \tan\theta = 1$ or $3\tan^2\theta = 1 \Leftrightarrow \tan\theta = 1$ or
$\tan\theta = \pm\frac{1}{\sqrt{3}} \Leftrightarrow \theta = \frac{\pi}{6} + k\pi, \frac{\pi}{4} + k\pi, \frac{5\pi}{6} + k\pi$ for any integer k.

(b) The solutions in $[0, 2\pi)$ are $\theta = \frac{\pi}{6}, \frac{\pi}{4}, \frac{5\pi}{6}, \frac{7\pi}{6}, \frac{5\pi}{4}, \frac{11\pi}{6}$.

33. (a) $2\sin\theta\tan\theta - \tan\theta = 1 - 2\sin\theta \Leftrightarrow 2\sin\theta\tan\theta - \tan\theta + 2\sin\theta - 1 = 0 \Leftrightarrow (2\sin\theta - 1)(\tan\theta + 1) = 0 \Leftrightarrow$
$2\sin\theta - 1 = 0$ or $\tan\theta + 1 = 0 \Leftrightarrow \sin\theta = \frac{1}{2}$ or $\tan\theta = -1 \Leftrightarrow \theta = \frac{\pi}{6} + 2k\pi, \frac{5\pi}{6} + 2k\pi, \frac{3\pi}{4} + k\pi$ for any integer k.

(b) The solutions in $[0, 2\pi)$ are $\frac{\pi}{6}, \frac{3\pi}{4}, \frac{5\pi}{6}, \frac{7\pi}{4}$.

35. (a)

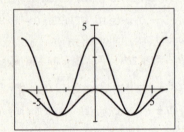

The points of intersection are
approximately $(\pm 3.14, -2)$.

(b) $f(x) = 3\cos x + 1$; $g(x) = \cos x - 1$. $f(x) = g(x)$ when
$3\cos x + 1 = \cos x - 1 \Leftrightarrow 2\cos x = -2 \Leftrightarrow \cos x = -1 \Leftrightarrow$
$x = \pi + 2k\pi = (2k + 1)\pi$. The points of intersection are
$((2k + 1)\pi, -2)$ for any integer k.

37. (a)

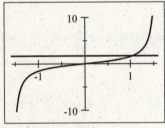

The point of intersection is
approximately $(1.04, 1.73)$.

(b) $f(x) = \tan x$; $g(x) = \sqrt{3}$. $f(x) = g(x)$ when $\tan x = \sqrt{3} \Leftrightarrow$
$x = \frac{\pi}{3} + k\pi$. The intersection points are $\left(\frac{\pi}{3} + k\pi, \sqrt{3}\right)$ for any integer k.

39. $\cos\theta\cos 3\theta - \sin\theta\sin 3\theta = 0 \Leftrightarrow \cos(\theta + 3\theta) = 0 \Leftrightarrow \cos 4\theta = 0 \Leftrightarrow 4\theta = \frac{\pi}{2}, \frac{3\pi}{2}, \frac{5\pi}{2}, \frac{7\pi}{2}, \frac{9\pi}{2}, \frac{11\pi}{2}, \frac{13\pi}{2}, \frac{15\pi}{2}$ in
$[0, 8\pi) \Leftrightarrow \theta = \frac{\pi}{8}, \frac{3\pi}{8}, \frac{5\pi}{8}, \frac{7\pi}{8}, \frac{9\pi}{8}, \frac{11\pi}{8}, \frac{13\pi}{8}, \frac{15\pi}{8}$ in $[0, 2\pi)$.

41. $\sin 2\theta\cos\theta - \cos 2\theta\sin\theta = \frac{\sqrt{3}}{2} \Leftrightarrow \sin(2\theta - \theta) = \frac{\sqrt{3}}{2} \Leftrightarrow \sin\theta = \frac{\sqrt{3}}{2} \Leftrightarrow \theta = \frac{\pi}{3}, \frac{2\pi}{3}$ in $[0, 2\pi)$.

43. $\sin 2\theta + \cos\theta = 0 \Leftrightarrow 2\sin\theta\cos\theta + \cos\theta = 0 \Leftrightarrow \cos\theta(2\sin\theta + 1) = 0 \Leftrightarrow \cos\theta = 0$ or $\sin\theta = -\frac{1}{2} \Leftrightarrow \theta = \frac{\pi}{2}, \frac{7\pi}{6},$
$\frac{3\pi}{2}, \frac{11\pi}{6}$ in $[0, 2\pi)$.

45. $\cos 2\theta + \cos\theta = 2 \Leftrightarrow 2\cos^2\theta - 1 + \cos\theta - 2 = 0 \Leftrightarrow 2\cos^2\theta + \cos\theta - 3 = 0 \Leftrightarrow (2\cos\theta + 3)(\cos\theta - 1) = 0 \Leftrightarrow$
$2\cos\theta + 3 = 0$ or $\cos\theta - 1 = 0 \Leftrightarrow \cos\theta = -\frac{3}{2}$ (which is impossible) or $\cos\theta = 1 \Leftrightarrow \theta = 0$ in $[0, 2\pi)$.

47. $\cos 2\theta - \cos^2\theta = 0 \Leftrightarrow 2\cos^2\theta - 1 - \cos^2\theta = 0 \Leftrightarrow \cos^2\theta = 1 \Leftrightarrow \theta = k\pi$ for any integer k. On $[0, 2\pi)$, the solutions
are $\theta = 0, \pi$.

49. $\cos 2\theta - \cos 4\theta = 0 \Leftrightarrow \cos 2\theta - \left(2\cos^2 2\theta - 1\right) = 0 \Leftrightarrow (\cos 2\theta - 1)(2\cos 2\theta + 1) = 0.$ The first factor has zeros at

$\theta = 0$, π and the second has zeros at $\theta = \frac{\pi}{3}, \frac{2\pi}{3}, \frac{4\pi}{3}, \frac{5\pi}{3}$. Thus, solutions of the original equation are are $\theta = 0, \frac{\pi}{3}, \frac{2\pi}{3},$

$\pi, \frac{4\pi}{3}, \frac{5\pi}{3}$ in $[0, 2\pi)$.

51. $\cos\theta - \sin\theta = \sqrt{2}\sin\frac{\theta}{2} \Leftrightarrow \cos\theta - \sin\theta = \sqrt{2}\left(\pm\sqrt{\dfrac{1-\cos\theta}{2}}\right)$. Squaring both sides, we have

$\cos^2\theta + \sin^2\theta - 2\sin\theta\cos\theta = 1 - \cos\theta \Leftrightarrow 1 - 2\sin\theta\cos\theta = 1 - \cos\theta \Leftrightarrow$ either $2\sin\theta = 1$ or $\cos\theta = 0 \Leftrightarrow \theta = \frac{\pi}{6}$,

$\frac{\pi}{2}, \frac{5\pi}{6}, \frac{3\pi}{2}$ in $[0, 2\pi)$. Of these, only $\frac{\pi}{6}$ and $\frac{3\pi}{2}$ satisfy the original equation.

53. $\sin\theta + \sin 3\theta = 0 \Leftrightarrow 2\sin 2\theta\cos(-\theta) = 0 \Leftrightarrow 2\sin 2\theta\cos\theta = 0 \Leftrightarrow \sin 2\theta = 0$ or $\cos\theta = 0 \Leftrightarrow 2\theta = k\pi$ or $\theta = k\frac{\pi}{2} \Leftrightarrow$

$\theta = \frac{1}{2}k\pi$ for any integer k.

55. $\cos 4\theta + \cos 2\theta = \cos\theta \Leftrightarrow 2\cos 3\theta\cos\theta = \cos\theta \Leftrightarrow \cos\theta(2\cos 3\theta - 1) = 0 \Leftrightarrow \cos\theta = 0$ or $\cos 3\theta = \frac{1}{2} \Leftrightarrow \theta = \frac{\pi}{2}$ or

$3\theta = \frac{\pi}{3} + 2k\pi, \frac{5\pi}{3} + 2k\pi, \frac{7\pi}{3} + 2k\pi, \frac{11\pi}{3} + 2k\pi, \frac{13\pi}{3} + 2k\pi, \frac{17\pi}{3} + 2k\pi \Leftrightarrow \theta = \frac{\pi}{2} + k\pi, \frac{\pi}{9} + \frac{2}{3}k\pi, \frac{5\pi}{9} + \frac{2}{3}k\pi$ for

any integer k.

57. $\sin 2x = x$

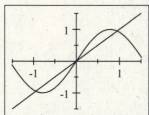

The three solutions are $x = 0$ and $x \approx \pm 0.95$.

59. $2^{\sin x} = x$

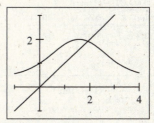

The only solution is $x \approx 1.92$.

61. $\dfrac{\cos x}{1 + x^2} = x^2$

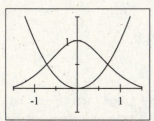

The two solutions are $x \approx \pm 0.71$.

63. We substitute $v_0 = 2200$ and $R(\theta) = 5000$ and solve for

θ. So $5000 = \dfrac{(2200)^2 \sin 2\theta}{32} \Leftrightarrow 5000 = 151250\sin 2\theta$

$\Leftrightarrow \sin 2\theta = 0.03308 \Rightarrow 2\theta = 1.89442°$ or

$2\theta = 180° - 1.89442° = 178.10558°$. If $2\theta = 1.89442°$,

then $\theta = 0.94721°$, and if $2\theta = 178.10558°$, then

$\theta = 89.05279°$.

65. (a) $10 = 12 + 2.83\sin\left(\frac{2\pi}{3}(t-80)\right) \Leftrightarrow 2.83\sin\left(\frac{2\pi}{3}(t-80)\right) = -2 \Leftrightarrow \sin\left(\frac{2\pi}{3}(t-80)\right) = -0.70671.$ Now

$\sin\theta = -0.70671$ and $\theta = -0.78484$. If $\frac{2\pi}{3}(t-80) = -0.78484 \Leftrightarrow t - 80 = 45.6 \Leftrightarrow t = 34.4$. Now in the interval

$[0, 2\pi)$, we have $\theta = \pi + 0.78484 \approx 3.92644$ and $\theta = 2\pi - 0.78484 \approx 5.49834$. If $\frac{2\pi}{3}(t-80) = 3.92644 \Leftrightarrow$

$t - 80 = 228.1 \Leftrightarrow t = 308.1$. And if $\frac{2\pi}{3}(t-80) = 5.49834 \Leftrightarrow t - 80 = 319.4 \Leftrightarrow t = 399.4$ $(399.4 - 365 = 34.4)$.

So according to this model, there should be 10 hours of sunshine on the 34th day (February 3) and on the 308th day

(November 4).

(b) Since $L(t) = 12 + 2.83\sin\left(\frac{2\pi}{3}(t-80)\right) \geq 10$ for $t \in [34, 308]$, the number of days with more than 10 hours of

daylight is $308 - 34 + 1 = 275$ days.

67. $\sin(\cos x)$ is a function of a function, that is, a composition of trigonometric functions (see Section 2.6). Most of the other equations involve sums, products, differences, or quotients of trigonometric functions.

$\sin(\cos x) = 0 \Leftrightarrow \cos x = 0$ or $\cos x = \pi$. However, since $|\cos x| \le 1$, the only solution is $\cos x = 0 \Rightarrow x = \frac{\pi}{2} + k\pi$. The graph of $f(x) = \sin(\cos x)$ is shown.

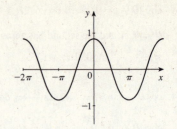

CHAPTER 7 REVIEW

1. $\sin\theta(\cot\theta + \tan\theta) = \sin\theta\left(\dfrac{\cos\theta}{\sin\theta} + \dfrac{\sin\theta}{\cos\theta}\right) = \cos\theta + \dfrac{\sin^2\theta}{\cos\theta} = \dfrac{\cos^2\theta + \sin^2\theta}{\cos\theta} = \dfrac{1}{\cos\theta} = \sec\theta$

3. $\cos^2 x \csc x - \csc x = \left(1 - \sin^2 x\right)\csc x - \csc x = \csc x - \sin^2 x \csc x - \csc x = -\sin^2 x \cdot \dfrac{1}{\sin x} = -\sin x$

5. $\dfrac{\cos^2 x - \tan^2 x}{\sin^2 x} = \dfrac{\cos^2 x}{\sin^2 x} - \dfrac{\tan^2 x}{\sin^2 x} = \cot^2 x - \dfrac{1}{\cos^2 x} = \cot^2 x - \sec^2 x$

7. $\dfrac{\cos^2 x}{1 - \sin x} = \dfrac{\cos x}{\dfrac{1}{\cos x}(1 - \sin x)} = \dfrac{\cos x}{\dfrac{1}{\cos x} - \dfrac{\sin x}{\cos x}} = \dfrac{\cos x}{\sec x - \tan x}$

9. $\sin^2 x \cot^2 x + \cos^2 x \tan^2 x = \sin^2 x \cdot \dfrac{\cos^2 x}{\sin^2 x} + \cos^2 x \cdot \dfrac{\sin^2 x}{\cos^2 x} = \cos^2 x + \sin^2 x = 1$

11. $\dfrac{\sin 2x}{1 + \cos 2x} = \dfrac{2\sin x \cos x}{1 + 2\cos^2 x - 1} = \dfrac{2\sin x \cos x}{2\cos^2 x} = \dfrac{2\sin x}{2\cos x} = \tan x$

13. $\tan\dfrac{x}{2} = \dfrac{1 - \cos x}{\sin x} = \dfrac{1}{\sin x} - \dfrac{\cos x}{\sin x} = \csc x - \cot x$

15. $\sin(x + y)\sin(x - y) = \frac{1}{2}\left[\cos((x+y) - (x-y)) - \cos((x+y) + (x-y))\right] = \frac{1}{2}\left(\cos 2y - \cos 2x\right)$
$= \frac{1}{2}\left[1 - 2\sin^2 y - \left(1 - 2\sin^2 x\right)\right] = \frac{1}{2}\left(2\sin^2 x - 2\sin^2 y\right) = \sin^2 x - \sin^2 y$

17. $1 + \tan x \tan\dfrac{x}{2} = 1 + \dfrac{\sin x}{\cos x} \cdot \dfrac{1 - \cos x}{\sin x} = 1 + \dfrac{1 - \cos x}{\cos x} = 1 + \dfrac{1}{\cos x} - 1 = \dfrac{1}{\cos x} = \sec x$

19. $\left(\cos\dfrac{x}{2} - \sin\dfrac{x}{2}\right)^2 = \cos^2\dfrac{x}{2} - 2\sin\dfrac{x}{2}\cos\dfrac{x}{2} + \sin^2\dfrac{x}{2} = \sin^2\dfrac{x}{2} + \cos^2\dfrac{x}{2} - 2\sin\dfrac{x}{2}\cos\dfrac{x}{2} = 1 - \sin\left(2\cdot\dfrac{x}{2}\right) = 1 - \sin x$

21. $\dfrac{\sin 2x}{\sin x} - \dfrac{\cos 2x}{\cos x} = \dfrac{2\sin x \cos x}{\sin x} - \dfrac{2\cos^2 x - 1}{\cos x} = 2\cos x - 2\cos x + \dfrac{1}{\cos x} = \sec x$

23. $\tan\left(x + \dfrac{\pi}{4}\right) = \dfrac{\tan x + \tan\dfrac{\pi}{4}}{1 - \tan x \tan\dfrac{\pi}{4}} = \dfrac{1 + \tan x}{1 - \tan x}$

25. (a) $f(x) = 1 - \left(\cos\frac{x}{2} - \sin\frac{x}{2}\right)^2$, $g(x) = \sin x$

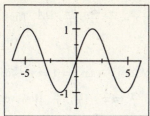

(b) The graphs suggest that $f(x) = g(x)$ is an identity. To prove this, expand $f(x)$ and simplify, using the double-angle formula for sine:

$$\begin{aligned} f(x) &= 1 - \left(\cos\frac{x}{2} - \sin\frac{x}{2}\right)^2 \\ &= 1 - \left(\cos^2\frac{x}{2} - 2\cos\frac{x}{2}\sin\frac{x}{2} + \sin^2\frac{x}{2}\right) \\ &= 1 + 2\cos\frac{x}{2}\sin\frac{x}{2} - \left(\cos^2\frac{x}{2} + \sin^2\frac{x}{2}\right) \\ &= 1 + \sin x - (1) = \sin x = g(x) \end{aligned}$$

27. (a) $f(x) = \tan x \tan\frac{x}{2}$, $g(x) = \dfrac{1}{\cos x}$

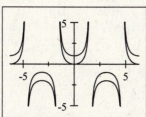

(b) The graphs suggest that $f(x) \neq g(x)$ in general. For example, choose $x = \frac{\pi}{3}$ and evaluate: $f\left(\frac{\pi}{3}\right) = \tan\frac{\pi}{3}$ $\tan\frac{\pi}{6} = \sqrt{3} \cdot \frac{1}{\sqrt{3}} = 1$, whereas $g\left(\frac{\pi}{3}\right) = \dfrac{1}{\frac{1}{2}} = 2$, so $f(x) \neq g(x)$.

29. (a) $f(x) = 2\sin^2 3x + \cos 6x$

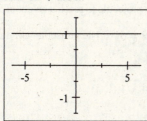

(b) The graph suggests that $f(x) = 1$ for all x. To prove this, we use the double angle formula to note that $\cos 6x = \cos(2(3x)) = 1 - 2\sin^2 3x$, so $f(x) = 2\sin^2 3x + \left(1 - 2\sin^2 3x\right) = 1$.

31. $4\sin\theta - 3 = 0 \Leftrightarrow 4\sin\theta = 3 \Leftrightarrow \sin\theta = \frac{3}{4} \Leftrightarrow \theta = \sin^{-1}\frac{3}{4} \approx 0.8481$ or $\theta = \pi - \sin^{-1}\frac{3}{4} \approx 2.2935$.

33. $\cos x \sin x - \sin x = 0 \Leftrightarrow \sin x (\cos x - 1) = 0 \Leftrightarrow \sin x = 0$ or $\cos x = 1 \Leftrightarrow x = 0, \pi$ or $x = 0$. Therefore, the solutions are $x = 0$ and π.

35. $2\sin^2 x - 5\sin x + 2 = 0 \Leftrightarrow (2\sin x - 1)(\sin x - 2) = 0 \Leftrightarrow \sin x = \frac{1}{2}$ or $\sin x = 2$ (which is inadmissible) $\Leftrightarrow x = \frac{\pi}{6}$, $\frac{5\pi}{6}$. Thus, the solutions in $[0, 2\pi)$ are $x = \frac{\pi}{6}$ and $\frac{5\pi}{6}$.

37. $2\cos^2 x - 7\cos x + 3 = 0 \Leftrightarrow (2\cos x - 1)(\cos x - 3) = 0 \Leftrightarrow \cos x = \frac{1}{2}$ or $\cos x = 3$ (which is inadmissible) $\Leftrightarrow x = \frac{\pi}{3}$, $\frac{5\pi}{3}$. Therefore, the solutions in $[0, 2\pi)$ are $x = \frac{\pi}{3}, \frac{5\pi}{3}$.

39. Note that $x = \pi$ is not a solution because the denominator is zero. $\dfrac{1 - \cos x}{1 + \cos x} = 3 \Leftrightarrow 1 - \cos x = 3 + 3\cos x \Leftrightarrow -4\cos x = 2 \Leftrightarrow \cos x = -\frac{1}{2} \Leftrightarrow x = \frac{2\pi}{3}, \frac{4\pi}{3}$ in $[0, 2\pi)$.

41. Factor by grouping: $\tan^3 x + \tan^2 x - 3\tan x - 3 = 0 \Leftrightarrow (\tan x + 1)\left(\tan^2 x - 3\right) = 0 \Leftrightarrow \tan x = -1$ or $\tan x = \pm\sqrt{3} \Leftrightarrow x = \frac{3\pi}{4}, \frac{7\pi}{4}$ or $x = \frac{\pi}{3}, \frac{2\pi}{3}, \frac{4\pi}{3}, \frac{5\pi}{3}$. Therefore, the solutions in $[0, 2\pi)$ are $x = \frac{\pi}{3}, \frac{2\pi}{3}, \frac{3\pi}{4}, \frac{4\pi}{3}, \frac{5\pi}{3}, \frac{7\pi}{4}$.

43. $\tan \frac{1}{2}x + 2\sin 2x = \csc x \Leftrightarrow \dfrac{1 - \cos x}{\sin x} + 4\sin x \cos x = \dfrac{1}{\sin x} \Leftrightarrow 1 - \cos x + 4\sin^2 x \cos x = 1 \Leftrightarrow 4\sin^2 x \cos x - \cos x = 0$

$\Leftrightarrow \cos x \left(4\sin^2 x - 1\right) = 0 \Leftrightarrow \cos x = 0$ or $\sin x = \pm\frac{1}{2} \Leftrightarrow x = \frac{\pi}{2}, \frac{3\pi}{2}$ or $x = \frac{\pi}{6}, \frac{5\pi}{6}, \frac{7\pi}{6}, \frac{11\pi}{6}$. Thus, the solutions in

$[0, 2\pi)$ are $x = \frac{\pi}{6}, \frac{\pi}{2}, \frac{5\pi}{6}, \frac{7\pi}{6}, \frac{3\pi}{2}, \frac{11\pi}{6}$.

45. $\tan x + \sec x = \sqrt{3} \Leftrightarrow \dfrac{\sin x}{\cos x} + \dfrac{1}{\cos x} = \sqrt{3} \Leftrightarrow \sin x + 1 = \sqrt{3}\cos x \Leftrightarrow \sqrt{3}\cos x - \sin x = 1 \Leftrightarrow \frac{\sqrt{3}}{2}\cos x - \frac{1}{2}\sin x = \frac{1}{2}$

$\Leftrightarrow \cos \frac{\pi}{6}\cos x - \sin \frac{\pi}{6}\sin x = \frac{1}{2} \Leftrightarrow \cos\left(x + \frac{\pi}{6}\right) = \frac{1}{2} \Leftrightarrow x + \frac{\pi}{6} = \frac{\pi}{3}, \frac{5\pi}{3} \Leftrightarrow x = \frac{\pi}{6}, \frac{3\pi}{2}$. However, $x = \frac{3\pi}{2}$ is

inadmissible because $\sec \frac{3\pi}{2}$ is undefined. Thus, the only solution in $[0, 2\pi)$ is $x = \frac{\pi}{6}$.

47. We graph $f(x) = \cos x$ and $g(x) = x^2 - 1$ in the viewing
rectangle $[0, 6.5]$ by $[-2, 2]$. The two functions intersect
at only one point, $x \approx 1.18$.

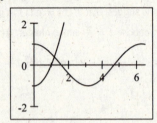

49. (a) $2000 = \dfrac{(400)^2 \sin^2 \theta}{64} \Leftrightarrow \sin^2 \theta = 0.8 \Leftrightarrow \sin \theta \approx 0.8944 \Leftrightarrow \theta \approx 63.4°$

(b) $\dfrac{(400)^2 \sin^2 \theta}{64} = 2500 \sin^2 \theta \le 2500$. Therefore it is impossible for the projectile to reach a height of 3000 ft.

(c) The function $M(\theta) = 2500 \sin^2 \theta$ is maximized when $\sin^2 \theta = 1$, so $\theta = 90°$. The projectile will travel the highest
when it is shot straight up.

51. Since $15°$ is in quadrant I, $\cos 15° = \sqrt{\dfrac{1 + \cos 30°}{2}} = \sqrt{\dfrac{2 + \sqrt{3}}{4}} = \frac{1}{2}\sqrt{2 + \sqrt{3}}$.

53. $\tan \frac{\pi}{8} = \dfrac{1 - \cos \frac{\pi}{4}}{\sin \frac{\pi}{4}} = \dfrac{1 - \frac{1}{\sqrt{2}}}{\frac{1}{\sqrt{2}}} = \left(1 - \frac{1}{\sqrt{2}}\right)\sqrt{2} = \sqrt{2} - 1$

55. $\sin 5° \cos 40° + \cos 5° \sin 40° = \sin(5° + 40°) = \sin 45° = \frac{1}{\sqrt{2}} = \frac{\sqrt{2}}{2}$

57. $\cos^2 \frac{\pi}{8} - \sin^2 \frac{\pi}{8} = \cos\left(2\left(\frac{\pi}{8}\right)\right) = \cos \frac{\pi}{4} = \frac{1}{\sqrt{2}} = \frac{\sqrt{2}}{2}$

59. We use a product-to-sum formula: $\cos 37.5° \cos 7.5° = \frac{1}{2}\left(\cos 45° + \cos 30°\right) = \frac{1}{2}\left(\frac{\sqrt{2}}{2} + \frac{\sqrt{3}}{2}\right) = \frac{1}{4}\left(\sqrt{2} + \sqrt{3}\right)$.

61. $\sin(x + y) = \sin x \cos y + \cos x \sin y = \frac{\sqrt{5}}{3} \cdot \frac{2\sqrt{2}}{3} + \frac{2}{3} \cdot \frac{1}{3} = \frac{2}{9}\left(1 + \sqrt{10}\right)$.

63. $\tan(x + y) = \dfrac{\tan x + \tan y}{1 - \tan x \tan y} = \dfrac{\frac{\sqrt{5}}{2} + \frac{\sqrt{2}}{4}}{1 - \left(\frac{\sqrt{5}}{2}\right)\left(\frac{\sqrt{2}}{4}\right)} = \dfrac{\frac{\sqrt{5}}{2} + \frac{\sqrt{2}}{4}}{1 - \left(\frac{\sqrt{5}}{2}\right)\left(\frac{\sqrt{2}}{4}\right)} \cdot \dfrac{8}{8} = \dfrac{2\left(2\sqrt{5} + \sqrt{2}\right)}{8 - \sqrt{10}} \cdot \dfrac{8 + \sqrt{10}}{8 + \sqrt{10}}$

$= \frac{2}{3}\left(\sqrt{2} + \sqrt{5}\right)$

65. $\cos \frac{y}{2} = \sqrt{\dfrac{1 + \cos y}{2}} = \sqrt{\dfrac{1 + \left(\frac{2\sqrt{2}}{3}\right)}{2}} = \sqrt{\dfrac{3 + 2\sqrt{2}}{6}}$ (since cosine is positive in quadrant I)

67. We sketch a triangle such that $\theta = \cos^{-1} \frac{3}{7}$. We see that $\tan \theta = \frac{2\sqrt{10}}{3}$, and the double-angle formula for tangent gives

$$\tan 2\theta = \frac{2\tan\theta}{1-\tan^2\theta} = \frac{2 \cdot \frac{2\sqrt{10}}{3}}{1-\left(\frac{2\sqrt{10}}{3}\right)^2} = \frac{\frac{4\sqrt{10}}{3}}{1-\frac{40}{9}} = -\frac{12\sqrt{10}}{31}.$$

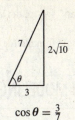

$$\cos\theta = \frac{3}{7}$$

69. The double-angle formula for tangent gives $\tan\left(2\tan^{-1}x\right) = \dfrac{2\tan\left(\tan^{-1}x\right)}{1-\tan^2\left(\tan^{-1}x\right)} = \dfrac{2x}{1-x^2}$.

71. (a) $\tan\theta = \dfrac{10}{x} \Leftrightarrow \theta = \tan^{-1}\left(\dfrac{10}{x}\right)$

(b) $\theta = \tan^{-1}\left(\dfrac{10}{x}\right)$, for $x > 0$. Since the road sign can first be seen when $\theta = 2°$,

we have $2° = \tan^{-1}\left(\dfrac{10}{x}\right) \Leftrightarrow x = \dfrac{10}{\tan 2°} \approx 286.4$ ft. Thus, the sign can first be

seen at a height of 286.4 ft.

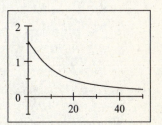

CHAPTER 7 TEST

1. (a) $\tan\theta\sin\theta + \cos\theta = \dfrac{\sin\theta}{\cos\theta}\sin\theta + \cos\theta = \dfrac{\sin^2\theta}{\cos\theta} + \dfrac{\cos^2\theta}{\cos\theta} = \dfrac{1}{\cos\theta} = \sec\theta$

(b) $\dfrac{\tan x}{1-\cos x} = \dfrac{\tan x}{1-\cos x} \cdot \dfrac{1+\cos x}{1+\cos x} = \dfrac{\tan x\,(1+\cos x)}{1-\cos^2 x} = \dfrac{\frac{\sin x}{\cos x}\,(1+\cos x)}{\sin^2 x} = \dfrac{1}{\sin x} \cdot \dfrac{1+\cos x}{\cos x} = \csc x\,(1+\sec x)$

(c) $\dfrac{2\tan x}{1+\tan^2 x} = \dfrac{2\tan x}{\sec^2 x} = \dfrac{2\sin x}{\cos x} \cdot \cos^2 x = 2\sin x\cos x = \sin 2x$

3. (a) $\sin 8° \cos 22° + \cos 8° \sin 22° = \sin\,(8° + 22°) = \sin 30° = \frac{1}{2}$

(b) $\sin 75° = \sin\,(45° + 30°) = \sin 45° \cos 30° + \cos 45° \sin 30° = \frac{\sqrt{2}}{2} \cdot \frac{\sqrt{3}}{2} + \frac{\sqrt{2}}{2} \cdot \frac{1}{2} = \frac{1}{4}\left(\sqrt{6} + \sqrt{2}\right)$

(c) $\sin\frac{\pi}{12} = \sin\left(\frac{\pi}{2}\right) = \sqrt{\dfrac{1-\cos\frac{\pi}{6}}{2}} = \sqrt{\dfrac{1-\frac{\sqrt{3}}{2}}{2}} = \sqrt{\dfrac{2-\sqrt{3}}{4}} = \frac{1}{2}\sqrt{2-\sqrt{3}}$

5. (a) $\sin 3x \cos 5x = \frac{1}{2}\left[\sin\,(3x+5x) + \sin\,(3x-5x)\right] = \frac{1}{2}\,(\sin 8x - \sin 2x)$

(b) $\sin 2x - \sin 5x = 2\cos\left(\dfrac{2x+5x}{2}\right)\sin\left(\dfrac{2x-5x}{2}\right) = -2\cos\dfrac{7x}{2}\sin\dfrac{3x}{2}$

7. (a) $3\sin\theta - 1 = 0 \Leftrightarrow 3\sin\theta = 1 \Leftrightarrow \sin\theta = \frac{1}{3} \Leftrightarrow \theta = \sin^{-1}\frac{1}{3} \approx 0.34$ or $\theta = \pi - \sin^{-1}\frac{1}{3} \approx 2.80$ on $[0, 2\pi)$.

(b) $(2\cos\theta - 1)\,(\sin\theta - 1) = 0 \Leftrightarrow \cos\theta = \frac{1}{2}$ or $\sin\theta = 1$. The first equation has solutions $\theta = \frac{\pi}{3} \approx 1.05$ and $\theta = \frac{5\pi}{3} \approx 5.24$ on $[0, 2\pi)$, while the second has the solution $\theta = \frac{\pi}{2} \approx 1.57$.

(c) $2\cos^2\theta + 5\cos\theta + 2 = 0 \Leftrightarrow (2\cos\theta + 1)\,(\cos\theta + 2) = 0 \Leftrightarrow \cos\theta = -\frac{1}{2}$ or $\cos\theta = -2$ (which is impossible). So in the interval $[0, 2\pi)$, the solutions are $\theta = \frac{2\pi}{3} \approx 2.09$, $\frac{4\pi}{3} \approx 4.19$.

(d) $\sin 2\theta - \cos\theta = 0 \Leftrightarrow 2\sin\theta\cos\theta - \cos\theta = 0 \Leftrightarrow \cos\theta\,(2\sin\theta - 1) = 0 \Leftrightarrow \cos\theta = 0$ or $\sin\theta = \frac{1}{2} \Leftrightarrow \theta = \frac{\pi}{2}, \frac{3\pi}{2}$ or $\theta = \frac{\pi}{6}, \frac{5\pi}{6}$. Therefore, the solutions in $[0, 2\pi)$ are $\theta = \frac{\pi}{6} \approx 0.52$, $\frac{\pi}{2} \approx 1.57$, $\frac{5\pi}{6} \approx 2.62$, $\frac{3\pi}{2} \approx 4.71$.

9. Let $u = \tan^{-1} \frac{9}{40}$ so $\tan u = \frac{9}{40}$. From the triangle, $\cos u = \frac{40}{41}$, so using a

double-angle formula for cosine, $\cos 2u = 2\cos^2 u - 1 = 2\left(\frac{40}{41}\right)^2 - 1 = \frac{1519}{1681}$.

FOCUS ON MODELING   Traveling and Standing Waves

1. (a) Substituting $x = 0$, we get $y(0, t) = 5\sin\left(2 \cdot 0 - \frac{\pi}{2}t\right) = 5\sin\left(-\frac{\pi}{2}t\right) = -5\sin\frac{\pi}{2}t$.

(b)

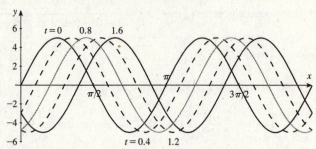

(c) We express the function in the standard form $y(x, t) = A\sin k(x - vt)$: $y(x, t) = 5\sin\left(2x - \frac{\pi}{2}t\right) = 5\sin 2\left(x - \frac{\pi}{4}t\right)$.

   Comparing this to the standard form, we see that the velocity of the wave is $v = \frac{\pi}{4}$.

3. From the graph, we see that the amplitude is $A = 2.7$ and the period is 9.2, so $k = \frac{2\pi}{9.2} \approx 0.68$. Since $v = 6$, we have

$kv = \frac{2\pi}{9.2} \cdot 6 \approx 4.10$, so the equation we seek is $y(x, t) = 2.7\sin(0.68x - 4.10t)$.

5. From the graphs, we see that the amplitude is $A = 0.6$. The nodes occur at $x = 0, 1, 2, 3$. Since $\sin \alpha x = 0$ when $\alpha x = k\pi$

(k any integer), we have $\alpha = \pi$. Then since the frequency is $\beta/2\pi$, we get $20 = \beta/2\pi \Leftrightarrow \beta = 40\pi$. Thus, an equation for

this model is $f(x, t) = 0.6\sin \pi x \cos 40\pi t$.

7. (a) The first standing wave has $\alpha = 1$, the second has $\alpha = 2$, the third has $\alpha = 3$, and the fourth has $\alpha = 4$.

(b) α is equal to the number of nodes minus 1. The first string has two nodes and $\alpha = 1$; the second string has three nodes and $\alpha = 2$, and so forth.

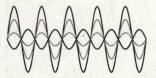 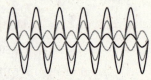

(c) Since the frequency is $\beta/2\pi$, we have $440 = \beta/2\pi \Leftrightarrow \beta = 880\pi$.

(d) The first standing wave has equation $y = \sin x \cos 880\pi t$, the second has equation $y = \sin 2x \cos 880\pi t$, the third has equation $y = \sin 3x \cos 880\pi t$, and the fourth has equation $y = \sin 4x \cos 880\pi t$.

CUMULATIVE REVIEW TEST: CHAPTERS 5, 6, and 7

1. Because P lies on the unit circle in the second quadrant and has y-coordinate $\frac{\sqrt{5}}{3}$, its x-coordinate is $-\sqrt{1 - \left(\frac{\sqrt{5}}{3}\right)^2} = -\frac{2}{3}$.

(a) $\sin t = y = \frac{\sqrt{5}}{3}$ **(b)** $\cos t = x = -\frac{2}{3}$ **(c)** $\tan t = \frac{y}{x} = -\frac{\sqrt{5}}{2}$ **(d)** $\csc t = \frac{1}{y} = \frac{3\sqrt{5}}{5}$

3. (a) The reference number for $\frac{7\pi}{6}$ is $\frac{\pi}{6}$, and cosine is negative in the third quadrant. Thus, $\cos\frac{7\pi}{6} = -\cos\frac{\pi}{6} = -\frac{\sqrt{3}}{2}$.

(b) The reference angle for $135°$ is $45°$, and tangent is negative in the second quadrant. Thus, $\tan 135° = -\tan 45° = -1$.

(c) The reference angle for $240°$ is $60°$, and sine (and thus cosecant) is negative in the third quadrant. Thus,

$\csc 240° = -\frac{1}{\sin 60°} = -\frac{2\sqrt{3}}{3}$.

(d) The reference number for $-\frac{9\pi}{2}$ is $\frac{\pi}{2}$, and sine is negative in the fourth quadrant. Thus, $\sin\left(-\frac{9\pi}{2}\right) = -\sin\frac{\pi}{2} = -1$.

5. (a) $f(x) = -2\sin\left(2x - \frac{\pi}{2}\right) = -2\sin 2\left(x - \frac{\pi}{4}\right)$

$\qquad = a\sin k(x - b)$

with $a = -2$, $k = 2$, and $b = \frac{\pi}{4}$.

Therefore, f has amplitude $|a| = |-2| = 2$, period

$2\pi/k = 2\pi/2 = \pi$, and phase shift $b = \frac{\pi}{4}$.

(b)

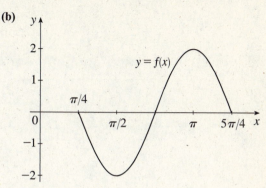

7. (a) We want to model the motion of the point P using a function h of the form $h(t) = a + b\cos kt$, where h is measured in cm and t in minutes. Because the center of the ferris wheel is 45 cm above the ground, we take $a = 45$. The motor turns the wheel at 4 rotations per minute, so the period of h is $\frac{2\pi}{k} = \frac{1}{4}$, and thus $k = \frac{\pi}{2}$. Because the amplitude of h is 40 cm and initially P is at its lowest point, we have $h(0) = 5 = 45 + b\cos 0$, so $b = -40$. Therefore, an equation that models the motion of P is

$h(t) = 45 - 40\cos\frac{\pi}{2}t$.

(b)

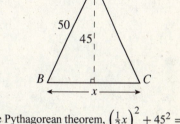

By the Pythagorean theorem, $\left(\frac{1}{2}x\right)^2 + 45^2 = 50^2$

$\Leftrightarrow \frac{1}{4}x^2 = 475 \Leftrightarrow x^2 = 1900 \Leftrightarrow$

$x = \sqrt{1900} \approx 43.6$ cm.

9. (a) $\dfrac{\sec\theta - 1}{\tan\theta} = \dfrac{\tan\theta}{\sec\theta + 1} \Leftrightarrow \tan^2\theta = (\sec\theta - 1)(\sec\theta + 1) = \sec^2\theta - 1$, a known identity.

(b) RHS $= 1 - \cos 4\theta = 1 - \left(1 - 2\sin^2 2\theta\right) = 2\sin^2 2\theta = 2(2\sin\theta\cos\theta)^2 = 8\sin^2\theta\cos^2\theta =$ LHS

11. (a) $f(x) = \cos^{-1}x$ has domain $[-1, 1]$ and range $[0, \pi]$.

(b) $\cos\frac{7\pi}{6} = -\frac{\sqrt{3}}{2}$, and the angle between 0 and π whose cosine is $-\frac{\sqrt{3}}{2}$ is $\frac{5\pi}{6}$. Therefore, $\cos^{-1}\left(\cos\frac{7\pi}{6}\right) = \frac{5\pi}{6}$.

(c)

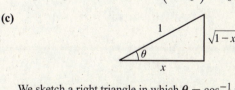

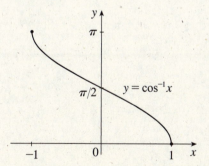

We sketch a right triangle in which $\theta = \cos^{-1}x$ and see that

$\tan\theta = \tan\left(\cos^{-1}x\right) = \dfrac{\sqrt{1-x^2}}{x}$.

8 POLAR COORDINATES AND PARAMETRIC EQUATIONS

8.1 POLAR COORDINATES

1. We can describe the location of a point in the plane using different *coordinate* systems. The point P shown in the figure has rectangular coordinates $(1, 1)$ and polar coordinates $\left(\sqrt{2}, \frac{\pi}{4}\right)$.

3.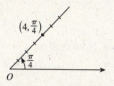

5.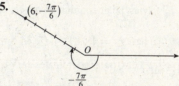

7.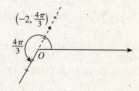

Answers to Exercises 9–14 will vary.

9. $\left(3, \frac{\pi}{2}\right)$ has polar coordinates $\left(3, \frac{5\pi}{2}\right)$ or $\left(-3, \frac{3\pi}{2}\right)$

11. $\left(-1, \frac{7\pi}{6}\right)$ has polar coordinates $\left(1, \frac{\pi}{6}\right)$ or $\left(-1, -\frac{5\pi}{6}\right)$.

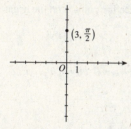

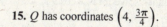

13. $(-5, 0)$ has polar coordinates $(5, \pi)$ or $(-5, 2\pi)$.

15. Q has coordinates $\left(4, \frac{3\pi}{4}\right)$.

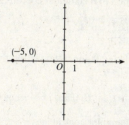

17. Q has coordinates $\left(-4, -\frac{\pi}{4}\right) = \left(4, \frac{3\pi}{4}\right)$.

19. P has coordinates $\left(4, -\frac{23\pi}{4}\right) = \left(4, \frac{\pi}{4}\right)$.

21. P has coordinates $\left(-4, \frac{101\pi}{4}\right) = \left(-4, \frac{5\pi}{4}\right) = \left(4, \frac{\pi}{4}\right)$.

23. $P = (-3, 3)$ in rectangular coordinates, so $r^2 = x^2 + y^2 = (-3)^2 + 3^2 = 18$ and we can take $r = 3\sqrt{2}$. $\tan \theta = \frac{y}{x} = \frac{3}{-3} = -1$, so since P is in quadrant 2 we take $\theta = \frac{3\pi}{4}$. Thus, polar coordinates for P are $\left(3\sqrt{2}, \frac{3\pi}{4}\right)$.

25. Here $r = 5$ and $\theta = -\frac{2\pi}{3}$, so $x = r \cos \theta = 5 \cos \left(-\frac{2\pi}{3}\right) = -\frac{5}{2}$ and $y = r \sin \theta = 5 \sin \left(-\frac{2\pi}{3}\right) = -\frac{5\sqrt{3}}{2}$. R has rectangular coordinates $\left(-\frac{5}{2}, -\frac{5\sqrt{3}}{2}\right)$.

27. $(r, \theta) = \left(4, \frac{\pi}{6}\right)$. So $x = r \cos \theta = 4 \cos \frac{\pi}{6} = 4 \cdot \frac{\sqrt{3}}{2} = 2\sqrt{3}$ and $y = r \sin \theta = 4 \sin \frac{\pi}{6} = 4 \cdot \frac{1}{2} = 2$. Thus, the rectangular coordinates are $\left(2\sqrt{3}, 2\right)$.

29. $(r, \theta) = \left(\sqrt{2}, -\frac{\pi}{4}\right)$. So $x = r\cos\theta = \sqrt{2}\cos\left(-\frac{\pi}{4}\right) = \sqrt{2} \cdot \frac{1}{\sqrt{2}} = 1$, and

$y = r\sin\theta = \sqrt{2}\sin\left(-\frac{\pi}{4}\right) = \sqrt{2}\left(-\frac{1}{\sqrt{2}}\right) = -1$. Thus, the rectangular coordinates are $(1, -1)$.

31. $(r, \theta) = (5, 5\pi)$. So $x = r\cos\theta = 5\cos 5\pi = -5$, and $y = r\sin\theta = 5\sin 5\pi = 0$. Thus, the rectangular coordinates are $(-5, 0)$.

33. $(r, \theta) = \left(6\sqrt{2}, \frac{11\pi}{6}\right)$. So $x = r\cos\theta = 6\sqrt{2}\cos\frac{11\pi}{6} = 3\sqrt{6}$ and $y = r\sin\theta = 6\sqrt{2}\sin\frac{11\pi}{6} = -3\sqrt{2}$. Thus, the rectangular coordinates are $\left(3\sqrt{6}, -3\sqrt{2}\right)$.

35. $(x, y) = (-1, 1)$. Since $r^2 = x^2 + y^2$, we have $r^2 = (-1)^2 + 1^2 = 2$, so $r = \sqrt{2}$. Now $\tan\theta = \frac{y}{x} = \frac{1}{-1} = -1$, so, since the point is in the second quadrant, $\theta = \frac{3\pi}{4}$. Thus, polar coordinates are $\left(\sqrt{2}, \frac{3\pi}{4}\right)$.

37. $(x, y) = \left(\sqrt{8}, \sqrt{8}\right)$. Since $r^2 = x^2 + y^2$, we have $r^2 = \left(\sqrt{8}\right)^2 + \left(\sqrt{8}\right)^2 = 16$, so $r = 4$. Now $\tan\theta = \frac{y}{x} = \frac{\sqrt{8}}{\sqrt{8}} = 1$, so, since the point is in the first quadrant, $\theta = \frac{\pi}{4}$. Thus, polar coordinates are $\left(4, \frac{\pi}{4}\right)$.

39. $(x, y) = (3, 4)$. Since $r^2 = x^2 + y^2$, we have $r^2 = 3^2 + 4^2 = 25$, so $r = 5$. Now $\tan\theta = \frac{y}{x} = \frac{4}{3}$, so, since the point is in the first quadrant, $\theta = \tan^{-1}\frac{4}{3}$. Thus, polar coordinates are $\left(5, \tan^{-1}\frac{4}{3}\right)$.

41. $(x, y) = (-6, 0)$. $r^2 = \left(-6^2\right) = 36$, so $r = 6$. Now $\tan\theta = \frac{y}{x} = 0$, so since the point is on the negative x-axis, $\theta = \pi$. Thus, polar coordinates are $(6, \pi)$.

43. $x = y \Leftrightarrow r\cos\theta = r\sin\theta \Leftrightarrow \tan\theta = 1$, and so $\theta = \frac{\pi}{4}$.

45. $y = x^2$. We substitute and then solve for r: $r\sin\theta = (r\cos\theta)^2 = r^2\cos^2\theta \Leftrightarrow \sin\theta = r\cos^2\theta \Leftrightarrow$

$r = \dfrac{\sin\theta}{\cos^2\theta} = \tan\theta\sec\theta$.

47. $x = 4$. We substitute and then solve for r: $r\cos\theta = 4 \Leftrightarrow r = \dfrac{4}{\cos\theta} = 4\sec\theta$.

49. $r = 7$. But $r^2 = x^2 + y^2$, so $x^2 + y^2 = r^2 = 49$. Hence, the equivalent equation in rectangular coordinates is $x^2 + y^2 = 49$.

51. $\theta = -\frac{\pi}{2} \Rightarrow \cos\theta = 0$, so an equivalent equation in rectangular coordinates is $x = 0$.

53. $r\cos\theta = 6$. But $x = r\cos\theta$, and so $x = 6$ is an equivalent rectangular equation.

55. $r = 4\sin\theta \Leftrightarrow r^2 = 4r\sin\theta$. Thus, $x^2 + y^2 = 4y$ is an equivalent rectangular equation. Completing the square, it can be written as $x^2 + (y - 2)^2 = 4$.

57. $r = 1 + \cos\theta$. If we multiply both sides of this equation by r we get $r^2 = r + r\cos\theta$. Thus $r^2 - r\cos\theta = r$, and squaring both sides gives $\left(r^2 - r\cos\theta\right)^2 = r^2$, or $\left(x^2 + y^2 - x\right)^2 = x^2 + y^2$ in rectangular coordinates.

59. $r = 1 + 2\sin\theta$. If we multiply both sides of this equation by r we get $r^2 = r + 2r\sin\theta$. Thus $r^2 - 2r\sin\theta = r$, and squaring both sides gives $\left(r^2 - 2r\sin\theta\right)^2 = r^2$, or $\left(x^2 + y^2 - 2y\right)^2 = x^2 + y^2$ in rectangular coordinates.

61. $r = \dfrac{1}{\sin\theta - \cos\theta} \Rightarrow r(\sin\theta - \cos\theta) = 1 \Leftrightarrow r\sin\theta - r\cos\theta = 1$, and since $r\cos\theta = x$ and $r\sin\theta = y$, we get $y - x = 1$.

63. $r = \dfrac{4}{1 + 2\sin\theta} \Leftrightarrow r(1 + 2\sin\theta) = 4 \Leftrightarrow r + 2r\sin\theta = 4$. Thus $r = 4 - 2r\sin\theta$. Squaring both sides, we get

$r^2 = (4 - 2r\sin\theta)^2$. Substituting, $x^2 + y^2 = (4 - 2y)^2 \Leftrightarrow x^2 + y^2 = 16 - 16y + 4y^2 \Leftrightarrow x^2 - 3y^2 + 16y - 16 = 0$.

65. $r^2 = \tan\theta$. Substituting $r^2 = x^2 + y^2$ and $\tan\theta = \frac{y}{x}$, we get $x^2 + y^2 = \frac{y}{x}$.

67. $\sec\theta = 2 \Leftrightarrow \cos\theta = \frac{1}{2} \Leftrightarrow \theta = \pm\frac{\pi}{3} \Leftrightarrow \tan\theta = \pm\sqrt{3} \Leftrightarrow \frac{y}{x} = \pm\sqrt{3} \Leftrightarrow y = \pm\sqrt{3}x$.

69. (a) In rectangular coordinates, the points (r_1, θ_1) and (r_2, θ_2) are $(x_1, y_1) = (r_1 \cos \theta_1, r_1 \sin \theta_1)$ and $(x_2, y_2) = (r_2 \cos \theta_2, r_2 \sin \theta_2)$. Then, the distance between the points is

$$D = \sqrt{(x_1 - x_2)^2 + (y_1 - y_2)^2} = \sqrt{(r_1 \cos \theta_1 - r_2 \cos \theta_2)^2 + (r_1 \sin \theta_1 - r_2 \sin \theta_2)^2}$$

$$= \sqrt{r_1^2 \left(\cos^2 \theta_1 + \sin^2 \theta_1\right) + r_2^2 \left(\cos^2 \theta_2 + \sin^2 \theta_2\right) - 2r_1 r_2 \left(\cos \theta_1 \cos \theta_2 + \sin \theta_1 \sin \theta_2\right)}$$

$$= \sqrt{r_1^2 + r_2^2 - 2r_1 r_2 \cos \left(\theta_2 - \theta_1\right)}$$

(b) The distance between the points $\left(3, \frac{3\pi}{4}\right)$ and $\left(-1, \frac{7\pi}{6}\right)$ is

$$D = \sqrt{3^2 + (-1)^2 - 2\,(3)\,(-1)\cos\left(\frac{7\pi}{6} - \frac{3\pi}{4}\right)} = \sqrt{9 + 1 + 6\cos\frac{5\pi}{12}} \approx 3.40$$

8.2 GRAPHS OF POLAR EQUATIONS

1. To plot points in polar coordinates we use a grid consisting of *circles* centered at the pole and *rays* emanating from the pole.

3. VI   **5.** II   **7.** I

9. Polar axis: $2 - \sin(-\theta) = 2 + \sin \theta \neq r$, so the graph is not symmetric about the polar axis.

Pole: $2 - \sin(\theta + \pi) = 2 - (\sin \pi \cos \theta + \cos \pi \sin \theta) = 2 - (-\sin \theta) = 2 + \sin \theta \neq r$, so the graph is not symmetric about the pole.

Line $\theta = \frac{\pi}{2}$: $2 - \sin(\pi - \theta) = 2 - (\sin \pi \cos \theta - \cos \pi \sin \theta) = 2 - \sin \theta = r$, so the graph is symmetric about $\theta = \frac{\pi}{2}$.

11. Polar axis: $3 \sec(-\theta) = 3 \sec \theta = r$, so the graph is symmetric about the polar axis.

Pole: $3 \sec(\theta + \pi) = \dfrac{3}{\cos(\theta + \pi)} = \dfrac{1}{\cos \pi \cos \theta - \sin \pi \sin \theta} = \dfrac{3}{-\cos \theta} = -3 \sec \theta \neq r$, so the graph is not symmetric about the pole.

Line $\theta = \frac{\pi}{2}$: $3 \sec(\pi - \theta) = \dfrac{3}{\cos(\pi - \theta)} = \dfrac{1}{\cos \pi \cos \theta + \sin \pi \sin \theta} = \dfrac{3}{-\cos \theta} = -3 \sec \theta \neq r$, so the graph is not symmetric about $\theta = \frac{\pi}{2}$.

13. Polar axis: $\dfrac{4}{3 - 2\sin(-\theta)} = \dfrac{4}{3 + 2\sin \theta} \neq r$, so the graph is not symmetric about the polar axis.

Pole: $\dfrac{4}{3 - 2\sin(\theta + \pi)} = \dfrac{4}{3 - 2(\sin \pi \cos \theta + \cos \pi \sin \theta)} = \dfrac{4}{3 - 2(-\sin \theta)} = \dfrac{4}{3 + 2\sin \theta} \neq r$, so the graph is not symmetric about the pole.

Line $\theta = \frac{\pi}{2}$: $\dfrac{4}{3 - 2\sin(\pi - \theta)} = \dfrac{4}{3 - 2(\sin \pi \cos \theta - \cos \pi \sin \theta)} = \dfrac{4}{3 - 2\sin \theta} = r$, so the graph is symmetric about $\theta = \frac{\pi}{2}$.

15. Polar axis: $4 \cos 2(-\theta) = 4 \cos 2\theta = r^2$, so the graph is symmetric about the polar axis.

Pole: $(-r)^2 = r^2$, so the graph is symmetric about the pole.

Line $\theta = \frac{\pi}{2}$: $4 \cos 2(\pi - \theta) = 4 \cos(2\pi - 2\theta) = 4 \cos(-2\theta) = 4 \cos 2\theta = r^2$, so the graph is symmetric about $\theta = \frac{\pi}{2}$.

17. $r = 2 \Rightarrow r^2 = 4 \Rightarrow x^2 + y^2 = 4$ is an equation of a circle with radius 2 centered at the origin.

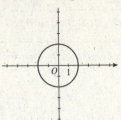

19. $\theta = -\frac{\pi}{2} \Rightarrow \cos\theta = 0 \Rightarrow x = 0$ is an equation of a vertical line.

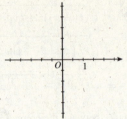

21. $r = 6\sin\theta \Rightarrow r^2 = 6r\sin\theta \Rightarrow x^2 + y^2 = 6y \Rightarrow$ $x^2 + (y-3)^2 = 9$, a circle of radius 3 centered at $(0,3)$.

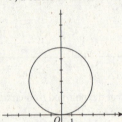

23. $r = -2\cos\theta$. Circle.

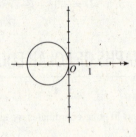

25. $r = 2 - 2\cos\theta$. Cardioid.

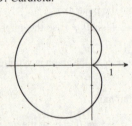

27. $r = -3(1 + \sin\theta)$. Cardioid.

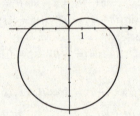

29. $r = \sin 2\theta$

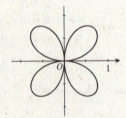

31. $r = -\cos 5\theta$

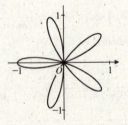

33. $r = \sqrt{3} - 2\sin\theta$

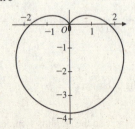

35. $r = \sqrt{3} + \cos\theta$

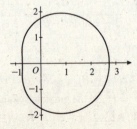

37. $r^2 = \cos 2\theta$

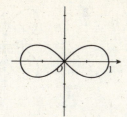

39. $r = \theta, \theta \geq 0$

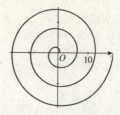

41. $r = 2 + \sec\theta$

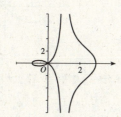

43. $r = \cos\left(\dfrac{\theta}{2}\right), \theta \in [0, 4\pi]$

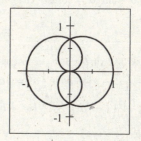

45. $r = 1 + 2\sin\left(\dfrac{\theta}{2}\right), \theta \in [0, 4\pi]$

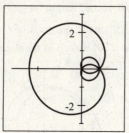

47. $r = 1 + \sin n\theta$. The number of loops is n.

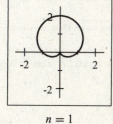 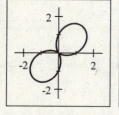 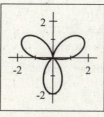 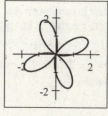 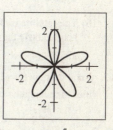

$n = 1$ $n = 2$ $n = 3$ $n = 4$ $n = 5$

49. The graph of $r = \sin\left(\dfrac{\theta}{2}\right)$ is IV, since the graph must contain the points $(0, 0)$, $\left(\dfrac{1}{\sqrt{2}}, \dfrac{\pi}{2}\right)$, $(1, \pi)$, and so on.

51. The graph of $r = \theta \sin\theta$ is III, since for $\theta = \dfrac{\pi}{2}, \dfrac{5\pi}{2}, \dfrac{7\pi}{2}, \ldots$ the values of r are also $\dfrac{\pi}{2}, \dfrac{5\pi}{2}, \dfrac{7\pi}{2}, \ldots$ Thus the graph must cross the vertical axis at an infinite number of points.

53. $\left(x^2 + y^2\right)^3 = 4x^2y^2 \Leftrightarrow \left(r^2\right)^3 = 4\,(r\cos\theta)^2\,(r\sin\theta)^2 \Leftrightarrow$

$r^6 = 4r^4 \cos^2\theta \sin^2\theta \Leftrightarrow r^2 = 4\cos^2\theta \sin^2\theta \Leftrightarrow r = 2\cos\theta\sin\theta = \sin 2\theta$. The equation is $r = \sin 2\theta$, a rose.

55. $\left(x^2 + y^2\right)^2 = x^2 - y^2 \Leftrightarrow \left(r^2\right)^2 = (r\cos\theta)^2 - (r\sin\theta)^2 \Leftrightarrow$

$r^4 = r^2 \cos^2\theta - r^2 \sin^2\theta \Leftrightarrow r^4 = r^2\left(\cos^2\theta - \sin^2\theta\right) \Leftrightarrow$

$r^2 = \cos^2\theta - \sin^2\theta = \cos 2\theta$. The graph is $r^2 = \cos 2\theta$, a leminiscate.

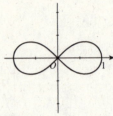

57. $r = a\cos\theta + b\sin\theta \Leftrightarrow r^2 = ar\cos\theta + br\sin\theta \Leftrightarrow x^2 + y^2 = ax + by \Leftrightarrow x^2 - ax + y^2 - by = 0 \Leftrightarrow$

$x^2 - ax + \frac{1}{4}a^2 + y^2 - by + \frac{1}{4}b^2 = \frac{1}{4}a^2 + \frac{1}{4}b^2 \quad \Leftrightarrow$

$\left(x - \frac{1}{2}a\right)^2 + \left(y - \frac{1}{2}b\right)^2 = \frac{1}{4}\left(a^2 + b^2\right)$. Thus, in rectangular coordinates the center is $\left(\frac{1}{2}a, \frac{1}{2}b\right)$ and the radius is

$\frac{1}{2}\sqrt{a^2 + b^2}$.

59. (a)

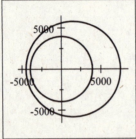

At $\theta = 0$, the satellite is at the "rightmost" point in its orbit, $(5625, 0)$. As θ increases, it travels counterclockwise. Note that it is moving fastest when $\theta = \pi$.

(b) The satellite is closest to earth when $\theta = \pi$. Its height above the earth's surface at this point is

$22500/\,(4 - \cos\pi) - 3960 = 4500 - 3960 = 540$ mi.

61. The graphs of $r = 1 + \sin\left(\theta - \frac{\pi}{6}\right)$ and

$r = 1 + \sin\left(\theta - \frac{\pi}{3}\right)$ have the same shape as

$r = 1 + \sin\theta$, rotated through angles of $\frac{\pi}{6}$ and $\frac{\pi}{3}$,

respectively. Similarly, the graph of $r = f\left(\theta - \alpha\right)$ is the

graph of $r = f\left(\theta\right)$ rotated by the angle α.

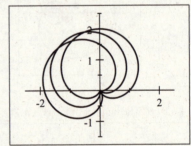

63. $y = 2 \Leftrightarrow r\sin\theta = 2 \Leftrightarrow r = 2\csc\theta$. The rectangular coordinate system gives the simpler equation here. It is easier to study lines in rectangular coordinates.

8.3   POLAR FORM OF COMPLEX NUMBERS; DE MOIVRE'S THEOREM

1. A complex number $z = a + bi$ has two parts: a is the *real* part and b is the *imaginary* part. To graph $a + bi$ we graph the ordered pair (a, b) in the complex plane.

3. (a) The complex number $z = -1 + i$ in polar form is $z = \sqrt{2}\left(\cos\frac{3\pi}{4} + i\sin\frac{3\pi}{4}\right)$. The complex number

$z = 2\left(\cos\frac{\pi}{6} + i\sin\frac{\pi}{6}\right)$ in rectangular form is $z = \sqrt{3} + i$.

(b) The complex number z can be expressed in rectangular form as $1 + i$ or in polar form as $\sqrt{2}\left(\cos\frac{\pi}{4} + i\sin\frac{\pi}{4}\right)$.

5. $|4i| = \sqrt{0^2 + 4^2} = 4$

7. $|-2| = \sqrt{4 + 0} = 2$

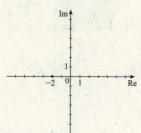

9. $|5 + 2i| = \sqrt{5^2 + 2^2} = \sqrt{29}$

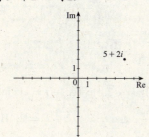

11. $\left|\sqrt{3} + i\right| = \sqrt{3 + 1} = 2$

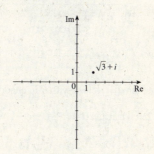

13. $\left|\dfrac{3 + 4i}{5}\right| = \sqrt{\dfrac{9}{25} + \dfrac{16}{25}} = 1$

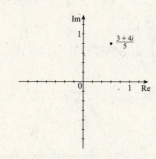

15. $z = 1 + i,\ 2z = 2 + 2i,\ -z = -1 - i,\ \frac{1}{2}z = \frac{1}{2} + \frac{1}{2}i$

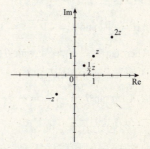

17. $z = 8 + 2i,\ \bar{z} = 8 - 2i$

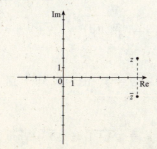

19. $z_1 = 2 - i,\ z_2 = 2 + i,\ z_1 + z_2 = 2 - i + 2 + i = 4,$

$z_1 z_2 = (2 - i)(2 + i) = 4 - i^2 = 5$

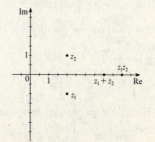

21. $\{z = a + bi \mid a \leq 0, b \geq 0\}$

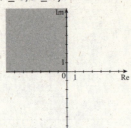

23. $\{z \mid |z| = 3\}$

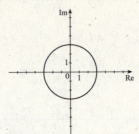

25. $\{z \mid |z| < 2\}$

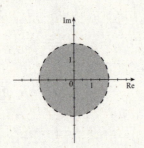

27. $\{z = a + bi \mid a + b < 2\}$

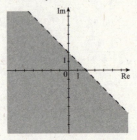

29. $1 + i$. Then $\tan \theta = \frac{1}{1} = 1$ with θ in quadrant I $\Rightarrow \theta = \frac{\pi}{4}$, and $r = \sqrt{1^2 + 1^2} = \sqrt{2}$. Hence, $1 + i = \sqrt{2}\left(\cos \frac{\pi}{4} + i \sin \frac{\pi}{4}\right)$.

31. $\sqrt{2} - \sqrt{2}i$. Then $\tan \theta = \frac{\sqrt{2}}{\sqrt{2}} = -1$ with θ in quadrant IV $\Rightarrow \theta = \frac{7\pi}{4}$, and $r = \sqrt{2 + 2} = 2$. Hence,

$\sqrt{2} - \sqrt{2}i = 2\left(\cos \frac{7\pi}{4} + i \sin \frac{7\pi}{4}\right)$.

33. $2\sqrt{3} - 2i$. Then $\tan \theta = \frac{-2}{2\sqrt{3}} = -\frac{1}{\sqrt{3}}$ with θ in quadrant IV $\Rightarrow \theta = \frac{11\pi}{6}$, and $r = \sqrt{12 + 4} = 4$. Hence,

$2\sqrt{3} - 2i = 4\left(\cos \frac{11\pi}{6} + i \sin \frac{11\pi}{6}\right)$.

35. $-3i$. Then $\theta = \frac{3\pi}{2}$, and $r = \sqrt{0 + 9} = 3$. Hence, $3i = 3\left(\cos \frac{3\pi}{2} + i \sin \frac{3\pi}{2}\right)$.

37. $5 + 5i$. Then $\tan \theta = \frac{5}{5} = 1$ with θ in quadrant I $\Rightarrow \theta = \frac{\pi}{4}$, and $r = \sqrt{25 + 25} = 5\sqrt{2}$. Hence,

$5 + 5i = 5\sqrt{2}\left(\cos \frac{\pi}{4} + i \sin \frac{\pi}{4}\right)$.

39. $4\sqrt{3} - 4i$. Then $\tan \theta = \frac{-4}{4\sqrt{3}} = -\frac{1}{\sqrt{3}}$ with θ in quadrant IV $\Rightarrow \theta = \frac{11\pi}{6}$, and $r = \sqrt{48 + 16} = 8$. Hence,

$4\sqrt{3} - 4i = 8\left(\cos \frac{11\pi}{6} + i \sin \frac{11\pi}{6}\right)$.

41. -20. Then $\theta = \pi$, and $r = 20$. Hence, $-20 = 20\left(\cos \pi + i \sin \pi\right)$.

43. $3 + 4i$. Then $\tan \theta = \frac{4}{3}$ with θ in quadrant I $\Rightarrow \theta = \tan^{-1} \frac{4}{3}$, and $r = \sqrt{9 + 16} = 5$. Hence,

$3 + 4i = 5\left[\cos\left(\tan^{-1} \frac{4}{3}\right) + i \sin\left(\tan^{-1} \frac{4}{3}\right)\right]$.

45. $3i(1 + i) = -3 + 3i$. Then $\tan \theta = \frac{3}{-3} = -1$ with θ in quadrant II $\Rightarrow \theta = \frac{3\pi}{4}$, and $r = \sqrt{9 + 9} = 3\sqrt{2}$. Hence,

$3i(1 + i) = 3\sqrt{2}\left(\cos \frac{3\pi}{4} + i \sin \frac{3\pi}{4}\right)$.

47. $4\left(\sqrt{3} + i\right) = 4\sqrt{3} + 4i$. Then $\tan \theta = \frac{4}{4\sqrt{3}} = \frac{1}{\sqrt{3}}$ with θ in quadrant I $\Rightarrow \theta = \frac{\pi}{6}$, and $r = \sqrt{48 + 16} = 8$. Hence,

$4\left(\sqrt{3} + i\right) = 8\left(\cos \frac{\pi}{6} + i \sin \frac{\pi}{6}\right)$.

49. $2 + i$. Then $\tan \theta = \frac{1}{2}$ with θ in quadrant I $\Rightarrow \theta = \tan^{-1} \frac{1}{2}$, and $r = \sqrt{4 + 1} = \sqrt{5}$. Hence,

$2 + i = \sqrt{5}\left[\cos\left(\tan^{-1} \frac{1}{2}\right) + i \sin\left(\tan^{-1} \frac{1}{2}\right)\right]$.

51. $\sqrt{2} + \sqrt{2}i$. Then $\tan\theta = \frac{\sqrt{2}}{\sqrt{2}} = 1$ with θ in quadrant I $\Rightarrow \theta = \frac{\pi}{4}$, and $r = \sqrt{2+2} = 2$. Hence,
$2 + \sqrt{2}i = 2\left(\cos\frac{\pi}{4} + i\sin\frac{\pi}{4}\right)$.

53. $z_1 = \cos\pi + i\sin\pi$, $z_2 = \cos\frac{\pi}{3} + i\sin\frac{\pi}{3}$, $z_1 z_2 = \cos\left(\pi + \frac{\pi}{3}\right) + i\sin\left(\pi + \frac{\pi}{3}\right) = \cos\frac{4\pi}{3} + i\sin\frac{4\pi}{3}$,
$z_1/z_2 = \cos\left(\pi - \frac{\pi}{3}\right) + i\sin\left(\pi - \frac{\pi}{3}\right) = \cos\frac{2\pi}{3} + i\sin\frac{2\pi}{3}$

55. $z_1 = 3\left(\cos\frac{\pi}{6} + i\sin\frac{\pi}{6}\right)$, $z_2 = 5\left(\cos\frac{4\pi}{3} + i\sin\frac{4\pi}{3}\right)$,
$z_1 z_2 = 3 \cdot 5\left[\cos\left(\frac{\pi}{6} + \frac{4\pi}{3}\right) + i\sin\left(\frac{\pi}{6} + \frac{4\pi}{3}\right)\right] = 15\left(\cos\frac{9\pi}{6} + i\sin\frac{9\pi}{6}\right) = 15\left(\cos\frac{3\pi}{2} + i\sin\frac{3\pi}{2}\right)$,
$z_1/z_2 = \frac{3}{5}\left[\cos\left(\frac{\pi}{6} - \frac{4\pi}{3}\right) + i\sin\left(\frac{\pi}{6} - \frac{4\pi}{3}\right)\right] = \frac{3}{5}\left[\cos\left(-\frac{7\pi}{6}\right) + i\sin\left(-\frac{7\pi}{6}\right)\right] = \frac{3}{5}\left[\cos\left(\frac{7\pi}{6}\right) - i\sin\left(\frac{7\pi}{6}\right)\right]$

57. $z_1 = 4(\cos 120° + i\sin 120°)$, $z_2 = 2(\cos 30° + i\sin 30°)$,
$z_1 z_2 = 4 \cdot 2\left[\cos(120° + 30°) + i\sin(120° + 30°)\right] = 8(\cos 150° + i\sin 150°)$,
$z_1/z_2 = \frac{4}{2}\left[\cos(120° - 30°) + i\sin(120° - 30°)\right] = 2(\cos 90° + i\sin 90°)$

59. $z_1 = 4(\cos 200° + i\sin 200°)$, $z_2 = 25(\cos 150° + i\sin 150°)$,
$z_1 z_2 = 4 \cdot 25\left[\cos(200° + 150°) + i\sin(200° + 150°)\right] = 100(\cos 350° + i\sin 350°)$,
$z_1/z_2 = \frac{4}{25}\left[\cos(200° - 150°) + i\sin(200° - 150°)\right] = \frac{4}{25}(\cos 50° + i\sin 50°)$

61. $z_1 = \sqrt{3} + i$, so $\tan\theta_1 = \frac{1}{\sqrt{3}}$ with θ_1 in quadrant I $\Rightarrow \theta_1 = \frac{\pi}{6}$, and $r_1 = \sqrt{3+1} = 2$.
$z_2 = 1 + \sqrt{3}i$, so $\tan\theta_2 = \sqrt{3}$ with θ_2 in quadrant I $\Rightarrow \theta_2 = \frac{\pi}{3}$, and $r_1 = \sqrt{1+3} = 2$.
Hence, $z_1 = 2\left(\cos\frac{\pi}{6} + i\sin\frac{\pi}{6}\right)$ and $z_2 = 2\left(\cos\frac{\pi}{3} + i\sin\frac{\pi}{3}\right)$.
Thus, $z_1 z_2 = 2 \cdot 2\left[\cos\left(\frac{\pi}{6} + \frac{\pi}{3}\right) + i\sin\left(\frac{\pi}{6} + \frac{\pi}{3}\right)\right] = 4\left(\cos\frac{\pi}{2} + i\sin\frac{\pi}{2}\right)$,
$z_1/z_2 = \frac{2}{2}\left[\cos\left(\frac{\pi}{6} - \frac{\pi}{3}\right) + i\sin\left(\frac{\pi}{6} - \frac{\pi}{3}\right)\right] = \cos\left(-\frac{\pi}{6}\right) + i\sin\left(-\frac{\pi}{6}\right) = \cos\frac{\pi}{6} - i\sin\frac{\pi}{6}$, and
$1/z_1 = \frac{1}{2}\left[\cos\left(-\frac{\pi}{6}\right) + i\sin\left(-\frac{\pi}{6}\right)\right] = \frac{1}{2}\left(\cos\frac{\pi}{6} - i\sin\frac{\pi}{6}\right)$.

63. $z_1 = 2\sqrt{3} - 2i$, so $\tan\theta_1 = \frac{-2}{2\sqrt{3}} = -\frac{1}{\sqrt{3}}$ with θ_1 in quadrant IV $\Rightarrow \theta_1 = \frac{11\pi}{6}$, and $r_1 = \sqrt{12+4} = 4$.
$z_2 = -1 + i$, so $\tan\theta_2 = -1$ with θ_2 in quadrant II $\Rightarrow \theta_2 = \frac{3\pi}{4}$, and $r_2 = \sqrt{1+1} = \sqrt{2}$.
Hence, $z_1 = 4\left(\cos\frac{11\pi}{6} + i\sin\frac{11\pi}{6}\right)$ and $z_2 = \sqrt{2}\left(\cos\frac{3\pi}{4} + i\sin\frac{3\pi}{4}\right)$.
Thus, $z_1 z_2 = 4 \cdot \sqrt{2}\left[\cos\left(\frac{11\pi}{6} + \frac{3\pi}{4}\right) + i\sin\left(\frac{11\pi}{6} + \frac{3\pi}{4}\right)\right] = 4\sqrt{2}\left(\cos\frac{7\pi}{12} + i\sin\frac{7\pi}{12}\right)$,
$z_1/z_2 = \frac{4}{\sqrt{2}}\left[\cos\left(\frac{11\pi}{6} - \frac{3\pi}{4}\right) + i\sin\left(\frac{11\pi}{6} - \frac{3\pi}{4}\right)\right] = 2\sqrt{2}\left(\cos\frac{13\pi}{12} + i\sin\frac{13\pi}{12}\right)$, and
$1/z_1 = \frac{1}{4}\left(\cos\left(-\frac{11\pi}{6}\right) + i\sin\left(-\frac{11\pi}{6}\right)\right) = \frac{1}{4}\left(\cos\frac{11\pi}{6} - i\sin\frac{11\pi}{6}\right)$.

65. $z_1 = 5 + 5i$, so $\tan\theta_1 = \frac{5}{5} = 1$ with θ_1 in quadrant I $\Rightarrow \theta_1 = \frac{\pi}{4}$, and $r_1 = \sqrt{25+25} = 5\sqrt{2}$.
$z_2 = 4$, so $\theta_2 = 0$, and $r_2 = 4$.
Hence, $z_1 = 5\sqrt{2}\left(\cos\frac{\pi}{4} + i\sin\frac{\pi}{4}\right)$ and $z_2 = 4(\cos 0 + i\sin 0)$.
Thus, $z_1 z_2 = 5\sqrt{2} \cdot 4\left[\cos\left(\frac{\pi}{4} + 0\right) + i\sin\left(\frac{\pi}{4} + 0\right)\right] = 20\sqrt{2}\left(\cos\frac{\pi}{4} + i\sin\frac{\pi}{4}\right)$, $z_1/z_2 = \frac{5\sqrt{2}}{4}\left(\cos\frac{\pi}{4} + i\sin\frac{\pi}{4}\right)$, and
$1/z_1 = \frac{1}{5\sqrt{2}}\left(\cos\left(-\frac{\pi}{4}\right) + i\sin\left(-\frac{\pi}{4}\right)\right) = \frac{\sqrt{2}}{10}\left(\cos\frac{\pi}{4} - i\sin\frac{\pi}{4}\right)$.

67. $z_1 = -20$, so $\theta_1 = \pi$, and $r_1 = 20$.

$z_2 = \sqrt{3} + i$, so $\tan\theta_2 = \frac{1}{\sqrt{3}}$ with θ_2 in quadrant I $\Rightarrow \theta_2 = \frac{\pi}{6}$, and $r_2 = \sqrt{3+1} = 2$.

Hence, $z_1 = 20\left(\cos\pi + i\sin\pi\right)$ and $z_2 = 2\left(\cos\frac{\pi}{6} + i\sin\frac{\pi}{6}\right)$.

Thus, $z_1 z_2 = 20 \cdot 2\left[\cos\left(\pi + \frac{\pi}{6}\right) + i\sin\left(\pi + \frac{\pi}{6}\right)\right] = 40\left(\cos\frac{7\pi}{6} + i\sin\frac{7\pi}{6}\right)$,

$z_1/z_2 = \frac{20}{2}\left[\cos\left(\pi - \frac{\pi}{6}\right) + i\sin\left(\pi - \frac{\pi}{6}\right)\right] = 10\left(\cos\frac{5\pi}{6} + i\sin\frac{5\pi}{6}\right)$, and

$1/z_1 = \frac{1}{20}\left[\cos\left(-\pi\right) + i\sin\left(-\pi\right)\right] = \frac{1}{20}\left(\cos\pi - i\sin\pi\right)$.

69. From Exercise 29, $1 + i = \sqrt{2}\left(\cos\frac{\pi}{4} + i\sin\frac{\pi}{4}\right)$. Thus,

$$(1+i)^{20} = \left(\sqrt{2}\right)^{20}\left[\cos 20\left(\frac{\pi}{4}\right) + i\sin 20\left(\frac{\pi}{4}\right)\right] = \left(2^{1/2}\right)^{20}\left(\cos 5\pi + i\sin 5\pi\right) = 2^{10}\left(-1 + 0i\right) = -1024.$$

71. $r = \sqrt{12+4} = 4$ and $\tan\theta = \frac{2}{2\sqrt{3}} = \frac{1}{\sqrt{3}} \Rightarrow \theta = \frac{\pi}{6}$. Thus, $2\sqrt{3} + 2i = 4\left(\cos\frac{\pi}{6} + i\sin\frac{\pi}{6}\right)$. So

$$\left(2\sqrt{3}+2i\right)^5 = 4^5\left(\cos\frac{5\pi}{6} + i\sin\frac{5\pi}{6}\right) = 1024\left(-\frac{\sqrt{3}}{2} + \frac{1}{2}i\right) = 512\left(-\sqrt{3}+i\right).$$

73. $r = \sqrt{\frac{1}{2} + \frac{1}{2}} = 1$ and $\tan\theta = 1 \Rightarrow \theta = \frac{\pi}{4}$. Thus $\frac{\sqrt{2}}{2} + \frac{\sqrt{2}}{2}i = \cos\frac{\pi}{4} + i\sin\frac{\pi}{4}$. Therefore,

$$\left(\frac{\sqrt{2}}{2} + \frac{\sqrt{2}}{2}i\right)^{12} = \cos 12\left(\frac{\pi}{4}\right) + i\sin 12\left(\frac{\pi}{4}\right) = \cos 3\pi + i\sin 3\pi = -1.$$

75. $r = \sqrt{4+4} = 4\sqrt{2}$ and $\tan\theta = -1$ with θ in quadrant IV $\Rightarrow \theta = \frac{7\pi}{4}$. Thus $2 - 2i = 2\sqrt{2}\left(\cos\frac{7\pi}{4} + i\sin\frac{7\pi}{4}\right)$, so

$$(2-2i)^8 = \left(2\sqrt{2}\right)^8\left(\cos 14\pi + i\sin 14\pi\right) = 4096\left(1 - 0i\right) = 4096.$$

77. $r = \sqrt{1+1} = \sqrt{2}$ and $\tan\theta = 1$ with θ in quadrant III $\Rightarrow \theta = \frac{5\pi}{4}$. Thus $-1 - i = \sqrt{2}\left(\cos\frac{5\pi}{4} + i\sin\frac{5\pi}{4}\right)$, so

$$(-1-i)^7 = \left(\sqrt{2}\right)^7\left(\cos\frac{35\pi}{4} + i\sin\frac{35\pi}{4}\right) = 8\sqrt{2}\left(\cos\frac{3\pi}{4} + i\sin\frac{3\pi}{4}\right) = 8\sqrt{2}\left(\frac{1}{\sqrt{2}} - i\frac{1}{\sqrt{2}}\right) = 8\left(-1+i\right).$$

79. $r = \sqrt{12+4} = 4$ and $\tan\theta = \frac{2}{2\sqrt{3}} = \frac{1}{\sqrt{3}} \Rightarrow \theta = \frac{\pi}{6}$. Thus $2\sqrt{3} + 2i = 4\left(\cos\frac{\pi}{6} + i\sin\frac{\pi}{6}\right)$, so

$$\left(2\sqrt{3}+2i\right)^{-5} = \left(\frac{1}{4}\right)^5\left(\cos\frac{-5\pi}{6} + i\sin\frac{-5\pi}{6}\right) = \frac{1}{1024}\left(-\frac{\sqrt{3}}{2} - \frac{1}{2}i\right) = \frac{1}{2048}\left(-\sqrt{3}-i\right)$$

81. $r = \sqrt{48+16} = 8$ and $\tan\theta = \frac{4}{4\sqrt{3}} = \frac{1}{\sqrt{3}} \Rightarrow \theta = \frac{\pi}{6}$. Thus

$4\sqrt{3} + 4i = 8\left(\cos\frac{\pi}{6} + i\sin\frac{\pi}{6}\right)$. So,

$$\left(4\sqrt{3}+4i\right)^{1/2} = \sqrt{8}\left[\cos\left(\frac{\pi/6 + 2k\pi}{2}\right) + i\sin\left(\frac{\pi/6 + 2k\pi}{2}\right)\right]\text{ for }k = 0, 1.$$

Thus the two roots are $w_0 = 2\sqrt{2}\left(\cos\frac{\pi}{12} + i\sin\frac{\pi}{12}\right)$ and

$w_1 = 2\sqrt{2}\left(\cos\frac{13\pi}{12} + i\sin\frac{13\pi}{12}\right)$.

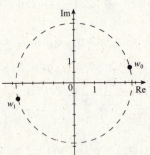

83. $-81i = 81\left(\cos\frac{3\pi}{2} + i\sin\frac{3\pi}{2}\right)$. Thus,

$$(-81i)^{1/4} = 81^{1/4}\left[\cos\left(\frac{3\pi/2 + 2k\pi}{4}\right) + i\sin\left(\frac{3\pi/2 + 2k\pi}{4}\right)\right]\text{ for }k = 0, 1,$$

2, 3. The four roots are $w_0 = 3\left(\cos\frac{3\pi}{8} + i\sin\frac{3\pi}{8}\right)$,

$w_1 = 3\left(\cos\frac{7\pi}{8} + i\sin\frac{7\pi}{8}\right)$, $w_2 = 3\left(\cos\frac{11\pi}{8} + i\sin\frac{11\pi}{8}\right)$, and

$w_3 = 3\left(\cos\frac{15\pi}{8} + i\sin\frac{15\pi}{8}\right)$.

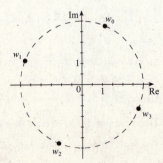

85. $1 = \cos 0 + i \sin 0$. Thus, $1^{1/8} = \cos \dfrac{2k\pi}{8} + i \sin \dfrac{2k\pi}{8}$, for $k = 0, 1, 2, 3, 4, 5, 6$,

7. So the eight roots are $w_0 = \cos 0 + i \sin 0 = 1$,

$w_1 = \cos \dfrac{\pi}{4} + i \sin \dfrac{\pi}{4} = \dfrac{\sqrt{2}}{2} + i \dfrac{\sqrt{2}}{2}$, $w_2 = \cos \dfrac{\pi}{2} + i \sin \dfrac{\pi}{2} = i$,

$w_3 = \cos \dfrac{3\pi}{4} + i \sin \dfrac{3\pi}{4} = -\dfrac{\sqrt{2}}{2} + i \dfrac{\sqrt{2}}{2}$, $w_4 = \cos \pi + i \sin \pi = -1$,

$w_5 = \cos \dfrac{5\pi}{4} + i \sin \dfrac{5\pi}{4} = -\dfrac{\sqrt{2}}{2} - i \dfrac{\sqrt{2}}{2}$, $w_6 = \cos \dfrac{3\pi}{2} + i \sin \dfrac{3\pi}{2} = -i$, and

$w_7 = \cos \dfrac{7\pi}{4} + i \sin \dfrac{7\pi}{4} = \dfrac{\sqrt{2}}{2} - i \dfrac{\sqrt{2}}{2}$.

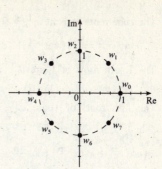

87. $i = \cos \dfrac{\pi}{2} + i \sin \dfrac{\pi}{2}$, so $i^{1/3} = \cos \left(\dfrac{\pi/2 + 2k\pi}{3} \right) + i \sin \left(\dfrac{\pi/2 + 2k\pi}{3} \right)$ for

$k = 0, 1, 2$. Thus the three roots are $w_0 = \cos \dfrac{\pi}{6} + i \sin \dfrac{\pi}{6} = \dfrac{\sqrt{3}}{2} + \dfrac{1}{2}i$,

$w_1 = \cos \dfrac{5\pi}{6} + i \sin \dfrac{5\pi}{6} = -\dfrac{\sqrt{3}}{2} + \dfrac{1}{2}i$, and $w_2 = \cos \dfrac{3\pi}{2} + i \sin \dfrac{3\pi}{2} = -i$.

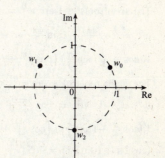

89. $-1 = \cos \pi + i \sin \pi$. Then $(-1)^{1/4} = \cos \left(\dfrac{\pi + 2k\pi}{4} \right) + i \sin \left(\dfrac{\pi + 2k\pi}{4} \right)$ for

$k = 0, 1, 2, 3$. So the four roots are $w_0 = \cos \dfrac{\pi}{4} + i \sin \dfrac{\pi}{4} = \dfrac{\sqrt{2}}{2} + i \dfrac{\sqrt{2}}{2}$,

$w_1 = \cos \dfrac{3\pi}{4} + i \sin \dfrac{3\pi}{4} = -\dfrac{\sqrt{2}}{2} + i \dfrac{\sqrt{2}}{2}$, $w_2 = \cos \dfrac{5\pi}{4} + i \sin \dfrac{5\pi}{4} = -\dfrac{\sqrt{2}}{2} - i \dfrac{\sqrt{2}}{2}$,

and $w_3 = \cos \dfrac{7\pi}{4} + i \sin \dfrac{7\pi}{4} = \dfrac{\sqrt{2}}{2} - i \dfrac{\sqrt{2}}{2}$.

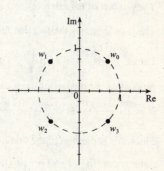

91. $z^4 + 1 = 0 \Leftrightarrow z = (-1)^{1/4} = \dfrac{\sqrt{2}}{2} (\pm 1 \pm i)$ (from Exercise 85)

93. $z^3 - 4\sqrt{3} - 4i = 0 \Leftrightarrow z = \left(4\sqrt{3} + 4i \right)^{1/3}$. Since $4\sqrt{3} + 4i = 8 \left(\cos \dfrac{\pi}{6} + i \sin \dfrac{\pi}{6} \right)$,

$\left(4\sqrt{3} + 4i \right)^{1/3} = 8^{1/3} \left[\cos \left(\dfrac{\pi/6 + 2k\pi}{3} \right) + i \sin \left(\dfrac{\pi/6 + 2k\pi}{3} \right) \right]$, for $k = 0, 1, 2$. Thus the three roots are

$z = 2 \left(\cos \dfrac{\pi}{18} + i \sin \dfrac{\pi}{18} \right)$, $z = 2 \left(\cos \dfrac{13\pi}{18} + i \sin \dfrac{13\pi}{8} \right)$, and $z = 2 \left(\cos \dfrac{25\pi}{18} + i \sin \dfrac{25\pi}{18} \right)$.

95. $z^3 + 1 = -i \Rightarrow z = (-1 - i)^{1/3}$. Since $-1 - i = \sqrt{2} \left(\cos \dfrac{5\pi}{4} + i \sin \dfrac{5\pi}{4} \right)$,

$z = (-1 - i)^{1/3} = 2^{1/6} \left[\cos \left(\dfrac{5\pi/4 + 2k\pi}{3} \right) + i \sin \left(\dfrac{5\pi/4 + 2k\pi}{3} \right) \right]$ for $k = 0, 1, 2$. Thus the three solutions to this

equation are $z = 2^{1/6} \left(\cos \dfrac{5\pi}{12} + i \sin \dfrac{5\pi}{12} \right)$, $2^{1/6} \left(\cos \dfrac{13\pi}{12} + i \sin \dfrac{13\pi}{12} \right)$, and $2^{1/6} \left(\cos \dfrac{21\pi}{12} + i \sin \dfrac{21\pi}{12} \right)$.

97. (a) $w = \cos \dfrac{2\pi}{n} + i \sin \dfrac{2\pi}{n}$ for a positive integer n. Then, $w^k = \cos \dfrac{2k\pi}{n} + i \sin \dfrac{2k\pi}{n}$. Now $w^0 = \cos 0 + i \sin 0 = 1$

and for $k \neq 0$, $\left(w^k \right)^n = \cos 2k\pi + i \sin 2k\pi = 1$. So the nth roots of 1 are $\cos \dfrac{2k\pi}{n} + i \sin \dfrac{2k\pi}{n} = w^k$ for

$k = 0, 1, 2, \ldots, n - 1$. In other words, the nth roots of 1 are $w^0, w^1, w^2, w^3, \ldots, w^{n-1}$ or $1, w, w^2, w^3, \ldots, w^{n-1}$.

(b) For $k = 0, 1, \ldots, n - 1$, we have $\left(sw^k \right)^n = s^n \left(w^k \right)^n = z \cdot 1 = z$, so sw^k are nth roots of z for $k = 0, 1, \ldots, n - 1$.

99. The cube roots of 1 are $w^0 = 1$, $w^1 = \cos\frac{2\pi}{3} + i\sin\frac{2\pi}{3}$, and $w^2 = \cos\frac{4\pi}{3} + i\sin\frac{4\pi}{3}$, so their product is

$$w^0 \cdot w^1 \cdot w^2 = (1)\left(\cos\frac{2\pi}{3} + i\sin\frac{2\pi}{3}\right)\left(\cos\frac{4\pi}{3} + i\sin\frac{4\pi}{3}\right) = \cos 2\pi + i\sin 2\pi = 1.$$

The fourth roots of 1 are $w^0 = 1$, $w^1 = i$, $w^2 = -1$, and $w^3 = -i$, so their product is

$$w^0 \cdot w^1 \cdot w^2 \cdot w^3 = (1) \cdot (i) \cdot (-1) \cdot (-i) = i^2 = -1.$$

The fifth roots of 1 are $w^0 = 1$, $w^1 = \cos\frac{2\pi}{5} + i\sin\frac{2\pi}{5}$, $w^2 = \cos\frac{4\pi}{5} +$

$i\sin\frac{4\pi}{5}$, $w^3 = \cos\frac{6\pi}{5} + i\sin\frac{6\pi}{5}$, and $w^4 = \cos\frac{8\pi}{5} + i\sin\frac{8\pi}{5}$, so their product is

$$1\left(\cos\frac{2\pi}{5} + i\sin\frac{2\pi}{5}\right)\left(\cos\frac{4\pi}{5} + i\sin\frac{4\pi}{5}\right)\left(\cos\frac{6\pi}{5} + i\sin\frac{6\pi}{5}\right)\left(\cos\frac{8\pi}{5} + i\sin\frac{8\pi}{5}\right) = \cos 4\pi + i\sin 4\pi = 1.$$

The sixth roots of 1 are $w^0 = 1$, $w^1 = \cos\frac{\pi}{3} + i\sin\frac{\pi}{3}$, $w^2 = \cos\frac{2\pi}{3} + i\sin\frac{2\pi}{3} = -\frac{1}{2} + \frac{\sqrt{3}}{2}i$, $w^3 = -1$,

$w^4 = \cos\frac{4\pi}{3} + i\sin\frac{4\pi}{3} = -\frac{1}{2} - \frac{\sqrt{3}}{2}i$, and $w^5 = \cos\frac{5\pi}{3} + i\sin\frac{5\pi}{3} = \frac{1}{2} - \frac{\sqrt{3}}{2}i$, so their product is

$$1\left(\cos\frac{\pi}{3} + i\sin\frac{\pi}{3}\right)\left(\cos\frac{2\pi}{3} + i\sin\frac{2\pi}{3}\right)(-1)\left(\cos\frac{4\pi}{3} + i\sin\frac{4\pi}{3}\right)\left(\cos\frac{5\pi}{3} + i\sin\frac{5\pi}{3}\right) = \cos 5\pi + i\sin 5\pi = -1.$$

The eight roots of 1 are $w^0 = 1$, $w^1 = \cos\frac{\pi}{4} + i\sin\frac{\pi}{4}$, $w^2 = i$, $w^3 = \cos\frac{3\pi}{4} + i\sin\frac{3\pi}{4}$,

$w^4 = -1$, $w^5 = \cos\frac{5\pi}{4} + i\sin\frac{5\pi}{4}$, $w^6 = -i$, $w^7 = \cos\frac{7\pi}{4} + i\sin\frac{7\pi}{4}$, so their product is

$$1\left(\cos\frac{\pi}{4} + i\sin\frac{\pi}{4}\right)i\left(\cos\frac{3\pi}{4} + i\sin\frac{3\pi}{4}\right)(-1)\left(\cos\frac{5\pi}{4} + i\sin\frac{5\pi}{4}\right)(-i)\left(\cos\frac{7\pi}{4} + i\sin\frac{7\pi}{4}\right) = i^2 \cdot$$

$$(\cos 2\pi + i\sin 2\pi) = -1.$$

The product of the nth roots of 1 is -1 if n is even and 1 if n is odd.

The proof requires the fact that the sum of the first m integers is $\dfrac{m(m+1)}{2}$.

Let $w = \cos\dfrac{2\pi}{n} + i\sin\dfrac{2\pi}{n}$. Then $w^k = \cos\dfrac{2k\pi}{n} + i\sin\dfrac{2k\pi}{n}$ for $k = 0, 1, 2, \ldots, n - 1$. The argument of the

product of the n roots of unity can be found by adding the arguments of each w^k. So the argument of the product is

$$\theta = 0 + \frac{2(1)\pi}{n} + \frac{2(2)\pi}{n} + \frac{2(3)\pi}{n} + \cdots + \frac{2(n-2)\pi}{n} + \frac{2(n-1)\pi}{n} = \frac{2\pi}{n}[0 + 1 + 2 + 3 + \cdots + (n-2) + (n-1)].$$

Since this is the sum of the first $n - 1$ integers, this sum is $\dfrac{2\pi}{n} \cdot \dfrac{(n-1)n}{2} = (n-1)\pi$. Thus the product of the n roots of

unity is $\cos((n-1)\pi) + i\sin((n-1)\pi) = -1$ if n is even and 1 if n is odd.

8.4   PLANE CURVES AND PARAMETRIC EQUATIONS

1. (a) The parametric equations $x = f(t)$ and $y = g(t)$ give the coordinates of a point $(x, y) = (f(t), g(t))$ for appropriate values of t. The variable t is called a *parameter*.

(b) When $t = 0$ the object is at $\left(0, 0^2\right) = (0, 0)$ and when $t = 1$ the object is at $\left(1, 1^2\right) = (1, 1)$.

(c) If we eliminate the parameter in part (b) we get the equation $y = x^2$. We see from this equation that the path of the moving object is a *parabola*.

3. (a) $x = 2t$, $y = t + 6$

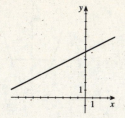

(b) Since $x = 2t$, $t = \dfrac{x}{2}$ and so $y = \dfrac{x}{2} + 6 \Leftrightarrow$

$x - 2y + 12 = 0$.

5. (a) $x = t^2$, $y = t - 2$, $2 \le t \le 4$

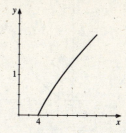

(b) Since $y = t - 2 \Leftrightarrow t = y + 2$, we have $x = t^2$

$\Leftrightarrow x = (y + 2)^2$, and since $2 \le t \le 4$, we have

$4 \le x \le 16$.

7. (a) $x = \sqrt{t}$, $y = 1 - t \Rightarrow t \ge 0$

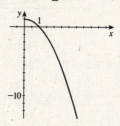

(b) Since $x = \sqrt{t}$, we have $x^2 = t$, and so $y = 1 - x^2$

with $x \ge 0$.

9. (a) $x = \dfrac{1}{t}$, $y = t + 1$

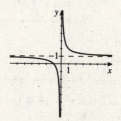

(b) Since $x = \dfrac{1}{t}$ we have $t = \dfrac{1}{x}$ and so $y = \dfrac{1}{x} + 1$.

11. (a) $x = 4t^2$, $y = 8t^3$

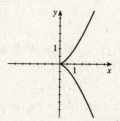

(b) Since $y = 8t^3 \Leftrightarrow y^2 = 64t^6 = \left(4t^2\right)^3 = x^3$, we

have $y^2 = x^3$.

13. (a) $x = 2\sin t$, $y = 2\cos t$, $0 \le t \le \pi$

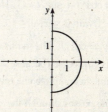

(b) $x^2 = (2\sin t)^2 = 4\sin^2 t$ and $y^2 = 4\cos^2 t$. Hence,

$x^2 + y^2 = 4\sin^2 t + 4\cos^2 t = 4 \Leftrightarrow x^2 + y^2 = 4$,

where $x \ge 0$.

15. (a) $x = \sin^2 t$, $y = \sin^4 t$

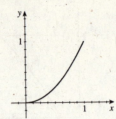

(b) Since $x = \sin^2 t$ we have $x^2 = \sin^4 t$ and so $y = x^2$. But since $0 \leq \sin^2 t \leq 1$ we only get the part of this parabola for which $0 \leq x \leq 1$.

17. (a) $x = \cos t$, $y = \cos 2t$

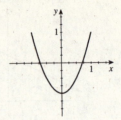

(b) Since $x = \cos t$ we have $x^2 = \cos^2 t$, so $2x^2 - 1 = 2\cos^2 t - 1 = \cos 2t = y$. Hence, the rectangular equation is $y = 2x^2 - 1$, $-1 \leq x \leq 1$.

19. (a) $x = \sec t$, $y = \tan t$, $0 \leq t < \frac{\pi}{2}$ $\Rightarrow$ $x \geq 1$ and $y \geq 0$.

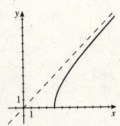

(b) $x^2 = \sec^2 t$, $y^2 = \tan^2 t$, and
$y^2 + 1 = \tan^2 t + 1 = \sec^2 t = x^2$. Therefore,
$y^2 + 1 = x^2 \Leftrightarrow x^2 - y^2 = 1$, $x \geq 1$, $y \geq 0$.

21. (a) $x = e^t$, $y = e^{-t}$ $\Rightarrow$ $x > 0$, $y > 0$.

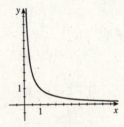

(b) $xy = e^t \cdot e^{-t} = e^0 = 1$. Hence, the equation is $xy = 1$, with $x > 0$, $y > 0$.

23. (a) $x = \cos^2 t$, $y = \sin^2 t$

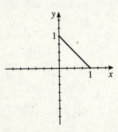

(b) $x + y = \cos^2 t + \sin^2 t = 1$. Hence, the equation is $x + y = 1$ with $0 \leq x, y \leq 1$.

25. $x = 3\cos t$, $y = 3\sin t$. The radius of the circle is 3, the position at time 0 is
$(x(0), y(0)) = (3\cos 0, 3\sin 0) = (3, 0)$ and the orientation is counterclockwise (because x is decreasing and y is increasing initially). $(x, y) = (3, 0)$ again when $t = 2\pi$, so it takes 2π units of time to complete one revolution.

27. $x = \sin 2t$, $y = \cos 2t$. The radius of the circle is 1, the position at time 0 is $(x(0), y(0)) = (\sin 0, \cos 0) = (0, 1)$ and the orientation is clockwise (because x is increasing and y is decreasing initially). $(x, y) = (0, 1)$ again when $t = \pi$, so it takes π units of time to complete one revolution.

29. Since the line passes through the point $(4, -1)$ and has slope $\frac{1}{2}$, parametric equations for the line are $x = 4 + t$, $y = -1 + \frac{1}{2}t$.

31. Since the line passes through the points $(6, 7)$ and $(7, 8)$, its slope is $\dfrac{8 - 7}{7 - 6} = 1$. Thus, parametric equations for the line are $x = 6 + t$, $y = 7 + t$.

33. Since $\cos^2 t + \sin^2 t = 1$, we have $a^2\cos^2 t + a^2\sin^2 t = a^2$. If we let $x = a\cos t$ and $y = a\sin t$, then $x^2 + y^2 = a^2$. Hence, parametric equations for the circle are $x = a\cos t$, $y = a\sin t$.

35. $x = a\tan\theta \Leftrightarrow \tan\theta = \dfrac{x}{a} \;\Rightarrow\; \tan^2\theta = \dfrac{x^2}{a^2}$. Also, $y = b\sec\theta \Leftrightarrow \sec\theta = \dfrac{y}{b} \;\Rightarrow\; \sec^2\theta = \dfrac{y^2}{b^2}$. Since

$\tan^2\theta = \sec^2\theta - 1$, we have $\dfrac{x^2}{a^2} = \dfrac{y^2}{b^2} - 1 \Leftrightarrow \dfrac{y^2}{b^2} - \dfrac{x^2}{a^2} = 1$, which is the equation of a hyperbola.

37. $x = t\cos t$, $y = t\sin t$, $t \ge 0$

t	x	y
0	0	0
$\dfrac{\pi}{4}$	$\dfrac{\pi\sqrt{2}}{8}$	$\dfrac{\pi\sqrt{2}}{8}$
$\dfrac{\pi}{2}$	0	$\dfrac{\pi}{2}$
$\dfrac{3\pi}{4}$	$-\dfrac{3\pi\sqrt{2}}{8}$	$\dfrac{3\pi\sqrt{2}}{8}$
π	$-\pi$	0

t	x	y
$\dfrac{5\pi}{4}$	$-\dfrac{5\pi\sqrt{2}}{8}$	$-\dfrac{5\pi\sqrt{2}}{8}$
$\dfrac{3\pi}{2}$	0	$-\dfrac{3\pi}{2}$
$\dfrac{7\pi}{4}$	$\dfrac{7\pi\sqrt{2}}{8}$	$-\dfrac{7\pi\sqrt{2}}{8}$
2π	2π	0

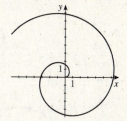

39. $x = \dfrac{3t}{1+t^3}$, $y = \dfrac{3t^2}{1+t^3}$, $t \ne -1$

t	x	y
-0.9	-9.96	8.97
-0.75	-3.89	2.92
-0.5	-1.71	0.86
0	0	0
0.5	1.33	0.67
1	1.5	1.5
1.5	1.03	1.54

t	x	y
2	0.67	1.33
2.5	0.45	1.13
3	0.32	0.96
4	0.18	0.74
5	0.12	0.60
6	0.08	0.50

t	x	y
-1.1	9.97	-10.97
-1.25	3.93	-4.92
-1.5	1.89	-2.84
-2	0.86	-1.71
-2.5	0.51	-1.28
-3	0.35	-1.04
-3.5	0.25	-0.87

t	x	y
-4	0.19	-0.76
-4.5	0.15	-0.67
-5	0.12	-0.60
-6	0.08	-0.50
-7	0.06	-0.43
-8	0.05	-0.38

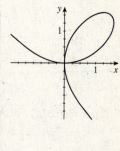

As $t \to -1^-$ we have $x \to \infty$ and $y \to -\infty$. As $t \to -1^+$ we have $x \to -\infty$ and $y \to \infty$. As $t \to \infty$ we have $x \to 0^+$ and $y \to 0^+$. As $t \to -\infty$ we have $x \to 0^+$ and $y \to 0^-$.

41. $x = (v_0\cos\alpha)\,t$, $y = (v_0\sin\alpha)\,t - 16t^2$. From the equation for x, $t = \dfrac{x}{v_0\cos\alpha}$. Substituting into the equation for y gives

$y = (v_0\sin\alpha)\dfrac{x}{v_0\cos\alpha} - 16\left(\dfrac{x}{v_0\cos\alpha}\right)^2 = x\tan\alpha - \dfrac{16x^2}{v_0^2\cos^2\alpha}$. Thus the equation is of the form $y = c_1 x - c_2 x^2$,

where c_1 and c_2 are constants, so its graph is a parabola.

43. $x = \sin t$, $y = 2\cos 3t$

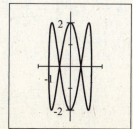

45. $x = 3\sin 5t$, $y = 5\cos 3t$

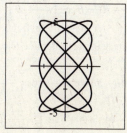

47. $x = \sin(\cos t)$, $y = \cos t^{3/2}$, $0 \le t \le 2\pi$

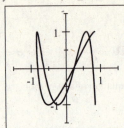

49. (a) $r = 2^{\theta/12}$, $0 \le \theta \le 4\pi \Rightarrow x = 2^{t/12}\cos t$, $y = 2^{t/12}\sin t$

(b)

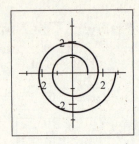

51. (a) $r = \dfrac{4}{2 - \cos\theta} \Leftrightarrow x = \dfrac{4\cos t}{2 - \cos t}$, $y = \dfrac{4\sin t}{2 - \cos t}$

(b)

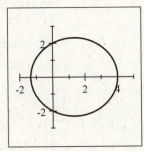

53. $x = t^3 - 2t$, $y = t^2 - t$ is Graph III, since

$$y = t^2 - t = \left(t^2 - t + \tfrac{1}{4}\right) - \tfrac{1}{4} = \left(t - \tfrac{1}{2}\right)^2 - \tfrac{1}{4}, \text{ and so}$$

$y \ge -\tfrac{1}{4}$ on this curve, while x is unbounded.

55. $x = t + \sin 2t$, $y = t + \sin 3t$ is Graph II, since the values of x and y oscillate about their values on the line $x = t$, $y = t \Leftrightarrow y = x$.

57. (a) If we modify Figure 8 so that $|PC| = b$, then by the same reasoning as in Example 6, we see that

$x = |OT| - |PQ| = a\theta - b\sin\theta$ and

$y = |TC| - |CQ| = a - b\cos\theta$.

We graph the case where $a = 3$ and $b = 2$.

(b)

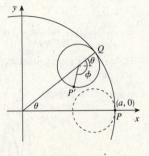

59. (a) We first note that the center of circle C (the small circle) has coordinates

$([a - b]\cos\theta, [a - b]\sin\theta)$. Now the arc PQ has the same length as the arc

$P'Q$, so $b\phi = a\theta \Leftrightarrow \phi = \dfrac{a}{b}\theta$, and so $\phi - \theta = \dfrac{a}{b}\theta - \theta = \dfrac{a - b}{b}\theta$. Thus the

x-coordinate of P is the x-coordinate of the center of circle C plus

$b\cos(\phi - \theta) = b\cos\left(\dfrac{a - b}{b}\theta\right)$, and the y-coordinate of P is the

y-coordinate of the center of circle C minus $b \cdot \sin(\phi - \theta) = b\sin\left(\dfrac{a - b}{b}\theta\right)$.

So $x = (a - b)\cos\theta + b\cos\left(\dfrac{a - b}{b}\theta\right)$ and

$y = (a - b)\sin\theta - b\sin\left(\dfrac{a - b}{b}\theta\right)$.

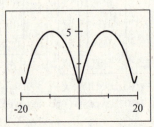

(b) If $a = 4b$, $b = \dfrac{a}{4}$, and $x = \frac{3}{4}a\cos\theta + \frac{1}{4}a\cos 3\theta$, $y = \frac{3}{4}a\sin\theta - \frac{1}{4}a\sin 3\theta$.

From Example 2 in Section 7.3, $\cos 3\theta = 4\cos^3\theta - 3\cos\theta$. Similarly, one can

prove that $\sin 3\theta = 3\sin\theta - 4\sin^3\theta$. Substituting, we get

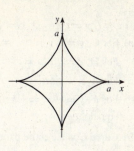

$x = \frac{3}{4}a\cos\theta + \frac{1}{4}a\left(4\cos^3\theta - 3\cos\theta\right) = a\cos^3\theta$

$y = \frac{3}{4}a\sin\theta - \frac{1}{4}a\left(3\sin\theta - 4\sin^3\theta\right) = a\sin^3\theta$. Thus,

$x^{2/3} + y^{2/3} = a^{2/3}\cos^2\theta + a^{2/3}\sin^2\theta = a^{2/3}$, so $x^{2/3} + y^{2/3} = a^{2/3}$.

61. A polar equation for the circle is $r = 2a\sin\theta$. Thus the coordinates of Q are $x = r\cos\theta = 2a\sin\theta\cos\theta$ and
$y = r\sin\theta = 2a\sin^2\theta$. The coordinates of R are $x = 2a\cot\theta$ and $y = 2a$. Since P is the midpoint of QR, we use the
midpoint formula to get $x = a\left(\sin\theta\cos\theta + \cot\theta\right)$ and $y = a\left(1 + \sin^2\theta\right)$.

63. (a) A has coordinates $(a\cos\theta, a\sin\theta)$. Since OA is perpendicular to AB,
$\triangle OAB$ is a right triangle and B has coordinates $(a\sec\theta, 0)$. It follows
that P has coordinates $(a\sec\theta, b\sin\theta)$. Thus, the parametric equations
are $x = a\sec\theta$, $y = b\sin\theta$.

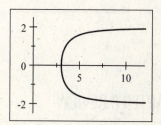

(b) The right half of the curve is graphed with $a = 3$ and $b = 2$.

65. We use the equation for y from Example 6 and solve for θ. Thus for $0 \le \theta \le \pi$, $y = a\left(1 - \cos\theta\right) \Leftrightarrow \dfrac{a - y}{a} = \cos\theta \Leftrightarrow$

$\theta = \cos^{-1}\left(\dfrac{a - y}{a}\right)$. Substituting into the equation for x, we get $x = a\left[\cos^{-1}\left(\dfrac{a - y}{a}\right) - \sin\left(\cos^{-1}\left(\dfrac{a - y}{a}\right)\right)\right]$.

However, $\sin\left(\cos^{-1}\left(\dfrac{a - y}{a}\right)\right) = \sqrt{1 - \left(\dfrac{a - y}{a}\right)^2} = \dfrac{\sqrt{2ay - y^2}}{a}$. Thus, $x = a\left[\cos^{-1}\left(\dfrac{a - y}{a}\right) - \dfrac{\sqrt{2ay - y^2}}{a}\right]$,

and we have $\dfrac{\sqrt{2ay - y^2} + x}{a} = \cos^{-1}\left(\dfrac{a - y}{a}\right) \Rightarrow 1 - \dfrac{y}{a} = \cos\left(\dfrac{\sqrt{2ay - y^2} + x}{a}\right) \Rightarrow$

$y = a\left[1 - \cos\left(\dfrac{\sqrt{2ay - y^2} + x}{a}\right)\right]$.

67. (a) In the figure, since OQ and QT are perpendicular and OT and TD are perpendicular, the angles formed by
their intersections are equal, that is, $\theta = \angle DTQ$. Now the coordinates of T are $(\cos\theta, \sin\theta)$. Since $|TD|$ is the
length of the string that has been unwound from the circle, it must also have arc length θ, so $|TD| = \theta$. Thus the
x-displacement from T to D is $\theta \cdot \sin\theta$ while the y-displacement from T to D is $\theta \cdot \cos\theta$. So the coordinates of D are
$x = \cos\theta + \theta\sin\theta$ and $y = \sin\theta - \theta\cos\theta$.

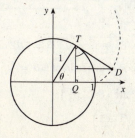

(b)

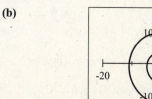

69. $C: x = t, y = t^2;$ $D: x = \sqrt{t}, y = t, t \geq 0$ $E: x = \sin t, y = 1 - \cos^2 t$ $F: x = e^t, y = e^{2t}$

  (a) For C, $x = t$, $y = t^2$ $\Rightarrow$ $y = x^2$.

    For D, $x = \sqrt{t}$, $y = t$ $\Rightarrow$ $y = x^2$.

    For E, $x = \sin t$ $\Rightarrow$ $x^2 = \sin^2 t = 1 - \cos^2 t = y$ and so $y = x^2$.

    For F, $x = e^t$ $\Rightarrow$ $x^2 = e^{2t} = y$ and so $y = x^2$. Therefore, the points on all four curves satisfy the same rectangular equation.

  (b) Curve C is the entire parabola $y = x^2$. Curve D is the right half of the parabola because $t \geq 0$ and so $x \geq 0$. Curve E is the portion of the parabola for $-1 \leq x \leq 1$. Curve F is the portion of the parabola where $x > 0$, since $e^t > 0$ for all t.

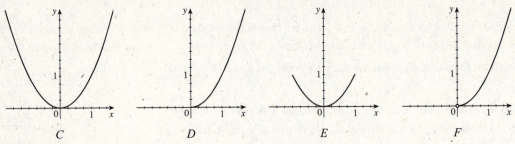

C D E F

CHAPTER 8 REVIEW

1. (a)

  (b) $x = 12 \cos \frac{\pi}{6} = 12 \cdot \frac{\sqrt{3}}{2} = 6\sqrt{3}$,

    $y = 12 \sin \frac{\pi}{6} = 12 \cdot \frac{1}{2} = 6$. Thus, the rectangular

    coordinates of P are $\left(6\sqrt{3}, 6\right)$.

3. (a)

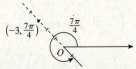

  (b) $x = -3 \cos \frac{7\pi}{4} = -3\left(\frac{\sqrt{2}}{2}\right) = -\frac{3\sqrt{2}}{2}$,

    $y = -3 \sin \frac{7\pi}{4} = -3\left(-\frac{\sqrt{2}}{2}\right) = \frac{3\sqrt{2}}{2}$. Thus, the

    rectangular coordinates of P are $\left(-\frac{3\sqrt{2}}{2}, \frac{3\sqrt{2}}{2}\right)$.

5. (a)

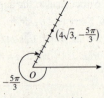

  (b) $x = 4\sqrt{3} \cos\left(-\frac{5\pi}{3}\right) = 4\sqrt{3}\left(\frac{1}{2}\right) = 2\sqrt{3}$,

    $y = 4\sqrt{3} \sin\left(-\frac{5\pi}{3}\right) = 4\sqrt{3}\left(\frac{\sqrt{3}}{2}\right) = 6$. Thus, the

    rectangular coordinates of P are $\left(2\sqrt{3}, 6\right)$.

7. (a)

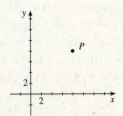

  (b) $r = \sqrt{8^2 + 8^2} = \sqrt{128} = 8\sqrt{2}$ and $\overline{\theta} = \tan^{-1} \frac{8}{8}$.

    Since P is in quadrant I, $\theta = \frac{\pi}{4}$. Polar coordinates

    for P are $\left(8\sqrt{2}, \frac{\pi}{4}\right)$.

  (c) $\left(-8\sqrt{2}, \frac{5\pi}{4}\right)$

9. (a)

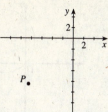

(b) $r = \sqrt{\left(-6\sqrt{2}\right)^2 + \left(-6\sqrt{2}\right)^2} = \sqrt{144} = 12$ and

$\bar{\theta} = \tan^{-1}\dfrac{-6\sqrt{2}}{-6\sqrt{2}} = \dfrac{\pi}{4}$. Since P is in quadrant III,

$\theta = \dfrac{5\pi}{4}$. Polar coordinates for P are $\left(12, \dfrac{5\pi}{4}\right)$.

(c) $\left(-12, \dfrac{\pi}{4}\right)$

11. (a)

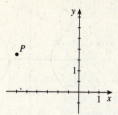

(b) $r = \sqrt{(-3)^2 + \left(\sqrt{3}\right)^2} = \sqrt{12} = 2\sqrt{3}$ and

$\bar{\theta} = \tan^{-1}\dfrac{\sqrt{3}}{-3}$. Since P is in quadrant II, $\theta = \dfrac{5\pi}{6}$.

Polar coordinates for P are $\left(2\sqrt{3}, \dfrac{5\pi}{6}\right)$.

(c) $\left(-2\sqrt{3}, -\dfrac{\pi}{6}\right)$

13. (a) $x + y = 4 \Leftrightarrow r\cos\theta + r\sin\theta = 4 \Leftrightarrow$

$r\left(\cos\theta + \sin\theta\right) = 4 \Leftrightarrow r = \dfrac{4}{\cos\theta + \sin\theta}$

(b) The rectangular equation is easier to graph.

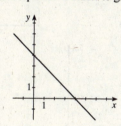

15. (a) $x^2 + y^2 = 4x + 4y \Leftrightarrow r^2 = 4r\cos\theta + 4r\sin\theta \Leftrightarrow$

$r^2 = r\left(4\cos\theta + 4\sin\theta\right) \Leftrightarrow r = 4\cos\theta + 4\sin\theta$

(b) The polar equation is easier to graph.

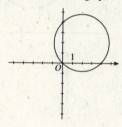

17. (a)

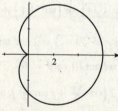

(b) $r = 3 + 3\cos\theta \Leftrightarrow r^2 = 3r + 3r\cos\theta$, which gives

$x^2 + y^2 = 3\sqrt{x^2 + y^2} + 3x \Leftrightarrow$

$x^2 - 3x + y^2 = 3\sqrt{x^2 + y^2}$. Squaring both sides

gives $\left(x^2 - 3x + y^2\right)^2 = 9\left(x^2 + y^2\right)$.

19. (a)

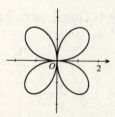

(b) $r = 2\sin 2\theta \Leftrightarrow r = 2 \cdot 2\sin\theta\cos\theta \Leftrightarrow$

$r^3 = 4r^2\sin\theta\cos\theta \Leftrightarrow$

$\left(r^2\right)^{3/2} = 4\left(r\sin\theta\right)\left(r\cos\theta\right)$ and so, since

$x = r\cos\theta$ and $y = r\sin\theta$, we get

$\left(x^2 + y^2\right)^3 = 16x^2y^2$.

21. (a)

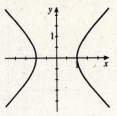

(b) $r^2 = \sec 2\theta = \dfrac{1}{\cos 2\theta} = \dfrac{1}{\cos^2 \theta - \sin^2 \theta} \Leftrightarrow$

$\qquad r^2 \left(\cos^2 \theta - \sin^2 \theta\right) = 1 \Leftrightarrow$

$\qquad r^2 \cos^2 \theta - r^2 \sin^2 \theta = 1 \Leftrightarrow$

$\qquad (r \cos \theta)^2 - (r \sin \theta)^2 = 1 \Leftrightarrow x^2 - y^2 = 1.$

25. $r = \cos(\theta/3)$, $\theta \in [0, 3\pi]$.

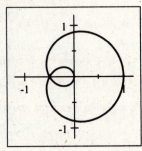

29. (a)

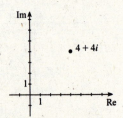

(b) $4 + 4i$ has $r = \sqrt{16 + 16} = 4\sqrt{2}$, and

$\qquad \theta = \tan^{-1} \frac{4}{4} = \frac{\pi}{4}$ (in quadrant I).

(c) $4 + 4i = 4\sqrt{2}\left(\cos \frac{\pi}{4} + i \sin \frac{\pi}{4}\right)$

33. (a)

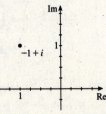

(b) $-1 + i$ has $r = \sqrt{1 + 1} = \sqrt{2}$ and $\tan \theta = \dfrac{1}{-1}$ with

$\qquad \theta$ in quadrant II $\Leftrightarrow \theta = \frac{3\pi}{4}$.

(c) $-1 + i = \sqrt{2}\left(\cos \frac{3\pi}{4} + i \sin \frac{3\pi}{4}\right)$

23. (a)

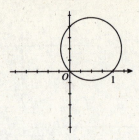

(b) $r = \sin \theta + \cos \theta \Leftrightarrow r^2 = r \sin \theta + r \cos \theta$, so

$\qquad x^2 + y^2 = y + x \quad \Leftrightarrow$

$\qquad \left(x^2 - x + \frac{1}{4}\right) + \left(y^2 - y + \frac{1}{4}\right) = \frac{1}{2} \Leftrightarrow$

$\qquad \left(x - \frac{1}{2}\right)^2 + \left(y - \frac{1}{2}\right)^2 = \frac{1}{2}.$

27. $r = 1 + 4\cos(\theta/3)$, $\theta \in [0, 6\pi]$.

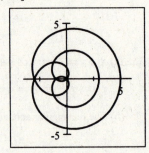

31. (a)

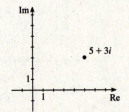

(b) $5 + 3i$. Then $r = \sqrt{25 + 9} = \sqrt{34}$, and

$\qquad \theta = \tan^{-1} \frac{3}{5}.$

(c) $5 + 3i = \sqrt{34}\left[\cos\left(\tan^{-1} \frac{3}{5}\right) + i \sin\left(\tan^{-1} \frac{3}{5}\right)\right]$

35. $1 - \sqrt{3}i$ has $r = \sqrt{1 + 3} = 2$ and $\tan \theta = \dfrac{-\sqrt{3}}{1} = -\sqrt{3}$

with θ in quadrant III $\Leftrightarrow \theta = \frac{5\pi}{3}$. Therefore,

$1 - \sqrt{3}i = 2\left(\cos \frac{5\pi}{3} + i \sin \frac{5\pi}{3}\right)$, and so

$\left(1 - \sqrt{3}i\right)^4 = 2^4 \left(\cos \frac{20\pi}{3} + i \sin \frac{20\pi}{3}\right)$

$\qquad = 16\left(\cos \frac{2\pi}{3} + i \sin \frac{2\pi}{3}\right)$

$\qquad = 16\left(-\frac{1}{2} + i \frac{\sqrt{3}}{2}\right) = 8\left(-1 + i\sqrt{3}\right).$

37. $\sqrt{3} + i$ has $r = \sqrt{3+1} = 2$ and $\tan\theta = \frac{1}{\sqrt{3}}$ with θ in quadrant I $\Leftrightarrow \theta = \frac{\pi}{6}$. Therefore, $\sqrt{3} + i = 2\left(\cos\frac{\pi}{6} + i\sin\frac{\pi}{6}\right)$, and so

$$\left(\sqrt{3}+i\right)^{-4} = 2^{-4}\left(\cos\frac{-4\pi}{6} + i\sin\frac{-4\pi}{6}\right) = \frac{1}{16}\left(\cos\frac{2\pi}{3} - i\sin\frac{2\pi}{3}\right) = \frac{1}{16}\left(-\frac{1}{2} - i\frac{\sqrt{3}}{2}\right)$$
$$= \frac{1}{32}\left(-1 - i\sqrt{3}\right) = -\frac{1}{32}\left(1 + i\sqrt{3}\right).$$

39. $-16i$ has $r = 16$ and $\theta = \frac{3\pi}{2}$. Thus, $-16i = 16\left(\cos\frac{3\pi}{2} + i\sin\frac{3\pi}{2}\right)$ and so

$$(-16i)^{1/2} = 16^{1/2}\left[\cos\left(\frac{3\pi + 4k\pi}{2}\right) + i\sin\left(\frac{3\pi + 4k\pi}{2}\right)\right] \text{ for } k = 0, 1. \text{ Thus}$$

the roots are $w_0 = 4\left(\cos\frac{3\pi}{4} + i\sin\frac{3\pi}{4}\right) = 4\left(-\frac{1}{\sqrt{2}} + i\frac{1}{\sqrt{2}}\right) = 2\sqrt{2}(-1+i)$ and

$w_1 = 4\left(\cos\frac{7\pi}{4} + i\sin\frac{7\pi}{4}\right) = 4\left(\frac{1}{\sqrt{2}} - i\frac{1}{\sqrt{2}}\right) = 2\sqrt{2}(1 - i).$

41. $1 = \cos 0 + i\sin 0$. Then $1^{1/6} = 1\left(\cos\frac{2k\pi}{6} + i\sin\frac{2k\pi}{6}\right)$ for $k = 0, 1, 2, 3, 4, 5$. Thus the six roots are

$w_0 = 1(\cos 0 + i\sin 0) = 1$, $w_1 = 1\left(\cos\frac{\pi}{3} + i\sin\frac{\pi}{3}\right) = \frac{1}{2} + i\frac{\sqrt{3}}{2}$, $w_2 = 1\left(\cos\frac{2\pi}{3} + i\sin\frac{2\pi}{3}\right) = -\frac{1}{2} + i\frac{\sqrt{3}}{2}$,

$w_3 = 1(\cos\pi + i\sin\pi) = -1$, $w_4 = 1\left(\cos\frac{4\pi}{3} + i\sin\frac{4\pi}{3}\right) = -\frac{1}{2} - i\frac{\sqrt{3}}{2}$, and $w_5 = 1\left(\cos\frac{5\pi}{3} + i\sin\frac{5\pi}{3}\right) = \frac{1}{2} - i\frac{\sqrt{3}}{2}$.

43. (a)

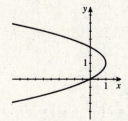

(b) $x = 1 - t^2$, $y = 1 + t \Leftrightarrow t = y - 1$. Substituting for t gives $x = 1 - (y - 1)^2 \Leftrightarrow (y - 1)^2 = 1 - x$ in rectangular coordinates.

45. (a)

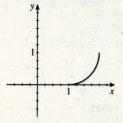

(b) $x = 1 + \cos t \Leftrightarrow \cos t = x - 1$, and $y = 1 - \sin t \Leftrightarrow \sin t = 1 - y$. Since $\cos^2 t + \sin^2 t = 1$, it follows that $(x - 1)^2 + (1 - y)^2 = 1 \Leftrightarrow (x - 1)^2 + (y - 1)^2 = 1$. Since t is restricted by $0 \le t \le \frac{\pi}{2}$, $1 + \cos 0 \le x \le 1 + \cos\frac{\pi}{2}$ $\Leftrightarrow 1 \le x \le 2$, and similarly, $0 \le y \le 1$. (This is the lower right quarter of the circle.)

47. $x = \cos 2t$, $y = \sin 3t$

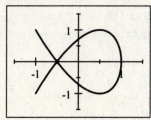

49. The coordinates of Q are $x = \cos\theta$ and $y = \sin\theta$. The coordinates of R are $x = 1$ and $y = \tan\theta$. Hence, the midpoint P is $\left(\dfrac{1 + \cos\theta}{2}, \dfrac{\sin\theta + \tan\theta}{2}\right)$, so parametric equations for the curve are $x = \dfrac{1 + \cos\theta}{2}$ and $y = \dfrac{\sin\theta + \tan\theta}{2}$.

CHAPTER 8 TEST

1. (a) $x = 8 \cos \frac{5\pi}{4} = 8 \left(-\frac{\sqrt{2}}{2} \right) = -4\sqrt{2}$, $y = 8 \sin \frac{5\pi}{4} = 8 \left(-\frac{\sqrt{2}}{2} \right) = -4\sqrt{2}$. So the point has rectangular coordinates $\left(-4\sqrt{2}, -4\sqrt{2} \right)$.

(b) $P = \left(-6, 2\sqrt{3} \right)$ in rectangular coordinates. So $\tan \theta = \frac{2\sqrt{3}}{-6}$ and the reference angle is $\bar{\theta} = \frac{\pi}{6}$. Since P is in quadrant II, we have $\theta = \frac{5\pi}{6}$. Next, $r^2 = (-6)^2 + \left(2\sqrt{3} \right)^2 = 36 + 12 = 48$, so $r = 4\sqrt{3}$. Thus, polar coordinates for the point are $\left(4\sqrt{3}, \frac{5\pi}{6} \right)$ or $\left(-4\sqrt{3}, \frac{11\pi}{6} \right)$.

3.

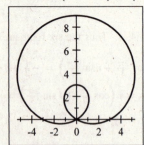

The curve is a limaçon.

5. $z_1 = 4 \left(\cos \frac{7\pi}{12} + i \sin \frac{7\pi}{12} \right)$ and $z_2 = 2 \left(\cos \frac{5\pi}{12} + i \sin \frac{5\pi}{12} \right)$.

Then $z_1 z_2 = 4 \cdot 2 \left[\cos \left(\frac{7\pi + 5\pi}{12} \right) + i \sin \left(\frac{7\pi + 5\pi}{12} \right) \right] = 8 \left(\cos \pi + i \sin \pi \right) = -8$ and

$z_1/z_2 = \frac{4}{2} \left[\cos \left(\frac{7\pi - 5\pi}{12} \right) + i \sin \left(\frac{7\pi - 5\pi}{12} \right) \right] = 2 \left(\cos \frac{\pi}{6} + i \sin \frac{\pi}{6} \right) = 2 \left(\frac{\sqrt{3}}{2} + \frac{1}{2}i \right) = \sqrt{3} + i$.

7. (a) $x = 3 \sin t + 3$, $y = 2 \cos t$, $0 \le t \le \pi$. From the work of part (b), we see that this is the half-ellipse shown.

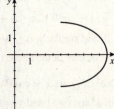

(b) $x = 3 \sin t + 3 \Leftrightarrow x - 3 = 3 \sin t \Leftrightarrow \frac{x - 3}{3} = \sin t$.

Squaring both sides gives $\frac{(x - 3)^2}{9} = \sin^2 t$. Similarly, $y = 2 \cos t \Leftrightarrow \frac{y}{2} = \cos t$, and squaring both sides gives $\frac{y^2}{4} = \cos^2 t$. Since $\sin^2 t + \cos^2 t = 1$, it follows that $\frac{(x - 3)^2}{9} + \frac{y^2}{4} = 1$. Since $0 \le t \le \pi$, $\sin t \ge 0$, so $3 \sin t \ge 0 \Rightarrow 3 \sin t + 3 \ge 3$, and so $x \ge 3$. Thus the curve consists of only the right half of the ellipse.

FOCUS ON MODELING   The Path of a Projectile

1. From $x = (v_0 \cos \theta) t$, we get $t = \frac{x}{v_0 \cos \theta}$. Substituting this value for t into the equation for y, we get

$y = (v_0 \sin \theta) t - \frac{1}{2} g t^2 \Leftrightarrow y = (v_0 \sin \theta) \left(\frac{x}{v_0 \cos \theta} \right) - \frac{1}{2} g \left(\frac{x}{v_0 \cos \theta} \right)^2 \Leftrightarrow y = (\tan \theta) x - \frac{g}{2 v_0^2 \cos^2 \theta} x^2$.

This shows that y is a quadratic function of x, so its graph is a parabola as long as $\theta \ne 90°$. When $\theta = 90°$, the path of the projectile is a straight line up (then down).

3. (a) We use the equation $t = \frac{2v_0 \sin\theta}{g}$. Substituting $g \approx 32$ ft/s^2, $\theta = 5°$, and $v_0 = 1000$, we get

$t = \frac{2 \cdot 1000 \cdot \sin 5°}{32} \approx 5.447$ seconds.

(b) Substituting the given values into $y = (v_0 \sin\theta)\, t - \frac{1}{2}gt^2$, we get

(d)

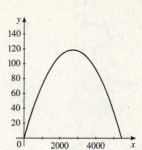

$y = 87.2t - 16t^2$. The maximum value of y is attained at the vertex of the

parabola; thus $y = 87.2t - 16t^2 = -16\left(t^2 - 5.45t\right)$ $\Leftrightarrow$

$y = -16\left[t^2 - 2\,(2.725)\,t + 7.425625\right] + 118.7$. Thus the greatest height

is 118.7 ft.

(c) The rocket hits the ground after 5.447 s, so substituting this into the

expression for the horizontal distance gives

$x = (1000\cos 5°)\,5.447 = 5426$ ft.

5. We use the equation of the parabola from Exercise 1 and find its vertex:

$$y = (\tan\theta)\, x - \frac{g}{2v_0^2\cos^2\theta}x^2 \Leftrightarrow y = -\frac{g}{2v_0^2\cos^2\theta}\left[x^2 - \frac{2v_0^2\sin\theta\cos\theta\, x}{g}\right] \Leftrightarrow$$

$$y = -\frac{g}{2v_0^2\cos^2\theta}\left[x^2 - \frac{2v_0^2\sin\theta\cos\theta\, x}{g} + \left(\frac{v_0^2\sin\theta\cos\theta}{g}\right)^2\right] + \frac{g}{2v_0^2\cos^2\theta}\cdot\left(\frac{v_0^2\sin\theta\cos\theta}{g}\right)^2 \Leftrightarrow$$

$$y = -\frac{g}{2v_0^2\cos^2\theta}\left[x - \frac{v_0^2\sin\theta\cos\theta}{g}\right]^2 + \frac{v_0^2\sin^2\theta}{2g}.$$ Thus the vertex is at $\left(\frac{v_0^2\sin\theta\cos\theta}{g}, \frac{v_0^2\sin^2\theta}{2g}\right)$, so the maximum

height is $\frac{v_0^2\sin^2\theta}{2g}$.

7. In Exercise 6 we derived the equations $x = (v_0\cos\theta - w)\,t$,

$y = (v_0\sin\theta)\,t - \frac{1}{2}gt^2$. We plot the graphs for the given values of

v_0, w, and θ in the figure to the right. The projectile will be blown

backwards if the horizontal component of its velocity is less than the

speed of the wind, that is, $32\cos\theta < 24$ $\Leftrightarrow$ $\cos\theta < \frac{3}{4}$ $\Rightarrow$

$\theta > 41.4°$.

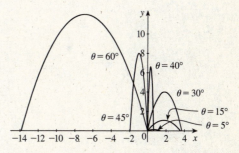

The optimal firing angle appears to be between 15° and 30°. We

graph the trajectory for $\theta = 20°$, $\theta = 23°$, and $\theta = 25°$. The solution

appears to be close to 23°.

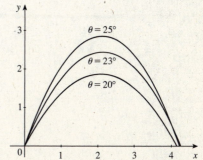

9 VECTORS IN TWO AND THREE DIMENSIONS

9.1 VECTORS IN TWO DIMENSIONS

1. (a) The vector **u** has initial point A and terminal point B.

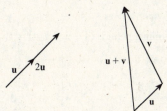

(b) The vector **u** has initial point $(2, 1)$ and terminal point $(4, 3)$. In component form we write $\mathbf{u} = \langle 2, 2 \rangle$ and $\mathbf{v} = \langle -3, 6 \rangle$. Then $2\mathbf{u} = \langle 4, 4 \rangle$ and $\mathbf{u} + \mathbf{v} = \langle -1, 8 \rangle$.

3. $2\mathbf{u} = 2 \langle -2, 3 \rangle = \langle -4, 6 \rangle$

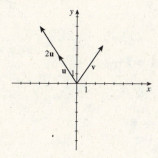

5. $\mathbf{u} + \mathbf{v} = \langle -2, 3 \rangle + \langle 3, 4 \rangle = \langle -2 + 3, 3 + 4 \rangle = \langle 1, 7 \rangle$

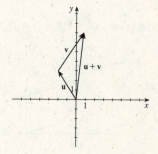

7. $\mathbf{v} - 2\mathbf{u} = \langle 3, 4 \rangle - 2 \langle -2, 3 \rangle = \langle 3 - 2(-2), 4 - 2(3) \rangle$

$\qquad = \langle 7, -2 \rangle$

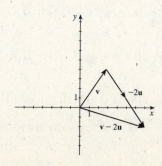

In Solutions 9–17, v represents the vector with initial point P and terminal point Q.

9. $P(2, 1)$, $Q(5, 4)$. $\mathbf{v} = \langle 5 - 2, 4 - 1 \rangle = \langle 3, 3 \rangle$

11. $P(1, 2)$, $Q(4, 1)$. $\mathbf{v} = \langle 4 - 1, 1 - 2 \rangle = \langle 3, -1 \rangle$

13. $P(3, 2)$, $Q(8, 9)$. $\mathbf{v} = \langle 8 - 3, 9 - 2 \rangle = \langle 5, 7 \rangle$

15. $P(5, 3)$, $Q(1, 0)$. $\mathbf{v} = \langle 1 - 5, 0 - 3 \rangle = \langle -4, -3 \rangle$

17. $P(-1, -1)$, $Q(-1, 1)$.
$\mathbf{v} = \langle -1 - (-1), 1 - (-1) \rangle = \langle 0, 2 \rangle$

19.

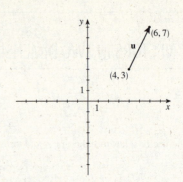

The terminal point is $(4 + 2, 3 + 4) = (6, 7)$.

21.

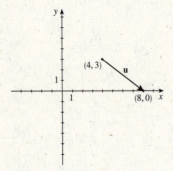

The terminal point is $(4 + 4, 3 - 3) = (8, 0)$.

23.

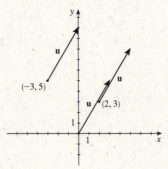

25.

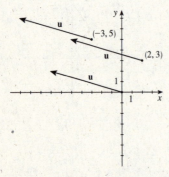

27. $\mathbf{u} = \langle 1, 4 \rangle = \mathbf{i} + 4\mathbf{j}$

29. $\mathbf{u} = \langle 3, 0 \rangle = 3\mathbf{i}$

31. $\mathbf{u} = \langle 2, 7 \rangle$, $\mathbf{v} = \langle 3, 1 \rangle$. $2\mathbf{u} = 2 \cdot \langle 2, 7 \rangle = \langle 4, 14 \rangle$; $-3\mathbf{v} = -3 \cdot \langle 3, 1 \rangle = \langle -9, -3 \rangle$; $\mathbf{u} + \mathbf{v} = \langle 2, 7 \rangle + \langle 3, 1 \rangle = \langle 5, 8 \rangle$; $3\mathbf{u} - 4\mathbf{v} = \langle 6, 21 \rangle - \langle 12, 4 \rangle = \langle -6, 17 \rangle$

33. $\mathbf{u} = \langle 0, -1 \rangle$, $\mathbf{v} = \langle -2, 0 \rangle$. $2\mathbf{u} = 2 \cdot \langle 0, -1 \rangle = \langle 0, -2 \rangle$; $-3\mathbf{v} = -3 \cdot \langle -2, 0 \rangle = \langle 6, 0 \rangle$; $\mathbf{u} + \mathbf{v} = \langle 0, -1 \rangle + \langle -2, 0 \rangle = \langle -2, -1 \rangle$; $3\mathbf{u} - 4\mathbf{v} = \langle 0, -3 \rangle - \langle -8, 0 \rangle = \langle 8, -3 \rangle$

35. $\mathbf{u} = 2\mathbf{i}$, $\mathbf{v} = 3\mathbf{i} - 2\mathbf{j}$. $2\mathbf{u} = 2 \cdot 2\mathbf{i} = 4\mathbf{i}$; $-3\mathbf{v} = -3(3\mathbf{i} - 2\mathbf{j}) = -9\mathbf{i} + 6\mathbf{j}$; $\mathbf{u} + \mathbf{v} = 2\mathbf{i} + 3\mathbf{i} - 2\mathbf{j} = 5\mathbf{i} - 2\mathbf{j}$; $3\mathbf{u} - 4\mathbf{v} = 3 \cdot 2\mathbf{i} - 4(3\mathbf{i} - 2\mathbf{j}) = -6\mathbf{i} + 8\mathbf{j}$

37. $\mathbf{u} = 2\mathbf{i} + \mathbf{j}$, $\mathbf{v} = 3\mathbf{i} - 2\mathbf{j}$. Then $|\mathbf{u}| = \sqrt{2^2 + 1^2} = \sqrt{5}$; $|\mathbf{v}| = \sqrt{3^2 + 2^2} = \sqrt{13}$; $2\mathbf{u} = 4\mathbf{i} + 2\mathbf{j}$; $|2\mathbf{u}| = \sqrt{4^2 + 2^2} = 2\sqrt{5}$; $\frac{1}{2}\mathbf{v} = \frac{3}{2}\mathbf{i} - \mathbf{j}$; $\left|\frac{1}{2}\mathbf{v}\right| = \sqrt{\left(\frac{3}{2}\right)^2 + 1^2} = \frac{1}{2}\sqrt{13}$; $\mathbf{u} + \mathbf{v} = 5\mathbf{i} - \mathbf{j}$; $|\mathbf{u} + \mathbf{v}| = \sqrt{5^2 + 1^2} = \sqrt{26}$; $\mathbf{u} - \mathbf{v} = 2\mathbf{i} + \mathbf{j} - 3\mathbf{i} + 2\mathbf{j} = -\mathbf{i} + 3\mathbf{j}$;
$|\mathbf{u} - \mathbf{v}| = \sqrt{1^2 + 3^2} = \sqrt{10}$; $|\mathbf{u}| - |\mathbf{v}| = \sqrt{5} - \sqrt{13}$

39. $\mathbf{u} = \langle 10, -1 \rangle$, $\mathbf{v} = \langle -2, -2 \rangle$. Then $|\mathbf{u}| = \sqrt{10^2 + 1^2} = \sqrt{101}$; $|\mathbf{v}| = \sqrt{(-2)^2 + (-2)^2} = 2\sqrt{2}$; $2\mathbf{u} = \langle 20, -2 \rangle$;

$|2\mathbf{u}| = \sqrt{20^2 + 2^2} = \sqrt{404} = 2\sqrt{101}$; $\frac{1}{2}\mathbf{v} = \langle -1, -1 \rangle$; $\left|\frac{1}{2}\mathbf{v}\right| = \sqrt{(-1)^2 + (-1)^2} = \sqrt{2}$; $\mathbf{u} + \mathbf{v} = \langle 8, -3 \rangle$;

$|\mathbf{u} + \mathbf{v}| = \sqrt{8^2 + 3^2} = \sqrt{73}$; $\mathbf{u} - \mathbf{v} = \langle 12, 1 \rangle$; $|\mathbf{u} - \mathbf{v}| = \sqrt{12^2 + 1^2} = \sqrt{145}$; $|\mathbf{u}| - |\mathbf{v}| = \sqrt{101} - 2\sqrt{2}$

41. $|\mathbf{v}| = 40$, direction $\theta = 30°$. $x = 40 \cos 30° = 20\sqrt{3}$ and $y = 40 \sin 30° = 20$. Thus, $\mathbf{v} = x\mathbf{i} + y\mathbf{j} = 20\sqrt{3}\mathbf{i} + 20\mathbf{j}$.

43. $|\mathbf{v}| = 1$, direction $\theta = 225°$. $x = \cos 225° = -\frac{1}{\sqrt{2}}$ and $y = \sin 225° = -\frac{1}{\sqrt{2}}$. Thus,

$\mathbf{v} = x\mathbf{i} + y\mathbf{j} = -\frac{1}{\sqrt{2}}\mathbf{i} - \frac{1}{\sqrt{2}}\mathbf{j} = -\frac{\sqrt{2}}{2}\mathbf{i} - \frac{\sqrt{2}}{2}\mathbf{j}$.

45. $|\mathbf{v}| = 4$, direction $\theta = 10°$. $x = 4 \cos 10° \approx 3.94$ and $y = 4 \sin 10° \approx 0.69$. Thus,

$\mathbf{v} = x\mathbf{i} + y\mathbf{j} = (4 \cos 10°)\mathbf{i} + (4 \sin 10°)\mathbf{j} \approx 3.94\mathbf{i} + 0.69\mathbf{j}$.

47. $\mathbf{v} = \langle 3, 4 \rangle$. The magnitude is $|\mathbf{v}| = \sqrt{3^2 + 4^2} = 5$. The direction is θ, where $\tan \theta = \frac{4}{3} \Leftrightarrow \theta = \tan^{-1}\left(\frac{4}{3}\right) \approx 53.13°$.

49. $\mathbf{v} = \langle -12, 5 \rangle$. The magnitude is $|\mathbf{v}| = \sqrt{(-12)^2 + 5^2} = \sqrt{169} = 13$. The direction is θ, where $\tan \theta = -\frac{5}{12}$ with θ in

quadrant II $\Leftrightarrow \theta = \pi + \tan^{-1}\left(-\frac{5}{12}\right) \approx 157.38°$.

51. $\mathbf{v} = \mathbf{i} + \sqrt{3}\mathbf{j}$. The magnitude is $|\mathbf{v}| = \sqrt{1^2 + \left(\sqrt{3}\right)^2} = 2$. The direction is θ, where $\tan \theta = \sqrt{3}$ with θ in quadrant I $\Leftrightarrow$

$\theta = \tan^{-1}\sqrt{3} = 60°$.

53. $|\mathbf{v}| = 30$, direction $\theta = 30°$. $x = 30 \cos 30° = 30 \cdot \frac{\sqrt{3}}{2} \approx 25.98$, $y = 30 \sin 30° = 15$. So the horizontal component of

force is $15\sqrt{3}$ lb and the vertical component is -15 lb.

55. The flow of the river can be represented by the vector $\mathbf{v} = -3\mathbf{j}$ and the swimmer can be represented by the vector $\mathbf{u} = 2\mathbf{i}$.

Therefore the true velocity is $\mathbf{u} + \mathbf{v} = 2\mathbf{i} - 3\mathbf{j}$.

57. The speed of the airplane is 300 mi/h, so its velocity relative to the air is

$\mathbf{v} = (-300 \cos \theta)\mathbf{i} + (-300 \sin \theta)\mathbf{j}$. The wind has velocity $\mathbf{w} = 30\mathbf{j}$, so

the true course of the airplane is given by

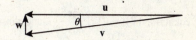

$\mathbf{u} = \mathbf{v} + \mathbf{w} = (-300 \cos \theta)\mathbf{i} + (-300 \sin \theta + 30)\mathbf{j}$. We want the y-component of the airplane's velocity to be 0, so we

solve $-300 \sin \theta + 30 = 0 \Leftrightarrow \sin \theta = \frac{1}{10} \Leftrightarrow \theta \approx 5.74°$. Therefore, the airplane should head in the direction 185.74° (or

S 84.26° W).

59. **(a)** The velocity of the wind is $40\mathbf{j}$.

(b) The velocity of the jet relative to the air is $425\mathbf{i}$.

(c) The true velocity of the jet is $\mathbf{v} = 425\mathbf{i} + 40\mathbf{j} = \langle 425, 40 \rangle$.

(d) The true speed of the jet is $|\mathbf{v}| = \sqrt{425^2 + 40^2} \approx 427$ mi/h, and the true direction is $\theta = \tan^{-1}\left(\frac{40}{425}\right) \approx 5.4° \Rightarrow \theta$ is

N 84.6° E.

61. If the direction of the plane is N 30° W, the airplane's velocity is $\mathbf{u} = \langle \mathbf{u}_x, \mathbf{u}_y \rangle$ where $\mathbf{u}_x = -765 \cos 60° = -382.5$,

and $\mathbf{u}_y = 765 \sin 60° \approx 662.51$. If the direction of the wind is N 30° E, the wind velocity is

$\mathbf{w} = \langle w_x, w_y \rangle$ where $w_x = 55 \cos 60° = 27.5$, and $w_y = 55 \sin 60° \approx 47.63$. Thus, the actual flight

path is $\mathbf{v} = \mathbf{u} + \mathbf{w} = \langle -382.5 + 27.5, 662.51 + 47.63 \rangle = \langle -355, 710.14 \rangle$, and so the true speed is

$|\mathbf{v}| = \sqrt{355^2 + 710.14^2} \approx 794$ mi/h, and the true direction is $\theta = \tan^{-1}\left(-\frac{710.14}{355}\right) \approx 116.6°$ so θ is N 26.6° W.

63. **(a)** The velocity of the river is represented by the vector $\mathbf{r} = \langle 10, 0 \rangle$.

(b) Since the boater direction is 60° from the shore at 20 mi/h, the velocity of the boat is represented by the vector

$\mathbf{b} = \langle 20 \cos 60°, 20 \sin 60° \rangle \approx \langle 10, 17.32 \rangle$.

(c) $\mathbf{w} = \mathbf{r} + \mathbf{b} = \langle 10 + 10, 0 + 17.32 \rangle = \langle 20, 17.32 \rangle$

(d) The true speed of the boat is $|\mathbf{w}| = \sqrt{20^2 + 17.32^2} \approx 26.5$ mi/h, and the true direction is

$\theta = \tan^{-1}\left(\frac{17.32}{20}\right) \approx 40.9° \approx$ N 49.1° E.

65. (a) Let $\mathbf{b} = \langle b_x, b_y \rangle$ represent the velocity of the boat relative to the water. Then $\mathbf{b} = \langle 24\cos 18°, 24\sin 18° \rangle$.

(b) Let $\mathbf{w} = \langle w_x, w_y \rangle$ represent the velocity of the water. Then $\mathbf{w} = \langle 0, w \rangle$ where w is the speed of the water. So the true velocity of the boat is $\mathbf{b} + \mathbf{w} = \langle 24\cos 18°, 24\sin 18° - w \rangle$. For the direction to be due east, we must have $24\sin 18° - w = 0 \Leftrightarrow w = 7.42$ mi/h. Therefore, the true speed of the water is 7.4 mi/h. Since $\mathbf{b} + \mathbf{w} = \langle 24\cos 18°, 0 \rangle$, the true speed of the boat is $|\mathbf{b} + \mathbf{w}| = 24\cos 18° \approx 22.8$ mi/h.

67. $\mathbf{F}_1 = \langle 2, 5 \rangle$ and $\mathbf{F}_2 = \langle 3, -8 \rangle$.

(a) $\mathbf{F}_1 + \mathbf{F}_2 = \langle 2 + 3, 5 - 8 \rangle = \langle 5, -3 \rangle$

(b) The additional force required is $\mathbf{F}_3 = \langle 0, 0 \rangle - \langle 5, -3 \rangle = \langle -5, 3 \rangle$.

69. $\mathbf{F}_1 = 4\mathbf{i} - \mathbf{j}$, $\mathbf{F}_2 = 3\mathbf{i} - 7\mathbf{j}$, $\mathbf{F}_3 = -8\mathbf{i} + 3\mathbf{j}$, and $\mathbf{F}_4 = \mathbf{i} + \mathbf{j}$.

(a) $\mathbf{F}_1 + \mathbf{F}_2 + \mathbf{F}_3 + \mathbf{F}_4 = (4 + 3 - 8 + 1)\mathbf{i} + (-1 - 7 + 3 + 1)\mathbf{j} = 0\mathbf{i} - 4\mathbf{j}$

(b) The additional force required is $\mathbf{F}_5 = 0\mathbf{i} + 0\mathbf{j} - (0\mathbf{i} - 4\mathbf{j}) = 4\mathbf{j}$.

71. $\mathbf{F}_1 = \langle 10\cos 60°, 10\sin 60° \rangle = \langle 5, 5\sqrt{3} \rangle$, $\mathbf{F}_2 = \langle -8\cos 30°, 8\sin 30° \rangle = \langle -4\sqrt{3}, 4 \rangle$, and

$\mathbf{F}_3 = \langle -6\cos 20°, -6\sin 20° \rangle \approx \langle -5.638, -2.052 \rangle$.

(a) $\mathbf{F}_1 + \mathbf{F}_2 + \mathbf{F}_3 = \langle 5 - 4\sqrt{3} - 5.638, 5\sqrt{3} + 4 - 2.052 \rangle \approx \langle -7.57, 10.61 \rangle$.

(b) The additional force required is $\mathbf{F}_4 = \langle 0, 0 \rangle - \langle -7.57, 10.61 \rangle = \langle 7.57, -10.61 \rangle$.

73. From the figure we see that $\mathbf{T}_1 = -|\mathbf{T}_1|\cos 50°\mathbf{i} + |\mathbf{T}_1|\sin 50°\mathbf{j}$ and $\mathbf{T}_2 = |\mathbf{T}_2|\cos 30°\mathbf{i} + |\mathbf{T}_2|\sin 30°\mathbf{j}$. Since $\mathbf{T}_1 + \mathbf{T}_2 = 100\mathbf{j}$ we get $-|\mathbf{T}_1|\cos 50° + |\mathbf{T}_2|\cos 30° = 0$ and $|\mathbf{T}_1|\sin 50° + |\mathbf{T}_2|\sin 30° = 100$. From the first equation, $|\mathbf{T}_2| = |\mathbf{T}_1|\dfrac{\cos 50°}{\cos 30°}$, and substituting into the second equation gives $|\mathbf{T}_1|\sin 50° + |\mathbf{T}_1|\dfrac{\cos 50°\sin 30°}{\cos 30°} = 100$

$\Leftrightarrow |\mathbf{T}_1|(\sin 50°\cos 30° + \cos 50°\sin 30°) = 100\cos 30° \Leftrightarrow |\mathbf{T}_1|\sin(50° + 30°) = 100\cos 30° \Leftrightarrow$

$|\mathbf{T}_1| = 100\dfrac{\cos 30°}{\sin 80°} \approx 87.9385$.

Similarly, solving for $|\mathbf{T}_1|$ in the first equation gives $|\mathbf{T}_1| = |\mathbf{T}_2|\dfrac{\cos 30°}{\cos 50°}$ and substituting gives

$|\mathbf{T}_2|\dfrac{\cos 30°\sin 50°}{\cos 50°} + |\mathbf{T}_2|\sin 30° = 100 \Leftrightarrow |\mathbf{T}_2|(\cos 30°\sin 50° + \cos 50°\sin 30°) = 100\cos 50° \Leftrightarrow$

$|\mathbf{T}_2| = \dfrac{100\cos 50°}{\sin 80°} \approx 65.2704$. Thus, $\mathbf{T}_1 \approx (-87.9416\cos 50°)\mathbf{i} + (87.9416\sin 50°)\mathbf{j} \approx -56.5\mathbf{i} + 67.4\mathbf{j}$ and

$\mathbf{T}_2 \approx (65.2704\cos 30°)\mathbf{i} + (65.2704\sin 30°)\mathbf{j} \approx 56.5\mathbf{i} + 32.6\mathbf{j}$.

75. When we add two (or more vectors), the resultant vector can be found by first placing the initial point of the second vector at the terminal point of the first vector. The resultant vector can then found by using the new terminal point of the second vector and the initial point of the first vector. When the n vectors are placed head to tail in the plane so that they form a polygon, the initial point and the terminal point are the same. Thus the sum of these n vectors is the zero vector.

9.2 THE DOT PRODUCT

1. The dot product of $\mathbf{a} = \langle a_1, a_2 \rangle$ and $\mathbf{b} = \langle b_1, b_2 \rangle$ is defined by $\mathbf{a} \cdot \mathbf{b} = a_1 a_2 + b_1 b_2$. The dot product of two vectors is a *real number, or scalar*, not a vector.

3. (a) The component of $\mathbf{a}$ along $\mathbf{b}$ is the scalar $|\mathbf{a}|\cos\theta$ and can be expressed in terms of the dot product as $\dfrac{\mathbf{a}\cdot\mathbf{b}}{|\mathbf{b}|}$.

(b) The projection of $\mathbf{a}$ onto $\mathbf{b}$ is the vector $\text{proj}_{\mathbf{b}}\,\mathbf{a}=\left(\dfrac{\mathbf{a}\cdot\mathbf{b}}{|\mathbf{b}|^2}\right)\mathbf{b}$.

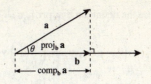

5. (a) $\mathbf{u}\cdot\mathbf{v}=\langle 2,0\rangle\cdot\langle 1,1\rangle=2+0=2$

(b) $\cos\theta=\dfrac{\mathbf{u}\cdot\mathbf{v}}{|\mathbf{u}|\,|\mathbf{v}|}=\dfrac{2}{2\cdot\sqrt{2}}=\dfrac{1}{\sqrt{2}}\Rightarrow\theta=45°$

7. (a) $\mathbf{u}\cdot\mathbf{v}=\langle 2,7\rangle\cdot\langle 3,1\rangle=6+7=13$

(b) $\cos\theta=\dfrac{\mathbf{u}\cdot\mathbf{v}}{|\mathbf{u}|\,|\mathbf{v}|}=\dfrac{13}{\sqrt{53}\cdot\sqrt{10}}\Rightarrow\theta\approx 56°$

9. (a) $\mathbf{u}\cdot\mathbf{v}=\langle 3,-2\rangle\cdot\langle 1,2\rangle=3+(-4)=-1$

(b) $\cos\theta=\dfrac{\mathbf{u}\cdot\mathbf{v}}{|\mathbf{u}|\,|\mathbf{v}|}=\dfrac{-1}{\sqrt{13}\cdot\sqrt{5}}\Rightarrow\theta\approx 97°$

11. (a) $\mathbf{u}\cdot\mathbf{v}=\langle 0,-5\rangle\cdot\left\langle -1,-\sqrt{3}\right\rangle=0+5\sqrt{3}=5\sqrt{3}$

(b) $\cos\theta=\dfrac{\mathbf{u}\cdot\mathbf{v}}{|\mathbf{u}|\,|\mathbf{v}|}=\dfrac{5\sqrt{3}}{5\cdot 2}=\dfrac{\sqrt{3}}{2}\Rightarrow\theta=30°$

13. (a) $\mathbf{u}\cdot\mathbf{v}=(\mathbf{i}+3\mathbf{j})\cdot(4\mathbf{i}-\mathbf{j})=4-3=1$

(b) $\cos\theta=\dfrac{\mathbf{u}\cdot\mathbf{v}}{|\mathbf{u}|\,|\mathbf{v}|}=\dfrac{1}{\sqrt{10}\cdot\sqrt{17}}\Rightarrow\theta\approx 85.6°$

15. $\mathbf{u}\cdot\mathbf{v}=-12+12=0\Rightarrow$ vectors are orthogonal

17. $\mathbf{u}\cdot\mathbf{v}=-8+12=4\neq 0\Rightarrow$ vectors are not orthogonal

19. $\mathbf{u}\cdot\mathbf{v}=-24+24=0\Rightarrow$ vectors are orthogonal

21. $\mathbf{u}\cdot\mathbf{v}+\mathbf{u}\cdot\mathbf{w}=\langle 2,1\rangle\cdot\langle 1,-3\rangle+\langle 2,1\rangle\cdot\langle 3,4\rangle$
$$=2-3+6+4=9$$

23. $(\mathbf{u}+\mathbf{v})\cdot(\mathbf{u}-\mathbf{v})=[\langle 2,1\rangle+\langle 1,-3\rangle]\cdot[\langle 2,1\rangle-\langle 1,-3\rangle]$
$$=\langle 3,-2\rangle\cdot\langle 1,4\rangle=3-8=-5$$

25. $x=\dfrac{\mathbf{u}\cdot\mathbf{v}}{|\mathbf{v}|}=\dfrac{12-24}{5}=-\dfrac{12}{5}$

27. $x=\dfrac{\mathbf{u}\cdot\mathbf{v}}{|\mathbf{v}|}=\dfrac{0-24}{1}=-24$

29. (a) $\mathbf{u}_1=\text{proj}_{\mathbf{v}}\,\mathbf{u}=\left(\dfrac{\mathbf{u}\cdot\mathbf{v}}{|\mathbf{v}|^2}\right)\mathbf{v}=\left(\dfrac{\langle -2,4\rangle\cdot\langle 1,1\rangle}{1^2+1^2}\right)\langle 1,1\rangle=\langle 1,1\rangle.$

(b) $\mathbf{u}_2=\mathbf{u}-\mathbf{u}_1=\langle -2,4\rangle-\langle 1,1\rangle=\langle -3,3\rangle$

31. (a) $\mathbf{u}_1=\text{proj}_{\mathbf{v}}\,\mathbf{u}=\left(\dfrac{\mathbf{u}\cdot\mathbf{v}}{|\mathbf{v}|^2}\right)\mathbf{v}=\left(\dfrac{\langle 1,2\rangle\cdot\langle 1,-3\rangle}{1^2+(-3)^2}\right)\langle 1,-3\rangle=-\dfrac{1}{2}\langle 1,-3\rangle=\left\langle -\dfrac{1}{2},\dfrac{3}{2}\right\rangle$

(b) $\mathbf{u}_2=\mathbf{u}-\mathbf{u}_1=\langle 1,2\rangle-\left\langle -\dfrac{1}{2},\dfrac{3}{2}\right\rangle=\left\langle \dfrac{3}{2},\dfrac{1}{2}\right\rangle$

33. (a) $\mathbf{u}_1=\text{proj}_{\mathbf{v}}\,\mathbf{u}=\left(\dfrac{\mathbf{u}\cdot\mathbf{v}}{|\mathbf{v}|^2}\right)\mathbf{v}=\left(\dfrac{\langle 2,9\rangle\cdot\langle -3,4\rangle}{(-3)^3+4^2}\right)\langle -3,4\rangle=\dfrac{6}{5}\langle -3,4\rangle=\left\langle -\dfrac{18}{5},\dfrac{24}{5}\right\rangle$

(b) $\mathbf{u}_2=\mathbf{u}-\mathbf{u}_1=\langle 2,9\rangle-\left\langle -\dfrac{18}{5},\dfrac{24}{5}\right\rangle=\left\langle \dfrac{28}{5},\dfrac{21}{5}\right\rangle$

35. $W=\mathbf{F}\cdot\mathbf{d}=\langle 4,-5\rangle\cdot\langle 3,8\rangle=-28$

37. $W=\mathbf{F}\cdot\mathbf{d}=\langle 10,3\rangle\cdot\langle 4,-5\rangle=25$

39. Let $\mathbf{u}=\langle u_1,u_2\rangle$ and $\mathbf{v}=\langle v_1,v_2\rangle$. Then
$$\mathbf{u}\cdot\mathbf{v}=\langle u_1,u_2\rangle\cdot\langle v_1,v_2\rangle=u_1v_1+u_2v_2=v_1u_1+v_2u_2=\langle v_1,v_2\rangle\cdot\langle u_1,u_2\rangle=\mathbf{v}\cdot\mathbf{u}$$

41. Let $\mathbf{u}=\langle u_1,u_2\rangle$, $\mathbf{v}=\langle v_1,v_2\rangle$, and $\mathbf{w}=\langle w_1,w_2\rangle$. Then
$$(\mathbf{u}+\mathbf{v})\cdot\mathbf{w}=(\langle u_1,u_2\rangle+\langle v_1,v_2\rangle)\cdot\langle w_1,w_2\rangle=\langle u_1+v_1,u_2+v_2\rangle\cdot\langle w_1,w_2\rangle$$
$$=u_1w_1+v_1w_1+u_2w_2+v_2w_2=u_1w_1+u_2w_2+v_1w_1+v_2w_2$$
$$=\langle u_1,u_2\rangle\cdot\langle w_1,w_2\rangle+\langle v_1,v_2\rangle\cdot\langle w_1,w_2\rangle=\mathbf{u}\cdot\mathbf{w}+\mathbf{v}\cdot\mathbf{w}$$

43. We use the definition that $\text{proj}_\mathbf{v}\,\mathbf{u} = \left(\dfrac{\mathbf{u} \cdot \mathbf{v}}{|\mathbf{v}|^2}\right)\mathbf{v}$. Then

$$\text{proj}_\mathbf{v}\,\mathbf{u} \cdot (\mathbf{u} - \text{proj}_\mathbf{v}\,\mathbf{u}) = \left(\dfrac{\mathbf{u} \cdot \mathbf{v}}{|\mathbf{v}|^2}\right)\mathbf{v} \cdot \left[\mathbf{u} - \left(\dfrac{\mathbf{u} \cdot \mathbf{v}}{|\mathbf{v}|^2}\right)\mathbf{v}\right] = \left(\dfrac{\mathbf{u} \cdot \mathbf{v}}{|\mathbf{v}|^2}\right)(\mathbf{v} \cdot \mathbf{u}) - \left(\dfrac{\mathbf{u} \cdot \mathbf{v}}{|\mathbf{v}|^2}\right)\mathbf{v} \cdot \left(\dfrac{\mathbf{u} \cdot \mathbf{v}}{|\mathbf{v}|^2}\right)\mathbf{v}$$

$$= \dfrac{(\mathbf{u} \cdot \mathbf{v})^2}{|\mathbf{v}|^2} - \dfrac{(\mathbf{u} \cdot \mathbf{v})^2}{|\mathbf{v}|^4}\,|\mathbf{v}|^2 = \dfrac{(\mathbf{u} \cdot \mathbf{v})^2}{|\mathbf{v}|^2} - \dfrac{(\mathbf{u} \cdot \mathbf{v})^2}{|\mathbf{v}|^2} = 0$$

Thus $\mathbf{u}$ and $\mathbf{u} - \text{proj}_\mathbf{v}\,\mathbf{u}$ are orthogonal.

45. $W = \mathbf{F} \cdot \mathbf{d} = \langle 4, -7 \rangle \cdot \langle 4, 0 \rangle = 16$ ft-lb

47. The distance vector is $\mathbf{D} = \langle 200, 0 \rangle$ and the force vector is $\mathbf{F} = \langle 50\cos 30°, 50\sin 30° \rangle$. Hence, the work done is
$W = \mathbf{F} \cdot \mathbf{D} = \langle 200, 0 \rangle \cdot \langle 50\cos 30°, 50\sin 30° \rangle = 200 \cdot 50\cos 30° \approx 8660$ ft-lb.

49. Since the weight of the car is 2755 lb, the force exerted perpendicular to the earth is 2755 lb. Resolving this into a force $\mathbf{u}$ perpendicular to the driveway gives $|\mathbf{u}| = 2766\cos 65° \approx 1164$ lb. Thus, a force of about 1164 lb is required.

51. Since the force required parallel to the plane is 80 lb and the weight of the package is 200 lb, it follows that $80 = 200\sin\theta$, where θ is the angle of inclination of the plane. Then $\theta = \sin^{-1}\left(\frac{80}{200}\right) \approx 23.58°$, and so the angle of inclination is approximately $23.6°$.

53. (a) $2(0) + 4(2) = 8$, so $Q(0, 2)$ lies on L. $2(2) + 4(1) = 4 + 4 = 8$, so $R(2, 1)$ lies on L.

(b) $\mathbf{u} = \overrightarrow{QP} = \langle 0, 2 \rangle - \langle 3, 4 \rangle = \langle -3, -2 \rangle$.

$\mathbf{v} = \overrightarrow{QR} = \langle 0, 2 \rangle - \langle 2, 1 \rangle = \langle -2, 1 \rangle$.

$\mathbf{w} = \text{proj}_\mathbf{v}\,\mathbf{u} = \left(\dfrac{\mathbf{u} \cdot \mathbf{v}}{|\mathbf{v}|^2}\right)\mathbf{v} = \dfrac{\langle -3, -2 \rangle \cdot \langle 2, 1 \rangle}{(-2)^2 + 1^2}\langle -2, 1 \rangle$

$= -\frac{8}{5}\langle -2, 1 \rangle = \left\langle \frac{16}{5}, -\frac{8}{5} \right\rangle$

(c) From the graph, we can see that $\mathbf{u} - \mathbf{w}$ is orthogonal to $\mathbf{v}$ (and thus to L). Thus, the distance from P to L is $|\mathbf{u} - \mathbf{w}|$.

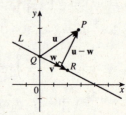

9.3   THREE-DIMENSIONAL COORDINATE GEOMETRY

1. In a three-dimensional coordinate system the three mutually perpendicular axes are called the x-axis, the y-axis, and the z-axis. The point P has coordinates $(5, 2, 3)$. The equation of the plane passing through P and parallel to the xz-plane is $y = 2$.

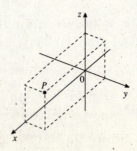

3. (a)

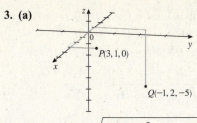

(b) $d(P, Q) = \sqrt{(-1-3)^2 + (2-1)^2 + (-5-0)^2}$

$\qquad = \sqrt{42}$

5. (a)

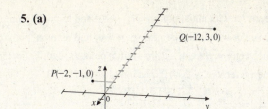

(b) $d(P, Q) = \sqrt{(-12+2)^2 + (3+1)^2 + (0-0)^2}$

$\qquad = 2\sqrt{29}$

7. $x = 4$ is a plane parallel to the yz-plane.

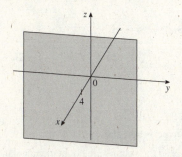

9. $z = 8$ is a plane parallel to the xy-plane.

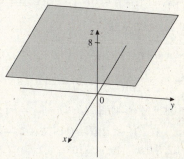

11. A sphere with radius $r = 5$ and center $C(2, -5, 3)$ has equation $(x-2)^2 + [y-(-5)]^2 + (z-3)^2 = 5^2$, or $(x-2)^2 + (y+5)^2 + (z-3)^2 = 25$.

13. A sphere with radius $r = \sqrt{6}$ and center $C(3, -1, 0)$ has equation $(x-3)^2 + (y+1)^2 + z^2 = 6$.

15. We complete the squares in x, y, and z: $x^2 + y^2 + z^2 - 10x + 2y + 8z = 9 \Leftrightarrow$

$\left(x^2 - 10x + 25\right) + \left(y^2 + 2y + 1\right) + \left(z^2 + 8z + 16\right) = 9 + 25 + 1 + 16 \Leftrightarrow (x-5)^2 + (y+1)^2 + (z+4)^2 = 51$. This

is an equation of a sphere with center $(5, -1, -4)$ and radius $\sqrt{51}$.

17. We complete the squares in x, y, and z: $x^2 + y^2 + z^2 = 12x + 2y \Leftrightarrow \left(x^2 - 12x + 36\right) + \left(y^2 - 2y + 1\right) + z^2 = 36 + 1$

$\Leftrightarrow (x-6)^2 + (y-1)^2 + z^2 = 37$. This is an equation of a sphere with center $(6, 1, 0)$ and radius $\sqrt{37}$.

19. (a) To find the trace in the yz-plane, we set $x = 0$: $(0+1)^2 + (y-2)^2 + (z+10)^2 = 100 \Leftrightarrow (y-2)^2 + (z+10)^2 = 99$. This represents a circle with center $(0, 2, -10)$ and radius $3\sqrt{11}$.

(b) We set $x = 4$ and find $(4+1)^2 + (y-2)^2 + (z+10)^2 = 100 \Leftrightarrow (y-2)^2 + (z+10)^2 = 75$. This represents a circle with center $(4, 2, -10)$ and radius $5\sqrt{3}$.

21. With the origin at its center, an equation of the tank is $x^2 + y^2 + z^2 = 25$. The metal circle is the trace in the plane $z = -4$, so its equation is $x^2 + y^2 + (-4)^2 = 25$ or $x^2 + y^2 = 9$. Therefore, its radius is 3.

23. We use the Distance Formula to write the condition that $X(x, y, z)$ is equidistant from $P(0, 0, 0)$ and $Q(0, 3, 0)$. (It is easier to equate the *squares* of the distances.) $|XP|^2 = |XQ|^2 \Leftrightarrow (x-0)^2 + (y-0)^2 + (z-0)^2 = (x-0)^2 + (y-3)^2 + (z-0)^2$

$\Leftrightarrow y^2 = (y-3)^2 \Leftrightarrow y^2 = y^2 - 6y + 9 \Leftrightarrow 6y = 9 \Leftrightarrow y = \frac{3}{2}$. This is an equation of a plane parallel to the xz-plane.

9.4 VECTORS IN THREE DIMENSIONS

1. A vector in three dimensions can be written in terms of the *unit* vectors $\mathbf{i}$, $\mathbf{j}$, and $\mathbf{k}$ as $a = a_1\mathbf{i} + a_2\mathbf{j} + a_3\mathbf{k}$. The magnitude of the vector $\mathbf{a}$ is $|a| = \sqrt{a_1^2 + a_2^2 + a_3^2}$. So $\langle 4, -2, 4 \rangle = 4\mathbf{i} + (-2)\mathbf{j} + 4\mathbf{k}$ and $7\mathbf{j} - 24\mathbf{k} = \langle 0, 7, -24 \rangle$.

3. The vector with initial point $P(1, -1, 0)$ and terminal point $Q(0, -2, 5)$ is $\mathbf{v} = \langle 0 - 1, -2 - (-1), 5 - 0 \rangle = \langle -1, -1, 5 \rangle$.

5. The vector with initial point $P(6, -1, 0)$ and terminal point $Q(0, -3, 0)$ is $\mathbf{v} = \langle 0 - 6, -3 - (-1), 0 - 0 \rangle = \langle -6, -2, 0 \rangle$.

7. If the vector $\mathbf{v} = \langle 3, 4, -2 \rangle$ has initial point $P(2, 0, 1)$, its terminal point is $(2 + 3, 0 + 4, 1 - 2) = (5, 4, -1)$.

9. If the vector $\mathbf{v} = \langle -2, 0, 2 \rangle$ has initial point $P(3, 0, -3)$, its terminal point is $(3 - 2, 0 + 0, -3 + 2) = (1, 0, -1)$.

11. $|\langle -2, 1, 2 \rangle| = \sqrt{(-2)^2 + 1^2 + 2^2} = 3$ **13.** $|\langle 3, 5, -4 \rangle| = \sqrt{3^2 + 5^2 + (-4)^2} = 5\sqrt{2}$

15. If $\mathbf{u} = \langle 2, -7, 3 \rangle$ and $\mathbf{v} = \langle 0, 4, -1 \rangle$, then $\mathbf{u} + \mathbf{v} = \langle 2 + 0, -7 + 4, 3 - 1 \rangle = \langle 2, -3, 2 \rangle$,

$\mathbf{u} - \mathbf{v} = \langle 2 - 0, -7 - 4, 3 - (-1) \rangle = \langle 2, -11, 4 \rangle$, and

$3\mathbf{u} - \frac{1}{2}\mathbf{v} = \left\langle 3(2) - \frac{1}{2}(0), 3(-7) - \frac{1}{2}(4), 3(3) - \frac{1}{2}(-1) \right\rangle = \left\langle 6, -23, \frac{19}{2} \right\rangle$.

17. If $\mathbf{u} = \mathbf{i} + \mathbf{j}$ and $\mathbf{v} = -\mathbf{j} - 2\mathbf{k}$, then $\mathbf{u} + \mathbf{v} = \mathbf{i} + \mathbf{j} - \mathbf{j} - 2\mathbf{k} = \mathbf{i} - 2\mathbf{k}$, $\mathbf{u} - \mathbf{v} = \mathbf{i} + \mathbf{j} - (-\mathbf{j} - 2\mathbf{k}) = \mathbf{i} + 2\mathbf{j} + 2\mathbf{k}$, and

$3\mathbf{u} - \frac{1}{2}\mathbf{v} = 3(\mathbf{i} + \mathbf{j}) - \frac{1}{2}(-\mathbf{j} - 2\mathbf{k}) = 3\mathbf{i} + \frac{7}{2}\mathbf{j} + \mathbf{k}$.

19. $\langle 12, 0, 2 \rangle = 12\mathbf{i} + 2\mathbf{k}$ **21.** $\langle 3, -3, 0 \rangle = 3\mathbf{i} - 3\mathbf{j}$

23. (a) $-2\mathbf{u} + 3\mathbf{v} = -2\langle 0, -2, 1 \rangle + 3\langle 1, -1, 0 \rangle = \langle 3, 1, -2 \rangle$ **(b)** $-2\mathbf{u} + 3\mathbf{v} = 3\mathbf{i} + \mathbf{j} - 2\mathbf{k}$

25. $\mathbf{u} \cdot \mathbf{v} = \langle 2, 5, 0 \rangle \cdot \left\langle \frac{1}{2}, -1, 10 \right\rangle = 2\left(\frac{1}{2}\right) + 5(-1) + 0(10) = -4$

27. $\mathbf{u} \cdot \mathbf{v} = (6\mathbf{i} - 4\mathbf{j} - 2\mathbf{k}) \cdot \left(\frac{5}{6}\mathbf{i} + \frac{3}{2}\mathbf{j} - \mathbf{k}\right) = 6\left(\frac{5}{6}\right) - 4\left(\frac{3}{2}\right) - 2(-1) = 1$

29. $\langle 4, -2, -4 \rangle \cdot \langle 1, -2, 2 \rangle = 4(1) - 2(-2) - 4(2) = 0$, so the vectors are perpendicular.

31. $\langle 0.3, 1.2, -0.9 \rangle \cdot \langle 10, -5, 10 \rangle = 0.3(10) + 1.2(-5) - 0.9(10) = -12$, so the vectors are not perpendicular.

33. $\cos\theta = \dfrac{\mathbf{u} \cdot \mathbf{v}}{|\mathbf{u}|\,|\mathbf{v}|} = \dfrac{\langle 2, -2, -1 \rangle \cdot \langle 1, 2, 2 \rangle}{|\langle 2, -2, -1 \rangle|\,|\langle 1, 2, 2 \rangle|} = \dfrac{2(1) - 2(2) - 1(2)}{\sqrt{2^2 + (-2)^2 + (-1)^2}\sqrt{1^2 + 2^2 + 2^2}} = -\dfrac{4}{9}$, so

$\theta = \cos^{-1}\left(-\dfrac{4}{9}\right) \approx 116.4°$.

35. $\cos\theta = \dfrac{\mathbf{u} \cdot \mathbf{v}}{|\mathbf{u}|\,|\mathbf{v}|} = \dfrac{(\mathbf{j} + \mathbf{k}) \cdot (\mathbf{i} + 2\mathbf{j} - 3\mathbf{k})}{|\mathbf{j} + \mathbf{k}|\,|\mathbf{i} + 2\mathbf{j} - 3\mathbf{k}|} = \dfrac{1(2) + 1(-3)}{\sqrt{1^2 + 1^2}\sqrt{1^2 + 2^2 + (-3)^2}} = -\dfrac{\sqrt{28}}{28}$, so $\theta = \cos^{-1}\left(-\dfrac{\sqrt{28}}{28}\right) \approx 100.9°$.

37. The length of the vector $3\mathbf{i} + 4\mathbf{j} + 5\mathbf{k}$ is $\sqrt{3^2 + 4^2 + 5^2} = 5\sqrt{2}$, so by definition, its direction angles satisfy $\cos\alpha = \dfrac{3}{5\sqrt{2}}$,

$\cos\beta = \dfrac{4}{5\sqrt{2}}$, and $\cos\gamma = \dfrac{1}{\sqrt{2}}$. Thus, $\alpha = \cos^{-1}\dfrac{3}{5\sqrt{2}} \approx 65°$, $\beta = \cos^{-1}\dfrac{2\sqrt{2}}{5} \approx 56°$, and $\gamma = \cos^{-1}\dfrac{1}{\sqrt{2}} = 45°$.

39. $|\langle 2, 3, -6 \rangle| = \sqrt{2^2 + 3^2 + (-6)^2} = 7$, so $\cos\alpha = \dfrac{2}{7}$, $\cos\beta = \dfrac{3}{7}$, and $\cos\gamma = \dfrac{-6}{7} \Leftrightarrow \alpha = \cos^{-1}\dfrac{2}{7} \approx 73°$,

$\beta = \cos^{-1}\dfrac{3}{7} \approx 65°$, and $\gamma = \cos^{-1}\left(-\dfrac{6}{7}\right) \approx 149°$.

41. We are given that $\alpha = \dfrac{\pi}{3}$, $\gamma = \dfrac{2\pi}{3}$, and β is acute. Using the property of direction cosines $\cos^2\alpha + \cos^2\beta + \cos^2\gamma = 1$,

we have $\left(\cos\dfrac{\pi}{3}\right)^2 + \cos^2\beta + \left(\cos\dfrac{2\pi}{3}\right)^2 = 1 \Leftrightarrow \left(\dfrac{1}{2}\right)^2 + \cos^2\beta + \left(-\dfrac{1}{2}\right)^2 = 1 \Leftrightarrow \cos^2\beta = \dfrac{1}{2} \Leftrightarrow \cos\beta = \pm\dfrac{1}{\sqrt{2}}$. Because

β is acute, we have $\beta = \cos^{-1}\dfrac{1}{\sqrt{2}} = 45°$.

43. We are given that $\alpha = 60°$, $\beta = 50°$, and γ is obtuse, so $\cos^2 60° + \cos^2 50° + \cos^2\gamma = 1 \Leftrightarrow$

$\cos^2\gamma = 1 - \cos^2 60° - \cos^2 50° \approx 0.337$. Because γ is obtuse, $\gamma \approx \cos^{-1}\left(-\sqrt{0.337}\right) \approx 125°$.

45. Here $\cos^2\alpha + \cos^2\beta = \cos^2 20° + \cos^2 45° \approx 1.38 > 1$, so there is no angle γ satisfying the property of direction cosines $\cos^2\alpha + \cos^2\beta + \cos^2\gamma = 1$.

47. (a) The second and third forces are $\mathbf{F}_2 = 24\mathbf{j}$ and $\mathbf{F}_3 = -25\mathbf{k}$. Therefore, $\mathbf{F}_1 + \mathbf{F}_2 + \mathbf{F}_3 + \mathbf{F}_4 = 0 \Leftrightarrow$

$7\mathbf{i} + 24\mathbf{j} - 25\mathbf{k} + \mathbf{F}_4 = 0 \Leftrightarrow \mathbf{F}_4 = -7\mathbf{i} - 24\mathbf{j} + 25\mathbf{k}$.

(b) $|\mathbf{F}_4| = \sqrt{(-7)^2 + (-24)^2 + (-25)^2} = 25\sqrt{2}$

49. (a) We solve $\mathbf{v} = a\mathbf{u} \Leftrightarrow \langle -6, 4, -8 \rangle = a\langle 3, -2, 4 \rangle \Leftrightarrow a = -2$. Therefore, the vectors are parallel and $\mathbf{v} = -2\mathbf{u}$.

(b) $\mathbf{v} = a\mathbf{u} \Leftrightarrow \langle 12, 8, -16 \rangle = a\langle -9, -6, 12 \rangle \Leftrightarrow a = -\frac{4}{3}$, so the vectors are parallel and $\mathbf{v} = -\frac{4}{3}\mathbf{u}$.

(c) $\mathbf{v} = a\mathbf{u} \Leftrightarrow 2\mathbf{i} + 2\mathbf{j} - 2\mathbf{k} = a(\mathbf{i} + \mathbf{j} + \mathbf{k})$ has no solution, so the vectors are not parallel.

51. (a) $(\mathbf{r} - \mathbf{a}) \cdot (\mathbf{r} - \mathbf{b}) = 0 \Leftrightarrow \langle x - 2, y - 2, z - 2 \rangle \cdot \langle x - (-2), y - (-2), z - 0 \rangle = 0 \Leftrightarrow (x - 2)(x + 2) +$
$(y - 2)(y + 2) + (z - 2)z = 0 \Leftrightarrow x^2 - 4 + y^2 - 4 + z^2 - 2z = 0 \Leftrightarrow x^2 + y^2 + (z - 1)^2 = 4 + 4 + 1 = 9.$

(b) The sphere with equation $x^2 + y^2 + (z - 1)^2 = 9$ has center $(0, 0, 1)$ and radius 3.

(c) The diagram shows the plane determined by $\mathbf{a}$, $\mathbf{b}$, and $\mathbf{r}$, along with the trace of the sphere in that plane. We see that the equation $(\mathbf{r} - \mathbf{a}) \cdot (\mathbf{r} - \mathbf{b}) = 0$ states the fact that lines from the ends of a diameter of a circle to any point on its surface meet at right angles.

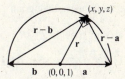

(d) Let $\mathbf{a} = \langle 0, 1, 3 \rangle$ and $\mathbf{b} = \langle 2, -1, 4 \rangle$. Then $(\mathbf{r} - \mathbf{a}) \cdot (\mathbf{r} - \mathbf{b}) = 0 \Leftrightarrow \langle x, y - 1, z - 3 \rangle \cdot \langle x - 2, y + 1, z - 4 \rangle = 0$
$\Leftrightarrow x(x - 2) + (y - 1)(y + 1) + (z - 3)(z - 4) = 0 \Leftrightarrow x^2 - 2x + y^2 - 1 + z^2 - 7x + 12 = 0 \Leftrightarrow$
$(x - 1)^2 + y^2 + \left(z - \frac{7}{2}\right)^2 = 1 - 12 - 1 + \frac{49}{4} = \frac{1}{4}$, an equation of a circle with center $\left(1, 0, \frac{7}{2}\right)$ and radius $\frac{1}{2}$.

9.5   THE CROSS PRODUCT

1. The cross product of the vectors $\mathbf{a} = \langle a_1, a_2, a_3 \rangle$ and $\mathbf{b} = \langle b_1, b_2, b_3 \rangle$ is the vector

$$\mathbf{a} \times \mathbf{b} = \begin{vmatrix} \mathbf{i} & \mathbf{j} & \mathbf{k} \\ a_1 & a_2 & a_3 \\ b_1 & b_2 & b_3 \end{vmatrix} = (a_2 b_3 - a_3 b_2)\mathbf{i} + (a_3 b_1 - a_1 b_3)\mathbf{j} + (a_1 b_2 - a_2 b_1)\mathbf{k}.$$ So the cross product of $\mathbf{a} = \langle 1, 0, 1 \rangle$ and

$\mathbf{b} = \langle 2, 3, 0 \rangle$ is $\mathbf{a} \times \mathbf{b} = \begin{vmatrix} \mathbf{i} & \mathbf{j} & \mathbf{k} \\ 1 & 0 & 1 \\ 2 & 3 & 0 \end{vmatrix} = -3\mathbf{i} + 2\mathbf{j} + 3\mathbf{k}.$

3. $\mathbf{a} \times \mathbf{b} = \begin{vmatrix} \mathbf{i} & \mathbf{j} & \mathbf{k} \\ 1 & 0 & -3 \\ 2 & 3 & 0 \end{vmatrix} = 9\mathbf{i} - 6\mathbf{j} + 3\mathbf{k}$ $\qquad$ **5.** $\mathbf{a} \times \mathbf{b} = \begin{vmatrix} \mathbf{i} & \mathbf{j} & \mathbf{k} \\ 6 & -2 & 8 \\ -9 & 3 & -12 \end{vmatrix} = \mathbf{0}$

7. $\mathbf{a} \times \mathbf{b} = \begin{vmatrix} \mathbf{i} & \mathbf{j} & \mathbf{k} \\ 1 & 1 & 1 \\ 3 & 0 & -4 \end{vmatrix} = -4\mathbf{i} + 7\mathbf{j} - 3\mathbf{k}$

9. (a) $\mathbf{a} \times \mathbf{b} = \begin{vmatrix} \mathbf{i} & \mathbf{j} & \mathbf{k} \\ 1 & 1 & -1 \\ -1 & 1 & -1 \end{vmatrix} = \langle 0, 2, 2 \rangle$ is perpendicular to both $\mathbf{a}$ and $\mathbf{b}$.

(b) $\dfrac{\mathbf{a} \times \mathbf{b}}{|\mathbf{a} \times \mathbf{b}|} = \dfrac{\langle 0, 2, 2 \rangle}{\sqrt{2^2 + 2^2}} = \left\langle 0, \dfrac{\sqrt{2}}{2}, \dfrac{\sqrt{2}}{2} \right\rangle$ is a unit vector perpendicular to both $\mathbf{a}$ and $\mathbf{b}$.

11. (a) $\mathbf{a} \times \mathbf{b} = \begin{vmatrix} \mathbf{i} & \mathbf{j} & \mathbf{k} \\ \frac{1}{2} & -1 & \frac{2}{3} \\ 6 & -12 & -6 \end{vmatrix} = \langle 14, 7, 0 \rangle$ is perpendicular to both $\mathbf{a}$ and $\mathbf{b}$.

(b) $\dfrac{\mathbf{a} \times \mathbf{b}}{|\mathbf{a} \times \mathbf{b}|} = \dfrac{\langle 14, 7, 0\rangle}{\sqrt{14^2 + 7^2}} = \left\langle \frac{2\sqrt{5}}{5}, \frac{\sqrt{5}}{5}, 0\right\rangle$ is a unit vector perpendicular to both $\mathbf{a}$ and $\mathbf{b}$.

13. $|\mathbf{a} \times \mathbf{b}| = |\mathbf{a}|\,|\mathbf{b}| \sin\theta = 6\left(\frac{1}{2}\right)\sin 60^\circ = \frac{3\sqrt{3}}{2}$ **15.** $|\mathbf{a} \times \mathbf{b}| = |\mathbf{a}|\,|\mathbf{b}| \sin\theta = 10\,(10) \sin 90^\circ = 100$

17. $\overrightarrow{PQ} \times \overrightarrow{PR} = \langle 1, 1, -1\rangle \times \langle -2, 0, 0\rangle = \langle 0, 2, 2\rangle$ is perpendicular to the plane passing through P, Q, and R.

19. $\overrightarrow{PQ} \times \overrightarrow{PR} = \langle 1, 1, 5\rangle \times \langle -1, -1, 5\rangle = \langle 10, -10, 0\rangle$ is perpendicular to the plane passing through P, Q, and R.

21. The area of the parallelogram determined by $\mathbf{u} = \langle 3, 2, 1\rangle$ and $\mathbf{v} = \langle 1, 2, 3\rangle$ is $|\mathbf{u} \times \mathbf{v}| = |\langle 4, -8, 4\rangle| = 4\sqrt{6}$.

23. The area of the parallelogram determined by $\mathbf{u} = 2\mathbf{i} - \mathbf{j} + 4\mathbf{k}$ and $\mathbf{v} = \frac{1}{2}\mathbf{i} + 2\mathbf{j} - \frac{3}{2}\mathbf{k}$ is $|\mathbf{u} \times \mathbf{v}| = \left|\left\langle -\frac{13}{2}, 5, \frac{9}{2}\right\rangle\right| = \frac{5\sqrt{14}}{2}$.

25. The area of triangle PQR is one-half the area of the parallelogram determined by $\overrightarrow{PQ}$ and $\overrightarrow{PR}$, that is,

$$\tfrac{1}{2}\left|\overrightarrow{PQ} \times \overrightarrow{PR}\right| = \tfrac{1}{2}\,|\langle -1, 1, -1\rangle \times \langle 1, 3, 3\rangle| = \tfrac{1}{2}\sqrt{6^2 + 2^2 + (-4)^2} = \sqrt{14}.$$

27. The area of triangle PQR is one-half the area of the parallelogram determined by $\overrightarrow{PQ}$ and $\overrightarrow{PR}$, that is,

$$\tfrac{1}{2}\left|\overrightarrow{PQ} \times \overrightarrow{PR}\right| = \tfrac{1}{2}\,|\langle -6, -6, 0\rangle \times \langle -6, 0, -6\rangle| = \tfrac{1}{2}\sqrt{36^2 + (-36)^2 + (-36)^2} = 18\sqrt{3}.$$

29. (a) $\mathbf{a} \cdot (\mathbf{b} \times \mathbf{c}) = \langle 1, 2, 3\rangle \cdot (\langle -3, 2, 1\rangle \times \langle 0, 8, 10\rangle) = \langle 1, 2, 3\rangle \cdot \langle 12, 30, -24\rangle = 0$

    (b) Because their scalar triple product is 0, the vectors are coplanar.

31. (a) $\mathbf{a} \cdot (\mathbf{b} \times \mathbf{c}) = \langle 2, 3, -2\rangle \cdot (\langle -1, 4, 0\rangle \times \langle 3, -1, 3\rangle) = \langle 2, 3, -2\rangle \cdot \langle 12, 3, -11\rangle = 55$

    (b) Because their scalar triple product is nonzero, the vectors are not coplanar. The volume of the parallelepiped that they determine is $|\mathbf{a} \cdot (\mathbf{b} \times \mathbf{c})| = 55$.

33. (a) $\mathbf{a} \cdot (\mathbf{b} \times \mathbf{c}) = \langle 1, -1, 1\rangle \cdot (\langle 0, -1, 1\rangle \times \langle 1, 1, 1\rangle) = \langle 1, -1, 1\rangle \cdot \langle -2, 1, 1\rangle = -2$

    (b) Because their scalar triple product is nonzero, the vectors are not coplanar. The volume of the parallelepiped that they determine is $|\mathbf{a} \cdot (\mathbf{b} \times \mathbf{c})| = 2$.

35. (a) We have $|\mathbf{a}| = 120$ cm, $|\mathbf{b}| = 150$ cm, $|\mathbf{c}| = 300$ cm, the angle between $\mathbf{b}$ and $\mathbf{c}$ is $90^\circ - 30^\circ = 60^\circ$, and the angle between $\mathbf{a}$ and $\mathbf{b} \times \mathbf{c}$ is 0° (because $\mathbf{a}$ is perpendicular to both $\mathbf{b}$ and $\mathbf{c}$). Therefore,

    $\mathbf{a} \cdot (\mathbf{b} \times \mathbf{c}) = |\mathbf{a}|\,|\mathbf{b} \times \mathbf{c}|\cos 0^\circ = 120\,(150 \cdot 300 \cdot \sin 60^\circ) = 2{,}700{,}000\sqrt{3} \approx 4{,}676{,}537.$

    (b) The capacity in liters is approximately $\dfrac{4{,}676{,}537 \text{ cm}^3}{1000 \text{ cm}^3/\text{L}} \approx 4677$ liters.

37. (a) $\mathbf{u} \cdot (\mathbf{v} \times \mathbf{w}) = \langle 0, 1, 1\rangle \cdot (\langle 1, 0, 1\rangle \times \langle 1, 1, 0\rangle) = \langle 0, 1, 1\rangle \cdot \langle -1, 1, 1\rangle = 2,\ \mathbf{u} \cdot (\mathbf{w} \times \mathbf{v}) = \langle 0, 1, 1\rangle \cdot$
    $(\langle 1, 1, 0\rangle \times \langle 1, 0, 1\rangle) = \langle 0, 1, 1\rangle \cdot \langle 1, -1, -1\rangle = -2,\ \mathbf{v} \cdot (\mathbf{u} \times \mathbf{w}) = \langle 1, 0, 1\rangle \cdot (\langle 0, 1, 1\rangle \times \langle 1, 1, 0\rangle) = \langle 1, 0, 1\rangle \cdot$
    $\langle -1, 1, -1\rangle = -2,\ \mathbf{v} \cdot (\mathbf{w} \times \mathbf{u}) = \langle 1, 0, 1\rangle \cdot (\langle 1, 1, 0\rangle \times \langle 0, 1, 1\rangle) = \langle 1, 0, 1\rangle \cdot \langle 1, -1, 1\rangle = 2,$
    $\mathbf{w} \cdot (\mathbf{u} \times \mathbf{v}) = \langle 1, 1, 0\rangle \cdot (\langle 0, 1, 1\rangle \times \langle 1, 0, 1\rangle) = \langle 1, 1, 0\rangle \cdot \langle 1, 1, -1\rangle = 2,$ and $\mathbf{w} \cdot (\mathbf{v} \times \mathbf{u}) = \langle 1, 1, 0\rangle \cdot$
    $(\langle 1, 0, 1\rangle \times \langle 0, 1, 1\rangle) = \langle 1, 1, 0\rangle \cdot \langle -1, -1, 1\rangle = -2.$

    (b) It appears that $\mathbf{u} \cdot (\mathbf{v} \times \mathbf{w}) = \mathbf{v} \cdot (\mathbf{w} \times \mathbf{u}) = \mathbf{w} \cdot (\mathbf{u} \times \mathbf{v}) = -\mathbf{u} \cdot (\mathbf{w} \times \mathbf{v}) = -\mathbf{v} \cdot (\mathbf{u} \times \mathbf{w}) = -\mathbf{w} \cdot (\mathbf{v} \times \mathbf{u}).$

    (c) We know that the absolute values of the six scalar triple products must be equal because they all represent the volume of the parallelepiped determined by $\mathbf{u}$, $\mathbf{v}$, and $\mathbf{w}$. The fact that $\mathbf{a} \times \mathbf{b} = -(\mathbf{b} \times \mathbf{a})$ completes the proof.

9.6   EQUATIONS OF LINES AND PLANES

1. A line in space is described algebraically by using *parametric* equations. The line that passes through the point $P(x_0, y_0, z_0)$ and is parallel to the vector $\mathbf{v} = \langle a, b, c\rangle$ is described by the equations $x = x_0 + at$, $y = y_0 + bt$, $z = z_0 + ct$.

3. The line passing through $P(1, 0, -2)$ parallel to $\mathbf{v} = \langle 3, 2, -3\rangle$ has parametric equations $x = 1 + 3t$, $y = 2t$, $z = -2 - 3t$.

5. $x = 3,\ y = 2 - 4t,\ z = 1 + 2t$ **7.** $x = 1 + 2t,\ y = 0,\ z = -2 - 5t$

9. We first find a vector determined by $P(1, -3, 2)$ and $Q(2, 1, -1)$: $\mathbf{v} = \langle 2 - 1, 1 - (-3), -1 - 2\rangle = \langle 1, 4, -3\rangle$. Now we use $\mathbf{v}$ and the point $(1, -3, 2)$ to find parametric equations: $x = 1 + t$, $y = -3 + 4t$, $z = 2 - 3t$ where t is any real number.

11. A vector determined by $P(1, 1, 0)$ and $Q(0, 2, 2)$ is $\langle -1, 1, 2 \rangle$, so parametric equations are $x = 1 - t$, $y = 1 + t$, $z = 2t$.

13. A vector determined by $P(3, 7, -5)$ and $Q(7, 3, -5)$ is $\langle 4, -4, 0 \rangle$, so parametric equations are $x = 3 + 4t$, $y = 7 - 4t$, $z = -5$.

15. (a) An equation of the plane with normal vector
$\mathbf{n} = \langle 1, 1, -1 \rangle$ that passes through $P(0, 2, -3)$ is
$1(x - 0) + 1(y - 2) + (-1)[z - (-3)] = 0$ or
$x + y - z = 5$.

(b) Setting $y = z = 0$, we find $x = 5$, so the x-intercept
is 5. Similarly, the y-intercept is 5 and the z-intercept
is -5.

17. (a) $3(x - 2) - \frac{1}{2}(z - 8) = 0 \Leftrightarrow 6x - z = 4$

(b) x-intercept $\frac{2}{3}$, no y-intercept, z-intercept -4

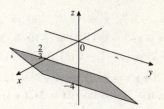

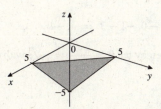

19. (a) $3x - (y - 2) + 2(z + 3) = 0 \Leftrightarrow 3x - y + 2z = -8$

(b) x-intercept $-\frac{8}{3}$, y-intercept 8, z-intercept -4

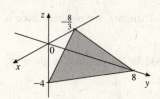

21. The vector
$\overrightarrow{PQ} \times \overrightarrow{PR} = \langle -1, -1, -2 \rangle \times \langle 1, 2, -1 \rangle = \langle 5, -3, -1 \rangle$ is
perpendicular to both $\overrightarrow{PQ}$ and $\overrightarrow{PR}$ and is therefore
perpendicular to the plane through P, Q, and R. Using the
formula for an equation of a plane with the point P, we
have $5(x - 6) - 3(y + 2) - (z - 1) = 0 \Leftrightarrow$
$5x - 3y - z = 35$.

23. $\overrightarrow{PQ} \times \overrightarrow{PR} = \left\langle 1, \frac{1}{3}, 2 \right\rangle \times \left\langle -1, -\frac{1}{3}, 6 \right\rangle = \left\langle \frac{8}{3}, -8, 0 \right\rangle$, so an equation is $\frac{8}{3}(x - 3) - 8\left(y + \frac{1}{3}\right) = 0 \Leftrightarrow x - 3y = 2$.

25. $\overrightarrow{PQ} \times \overrightarrow{PR} = \langle -3, 1, -1 \rangle \times \langle -6, -1, -1 \rangle = \langle -2, 3, 9 \rangle$, so an equation is $-2(x - 6) + 3(y - 1) + 9(z - 1) = 0 \Leftrightarrow$
$2x - 3y - 9z = 0$

27. The line passes through $(0, 0, 4)$ and $(2, 5, 0)$. A vector determined by these two points is
$\mathbf{v} = \langle 2 - 0, 5 - 0, 0 - 4 \rangle = \langle 2, 5, -4 \rangle$. Now we use $\mathbf{v}$ and the point $(0, 0, 4)$ to find parametric equations: $x = 2t$, $y = 5t$,
$z = 4 - 4t$, where t is any real number.

29. The line passes through $(2, -1, 5)$ and is parallel to $\mathbf{j}$, so equations are $x = 2$, $y = -1 + t$, $z = 5$, where t is any real
number.

31. The plane passes through $P(1, 0, 0)$, $Q(0, 3, 0)$, and $R(0, 0, 4)$. The vector $\overrightarrow{PQ} \times \overrightarrow{PR} = \langle -1, 3, 0 \rangle \times \langle -1, 0, 4 \rangle = \langle 12, 4, 3 \rangle$
is perpendicular to both $\overrightarrow{PQ}$ and $\overrightarrow{PR}$ and is therefore perpendicular to the plane through P, Q, and R. Using the formula
for an equation of a plane, we have $12(x - 1) + 4y + 3z = 0 \Leftrightarrow 12x + 4y + 3z = 12$.

33. This plane has the same normal vector as $x - 2y + 4z = 6$, and because it contains the origin, its equation is $x - 2y + 4z = 0$.

35. (a) To find the point of intersection, we substitute the parametric equations of the line into the equation of the plane:
$5(2 + t) - 2(3t) - 2(5 - t) = 1 \Leftrightarrow 10 + 5t - 6t - 10 + 2t = 1 \Leftrightarrow t = 1$.

(b) The parameter value $t = 1$ corresponds to the point $(3, 3, 4)$.

37. (a) Setting $t = 0$ in the equation for Line 1 gives the point $P(1, 0, -6)$. Setting $t = 1$ gives $Q(0, 3, -1)$. If we set $t = 1$ in
the equation for Line 2, we have the point $P'(1, 0, -6) = P$, and if we set $t = \frac{1}{2}$, we get $Q'(0, 3, -1) = Q$.

(b) Setting $t = 0$ in the equation for Line 1 gives the point $(0, 3, -5)$. But if a point on Line 4 has x-coordinate 0, it must have $x = 8 - 2t = 0 \Leftrightarrow t = 4$. But this parameter value gives the point $(0, 3, 2)$ on Line 2, so the two lines are not the same.

CHAPTER 9 REVIEW

1. $\mathbf{u} = \langle -2, 3 \rangle$, $\mathbf{v} = \langle 8, 1 \rangle$. $|\mathbf{u}| = \sqrt{(-2)^2 + 3^2} = \sqrt{13}$, $\mathbf{u} + \mathbf{v} = \langle -2 + 8, 3 + 1 \rangle = \langle 6, 4 \rangle$, $\mathbf{u} - \mathbf{v} = \langle -2 - 8, 3 - 1 \rangle = \langle -10, 2 \rangle$, $2\mathbf{u} = \langle 2(-2), 2(3) \rangle = \langle -4, 6 \rangle$, and $3\mathbf{u} - 2\mathbf{v} = \langle 3(-2) - 2(8), 3(3) - 2(1) \rangle = \langle -22, 7 \rangle$.

3. $\mathbf{u} = 2\mathbf{i} + \mathbf{j}$, $\mathbf{v} = \mathbf{i} - 2\mathbf{j}$. $|\mathbf{u}| = \sqrt{2^2 + 1^2} = \sqrt{5}$, $\mathbf{u} + \mathbf{v} = (2 + 1)\mathbf{i} + (1 - 2)\mathbf{j} = 3\mathbf{i} - \mathbf{j}$, $\mathbf{u} - \mathbf{v} = (2 - 1)\mathbf{i} + (1 + 2)\mathbf{j} = \mathbf{i} + 3\mathbf{j}$, $2\mathbf{u} = 4\mathbf{i} + 2\mathbf{j}$, and $3\mathbf{u} - 2\mathbf{v} = 3(2\mathbf{i} + \mathbf{j}) - 2(\mathbf{i} - 2\mathbf{j}) = 4\mathbf{i} + 7\mathbf{j}$.

5. The vector with initial point $P(0, 3)$ and terminal point $Q(3, -1)$ is $\langle 3 - 0, -1 - 3 \rangle = \langle 3, -4 \rangle$.

7. $\mathbf{u} = \langle -2, 2\sqrt{3} \rangle$ has length $\sqrt{(-2)^2 + \left(2\sqrt{3}\right)^2} = 4$. Its direction is given by $\tan\theta = \dfrac{2\sqrt{3}}{-2} = -\sqrt{3}$ with θ in quadrant II, so $\theta = \pi + \tan^{-1}\left(-\sqrt{3}\right) = 120°$.

9. $\mathbf{u} = \langle |\mathbf{u}| \cos\theta, |\mathbf{u}| \sin\theta \rangle = \langle 20\cos 60°, 20\sin 60° \rangle = \left(10, 10\sqrt{3} \right)$.

11. (a) The force exerted by the first tugboat can be expressed in component form as
$$\mathbf{u} = \left(2.0 \times 10^4 \cos 40°, 2.0 \times 10^4 \sin 40° \right) \approx \langle 15321, 12856 \rangle, \text{ and that of the second tugboat is}$$
$$\left(3.4 \times 10^4 \cos(-15°), 3.4 \times 10^4 \sin(-15°) \right) \approx \langle 32841, -8800 \rangle. \text{ Therefore, the resultant force is}$$
$$\mathbf{w} = \mathbf{u} + \mathbf{v} \approx \langle 15321 + 32841, 12856 - 8800 \rangle = \langle 48162, 4056 \rangle.$$

(b) The magnitude of the resultant force is $\sqrt{48162^2 + 4056^2} \approx 48{,}332$ lb. Its direction is given by
$$\tan\theta = \frac{4056}{48{,}162} \approx 0.084, \text{ so } \theta \approx \tan^{-1} 0.084 \approx 4.8° \text{ or N } 85.2° \text{ E.}$$

13. $\mathbf{u} = \langle 4, -3 \rangle$, $\mathbf{v} = \langle 9, -8 \rangle$. $|\mathbf{u}| = \sqrt{4^2 + (-3)^2} = 5$, $\mathbf{u} \cdot \mathbf{u} = 4^2 + (-3)^2 = 25$, and $\mathbf{u} \cdot \mathbf{v} = 4(9) + (-3)(-8) = 60$.

15. $\mathbf{u} = -2\mathbf{i} + 2\mathbf{j}$, $\mathbf{v} = \mathbf{i} + \mathbf{j}$. $|\mathbf{u}| = \sqrt{(-2)^2 + 2^2} = 2\sqrt{2}$, $\mathbf{u} \cdot \mathbf{u} = (-2)^2 + 2^2 = 8$, and $\mathbf{u} \cdot \mathbf{v} = -2(1) + 2(1) = 0$.

17. $\mathbf{u} \cdot \mathbf{v} = \langle -4, 2 \rangle \cdot \langle 3, 6 \rangle = -4(3) + 2(6) = 0$, so the vectors are perpendicular.

19. $\mathbf{u} \cdot \mathbf{v} = (2\mathbf{i} + \mathbf{j}) \cdot (\mathbf{i} + 3\mathbf{j}) = 2(1) + 1(3) = 5$, so the vectors are not perpendicular. The angle between them is given by
$$\cos\theta = \frac{\mathbf{u} \cdot \mathbf{v}}{|\mathbf{u}||\mathbf{v}|} = \frac{5}{\sqrt{2^2 + 1^2}\sqrt{1^2 + 3^2}} = \frac{\sqrt{2}}{2}, \text{ so } \theta = \cos^{-1}\frac{\sqrt{2}}{2} = 45°.$$

21. (a) $\mathbf{u} = \langle 3, 1 \rangle$, $\mathbf{v} = \langle 6, -1 \rangle$. The component of $\mathbf{u}$ along $\mathbf{v}$ is $\dfrac{\mathbf{u} \cdot \mathbf{v}}{|\mathbf{v}|} = \dfrac{3(6) + 1(-1)}{\sqrt{6^2 + 1^2}} = \dfrac{17\sqrt{37}}{37}$.

(b) $\text{proj}_{\mathbf{v}}\,\mathbf{u} = \left(\dfrac{\mathbf{u} \cdot \mathbf{v}}{|\mathbf{v}|^2} \right)\mathbf{v} = \dfrac{17}{37}\langle 6, -1 \rangle = \left\langle \dfrac{102}{37}, -\dfrac{17}{37} \right\rangle$

(c) $\mathbf{u}_1 = \text{proj}_{\mathbf{v}}\,\mathbf{u} = \left\langle \dfrac{102}{37}, -\dfrac{17}{37} \right\rangle$ and $\mathbf{u}_2 = \mathbf{u} - \text{proj}_{\mathbf{v}}\,\mathbf{u} = \langle 3, 1 \rangle - \left\langle \dfrac{102}{37}, -\dfrac{17}{37} \right\rangle = \left\langle \dfrac{9}{37}, \dfrac{54}{37} \right\rangle$.

23. (a) $\mathbf{u} = \mathbf{i} + 2\mathbf{j}$, $\mathbf{v} = 4\mathbf{i} - 9\mathbf{j}$. The component of $\mathbf{u}$ along $\mathbf{v}$ is $\dfrac{\mathbf{u} \cdot \mathbf{v}}{|\mathbf{v}|} = \dfrac{1(4) + 2(-9)}{\sqrt{4^2 + (-9)^2}} = -\dfrac{14\sqrt{97}}{97}$.

(b) $\text{proj}_{\mathbf{v}}\,\mathbf{u} = \left(\dfrac{\mathbf{u} \cdot \mathbf{v}}{|\mathbf{v}|^2} \right)\mathbf{v} = \dfrac{-14}{97}(4\mathbf{i} - 9\mathbf{j}) = -\dfrac{56}{97}\mathbf{i} + \dfrac{126}{97}\mathbf{j}$

(c) $\mathbf{u}_1 = \text{proj}_{\mathbf{v}}\,\mathbf{u} = -\dfrac{56}{97}\mathbf{i} + \dfrac{126}{97}\mathbf{j}$ and $\mathbf{u}_2 = \mathbf{u} - \text{proj}_{\mathbf{v}}\,\mathbf{u} = \mathbf{i} + 2\mathbf{j} - \left(-\dfrac{56}{97}\mathbf{i} + \dfrac{126}{97}\mathbf{j} \right) = \dfrac{153}{97}\mathbf{i} + \dfrac{68}{97}\mathbf{j}$.

25.

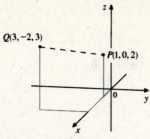

The distance between P and Q is

$\sqrt{(3-1)^2 + (-2-0)^2 + (3-2)^2} = 3.$

27. The sphere with radius $r = 6$ and center $C(0, 0, 0)$ has

equation $(x - 0)^2 + (y - 0)^2 + (z - 0)^2 = 6^2 \Leftrightarrow$

$x^2 + y^2 + z^2 = 36.$

29. We complete the squares to find

$x^2 + y^2 + z^2 - 2x - 6y + 4z = 2 \Leftrightarrow$

$\left(x^2 - 2x + 1\right) + \left(y^2 - 6y + 9\right) + \left(z^2 + 4z + 4\right)$

$= 2 + 1 + 9 + 4$

$\Leftrightarrow (x - 1)^2 + (y - 3)^2 + (z + 2)^2 = 16$, an equation of

the sphere with center $(1, 3, -2)$ and radius 4.

31. $\mathbf{u} = \langle 4, -2, 4 \rangle$ and $\mathbf{v} = \langle 2, 3, -1 \rangle$, so $|\mathbf{u}| = \sqrt{4^2 + (-2)^2 + (4)^2} = 6$, $\mathbf{u} + \mathbf{v} = \langle 4 + 2, -2 + 3, 4 + (-1) \rangle = \langle 6, 1, 3 \rangle$,

$\mathbf{u} - \mathbf{v} = \langle 4 - 2, -2 - 3, 4 - (-1) \rangle = \langle 2, -5, 5 \rangle$, and $\frac{3}{4}\mathbf{u} - 2\mathbf{v} = \frac{3}{4}\langle 4, -2, 4 \rangle - 2\langle 2, 3, -1 \rangle = \left\langle -1, -\frac{15}{2}, 5 \right\rangle$.

33. (a) $\mathbf{u} \cdot \mathbf{v} = \langle 3, -2, 4 \rangle \cdot \langle 3, 1, -2 \rangle = 3(3) - 2(1) + 4(-2) = -1$

(b) $\mathbf{u} \cdot \mathbf{v} \neq 0$, so the vectors are not perpendicular. The angle between them is given by

$\cos\theta = \dfrac{\mathbf{u} \cdot \mathbf{v}}{|\mathbf{u}||\mathbf{v}|} = \dfrac{-1}{\sqrt{3^2 + (-2)^2 + 4^2}\sqrt{3^2 + 1^2 + (-2)^2}} = -\dfrac{\sqrt{406}}{406}$, so $\theta = \cos^{-1}\left(-\dfrac{\sqrt{406}}{406}\right) \approx 92.8°$.

35. (a) $\mathbf{u} \cdot \mathbf{v} = (2\mathbf{i} - \mathbf{j} + 4\mathbf{k}) \cdot (3\mathbf{i} + 2\mathbf{j} - \mathbf{k}) = 2(3) - 1(2) + 4(-1) = 0$

(b) $\mathbf{u} \cdot \mathbf{v} = 0$, so the vectors are perpendicular.

37. (a) $\mathbf{u} \times \mathbf{v} = \langle 1, 1, 3 \rangle \times \langle 5, 0, -2 \rangle = (-2 - 0)\mathbf{i} - (-2 - 15)\mathbf{j} + (0 - 5)\mathbf{k} = \langle -2, 17, -5 \rangle$

(b) A unit vector perpendicular to $\mathbf{u}$ and $\mathbf{v}$ is $\dfrac{\mathbf{u} \times \mathbf{v}}{|\mathbf{u} \times \mathbf{v}|} = \dfrac{\langle -2, 17, -5 \rangle}{\sqrt{(-2)^2 + 17^2 + (-5)^2}} = \left\langle -\dfrac{\sqrt{318}}{159}, \dfrac{17\sqrt{318}}{318}, -\dfrac{5\sqrt{318}}{318} \right\rangle$.

39. (a) $\mathbf{u} \times \mathbf{v} = (\mathbf{i} - \mathbf{j}) \times (2\mathbf{j} - \mathbf{k}) = (1 - 0)\mathbf{i} - (-1 - 0)\mathbf{j} + (2 - 0)\mathbf{k} = \mathbf{i} + \mathbf{j} + 2\mathbf{k}$

(b) A unit vector perpendicular to $\mathbf{u}$ and $\mathbf{v}$ is $\dfrac{\mathbf{u} \times \mathbf{v}}{|\mathbf{u} \times \mathbf{v}|} = \dfrac{\mathbf{i} + \mathbf{j} + 2\mathbf{k}}{\sqrt{1^2 + 1^2 + 2^2}} = \dfrac{\sqrt{6}}{6}\mathbf{i} + \dfrac{\sqrt{6}}{6}\mathbf{j} + \dfrac{\sqrt{6}}{3}\mathbf{k}$.

41. The area of triangle PQR is one-half the area of the parallelogram determined by $\overrightarrow{PQ}$ and $\overrightarrow{PR}$, that is,

$\frac{1}{2}\left|\overrightarrow{PQ} \times \overrightarrow{PR}\right| = \frac{1}{2}|(1 - 6)\mathbf{i} - (2 + 8)\mathbf{j} + (-6 - 4)\mathbf{k}| = \frac{1}{2}\sqrt{(-5)^2 + (-10)^2 + (-10)^2} = \frac{1}{2}\sqrt{225} = \frac{15}{2}$.

43. The volume of the parallelepiped determined by $\mathbf{a} = 2\mathbf{i} - \mathbf{j}$, $\mathbf{b} = 2\mathbf{j} + \mathbf{k}$, and $\mathbf{c} = 3\mathbf{i} + \mathbf{j} - \mathbf{k}$ is the absolute value of their scalar triple product: $V = |\mathbf{a} \cdot (\mathbf{b} \times \mathbf{c})| = |(2\mathbf{i} - \mathbf{j}) \cdot (-3\mathbf{i} + 3\mathbf{j} - 6\mathbf{k})| = |-6 - 3| = 9$.

45. The line that passes through $P(2, 0, -6)$ and is parallel to $\mathbf{v} = \langle 3, 1, 0 \rangle$ has parametric equations $x = 2 + 3t$, $y = t$, $z = -6$.

47. A vector determined by $P(6, -2, -3)$ and $Q(4, 1, -2)$ is $\langle -2, 3, 1 \rangle$, so parametric equations are $x = 6 - 2t$, $y = -2 + 3t$, $z = -3 + t$.

49. Using the formula for an equation of a plane, the plane with normal vector $\mathbf{n} = \langle 2, 3, -5 \rangle$ passing through $P(2, 1, 1)$ has equation $2(x - 2) + 3(y - 1) - 5(z - 1) = 0 \Leftrightarrow 2x + 3y - 5z = 2$.

51. The plane passes through $P(1, 1, 1)$, $Q(3, -4, 2)$, and $R(6, -1, 0)$. $\overrightarrow{PQ} \times \overrightarrow{PR} = \langle 2, -5, 1 \rangle \times \langle 5, -2, -1 \rangle = \langle 7, 7, 6 \rangle$, so an equation is $7(x - 1) + 7(y - 1) + 6(z - 1) = 0 \Leftrightarrow 7x + 7y + 6z = 20$.

53. The line passes through the points $(2, 0, 0)$ and $(0, 0, -4)$. A vector determined by these points is $\langle -2, 0, -4 \rangle$ or $\langle 1, 0, 2 \rangle$, so parametric equations are $x = 2 + t$, $y = 0$, $z = 2t$.

CHAPTER 9 TEST

1. (a)

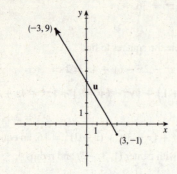

(b) $\mathbf{u} = (-3 - 3)\,\mathbf{i} + [9 - (-1)]\,\mathbf{j} = -6\mathbf{i} + 10\mathbf{j}$

(c) $|\mathbf{u}| = \sqrt{(-6)^2 + 10^2} = 2\sqrt{34}$

3. (a)

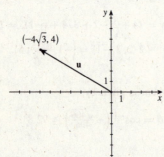

(b) The length of $\mathbf{u}$ is $|\mathbf{u}| = \sqrt{\left(-4\sqrt{3}\right)^2 + 4^2} = 8$. Its direction is given by

$\tan\theta = \dfrac{4}{-4\sqrt{3}} = -\dfrac{\sqrt{3}}{3}$ with θ in quadrant II, so

$\theta = 180° - \tan^{-1}\left(\dfrac{\sqrt{3}}{3}\right) = 150°$.

5. (a) $\cos\theta = \dfrac{\mathbf{u} \cdot \mathbf{v}}{|\mathbf{u}|\,|\mathbf{v}|} = \dfrac{3\,(5) + 2\,(-1)}{\sqrt{3^2 + 2^2}\sqrt{5^2 + (-1)^2}} = \dfrac{\sqrt{338}}{26}$, so $\theta = \cos^{-1}\dfrac{\sqrt{338}}{26} \approx 45.0°$.

(b) The component of $\mathbf{u}$ along $\mathbf{v}$ is $\dfrac{\mathbf{u} \cdot \mathbf{v}}{|\mathbf{v}|} = \dfrac{13}{\sqrt{26}} = \dfrac{\sqrt{26}}{2}$.

(c) $\text{proj}_{\mathbf{v}}\,\mathbf{u} = \left(\dfrac{\mathbf{u} \cdot \mathbf{v}}{|\mathbf{v}|^2}\right)\mathbf{v} = \dfrac{13}{26}\,(5\mathbf{i} - \mathbf{j}) = \dfrac{5}{2}\mathbf{i} - \dfrac{1}{2}\mathbf{j}$

7. (a) The distance between $P\,(4, 3, -1)$ and $Q\,(6, -1, 3)$ is $d = \sqrt{(6 - 4)^2 + (-1 - 3)^2 + [3 - (-1)]^2} = 6$.

(b) An equation is $(x - 4)^2 + (y - 3)^2 + [z - (-1)]^2 = 6^2 \Leftrightarrow (x - 4)^2 + (y - 3)^2 + (z + 1)^2 = 36$.

(c) $\mathbf{u} = \langle 6 - 4, -1 - 3, 3 - (-1)\rangle = \langle 2, -4, 4\rangle = 2\mathbf{i} - 4\mathbf{j} + 4\mathbf{k}$

9. A vector perpendicular to both $\mathbf{u} = \mathbf{j} + 2\mathbf{k}$ and $\mathbf{v} = \mathbf{i} - 2\mathbf{j} + 3\mathbf{k}$ is $\mathbf{u} \times \mathbf{v} = \langle 0, 1, 2\rangle \times \langle 1, -2, 3\rangle = \langle 7, 2, -1\rangle$, so two unit

vectors perpendicular to $\mathbf{u}$ and $\mathbf{v}$ are $\dfrac{\mathbf{u} \times \mathbf{v}}{|\mathbf{u} \times \mathbf{v}|} = \dfrac{\langle 7, 2, -1\rangle}{\sqrt{7^2 + 2^2 + (-1)^2}} = \left\langle \dfrac{7\sqrt{6}}{18}, \dfrac{\sqrt{6}}{9}, -\dfrac{\sqrt{6}}{18} \right\rangle$ and $\left\langle -\dfrac{7\sqrt{6}}{18}, -\dfrac{\sqrt{6}}{9}, \dfrac{\sqrt{6}}{18} \right\rangle$.

11. A vector determined by the two points is $\overrightarrow{PQ} = \langle -2, 1, -2\rangle$, so parametric equations are $x = 2 - 2t$, $y = -4 + t$, $z = 7 - 2t$.

FOCUS ON MODELING   Vector Fields

1. $\mathbf{F}(x, y) = \frac{1}{2}\mathbf{i} + \frac{1}{2}\mathbf{j}$

All vectors point in the same direction and have length $\frac{\sqrt{2}}{2}$.

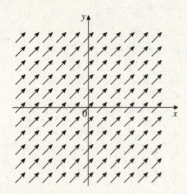

3. $\mathbf{F}(x, y) = y\mathbf{i} + \frac{1}{2}\mathbf{j}$

The vectors point to the left for $y < 0$ and to the right for $y > 0$.

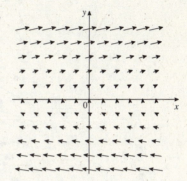

5. $\mathbf{F}(x, y) = \dfrac{y\mathbf{i} + x\mathbf{j}}{\sqrt{x^2 + y^2}}$

The length of the vector $\dfrac{y\mathbf{i} + x\mathbf{j}}{\sqrt{x^2 + y^2}}$ is 1.

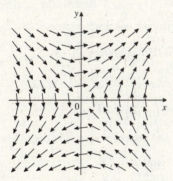

7. $\mathbf{F}(x, y, z) = \mathbf{j}$

All vectors in this field are parallel to the y-axis and have length 1.

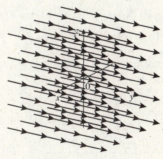

9. $\mathbf{F}(x, y, z) = z\mathbf{j}$

At each point (x, y, z), $\mathbf{F}(x, y, z)$ is a vector of length $|z|$. For $z > 0$, all point in the direction of the positive y-axis while for $z < 0$, all are in the direction of the negative y-axis.

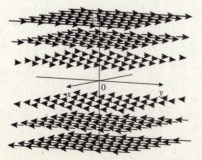

11. $\mathbf{F}(x, y) = \langle y, x \rangle$ corresponds to graph II, because in the first quadrant all the vectors have positive x- and y-components, in the second quadrant all vectors have positive x-components and negative y-components, in the third quadrant all vectors have negative x- and y-components, and in the fourth quadrant all vectors have negative x-components and positive y-components.

13. $\mathbf{F}(x, y) = \langle x - 2, y + 1 \rangle$ corresponds to graph I because the vectors are independent of y (vectors along vertical lines are identical) and, as we move to the right, both the x- and the y-components get larger.

15. $\mathbf{F}(x, y, z) = \mathbf{i} + 2\mathbf{j} + 3\mathbf{k}$ corresponds to graph IV, since all vectors have identical length and direction.

17. $\mathbf{F}(x, y, z) = x\mathbf{i} + y\mathbf{j} + 3\mathbf{k}$ corresponds to graph III; the projection of each vector onto the xy-plane is $x\mathbf{i} + y\mathbf{j}$, which points away from the origin, and the vectors point generally upward because their z-components are all 3.

19.

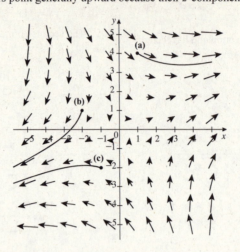

CUMULATIVE REVIEW TEST: CHAPTERS 8 and 9

1. For the point $(8, -8)$, $r^2 = x^2 + y^2 = 8^2 + (-8)^2 = 128$, so $r = \pm\sqrt{128} = \pm 8\sqrt{2}$. Because the point lies on the line $y = -x$ (with polar representations $\theta = \frac{3\pi}{4}$ or $\theta = \frac{7\pi}{4}$) in the fourth quadrant, one representation is $\left(8\sqrt{2}, \frac{7\pi}{4}\right)$. To find a representation with $r < 0$ and $0 \le \theta \le 2\pi$, we replace r by $-r$ and add $-\pi$ to θ, obtaining $\left(-8\sqrt{2}, \frac{3\pi}{4}\right)$.

3. (a)

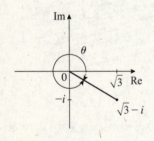

An argument for $z = \sqrt{3} - i$ is $\theta = \frac{11\pi}{6}$ (or we could use $\theta = -\frac{\pi}{6}$) and $r = \sqrt{3 + 1} = 2$. Thus, $z = 2\left(\cos\frac{11\pi}{6} + i\sin\frac{11\pi}{6}\right)$.

(b) $w = 6\left(\cos\frac{5\pi}{12} + i\sin\frac{5\pi}{12}\right)$ and from part (a), $z = 2\left(\cos\frac{11\pi}{6} + i\sin\frac{11\pi}{6}\right)$.

Thus, by the Multiplication Formula,

$$zw = 2 \cdot 6\left[\cos\left(\frac{11\pi}{6} + \frac{5\pi}{12}\right) + i\sin\left(\frac{11\pi}{6} + \frac{5\pi}{12}\right)\right]$$
$$= 12\left(\cos\frac{9\pi}{4} + i\sin\frac{9\pi}{4}\right) = 12\left(\cos\frac{\pi}{4} + i\sin\frac{\pi}{4}\right) = 6\sqrt{2} + 6\sqrt{2}i.$$

By the Division Formula,

$$z/w = \frac{2}{6}\left[\cos\left(\frac{11\pi}{6} - \frac{5\pi}{12}\right) + i\sin\left(\frac{11\pi}{6} - \frac{5\pi}{12}\right)\right] = \frac{1}{3}\left(\cos\frac{17\pi}{12} + i\sin\frac{17\pi}{12}\right) = -\frac{\sqrt{6}-\sqrt{2}}{12} - \frac{\sqrt{6}+\sqrt{2}}{12}i.$$

(c) By De Moivre's Theorem,

$$z^{10} = 2^{10}\left[\cos\left(10 \cdot \frac{11\pi}{6}\right) + i\sin\left(10 \cdot \frac{11\pi}{6}\right)\right] = 1024\left(\cos\frac{110\pi}{6} + i\sin\frac{110\pi}{6}\right) = 1024\left(\cos\frac{\pi}{3} + i\sin\frac{\pi}{3}\right)$$
$$= 512 + 512\sqrt{3}i.$$

(d) Applying the formula for nth roots with $n = 3$, we have $z_k = 2^{1/3}\left[\cos\left(\frac{\frac{11\pi}{6} + 2k\pi}{3}\right) + i\sin\left(\frac{\frac{11\pi}{6} + 2k\pi}{3}\right)\right]$ for

$k = 0, 1, 2$. Thus, the cube roots of z are $z_0 = \sqrt[3]{2}\left(\cos\frac{11\pi}{18} + i\sin\frac{11\pi}{18}\right)$, $z_1 = \sqrt[3]{2}\left(\cos\frac{23\pi}{18} + i\sin\frac{23\pi}{18}\right)$, and

$z_3 = \sqrt[3]{2}\left(\cos\frac{35\pi}{18} + i\sin\frac{35\pi}{18}\right)$.

5. (a)

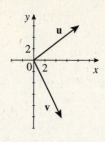

(b) $\mathbf{u} = \langle 8, 6 \rangle$ and $\mathbf{v} = 5\mathbf{i} - 10\mathbf{j} = \langle 5, -10 \rangle$, so $\mathbf{u} + \mathbf{v} = \langle 8 + 5, 6 - 10 \rangle = \langle 13, -4 \rangle$ and $2\mathbf{u} - \mathbf{v} = \langle 2 \cdot 8 - 5, 2 \cdot 6 - (-10) \rangle = \langle 11, 22 \rangle$. The angle θ between $\mathbf{u}$ and

$\mathbf{v}$ satisfies $\cos \theta = \dfrac{\mathbf{u} \cdot \mathbf{v}}{|\mathbf{u}|\,|\mathbf{v}|} = \dfrac{8\,(5) + 6\,(-10)}{\sqrt{8^2 + 6^2}\sqrt{5^2 + (-10)^2}} = -\dfrac{2\sqrt{5}}{25}$, so

$\theta = \cos^{-1}\left(-\dfrac{2\sqrt{5}}{25}\right) \approx 100.3°$.

$\operatorname{proj}_{\mathbf{v}} \mathbf{u} = \left(\dfrac{\mathbf{u} \cdot \mathbf{v}}{|\mathbf{v}|^2}\right)\mathbf{v} = \dfrac{8\,(5) + 6\,(-10)}{5^2 + (-10)^2}\,\langle 5, -10 \rangle = -\dfrac{4}{25}\,\langle 5, -10 \rangle = \left\langle -\dfrac{4}{5}, \dfrac{8}{5}\right\rangle.$

(c) The distance vector from $(2, 0)$ to $(10, 3)$ is $\mathbf{d} = \langle 10 - 2, 3 - 0 \rangle = \langle 8, 3 \rangle$, so the work done is
$W = \mathbf{u} \cdot \mathbf{d} = \langle 8, 6 \rangle \cdot \langle 8, 3 \rangle = 82$.

7. (a) $\mathbf{a} = \langle 2, 1, -3 \rangle$ and $\mathbf{b} = 3\mathbf{i} + 2\mathbf{k} = \langle 3, 0, 2 \rangle$, so $\mathbf{a} \cdot \mathbf{b} = 2\,(3) + 1\,(0) + (-3)\,(2) = 0$ and
$\mathbf{a} \times \mathbf{b} = \langle 1\,(2) - (-3)\,(0),\, -3\,(3) - 2\,(2),\, 2\,(0) - 1\,(3) \rangle = \langle 2, -13, -3 \rangle$. Because $\mathbf{a} \cdot \mathbf{b} = 0$, $\mathbf{a}$ and $\mathbf{b}$ are perpendicular.

(b) $\mathbf{a} \times \mathbf{b} = \langle 2, -13, -3 \rangle$ is a normal vector to any plane parallel to $\mathbf{a}$ and $\mathbf{b}$, so such a plane containing the point $\langle 3, 0, -5 \rangle$ has equation $2\,(x - 3) - 13\,(y - 0) - 3\,[z - (-5)] = 0 \Leftrightarrow 2x - 13y - 3z = 21$.

10 SYSTEMS OF EQUATIONS AND INEQUALITIES

10.1 SYSTEMS OF LINEAR EQUATIONS IN TWO VARIABLES

1. The given system is a system of two equations in the two variables x and y. To check if $(5, -1)$ is a solution of this system, we check if $x = 5$ and $y = -1$ satisfy each *equation* in the system. The only solution of the given system is $(2, 1)$.

3. A system of two linear equations in two variables can have one solution, *no* solution, or *infinitely many* solutions.

5. $\begin{cases} x - y = 1 \\ 4x + 3y = 18 \end{cases}$ Solving the first equation for x, we get $x = y + 1$, and substituting this into the second equation gives
$4(y + 1) + 3y = 18 \Leftrightarrow 7y + 4 = 18 \Leftrightarrow 7y = 14 \Leftrightarrow y = 2$. Substituting for y we get $x = y + 1 = 2 + 1 = 3$. Thus, the solution is $(3, 2)$.

7. $\begin{cases} x - y = 2 \\ 2x + 3y = 9 \end{cases}$ Solving the first equation for x, we get $x = y + 2$, and substituting this into the second equation gives
$2(y + 2) + 3y = 9 \Leftrightarrow 5y + 4 = 9 \Leftrightarrow 5y = 5 \Leftrightarrow y = 1$. Substituting for y we get $x = y + 2 = (1) + 2 = 3$. Thus, the solution is $(3, 1)$.

9. $\begin{cases} 3x + 4y = 10 \\ x - 4y = -2 \end{cases}$ Adding the two equations, we get $4x = 8 \Leftrightarrow x = 2$, and substituting into the first equation in the
original system gives $3(2) + 4y = 10 \Leftrightarrow 4y = 4 \Leftrightarrow y = 1$. Thus, the solution is $(2, 1)$.

11. $\begin{cases} x + 2y = 5 \\ 2x + 3y = 8 \end{cases}$ Multiplying the first equation by 2 and the second by -1 gives the system $\begin{cases} 2x + 4y = 10 \\ -2x - 3y = -8 \end{cases}$ Adding,
we get $y = 2$, and substituting into the first equation in the original system gives $x + 2(2) = 5 \Leftrightarrow x + 4 = 5 \Leftrightarrow x = 1$. The solution is $(1, 2)$.

13. $\begin{cases} 2x + y = -1 \\ x - 2y = -8 \end{cases}$ By inspection of the graph, it appears that $(-2, 3)$ is the solution to the system. We check this in both
equations to verify that it is a solution. $2(-2) + 3 = -4 + 3 = -1$ and $-2 - 2(3) = -2 - 6 = -8$. Since both equations are satisfied, the solution is $(-2, 3)$.

15. $\begin{cases} x - y = 4 \\ 2x + y = 2 \end{cases}$

The solution is $x = 2$, $y = -2$.

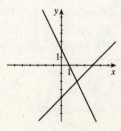

17. $\begin{cases} 2x - 3y = 12 \\ -x + \frac{3}{2}y = 4 \end{cases}$

The lines are parallel, so there is no intersection and hence no solution.

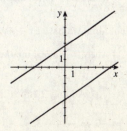

19. $\begin{cases} -x + \frac{1}{2}y = -5 \\ 2x - \ y = 10 \end{cases}$

There are infinitely many solutions.

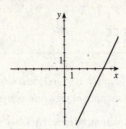

21. $\begin{cases} x + y = 4 \\ -x + y = 0 \end{cases}$ Adding the two equations gives

$2y = 4 \Leftrightarrow y = 2$. Substituting for y in the first equation gives $x + 2 = 4 \Leftrightarrow x = 2$. Hence, the solution is $(2, 2)$.

23. $\begin{cases} 2x - 3y = 9 \\ 4x + 3y = 9 \end{cases}$ Adding the two equations gives

$6x = 18 \Leftrightarrow x = 3$. Substituting for x in the second equation gives $4(3) + 3y = 9 \Leftrightarrow 12 + 3y = 9 \Leftrightarrow 3y = -3 \Leftrightarrow x = -1$. Hence, the solution is $(3, -1)$.

25. $\begin{cases} x + 3y = 5 \\ 2x - \ y = 3 \end{cases}$ Solving the first equation for x gives $x = -3y + 5$. Substituting for x in the second equation gives $2(-3y + 5) - y = 3 \Leftrightarrow -6y + 10 - y = 3 \Leftrightarrow -7y = -7 \Leftrightarrow y = 1$. Then $x = -3(1) + 5 = 2$. Hence, the solution is $(2, 1)$.

27. $-x + y = 2 \Leftrightarrow y = x + 2$. Substituting for y into $4x - 3y = -3$ gives $4x - 3(x + 2) = -3 \Leftrightarrow 4x - 3x - 6 = -3 \Leftrightarrow x = 3$, and so $y = (3) + 2 = 5$. Hence, the solution is $(3, 5)$.

29. $x + 2y = 7 \Leftrightarrow x = 7 - 2y$. Substituting for x into $5x - y = 2$ gives $5(7 - 2y) - y = 2 \Leftrightarrow 35 - 10y - y = 2 \Leftrightarrow -11y = -33 \Leftrightarrow y = 3$, and so $x = 7 - 2(3) = 1$. Hence, the solution is $(1, 3)$.

31. $\frac{1}{2}x + \frac{1}{3}y = 2 \Leftrightarrow x + \frac{2}{3}y = 4 \Leftrightarrow x = 4 - \frac{2}{3}y$. Substituting for x into $\frac{1}{5}x - \frac{2}{3}y = 8$ gives $\frac{1}{5}\left(4 - \frac{2}{3}y\right) - \frac{2}{3}y = 8 \Leftrightarrow \frac{4}{5} - \frac{2}{15}y - \frac{10}{15}y = 8 \Leftrightarrow 12 - 2y - 10y = 120 \Leftrightarrow y = -9$, and so $x = 4 - \frac{2}{3}(-9) = 10$. Hence, the solution is $(10, -9)$.

33. $\begin{cases} 3x + 2y = 8 \\ x - 2y = 0 \end{cases}$ Multiplying the second equation by 3 gives the system $\begin{cases} 3x + 2y = 8 \\ 3x - 6y = 0 \end{cases}$ Subtracting the second equation from the first gives $8y = 8 \Leftrightarrow y = 1$. Substituting into the first equation we get $3x + 2(1) = 8 \Leftrightarrow 3x = 6 \Leftrightarrow x = 2$. Thus, the solution is $(2, 1)$.

35. $\begin{cases} x + 4y = 8 \\ 3x + 12y = 2 \end{cases}$ Adding -3 times the first equation to the second equation gives $0 = -22$, which is never true. Thus, the system has no solution.

37. $\begin{cases} 2x - 6y = \ 10 \\ -3x + 9y = -15 \end{cases}$ Adding 3 times the first equation to 2 times the second equation gives $0 = 0$. Writing the equation in slope-intercept form, we have $2x - 6y = 10 \Leftrightarrow -6y = -2x + 10 \Leftrightarrow y = \frac{1}{3}x - \frac{5}{3}$, so the solutions are all pairs of the form $\left(x, \frac{1}{3}x - \frac{5}{3}\right)$ where x is a real number.

39. $\begin{cases} 6x + 4y = 12 \\ 9x + 6y = 18 \end{cases}$ Adding 3 times the first equation to -2 times the second equation gives $0 = 0$. Writing the equation in slope-intercept form, we have $6x + 4y = 12 \Leftrightarrow 4y = -6x + 12 \Leftrightarrow y = -\frac{3}{2}x + 3$, so the solutions are all pairs of the form $\left(x, -\frac{3}{2}x + 3\right)$ where x is a real number.

41. $\begin{cases} 8s - 3t = -3 \\ 5s - 2t = -1 \end{cases}$ Adding 2 times the first equation to 3 times the second equation gives $s = -3$, so $8(-3) - 3t = -3 \Leftrightarrow -24 - 3t = -3 \Leftrightarrow t = -7$. Thus, the solution is $(-3, -7)$.

43. $\begin{cases} \frac{1}{2}x + \frac{3}{5}y = 3 \\ \frac{5}{3}x + 2y = 10 \end{cases}$ Adding 10 times the first equation to -3 times the second equation gives $0 = 0$. Writing the equation

in slope-intercept form, we have $\frac{1}{2}x + \frac{3}{5}y = 3 \Leftrightarrow \frac{3}{5}y = -\frac{1}{2}x + 3 \Leftrightarrow y = -\frac{5}{6}x + 5$, so the solutions are all pairs of the

form $\left(x, -\frac{5}{6}x + 5\right)$ where x is a real number.

45. $\begin{cases} 0.4x + 1.2y = 14 \\ 12x - 5y = 10 \end{cases}$ Adding 30 times the first equation to -1 times the second equation gives $41y = 410 \Leftrightarrow y = 10$, so

$12x - 5(10) = 10 \Leftrightarrow 12x = 60 \Leftrightarrow x = 5$. Thus, the solution is $(5, 10)$.

47. $\begin{cases} \frac{1}{3}x - \frac{1}{4}y = 2 \\ -8x + 6y = 10 \end{cases}$ Adding 24 times the first equation to the second equation gives $0 = 58$, which is never true. Thus,

the system has no solution.

49. $\begin{cases} 0.21x + 3.17y = 9.51 \\ 2.35x - 1.17y = 5.89 \end{cases}$ **51.** $\begin{cases} 2371x - 6552y = 13{,}591 \\ 9815x + 992y = 618{,}555 \end{cases}$

The solution is approximately $(3.87, 2.74)$.   The solution is approximately $(61.00, 20.00)$.

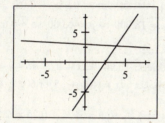 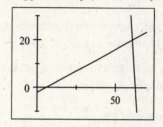

53. Subtracting the first equation from the second, we get $ay - y = 1 \Leftrightarrow y(a - 1) = 1 \Leftrightarrow y = \dfrac{1}{a - 1}, a \neq 1$. So

$x + \left(\dfrac{1}{a - 1}\right) = 0 \Leftrightarrow x = \dfrac{1}{1 - a} = -\dfrac{1}{a - 1}$. Thus, the solution is $\left(-\dfrac{1}{a - 1}, \dfrac{1}{a - 1}\right)$.

55. Subtracting b times the first equation from a times the second, we get $\left(a^2 - b^2\right)y = a - b \Leftrightarrow y = \dfrac{a - b}{a^2 - b^2} = \dfrac{1}{a + b}$,

$a^2 - b^2 \neq 0$. So $ax + \dfrac{b}{a + b} = 1 \Leftrightarrow ax = \dfrac{a}{a + b} \Leftrightarrow x = \dfrac{1}{a + b}$. Thus, the solution is $\left(\dfrac{1}{a + b}, \dfrac{1}{a + b}\right)$.

57. Let the two numbers be x and y. Then $\begin{cases} x + y = 34 \\ x - y = 10 \end{cases}$ Adding these two equations gives $2x = 44 \Leftrightarrow x = 22$. So

$22 + y = 34 \Leftrightarrow y = 12$. Therefore, the two numbers are 22 and 12.

59. Let d be the number of dimes and q be the number of quarters. This gives $\begin{cases} d + q = 14 \\ 0.10d + 0.25q = 2.75 \end{cases}$ Subtracting the

first equation from 10 times the second gives $1.5q = 13.5 \Leftrightarrow q = 9$. So $d + 9 = 14 \Leftrightarrow d = 5$. Thus, the number of dimes

is 5 and the number of quarters is 9.

61. Let r be the amount of regular gas sold and p the amount of premium gas sold. Then $\begin{cases} r + p = 280 \\ 2.20r + 3.00p = 680 \end{cases}$ Subtracting

the second equation from three times the first equation gives $3r - 2.2 = 3(280) - 680 \Leftrightarrow 0.8r = 160 \Leftrightarrow r = 200$.

Substituting this value of r into the original first equation gives $200 + p = 280 \Leftrightarrow p = 80$. Thus, 200 gallons of regular gas

and 80 gallons of premium were sold.

63. Let x be the speed of the plane in still air and y be the speed of the wind. This gives $\begin{cases} 2x - 2y = 180 \\ 1.2x + 1.2y = 180 \end{cases}$ Subtracting

6 times the first equation from 10 times the second gives $24x = 2880 \Leftrightarrow x = 120$, so $2(120) - 2y = 180 \Leftrightarrow -2y = -60$
$\Leftrightarrow y = 30$. Therefore, the speed of the plane is 120 mi/h and the wind speed is 30 mi/h.

65. Let a and b be the number of grams of food A and food B. Then $\begin{cases} 0.12a + 0.20b = 32 \\ 100a + 50b = 22{,}000 \end{cases}$ Subtracting 250 times the

first equation from the second, we get $70a = 14{,}000 \Leftrightarrow a = 200$, so $0.12(200) + 0.20b = 32 \Leftrightarrow 0.20b = 8 \Leftrightarrow b = 40$.
Thus, she should use 200 grams of food A and 40 grams of food B.

67. Let x and y be the sulfuric acid concentrations in the first and second containers.

$\begin{cases} 300x + 600y = 900(0.15) \\ 100x + 500y = 600(0.125) \end{cases}$ Subtracting the first equation from 3 times the second gives $900y = 90 \Leftrightarrow y = 0.10$, so

$100x + 500(0.10) = 75 \Leftrightarrow x = 0.25$. Thus, the concentrations of sulfuric acid are 25% in the first container and 10% in
the second.

69. Let x be the amount invested at 5% and y the amount invested at 8%. $\begin{cases} \text{Total invested:} \quad x + y = 20{,}000 \\ \text{Interest earned:} \ 0.05x + 0.08y = 1180 \end{cases}$

Subtracting 5 times the first equation from 100 times the second gives $3y = 18{,}000 \Leftrightarrow y = 6{,}000$, so $x + 6{,}000 = 20{,}000$
$\Leftrightarrow x = 14{,}000$. She invests \$14,000 at 5% and \$6,000 at 8%.

71. Let x be the length of time John drives and y be the length of time Mary drives. Then $y = x + 0.25$, so $-x + y = 0.25$, and
multiplying by 40, we get $-40x + 40y = 10$. Comparing the distances, we get $60x = 40y + 35$, or $60x - 40y = 35$. This

gives the system $\begin{cases} -40x + 40y = 10 \\ 60x - 40y = 35 \end{cases}$ Adding, we get $20x = 45 \Leftrightarrow x = 2.25$, so $y = 2.25 + 0.25 = 2.5$. Thus, John

drives for $2\frac{1}{4}$ hours and Mary drives for $2\frac{1}{2}$ hours.

73. Let x be the tens digit and y be the ones digit of the number. $\begin{cases} x + y = 7 \\ 10y + x = 27 + 10x + y \end{cases}$ Adding 9 times the first

equation to the second gives $18x = 36 \Leftrightarrow x = 2$, so $2 + y = 7 \Leftrightarrow y = 5$. Thus, the number is 25.

75. $n = 5$, so $\sum_{k=1}^{n} x_k = 1 + 2 + 3 + 5 + 7 = 18$, $\sum_{k=1}^{n} y_k = 3 + 5 + 6 + 6 + 9 = 29$,

$\sum_{k=1}^{n} x_k y_k = 1(3) + 2(5) + 3(6) + 5(6) + 7(9) = 124$, and

$\sum_{k=1}^{n} x_k^2 = 1^2 + 2^2 + 3^2 + 5^2 + 7^2 = 88$. Thus we get the system

$\begin{cases} 18a + 5b = 29 \\ 88a + 18b = 124 \end{cases}$ Subtracting 18 times the first equation from 5 times the

second, we get $116a = 98 \Leftrightarrow a \approx 0.845$. Then

$b = \frac{1}{5}[-18(0.845) + 29] \approx 2.758$. So the regression line is $y = 0.845x + 2.758$.

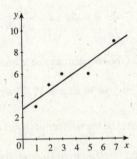

10.2 SYSTEMS OF LINEAR EQUATIONS IN SEVERAL VARIABLES

1. If we add 2 times the first equation to the second equation, the second equation becomes $x + 3z = 1$.

3. The equation $6x - \sqrt{3}y + \frac{1}{2}z = 0$ is linear.

5. The system $\begin{cases} xy - 3y + z = 5 \\ x - y^2 + 5z = 0 \\ 2x + yz = 3 \end{cases}$ is not a linear system, since the first equation contains a product of variables. In fact both the second and the third equation are not linear.

7. $\begin{cases} x - 2y + 4z = 3 \\ y + 2z = 7 \\ z = 2 \end{cases}$ Substituting $z = 2$ into the second equation gives $y + 2(2) = 7 \Leftrightarrow y = 3$. Substituting $z = 2$ and $y = 3$ into the first equation gives $x - 2(3) + 4(2) = 3 \Leftrightarrow x = 1$. Thus, the solution is $(1, 3, 2)$.

9. $\begin{cases} x + 2y + z = 7 \\ -y + 3z = 9 \\ 2z = 6 \end{cases}$ Solving we get $2z = 6 \Leftrightarrow z = 3$. Substituting $z = 3$ into the second equation gives $-y + 3(3) = 9 \Leftrightarrow y = 0$. Substituting $z = 3$ and $y = 0$ into the first equation gives $x + 2(0) + 3 = 7 \Leftrightarrow x = 4$. Thus, the solution is $(4, 0, 3)$.

11. $\begin{cases} 2x - y + 6z = 5 \\ y + 4z = 0 \\ -2z = 1 \end{cases}$ Solving we get $-2z = 1 \Leftrightarrow z = -\frac{1}{2}$. Substituting $z = -\frac{1}{2}$ into the second equation gives $y + 4\left(-\frac{1}{2}\right) = 0 \Leftrightarrow y = 2$. Substituting $z = -\frac{1}{2}$ and $y = 2$ into the first equation gives $2x - (2) + 6\left(-\frac{1}{2}\right) = 5 \Leftrightarrow x = 5$. Thus, the solution is $\left(5, 2, -\frac{1}{2}\right)$.

13. $\begin{cases} x - 2y - z = 4 \\ x - y + 3z = 0 \\ 2x + y + z = 0 \end{cases}$ Subtract the first equation from the second equation: $\begin{cases} x - 2y - z = 4 \\ y + 4z = -4 \\ 2x + y + z = 0 \end{cases}$

Or, subtract $\frac{1}{2}$ times the third equation from the second equation: $\begin{cases} x - 2y - z = 4 \\ -\frac{3}{2}y + \frac{5}{2}z = 0 \\ 2x + y + z = 0 \end{cases}$

15. $\begin{cases} 2x - y + 3z = 2 \\ x + 2y - z = 4 \\ -4x + 5y + z = 10 \end{cases}$ Add 2 times the first equation to the third equation: $\begin{cases} 2x - y + 3z = 2 \\ x + 2y - z = 4 \\ 3y + 7z = 14 \end{cases}$

Or, add 4 times the second equation to the third equation: $\begin{cases} 2x - y + 3z = 2 \\ x + 2y - z = 4 \\ 13y - 3z = 26 \end{cases}$

17. $\begin{cases} x - y - z = 4 \\ 2y + z = -1 \\ -x + y - 2z = 5 \end{cases} \Leftrightarrow \begin{cases} x - y - z = 4 \\ 2y + z = -1 \\ -3z = 9 \end{cases}$ Eq. 1 + Eq. 3

So $z = -3$ and $2y + (-3) = -1 \Leftrightarrow 2y = 2 \Leftrightarrow y = 1$. Thus, $x - 1 - (-3) = 4 \Leftrightarrow x = 2$. So the solution is $(2, 1, -3)$.

19. $\begin{cases} x + y + z = 4 \\ x + 3y + 3z = 10 \\ 2x + y - z = 3 \end{cases} \Leftrightarrow \begin{cases} x + y + z = 4 \\ 2y + 2z = 6 \\ y + 3z = 5 \end{cases} \begin{matrix} \\ (-1) \times \text{Eq. 1} + \text{Eq. 2} \\ 2 \times \text{Eq. 1} + (-1) \times \text{Eq. 3} \end{matrix} \Leftrightarrow \begin{cases} x + y + z = 4 \\ y + 3z = 5 \\ 2y + 2z = 6 \end{cases} \begin{matrix} \\ \text{Eq. 3} \\ \text{Eq. 2} \end{matrix} \Leftrightarrow$

$\begin{cases} x + y + z = 4 \\ y + 3z = 5 \\ -4z = -4 \end{cases} \begin{matrix} \\ \\ (-2) \times \text{Eq. 2} + \text{Eq. 3} \end{matrix}$

So $z = 1$ and $y + 3(1) = 8 \Leftrightarrow y = 2$. Then $x + 2 + 1 = 4 \Leftrightarrow x = 1$. So the solution is $(1, 2, 1)$.

21. $\begin{cases} x - 4z = 1 \\ 2x - y - 6z = 4 \\ 2x + 3y - 2z = 8 \end{cases} \Leftrightarrow \begin{cases} x - 4z = 1 \\ -y + 2z = 2 \\ 3y + 6z = 6 \end{cases} \begin{matrix} \\ (-2) \times \text{Eq. 1} + \text{Eq. 2} \\ (-2) \times \text{Eq. 1} + \text{Eq. 3} \end{matrix} \Leftrightarrow \begin{cases} x - 4z = 1 \\ -y + 2z = 2 \\ 12z = 12 \end{cases} \begin{matrix} \\ \\ 3 \times \text{Eq. 2} + \text{Eq. 3} \end{matrix}$

So $z = 1$ and $-y + 2(1) = 2 \Leftrightarrow y = 0$. Then $x - 4(1) = 1 \Leftrightarrow x = 5$. So the solution is $(5, 0, 1)$.

23. $\begin{cases} 2x + 4y - z = 2 \\ x + 2y - 3z = -4 \\ 3x - y + z = 1 \end{cases} \Leftrightarrow \begin{cases} 2x + 4y - z = 2 \\ 5z = 10 \\ 14y - 5z = 4 \end{cases} \begin{matrix} \\ \text{Eq. 1} + (-2) \times \text{Eq. 2} \\ 3 \times \text{Eq. 1} + (-2) \times \text{Eq. 3} \end{matrix} \Leftrightarrow \begin{cases} x + 2y - 3z = -4 \\ 14y - 5z = 4 \\ 5z = 10 \end{cases} \begin{matrix} \\ \text{Eq. 2} \leftrightarrow \text{Eq. 3} \\ \text{Eq. 2} \leftrightarrow \text{Eq. 3} \end{matrix}$

So $z = 2$ and $14y - 5(2) = 4 \Leftrightarrow y = 1$. Then $x + 2(1) - 3(2) = -4 \Leftrightarrow x = 0$. So the solution is $(0, 1, 2)$.

25. $\begin{cases} y - 2z = 0 \\ 2x + 3y = 2 \\ -x - 2y + z = -1 \end{cases} \Leftrightarrow \begin{cases} -x - 2y + z = -1 \\ y - 2z = 0 \\ 2x + 3y = 2 \end{cases} \begin{matrix} \text{Eq. 3} \\ \text{Eq. 1} \\ \text{Eq. 2} \end{matrix} \Leftrightarrow \begin{cases} -x - 2y + z = -1 \\ y - 2z = 0 \\ -y + 2z = 0 \end{cases} \begin{matrix} \\ \\ 2 \times \text{Eq. 1} + \text{Eq. 3} \end{matrix} \Leftrightarrow$

$\begin{cases} -x - 2y + z = -1 \\ y - 2z = 0 \\ 0 = 0 \end{cases} \begin{matrix} \\ \\ \text{Eq. 2} + \text{Eq. 3} \end{matrix}$

The system is dependent, so when $z = t$ we solve for y to get $y - 2t = 0 \Leftrightarrow y = 2t$. Then $-x - 2(2t) + t = -1 \Leftrightarrow -x - 3t = -1 \Leftrightarrow x = -3t + 1$. So the solutions are $(-3t + 1, 2t, t)$, where t is any real number.

27. $\begin{cases} x + 2y - z = 1 \\ 2x + 3y - 4z = -3 \\ 3x + 6y - 3z = 4 \end{cases} \Leftrightarrow \begin{cases} x + 2y - z = 1 \\ -y - 2z = -5 \\ 0 = 1 \end{cases} \begin{matrix} \\ (-2) \times \text{Eq. 1} + \text{Eq. 2} \\ (-3) \times \text{Eq. 1} + \text{Eq. 3} \end{matrix}$ Since $0 = 1$ is false, this system is inconsistent.

29. $\begin{cases} 2x + 3y - z = 1 \\ x + 2y = 3 \\ x + 3y + z = 4 \end{cases} \Leftrightarrow \begin{cases} x + 2y = 3 \\ 2x + 3y - z = 1 \\ x + 3y + z = 4 \end{cases} \begin{matrix} \text{Eq. 2} \\ \text{Eq. 1} \\ \end{matrix} \Leftrightarrow \begin{cases} x + 2y = 3 \\ -y - z = -5 \\ y + z = 1 \end{cases} \begin{matrix} \\ \text{Eq. 2} + (-2) \times \text{Eq. 1} \\ \text{Eq. 3} - \text{Eq. 1} \end{matrix} \Leftrightarrow$

$\begin{cases} x + 2y = 3 \\ -y - z = -5 \\ 0 = -4 \end{cases} \begin{matrix} \\ \\ \text{Eq. 2} + \text{Eq. 3} \end{matrix}$ Since $0 = -4$ is false, this system is inconsistent.

31. $\begin{cases} x + y - z = 0 \\ x + 2y - 3z = -3 \\ 2x + 3y - 4z = -3 \end{cases} \Leftrightarrow \begin{cases} x + y - z = 0 \\ y - 2z = -3 \\ y - 2z = -3 \end{cases} \begin{matrix} \\ \text{Eq. 2} - \text{Eq. 1} \\ (-2) \times \text{Eq. 1} + \text{Eq. 3} \end{matrix} \Leftrightarrow \begin{cases} x + y - z = 0 \\ y - 2z = -3 \\ 0 = 0 \end{cases} \begin{matrix} \\ \\ \text{Eq. 2} - \text{Eq. 3} \end{matrix}$

So $z = t$ and $y - 2t = -3 \Leftrightarrow y = 2t - 3$. Then $x + (2t - 3) - t = 0 \Leftrightarrow x = -t + 3$. So the solutions are $(-t + 3, 2t - 3, t)$, where t is any real number.

33.
$$\begin{cases} x + 3y - 2z = 0 \\ 2x \qquad + 4z = 4 \\ 4x + 6y \qquad = 4 \end{cases} \Leftrightarrow \begin{cases} x + 3y - 2z = 0 \\ -6y + 8z = 4 \quad \text{Eq. 2} + (-2) \times \text{Eq. 1} \\ -6y + 8z = 4 \quad \text{Eq. 3} + (-4) \times \text{Eq. 1} \end{cases} \Leftrightarrow \begin{cases} x + 3y - 2z = 0 \\ -6y + 8z = 4 \\ 0 = 0 \quad \text{Eq. 2} - \text{Eq. 3} \end{cases}$$

So $z = t$ and $-6y + 8t = 4 \Leftrightarrow -6y = -8t + 4 \Leftrightarrow y = \frac{4}{3}t - \frac{2}{3}$. Then $x + 3\left(\frac{4}{3}t - \frac{2}{3}\right) - 2t = 0 \Leftrightarrow x = -2t + 2$. So the

solutions are $\left(-2t + 2, \frac{4}{3}t - \frac{2}{3}, t\right)$, where t is any real number.

35.
$$\begin{cases} x \qquad + z + 2w = 6 \\ y - 2z \qquad = -3 \\ x + 2y - z \qquad = -2 \\ 2x + y + 3z - 2w = 0 \end{cases} \Leftrightarrow \begin{cases} x \qquad + z + 2w = 6 \\ y - 2z \qquad = -3 \\ 2y - 2z - 2w = -8 \quad \text{Eq. 3} - \text{Eq. 1} \\ y + z - 6w = -12 \quad \text{Eq. 4} + (-2) \times \text{Eq. 1} \end{cases} \Leftrightarrow$$

$$\begin{cases} x \qquad + z + 2w = 6 \\ y - 2z \qquad = -3 \\ 2z - 2w = -2 \quad \text{Eq. 3} + (-2) \times \text{Eq. 2} \\ 3z - 6w = -9 \quad \text{Eq. 4} - \text{Eq. 2} \end{cases} \Leftrightarrow \begin{cases} x \qquad + z + 2w = 6 \\ y - 2z \qquad = -3 \\ z - w = -1 \quad \frac{1}{2}\text{Eq. 3} \\ 6w = 12 \quad 3 \times \text{Eq. 3} + (-2) \times \text{Eq. 4} \end{cases}$$

So $w = 2$ and $z - 2 = -1 \Leftrightarrow z = 1$. Then $y - 2(1) = -3 \Leftrightarrow y = -1$ and $x + 1 + 2(2) = 6 \Leftrightarrow x = 1$. Thus, the solution
is $(1, -1, 1, 2)$.

37. Let x be the amount invested at 4%, y the amount invested at 5%, and z the amount invested at 6%. We set

up a model and get the following equations: $\begin{cases} \text{Total money:} & x + y + z = 100{,}000 \\ \text{Annual income:} & 0.04x + 0.05y + 0.06z = 0.051\,(100{,}000) \\ \text{Equal amounts:} & x = y \end{cases}$

$$\Leftrightarrow \begin{cases} x + y + z = 100{,}000 \\ 4x + 5y + 6z = 510{,}000 \\ x - y = 0 \end{cases} \Leftrightarrow \begin{cases} x + y + z = 100{,}000 \\ y + 2z = 110{,}000 \quad \text{Eq. 2} + (-4) \times \text{Eq. 1} \\ -2y - z = -100{,}000 \quad \text{Eq. 3} - \text{Eq. 1} \end{cases} \Leftrightarrow$$

$$\begin{cases} x + y + z = 100{,}000 \\ y + 2z = 110{,}000 \\ 3z = 120{,}000 \quad 2 \times \text{Eq. 2} + \text{Eq. 3} \end{cases}$$

So $z = 40{,}000$ and $y + 2(40{,}000) = 110{,}000 \Leftrightarrow y = 30{,}000$. Since $x = y$, $x = 30{,}000$. She must invest \$30,000 in
short-term bonds, \$30,000 in intermediate-term bonds, and \$40,000 in long-term bonds.

39. Let x, y, and z be the number of acres of land planted with corn, wheat, and soybeans. We set up a model and

get the following equations: $\begin{cases} \text{Total acres:} & x + y + z = 1200 \\ \text{Market demand:} & 2x = y \\ \text{Total cost:} & 45x + 60y + 50z = 63{,}750 \end{cases}$ Substituting $2x$ for y, we get

$$\begin{cases} x + 2x + z = 1200 \\ 2x = y \\ 45x + 60\,(2x) + 50z = 63{,}750 \end{cases} \Leftrightarrow \begin{cases} 3x + z = 1200 \\ 2x - y = 0 \\ 165x + 50z = 63{,}750 \end{cases} \Leftrightarrow \begin{cases} 3x + z = 1200 \\ 2x - y = 0 \\ 15x = 3750 \quad \text{Eq. 3} + (-50) \times \text{Eq. 1} \end{cases}$$

So $15x = 3{,}750 \Leftrightarrow x = 250$ and $y = 2(250) = 500$. Substituting into the original equation, we have $250 + 500 + z = 1200$
$\Leftrightarrow z = 450$. Thus the farmer should plant 250 acres of corn, 500 acres of wheat, and 450 acres of soybeans.

41. Let a, b, and c be the number of ounces of Type A, Type B, and Type C pellets used. The requirements for the different vitamins gives the following system:
$$\begin{cases} 2a + 3b + c = 9 \\ 3a + b + 3c = 14 \\ 8a + 5b + 7c = 32 \end{cases} \Leftrightarrow$$

$$\begin{cases} 2a + 3b + c = 9 \\ -7b + 3c = 1 \quad \text{2 × Eq. 2 + (−3) × Eq. 1} \\ -7b + 3c = -4 \quad \text{Eq. 3 + (−4) × Eq. 1} \end{cases} \qquad \text{Equations 2 and 3 are inconsistent, so there is no solution.}$$

43. Let a, b, and c represent the number of Midnight Mango, Tropical Torrent, and Pineapple Power smoothies sold. The given information leads to the system
$$\begin{cases} 8a + 6b + 2c = 820 \\ 3a + 5b + 8c = 690 \\ 3a + 3b + 4c = 450 \end{cases} \Leftrightarrow \begin{cases} 8a + 6b + 2c = 820 \\ 22b + 58c = 3060 \quad \text{8 × Eq. 2 + (−3) × Eq. 1} \\ 2b + 4c = 240 \quad \text{Eq. 2 − Eq. 3} \end{cases} \Leftrightarrow$$

$$\begin{cases} 8a + 6b + 2c = 820 \\ 22b + 58c = 3060 \\ 14c = 420 \quad \text{Eq. 2 + (−11) × Eq. 3} \end{cases}$$

Thus, $c = 30$, so $22b + 58(30) = 3060 \Leftrightarrow 22b = 1320 \Leftrightarrow b = 60$ and $8a + 6(60) + 2(30) = 820 \Leftrightarrow a = 50$. Thus, The Juice Company sold 50 Midnight Mango, 60 Tropical Torrent, and 30 Pineapple Power smoothies on that particular day.

45. Let a, b, and c be the number of shares of Stock A, Stock B, and Stock C in the investor's portfolio. Since the total value remains unchanged, we get the following system:

$$\begin{cases} 10a + 25b + 29c = 74{,}000 \\ 12a + 20b + 32c = 74{,}000 \\ 16a + 15b + 32c = 74{,}000 \end{cases} \Leftrightarrow \begin{cases} 10a + 25b + 29c = 74{,}000 \\ 50b + 14c = 74{,}000 \quad \text{6 × Eq. 1 + (−5) × Eq. 2} \\ 125b + 72c = 222{,}000 \quad \text{8 × Eq. 1 + (−5) × Eq. 3} \end{cases} \Leftrightarrow$$

$$\begin{cases} 10a + 25b + 29c = 74{,}000 \\ 50b + 14c = 74{,}000 \\ 74c = 74{,}000 \quad \text{(−5) × Eq. 2 + 2 × Eq. 3} \end{cases}$$

So $c = 1{,}000$. Back-substituting we have $50b + 14(1000) = 74{,}000 \Leftrightarrow 50b = 60{,}000 \Leftrightarrow b = 1{,}200$. And finally $10a + 25(1200) + 29(1000) = 74{,}000 \quad 10a + 30{,}000 + 29{,}000 = 74{,}000 \Leftrightarrow 10a = 15{,}000 \Leftrightarrow a = 1{,}500$. Thus the portfolio consists of 1,500 shares of Stock A, 1,200 shares of Stock B, and 1,000 shares of Stock C.

47. (a) We begin by substituting $\dfrac{x_0 + x_1}{2}$, $\dfrac{y_0 + y_1}{2}$, and $\dfrac{z_0 + z_1}{2}$ into the left-hand side of the first equation:

$$a_1\left(\frac{x_0 + x_1}{2}\right) + b_1\left(\frac{y_0 + y_1}{2}\right) + c_1\left(\frac{z_0 + z_1}{2}\right) = \tfrac{1}{2}\left[(a_1 x_0 + b_1 y_0 + c_1 z_0) + (a_1 x_1 + b_1 y_1 + c_1 z_1)\right]$$
$$= \tfrac{1}{2}\left[d_1 + d_1\right] = d_1$$

Thus the given ordered triple satisfies the first equation. We can show that it satisfies the second and the third in exactly the same way. Thus it is a solution of the system.

(b) We have shown in part (a) that if the system has two different solutions, we can find a third one by averaging the two solutions. But then we can find a fourth and a fifth solution by averaging the new one with each of the previous two. Then we can find four more by repeating this process with these new solutions, and so on. Clearly this process can continue indefinitely, so there are infinitely many solutions.

10.3 MATRICES AND SYSTEMS OF LINEAR EQUATIONS

1. A system of linear equations with infinitely many solutions is called *dependent*. A system of linear equations with no solution is called *inconsistent*.

3. (a) The leading variables are x *and* y.

(b) The system is dependent.

(c) The solution of the system is $x = 3 + t$, $y = 5 - 2t$, $z = t$.

5. 3×2 **7.** 2×1 **9.** 1×3

11. (a) Yes, this matrix is in row-echelon form.

(b) Yes, this matrix is in reduced row-echelon form.

(c) $\begin{cases} x = -3 \\ y = 5 \end{cases}$

13. (a) Yes, this matrix is in row-echelon form.

(b) No, this matrix is not in reduced row-echelon form, since the leading 1 in the second row does not have a zero above it.

(c) $\begin{cases} x + 2y + 8z = 0 \\ y + 3z = 2 \\ 0 = 0 \end{cases}$

15. (a) No, this matrix is not in row-echelon form, since the row of zeros is not at the bottom.

(b) No, this matrix is not in reduced row-echelon form.

(c) $\begin{cases} x = 0 \\ 0 = 0 \\ y + 5z = 1 \end{cases}$

17. (a) Yes, this matrix is in row-echelon form.

(b) Yes, this matrix is in reduced row-echelon form.

(c) $\begin{cases} x + 3y - w = 0 \\ z + 2w = 0 \\ 0 = 1 \\ 0 = 0 \end{cases}$

Notice that this system has no solution.

19. $\begin{bmatrix} 1 & -2 & 1 & 1 \\ 0 & 1 & 2 & 5 \\ 1 & 1 & 3 & 8 \end{bmatrix} \xrightarrow{R_3 - R_1 \to R_3} \begin{bmatrix} 1 & -2 & 1 & 1 \\ 0 & 1 & 2 & 5 \\ 0 & 3 & 2 & 7 \end{bmatrix} \xrightarrow{R_3 - 3R_2 \to R_3} \begin{bmatrix} 1 & -2 & 1 & 1 \\ 0 & 1 & 2 & 5 \\ 0 & 0 & -4 & -8 \end{bmatrix}$. Thus, $-4z = -8 \Leftrightarrow$

$z = 2$; $y + 2(2) = 5 \Leftrightarrow y = 1$; and $x - 2(1) + (2) = 1 \Leftrightarrow x = 1$. Therefore, the solution is $(1, 1, 2)$.

21. $\begin{bmatrix} 1 & 1 & 1 & 2 \\ 2 & -3 & 2 & 4 \\ 4 & 1 & -3 & 1 \end{bmatrix} \xrightarrow[R_3 - 4R_1 \to R_3]{R_2 - 2R_1 \to R_2} \begin{bmatrix} 1 & 1 & 1 & 2 \\ 0 & -5 & 0 & 0 \\ 0 & -3 & -7 & -7 \end{bmatrix} \xrightarrow{R_3 - \frac{3}{5}R_2 \to R_3} \begin{bmatrix} 1 & 1 & 1 & 2 \\ 0 & -5 & 0 & 0 \\ 0 & 0 & -7 & -7 \end{bmatrix}$. Thus, $-7z = -7$

$\Leftrightarrow z = 1$; $-5y = 0 \Leftrightarrow y = 0$; and $x + 0 + 1 = 2 \Leftrightarrow x = 1$. Therefore, the solution is $(1, 0, 1)$.

23. $\begin{bmatrix} 1 & 2 & -1 & -2 \\ 1 & 0 & 1 & 0 \\ 2 & -1 & -1 & -3 \end{bmatrix} \xrightarrow[R_3 - 2R_1 \to R_3]{R_2 - R_1 \to R_2} \begin{bmatrix} 1 & 2 & -1 & -2 \\ 0 & -2 & 2 & 2 \\ 0 & -5 & 1 & 1 \end{bmatrix} \xrightarrow{-\frac{1}{2}R_2} \begin{bmatrix} 1 & 2 & -1 & -2 \\ 0 & 1 & 1 & 1 \\ 0 & -5 & 1 & 1 \end{bmatrix} \xrightarrow{R_3 + 5R_2 \to R_3}$

$\begin{bmatrix} 1 & 2 & -1 & -2 \\ 0 & 1 & 1 & 1 \\ 0 & 0 & 6 & 6 \end{bmatrix}$. Thus, $6z = 6 \Leftrightarrow z = 1$; $y + (1) = 1 \Leftrightarrow y = 0$; and $x + 2(0) - (1) = -2 \Leftrightarrow x = -1$. Therefore, the

solution is $(-1, 0, 1)$.

25. $\begin{bmatrix} 1 & 2 & -1 & 9 \\ 2 & 0 & -1 & -2 \\ 3 & 5 & 2 & 22 \end{bmatrix}$ $\xrightarrow[R_3 - 3R_1 \to R_3]{R_2 - 2R_1 \to R_2}$ $\begin{bmatrix} 1 & 2 & -1 & 9 \\ 0 & -4 & 1 & -20 \\ 0 & -1 & 5 & -5 \end{bmatrix}$ $\xrightarrow{4R_3 - R_2 \to R_3}$ $\begin{bmatrix} 1 & 2 & -1 & 9 \\ 0 & -4 & 1 & -20 \\ 0 & 0 & 19 & 0 \end{bmatrix}$ Thus, $19x_3 = 0$

$\Leftrightarrow x_3 = 0$; $-4x_2 = -20 \Leftrightarrow x_2 = 5$; and $x_1 + 2(5) = 9 \Leftrightarrow x_1 = -1$. Therefore, the solution is $(-1, 5, 0)$.

27. $\begin{bmatrix} 2 & -3 & -1 & 13 \\ -1 & 2 & -5 & 6 \\ 5 & -1 & -1 & 49 \end{bmatrix}$ $\xrightarrow[2R_3 - 5R_1 \to R_3]{2R_2 + R_1 \to R_2}$ $\begin{bmatrix} 2 & -3 & -1 & 13 \\ 0 & 1 & -11 & 25 \\ 0 & 13 & 3 & 33 \end{bmatrix}$ $\xrightarrow{R_3 - 13R_2 \to R_3}$ $\begin{bmatrix} 2 & -3 & -1 & 13 \\ 0 & 1 & -11 & 25 \\ 0 & 0 & 146 & -292 \end{bmatrix}$ Thus,

$146z = -292 \Leftrightarrow z = -2$; $y - 11(-2) = 25 \Leftrightarrow y = 3$; and $2x - 3 \cdot 3 + 2 = 13 \Leftrightarrow x = 10$. Therefore, the solution is $(10, 3, -2)$.

29. $\begin{bmatrix} 1 & 1 & 1 & 2 \\ 0 & 1 & -3 & 1 \\ 2 & 1 & 5 & 0 \end{bmatrix}$ $\xrightarrow{R_3 - 2R_1 \to R_3}$ $\begin{bmatrix} 1 & 1 & 1 & 2 \\ 0 & 1 & -3 & 1 \\ 0 & -1 & 3 & -4 \end{bmatrix}$ $\xrightarrow{R_3 + R_2 \to R_3}$ $\begin{bmatrix} 1 & 1 & 1 & 3 \\ 0 & 1 & -3 & 1 \\ 0 & 0 & 0 & -3 \end{bmatrix}$. The third row of the

matrix states $0 = -3$, which is impossible. Hence, the system is inconsistent, and there is no solution.

31. $\begin{bmatrix} 2 & -3 & -9 & -5 \\ 1 & 0 & 3 & 2 \\ -3 & 1 & -4 & -3 \end{bmatrix}$ $\xrightarrow{R_1 \leftrightarrow R_2}$ $\begin{bmatrix} 1 & 0 & 3 & 2 \\ 2 & -3 & -9 & -5 \\ -3 & 1 & -4 & -3 \end{bmatrix}$ $\xrightarrow[R_3 + 3R_1 \to R_3]{R_2 - 2R_1 \to R_2}$ $\begin{bmatrix} 1 & 0 & 3 & 2 \\ 0 & -3 & -15 & -9 \\ 0 & 1 & 5 & 3 \end{bmatrix}$ $\xrightarrow{-\frac{1}{3}R_2}$

$\begin{bmatrix} 1 & 0 & 3 & 2 \\ 0 & 1 & 5 & 3 \\ 0 & 1 & 5 & 3 \end{bmatrix}$ $\xrightarrow{R_3 - R_2 \to R_3}$ $\begin{bmatrix} 1 & 0 & 3 & 2 \\ 0 & 1 & 5 & 3 \\ 0 & 0 & 0 & 0 \end{bmatrix}$. Therefore, this system has infinitely many solutions, given by $x + 3t = 2$

$\Leftrightarrow x = 2 - 3t$, and $y + 5t = 3 \Leftrightarrow y = 3 - 5t$. Hence, the solutions are $(2 - 3t, 3 - 5t, t)$, where t is any real number.

33. $\begin{bmatrix} 1 & -1 & 3 & 3 \\ 4 & -8 & 32 & 24 \\ 2 & -3 & 11 & 4 \end{bmatrix}$ $\xrightarrow[R_3 - 2R_1 \to R_3]{R_2 - 4R_1 \to R_2}$ $\begin{bmatrix} 1 & -1 & 3 & 3 \\ 0 & -4 & 20 & 12 \\ 0 & -1 & 5 & -2 \end{bmatrix}$ $\xrightarrow[R_3 + R_2 \to R_3]{-\frac{1}{4}R_2}$ $\begin{bmatrix} 1 & -1 & 3 & 3 \\ 0 & 1 & -5 & -3 \\ 0 & 0 & 0 & -5 \end{bmatrix}$ The third row of the

matrix states $0 = -5$, which is impossible. Hence, the system is inconsistent, and there is no solution.

35. $\begin{bmatrix} 1 & 4 & -2 & -3 \\ 2 & -1 & 5 & 12 \\ 8 & 5 & 11 & 30 \end{bmatrix}$ $\xrightarrow[R_3 - 8R_1 \to R_3]{R_2 - 2R_1 \to R_2}$ $\begin{bmatrix} 1 & 4 & -2 & -3 \\ 0 & -9 & 9 & 18 \\ 0 & -27 & 27 & 54 \end{bmatrix}$ $\xrightarrow{R_3 - 3R_2 \to R_3}$ $\begin{bmatrix} 1 & 4 & -2 & -3 \\ 0 & -9 & 9 & 18 \\ 0 & 0 & 0 & 0 \end{bmatrix}$. Therefore, this

system has infinitely many solutions, given by $-9y + 9t = 18 \Leftrightarrow y = -2 + t$, and $x + 4(-2 + t) - 2t = -3 \Leftrightarrow x = 5 - 2t$. Hence, the solutions are $(5 - 2t, -2 + t, t)$, where t is any real number.

37. $\begin{bmatrix} 2 & 1 & -2 & 12 \\ -1 & -\frac{1}{2} & 1 & -6 \\ 3 & \frac{3}{2} & -3 & 18 \end{bmatrix}$ $\xrightarrow[-R_1]{R_1 \leftrightarrow R_2}$ $\begin{bmatrix} 1 & \frac{1}{2} & -1 & 6 \\ 2 & 1 & -2 & 12 \\ 3 & \frac{3}{2} & -3 & 18 \end{bmatrix}$ $\xrightarrow[R_3 - 3R_1 \to R_3]{R_2 - 2R_1 \to R_2}$ $\begin{bmatrix} 1 & \frac{1}{2} & -1 & 6 \\ 0 & 0 & 0 & 0 \\ 0 & 0 & 0 & 0 \end{bmatrix}$ Therefore, this system

has infinitely many solutions, given by $x + \frac{1}{2}s - t = 6 \Leftrightarrow x = 6 - \frac{1}{2}s + t$. Hence, the solutions are $\left(6 - \frac{1}{2}s + t, s, t\right)$, where s and t are any real numbers.

39. $\begin{bmatrix} 4 & -3 & 1 & -8 \\ -2 & 1 & -3 & -4 \\ 1 & -1 & 2 & 3 \end{bmatrix} \xrightarrow{R_1 \leftrightarrow R_3} \begin{bmatrix} 1 & -1 & 2 & 3 \\ -2 & 1 & -3 & -4 \\ 4 & -3 & 1 & -8 \end{bmatrix} \xrightarrow[R_3 - 4R_1 \to R_3]{R_2 + 2R_1 \to R_2} \begin{bmatrix} 1 & -1 & 2 & 3 \\ 0 & -1 & 1 & 2 \\ 0 & 1 & -7 & -20 \end{bmatrix} \xrightarrow{-R_2}$

$\begin{bmatrix} 1 & -1 & 2 & 3 \\ 0 & 1 & -1 & -2 \\ 0 & 1 & -7 & -20 \end{bmatrix} \xrightarrow{R_3 - R_2 \to R_3} \begin{bmatrix} 1 & -1 & 2 & 3 \\ 0 & 1 & -1 & -2 \\ 0 & 0 & -6 & -18 \end{bmatrix}$. Therefore, $-6z = -18 \Leftrightarrow z = 3$; $y - (3) = -2 \Leftrightarrow$

$y = 1$; and $x - (1) + 2(3) = 3 \Leftrightarrow x = -2$. Hence, the solution is $(-2, 1, 3)$.

41. $\begin{bmatrix} 2 & 1 & 3 & 9 \\ -1 & 0 & -7 & 10 \\ 3 & 2 & -1 & 4 \end{bmatrix} \xrightarrow[R_3 + 3R_2 \to R_3]{2R_2 + R_1 \to R_2} \begin{bmatrix} 2 & 1 & 3 & 9 \\ 0 & 1 & -11 & 29 \\ 0 & 2 & -22 & 34 \end{bmatrix} \xrightarrow{2R_2 - R_3 \to R_3} \begin{bmatrix} 2 & 1 & 3 & 9 \\ 0 & 1 & -11 & 29 \\ 0 & 0 & 0 & 24 \end{bmatrix}$. Therefore, the

system is inconsistent and there is no solution.

43. $\begin{bmatrix} 1 & 2 & -3 & -5 \\ -2 & -4 & -6 & 10 \\ 3 & 7 & -2 & -13 \end{bmatrix} \xrightarrow[R_3 - 3R_1 \to R_3]{R_2 + 2R_1 \to R_2} \begin{bmatrix} 1 & 2 & -3 & -5 \\ 0 & 0 & -12 & 0 \\ 0 & 1 & 7 & 2 \end{bmatrix} \xrightarrow{R_2 \leftrightarrow R_3} \begin{bmatrix} 1 & 2 & -3 & -5 \\ 0 & 1 & 7 & 2 \\ 0 & 0 & -12 & 0 \end{bmatrix}$. Therefore,

$-12z = 0 \Leftrightarrow z = 0$; $y + 7(0) = 2 \Leftrightarrow y = 2$; and $x + 2(2) - 3(0) = -5 \Leftrightarrow x = -9$. Hence, the solution is $(-9, 2, 0)$.

45. $\begin{bmatrix} 1 & -1 & 6 & 8 \\ 1 & 0 & 1 & 5 \\ 1 & 3 & -14 & -4 \end{bmatrix} \xrightarrow[R_3 - R_1 \to R_3]{R_2 - R_1 \to R_2} \begin{bmatrix} 1 & -1 & 6 & 8 \\ 0 & 1 & -5 & -3 \\ 0 & 4 & -20 & -12 \end{bmatrix} \xrightarrow{R_3 - 4R_2 \to R_3} \begin{bmatrix} 1 & -1 & 6 & 8 \\ 0 & 1 & -5 & -3 \\ 0 & 0 & 0 & 0 \end{bmatrix}$. Therefore,

the system is dependent. Let $z = t$. Then $y - 5t = -3 \Rightarrow y = -3 + 5t$ and $x - y + 6t = 8 \Leftrightarrow x = 8 + (-3 + 5t) - 6t = 5 - t$.
The solutions are $(5 - t, -3 + 5t, t)$, where t is any real number.

47. $\begin{bmatrix} -1 & 2 & 1 & -3 & 3 \\ 3 & -4 & 1 & 1 & 9 \\ -1 & -1 & 1 & 1 & 0 \\ 2 & 1 & 4 & -2 & 3 \end{bmatrix} \xrightarrow{-R_1} \begin{bmatrix} 1 & -2 & -1 & 3 & -3 \\ 3 & -4 & 1 & 1 & 9 \\ -1 & -1 & 1 & 1 & 0 \\ 2 & 1 & 4 & -2 & 3 \end{bmatrix} \xrightarrow[\substack{R_3 + R_1 \to R_3 \\ R_4 - 2R_1 \to R_4}]{R_2 - 3R_1 \to R_2} \begin{bmatrix} 1 & -2 & -1 & 3 & -3 \\ 0 & 2 & 4 & -8 & 18 \\ 0 & -3 & 0 & 4 & -3 \\ 0 & 5 & 6 & -8 & 9 \end{bmatrix}$

$\xrightarrow{\frac{1}{2}R_2} \begin{bmatrix} 1 & -2 & -1 & 3 & -3 \\ 0 & 1 & 2 & -4 & 9 \\ 0 & -3 & 0 & 4 & -3 \\ 0 & 5 & 6 & -8 & 9 \end{bmatrix} \xrightarrow[R_4 - 5R_2 \to R_4]{R_3 + 3R_2 \to R_3} \begin{bmatrix} 1 & -2 & -1 & 3 & -3 \\ 0 & 1 & 2 & -4 & 9 \\ 0 & 0 & 6 & -8 & 24 \\ 0 & 0 & -4 & 12 & -36 \end{bmatrix} \xrightarrow{3R_4 + 2R_3 \to R_4}$

$\begin{bmatrix} 1 & -2 & -1 & 3 & -3 \\ 0 & 1 & 2 & -4 & 9 \\ 0 & 0 & 6 & -8 & 24 \\ 0 & 0 & 0 & 20 & -60 \end{bmatrix}$. Therefore, $20w = -60 \Leftrightarrow w = -3$; $6z + 24 = 24 \Leftrightarrow z = 0$. Then $y + 12 = 9 \Leftrightarrow y = -3$ and

$x + 6 - 9 = -3 \Leftrightarrow x = 0$. Hence, the solution is $(0, -3, 0, -3)$.

49.
$$\begin{bmatrix} 1 & 1 & 2 & -1 & -2 \\ 0 & 3 & 1 & 2 & 2 \\ 1 & 1 & 0 & 3 & 2 \\ -3 & 0 & 1 & 2 & 5 \end{bmatrix} \xrightarrow[R_4+3R_1 \to R_4]{R_3-R_1 \to R_3} \begin{bmatrix} 1 & 1 & 2 & -1 & -2 \\ 0 & 3 & 1 & 2 & 2 \\ 0 & 0 & -2 & 4 & 4 \\ 0 & 3 & 7 & -1 & -1 \end{bmatrix} \xrightarrow{R_4-R_2 \to R_4} \begin{bmatrix} 1 & 1 & 2 & -1 & -2 \\ 0 & 3 & 1 & 2 & 2 \\ 0 & 0 & -2 & 4 & 4 \\ 0 & 0 & 6 & -3 & -3 \end{bmatrix}$$

$$\xrightarrow{R_4+3R_3 \to R_4} \begin{bmatrix} 1 & 1 & 2 & -1 & -2 \\ 0 & 3 & 1 & 2 & 2 \\ 0 & 0 & -2 & 4 & 4 \\ 0 & 0 & 0 & 9 & 9 \end{bmatrix}.$$ Therefore, $9w = 9 \Leftrightarrow w = 1$; $-2z + 4(1) = 4 \Leftrightarrow z = 0$. Then

$3y + (0) + 2(1) = 2 \Leftrightarrow y = 0$ and $x + (0) + 2(0) - (1) = -2 \Leftrightarrow x = -1$. Hence, the solution is $(-1, 0, 0, 1)$.

51.
$$\begin{bmatrix} 1 & -1 & 0 & 1 & 0 \\ 3 & 0 & -1 & 2 & 0 \\ 1 & -4 & 1 & 2 & 0 \end{bmatrix} \xrightarrow[R_3-R_1 \to R_3]{R_2-3R_1 \to R_2} \begin{bmatrix} 1 & -1 & 0 & 1 & 0 \\ 0 & 3 & -1 & -1 & 0 \\ 0 & -3 & 1 & 1 & 0 \end{bmatrix} \xrightarrow{R_3+R_2 \to R_3} \begin{bmatrix} 1 & -1 & 0 & 1 & 0 \\ 0 & 3 & -1 & -1 & 0 \\ 0 & 0 & 0 & 0 & 0 \end{bmatrix}.$$

Therefore, the system has infinitely many solutions, given by $3y - s - t = 0 \Leftrightarrow y = \frac{1}{3}(s+t)$ and $x - \frac{1}{3}(s+t) + t = 0$

$\Leftrightarrow x = \frac{1}{3}(s - 2t)$. So the solutions are $\left(\frac{1}{3}(s - 2t), \frac{1}{3}(s+t), s, t\right)$, where s and t are any real numbers.

53.
$$\begin{bmatrix} 1 & 0 & 1 & 1 & 4 \\ 0 & 1 & -1 & 0 & -4 \\ 1 & -2 & 3 & 1 & 12 \\ 2 & 0 & -2 & 5 & -1 \end{bmatrix} \xrightarrow[R_4-2R_1 \to R_4]{R_3-R_1 \to R_3} \begin{bmatrix} 1 & 0 & 1 & 1 & 4 \\ 0 & 1 & -1 & 0 & -4 \\ 0 & -2 & 2 & 0 & 8 \\ 0 & 0 & -4 & 3 & -9 \end{bmatrix} \xrightarrow{R_3+2R_2 \to R_3} \begin{bmatrix} 1 & 0 & 1 & 1 & 4 \\ 0 & 1 & -1 & 0 & -4 \\ 0 & 0 & 0 & 0 & 0 \\ 0 & 0 & -4 & 3 & -9 \end{bmatrix}$$

$$\xrightarrow{R_3 \leftrightarrow -R_4} \begin{bmatrix} 1 & 0 & 1 & 1 & 4 \\ 0 & 1 & -1 & 0 & -4 \\ 0 & 0 & 4 & -3 & 9 \\ 0 & 0 & 0 & 0 & 0 \end{bmatrix}.$$ Therefore, $4z - 3t = 9 \Leftrightarrow 4z = 9 + 3t \Leftrightarrow z = \frac{9}{4} + \frac{3}{4}t$. Then we have

$y - \left(\frac{9}{4} + \frac{3}{4}t\right) = -4 \Leftrightarrow y = \frac{-7}{4} + \frac{3}{4}t$ and $x + \left(\frac{9}{4} + \frac{3}{4}t\right) + t = 4 \Leftrightarrow x = \frac{7}{4} - \frac{7}{4}t$. Hence, the solutions are

$\left(\frac{7}{4} - \frac{7}{4}t, -\frac{7}{4} + \frac{3}{4}t, \frac{9}{4} + \frac{3}{4}t, t\right)$, where t is any real number.

55. Let x, y, z represent the number of VitaMax, Vitron, and VitaPlus pills taken daily. The matrix representation for the system of equations is

$$\begin{bmatrix} 5 & 10 & 15 & 50 \\ 15 & 20 & 0 & 50 \\ 10 & 10 & 10 & 50 \end{bmatrix} \xrightarrow[\frac{1}{5}R_3]{\substack{\frac{1}{5}R_1 \\ \frac{1}{5}R_2}} \begin{bmatrix} 1 & 2 & 3 & 10 \\ 3 & 4 & 0 & 10 \\ 2 & 2 & 2 & 10 \end{bmatrix} \xrightarrow[R_3-2R_1 \to R_3]{R_2-3R_1 \to R_2} \begin{bmatrix} 1 & 2 & 3 & 10 \\ 0 & -2 & -9 & -20 \\ 0 & -2 & -4 & -10 \end{bmatrix} \xrightarrow{R_3-R_2 \to R_3} \begin{bmatrix} 1 & 2 & 3 & 10 \\ 0 & -2 & -9 & -20 \\ 0 & 0 & 5 & 10 \end{bmatrix}.$$

Thus, $5z = 10 \Leftrightarrow z = 2$; $-2y - 18 = -20 \Leftrightarrow y = 1$; and $x + 2 + 6 = 10 \Leftrightarrow x = 2$. Hence, he should take 2 VitaMax, 1 Vitron, and 2 VitaPlus pills daily.

57. Let x, y, and z represent the distance, in miles, of the run, swim, and cycle parts of the race respectively. Then, since time $= \dfrac{\text{distance}}{\text{speed}}$, we get the following equations from the three contestants' race times:

$$\begin{cases} \left(\frac{x}{10}\right) + \left(\frac{y}{4}\right) + \left(\frac{z}{20}\right) = 2.5 \\ \left(\frac{x}{7.5}\right) + \left(\frac{y}{6}\right) + \left(\frac{z}{15}\right) = 3 \\ \left(\frac{x}{15}\right) + \left(\frac{y}{3}\right) + \left(\frac{z}{40}\right) = 1.75 \end{cases} \Leftrightarrow \begin{cases} 2x + 5y + z = 50 \\ 4x + 5y + 2z = 90 \\ 8x + 40y + 3z = 210 \end{cases}$$ which has the following matrix representation:

$$\begin{bmatrix} 2 & 5 & 1 & 50 \\ 4 & 5 & 2 & 90 \\ 8 & 40 & 3 & 210 \end{bmatrix} \xrightarrow[R_3 - 4R_1 \to R_3]{R_2 - 2R_1 \to R_2} \begin{bmatrix} 2 & 5 & 1 & 50 \\ 0 & -5 & 0 & -10 \\ 0 & 20 & -1 & 10 \end{bmatrix} \xrightarrow{R_3 + 4R_2 \to R_3} \begin{bmatrix} 2 & 5 & 1 & 50 \\ 0 & -5 & 0 & -10 \\ 0 & 0 & -1 & -30 \end{bmatrix}.$$

Thus, $-z = -30 \Leftrightarrow z = 30$; $-5y = -10 \Leftrightarrow y = 2$; and $2x + 10 + 30 = 50 \Leftrightarrow x = 5$. So the race has a 5 mile run, 2 mile swim, and 30 mile cycle.

59. Let t be the number of tables produced, c the number of chairs, and a the number of armoires. Then, the system of equations

is $\begin{cases} \frac{1}{2}t + c + a = 300 \\ \frac{1}{2}t + \frac{3}{2}c + a = 400 \\ t + \frac{3}{2}c + 2a = 590 \end{cases} \Leftrightarrow \begin{cases} t + 2c + 2a = 600 \\ t + 3c + 2a = 800 \\ 2t + 3c + 4a = 1180 \end{cases}$ and a matrix representation is

$$\begin{bmatrix} 1 & 2 & 2 & 600 \\ 1 & 3 & 2 & 800 \\ 2 & 3 & 4 & 1180 \end{bmatrix} \xrightarrow[R_3 - 2R_1 \to R_3]{R_2 - R_1 \to R_2} \begin{bmatrix} 1 & 2 & 2 & 600 \\ 0 & 1 & 0 & 200 \\ 0 & -1 & 0 & -20 \end{bmatrix} \xrightarrow{R_3 + R_2 \to R_3} \begin{bmatrix} 1 & 2 & 2 & 600 \\ 0 & 1 & 0 & 200 \\ 0 & 0 & 0 & 180 \end{bmatrix}.$$ The third row states

$0 = 180$, which is impossible, and so the system is inconsistent. Therefore, it is impossible to use all of the available labor-hours.

61. *Line containing the points* $(0, 0)$ *and* $(1, 12)$*:* Using the general form of a line, $y = ax + b$, we substitute for x and y and solve for a and b. The point $(0, 0)$ gives $0 = a(0) + b \Rightarrow b = 0$; the point $(1, 12)$ gives $12 = a(1) + b \Rightarrow a = 12$. Since $a = 12$ and $b = 0$, the equation of the line is $y = 12x$.

Quadratic containing the points $(0, 0)$*,* $(1, 12)$*, and* $(3, 6)$*:* Using the general form of a quadratic, $y = ax^2 + bx + c$, we substitute for x and y and solve for a, b, and c. The point $(0, 0)$ gives $0 = a(0)^2 + b(0) + c \Rightarrow c = 0$; the point $(1, 12)$ gives $12 = a(1)^2 + b(1) + c \Rightarrow a + b = 12$; the point $(3, 6)$ gives $6 = a(3)^2 + b(3) + c \Rightarrow 9a + 3b = 6$. Subtracting the third equation from -3 times the third gives $6a = -30 \Leftrightarrow a = -5$. So $a + b = 12 \Leftrightarrow b = 12 - a \Rightarrow b = 17$. Since $a = -5$, $b = 17$, and $c = 0$, the equation of the quadratic is $y = -5x^2 + 17x$. *Cubic containing the points* $(0, 0)$*,* $(1, 12)$*,* $(2, 40)$*, and* $(3, 6)$*:* Using the general form of a cubic, $y = ax^3 + bx^2 + cx + d$, we substitute for x and y and solve for a, b, c, and d. The point $(0, 0)$ gives $0 = a(0)^3 + b(0)^2 + c(0) + d \Rightarrow d = 0$; the point the point $(1, 12)$ gives $12 = a(1)^3 + b(1)^2 + c(1) + d \Rightarrow a + b + c + d = 12$; the point $(2, 40)$ gives $40 = a(2)^3 + b(2)^2 + c(2) + d \Rightarrow 8a + 4b + 2c + d = 40$; the point $(3, 6)$ gives $6 = a(3)^3 + b(3)^2 + c(3) + d \Rightarrow 27a + 9b + 3c + d = 6$. Since $d = 0$, the

system reduces to $\begin{cases} a + b + c = 12 \\ 8a + 4b + 2c = 40 \\ 27a + 9b + 3c = 6 \end{cases}$ which has representation

$$\begin{bmatrix} 1 & 1 & 1 & 12 \\ 8 & 4 & 2 & 40 \\ 27 & 9 & 3 & 6 \end{bmatrix} \xrightarrow[R_3 - 27R_1 \to R_3]{R_2 - 8R_1 \to R_2} \begin{bmatrix} 1 & 1 & 1 & 12 \\ 0 & -4 & -6 & -56 \\ 0 & -18 & -24 & -318 \end{bmatrix} \xrightarrow[-\frac{1}{6}R_3]{-\frac{1}{2}R_2} \begin{bmatrix} 1 & 1 & 1 & 12 \\ 0 & 2 & 3 & 28 \\ 0 & 3 & 4 & 53 \end{bmatrix} \xrightarrow{2R_3 - 3R_2 \to R_3} \begin{bmatrix} 1 & 1 & 1 & 12 \\ 0 & 2 & 3 & 28 \\ 0 & 0 & -1 & 22 \end{bmatrix}.$$

So $c = -22$ and back-substituting we have $2b + 3(-22) = 28 \Leftrightarrow b = 47$ and $a + 47 + (-22) = 0 \Leftrightarrow a = -13$. So the cubic is $y = -13x^3 + 47x^2 - 22x$.

Fourth-degree polynomial containing the points $(0, 0)$, $(1, 12)$, $(2, 40)$, $(3, 6)$, *and* $(-1, -14)$: Using the general form of a fourth-degree polynomial, $y = ax^4 + bx^3 + cx^2 + dx + e$, we substitute for x and y and solve for a, b, c, d, and e. The point $(0, 0)$ gives $0 = a(0)^4 + b(0)^3 + c(0)^2 + d(0) + e \Rightarrow e = 0$; the point $(1, 12)$ gives $12 = a(1)^4 + b(1)^3 + c(1)^2 + d(1) + e$; the point $(2, 40)$ gives $40 = a(2)^4 + b(2)^3 + c(2)^2 + d(2) + e$; the point $(3, 6)$ gives $6 = a(3)^4 + b(3)^3 + c(3)^2 + d(3) + e$; the point $(-1, -14)$ gives $-14 = a(-1)^4 + b(-1)^3 + c(-1)^2 + d(-1) + e$.

Because the first equation is $e = 0$, we eliminate e from the other equations to get

$$\begin{cases} a + b + c + d = 12 \\ 16a + 8b + 4c + 2d = 40 \\ 81a + 27b + 9c + 3d = 6 \\ a - b + c - d = -14 \end{cases} \Leftrightarrow \left[\begin{array}{cccc|c} 1 & 1 & 1 & 1 & 12 \\ 16 & 8 & 4 & 2 & 40 \\ 81 & 27 & 9 & 3 & 6 \\ 1 & -1 & 1 & -1 & -14 \end{array}\right] \xrightarrow[\substack{R_2 - 16R_1 \to R_2 \\ R_3 - 81R_1 \to R_3 \\ R_4 - R_1 \to R_4}]{} \left[\begin{array}{cccc|c} 1 & 1 & 1 & 1 & 12 \\ 0 & -8 & -12 & -14 & -152 \\ 0 & -54 & -72 & -78 & -966 \\ 0 & -2 & 0 & -2 & -26 \end{array}\right] \xrightarrow[\substack{-\frac{1}{8}R_4 \to R_2 \\ R_2 \to R_3 \\ R_3 \to R_4}]{}$$

$$\left[\begin{array}{cccc|c} 1 & 1 & 1 & 1 & 12 \\ 0 & 1 & 0 & 1 & 13 \\ 0 & -8 & -12 & -14 & -152 \\ 0 & -54 & -72 & -78 & -966 \end{array}\right] \xrightarrow[\substack{R_3 + 8R_2 \to R_3 \\ R_4 + 54R_2 \to R_4}]{} \left[\begin{array}{cccc|c} 1 & 1 & 1 & 1 & 12 \\ 0 & 1 & 0 & 1 & 13 \\ 0 & 0 & -12 & -6 & -48 \\ 0 & 0 & -72 & -24 & -264 \end{array}\right] \xrightarrow[\substack{R_4 - 6R_3 \to R_4}]{} \left[\begin{array}{cccc|c} 1 & 1 & 1 & 1 & 12 \\ 0 & 1 & 0 & 1 & 13 \\ 0 & 0 & -12 & -6 & -48 \\ 0 & 0 & 0 & 12 & 24 \end{array}\right].$$

So $d = 2$. Then $-12c - 6(2) = -48 \Leftrightarrow c = 3$ and $b + 2 = 13 \Leftrightarrow b = 11$. Finally, $a + 11 + 3 + 2 = 12 \Leftrightarrow a = -4$. So the fourth-degree polynomial containing these points is $y = -4x^4 + 11x^3 + 3x^2 + 2x$.

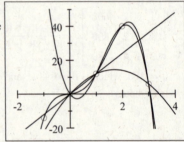

10.4 THE ALGEBRA OF MATRICES

1. We can add (or subtract) two matrices only if they have the same *dimension*.

3. (i) $A + A$ and (ii) $2A$ exist for all matrices A, but (iii) $A \cdot A$ is not defined when A is not square.

5. The matrices have different dimensions, so they cannot be equal.

7. $\begin{bmatrix} 2 & 6 \\ -5 & 3 \end{bmatrix} + \begin{bmatrix} -1 & -3 \\ 6 & 2 \end{bmatrix} = \begin{bmatrix} 1 & 3 \\ 1 & 5 \end{bmatrix}$

9. $3\begin{bmatrix} 1 & 2 \\ 4 & -1 \\ 1 & 0 \end{bmatrix} = \begin{bmatrix} 3 & 6 \\ 12 & -3 \\ 3 & 0 \end{bmatrix}$

11. $\begin{bmatrix} 2 & 6 \\ 1 & 3 \\ 2 & 4 \end{bmatrix}\begin{bmatrix} 1 & -2 \\ 3 & 6 \\ -2 & 0 \end{bmatrix}$ is undefined because these matrices have incompatible dimensions.

13. $\begin{bmatrix} 1 & 2 \\ -1 & 4 \end{bmatrix}\begin{bmatrix} 1 & -2 & 3 \\ 2 & 2 & -1 \end{bmatrix} = \begin{bmatrix} 5 & 2 & 1 \\ 7 & 10 & -7 \end{bmatrix}$

15. $2X + A = B \Leftrightarrow X = \frac{1}{2}(B - A) = \frac{1}{2}\left(\begin{bmatrix} 2 & 5 \\ 3 & 7 \end{bmatrix} - \begin{bmatrix} 4 & 6 \\ 1 & 3 \end{bmatrix}\right) = \frac{1}{2}\begin{bmatrix} -2 & -1 \\ 2 & 4 \end{bmatrix} = \begin{bmatrix} -1 & -\frac{1}{2} \\ 1 & 2 \end{bmatrix}$.

17. $2(B - X) = D$. Since B is a 2×2 matrix, $B - X$ is defined only when X is a 2×2 matrix, so $2(B - X)$ is a 2×2 matrix. But D is a 3×2 matrix. Thus, there is no solution.

19. $\frac{1}{5}(X + D) = C \Leftrightarrow X + D = 5C \Leftrightarrow$

$$X = 5C - D = 5 \begin{bmatrix} 2 & 3 \\ 1 & 0 \\ 0 & 2 \end{bmatrix} - \begin{bmatrix} 10 & 20 \\ 30 & 20 \\ 10 & 0 \end{bmatrix} = \begin{bmatrix} 10 & 15 \\ 5 & 0 \\ 0 & 10 \end{bmatrix} - \begin{bmatrix} 10 & 20 \\ 30 & 20 \\ 10 & 0 \end{bmatrix} = \begin{bmatrix} 0 & -5 \\ -25 & -20 \\ -10 & 10 \end{bmatrix}.$$

In Solutions 21–33, the matrices A, B, C, D, E, F, and G are defined as follows:

$$A = \begin{bmatrix} 2 & -5 \\ 0 & 7 \end{bmatrix} \qquad B = \begin{bmatrix} 3 & \frac{1}{2} & 5 \\ 1 & -1 & 3 \end{bmatrix} \qquad C = \begin{bmatrix} 2 & -\frac{5}{2} & 0 \\ 0 & 2 & -3 \end{bmatrix} \qquad D = \begin{bmatrix} 7 & 3 \end{bmatrix}$$

$$E = \begin{bmatrix} 1 \\ 2 \\ 0 \end{bmatrix} \qquad F = \begin{bmatrix} 1 & 0 & 0 \\ 0 & 1 & 0 \\ 0 & 0 & 1 \end{bmatrix} \qquad G = \begin{bmatrix} 5 & -3 & 10 \\ 6 & 1 & 0 \\ -5 & 2 & 2 \end{bmatrix} \qquad H = \begin{bmatrix} 3 & 1 \\ 2 & -1 \end{bmatrix}$$

21. (a) $B + C = \begin{bmatrix} 3 & \frac{1}{2} & 5 \\ 1 & -1 & 3 \end{bmatrix} + \begin{bmatrix} 2 & -\frac{5}{2} & 0 \\ 0 & 2 & -3 \end{bmatrix} = \begin{bmatrix} 5 & -2 & 5 \\ 1 & 1 & 0 \end{bmatrix}$

(b) $B + F$ is undefined because B (2×3) and F (3×3) don't have the same dimensions.

23. (a) $5A = 5 \begin{bmatrix} 2 & -5 \\ 0 & 7 \end{bmatrix} = \begin{bmatrix} 10 & -25 \\ 0 & 35 \end{bmatrix}$

(b) $C - 5A$ is undefined because C (2×3) and A (2×2) don't have the same dimensions.

25. (a) AD is undefined because A (2×2) and D (1×2) have incompatible dimensions.

(b) $DA = \begin{bmatrix} 7 & 3 \end{bmatrix} \begin{bmatrix} 2 & -5 \\ 0 & 7 \end{bmatrix} = \begin{bmatrix} 14 & -14 \end{bmatrix}$

27. (a) $AH = \begin{bmatrix} 2 & -5 \\ 0 & 7 \end{bmatrix} \begin{bmatrix} 3 & 1 \\ 2 & -1 \end{bmatrix} = \begin{bmatrix} -4 & 7 \\ 14 & -7 \end{bmatrix}$

(b) $\begin{bmatrix} 3 & 1 \\ 2 & -1 \end{bmatrix} \begin{bmatrix} 2 & -5 \\ 0 & 7 \end{bmatrix} = \begin{bmatrix} 6 & -8 \\ 4 & -17 \end{bmatrix}$

29. (a) $GF = \begin{bmatrix} 5 & -3 & 10 \\ 6 & 1 & 0 \\ -5 & 2 & 2 \end{bmatrix} \begin{bmatrix} 1 & 0 & 0 \\ 0 & 1 & 0 \\ 0 & 0 & 1 \end{bmatrix} = \begin{bmatrix} 5 & -3 & 10 \\ 6 & 1 & 0 \\ -5 & 2 & 2 \end{bmatrix}$

(b) $GE = \begin{bmatrix} 5 & -3 & 10 \\ 6 & 1 & 0 \\ -5 & 2 & 2 \end{bmatrix} \begin{bmatrix} 1 \\ 2 \\ 0 \end{bmatrix} = \begin{bmatrix} -1 \\ 8 \\ -1 \end{bmatrix}$

31. (a) $A^2 = \begin{bmatrix} 2 & -5 \\ 0 & 7 \end{bmatrix} \begin{bmatrix} 2 & -5 \\ 0 & 7 \end{bmatrix} = \begin{bmatrix} 4 & -45 \\ 0 & 49 \end{bmatrix}$

(b) $A^3 = \begin{bmatrix} 2 & -5 \\ 0 & 7 \end{bmatrix} \begin{bmatrix} 2 & -5 \\ 0 & 7 \end{bmatrix} \begin{bmatrix} 2 & -5 \\ 0 & 7 \end{bmatrix} = \begin{bmatrix} 4 & -45 \\ 0 & 49 \end{bmatrix} \begin{bmatrix} 2 & -5 \\ 0 & 7 \end{bmatrix} = \begin{bmatrix} 8 & -335 \\ 0 & 343 \end{bmatrix}$

33. (a) $ABE = \begin{bmatrix} 2 & -5 \\ 0 & 7 \end{bmatrix} \begin{bmatrix} 3 & \frac{1}{2} & 5 \\ 1 & -1 & 3 \end{bmatrix} \begin{bmatrix} 1 \\ 2 \\ 0 \end{bmatrix} = \begin{bmatrix} 1 & 6 & -5 \\ 7 & -7 & 21 \end{bmatrix} \begin{bmatrix} 1 \\ 2 \\ 0 \end{bmatrix} = \begin{bmatrix} 13 \\ -7 \end{bmatrix}$

(b) AHE is undefined because the dimensions of AH (2×2) and E (3×1) are incompatible.

35. $\begin{bmatrix} x & 2y \\ 4 & 6 \end{bmatrix} = \begin{bmatrix} 2 & -2 \\ 2x & -6y \end{bmatrix}$. Thus we must solve the system $\begin{cases} x = 2 \\ 2y = -2 \\ 4 = 2x \\ 6 = -6y \end{cases}$ So $x = 2$ and $2y = -2 \Leftrightarrow y = -1$. Since these values for x and y also satisfy the last two equations, the solution is $x = 2$, $y = -1$.

37. $2 \begin{bmatrix} x & y \\ x+y & x-y \end{bmatrix} = \begin{bmatrix} 2 & -4 \\ -2 & 6 \end{bmatrix}$. Since $2 \begin{bmatrix} x & y \\ x+y & x-y \end{bmatrix} = \begin{bmatrix} 2x & 2y \\ 2(x+y) & 2(x-y) \end{bmatrix}$, Thus we must solve the

system $\begin{cases} 2x = 2 \\ 2y = -4 \\ 2(x+y) = -2 \\ 2(x-y) = 6 \end{cases}$ So $x = 1$ and $y = -2$. Since these values for x and y also satisfy the last two equations, the solution is $x = 1$, $y = -2$.

39. $\begin{cases} 2x - 5y = 7 \\ 3x + 2y = 4 \end{cases}$ written as a matrix equation is $\begin{bmatrix} 2 & -5 \\ 3 & 2 \end{bmatrix} \begin{bmatrix} x \\ y \end{bmatrix} = \begin{bmatrix} 7 \\ 4 \end{bmatrix}$.

41. $\begin{cases} 3x_1 + 2x_2 - x_3 + x_4 = 0 \\ x_1 \quad - x_3 \quad = 5 \\ 3x_2 + x_3 - x_4 = 4 \end{cases}$ written as a matrix equation is $\begin{bmatrix} 3 & 2 & -1 & 1 \\ 1 & 0 & -1 & 0 \\ 0 & 3 & 1 & -1 \end{bmatrix} \begin{bmatrix} x_1 \\ x_2 \\ x_3 \\ x_4 \end{bmatrix} = \begin{bmatrix} 0 \\ 5 \\ 4 \end{bmatrix}$.

43. $A = \begin{bmatrix} 1 & 0 & 6 & -1 \\ 2 & \frac{1}{2} & 4 & 0 \end{bmatrix}$, $B = \begin{bmatrix} 1 & 7 & -9 & 2 \end{bmatrix}$, and $C = \begin{bmatrix} 1 \\ 0 \\ -1 \\ -2 \end{bmatrix}$. ABC is undefined because the dimensions of A (2×4)

and B (1×4) are not compatible. $ACB = \begin{bmatrix} -3 \\ -2 \end{bmatrix} \begin{bmatrix} 1 & 7 & -9 & 2 \end{bmatrix} = \begin{bmatrix} -3 & -21 & 27 & -6 \\ -2 & -14 & 18 & -4 \end{bmatrix}$. BAC is undefined because

the dimensions of B (1×4) and A (2×4) are not compatible. BCA is undefined because the dimensions of C (4×1) and A (2×4) are not compatible. CAB is undefined because the dimensions of C (4×1) and A (2×4) are not compatible. CBA is undefined because the dimensions of B (1×4) and A (2×4) are not compatible.

45. (a) $BA = \begin{bmatrix} \$0.90 & \$0.80 & \$1.10 \end{bmatrix} \begin{bmatrix} 4000 & 1000 & 3500 \\ 400 & 300 & 200 \\ 700 & 500 & 9000 \end{bmatrix} = \begin{bmatrix} \$4690 & \$1690 & \$13,210 \end{bmatrix}$

(b) The entries in the product matrix represent the total food sales in Santa Monica, Long Beach, and Anaheim, respectively.

47. (a) $AB = \begin{bmatrix} 6 & 10 & 14 & 28 \end{bmatrix} \begin{bmatrix} 2000 & 2500 \\ 3000 & 1500 \\ 2500 & 1000 \\ 1000 & 500 \end{bmatrix} = \begin{bmatrix} 105{,}000 & 58{,}000 \end{bmatrix}$

(b) That day they canned 105,000 ounces of tomato sauce and 58,000 ounces of tomato paste.

49. (a) $\begin{bmatrix} 1 & 0 & 1 & 0 & 1 & 1 \\ 0 & 3 & 0 & 1 & 2 & 1 \\ 1 & 2 & 0 & 0 & 3 & 0 \\ 1 & 3 & 2 & 3 & 2 & 0 \\ 0 & 3 & 0 & 0 & 2 & 1 \\ 1 & 2 & 0 & 1 & 3 & 1 \end{bmatrix}$
(b) $\begin{bmatrix} 2 & 1 & 2 & 1 & 2 & 2 \\ 1 & 3 & 1 & 2 & 3 & 2 \\ 2 & 3 & 1 & 1 & 3 & 1 \\ 2 & 3 & 3 & 3 & 3 & 1 \\ 1 & 3 & 1 & 1 & 3 & 2 \\ 2 & 3 & 1 & 2 & 3 & 2 \end{bmatrix}$
(c) $\begin{bmatrix} 2 & 3 & 2 & 3 & 2 & 2 \\ 3 & 0 & 3 & 2 & 1 & 2 \\ 2 & 1 & 3 & 3 & 0 & 3 \\ 2 & 0 & 1 & 0 & 1 & 3 \\ 3 & 0 & 3 & 3 & 1 & 2 \\ 2 & 1 & 3 & 2 & 0 & 2 \end{bmatrix}$

(d) $\begin{bmatrix} 3 & 3 & 3 & 3 & 3 & 3 \\ 3 & 0 & 3 & 3 & 0 & 3 \\ 3 & 0 & 3 & 3 & 0 & 3 \\ 3 & 0 & 0 & 0 & 0 & 3 \\ 3 & 0 & 3 & 3 & 0 & 3 \\ 3 & 0 & 3 & 3 & 0 & 3 \end{bmatrix}$

(e)

51. $A = \begin{bmatrix} 1 & 1 \\ 0 & 1 \end{bmatrix}$; $A^2 = \begin{bmatrix} 1 & 1 \\ 0 & 1 \end{bmatrix}\begin{bmatrix} 1 & 1 \\ 0 & 1 \end{bmatrix} = \begin{bmatrix} 1 & 2 \\ 0 & 1 \end{bmatrix}$; $A^3 = A \cdot A^2 = \begin{bmatrix} 1 & 1 \\ 0 & 1 \end{bmatrix}\begin{bmatrix} 1 & 2 \\ 0 & 1 \end{bmatrix} = \begin{bmatrix} 1 & 3 \\ 0 & 1 \end{bmatrix}$;

$A^4 = A \cdot A^3 = \begin{bmatrix} 1 & 1 \\ 0 & 1 \end{bmatrix}\begin{bmatrix} 1 & 3 \\ 0 & 1 \end{bmatrix} = \begin{bmatrix} 1 & 4 \\ 0 & 1 \end{bmatrix}$. Therefore, it seems that $A^n = \begin{bmatrix} 1 & n \\ 0 & 1 \end{bmatrix}$.

53. Let $A = \begin{bmatrix} a & b \\ c & d \end{bmatrix}$.   For the first matrix, we have $A^2 = \begin{bmatrix} a & b \\ c & d \end{bmatrix}$.

$\begin{bmatrix} a & b \\ c & d \end{bmatrix} = \begin{bmatrix} a^2 + bc & ab + bd \\ ac + cd & bc + d^2 \end{bmatrix} = \begin{bmatrix} a^2 + bc & b(a+d) \\ c(a+d) & bc + d^2 \end{bmatrix}$. So $A^2 = \begin{bmatrix} 4 & 0 \\ 0 & 9 \end{bmatrix} \Leftrightarrow \begin{cases} a^2 + bc = 4 \\ b(a+d) = 0 \\ c(a+d) = 0 \\ bc + d^2 = 9 \end{cases}$

If $a + d = 0$, then $a = -d$, so $4 = a^2 + bc = (-d)^2 + bc = d^2 + bc = 9$, which is a contradiction. Thus $a + d \neq 0$. Since $b(a+d) = 0$ and $c(a+d) = 0$, we must have $b = 0$ and $c = 0$. So the first equation becomes $a^2 = 4 \Rightarrow a = \pm 2$, and the fourth equation becomes $d^2 = 9 \Rightarrow d = \pm 3$.

Thus the square roots of $\begin{bmatrix} 4 & 0 \\ 0 & 9 \end{bmatrix}$ are $A_1 = \begin{bmatrix} 2 & 0 \\ 0 & 3 \end{bmatrix}$, $A_2 = \begin{bmatrix} 2 & 0 \\ 0 & -3 \end{bmatrix}$, $A_3 = \begin{bmatrix} -2 & 0 \\ 0 & 3 \end{bmatrix}$, and $A_4 = \begin{bmatrix} -2 & 0 \\ 0 & -3 \end{bmatrix}$.

For the second matrix, we have $A^2 = \begin{bmatrix} 1 & 5 \\ 0 & 9 \end{bmatrix} \Leftrightarrow \begin{cases} a^2 + bc = 1 \\ b(a+d) = 5 \\ c(a+d) = 0 \\ bc + d^2 = 9 \end{cases}$ Since $a + d \neq 0$ and $c(a+d) = 0$, we must have

$c = 0$. The equations then simplify into the system $\begin{cases} a^2 = 1 \\ b(a+d) = 5 \\ d^2 = 9 \end{cases} \Rightarrow \begin{cases} a = \pm 1 \\ b(a+d) = 5 \\ d = \pm 3 \end{cases}$

We consider the four possible values of a and d. If $a = 1$ and $d = 3$, then $b(a+d) = 5 \Rightarrow b(4) = 5 \Rightarrow b = \frac{5}{4}$. If $a = 1$ and $d = -3$, then $b(a+d) = 5 \Rightarrow b(-2) = 5 \Rightarrow b = -\frac{5}{2}$. If $a = -1$ and $d = 3$, then $b(a+d) = 5 \Rightarrow b(2) = 5 \Rightarrow b = \frac{5}{2}$. If $a = -1$ and $d = -3$, then $b(a+d) = 5 \Rightarrow b(-4) = 5 \Rightarrow b = -\frac{5}{4}$. Thus, the square roots of $\begin{bmatrix} 1 & 5 \\ 0 & 9 \end{bmatrix}$ are

$A_1 = \begin{bmatrix} 1 & \frac{5}{4} \\ 0 & 3 \end{bmatrix}$, $A_2 = \begin{bmatrix} 1 & -\frac{5}{2} \\ 0 & -3 \end{bmatrix}$, $A_3 = \begin{bmatrix} -1 & \frac{5}{2} \\ 0 & 3 \end{bmatrix}$, and $A_4 = \begin{bmatrix} -1 & -\frac{5}{4} \\ 0 & -3 \end{bmatrix}$.

10.5 INVERSES OF MATRICES AND MATRIX EQUATIONS

1. (a) The matrix $I = \begin{bmatrix} 1 & 0 \\ 0 & 1 \end{bmatrix}$ is called an *identity* matrix.

(b) If A is a 2×2 matrix then $A \times I = A$ and $I \times A = A$.

(c) If A and B are 2×2 matrices with $AB = I$ then B is the *inverse* of A.

3. $A = \begin{bmatrix} 4 & 1 \\ 7 & 2 \end{bmatrix}$; $B = \begin{bmatrix} 2 & -1 \\ -7 & 4 \end{bmatrix}$.

$AB = \begin{bmatrix} 4 & 1 \\ 7 & 2 \end{bmatrix}\begin{bmatrix} 2 & -1 \\ -7 & 4 \end{bmatrix} = \begin{bmatrix} 1 & 0 \\ 0 & 1 \end{bmatrix}$ and $BA = \begin{bmatrix} 2 & -1 \\ -7 & 4 \end{bmatrix}\begin{bmatrix} 4 & 1 \\ 7 & 2 \end{bmatrix} = \begin{bmatrix} 1 & 0 \\ 0 & 1 \end{bmatrix}$.

5. $A = \begin{bmatrix} 1 & 3 & -1 \\ 1 & 4 & 0 \\ -1 & -3 & 2 \end{bmatrix}$; $B = \begin{bmatrix} 8 & -3 & 4 \\ -2 & 1 & -1 \\ 1 & 0 & 1 \end{bmatrix}$. $AB = \begin{bmatrix} 1 & 3 & -1 \\ 1 & 4 & 0 \\ -1 & -3 & 2 \end{bmatrix}\begin{bmatrix} 8 & -3 & 4 \\ -2 & 1 & -1 \\ 1 & 0 & 1 \end{bmatrix} = \begin{bmatrix} 1 & 0 & 0 \\ 0 & 1 & 0 \\ 0 & 0 & 1 \end{bmatrix}$ and

$BA = \begin{bmatrix} 8 & -3 & 4 \\ -2 & 1 & -1 \\ 1 & 0 & 1 \end{bmatrix}\begin{bmatrix} 1 & 3 & -1 \\ 1 & 4 & 0 \\ -1 & -3 & 2 \end{bmatrix} = \begin{bmatrix} 1 & 0 & 0 \\ 0 & 1 & 0 \\ 0 & 0 & 1 \end{bmatrix}$.

7. $A = \begin{bmatrix} 7 & 4 \\ 3 & 2 \end{bmatrix} \Leftrightarrow A^{-1} = \frac{1}{14-12}\begin{bmatrix} 2 & -4 \\ -3 & 7 \end{bmatrix} = \begin{bmatrix} 1 & -2 \\ -\frac{3}{2} & \frac{7}{2} \end{bmatrix}$. Then, $AA^{-1} = \begin{bmatrix} 7 & 4 \\ 3 & 2 \end{bmatrix}\begin{bmatrix} 1 & -2 \\ -\frac{3}{2} & \frac{7}{2} \end{bmatrix} = \begin{bmatrix} 1 & 0 \\ 0 & 1 \end{bmatrix}$

and $A^{-1}A = \begin{bmatrix} 1 & -2 \\ -\frac{3}{2} & \frac{7}{2} \end{bmatrix}\begin{bmatrix} 7 & 4 \\ 3 & 2 \end{bmatrix} = \begin{bmatrix} 1 & 0 \\ 0 & 1 \end{bmatrix}$.

9. $\begin{bmatrix} -3 & -5 \\ 2 & 3 \end{bmatrix}^{-1} = \frac{1}{-9+10}\begin{bmatrix} 3 & 5 \\ -2 & -3 \end{bmatrix} = \begin{bmatrix} 3 & 5 \\ -2 & -3 \end{bmatrix}$

11. $\begin{bmatrix} 2 & 5 \\ -5 & -13 \end{bmatrix}^{-1} = \dfrac{1}{-26+25} \begin{bmatrix} -13 & -5 \\ 5 & 2 \end{bmatrix} = \begin{bmatrix} 13 & 5 \\ -5 & -2 \end{bmatrix}$

13. $\begin{bmatrix} 6 & -3 \\ -8 & 4 \end{bmatrix}^{-1} = \dfrac{1}{24-24} \begin{bmatrix} 4 & 3 \\ 8 & 6 \end{bmatrix}$, which is not defined, and so there is no inverse.

15. $\begin{bmatrix} 0.4 & -1.2 \\ 0.3 & 0.6 \end{bmatrix}^{-1} = \dfrac{1}{0.24+0.36} \begin{bmatrix} 0.6 & 1.2 \\ -0.3 & 0.4 \end{bmatrix} = \begin{bmatrix} 1 & 2 \\ -\frac{1}{2} & \frac{2}{3} \end{bmatrix}$

17. $\begin{bmatrix} 2 & 4 & 1 & 1 & 0 & 0 \\ -1 & 1 & -1 & 0 & 1 & 0 \\ 1 & 4 & 0 & 0 & 0 & 1 \end{bmatrix} \xrightarrow[2R_3-R_1 \to R_3]{2R_2+R_1 \to R_2} \begin{bmatrix} 2 & 4 & 1 & 1 & 0 & 0 \\ 0 & 6 & -1 & 1 & 2 & 0 \\ 0 & 4 & -1 & -1 & 0 & 2 \end{bmatrix} \xrightarrow[3R_1-2R_2 \to R_1]{3R_3-2R_2 \to R_3} \begin{bmatrix} 6 & 0 & 5 & 1 & -4 & 0 \\ 0 & 6 & -1 & 1 & 2 & 0 \\ 0 & 0 & -1 & -5 & -4 & 6 \end{bmatrix}$

$\xrightarrow[R_2-R_3 \to R_2]{R_1+5R_3 \to R_1} \begin{bmatrix} 6 & 0 & 0 & -24 & -24 & 30 \\ 0 & 6 & 0 & 6 & 6 & -6 \\ 0 & 0 & -1 & -5 & -4 & 6 \end{bmatrix} \xrightarrow[\frac{1}{6}R_2,\, -R_3]{\frac{1}{6}R_1} \begin{bmatrix} 1 & 0 & 0 & -4 & -4 & 5 \\ 0 & 1 & 0 & 1 & 1 & -1 \\ 0 & 0 & 1 & 5 & 4 & -6 \end{bmatrix}$.

Therefore, the inverse matrix is $\begin{bmatrix} -4 & -4 & 5 \\ 1 & 1 & -1 \\ 5 & 4 & -6 \end{bmatrix}$.

19. $\begin{bmatrix} 1 & 2 & 3 & 1 & 0 & 0 \\ 4 & 5 & -1 & 0 & 1 & 0 \\ 1 & -1 & -10 & 0 & 0 & 1 \end{bmatrix} \xrightarrow[R_3-R_1 \to R_3]{R_2-4R_1 \to R_2} \begin{bmatrix} 1 & 2 & 3 & 1 & 0 & 0 \\ 0 & -3 & -13 & -4 & 1 & 0 \\ 0 & -3 & -13 & -1 & 0 & 1 \end{bmatrix} \xrightarrow{R_3-R_2 \to R_3} \begin{bmatrix} 1 & 2 & 3 & 1 & 0 & 0 \\ 0 & -3 & -13 & -4 & 1 & 0 \\ 0 & 0 & 0 & 3 & -1 & 1 \end{bmatrix}$.

Since the left half of the last row consists entirely of zeros, there is no inverse matrix.

21. $\begin{bmatrix} 0 & -2 & 2 & 1 & 0 & 0 \\ 3 & 1 & 3 & 0 & 1 & 0 \\ 1 & -2 & 3 & 0 & 0 & 1 \end{bmatrix} \xrightarrow{R_1 \leftrightarrow R_3} \begin{bmatrix} 1 & -2 & 3 & 0 & 0 & 1 \\ 3 & 1 & 3 & 0 & 1 & 0 \\ 0 & -2 & 2 & 1 & 0 & 0 \end{bmatrix} \xrightarrow{R_2-3R_1 \to R_2} \begin{bmatrix} 1 & -2 & 3 & 0 & 0 & 1 \\ 0 & 7 & -6 & 0 & 1 & -3 \\ 0 & -2 & 2 & 1 & 0 & 0 \end{bmatrix}$

$\xrightarrow[R_2+3R_3 \to R_2]{R_1-R_3 \to R_1} \begin{bmatrix} 1 & 0 & 1 & -1 & 0 & 1 \\ 0 & 1 & 0 & 3 & 1 & -3 \\ 0 & -2 & 2 & 1 & 0 & 0 \end{bmatrix} \xrightarrow{R_3+2R_2 \to R_3} \begin{bmatrix} 1 & 0 & 1 & -1 & 0 & 1 \\ 0 & 1 & 0 & 3 & 1 & -3 \\ 0 & 0 & 2 & 7 & 2 & -6 \end{bmatrix} \xrightarrow{\frac{1}{2}R_3}$

$\begin{bmatrix} 1 & 0 & 1 & -1 & 0 & 1 \\ 0 & 1 & 0 & 3 & 1 & -3 \\ 0 & 0 & 1 & \frac{7}{2} & 1 & -3 \end{bmatrix} \xrightarrow{R_1-R_3 \to R_1} \begin{bmatrix} 1 & 0 & 0 & -\frac{9}{2} & -1 & 4 \\ 0 & 1 & 0 & 3 & 1 & -3 \\ 0 & 0 & 1 & \frac{7}{2} & 1 & -3 \end{bmatrix}$. Therefore, the inverse matrix is

$\begin{bmatrix} -\frac{9}{2} & -1 & 4 \\ 3 & 1 & -3 \\ \frac{7}{2} & 1 & -3 \end{bmatrix}$.

23.
$$\begin{bmatrix} 1 & 2 & 0 & 3 & 1 & 0 & 0 & 0 \\ 0 & 1 & 1 & 1 & 0 & 1 & 0 & 0 \\ 0 & 1 & 0 & 1 & 0 & 0 & 1 & 0 \\ 1 & 2 & 0 & 2 & 0 & 0 & 0 & 1 \end{bmatrix} \xrightarrow[R_4 - R_1 \to R_4]{R_3 - R_2 \to R_3} \begin{bmatrix} 1 & 2 & 0 & 3 & 1 & 0 & 0 & 0 \\ 0 & 1 & 1 & 1 & 0 & 1 & 0 & 0 \\ 0 & 0 & -1 & 0 & 0 & -1 & 1 & 0 \\ 0 & 0 & 0 & -1 & -1 & 0 & 0 & 1 \end{bmatrix} \xrightarrow[-R_4]{-R_3}$$

$$\begin{bmatrix} 1 & 2 & 0 & 3 & 1 & 0 & 0 & 0 \\ 0 & 1 & 1 & 1 & 0 & 1 & 0 & 0 \\ 0 & 0 & 1 & 0 & 0 & 1 & -1 & 0 \\ 0 & 0 & 0 & 1 & 1 & 0 & 0 & -1 \end{bmatrix} \xrightarrow[R_2 - R_3 \to R_2]{R_1 - 2R_2 \to R_1} \begin{bmatrix} 1 & 0 & -2 & 1 & 1 & -2 & 0 & 0 \\ 0 & 1 & 0 & 1 & 0 & 0 & 1 & 0 \\ 0 & 0 & 1 & 0 & 0 & 1 & -1 & 0 \\ 0 & 0 & 0 & 1 & 1 & 0 & 0 & -1 \end{bmatrix} \xrightarrow[R_2 - R_4 \to R_2]{R_1 + 2R_3 \to R_1}$$

$$\begin{bmatrix} 1 & 0 & 0 & 1 & 1 & 0 & -2 & 0 \\ 0 & 1 & 0 & 0 & -1 & 0 & 1 & 1 \\ 0 & 0 & 1 & 0 & 0 & 1 & -1 & 0 \\ 0 & 0 & 0 & 1 & 1 & 0 & 0 & -1 \end{bmatrix} \xrightarrow{R_1 \to R_1 - R_4} \begin{bmatrix} 1 & 0 & 0 & 0 & 0 & 0 & -2 & 1 \\ 0 & 1 & 0 & 0 & -1 & 0 & 1 & 1 \\ 0 & 0 & 1 & 0 & 0 & 1 & -1 & 0 \\ 0 & 0 & 0 & 1 & 1 & 0 & 0 & -1 \end{bmatrix}.$$ Therefore, the inverse matrix is

$$\begin{bmatrix} 0 & 0 & -2 & 1 \\ -1 & 0 & 1 & 1 \\ 0 & 1 & -1 & 0 \\ 1 & 0 & 0 & -1 \end{bmatrix}.$$

25. $\begin{cases} -3x - 5y = 4 \\ 2x + 3y = 0 \end{cases}$ is equivalent to the matrix equation $\begin{bmatrix} -3 & -5 \\ 2 & 3 \end{bmatrix} \begin{bmatrix} x \\ y \end{bmatrix} = \begin{bmatrix} 4 \\ 0 \end{bmatrix}$. Using the inverse from Exercise 9,

$\begin{bmatrix} x \\ y \end{bmatrix} = \begin{bmatrix} 3 & 5 \\ -2 & -3 \end{bmatrix} \begin{bmatrix} 4 \\ 0 \end{bmatrix} = \begin{bmatrix} 12 \\ -8 \end{bmatrix}$. Therefore, $x = 12$ and $y = -8$.

27. $\begin{cases} 2x + 5y = 2 \\ -5x - 13y = 20 \end{cases}$ is equivalent to the matrix equation $\begin{bmatrix} 2 & 5 \\ -5 & -13 \end{bmatrix} \begin{bmatrix} x \\ y \end{bmatrix} = \begin{bmatrix} 2 \\ 20 \end{bmatrix}$. Using the inverse from

Exercise 11, $\begin{bmatrix} x \\ y \end{bmatrix} = \begin{bmatrix} 13 & 5 \\ -5 & -2 \end{bmatrix} \begin{bmatrix} 2 \\ 20 \end{bmatrix} = \begin{bmatrix} 126 \\ -50 \end{bmatrix}$. Therefore, $x = 126$ and $y = -50$.

29. $\begin{cases} 2x + 4y + z = 7 \\ -x + y - z = 0 \\ x + 4y = -2 \end{cases}$ is equivalent to the matrix equation $\begin{bmatrix} 2 & 4 & 1 \\ -1 & 1 & -1 \\ 1 & 4 & 0 \end{bmatrix} \begin{bmatrix} x \\ y \\ z \end{bmatrix} = \begin{bmatrix} 7 \\ 0 \\ -2 \end{bmatrix}$. Using the inverse from

Exercise 17, $\begin{bmatrix} x \\ y \\ z \end{bmatrix} = \begin{bmatrix} -4 & -4 & 5 \\ 1 & 1 & -1 \\ 5 & 4 & -6 \end{bmatrix} \begin{bmatrix} 7 \\ 0 \\ -2 \end{bmatrix} = \begin{bmatrix} -38 \\ 9 \\ 47 \end{bmatrix}$. Therefore, $x = -38$, $y = 9$, and $z = 47$.

31. $\begin{cases} -2y + 2z = 12 \\ 3x + y + 3z = -2 \\ x - 2y + 3z = 8 \end{cases}$ is equivalent to the matrix equation $\begin{bmatrix} 0 & -2 & 2 \\ 3 & 1 & 3 \\ 1 & -2 & 3 \end{bmatrix} \begin{bmatrix} x \\ y \\ z \end{bmatrix} = \begin{bmatrix} 12 \\ -2 \\ 8 \end{bmatrix}$. Using the inverse from

Exercise 21, $\begin{bmatrix} x \\ y \\ z \end{bmatrix} = \begin{bmatrix} -\frac{9}{2} & -1 & 4 \\ 3 & 1 & -3 \\ \frac{7}{2} & 1 & -3 \end{bmatrix} \begin{bmatrix} 12 \\ -2 \\ 8 \end{bmatrix} = \begin{bmatrix} -20 \\ 10 \\ 16 \end{bmatrix}$. Therefore, $x = -20$, $y = 10$, and $z = 16$.

33. Using a calculator, we get the result $(3, 2, 1)$.

35. Using a calculator, we get the result $(3, -2, 2)$.

37. Using a calculator, we get the result $(8, 1, 0, 3)$.

39. This has the form $MX = C$, so $M^{-1}(MX) = M^{-1}C$ and $M^{-1}(MX) = \left(M^{-1}M\right)X = X$.

Now $M^{-1} = \begin{bmatrix} 3 & -2 \\ -4 & 3 \end{bmatrix}^{-1} = \dfrac{1}{9-8}\begin{bmatrix} 3 & 2 \\ 4 & 3 \end{bmatrix} = \begin{bmatrix} 3 & 2 \\ 4 & 3 \end{bmatrix}$. Since $X = M^{-1}C$, we get

$\begin{bmatrix} x & y & z \\ u & v & w \end{bmatrix} = \begin{bmatrix} 3 & 2 \\ 4 & 3 \end{bmatrix}\begin{bmatrix} 1 & 0 & -1 \\ 2 & 1 & 3 \end{bmatrix} = \begin{bmatrix} 7 & 2 & 3 \\ 10 & 3 & 5 \end{bmatrix}$.

41. $\begin{bmatrix} a & -a \\ a & a \end{bmatrix}^{-1} = \dfrac{1}{a^2 - (-a^2)}\begin{bmatrix} a & a \\ -a & a \end{bmatrix} = \dfrac{1}{2a^2}\begin{bmatrix} a & a \\ -a & a \end{bmatrix} = \dfrac{1}{2a}\begin{bmatrix} 1 & 1 \\ -1 & 1 \end{bmatrix}$

43. $\begin{bmatrix} 2 & x \\ x & x^2 \end{bmatrix}^{-1} = \dfrac{1}{2x^2 - x^2}\begin{bmatrix} x^2 & -x \\ -x & 2 \end{bmatrix} = \dfrac{1}{x^2}\begin{bmatrix} x^2 & -x \\ -x & 2 \end{bmatrix} = \begin{bmatrix} 1 & -1/x \\ -1/x & 2/x^2 \end{bmatrix}$. The inverse does not exist when $x = 0$.

45. $\begin{bmatrix} 1 & e^x & 0 & 1 & 0 & 0 \\ e^x & -e^{2x} & 0 & 0 & 1 & 0 \\ 0 & 0 & 2 & 0 & 0 & 1 \end{bmatrix} \xrightarrow{R_2 - e^x R_1 \to R_2} \begin{bmatrix} 1 & e^x & 0 & 1 & 0 & 0 \\ 0 & -2e^{2x} & 0 & -e^x & 1 & 0 \\ 0 & 0 & 2 & 0 & 0 & 1 \end{bmatrix} \xrightarrow[\frac{1}{2}R_3]{-\frac{1}{2}e^{-2x}R_2}$

$\begin{bmatrix} 1 & e^x & 0 & 1 & 0 & 0 \\ 0 & 1 & 0 & \frac{1}{2}e^{-x} & -\frac{1}{2}e^{-2x} & 0 \\ 0 & 0 & 1 & 0 & 0 & \frac{1}{2} \end{bmatrix} \xrightarrow{R_1 - e^x R_2 \to R_1} \begin{bmatrix} 1 & 0 & 0 & \frac{1}{2} & \frac{1}{2}e^{-x} & 0 \\ 0 & 1 & 0 & \frac{1}{2}e^{-x} & -\frac{1}{2}e^{-2x} & 0 \\ 0 & 0 & 1 & 0 & 0 & \frac{1}{2} \end{bmatrix}$.

Therefore, the inverse matrix is $\begin{bmatrix} \frac{1}{2} & \frac{1}{2}e^{-x} & 0 \\ \frac{1}{2}e^{-x} & -\frac{1}{2}e^{-2x} & 0 \\ 0 & 0 & \frac{1}{2} \end{bmatrix}$. The inverse exists for all x.

47. (a) $\begin{bmatrix} 3 & 1 & 3 & 1 & 0 & 0 \\ 4 & 2 & 4 & 0 & 1 & 0 \\ 3 & 2 & 4 & 0 & 0 & 1 \end{bmatrix} \xrightarrow[R_1 \leftrightarrow R_2]{R_3 - R_1 \to R_3} \begin{bmatrix} 4 & 2 & 4 & 0 & 1 & 0 \\ 3 & 1 & 3 & 1 & 0 & 0 \\ 0 & 1 & 1 & -1 & 0 & 1 \end{bmatrix} \xrightarrow{R_1 - R_2 \to R_1} \begin{bmatrix} 1 & 1 & 1 & -1 & 1 & 0 \\ 3 & 1 & 3 & 1 & 0 & 0 \\ 0 & 1 & 1 & -1 & 0 & 1 \end{bmatrix}$

$\xrightarrow{R_2 - 3R_1 \to R_2} \begin{bmatrix} 1 & 1 & 1 & -1 & 1 & 0 \\ 0 & -2 & 0 & 4 & -3 & 0 \\ 0 & 1 & 1 & -1 & 0 & 1 \end{bmatrix} \xrightarrow[\substack{-\frac{1}{2}R_2 \\ R_3 + \frac{1}{2}R_2 \to R_3}]{R_1 + \frac{1}{2}R_2 \to R_1} \begin{bmatrix} 1 & 0 & 1 & 1 & -\frac{1}{2} & 0 \\ 0 & 1 & 0 & -2 & \frac{3}{2} & 0 \\ 0 & 0 & 1 & 1 & -\frac{3}{2} & 1 \end{bmatrix} \xrightarrow{R_1 - R_3 \to R_1}$

$\begin{bmatrix} 1 & 0 & 0 & 0 & 1 & -1 \\ 0 & 1 & 0 & -2 & \frac{3}{2} & 0 \\ 0 & 0 & 1 & 1 & -\frac{3}{2} & 1 \end{bmatrix}$. Therefore, the inverse of the matrix is $\begin{bmatrix} 0 & 1 & -1 \\ -2 & \frac{3}{2} & 0 \\ 1 & -\frac{3}{2} & 1 \end{bmatrix}$.

(b) $\begin{bmatrix} A \\ B \\ C \end{bmatrix} = \begin{bmatrix} 0 & 1 & -1 \\ -2 & \frac{3}{2} & 0 \\ 1 & -\frac{3}{2} & 1 \end{bmatrix}\begin{bmatrix} 10 \\ 14 \\ 13 \end{bmatrix} = \begin{bmatrix} 1 \\ 1 \\ 2 \end{bmatrix}$.

Therefore, he should feed the rats 1 oz of food A, 1 oz of food B, and 2 oz of food C.

(c) $\begin{bmatrix} A \\ B \\ C \end{bmatrix} = \begin{bmatrix} 0 & 1 & -1 \\ -2 & \frac{3}{2} & 0 \\ 1 & -\frac{3}{2} & 1 \end{bmatrix} \begin{bmatrix} 9 \\ 12 \\ 10 \end{bmatrix} = \begin{bmatrix} 2 \\ 0 \\ 1 \end{bmatrix}$.

Therefore, he should feed the rats 2 oz of food A, no food B, and 1 oz of food C.

(d) $\begin{bmatrix} A \\ B \\ C \end{bmatrix} = \begin{bmatrix} 0 & 1 & -1 \\ -2 & \frac{3}{2} & 0 \\ 1 & -\frac{3}{2} & 1 \end{bmatrix} \begin{bmatrix} 2 \\ 4 \\ 11 \end{bmatrix} = \begin{bmatrix} -7 \\ 2 \\ 7 \end{bmatrix}$.

Since $A < 0$, there is no combination of foods giving the required supply.

49. (a) $\begin{cases} x + y + 2z = 675 \\ 2x + y + z = 600 \\ x + 2y + z = 625 \end{cases}$ **(b)** $\begin{bmatrix} 1 & 1 & 2 \\ 2 & 1 & 1 \\ 1 & 2 & 1 \end{bmatrix} \begin{bmatrix} x \\ y \\ z \end{bmatrix} = \begin{bmatrix} 675 \\ 600 \\ 625 \end{bmatrix}$

(c) $\begin{bmatrix} 1 & 1 & 2 & 1 & 0 & 0 \\ 2 & 1 & 1 & 0 & 1 & 0 \\ 1 & 2 & 1 & 0 & 0 & 1 \end{bmatrix}$ $\xrightarrow[R_3 - R_1 \to R_3]{R_2 - 2R_1 \to R_2}$ $\begin{bmatrix} 1 & 1 & 2 & 1 & 0 & 0 \\ 0 & -1 & -3 & -2 & 1 & 0 \\ 0 & 1 & -1 & -1 & 0 & 1 \end{bmatrix}$ $\xrightarrow[R_3 + R_2 \to R_3]{R_1 + R_2 \to R_1}$ $\begin{bmatrix} 1 & 0 & -1 & -1 & 1 & 0 \\ 0 & -1 & -3 & -2 & 1 & 0 \\ 0 & 0 & -4 & -3 & 1 & 1 \end{bmatrix}$

$\xrightarrow[-\frac{1}{4}R_3]{-R_2}$ $\begin{bmatrix} 1 & 0 & -1 & -1 & 1 & 0 \\ 0 & 1 & 3 & 2 & -1 & 0 \\ 0 & 0 & 1 & \frac{3}{4} & -\frac{1}{4} & -\frac{1}{4} \end{bmatrix}$ $\xrightarrow[R_2 - 3R_3 \to R_2]{R_1 + R_3 \to R_1}$ $\begin{bmatrix} 1 & 0 & 0 & -\frac{1}{4} & \frac{3}{4} & -\frac{1}{4} \\ 0 & 1 & 0 & -\frac{1}{4} & -\frac{1}{4} & \frac{3}{4} \\ 0 & 0 & 1 & \frac{3}{4} & -\frac{1}{4} & -\frac{1}{4} \end{bmatrix}$. Therefore, the inverse of

the matrix is $\begin{bmatrix} -\frac{1}{4} & \frac{3}{4} & -\frac{1}{4} \\ -\frac{1}{4} & -\frac{1}{4} & \frac{3}{4} \\ \frac{3}{4} & -\frac{1}{4} & -\frac{1}{4} \end{bmatrix}$ and $\begin{bmatrix} x \\ y \\ z \end{bmatrix} = \begin{bmatrix} -\frac{1}{4} & \frac{3}{4} & -\frac{1}{4} \\ -\frac{1}{4} & -\frac{1}{4} & \frac{3}{4} \\ \frac{3}{4} & -\frac{1}{4} & -\frac{1}{4} \end{bmatrix} \begin{bmatrix} 675 \\ 600 \\ 625 \end{bmatrix} = \begin{bmatrix} 125 \\ 150 \\ 200 \end{bmatrix}$. Thus, he earns $125 on a

standard set, $150 on a deluxe set, and $200 on a leather-bound set.

10.6   DETERMINANTS AND CRAMER'S RULE

1. True. det (A) is defined only for a square matrix A.

3. True. If det $(A) = 0$ then A is not invertible.

5. The matrix $\begin{bmatrix} 2 & 0 \\ 0 & 3 \end{bmatrix}$ has determinant

$|D| = (2)(3) - (0)(0) = 6$.

7. The matrix $\begin{bmatrix} 4 & 5 \\ 0 & -1 \end{bmatrix}$ has determinant

$|D| = (4)(-1) - (5)(0) = -4$.

9. The matrix $\begin{bmatrix} 2 & 5 \end{bmatrix}$ does not have a determinant because the matrix is not square.

11. The matrix $\begin{bmatrix} \frac{1}{2} & \frac{1}{8} \\ 1 & \frac{1}{2} \end{bmatrix}$ has determinant

$|D| = \frac{1}{2} \cdot \frac{1}{2} - 1 \cdot \frac{1}{8} = \frac{1}{4} - \frac{1}{8} = \frac{1}{8}$.

In Solutions 13–17, $A = \begin{bmatrix} 1 & 0 & \frac{1}{2} \\ -3 & 5 & 2 \\ 0 & 0 & 4 \end{bmatrix}$.

13. $M_{11} = 5 \cdot 4 - 0 \cdot 2 = 20$, $A_{11} = (-1)^2 M_{11} = 20$ **15.** $M_{12} = -3 \cdot 4 - 0 \cdot 2 = -12$, $A_{12} = (-1)^3 M_{12} = 12$

17. $M_{23} = 1 \cdot 0 - 0 \cdot 0 = 0$, $A_{23} = (-1)^5 M_{23} = 0$

19. $M = \begin{bmatrix} 2 & 1 & 0 \\ 0 & -2 & 4 \\ 0 & 1 & -3 \end{bmatrix}$. Therefore, expanding by the first column, $|M| = 2 \begin{vmatrix} -2 & 4 \\ 1 & -3 \end{vmatrix} = 2(6-4) = 4$. Since $|M| \neq 0$, the matrix has an inverse.

21. $M = \begin{bmatrix} 30 & 0 & 20 \\ 0 & -10 & -20 \\ 40 & 0 & 10 \end{bmatrix}$. Therefore, expanding by the first row,

$|M| = 30 \begin{vmatrix} -10 & -20 \\ 0 & 10 \end{vmatrix} + 20 \begin{vmatrix} 0 & -10 \\ 40 & 0 \end{vmatrix} = 30(-100+0) + 20(0+400) = -3000 + 8000 = 5000$, and so M^{-1} exists.

23. $M = \begin{bmatrix} 1 & 3 & 7 \\ 2 & 0 & 8 \\ 0 & 2 & 2 \end{bmatrix}$. Therefore, expanding by the second row, $|M| = -2 \begin{vmatrix} 3 & 7 \\ 2 & 2 \end{vmatrix} - 8 \begin{vmatrix} 1 & 3 \\ 0 & 2 \end{vmatrix} = -2(6-14) - 16 = 0$. Since $|M| = 0$, the matrix does not have an inverse.

25. $M = \begin{bmatrix} 1 & 3 & 3 & 0 \\ 0 & 2 & 0 & 1 \\ -1 & 0 & 0 & 2 \\ 1 & 6 & 4 & 1 \end{bmatrix}$. Therefore, expanding by the third row,

$|M| = -1 \begin{vmatrix} 3 & 3 & 0 \\ 2 & 0 & 1 \\ 6 & 4 & 1 \end{vmatrix} - 2 \begin{vmatrix} 1 & 3 & 3 \\ 0 & 2 & 0 \\ 1 & 6 & 4 \end{vmatrix} = 1 \begin{vmatrix} 3 & 3 \\ 6 & 4 \end{vmatrix} - 1 \begin{vmatrix} 3 & 3 \\ 2 & 0 \end{vmatrix} - 4 \begin{vmatrix} 1 & 3 \\ 1 & 4 \end{vmatrix} = -6 + 6 - 4 = -4$, and so M^{-1} exists.

27. $|M| = \begin{vmatrix} 0 & 0 & 4 & 6 \\ 2 & 1 & 1 & 3 \\ 2 & 1 & 2 & 3 \\ 3 & 0 & 1 & 7 \end{vmatrix} = \begin{vmatrix} 0 & 0 & 4 & 6 \\ 2 & 1 & 1 & 3 \\ 0 & 0 & 1 & 0 \\ 3 & 0 & 1 & 7 \end{vmatrix}$, by replacing R_3 with $R_3 - R_2$. Then, expanding by the third row,

$|M| = 1 \begin{vmatrix} 0 & 0 & 6 \\ 2 & 1 & 3 \\ 3 & 0 & 7 \end{vmatrix} = 6 \begin{vmatrix} 2 & 1 \\ 3 & 0 \end{vmatrix} = 6(2 \cdot 0 - 3 \cdot 1) = -18$.

29. $M = \begin{bmatrix} 1 & 2 & 3 & 4 & 5 \\ 0 & 2 & 4 & 6 & 8 \\ 0 & 0 & 3 & 6 & 9 \\ 0 & 0 & 0 & 4 & 8 \\ 0 & 0 & 0 & 0 & 5 \end{bmatrix}$, so $|M| = 5 \begin{vmatrix} 1 & 2 & 3 & 4 \\ 0 & 2 & 4 & 6 \\ 0 & 0 & 3 & 6 \\ 0 & 0 & 0 & 4 \end{vmatrix} = 5 \cdot 4 \begin{vmatrix} 1 & 2 & 3 \\ 0 & 2 & 4 \\ 0 & 0 & 3 \end{vmatrix} = 20 \cdot 3 \begin{vmatrix} 1 & 2 \\ 0 & 2 \end{vmatrix} = 60 \cdot 2 = 120$.

31. $B = \begin{bmatrix} 4 & 1 & 0 \\ -2 & -1 & 1 \\ 4 & 0 & 3 \end{bmatrix}$

(a) $|B| = 2 \begin{vmatrix} 1 & 0 \\ 0 & 3 \end{vmatrix} - 1 \begin{vmatrix} 4 & 0 \\ 4 & 3 \end{vmatrix} - 1 \begin{vmatrix} 4 & 1 \\ 4 & 0 \end{vmatrix} = 6 - 12 + 4 = -2$

(b) $|B| = -1 \begin{vmatrix} 4 & 1 \\ 4 & 0 \end{vmatrix} + 3 \begin{vmatrix} 4 & 1 \\ -2 & -1 \end{vmatrix} = 4 - 6 = -2$

(c) Yes, as expected, the results agree.

33. $\begin{cases} 2x - y = -9 \\ x + 2y = 8 \end{cases}$ Then $|D| = \begin{vmatrix} 2 & -1 \\ 1 & 2 \end{vmatrix} = 5$, $|D_x| = \begin{vmatrix} -9 & -1 \\ 8 & 2 \end{vmatrix} = -10$, and $|D_y| = \begin{vmatrix} 2 & -9 \\ 1 & 8 \end{vmatrix} = 25$.

Hence, $x = \dfrac{|D_x|}{|D|} = \dfrac{-10}{5} = -2$, $y = \dfrac{|D_y|}{|D|} = \dfrac{25}{5} = 5$, and so the solution is $(-2, 5)$.

35. $\begin{cases} x - 6y = 3 \\ 3x + 2y = 1 \end{cases}$ Then, $|D| = \begin{vmatrix} 1 & -6 \\ 3 & 2 \end{vmatrix} = 20$, $|D_x| = \begin{vmatrix} 3 & -6 \\ 1 & 2 \end{vmatrix} = 12$, and $|D_y| = \begin{vmatrix} 1 & 3 \\ 3 & 1 \end{vmatrix} = -8$.

Hence, $x = \dfrac{|D_x|}{|D|} = \dfrac{12}{20} = 0.6$, $y = \dfrac{|D_y|}{|D|} = \dfrac{-8}{20} = -0.4$, and so the solution is $(0.6, -0.4)$.

37. $\begin{cases} 0.4x + 1.2y = 0.4 \\ 1.2x + 1.6y = 3.2 \end{cases}$ Then, $|D| = \begin{vmatrix} 0.4 & 1.2 \\ 1.2 & 1.6 \end{vmatrix} = -0.8$, $|D_x| = \begin{vmatrix} 0.4 & 1.2 \\ 3.2 & 1.6 \end{vmatrix} = -3.2$, and $|D_y| = \begin{vmatrix} 0.4 & 0.4 \\ 1.2 & 3.2 \end{vmatrix} = 0.8$.

Hence, $x = \dfrac{|D_x|}{|D|} = \dfrac{-3.2}{-0.8} = 4$, $y = \dfrac{|D_y|}{|D|} = \dfrac{0.8}{-0.8} = -1$, and so the solution is $(4, -1)$.

39. $\begin{cases} x - y + 2z = 0 \\ 3x \qquad + z = 11 \\ -x + 2y \qquad = 0 \end{cases}$ Then expanding by the second row,

$|D| = \begin{vmatrix} 1 & -1 & 2 \\ 3 & 0 & 1 \\ -1 & 2 & 0 \end{vmatrix} = -3 \begin{vmatrix} -1 & 2 \\ 2 & 0 \end{vmatrix} - 1 \begin{vmatrix} 1 & -1 \\ -1 & 2 \end{vmatrix} = 12 - 1 = 11$, $|D_x| = \begin{vmatrix} 0 & -1 & 2 \\ 11 & 0 & 1 \\ 0 & 2 & 0 \end{vmatrix} = -11 \begin{vmatrix} -1 & 2 \\ 2 & 0 \end{vmatrix} = 44$,

$|D_y| = \begin{vmatrix} 1 & 0 & 2 \\ 3 & 11 & 1 \\ -1 & 0 & 0 \end{vmatrix} = 11 \begin{vmatrix} 1 & 2 \\ -1 & 0 \end{vmatrix} = 22$, and $|D_z| = \begin{vmatrix} 1 & -1 & 0 \\ 3 & 0 & 11 \\ -1 & 2 & 0 \end{vmatrix} = -11 \begin{vmatrix} 1 & -1 \\ -1 & 2 \end{vmatrix} = -11$.

Therefore, $x = \dfrac{44}{11} = 4$, $y = \dfrac{22}{11} = 2$, $z = \dfrac{-11}{11} = -1$, and so the solution is $(4, 2, -1)$.

41. $\begin{cases} 2x_1 + 3x_2 - 5x_3 = 1 \\ x_1 + x_2 - x_3 = 2 \\ 2x_2 + x_3 = 8 \end{cases}$

Then, expanding by the third row,

$$|D| = \begin{vmatrix} 2 & 3 & -5 \\ 1 & 1 & -1 \\ 0 & 2 & 1 \end{vmatrix} = -2 \begin{vmatrix} 2 & -5 \\ 1 & -1 \end{vmatrix} + \begin{vmatrix} 2 & 3 \\ 1 & 1 \end{vmatrix} = -6 - 1 = -7,$$

$$|D_{x_1}| = \begin{vmatrix} 1 & 3 & -5 \\ 2 & 1 & -1 \\ 8 & 2 & 1 \end{vmatrix} = \begin{vmatrix} 1 & -1 \\ 2 & 1 \end{vmatrix} - 3 \begin{vmatrix} 2 & -1 \\ 8 & 1 \end{vmatrix} - 5 \begin{vmatrix} 2 & 1 \\ 8 & 2 \end{vmatrix} = 3 - 30 + 20 = -7,$$

$$|D_{x_2}| = \begin{vmatrix} 2 & 1 & -5 \\ 1 & 2 & -1 \\ 0 & 8 & 1 \end{vmatrix} = -8 \begin{vmatrix} 2 & -5 \\ 1 & -1 \end{vmatrix} + \begin{vmatrix} 2 & 1 \\ 1 & 2 \end{vmatrix} = -24 + 3 = -21, \text{ and}$$

$$|D_{x_3}| = \begin{vmatrix} 2 & 3 & 1 \\ 1 & 1 & 2 \\ 0 & 2 & 8 \end{vmatrix} = -2 \begin{vmatrix} 2 & 1 \\ 1 & 2 \end{vmatrix} + 8 \begin{vmatrix} 2 & 3 \\ 1 & 1 \end{vmatrix} = -6 - 8 = -14.$$

Thus, $x_1 = \frac{-7}{-7} = 1$, $x_2 = \frac{-21}{-7} = 3$, $x_3 = \frac{-14}{-7} = 2$, and so the solution is $(1, 3, 2)$.

43. $\begin{cases} \frac{1}{3}x - \frac{1}{5}y + \frac{1}{2}z = \frac{7}{10} \\ -\frac{2}{3}x + \frac{2}{5}y + \frac{3}{2}z = \frac{11}{10} \\ x - \frac{4}{5}y + z = \frac{9}{5} \end{cases} \Leftrightarrow \begin{cases} 10x - 6y + 15z = 21 \\ -20x + 12y + 45z = 33 \\ 5x - 4y + 5z = 9 \end{cases}$ Then

$$|D| = \begin{vmatrix} 10 & -6 & 15 \\ -20 & 12 & 45 \\ 5 & -4 & 5 \end{vmatrix} = 10 \begin{vmatrix} 12 & 45 \\ -4 & 5 \end{vmatrix} + 6 \begin{vmatrix} -20 & 45 \\ 5 & 5 \end{vmatrix} + 15 \begin{vmatrix} -20 & 12 \\ 5 & -4 \end{vmatrix} = 2400 - 1950 + 300 = 750,$$

$$|D_x| = \begin{vmatrix} 21 & -6 & 15 \\ 33 & 12 & 45 \\ 9 & -4 & 5 \end{vmatrix} = 21 \begin{vmatrix} 12 & 45 \\ -4 & 5 \end{vmatrix} + 6 \begin{vmatrix} 33 & 45 \\ 9 & 5 \end{vmatrix} + 15 \begin{vmatrix} 33 & 12 \\ 9 & -4 \end{vmatrix} = 5040 - 1440 - 3600 = 0,$$

$$|D_y| = \begin{vmatrix} 10 & 21 & 15 \\ -20 & 33 & 45 \\ 5 & 9 & 5 \end{vmatrix} = 10 \begin{vmatrix} 33 & 45 \\ 9 & 5 \end{vmatrix} - 21 \begin{vmatrix} -20 & 45 \\ 5 & 5 \end{vmatrix} + 15 \begin{vmatrix} -20 & 33 \\ 5 & 9 \end{vmatrix} = -2400 + 6825 - 5175 = -750, \text{ and}$$

$$|D_z| = \begin{vmatrix} 10 & -6 & 21 \\ -20 & 12 & 33 \\ 5 & -4 & 9 \end{vmatrix} = 10 \begin{vmatrix} 12 & 33 \\ -4 & 9 \end{vmatrix} + 6 \begin{vmatrix} -20 & 33 \\ 5 & 9 \end{vmatrix} + 21 \begin{vmatrix} -20 & 12 \\ 5 & -4 \end{vmatrix} = 2400 - 2070 + 420 = 750.$$

Therefore, $x = 0$, $y = -1$, $z = 1$, and so the solution is $(0, -1, 1)$.

45. $\begin{cases} 3y + 5z = 4 \\ 2x - z = 10 \\ 4x + 7y = 0 \end{cases}$ Then $|D| = \begin{vmatrix} 0 & 3 & 5 \\ 2 & 0 & -1 \\ 4 & 7 & 0 \end{vmatrix} = -3\begin{vmatrix} 2 & -1 \\ 4 & 0 \end{vmatrix} + 5\begin{vmatrix} 2 & 0 \\ 4 & 7 \end{vmatrix} = -12 + 70 = 58$,

$|D_x| = \begin{vmatrix} 4 & 3 & 5 \\ 10 & 0 & -1 \\ 0 & 7 & 0 \end{vmatrix} = -7\begin{vmatrix} 4 & 5 \\ 10 & -1 \end{vmatrix} = 378$, $|D_y| = \begin{vmatrix} 0 & 4 & 5 \\ 2 & 10 & -1 \\ 4 & 0 & 0 \end{vmatrix} = 4\begin{vmatrix} 4 & 5 \\ 10 & -1 \end{vmatrix} = -216$, and

$|D_z| = \begin{vmatrix} 0 & 3 & 4 \\ 2 & 0 & 10 \\ 4 & 7 & 0 \end{vmatrix} = 4\begin{vmatrix} 3 & 4 \\ 0 & 10 \end{vmatrix} - 7\begin{vmatrix} 0 & 4 \\ 2 & 10 \end{vmatrix} = 120 + 56 = 176$.

Thus, $x = \frac{189}{29}$, $y = -\frac{108}{29}$, and $z = \frac{88}{29}$, and so the solution is $\left(\frac{189}{29}, -\frac{108}{29}, \frac{88}{29}\right)$.

47. $\begin{cases} x + y + z + w = 0 \\ 2z + w = 0 \\ y - z = 0 \\ x + 2z = 1 \end{cases}$ Then

$|D| = \begin{vmatrix} 1 & 1 & 1 & 1 \\ 2 & 0 & 0 & 1 \\ 0 & 1 & -1 & 0 \\ 1 & 0 & 2 & 0 \end{vmatrix} = -1\begin{vmatrix} 2 & 0 & 1 \\ 0 & -1 & 0 \\ 1 & 2 & 0 \end{vmatrix} - 1\begin{vmatrix} 1 & 1 & 1 \\ 2 & 0 & 1 \\ 1 & 2 & 0 \end{vmatrix} = -\left(2\begin{vmatrix} -1 & 0 \\ 2 & 0 \end{vmatrix} + 1\begin{vmatrix} 0 & 1 \\ -1 & 0 \end{vmatrix}\right) - \left(-1\begin{vmatrix} 2 & 1 \\ 1 & 0 \end{vmatrix} - 2\begin{vmatrix} 1 & 1 \\ 2 & 1 \end{vmatrix}\right)$

$= -2(0) - 1(1) + 1(-1) + 2(-1) = -4$,

$|D_x| = \begin{vmatrix} 0 & 1 & 1 & 1 \\ 0 & 0 & 0 & 1 \\ 0 & 1 & -1 & 0 \\ 1 & 0 & 2 & 0 \end{vmatrix} = -1\begin{vmatrix} 1 & 1 & 1 \\ 0 & 0 & 1 \\ 1 & -1 & 0 \end{vmatrix} = -1(-1)\begin{vmatrix} 1 & 1 \\ 1 & -1 \end{vmatrix} = -2$,

$|D_y| = \begin{vmatrix} 1 & 0 & 1 & 1 \\ 2 & 0 & 0 & 1 \\ 0 & 0 & -1 & 0 \\ 1 & 1 & 2 & 0 \end{vmatrix} = 1\begin{vmatrix} 1 & 1 & 1 \\ 2 & 0 & 1 \\ 0 & -1 & 0 \end{vmatrix} = 1\begin{vmatrix} 0 & 1 \\ -1 & 0 \end{vmatrix} - 2\begin{vmatrix} 1 & 1 \\ -1 & 0 \end{vmatrix} = 1 - 2(1) = -1$,

$|D_z| = \begin{vmatrix} 1 & 1 & 0 & 1 \\ 2 & 0 & 0 & 1 \\ 0 & 1 & 0 & 0 \\ 1 & 0 & 1 & 0 \end{vmatrix} = -1\begin{vmatrix} 1 & 1 & 1 \\ 2 & 0 & 1 \\ 0 & 1 & 0 \end{vmatrix} = -1\begin{vmatrix} 0 & 1 \\ 1 & 0 \end{vmatrix} + 2\begin{vmatrix} 1 & 1 \\ 1 & 0 \end{vmatrix} = -1(-1) + 2(-1) = -1$, and

$|D_w| = \begin{vmatrix} 1 & 1 & 1 & 0 \\ 2 & 0 & 0 & 0 \\ 0 & 1 & -1 & 0 \\ 1 & 0 & 2 & 1 \end{vmatrix} = 1\begin{vmatrix} 1 & 1 & 1 \\ 2 & 0 & 0 \\ 0 & 1 & -1 \end{vmatrix} = -2\begin{vmatrix} 1 & 1 \\ 1 & -1 \end{vmatrix} = -2(-2) = 4$. Hence, we have $x = \frac{|D_x|}{|D|} = \frac{-2}{-4} = \frac{1}{2}$,

$y = \frac{|D_y|}{|D|} = \frac{-1}{-4} = \frac{1}{4}$, $z = \frac{|D_z|}{|D|} = \frac{-1}{-4} = \frac{1}{4}$, and $w = \frac{|D_w|}{|D|} = \frac{4}{-4} = -1$, and the solution is $\left(\frac{1}{2}, \frac{1}{4}, \frac{1}{4}, -1\right)$.

49.
$$\begin{vmatrix} a & 0 & 0 & 0 & 0 \\ 0 & b & 0 & 0 & 0 \\ 0 & 0 & c & 0 & 0 \\ 0 & 0 & 0 & d & 0 \\ 0 & 0 & 0 & 0 & e \end{vmatrix} = a\begin{vmatrix} b & 0 & 0 & 0 \\ 0 & c & 0 & 0 \\ 0 & 0 & d & 0 \\ 0 & 0 & 0 & e \end{vmatrix} = ab\begin{vmatrix} c & 0 & 0 \\ 0 & d & 0 \\ 0 & 0 & e \end{vmatrix} = abc\begin{vmatrix} d & 0 \\ 0 & e \end{vmatrix} = abcde$$

51.
$$\begin{vmatrix} x & 12 & 13 \\ 0 & x-1 & 23 \\ 0 & 0 & x-2 \end{vmatrix} = 0 \Leftrightarrow (x-2)\begin{vmatrix} x & 12 \\ 0 & x-1 \end{vmatrix} = 0 \Leftrightarrow (x-2)\cdot x(x-1) = 0 \Leftrightarrow x = 0, 1, \text{ or } 2$$

53.
$$\begin{vmatrix} 1 & 0 & x \\ x^2 & 1 & 0 \\ x & 0 & 1 \end{vmatrix} = 0 \Leftrightarrow 1\begin{vmatrix} 1 & 0 \\ 0 & 1 \end{vmatrix} + x\begin{vmatrix} x^2 & 1 \\ x & 0 \end{vmatrix} = 0 \Leftrightarrow 1 - x^2 = 0 \Leftrightarrow x^2 = 1 \Leftrightarrow x = \pm 1$$

55. Area $= \pm\dfrac{1}{2}\begin{vmatrix} 0 & 0 & 1 \\ 6 & 2 & 1 \\ 3 & 8 & 1 \end{vmatrix} = \pm\dfrac{1}{2}\begin{vmatrix} 6 & 2 \\ 3 & 8 \end{vmatrix} = \pm\dfrac{1}{2}(48-6) = \dfrac{1}{2}(42) = 21$

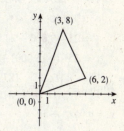

57. Area $= \pm\dfrac{1}{2}\begin{vmatrix} -1 & 3 & 1 \\ 2 & 9 & 1 \\ 5 & -6 & 1 \end{vmatrix} = \pm\dfrac{1}{2}\left[-1\begin{vmatrix} 9 & 1 \\ -6 & 1 \end{vmatrix} - 3\begin{vmatrix} 2 & 1 \\ 5 & 1 \end{vmatrix} + 1\begin{vmatrix} 2 & 9 \\ 5 & -6 \end{vmatrix} \right]$

$= \pm\dfrac{1}{2}[-1(9+6) - 3(2-5) + 1(-12-45)]$

$= \pm\dfrac{1}{2}[-15 - 3(-3) + (-57)] = \pm\dfrac{1}{2}(-63) = \dfrac{63}{2}$

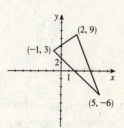

59.
$$\begin{vmatrix} 1 & x & x^2 \\ 1 & y & y^2 \\ 1 & z & z^2 \end{vmatrix} = 1\begin{vmatrix} y & y^2 \\ z & z^2 \end{vmatrix} - 1\begin{vmatrix} x & x^2 \\ z & z^2 \end{vmatrix} + 1\begin{vmatrix} x & x^2 \\ y & y^2 \end{vmatrix} = yz^2 - y^2z - \left(xz^2 - x^2z\right) + \left(xy^2 - xy^2\right)$$

$= yz^2 - y^2z - xz^2 - x^2z + xy^2 - xy^2 + xyz - xyz = xyz - xz^2 - y^2z + yz^2 - x^2y + x^2z + zy^2 - xyz$

$= z\left(xy - xz - y^2 + yz\right) - x\left(xy - xz - y^2 + yz\right) = (z-x)\left(xy - xz - y^2 + yz\right)$

$= (z-x)\left[x(y-z) - y(y-z)\right] = (z-x)(x-y)(y-z)$

61. (a) Using the points $(10, 25)$, $(15, 33.75)$, and $(40, 40)$, we substitute for x and y and get the system

$$\begin{cases} 100a + 10b + c = 25 \\ 225a + 15b + c = 33.75 \\ 1600a + 40b + c = 40 \end{cases}$$

(b) $|D| = \begin{vmatrix} 100 & 10 & 1 \\ 225 & 15 & 1 \\ 1600 & 40 & 1 \end{vmatrix} = 1 \cdot \begin{vmatrix} 225 & 15 \\ 1600 & 40 \end{vmatrix} - 1 \cdot \begin{vmatrix} 100 & 10 \\ 1600 & 40 \end{vmatrix} + 1 \cdot \begin{vmatrix} 100 & 10 \\ 225 & 15 \end{vmatrix}$

$= (9000 - 24{,}000) - (4000 - 16{,}000) + (1500 - 2250) = -15{,}000 + 12{,}000 - 750 = -3750,$

$|D_a| = \begin{vmatrix} 25 & 10 & 1 \\ 33.75 & 15 & 1 \\ 40 & 40 & 1 \end{vmatrix} = 1 \cdot \begin{vmatrix} 33.75 & 15 \\ 40 & 40 \end{vmatrix} - 1 \cdot \begin{vmatrix} 25 & 10 \\ 40 & 40 \end{vmatrix} + 1 \cdot \begin{vmatrix} 25 & 10 \\ 33.75 & 15 \end{vmatrix}$

$= (1350 - 600) - (1000 - 400) + (375 - 337.5) = 750 - 600 + 37.5 = 187.5,$

$|D_b| = \begin{vmatrix} 100 & 25 & 1 \\ 225 & 33.75 & 1 \\ 1600 & 40 & 1 \end{vmatrix} = 1 \cdot \begin{vmatrix} 225 & 33.75 \\ 1600 & 40 \end{vmatrix} - 1 \cdot \begin{vmatrix} 100 & 25 \\ 1600 & 40 \end{vmatrix} + 1 \cdot \begin{vmatrix} 100 & 25 \\ 225 & 33.75 \end{vmatrix}$

$= (9000 - 54{,}000) - (4000 - 40{,}000) + (3375 - 5625) = -45{,}000 + 36{,}000 - 2250 = -11{,}250,$ and

$|D_c| = \begin{vmatrix} 100 & 10 & 25 \\ 225 & 15 & 33.75 \\ 1600 & 40 & 40 \end{vmatrix} = 25 \cdot \begin{vmatrix} 225 & 15 \\ 1600 & 40 \end{vmatrix} - 33.75 \cdot \begin{vmatrix} 100 & 10 \\ 1600 & 40 \end{vmatrix} + 40 \cdot \begin{vmatrix} 100 & 10 \\ 225 & 15 \end{vmatrix}$

$= 25 \cdot (9{,}000 - 24{,}000) - 33.75 \cdot (4{,}000 - 16{,}000) + 40 \cdot (1{,}500 - 2{,}250)$

$= 25 \cdot (-15{,}000) + 33.75 \cdot 12{,}000 + 40 \cdot (-750) = -375{,}000 + 405{,}000 - 30{,}000 = 0.$

Thus, $a = \dfrac{|D_a|}{|D|} = \dfrac{187.5}{-3750} = 0.05$, $b = \dfrac{|D_b|}{|D|} = \dfrac{-11{,}250}{-3{,}750} = 3$, and $c = \dfrac{|D_c|}{|D|} = \dfrac{0}{-3{,}750} = 0$. Thus, the model is

$y = 0.05x^2 + 3x.$

63. (a) The coordinates of the vertices of the surrounding rectangle are (a_1, b_1), (a_2, b_1), (a_2, b_3), and (a_1, b_3). The area of the surrounding rectangle is given by $(a_2 - a_1) \cdot (b_3 - b_1) = a_2 b_3 + a_1 b_1 - a_2 b_1 - a_1 b_3 = a_1 b_1 + a_2 b_3 - a_1 b_3 - a_2 b_1$.

(b) The area of the three blue triangles are as follows:

Area of $\triangle ((a_1, b_1), (a_2, b_1), (a_2, b_2))$: $\frac{1}{2}(a_2 - a_1) \cdot (b_2 - b_1) = \frac{1}{2}(a_2 b_2 + a_1 b_1 - a_2 b_1 - a_1 b_2)$

Area of $\triangle ((a_2, b_2), (a_2, b_3), (a_3, b_3))$: $\frac{1}{2}(a_2 - a_3) \cdot (b_3 - b_2) = \frac{1}{2}(a_2 b_3 + a_3 b_2 - a_2 b_2 - a_3 b_3)$

Area of $\triangle ((a_1, b), (a_1, b_3), (a_3, b_3))$: $\frac{1}{2}(a_3 - a_1) \cdot (b_3 - b_1) = \frac{1}{2}(a_3 b_3 + a_1 b_1 - a_3 b_1 - a_1 b_3)$.

Thus the sum of the areas of the blue triangles, B, is

$B = \frac{1}{2}(a_2 b_2 + a_1 b_1 - a_2 b_1 - a_1 b_2) + \frac{1}{2}(a_2 b_3 + a_3 b_2 - a_2 b_2 - a_3 b_3) + \frac{1}{2}(a_3 b_3 + a_1 b_1 - a_3 b_1 - a_1 b_3)$

$= \frac{1}{2}(a_1 b_1 + a_1 b_1 + a_2 b_2 + a_2 b_3 + a_3 b_2 + a_3 b_3) - \frac{1}{2}(a_1 b_2 + a_1 b_3 + a_2 b_1 + a_2 b_2 + a_3 b_1 + a_3 b_3)$

$= a_1 b_1 + \frac{1}{2}(a_2 b_3 + a_3 b_2) - \frac{1}{2}(a_1 b_2 + a_1 b_3 + a_2 b_1 + a_3 b_1)$

So the area of the red triangle A is the area of the rectangle minus the sum of the areas of the blue triangles, that is,

$A = (a_1 b_1 + a_2 b_3 - a_1 b_3 - a_2 b_1) - \left[a_1 b_1 + \frac{1}{2}(a_2 b_3 + a_3 b_2) - \frac{1}{2}(a_1 b_2 + a_1 b_3 + a_2 b_1 + a_3 b_1) \right]$

$= a_1 b_1 + a_2 b_3 - a_1 b_3 - a_2 b_1 - a_1 b_1 - \frac{1}{2}(a_2 b_3 + a_3 b_2) + \frac{1}{2}(a_1 b_2 + a_1 b_3 + a_2 b_1 + a_3 b_1)$

$= \frac{1}{2}(a_1 b_2 + a_2 b_3 + a_3 b_1) - \frac{1}{2}(a_1 b_3 + a_2 b_1 + a_3 b_2)$

(c) We first find $Q = \begin{vmatrix} a_1 & b_1 & 1 \\ a_2 & b_2 & 1 \\ a_3 & b_3 & 1 \end{vmatrix}$ by expanding about the third column.

$Q = 1 \begin{vmatrix} a_2 & b_2 \\ a_3 & b_3 \end{vmatrix} - 1 \begin{vmatrix} a_1 & b_1 \\ a_3 & b_3 \end{vmatrix} + 1 \begin{vmatrix} a_1 & b_1 \\ a_2 & b_2 \end{vmatrix} = a_2 b_3 - a_3 b_2 - (a_1 b_3 - a_3 b_1) + a_1 b_2 - a_2 b_1$

$\quad = a_1 b_2 + a_2 b_3 + a_3 b_1 - a_1 b_3 - a_2 b_1 - a_3 b_2$

So $\frac{1}{2} Q = \frac{1}{2} (a_1 b_2 + a_2 b_3 + a_3 b_1) - \frac{1}{2} (a_1 b_3 - a_2 b_1 - a_3 b_2)$, the area of the red triangle. Since $\frac{1}{2} Q$ is not always positive, the area is $\pm \frac{1}{2} Q$.

65. (a) Let $|M| = \begin{vmatrix} x & y & 1 \\ x_1 & y_1 & 1 \\ x_2 & y_2 & 1 \end{vmatrix}$. Then, expanding by the third column,

$|M| = \begin{vmatrix} x_1 & y_1 \\ x_2 & y_2 \end{vmatrix} - \begin{vmatrix} x & y \\ x_2 & y_2 \end{vmatrix} + \begin{vmatrix} x & y \\ x_1 & y_1 \end{vmatrix} = (x_1 y_2 - x_2 y_1) - (x y_2 - x_2 y) + (x y_1 - x_1 y)$

$\quad = x_1 y_2 - x_2 y_1 - x y_2 + x_2 y + x y_1 - x_1 y = x_2 y - x_1 y - x y_2 + x y_1 + x_1 y_2 - x_2 y_1$

$\quad = (x_2 - x_1) y - (y_2 - y_1) x + x_1 y_2 - x_2 y_1$

So $|M| = 0 \Leftrightarrow (x_2 - x_1) y - (y_2 - y_1) x + x_1 y_2 - x_2 y_1 = 0 \Leftrightarrow (x_2 - x_1) y = (y_2 - y_1) x - x_1 y_2 + x_2 y_1 \Leftrightarrow$

$(x_2 - x_1) y = (y_2 - y_1) x - x_1 y_2 + x_1 y_1 - x_1 y_1 + x_2 y_1 \Leftrightarrow y = \dfrac{y_2 - y_1}{x_2 - x_1} x - \dfrac{x_1 (y_2 - y_1)}{x_2 - x_1} + \dfrac{y_1 (x_2 - x_1)}{x_2 - x_1} \Leftrightarrow$

$y = \dfrac{y_2 - y_1}{x_2 - x_1} (x - x_1) + y_1 \Leftrightarrow y - y_1 = \dfrac{y_2 - y_1}{x_2 - x_1} (x - x_1)$, which is the "two-point" form of the equation for the line

passing through the points (x_1, y_1) and (x_2, y_2).

(b) Using the result of part (a), the line has equation

$\begin{vmatrix} x & y & 1 \\ 20 & 50 & 1 \\ -10 & 25 & 1 \end{vmatrix} = 0 \Leftrightarrow \begin{vmatrix} 20 & 50 \\ -10 & 25 \end{vmatrix} - \begin{vmatrix} x & y \\ -10 & 25 \end{vmatrix} + \begin{vmatrix} x & y \\ 20 & 50 \end{vmatrix} = 0 \Leftrightarrow$

$(500 + 500) - (25x + 10y) + (50x - 20y) = 0 \Leftrightarrow 25x - 30y + 1000 = 0 \Leftrightarrow 5x - 6y + 200 = 0.$

67. Gaussian elimination is superior, since it takes much longer to evaluate six 5×5 determinants than it does to perform one five-equation Gaussian elimination.

10.7 PARTIAL FRACTIONS

1. (iii): $r(x) = \dfrac{4}{x (x - 2)^2} = \dfrac{A}{x} + \dfrac{B}{x - 2} + \dfrac{C}{(x - 2)^2}$

3. $\dfrac{1}{(x - 1)(x + 2)} = \dfrac{A}{x - 1} + \dfrac{B}{x + 2}$

5. $\dfrac{x^2 - 3x + 5}{(x - 2)^2 (x + 4)} = \dfrac{A}{x - 2} + \dfrac{B}{(x - 2)^2} + \dfrac{C}{x + 4}$

7. $\dfrac{x^2}{(x - 3)(x^2 + 4)} = \dfrac{A}{x - 3} + \dfrac{Bx + C}{x^2 + 4}$

9. $\dfrac{x^3 - 4x^2 + 2}{(x^2 + 1)(x^2 + 2)} = \dfrac{Ax + B}{x^2 + 1} + \dfrac{Cx + D}{x^2 + 2}$

11. $\dfrac{x^3 + x + 1}{x (2x - 5)^3 (x^2 + 2x + 5)^2} = \dfrac{A}{x} + \dfrac{B}{2x - 5} + \dfrac{C}{(2x - 5)^2} + \dfrac{D}{(2x - 5)^3} + \dfrac{Ex + F}{x^2 + 2x + 5} + \dfrac{Gx + H}{(x^2 + 2x + 5)^2}$

13. $\dfrac{2}{(x-1)(x+1)} = \dfrac{A}{x-1} + \dfrac{B}{x+1}$. Multiplying by $(x-1)(x+1)$, we get $2 = A(x+1) + B(x-1) \Leftrightarrow$

$2 = Ax + A + Bx - B$. Thus $\begin{cases} A+B = 0 \\ A-B = 2 \end{cases}$ Adding we get $2A = 2 \Leftrightarrow A = 1$. Now $A + B = 0 \Leftrightarrow B = -A$, so

$B = -1$. Thus, the required partial fraction decomposition is $\dfrac{2}{(x-1)(x+1)} = \dfrac{1}{x-1} - \dfrac{1}{x+1}$.

15. $\dfrac{5}{(x-1)(x+4)} = \dfrac{A}{x-1} + \dfrac{B}{x+4}$. Multiplying by $(x-1)(x+4)$, we get $5 = A(x+4) + B(x-1) \Leftrightarrow$

$5 = Ax + 4A + Bx - B$. Thus $\begin{cases} A+B = 0 \\ 4A-B = 5 \end{cases}$ Now $A + B = 0 \Leftrightarrow B = -A$, so substituting, we get $4A - (-A) = 5 \Leftrightarrow$

$5A = 5 \Leftrightarrow A = 1$ and $B = -1$. The required partial fraction decomposition is $\dfrac{5}{(x-1)(x+4)} = \dfrac{1}{x-1} - \dfrac{1}{x+4}$.

17. $\dfrac{12}{x^2-9} = \dfrac{12}{(x-3)(x+3)} = \dfrac{A}{x-3} + \dfrac{B}{x+3}$. Multiplying by $(x-3)(x+3)$, we get $12 = A(x+3) + B(x-3) \Leftrightarrow$

$12 = Ax + 3A + Bx - 3B$. Thus $\begin{cases} A+B = 0 \\ 3A-3B = 12 \end{cases} \Leftrightarrow \begin{cases} A+B = 0 \\ A-B = 4 \end{cases}$ Adding, we get $2A = 4 \Leftrightarrow A = 2$. So $2 + B = 0$

$\Leftrightarrow B = -2$. The required partial fraction decomposition is $\dfrac{12}{x^2-9} = \dfrac{2}{x-3} - \dfrac{2}{x+3}$.

19. $\dfrac{4}{x^2-4} = \dfrac{4}{(x-2)(x+2)} = \dfrac{A}{x-2} + \dfrac{B}{x+2}$. Multiplying by $x^2 - 4$, we get

$4 = A(x+2) + B(x-2) = (A+B)x + (2A - 2B)$, and so $\begin{cases} A+B = 0 \\ 2A-2B = 4 \end{cases} \Leftrightarrow \begin{cases} A+B = 0 \\ A-B = 2 \end{cases}$ Adding we get $2A = 2$

$\Leftrightarrow A = 1$, and $B = -1$. Therefore, $\dfrac{4}{x^2-4} = \dfrac{1}{x-2} - \dfrac{1}{x+2}$.

21. $\dfrac{x+14}{x^2-2x-8} = \dfrac{x+14}{(x-4)(x+2)} = \dfrac{A}{x-4} + \dfrac{B}{x+2}$. Hence, $x + 14 = A(x+2) + B(x-4) = (A+B)x + (2A - 4B)$,

and so $\begin{cases} A+B = 1 \\ 2A-4B = 14 \end{cases} \Leftrightarrow \begin{cases} 2A+2B = 2 \\ A-2B = 7 \end{cases}$ Adding, we get $3A = 9 \Leftrightarrow A = 3$. So $(3) + B = 1 \Leftrightarrow B = -2$.

Therefore, $\dfrac{x+14}{x^2-2x-8} = \dfrac{3}{x-4} - \dfrac{2}{x+2}$.

23. $\dfrac{x}{8x^2-10x+3} = \dfrac{x}{(4x-3)(2x-1)} = \dfrac{A}{4x-3} + \dfrac{B}{2x-1}$. Hence,

$x = A(2x-1) + B(4x-3) = (2A+4B)x + (-A - 3B)$, and so $\begin{cases} 2A+4B = 1 \\ -A-3B = 0 \end{cases} \Leftrightarrow \begin{cases} 2A+4B = 1 \\ -2A-6B = 0 \end{cases}$

Adding, we get $-2B = 1 \Leftrightarrow B = -\frac{1}{2}$, and $A = \frac{3}{2}$. Therefore, $\dfrac{x}{8x^2-10x+3} = \dfrac{\frac{3}{2}}{4x-3} - \dfrac{\frac{1}{2}}{2x-1}$.

25. $\dfrac{9x^2-9x+6}{2x^3-x^2-8x+4} = \dfrac{9x^2-9x+6}{(x-2)(x+2)(2x-1)} = \dfrac{A}{x-2} + \dfrac{B}{x+2} + \dfrac{C}{2x-1}$. Thus,

$$9x^2 - 9x + 6 = A(x+2)(2x-1) + B(x-2)(2x-1) + C(x-2)(x+2)$$
$$= A\left(2x^2 + 3x - 2\right) + B\left(2x^2 - 5x + 2\right) + C\left(x^2 - 4\right)$$
$$= (2A + 2B + C)x^2 + (3A - 5B)x + (-2A + 2B - 4C)$$

This leads to the system $\begin{cases} 2A + 2B + C = 9 \\ 3A - 5B = -9 \\ -2A + 2B - 4C = 6 \end{cases}$ $\begin{matrix} \text{Coefficients of } x^2 \\ \text{Coefficients of } x \\ \text{Constant terms} \end{matrix}$ $\Leftrightarrow$ $\begin{cases} 2A + 2B + C = 9 \\ 16B + 3C = 45 \\ 4B - 3C = 15 \end{cases}$ $\Leftrightarrow$

$\begin{cases} 2A + 2B + C = 9 \\ 16B + 3C = 45 \\ 15C = -15 \end{cases}$ Hence, $-15C = 15 \Leftrightarrow C = -1$; $16B - 3 = 45 \Leftrightarrow B = 3$; and $2A + 6 - 1 = 9 \Leftrightarrow A = 2$.

Therefore, $\dfrac{9x^2 - 9x + 6}{2x^3 - x^2 - 8x + 4} = \dfrac{2}{x-2} + \dfrac{3}{x+2} - \dfrac{1}{2x-1}$.

27. $\dfrac{x^2+1}{x^3+x^2} = \dfrac{x^2+1}{x^2(x+1)} = \dfrac{A}{x} + \dfrac{B}{x^2} + \dfrac{C}{x+1}$. Hence,

$x^2 + 1 = Ax(x+1) + B(x+1) + Cx^2 = (A+C)x^2 + (A+B)x + B$, and so $B = 1$; $A + 1 = 0 \Leftrightarrow A = -1$; and

$-1 + C = 1 \Leftrightarrow C = 2$. Therefore, $\dfrac{x^2+1}{x^3+x^2} = \dfrac{-1}{x} + \dfrac{1}{x^2} + \dfrac{2}{x+1}$.

29. $\dfrac{2x}{4x^2+12x+9} = \dfrac{2x}{(2x+3)^2} = \dfrac{A}{2x+3} + \dfrac{B}{(2x+3)^2}$. Hence, $2x = A(2x+3) + B = 2Ax + (3A+B)$. So $2A = 2 \Leftrightarrow$

$A = 1$; and $3(1) + B = 0 \Leftrightarrow B = -3$. Therefore, $\dfrac{2x}{4x^2+12x+9} = \dfrac{1}{2x+3} - \dfrac{3}{(2x+3)^2}$.

31. $\dfrac{4x^2 - x - 2}{x^4 + 2x^3} = \dfrac{4x^2 - x - 2}{x^3(x+2)} = \dfrac{A}{x} + \dfrac{B}{x^2} + \dfrac{C}{x^3} + \dfrac{D}{x+2}$. Hence,

$$4x^2 - x - 2 = Ax^2(x+2) + Bx(x+2) + C(x+2) + Dx^3$$
$$= (A+D)x^3 + (2A+B)x^2 + (2B+C)x + 2C$$

So $2C = -2 \Leftrightarrow C = -1$; $2B - 1 = -1 \Leftrightarrow B = 0$; $2A + 0 = 4 \Leftrightarrow A = 2$; and $2 + D = 0 \Leftrightarrow D = -2$. Therefore,

$\dfrac{4x^2 - x - 2}{x^4 + 2x^3} = \dfrac{2}{x} - \dfrac{1}{x^3} - \dfrac{2}{x+2}$.

33. $\dfrac{-10x^2 + 27x - 14}{(x-1)^3(x+2)} = \dfrac{A}{x+2} + \dfrac{B}{x-1} + \dfrac{C}{(x-1)^2} + \dfrac{D}{(x-1)^3}$. Thus,

$$-10x^2 + 27x - 14 = A(x-1)^3 + B(x+2)(x-1)^2 + C(x+2)(x-1) + D(x+2)$$
$$= A(x^3 - 3x^2 + 3x - 1) + B(x+2)(x^2 - 2x + 1) + C(x^2 + x - 2) + D(x+2)$$
$$= A(x^3 - 3x^2 + 3x - 1) + B(x^3 - 3x + 2) + C(x^2 + x - 2) + D(x+2)$$
$$= (A+B)x^3 + (-3A+C)x^2 + (3A - 3B + C + D)x + (-A + 2B - 2C + 2D)$$

which leads to the system

$$\begin{cases} A + B & = & 0 \\ -3A & + C & = -10 \\ 3A - 3B + C + D = & 27 \\ -A + 2B - 2C + 2D = -14 \end{cases}$$
Coefficients of x^3
Coefficients of x^2
Coefficients of x
Constant terms
$\Leftrightarrow$
$$\begin{cases} A + B & = & 0 \\ 3B + C & = -10 \\ -3B + 2C + D = & 17 \\ 3B - 5C + 7D = -15 \end{cases} \Leftrightarrow$$

$$\begin{cases} A + B & = & 0 \\ 3B + C & = -10 \\ 3C + D = & 7 \\ -3C + 8D = & 2 \end{cases} \Leftrightarrow \begin{cases} A + B & = & 0 \\ 3B + C & = -10 \\ 3C + D = & 7 \\ 9D = & 9 \end{cases}$$

Hence, $9D = 9 \Leftrightarrow D = 1$, $3C + 1 = 7 \Leftrightarrow C = 2$, $3B + 2 = -10 \Leftrightarrow B = -4$, and $A - 4 = 0 \Leftrightarrow A = 4$. Therefore,

$$\frac{-10x^2 + 27x - 14}{(x-1)^3 (x+2)} = \frac{4}{x+2} - \frac{4}{x-1} + \frac{2}{(x-1)^2} + \frac{1}{(x-1)^3}.$$

35. $\dfrac{3x^3 + 22x^2 + 53x + 41}{(x+2)^2 (x+3)^2} = \dfrac{A}{x+2} + \dfrac{B}{(x+2)^2} + \dfrac{C}{x+3} + \dfrac{D}{(x+3)^2}$. Thus,

$$3x^3 + 22x^2 + 53x + 41 = A(x+2)(x+3)^2 + B(x+3)^2 + C(x+2)^2(x+3) + D(x+2)^2$$

$$= A\left(x^3 + 8x^2 + 21x + 18\right) + B\left(x^2 + 6x + 9\right)$$

$$+ C\left(x^3 + 7x^2 + 16x + 12\right) + D\left(x^2 + 4x + 4\right)$$

$$= (A+C)x^3 + (8A+B+7C+D)x^2$$

$$+ (21A+6B+16C+4D)x + (18A+9B+12C+4D)$$

so we must solve the system $\begin{cases} A & + C & = 3 \\ 8A + B + 7C + D = 22 \\ 21A + 6B + 16C + 4D = 53 \\ 18A + 9B + 12C + 4D = 41 \end{cases}$
Coefficients of x^3
Coefficients of x^2
Coefficients of x
Constant terms
$\Leftrightarrow$
$\begin{cases} A & + C & = 3 \\ B - C + D = -2 \\ 6B - 5C + 4D = -10 \\ 9B - 6C + 4D = -13 \end{cases}$

$\Leftrightarrow \begin{cases} A + C & = 3 \\ B - C + D = -2 \\ C - 2D = 2 \\ 3C - 5D = 5 \end{cases} \Leftrightarrow \begin{cases} A + C & = 3 \\ B - C + D = -2 \\ C - 2D = 2 \\ D = -1 \end{cases}$ Hence, $D = -1$, $C + 2 = 2 \Leftrightarrow C = 0$, $B - 0 - 1 = -2$

$\Leftrightarrow B = -1$, and $A + 0 = 3 \Leftrightarrow A = 3$. Therefore, $\dfrac{3x^3 + 22x^2 + 53x + 41}{(x+2)^2 (x+3)^2} = \dfrac{3}{x+2} - \dfrac{1}{(x+2)^2} - \dfrac{1}{(x+3)^2}$.

37. $\dfrac{x-3}{x^3 + 3x} = \dfrac{x-3}{x(x^2+3)} = \dfrac{A}{x} + \dfrac{Bx+C}{x^2+3}$. Hence, $x - 3 = A\left(x^2 + 3\right) + Bx^2 + Cx = (A+B)x^2 + Cx + 3A$. So

$3A = -3 \Leftrightarrow A = -1$; $C = 1$; and $-1 + B = 0 \Leftrightarrow B = 1$. Therefore, $\dfrac{x-3}{x^3 + 3x} = -\dfrac{1}{x} + \dfrac{x+1}{x^2+3}$.

39. $\dfrac{2x^3 + 7x + 5}{\left(x^2 + x + 2\right)\left(x^2 + 1\right)} = \dfrac{Ax + B}{x^2 + x + 2} + \dfrac{Cx + D}{x^2 + 1}$. Thus,

$$2x^3 + 7x + 5 = (Ax + B)\left(x^2 + 1\right) + (Cx + D)\left(x^2 + x + 2\right)$$
$$= Ax^3 + Ax + Bx^2 + B + Cx^3 + Cx^2 + 2Cx + Dx^2 + Dx + 2D$$
$$= (A + C)x^3 + (B + C + D)x^2 + (A + 2C + D)x + (B + 2D)$$

We must solve the system

$$\begin{cases} A + C = 2 \\ B + C + D = 0 \\ A + 2C + D = 7 \\ B + 2D = 5 \end{cases} \begin{array}{l} \text{Coefficients of } x^3 \\ \text{Coefficients of } x^2 \\ \text{Coefficients of } x \\ \text{Constant terms} \end{array} \Leftrightarrow \begin{cases} A + C = 2 \\ B + C + D = 0 \\ C + D = 5 \\ C - D = -5 \end{cases} \Leftrightarrow \begin{cases} A + C = 2 \\ B + C + D = 0 \\ C + D = 5 \\ 2D = 10 \end{cases}$$

Hence, $2D = 10 \Leftrightarrow D = 5$, $C + 5 = 5 \Leftrightarrow C = 0$, $B + 0 + 5 = 0 \Leftrightarrow B = -5$, and $A + 0 = 2 \Leftrightarrow A = 2$. Therefore,

$$\dfrac{2x^3 + 7x + 5}{\left(x^2 + x + 2\right)\left(x^2 + 1\right)} = \dfrac{2x - 5}{x^2 + x + 2} + \dfrac{5}{x^2 + 1}.$$

41. $\dfrac{x^4 + x^3 + x^2 - x + 1}{x\left(x^2 + 1\right)^2} = \dfrac{A}{x} + \dfrac{Bx + C}{x^2 + 1} + \dfrac{Dx + E}{\left(x^2 + 1\right)^2}$. Hence,

$$x^4 + x^3 + x^2 - x + 1 = A\left(x^2 + 1\right)^2 + (Bx + C)x\left(x^2 + 1\right) + x(Dx + E)$$
$$= A\left(x^4 + 2x^2 + 1\right) + \left(Bx^2 + Cx\right)\left(x^2 + 1\right) + Dx^2 + Ex$$
$$= A\left(x^4 + 2x^2 + 1\right) + Bx^4 + Bx^2 + Cx^3 + Cx + Dx^2 + Ex$$
$$= (A + B)x^4 + Cx^3 + (2A + B + D)x^2 + (C + E)x + A$$

So $A = 1$, $1 + B = 1 \Leftrightarrow B = 0$; $C = 1$; $2 + 0 + D = 1 \Leftrightarrow D = -1$; and $1 + E = -1 \Leftrightarrow E = -2$. Therefore,

$$\dfrac{x^4 + x^3 + x^2 - x + 1}{x\left(x^2 + 1\right)^2} = \dfrac{1}{x} + \dfrac{1}{x^2 + 1} - \dfrac{x + 2}{\left(x^2 + 1\right)^2}.$$

43. We must first get a proper rational function. Using long division, we find that $\dfrac{x^5 - 2x^4 + x^3 + x + 5}{x^3 - 2x^2 + x - 2} = x^2 +$

$\dfrac{2x^2 + x + 5}{x^3 - 2x^2 + x - 2} = x^2 + \dfrac{2x^2 + x + 5}{(x - 2)\left(x^2 + 1\right)} = x^2 + \dfrac{A}{x - 2} + \dfrac{Bx + C}{x^2 + 1}$. Hence,

$$2x^2 + x + 5 = A\left(x^2 + 1\right) + (Bx + C)(x - 2) = Ax^2 + A + Bx^2 + Cx - 2Bx - 2C$$
$$= (A + B)x^2 + (C - 2B)x + (A - 2C)$$

Equating coefficients, we get the system

$$\begin{cases} A + B = 2 & \text{Coefficients of } x^2 \\ -2B + C = 1 & \text{Coefficients of } x \\ A - 2C = 5 & \text{Constant terms} \end{cases} \Leftrightarrow \begin{cases} A + B = 2 \\ -2B + C = 1 \\ B + 2C = -3 \end{cases} \Leftrightarrow \begin{cases} A + B = 2 \\ -2B + C = 1 \\ 5C = -5 \end{cases}$$

Therefore, $5C = -5 \Leftrightarrow C = -1$, $-2B - 1 = 1 \Leftrightarrow B = -1$, and $A - 1 = 2 \Leftrightarrow A = 3$, so

$$\frac{x^5 - 2x^4 + x^3 + x + 5}{x^3 - 2x^2 + x - 2} = x^2 + \frac{3}{x - 2} - \frac{x + 1}{x^2 + 1}.$$

45. $\dfrac{ax + b}{x^2 - 1} = \dfrac{A}{x - 1} + \dfrac{B}{x + 1}$. Hence, $ax + b = A(x + 1) + B(x - 1) = (A + B)x + (A - B)$.

So $\begin{cases} A + B = a \\ A - B = b \end{cases}$ Adding, we get $2A = a + b \Leftrightarrow A = \dfrac{a + b}{2}$.

Substituting, we get $B = a - A = \dfrac{2a}{2} - \dfrac{a + b}{2} = \dfrac{a - b}{2}$. Therefore, $A = \dfrac{a + b}{2}$ and $B = \dfrac{a - b}{2}$.

47. (a) The expression $\dfrac{x}{x^2 + 1} + \dfrac{1}{x + 1}$ is already a partial fraction decomposition. The denominator in the first term is a quadratic which cannot be factored and the degree of the numerator is less than 2. The denominator of the second term is linear and the numerator is a constant.

(b) The term $\dfrac{x}{(x + 1)^2}$ can be decomposed further, since the numerator and denominator both have linear factors.

$\dfrac{x}{(x + 1)^2} = \dfrac{A}{x + 1} + \dfrac{B}{(x + 1)^2}$. Hence, $x = A(x + 1) + B = Ax + (A + B)$. So $A = 1$, $B = -1$, and

$\dfrac{x}{(x + 1)^2} = \dfrac{1}{x + 1} + \dfrac{-1}{(x + 1)^2}$.

(c) The expression $\dfrac{1}{x + 1} + \dfrac{2}{(x + 1)^2}$ is already a partial fraction decomposition, since each numerator is constant.

(d) The expression $\dfrac{x + 2}{\left(x^2 + 1\right)^2}$ is already a partial fraction decomposition, since the denominator is the square of a quadratic which cannot be factored, and the degree of the numerator is less than 2.

10.8 SYSTEMS OF NONLINEAR EQUATIONS

1. The solutions of the system are the points of intersection of the two graphs, namely $(-2, 2)$ and $(4, 8)$.

3. $\begin{cases} y = x^2 \\ y = x + 12 \end{cases}$ Substituting $y = x^2$ into the second equation gives $x^2 = x + 12 \Leftrightarrow$

$0 = x^2 - x - 12 = (x - 4)(x + 3) \Rightarrow x = 4$ or $x = -3$. So since $y = x^2$, the solutions are $(-3, 9)$ and $(4, 16)$.

5. $\begin{cases} x^2 + y^2 = 8 \\ x + y = 0 \end{cases}$ Solving the second equation for y gives $y = -x$, and substituting this into the first equation gives

$x^2 + (-x)^2 = 8 \Leftrightarrow 2x^2 = 8 \Leftrightarrow x = \pm 2$. So since $y = -x$, the solutions are $(2, -2)$ and $(-2, 2)$.

7. $\begin{cases} x + y^2 = 0 \\ 2x + 5y^2 = 75 \end{cases}$ Solving the first equation for x gives $x = -y^2$, and substituting this into the second equation gives

$2\left(-y^2\right) + 5y^2 = 75 \Leftrightarrow 3y^2 = 75 \Leftrightarrow y^2 = 25 \Leftrightarrow y = \pm 5$. So since $x = -y^2$, the solutions are $(-25, -5)$ and $(-25, 5)$.

9. $\begin{cases} x^2 - 2y = 1 \\ x^2 + 5y = 29 \end{cases}$ Subtracting the first equation from the second equation gives $7y = 28 \Rightarrow y = 4$. Substituting $y = 4$ into

the first equation of the original system gives $x^2 - 2(4) = 1 \Leftrightarrow x^2 = 9 \Leftrightarrow x = \pm 3$. The solutions are $(3, 4)$ and $(-3, 4)$.

11. $\begin{cases} 3x^2 - y^2 = 11 \\ x^2 + 4y^2 = 8 \end{cases}$ Multiplying the first equation by 4 gives the system $\begin{cases} 12x^2 - 4y^2 = 44 \\ x^2 + 4y^2 = 8 \end{cases}$ Adding the equations

gives $13x^2 = 52 \Leftrightarrow x = \pm 2$. Substituting into the first equation we get $3(4) - y^2 = 11 \Leftrightarrow y = \pm 1$. Thus, the solutions are $(2, 1)$, $(2, -1)$, $(-2, 1)$, and $(-2, -1)$.

13. $\begin{cases} x - y^2 + 3 = 0 \\ 2x^2 + y^2 - 4 = 0 \end{cases}$ Adding the two equations gives $2x^2 + x - 1 = 0$. Using the quadratic formula we have

$x = \dfrac{-1 \pm \sqrt{1 - 4(2)(-1)}}{2(2)} = \dfrac{-1 \pm \sqrt{9}}{4} = \dfrac{-1 \pm 3}{4}$. So $x = \dfrac{-1 - 3}{4} = -1$ or $x = \dfrac{-1 + 3}{4} = \frac{1}{2}$. Substituting $x = -1$

into the first equation gives $-1 - y^2 + 3 = 0 \Leftrightarrow y^2 = 2 \Leftrightarrow y = \pm\sqrt{2}$. Substituting $x = \frac{1}{2}$ into the first equation gives

$\frac{1}{2} - y^2 + 3 = 0 \Leftrightarrow y^2 = \frac{7}{2} \Leftrightarrow y = \pm\sqrt{\frac{7}{2}}$. Thus the solutions are $\left(-1, \pm\sqrt{2}\right)$ and $\left(\frac{1}{2}, \pm\sqrt{\frac{7}{2}}\right)$.

15. $\begin{cases} x^2 + y = 8 \\ x - 2y = -6 \end{cases}$ By inspection of the graph, it appears that $(2, 4)$ is a solution, but is difficult to get accurate values

for the other point. Multiplying the first equation by 2 gives the system $\begin{cases} 2x^2 + 2y = 16 \\ x - 2y = -6 \end{cases}$ Adding the equations gives

$2x^2 + x = 10 \Leftrightarrow 2x^2 + x - 10 = 0 \Leftrightarrow (2x + 5)(x - 2) = 0$. So $x = -\frac{5}{2}$ or $x = 2$. If $x = -\frac{5}{2}$, then $-\frac{5}{2} - 2y = -6 \Leftrightarrow$

$-2y = -\frac{7}{2} \Leftrightarrow y = \frac{7}{4}$, and if $x = 2$, then $2 - 2y = -6 \Leftrightarrow -2y = -8 \Leftrightarrow y = 4$. Hence, the solutions are $\left(-\frac{5}{2}, \frac{7}{4}\right)$ and $(2, 4)$.

17. $\begin{cases} x^2 + y = 0 \\ x^3 - 2x - y = 0 \end{cases}$ By inspection of the graph, it appears that $(-2, -4)$, $(0, 0)$, and $(1, -1)$ are solutions to the system.

We check each point in both equations to verify that it is a solution.

For $(-2, -4)$: $(-2)^2 + (-4) = 4 - 4 = 0$ and $(-2)^3 - 2(-2) - (-4) = -8 + 4 + 4 = 0$.

For $(0, 0)$: $(0)^2 + (0) = 0$ and $(0)^3 - 2(0) - (0) = 0$.

For $(1, -1)$: $(1)^2 + (-1) = 1 - 1 = 0$ and $(1)^3 - 2(1) - (-1) = 1 - 2 + 1 = 0$.

Thus, the solutions are $(-2, -4)$, $(0, 0)$, and $(1, -1)$.

19. $\begin{cases} y + x^2 = 4x \\ y + 4x = 16 \end{cases}$ Subtracting the second equation from the first equation gives $x^2 - 4x = 4x - 16 \Leftrightarrow$

$x^2 - 8x + 16 = 0 \Leftrightarrow (x - 4)^2 = 0 \Leftrightarrow x = 4$. Substituting this value for x into either of the original equations gives $y = 0$.

Therefore, the solution is $(4, 0)$.

21. $\begin{cases} x - 2y = 2 \\ y^2 - x^2 = 2x + 4 \end{cases}$ Now $x - 2y = 2 \Leftrightarrow x = 2y + 2$. Substituting for x gives $y^2 - x^2 = 2x + 4 \Leftrightarrow$

$y^2 - (2y + 2)^2 = 2(2y + 2) + 4 \Leftrightarrow y^2 - 4y^2 - 8y - 4 = 4y + 4 + 4 \Leftrightarrow y^2 + 4y + 4 = 0 \Leftrightarrow (y + 2)^2 = 0 \Leftrightarrow y = -2$.

Since $x = 2y + 2$, we have $x = 2(-2) + 2 = -2$. Thus, the solution is $(-2, -2)$.

23. $\begin{cases} x - y = 4 \\ xy = 12 \end{cases}$ Now $x - y = 4 \Leftrightarrow x = 4 + y$. Substituting for x gives $xy = 12 \Leftrightarrow (4 + y) y = 12 \Leftrightarrow y^2 + 4y - 12 = 0$

$\Leftrightarrow (y + 6)(y - 2) = 0 \Leftrightarrow y = -6, y = 2$. Since $x = 4 + y$, the solutions are $(-2, -6)$ and $(6, 2)$.

25. $\begin{cases} x^2 y = 16 \\ x^2 + 4y + 16 = 0 \end{cases}$ Now $x^2 y = 16 \Leftrightarrow x^2 = \dfrac{16}{y}$. Substituting for x^2 gives $\dfrac{16}{y} + 4y + 16 = 0 \Rightarrow 4y^2 + 16y + 16 = 0$

$\Leftrightarrow y^2 + 4y + 4 = 0 \Leftrightarrow (y + 2)^2 = 0 \Leftrightarrow y = -2$. Therefore, $x^2 = \dfrac{16}{-2} = -8$, which has no real solution, and so the system

has no solution.

27. $\begin{cases} x^2 + y^2 = 9 \\ x^2 - y^2 = 1 \end{cases}$ Adding the equations gives $2x^2 = 10 \Leftrightarrow x^2 = 5 \Leftrightarrow x = \pm\sqrt{5}$. Now $x = \pm\sqrt{5} \Rightarrow y^2 = 9 - 5 = 4 \Leftrightarrow$

$y = \pm 2$, and so the solutions are $\left(\sqrt{5}, 2\right), \left(\sqrt{5}, -2\right), \left(-\sqrt{5}, 2\right)$, and $\left(-\sqrt{5}, -2\right)$.

29. $\begin{cases} 2x^2 - 8y^3 = 19 \\ 4x^2 + 16y^3 = 34 \end{cases}$ Multiplying the first equation by 2 gives the system $\begin{cases} 4x^2 - 16y^3 = 38 \\ 4x^2 + 16y^3 = 34 \end{cases}$ Adding the two

equations gives $8x^2 = 72 \Leftrightarrow x = \pm 3$, and then substituting into the first equation we have $2(9) - 8y^3 = 19 \Leftrightarrow$

$y^3 = -\frac{1}{8} \Leftrightarrow y = -\frac{1}{2}$. Therefore, the solutions are $\left(3, -\frac{1}{2}\right)$ and $\left(-3, -\frac{1}{2}\right)$.

31. $\begin{cases} \dfrac{2}{x} - \dfrac{3}{y} = 1 \\ -\dfrac{4}{x} + \dfrac{7}{y} = 1 \end{cases}$ If we let $u = \dfrac{1}{x}$ and $v = \dfrac{1}{y}$, the system is equivalent to $\begin{cases} 2u - 3v = 1 \\ -4u + 7v = 1 \end{cases}$ Multiplying the first

equation by 4 gives the system $\begin{cases} 4u - 6v = 2 \\ -4u + 7v = 1 \end{cases}$ Adding the equations gives $v = 3$, and then substituting into the first

equation gives $2u - 9 = 1 \Leftrightarrow u = 5$. Thus, the solution is $\left(\frac{1}{5}, \frac{1}{3}\right)$.

33. $\begin{cases} y = x^2 + 8x \\ y = 2x + 16 \end{cases}$

The solutions are $(-8, 0)$ and $(2, 20)$.

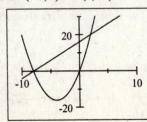

35. $\begin{cases} x^2 + y^2 = 25 \\ x + 3y = 2 \end{cases} \Leftrightarrow \begin{cases} y = \pm\sqrt{25 - x^2} \\ y = -\frac{1}{3}x + \frac{2}{3} \end{cases}$

The solutions are $(-4.51, 2.17)$ and $(4.91, -0.97)$.

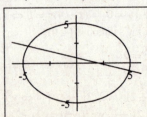

37. $\begin{cases} \dfrac{x^2}{9} + \dfrac{y^2}{18} = 1 \\ y = -x^2 + 6x - 2 \end{cases} \Leftrightarrow \begin{cases} y = \pm\sqrt{18 - 2x^2} \\ y = -x^2 + 6x - 2 \end{cases}$ **39.** $\begin{cases} x^4 + 16y^4 = 32 \\ x^2 + 2x + y = 0 \end{cases} \Leftrightarrow \begin{cases} y = \pm\dfrac{\sqrt[4]{32 - x^4}}{2} \\ y = -x^2 - 2x \end{cases}$

The solutions are $(1.23, 3.87)$ and $(-0.35, -4.21)$.   The solutions are $(-2.30, -0.70)$ and $(0.48, -1.19)$.

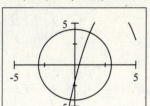   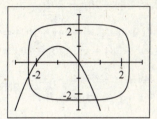

41. Let w and l be the lengths of the sides, in cm. Then we have the system $\begin{cases} lw = 180 \\ 2l + 2w = 54 \end{cases}$ We solve the second equation

for w giving, $w = 27 - l$, and substitute into the first equation to get $l\,(27 - l) = 180 \Leftrightarrow l^2 - 27l + 180 = 0 \Leftrightarrow$
$(l - 15)\,(l - 12) = 0 \Rightarrow l = 15$ or $l = 12$. If $l = 15$, then $w = 27 - 15 = 12$, and if $l = 12$, then $w = 27 - 12 = 15$.
Therefore, the dimensions of the rectangle are 12 cm by 15 cm.

43. Let l and w be the length and width, respectively, of the rectangle. Then, the system of equations is
$\begin{cases} 2l + 2w = 70 \\ \sqrt{l^2 + w^2} = 25 \end{cases}$ Solving the first equation for l, we have $l = 35 - w$, and substituting into the second gives

$\sqrt{l^2 + w^2} = 25 \Leftrightarrow l^2 + w^2 = 625 \Leftrightarrow (35 - w)^2 + w^2 = 625 \Leftrightarrow 1225 - 70w + w^2 + w^2 = 625 \Leftrightarrow 2w^2 - 70w + 600 = 0$
$\Leftrightarrow (w - 15)\,(w - 20) = 0 \Rightarrow w = 15$ or $w = 20$. So the dimensions of the rectangle are 15 and 20.

45. At the points where the rocket path and the hillside meet, we have $\begin{cases} y = \frac{1}{2}x \\ y = -x^2 + 401x \end{cases}$ Substituting for y in the second

equation gives $\frac{1}{2}x = -x^2 + 401x \Leftrightarrow x^2 - \frac{801}{2}x = 0 \Leftrightarrow x\left(x - \frac{801}{2}\right) = 0 \Rightarrow x = 0, x = \frac{801}{2}$. When $x = 0$, the rocket has

not left the pad. When $x = \frac{801}{2}$, then $y = \frac{1}{2}\left(\frac{801}{2}\right) = \frac{801}{4}$. So the rocket lands at the point $\left(\frac{801}{2}, \frac{801}{4}\right)$. The distance from

the base of the hill is $\sqrt{\left(\frac{801}{2}\right)^2 + \left(\frac{801}{4}\right)^2} \approx 447.77$ meters.

47. The point P is at an intersection of the circle of radius 26 centered at A $(22, 32)$ and the circle of radius 20 centered at B $(28, 20)$. We have the system

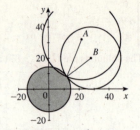

$$\begin{cases} (x - 22)^2 + (y - 32)^2 = 26^2 \\ (x - 28)^2 + (y - 20)^2 = 20^2 \end{cases} \Leftrightarrow$$

$$\begin{cases} x^2 - 44x + 484 + y^2 - 64y + 1024 = 676 \\ x^2 - 56x + 784 + y^2 - 40y + 400 = 400 \end{cases} \Leftrightarrow$$

$$\begin{cases} x^2 - 44x + y^2 - 64y = -832 \\ x^2 - 56x + y^2 - 40y = -784 \end{cases}$$ Subtracting the two equations, we get $12x - 24y = -48 \Leftrightarrow x - 2y = -4$,

which is the equation of a line. Solving for x, we have $x = 2y - 4$. Substituting into the first equation gives

$$(2y - 4)^2 - 44(2y - 4) + y^2 - 64y = -832 \Leftrightarrow 4y^2 - 16y + 16 - 88y + 176 + y^2 - 64y = -832 \Leftrightarrow 5y^2 - 168y + 192 = -832$$

$\Leftrightarrow 5y^2 - 168y + 1024 = 0$. Using the quadratic formula, we have $y = \dfrac{168 \pm \sqrt{168^2 - 4(5)(1024)}}{2(5)} = \dfrac{168 \pm \sqrt{7744}}{10} = \dfrac{168 \pm 88}{10}$

$\Leftrightarrow y = 8$ or $y = 25.60$. Since the y-coordinate of the point P must be less than that of point A, we have $y = 8$. Then $x = 2(8) - 4 = 12$. So the coordinates of P are $(12, 8)$.

To solve graphically, we must solve each equation for y. This gives $(x - 22)^2 + (y - 32)^2 = 26^2$

$\Leftrightarrow (y - 32)^2 = 26^2 - (x - 22)^2 \Rightarrow y - 32 = \pm\sqrt{676 - (x - 22)^2} \Leftrightarrow y = 32 \pm \sqrt{676 - (x - 22)^2}$. We use the function

$y = 32 - \sqrt{676 - (x - 22)^2}$ because the intersection we at interested in is below the point A. Likewise, solving the second

equation for y, we would get the function $y = 20 - \sqrt{400 - (x - 28)^2}$. In a three-dimensional situation, you would need a

minimum of three satellites, since a point on the earth can be uniquely specified as the intersection of three spheres centered at the satellites.

49. (a) $\begin{cases} \log x + \log y = \frac{3}{2} \\ 2 \log x - \log y = 0 \end{cases}$ Adding the two equations gives $3 \log x = \frac{3}{2} \Leftrightarrow \log x = \frac{1}{2} \Leftrightarrow x = \sqrt{10}$. Substituting into the

second equation we get $2 \log 10^{1/2} - \log y = 0 \Leftrightarrow \log 10 - \log y = 0 \Leftrightarrow \log y = 1 \Leftrightarrow y = 10$. Thus, the solution is

$\left(\sqrt{10}, 10\right)$.

(b) $\begin{cases} 2^x + 2^y = 10 \\ 4^x + 4^y = 68 \end{cases} \Leftrightarrow \begin{cases} 2^x + 2^y = 10 \\ 2^{2x} + 2^{2y} = 68 \end{cases}$ If we let $u = 2^x$ and $v = 2^y$, the system becomes $\begin{cases} u + v = 10 \\ u^2 + v^2 = 68 \end{cases}$

Solving the first equation for u, and substituting this into the second equation gives $u + v = 10 \Leftrightarrow u = 10 - v$, so

$(10 - v)^2 + v^2 = 68 \Leftrightarrow 100 - 20v + v^2 + v^2 = 68 \Leftrightarrow v^2 - 10v + 16 = 0 \Leftrightarrow (v - 8)(v - 2) = 0 \Rightarrow v = 2$ or $v = 8$.

If $v = 2$, then $u = 8$, and so $y = 1$ and $x = 3$. If $v = 8$, then $u = 2$, and so $y = 3$ and $x = 1$. Thus, the solutions are $(1, 3)$ and $(3, 1)$.

(c) $\begin{cases} x - y = 3 \\ x^3 - y^3 = 387 \end{cases}$ Solving the first equation for x gives $x = 3 + y$ and using the hint, $x^3 - y^3 = 387 \Leftrightarrow$

$(x - y)\left(x^2 + xy + y^2\right) = 387$. Next, substituting for x, we get $3\left[(3 + y)^2 + y(3 + y) + y^2\right] = 387 \Leftrightarrow 9 + 6y +$

$y^2 + 3y + y^2 + y^2 = 129 \Leftrightarrow 3y^2 + 9y + 9 = 129 \Leftrightarrow (y + 8)(y - 5) = 0 \Rightarrow y = -8$ or $y = 5$. If $y = -8$, then

$x = 3 + (-8) = -5$, and if $y = 5$, then $x = 3 + 5 = 8$. Thus the solutions are $(-5, -8)$ and $(8, 5)$.

(d) $\begin{cases} x^2 + xy = 1 \\ xy + y^2 = 3 \end{cases}$ Adding the equations gives $x^2 + xy + xy + y^2 = 4 \Leftrightarrow x^2 + 2xy + y^2 = 4 \Leftrightarrow$

$(x+y)^2 = 4 \Rightarrow x + y = \pm 2$. If $x + y = 2$, then from the first equation we get $x(x+y) = 1 \Rightarrow x \cdot 2 = 1 \Rightarrow x = \frac{1}{2}$,

and so $y = 2 - \frac{1}{2} = \frac{3}{2}$. If $x + y = -2$, then from the first equation we get $x(x+y) = 1 \Rightarrow x \cdot (-2) = 1 \Rightarrow x = -\frac{1}{2}$,

and so $y = -2 - \left(-\frac{1}{2}\right) = -\frac{3}{2}$. Thus the solutions are $\left(\frac{1}{2}, \frac{3}{2}\right)$ and $\left(-\frac{1}{2}, -\frac{3}{2}\right)$.

10.9   SYSTEMS OF INEQUALITIES

1. To graph an inequality we first graph the corresponding *equation*. So to graph $y \le x + 1$, we first graph the equation $y = x + 1$. To decide which side is the graph of the inequality we use *test* points.

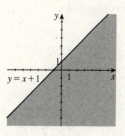

3. $x < 3$

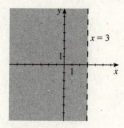

5. $y > x$

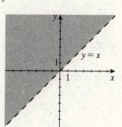

7. $y \le 2x + 2$

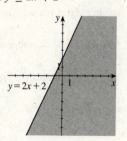

9. $2x - y \le 8$

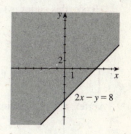

11. $4x + 5y < 20$

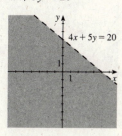

13. $y > x^2 + 1$

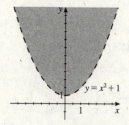

15. $x^2 + y^2 \le 25$

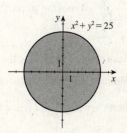

17. The boundary is a solid curve, so we have the inequality $y \le \frac{1}{2}x - 1$. We take the test point $(0, -2)$ and verify that it satisfies the inequality: $-2 \le \frac{1}{2}(0) - 1$.

19. The boundary is a broken curve, so we have the inequality $x^2 + y^2 > 4$. We take the test point $(0, 4)$ and verify that it satisfies the inequality: $0^2 + 4^2 > 4$.

21. $\begin{cases} x+y \leq 4 \\ \quad y \geq x \end{cases}$ The vertices occur where $\begin{cases} x+y=4 \\ \quad y=x \end{cases}$ Substituting, we have

$2x=4 \Leftrightarrow x=2$. Since $y=x$, the vertex is $(2,2)$, and the solution set is not bounded.

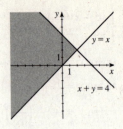

23. $\begin{cases} y < \frac{1}{4}x+2 \\ y \geq 2x-5 \end{cases}$ The vertex occurs where $\begin{cases} y=\frac{1}{4}x+2 \\ y=2x-5 \end{cases}$ Substituting for y

gives $\frac{1}{4}x+2=2x-5 \Leftrightarrow \frac{7}{4}x=7 \Leftrightarrow x=4$, so $y=3$. Hence, the vertex is $(4,3)$, and the solution is not bounded.

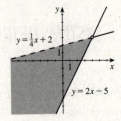

25. $\begin{cases} y \leq -2x+8 \\ y \leq -\frac{1}{2}x+5 \\ x \geq 0, y \geq 0 \end{cases}$ One vertex occurs where $\begin{cases} y \leq -2x+8 \\ y \leq -\frac{1}{2}x+5 \end{cases}$ Substituting for

y gives $-2x+8=-\frac{1}{2}x+5 \Leftrightarrow -\frac{3}{2}x=-3 \Leftrightarrow x=2$, so $y=-2(2)+8=4$.

Hence, this vertex is $(2,4)$. Another vertex occurs where $\begin{cases} y=-2x+8 \\ y=0 \end{cases} \Leftrightarrow$

$-2x+8=0 \Leftrightarrow x=4$; this vertex is $(4,0)$. Another occurs where

$\begin{cases} y=-\frac{1}{2}x+5 \\ x=0 \end{cases} \Leftrightarrow y=5$; this gives the vertex $(0,5)$. The origin is another

vertex, and the solution set is bounded.

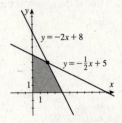

27. $\begin{cases} \quad x \geq 0 \\ \quad y \geq 0 \\ 3x+5y \leq 15 \\ 3x+2y \leq 9 \end{cases}$ From the graph, the points $(3,0)$, $(0,3)$ and $(0,0)$ are vertices,

and the fourth vertex occurs where the lines $3x+5y=15$ and $3x+2y=9$

intersect. Subtracting these two equations gives $3y=6 \Leftrightarrow y=2$, and so $x=\frac{5}{3}$.

Thus, the fourth vertex is $\left(\frac{5}{3},2\right)$, and the solution set is bounded.

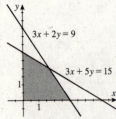

29. $\begin{cases} y \leq 9-x^2 \\ x \geq 0, y \geq 0 \end{cases}$ From the graph, the vertices occur at $(0,0)$, $(3,0)$, and $(0,9)$.

The solution set is bounded.

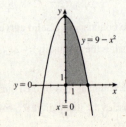

31. $\begin{cases} y < 9 - x^2 \\ y \geq x + 3 \end{cases}$ The vertices occur where $\begin{cases} y = 9 - x^2 \\ y = x + 3 \end{cases}$ Substituting for y

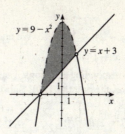

gives $9 - x^2 = x + 3 \Leftrightarrow x^2 + x - 6 = 0 \Leftrightarrow (x - 2)(x + 3) = 0 \Rightarrow x = -3$, $x = 2$. Therefore, the vertices are $(-3, 0)$ and $(2, 5)$, and the solution set is bounded.

33. $\begin{cases} x^2 + y^2 \leq 4 \\ x - y > 0 \end{cases}$ The vertices occur where $\begin{cases} x^2 + y^2 = 4 \\ x - y = 0 \end{cases}$ Since $x - y = 0$

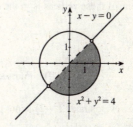

$\Leftrightarrow x = y$, substituting for x gives $y^2 + y^2 = 4 \Leftrightarrow y^2 = 2 \Rightarrow y = \pm\sqrt{2}$, and $x = \pm\sqrt{2}$. Therefore, the vertices are $\left(-\sqrt{2}, -\sqrt{2}\right)$ and $\left(\sqrt{2}, \sqrt{2}\right)$, and the solution set is bounded.

35. $\begin{cases} x^2 - y \leq 0 \\ 2x^2 + y \leq 12 \end{cases}$ The vertices occur where $\begin{cases} x^2 - y = 0 \\ 2x^2 + y = 12 \end{cases} \Leftrightarrow$

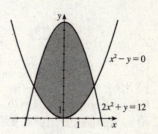

$\begin{cases} 2x^2 - 2y = 0 \\ 2x^2 + y = 12 \end{cases}$ Subtracting the equations gives $3y = 12 \Leftrightarrow y = 4$, and

$x = \pm 2$. Thus, the vertices are $(2, 4)$ and $(-2, 4)$, and the solution set is bounded.

37. $\begin{cases} x + 2y \leq 14 \\ 3x - y \geq 0 \\ x - y \geq 2 \end{cases}$ We find the vertices of the region by solving pairs of the

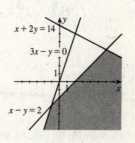

corresponding equations: $\begin{cases} x + 2y = 14 \\ x - y = 2 \end{cases} \Leftrightarrow \begin{cases} x + 2y = 14 \\ 3y = 12 \end{cases} \Leftrightarrow y = 4$ and $x = 6$.

$\begin{cases} 3x - y = 0 \\ x - y = 2 \end{cases} \Leftrightarrow \begin{cases} 3x - y = 0 \\ 2x - y = -2 \end{cases} \Leftrightarrow x = -1$ and $y = -3$. Therefore, the vertices

are $(6, 4)$ and $(-1, -3)$, and the solution set is not bounded.

39. $\begin{cases} x \geq 0, y \geq 0 \\ x \leq 5, x + y \leq 7 \end{cases}$ The points of intersection are $(0, 7)$, $(0, 0)$, $(7, 0)$, $(5, 2)$,

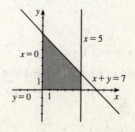

and $(5, 0)$. However, the point $(7, 0)$ is not in the solution set. Therefore, the vertices are $(0, 7)$, $(0, 0)$, $(5, 0)$, and $(5, 2)$, and the solution set is bounded.

41. $\begin{cases} y > x + 1 \\ x + 2y \le 12 \\ x + 1 > 0 \end{cases}$ We find the vertices of the region by solving pairs of the

corresponding equations. Using $x = -1$ and substituting for x in the line $y = x + 1$ gives the point $(-1, 0)$. Substituting for x in the line $x + 2y = 12$ gives the point

$\left(-1, \frac{13}{2}\right)$. $\begin{cases} y = x + 1 \\ x + 2y = 12 \end{cases} \Leftrightarrow x = y - 1$ and $y - 1 + 2y = 12 \Leftrightarrow 3y = 13 \Leftrightarrow$

$y = \frac{13}{3}$ and $x = \frac{10}{3}$. So the vertices are $(-1, 0)$, $\left(-1, \frac{13}{2}\right)$, and $\left(\frac{10}{3}, \frac{13}{3}\right)$, and none of these vertices is in the solution set. The solution set is bounded.

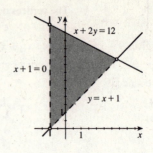

43. $\begin{cases} x^2 + y^2 \le 8 \\ x \ge 2, y \ge 0 \end{cases}$ The intersection points are $(2, \pm 2)$, $(2, 0)$, and $\left(2\sqrt{2}, 0\right)$.

However, since $(2, -2)$ is not part of the solution set, the vertices are $(2, 2)$, $(2, 0)$, and $\left(2\sqrt{2}, 0\right)$. The solution set is bounded.

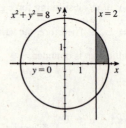

45. $\begin{cases} x^2 + y^2 < 9 \\ x + y > 0, x \le 0 \end{cases}$ Substituting $x = 0$ into the equations $x^2 + y^2 = 9$ and

$x + y = 0$ gives the vertices $(0, \pm 3)$ and $(0, 0)$. To find the points of intersection for the equations $x^2 + y^2 = 9$ and $x + y = 0$, we solve for $x = -y$ and substitute into the first equation. This gives $(-y)^2 + y^2 = 9 \Rightarrow y = \pm \frac{3\sqrt{2}}{2}$. The points

$(0, -3)$ and $\left(\frac{3\sqrt{2}}{2}, -\frac{3\sqrt{2}}{2}\right)$ lie away from the solution set, so the vertices are $(0, 0)$,

$(0, 3)$, and $\left(-\frac{3\sqrt{2}}{2}, \frac{3\sqrt{2}}{2}\right)$. Note that the vertices are not solutions in this case. The solution set is bounded.

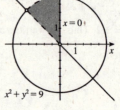

47. $\begin{cases} y \ge x - 3 \\ y \ge -2x + 6 \\ y \le 8 \end{cases}$ Using a graphing calculator, we find the region shown. The

vertices are $(3, 0)$, $(-1, 8)$, and $(11, 8)$.

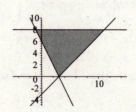

49. $\begin{cases} y \le 6x - x^2 \\ x + y \ge 4 \end{cases}$ Using a graphing calculator, we find the region shown. The

vertices are $(0.6, 3.4)$ and $(6.4, -2.4)$.

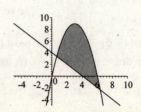

51. Let x be the number of fiction books published in a year and y the number of nonfiction books. Then the following system of inequalities holds:

$$\begin{cases} x \geq 0, \, y \geq 0 \\ x + y \leq 100 \\ y \geq 20, \, x \geq y \end{cases}$$ From the graph, we see that the vertices are $(50, 50)$, $(80, 20)$

and $(20, 20)$.

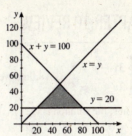

53. Let x be the number of Standard Blend packages and y be the number of Deluxe Blend packages. Since there are 16 ounces per pound, we get the following system of inequalities:

$$\begin{cases} x \geq 0 \\ y \geq 0 \\ \frac{1}{4}x + \frac{5}{8}y \leq 80 \\ \frac{3}{4}x + \frac{3}{8}y \leq 90 \end{cases}$$

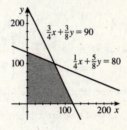

From the graph, we see that the vertices are $(0, 0)$, $(120, 0)$, $(70, 100)$ and $(0, 128)$.

55. $x + 2y > 4$, $-x + y < 1$, $x + 3y \leq 9$, $x < 3$.

Method 1:   We shade the solution to each inequality with lines perpendicular to the boundary. As you can see, as the number of inequalities in the system increases, it gets harder to locate the region where *all* of the shaded parts overlap.

Method 2: Here, if a region is shaded then it fails to satisfy at least one inequality. As a result, the region that is left unshaded satisfies each inequality, and is the solution to the system of inequalities. In this case, this method makes it easier to identify the solution set.

To finish, we find the vertices of the solution set. The line $x = 3$ intersects the line $x + 2y = 4$ at $\left(3, \frac{1}{2}\right)$ and the line $x + 3y = 9$ at $(3, 2)$. To find where the lines $-x + y = 1$ and $x + 2y = 4$ intersect, we add the two equations, which gives $3y = 5 \Leftrightarrow y = \frac{5}{3}$, and $x = \frac{2}{3}$. To find where the lines $-x + y = 1$ and $x + 3y = 9$ intersect, we add the two equations, which gives $4y = 10 \Leftrightarrow y = \frac{10}{4} = \frac{5}{2}$, and $x = \frac{3}{2}$. The vertices are $\left(3, \frac{1}{2}\right)$, $(3, 2)$, $\left(\frac{2}{3}, \frac{5}{3}\right)$, and $\left(\frac{3}{2}, \frac{5}{2}\right)$, and the solution set is bounded.

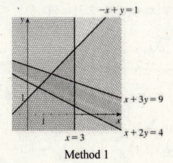

Method 1

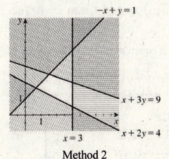

Method 2

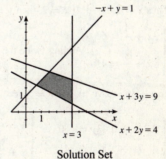

Solution Set

CHAPTER 10 REVIEW

1. $\begin{cases} 3x - y = 5 \\ 2x + y = 5 \end{cases}$ Adding, we get $5x = 10 \Leftrightarrow x = 2$. So $2\,(2) + y = 5 \Leftrightarrow y = 1$.

Thus, the solution is $(2, 1)$.

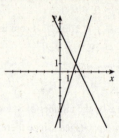

3. $\begin{cases} 2x - 7y = 28 \\ y = \frac{2}{7}x - 4 \end{cases} \Leftrightarrow \begin{cases} 2x - 7y = 28 \\ 2x - 7y = 28 \end{cases}$ Since these equations represent the

same line, any point on this line will satisfy the system. Thus the solution are

$\left(x, \frac{2}{7}x - 4\right)$, where x is any real number.

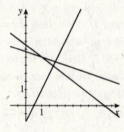

5. $\begin{cases} 2x - y = 1 \\ x + 3y = 10 \\ 3x + 4y = 15 \end{cases}$ Solving the first equation for y, we get $y = -2x + 1$.

Substituting into the second equation gives $x + 3\,(-2x + 1) = 10 \Leftrightarrow -5x = 7 \Leftrightarrow$

$x = -\frac{7}{5}$. So $y = -\left(-\frac{7}{5}\right) + 1 = \frac{12}{5}$. Checking the point $\left(-\frac{7}{5}, \frac{12}{5}\right)$ in the third

equation we have $3\left(-\frac{7}{5}\right) + 4\left(\frac{12}{5}\right) \overset{?}{=} 15$ but $-\frac{21}{5} + \frac{48}{5} \neq 15$. Thus, there is no

solution, and the lines do not intersect at one point.

7. $\begin{cases} y = x^2 + 2x \\ y = 6 + x \end{cases}$ Substituting for y gives $6 + x = x^2 + 2x \Leftrightarrow x^2 + x - 6 = 0$. Factoring, we have $(x - 2)(x + 3) = 0$.

Thus $x = 2$ or -3. If $x = 2$, then $y = 8$, and if $x = -3$, then $y = 3$. Thus the solutions are $(-3, 3)$ and $(2, 8)$.

9. $\begin{cases} 3x + \dfrac{4}{y} = 6 \\ x - \dfrac{8}{y} = 4 \end{cases}$ Adding twice the first equation to the second gives $7x = 16 \Leftrightarrow x = \frac{16}{7}$. So $\dfrac{16}{7} - \dfrac{8}{y} = 4 \Leftrightarrow$

$16y - 56 = 28y \Leftrightarrow -12y = 56 \Leftrightarrow y = -\frac{14}{3}$. Thus, the solution is $\left(\frac{16}{7}, -\frac{14}{3}\right)$.

11. $\begin{cases} 0.32x + 0.43y = 0 \\ 7x - 12y = 341 \end{cases} \Leftrightarrow \begin{cases} y = -\dfrac{32x}{43} \\ y = \dfrac{7x - 341}{12} \end{cases}$

The solution is approximately $(21.41, -15.93)$.

13. $\begin{cases} x - y^2 = 10 \\ x = \frac{1}{22}y + 12 \end{cases} \Leftrightarrow \begin{cases} y = \pm\sqrt{x - 10} \\ y = 22\,(x - 12) \end{cases}$

The solutions are $(11.94, -1.39)$ and $(12.07, 1.44)$.

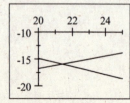

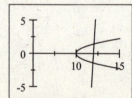

15. (a) 2×3

(b) Yes, this matrix is in row-echelon form.

(c) No, this matrix is not in reduced row-echelon form, since the leading 1 in the second row does not have a 0 above it.

(d) $\begin{cases} x + 2y = -5 \\ y = 3 \end{cases}$

17. (a) 3×4

(b) Yes, this matrix is in row-echelon form.

(c) Yes, this matrix is in reduced row-echelon form.

(d) $\begin{cases} x + 8z = 0 \\ y + 5z = -1 \\ 0 = 0 \end{cases}$

19. (a) 3×4

(b) No, this matrix is not in row-echelon form. The leading 1 in the second row is not to the left of the one above it.

(c) No, this matrix is not in reduced row-echelon form.

(d) $\begin{cases} y - 3z = 4 \\ x + y = 7 \\ x + 2y + z = 2 \end{cases}$

21. $\begin{cases} x + y + 2z = 6 \\ 2x + 5z = 12 \\ x + 2y + 3z = 9 \end{cases} \Leftrightarrow \begin{cases} x + y + 2z = 6 \\ 2y - z = 0 \\ 4y + z = 6 \end{cases} \Leftrightarrow$

$\begin{cases} x + y + 2z = 6 \\ 2y - z = 0 \\ 3z = 6 \end{cases}$ Therefore, $3z = 6 \Leftrightarrow z = 2$,

$2y - 2 = 0 \Leftrightarrow y = 1$, and $x + 1 + 2(2) = 6 \Leftrightarrow x = 1$. Hence, the solution is $(1, 1, 2)$.

23. $\begin{cases} x - 2y + 3z = 1 \\ 2x - y + z = 3 \\ 2x - 7y + 11z = 2 \end{cases} \Leftrightarrow \begin{cases} x - 2y + 3z = 1 \\ 3y - 5z = 1 \\ 6y - 10z = 1 \end{cases} \Leftrightarrow \begin{cases} x - 2y + 3z = 1 \\ 3y - 5z = 1 \\ 0 = -1 \end{cases}$ which is impossible. Therefore, the

system has no solution.

25. $\begin{bmatrix} 1 & 2 & 2 & 6 \\ 1 & -1 & 0 & -1 \\ 2 & 1 & 3 & 7 \end{bmatrix} \xrightarrow{R_2 \leftrightarrow R_1} \begin{bmatrix} 1 & -1 & 0 & -1 \\ 1 & 2 & 2 & 6 \\ 2 & 1 & 3 & 7 \end{bmatrix} \xrightarrow[R_3 - 2R_1 \to R_3]{R_2 - R_1 \to R_2} \begin{bmatrix} 1 & -1 & 0 & -1 \\ 0 & 3 & 2 & 7 \\ 0 & 3 & 3 & 9 \end{bmatrix} \xrightarrow{R_3 - R_2 \to R_3} \begin{bmatrix} 1 & -1 & 0 & -1 \\ 0 & 3 & 2 & 7 \\ 0 & 0 & 1 & 2 \end{bmatrix}.$

Thus, $z = 2$, $3y + 2(2) = 7 \Leftrightarrow 3y = 3 \Leftrightarrow y = 1$, and $x - (1) = -1 \Leftrightarrow x = 0$, and so the solution is $(0, 1, 2)$.

27. $\begin{bmatrix} 1 & -2 & 3 & -2 \\ 2 & -1 & 1 & 2 \\ 2 & -7 & 11 & -9 \end{bmatrix} \xrightarrow[R_3 - 2R_1 \to R_3]{R_2 - 2R_1 \to R_2} \begin{bmatrix} 1 & -2 & 3 & -2 \\ 0 & 3 & -5 & 6 \\ 0 & -3 & 5 & -5 \end{bmatrix} \xrightarrow{R_3 + R_2 \to R_3} \begin{bmatrix} 1 & -2 & 3 & -2 \\ 0 & 3 & -5 & 6 \\ 0 & 0 & 0 & 1 \end{bmatrix}.$

The last row corresponds to the equation $0 = 1$, which is always false. Thus, there is no solution.

29. $\begin{bmatrix} 1 & 1 & 1 & 1 & 0 \\ 1 & -1 & -4 & -1 & -1 \\ 1 & -2 & 0 & 4 & -7 \\ 2 & 2 & 3 & 4 & -3 \end{bmatrix} \xrightarrow[\substack{R_2 - R_1 \to R_2 \\ R_3 - R_1 \to R_3 \\ R_4 - 2R_1 \to R_4}]{} \begin{bmatrix} 1 & 1 & 1 & 1 & 0 \\ 0 & -2 & -5 & -2 & -1 \\ 0 & -3 & -1 & 3 & -7 \\ 0 & 0 & 1 & 2 & -3 \end{bmatrix} \xrightarrow{-R_3 + R_2 \to R_3} \begin{bmatrix} 1 & 1 & 1 & 1 & 0 \\ 0 & 1 & -4 & -5 & 6 \\ 0 & -3 & -1 & 3 & -7 \\ 0 & 0 & 1 & 2 & -3 \end{bmatrix}$

$\xrightarrow{R_3 + 3R_2 \to R_3} \begin{bmatrix} 1 & 1 & 1 & 1 & 0 \\ 0 & 1 & -4 & -5 & 6 \\ 0 & 0 & -13 & -12 & 11 \\ 0 & 0 & 1 & 2 & -3 \end{bmatrix} \xrightarrow{R_3 \leftrightarrow R_4} \begin{bmatrix} 1 & 1 & 1 & 1 & 0 \\ 0 & 1 & -4 & -5 & 6 \\ 0 & 0 & 1 & 2 & -3 \\ 0 & 0 & -13 & -12 & 11 \end{bmatrix} \xrightarrow{R_4 + 13R_3 \to R_4} \begin{bmatrix} 1 & 1 & 1 & 1 & 0 \\ 0 & 1 & -4 & -5 & 6 \\ 0 & 0 & 1 & 2 & -3 \\ 0 & 0 & 0 & 14 & -28 \end{bmatrix}.$

Therefore, $14w = -28 \Leftrightarrow w = -2$, $z + 2(-2) = -3 \Leftrightarrow z = 1$, $y - 4(1) - 5(-2) = 6 \Leftrightarrow y = 0$, and $x + 0 + 1 + (-2) = 0 \Leftrightarrow x = 1$. So the solution is $(1, 0, 1, -2)$.

31. $\begin{cases} x - 3y + z = 4 \\ 4x - y + 15z = 5 \end{cases} \Leftrightarrow \begin{cases} x - 3y + z = 4 \\ y + z = -1 \end{cases}$ Thus, the system has infinitely many solutions given by $z = t$,

$y + t = -1 \Leftrightarrow y = -1 - t$, and $x + 3(1 + t) + t = 4 \Leftrightarrow x = 1 - 4t$. Therefore, the solutions are $(1 - 4t, -1 - t, t)$, where t is any real number.

33. $\begin{cases} -x + 4y + z = 8 \\ 2x - 6y + z = -9 \\ x - 6y - 4z = -15 \end{cases} \Leftrightarrow \begin{cases} -x + 4y + z = 8 \\ 2y + 3z = 7 \\ 6y + 9z = 21 \end{cases} \Leftrightarrow \begin{cases} -x + 4y + z = 8 \\ 2y + 3z = 7 \\ 0 = 0 \end{cases}$

Thus, the system has infinitely many solutions. Letting $z = t$, we find $2y + 3t = 7 \Leftrightarrow y = \frac{7}{2} - \frac{3}{2}t$, and

$-x + 4\left(\frac{7}{2} - \frac{3}{2}t\right) + t = 8 \Leftrightarrow x = 6 - 5t$. Therefore, the solutions are $\left(6 - 5t, \frac{7}{2} - \frac{3}{2}t, t\right)$, where t is any real number.

35. $\begin{bmatrix} 1 & -1 & 3 & 2 \\ 2 & 1 & 1 & 2 \\ 3 & 0 & 4 & 4 \end{bmatrix} \xrightarrow[R_3 - 3R_1 \to R_3]{R_2 - 2R_1 \to R_2} \begin{bmatrix} 1 & -1 & 3 & 2 \\ 0 & 3 & -5 & -2 \\ 0 & 3 & -5 & -2 \end{bmatrix} \xrightarrow{R_3 - R_2 \to R_3} \begin{bmatrix} 1 & -1 & 3 & 2 \\ 0 & 3 & -5 & -2 \\ 0 & 0 & 0 & 0 \end{bmatrix} \xrightarrow{\frac{1}{3}R_2}$

$\begin{bmatrix} 1 & -1 & 3 & 2 \\ 0 & 1 & -\frac{5}{3} & -\frac{2}{3} \\ 0 & 0 & 0 & 0 \end{bmatrix} \xrightarrow{R_1 + R_2 \to R_1} \begin{bmatrix} 1 & 0 & \frac{4}{3} & \frac{4}{3} \\ 0 & 1 & -\frac{5}{3} & -\frac{2}{3} \\ 0 & 0 & 0 & 0 \end{bmatrix}$. The system is dependent, so let $z = t$: $y - \frac{5}{3}t = -\frac{2}{3} \Leftrightarrow$

$y = \frac{5}{3}t - \frac{2}{3}$ and $x + \frac{4}{3}t = \frac{4}{3} \Leftrightarrow x = -\frac{4}{3}t + \frac{4}{3}$. So the solution is $\left(-\frac{4}{3}t + \frac{4}{3}, \frac{5}{3}t - \frac{2}{3}, t\right)$, where t is any real number.

37. $\begin{bmatrix} 1 & -1 & 1 & -1 & 0 \\ 3 & -1 & -1 & -1 & 2 \end{bmatrix} \xrightarrow{R_2 - 3R_1 \to R_2} \begin{bmatrix} 1 & -1 & 1 & -1 & 0 \\ 0 & 2 & -4 & 2 & 2 \end{bmatrix} \xrightarrow{\frac{1}{2}R_4} \begin{bmatrix} 1 & -1 & 1 & -1 & 0 \\ 0 & 1 & -2 & 1 & 1 \end{bmatrix} \xrightarrow{R_1 + R_2 \to R_1}$

$\begin{bmatrix} 1 & 0 & -1 & 0 & 1 \\ 0 & 1 & -2 & 1 & 1 \end{bmatrix}$. Since the system is dependent, Let $z = s$ and $w = t$. Then $y - 2s + t = 1 \Leftrightarrow y = 2s - t + 1$ and

$x - s = 1 \Leftrightarrow x = s + 1$. So the solution is $(s + 1, 2s - t + 1, s, t)$, where s and t are any real numbers.

39. $\begin{bmatrix} 1 & -1 & 1 & 0 \\ 3 & 2 & -1 & 6 \\ 1 & 4 & -3 & 3 \end{bmatrix} \xrightarrow[R_3 - R_1 \to R_3]{R_2 - 3R_1 \to R_2} \begin{bmatrix} 1 & -1 & 1 & 0 \\ 0 & 5 & -4 & 6 \\ 0 & 5 & -4 & 3 \end{bmatrix} \xrightarrow{R_3 - R_2 \to R_3} \begin{bmatrix} 1 & -1 & 1 & 0 \\ 0 & 5 & -4 & 6 \\ 0 & 0 & 0 & 3 \end{bmatrix}$. The last row of this

matrix corresponds to the equation $0 = 3$, which is always false. Hence there is no solution.

41.
$$\begin{bmatrix} 1 & 1 & -1 & -1 & 2 \\ 1 & -1 & 1 & -1 & 0 \\ 2 & 0 & 0 & 2 & 2 \\ 2 & 4 & -4 & -2 & 6 \end{bmatrix} \xrightarrow{R_1 \leftrightarrow \frac{1}{2}R_3} \begin{bmatrix} 1 & 0 & 0 & 1 & 1 \\ 1 & -1 & 1 & -1 & 0 \\ 1 & 1 & -1 & -1 & 2 \\ 2 & 4 & -4 & -2 & 6 \end{bmatrix} \xrightarrow[\substack{R_2-R_1 \to R_2 \\ R_3-R_1 \to R_3 \\ R_4-2R_1 \to R_4}]{} \begin{bmatrix} 1 & 0 & 0 & 1 & 1 \\ 0 & -1 & 1 & -2 & -1 \\ 0 & 1 & -1 & -2 & 1 \\ 0 & 4 & -4 & -4 & 4 \end{bmatrix}$$

$$\xrightarrow[\substack{R_3+R_2 \to R_3 \\ R_4+4R_2 \to R_4}]{} \begin{bmatrix} 1 & 0 & 0 & 1 & 1 \\ 0 & -1 & 1 & -2 & -1 \\ 0 & 0 & 0 & -4 & 0 \\ 0 & 0 & 0 & -12 & 0 \end{bmatrix} \xrightarrow[\substack{-\frac{1}{4}R_3 \\ -\frac{1}{12}R_4}]{} \begin{bmatrix} 1 & 0 & 0 & 1 & 1 \\ 0 & -1 & 1 & -2 & -1 \\ 0 & 0 & 0 & 0 & 0 \\ 0 & 0 & 0 & 1 & 0 \end{bmatrix} \xrightarrow[\substack{R_1-R_3 \to R_1 \\ R_2+2R_3 \to R_2 \\ R_4-R_3 \to R_4}]{}$$

$$\begin{bmatrix} 1 & 0 & 0 & 0 & 1 \\ 0 & -1 & 1 & 0 & -1 \\ 0 & 0 & 0 & 1 & 0 \\ 0 & 0 & 0 & 0 & 0 \end{bmatrix}.$$ This system is dependent. Let $z = t$, so $-y + t = -1 \Leftrightarrow y = t + 1$; $x = 1 \Leftrightarrow x = 1$. So the

solution is $(1, t + 1, t, 0)$, where t is any real number.

43. Let x be the amount in the 6% account and y the amount in the 7% account. The system is $\begin{cases} y = 2x \\ 0.06x + 0.07y = 600 \end{cases}$

Substituting gives $0.06x + 0.07(2x) = 600 \Leftrightarrow 0.2x = 600 \Leftrightarrow x = 3000$, so $y = 2(3000) = 6000$. Hence, the man has $3,000 invested at 6% and $6,000 invested at 7%.

45. Let x be the amount invested in Bank A, y the amount invested in Bank B, and z the amount invested in Bank C.

We get the following system: $\begin{cases} x + y + z = 60,000 \\ 0.02x + 0.025y + 0.03z = 1575 \\ 2x + 2z = y \end{cases} \Leftrightarrow \begin{cases} x + y + z = 60,000 \\ 2x + 2.5y + 3z = 157,500 \\ 2x - y + 2z = 0 \end{cases}$ which

has matrix representation $\begin{bmatrix} 1 & 1 & 1 & 60,000 \\ 2 & 2.5 & 3 & 157,500 \\ 2 & -1 & 2 & 0 \end{bmatrix} \xrightarrow[\substack{R_2-2R_1 \to R_2 \\ R_3-2R_1 \to R_3}]{} \begin{bmatrix} 1 & 1 & 1 & 60,000 \\ 0 & 0.5 & 1 & 37,500 \\ 0 & -3 & 0 & -120,000 \end{bmatrix} \xrightarrow{R_2 \leftrightarrow -\frac{1}{3}R_3}$

$$\begin{bmatrix} 1 & 1 & 1 & 60,000 \\ 0 & 1 & 0 & 40,000 \\ 0 & 0.5 & 1 & 37,500 \end{bmatrix} \xrightarrow[\substack{R_1-R_2 \to R_1 \\ R_3-0.5R_2 \to R_3}]{} \begin{bmatrix} 1 & 0 & 1 & 20,000 \\ 0 & 1 & 0 & 40,000 \\ 0 & 0 & 1 & 17,500 \end{bmatrix} \xrightarrow{R_1-R_3 \to R_1} \begin{bmatrix} 1 & 0 & 1 & 2,500 \\ 0 & 1 & 0 & 40,000 \\ 0 & 0 & 1 & 17,500 \end{bmatrix}.$$ Thus, she invests

$2,500 in Bank A, $40,000 in Bank B, and $17,500 in Bank C.

In Solutions 51–61, the matrices $A, B, C, D, E, F,$ **and** G **are defined as follows:**

$$A = \begin{bmatrix} 2 & 0 & -1 \end{bmatrix} \qquad B = \begin{bmatrix} 1 & 2 & 4 \\ -2 & 1 & 0 \end{bmatrix} \qquad C = \begin{bmatrix} \frac{1}{2} & 3 \\ 2 & \frac{3}{2} \\ -2 & 1 \end{bmatrix}$$

$$D = \begin{bmatrix} 1 & 4 \\ 0 & -1 \\ 2 & 0 \end{bmatrix} \qquad E = \begin{bmatrix} 2 & -1 \\ -\frac{1}{2} & 1 \end{bmatrix} \qquad F = \begin{bmatrix} 4 & 0 & 2 \\ -1 & 1 & 0 \\ 7 & 5 & 0 \end{bmatrix} \qquad G = \begin{bmatrix} 5 \end{bmatrix}$$

47. $A + B$ is not defined because the matrix dimensions 1×3 and 2×3 are not compatible.

49. $2C + 3D = 2\begin{bmatrix} \frac{1}{2} & 3 \\ 2 & \frac{3}{2} \\ -2 & 1 \end{bmatrix} + 3\begin{bmatrix} 1 & 4 \\ 0 & -1 \\ 2 & 0 \end{bmatrix} = \begin{bmatrix} 1 & 6 \\ 4 & 3 \\ -4 & 2 \end{bmatrix} + \begin{bmatrix} 3 & 12 \\ 0 & -3 \\ 6 & 0 \end{bmatrix} = \begin{bmatrix} 4 & 18 \\ 4 & 0 \\ 2 & 2 \end{bmatrix}$

51. $GA = \begin{bmatrix} 5 \end{bmatrix}\begin{bmatrix} 2 & 0 & -1 \end{bmatrix} = \begin{bmatrix} 10 & 0 & -5 \end{bmatrix}$

53. $BC = \begin{bmatrix} 1 & 2 & 4 \\ -2 & 1 & 0 \end{bmatrix}\begin{bmatrix} \frac{1}{2} & 3 \\ 2 & \frac{3}{2} \\ -2 & 1 \end{bmatrix} = \begin{bmatrix} -\frac{7}{2} & 10 \\ 1 & -\frac{9}{2} \end{bmatrix}$ **55.** $BF = \begin{bmatrix} 1 & 2 & 4 \\ -2 & 1 & 0 \end{bmatrix}\begin{bmatrix} 4 & 0 & 2 \\ -1 & 1 & 0 \\ 7 & 5 & 0 \end{bmatrix} = \begin{bmatrix} 30 & 22 & 2 \\ -9 & 1 & -4 \end{bmatrix}$

57. $(C + D)E = \left(\begin{bmatrix} \frac{1}{2} & 3 \\ 2 & \frac{3}{2} \\ -2 & 1 \end{bmatrix} + \begin{bmatrix} 1 & 4 \\ 0 & -1 \\ 2 & 0 \end{bmatrix} \right)\begin{bmatrix} 2 & -1 \\ -\frac{1}{2} & 1 \end{bmatrix} = \begin{bmatrix} \frac{3}{2} & 7 \\ 2 & \frac{1}{2} \\ 0 & 1 \end{bmatrix}\begin{bmatrix} 2 & -1 \\ -\frac{1}{2} & 1 \end{bmatrix} = \begin{bmatrix} -\frac{1}{2} & \frac{11}{2} \\ \frac{15}{4} & -\frac{3}{2} \\ -\frac{1}{2} & 1 \end{bmatrix}$

59. $AB = \begin{bmatrix} 2 & -5 \\ -2 & 6 \end{bmatrix}\begin{bmatrix} 3 & \frac{5}{2} \\ 1 & 1 \end{bmatrix} = \begin{bmatrix} 1 & 0 \\ 0 & 1 \end{bmatrix}$ and $BA = \begin{bmatrix} 3 & \frac{5}{2} \\ 1 & 1 \end{bmatrix}\begin{bmatrix} 2 & -5 \\ -2 & 6 \end{bmatrix} = \begin{bmatrix} 1 & 0 \\ 0 & 1 \end{bmatrix}$.

In Solutions 65–69, $A = \begin{bmatrix} 2 & 1 \\ 3 & 2 \end{bmatrix}$, $B = \begin{bmatrix} 1 & -2 \\ -2 & 4 \end{bmatrix}$, **and** $C = \begin{bmatrix} 0 & 1 & 3 \\ -2 & 4 & 0 \end{bmatrix}$.

61. $A + 3X = B \Leftrightarrow 3X = B - A \Leftrightarrow X = \frac{1}{3}(B - A)$. Thus, $X = \frac{1}{3}\left(\begin{bmatrix} 1 & -2 \\ -2 & 4 \end{bmatrix} - \begin{bmatrix} 2 & 1 \\ 3 & 2 \end{bmatrix} \right) = \frac{1}{3}\begin{bmatrix} -1 & -3 \\ -5 & 2 \end{bmatrix}$.

63. $2(X - A) = 3B \Leftrightarrow X - A = \frac{3}{2}B \Leftrightarrow X = A + \frac{3}{2}B$. Thus,

$X = \begin{bmatrix} 2 & 1 \\ 3 & 2 \end{bmatrix} + \frac{3}{2}\begin{bmatrix} 1 & -2 \\ -2 & 4 \end{bmatrix} = \begin{bmatrix} 2 & 1 \\ 3 & 2 \end{bmatrix} + \begin{bmatrix} \frac{3}{2} & -3 \\ -3 & 6 \end{bmatrix} = \begin{bmatrix} \frac{7}{2} & -2 \\ 0 & 8 \end{bmatrix}$.

65. $AX = C \Leftrightarrow A^{-1}AX = X = A^{-1}C$. Now

$A^{-1} = \frac{1}{4-3}\begin{bmatrix} 2 & -1 \\ -3 & 2 \end{bmatrix} = \begin{bmatrix} 2 & -1 \\ -3 & 2 \end{bmatrix}$. Thus, $X = A^{-1}C = \begin{bmatrix} 2 & -1 \\ -3 & 2 \end{bmatrix}\begin{bmatrix} 0 & 1 & 3 \\ -2 & 4 & 0 \end{bmatrix} = \begin{bmatrix} 2 & -2 & 6 \\ -4 & 5 & -9 \end{bmatrix}$.

67. $D = \begin{bmatrix} 1 & 4 \\ 2 & 9 \end{bmatrix}$. Then $|D| = 1(9) - 2(4) = 1$, and so $D^{-1} = \begin{bmatrix} 9 & -4 \\ -2 & 1 \end{bmatrix}$.

69. $D = \begin{bmatrix} 4 & -12 \\ -2 & 6 \end{bmatrix}$. Then $|D| = 4(6) - 2(12) = 0$, and so D has no inverse.

71. $D = \begin{bmatrix} 3 & 0 & 1 \\ 2 & -3 & 0 \\ 4 & -2 & 1 \end{bmatrix}$. Then, $|D| = 1\begin{vmatrix} 2 & -3 \\ 4 & -2 \end{vmatrix} + 1\begin{vmatrix} 3 & 0 \\ 2 & -3 \end{vmatrix} = -4 + 12 - 9 = -1$. So D^{-1} exists.

$\begin{bmatrix} 3 & 0 & 1 & 1 & 0 & 0 \\ 2 & -3 & 0 & 0 & 1 & 0 \\ 4 & -2 & 1 & 0 & 0 & 1 \end{bmatrix} \xrightarrow{R_1 - R_2 \to R_1} \begin{bmatrix} 1 & 3 & 1 & 1 & -1 & 0 \\ 2 & -3 & 0 & 0 & 1 & 0 \\ 4 & -2 & 1 & 0 & 0 & 1 \end{bmatrix} \xrightarrow[R_3 - 4R_1 \to R_3]{R_2 - 2R_1 \to R_2} \begin{bmatrix} 1 & 3 & 1 & 1 & -1 & 0 \\ 0 & -9 & -2 & -2 & 3 & 0 \\ 0 & -14 & -3 & -4 & 4 & 1 \end{bmatrix} \xrightarrow[-2R_3]{-3R_2}$

$\begin{bmatrix} 1 & 3 & 1 & 1 & -1 & 0 \\ 0 & 27 & 6 & 6 & -9 & 0 \\ 0 & 28 & 6 & 8 & -8 & -2 \end{bmatrix} \xrightarrow{R_3 - R_2 \to R_3} \begin{bmatrix} 1 & 3 & 1 & 1 & -1 & 0 \\ 0 & 27 & 6 & 6 & -9 & 0 \\ 0 & 1 & 0 & 2 & 1 & -2 \end{bmatrix} \xrightarrow[\frac{1}{3}R_3]{R_3 \leftrightarrow R_2} \begin{bmatrix} 1 & 3 & 1 & 1 & -1 & 0 \\ 0 & 1 & 0 & 2 & 1 & -2 \\ 0 & 9 & 2 & 2 & -3 & 0 \end{bmatrix} \xrightarrow[R_1 - 3R_2 \to R_1]{R_3 - 9R_2 \to R_3}$

$\begin{bmatrix} 1 & 0 & 1 & -5 & -4 & 6 \\ 0 & 1 & 0 & 2 & 1 & -2 \\ 0 & 0 & 2 & -16 & -12 & 18 \end{bmatrix} \xrightarrow[R_1 - R_3 \to R_1]{\frac{1}{2}R_3} \begin{bmatrix} 1 & 0 & 0 & 3 & 2 & -3 \\ 0 & 1 & 0 & 2 & 1 & -2 \\ 0 & 0 & 1 & -8 & -6 & 9 \end{bmatrix}$. Thus, $D^{-1} = \begin{bmatrix} 3 & 2 & -3 \\ 2 & 1 & -2 \\ -8 & -6 & 9 \end{bmatrix}$.

73. $D = \begin{bmatrix} 1 & 0 & 0 & 1 \\ 0 & 2 & 0 & 2 \\ 0 & 0 & 3 & 3 \\ 0 & 0 & 0 & 4 \end{bmatrix}$. Thus, $|D| = \begin{vmatrix} 2 & 0 & 2 \\ 0 & 3 & 3 \\ 0 & 0 & 4 \end{vmatrix} = 2\begin{vmatrix} 3 & 3 \\ 0 & 4 \end{vmatrix} = 24$ and D^{-1} exists. $\begin{bmatrix} 1 & 0 & 0 & 1 & 1 & 0 & 0 & 0 \\ 0 & 2 & 0 & 2 & 0 & 1 & 0 & 0 \\ 0 & 0 & 3 & 3 & 0 & 0 & 1 & 0 \\ 0 & 0 & 0 & 4 & 0 & 0 & 0 & 1 \end{bmatrix} \xrightarrow[\frac{1}{4}R_4]{\substack{\frac{1}{2}R_2 \\ \frac{1}{3}R_3}}$

$\begin{bmatrix} 1 & 0 & 0 & 1 & 1 & 0 & 0 & 0 \\ 0 & 1 & 0 & 1 & 0 & \frac{1}{2} & 0 & 0 \\ 0 & 0 & 1 & 1 & 0 & 0 & \frac{1}{3} & 0 \\ 0 & 0 & 0 & 1 & 0 & 0 & 0 & \frac{1}{4} \end{bmatrix} \xrightarrow[R_3 - R_4 \to R_3]{\substack{R_1 - R_4 \to R_1 \\ R_2 - R_4 \to R_2}} \begin{bmatrix} 1 & 0 & 0 & 0 & 1 & 0 & 0 & -\frac{1}{4} \\ 0 & 1 & 0 & 0 & 0 & \frac{1}{2} & 0 & -\frac{1}{4} \\ 0 & 0 & 1 & 0 & 0 & 0 & \frac{1}{3} & -\frac{1}{4} \\ 0 & 0 & 0 & 1 & 0 & 0 & 0 & \frac{1}{4} \end{bmatrix}$. Therefore, $D^{-1} = \begin{bmatrix} 1 & 0 & 0 & -\frac{1}{4} \\ 0 & \frac{1}{2} & 0 & -\frac{1}{4} \\ 0 & 0 & \frac{1}{3} & -\frac{1}{4} \\ 0 & 0 & 0 & \frac{1}{4} \end{bmatrix}$.

75. $\begin{bmatrix} 12 & -5 \\ 5 & -2 \end{bmatrix}\begin{bmatrix} x \\ y \end{bmatrix} = \begin{bmatrix} 10 \\ 17 \end{bmatrix}$. If we let $A = \begin{bmatrix} 12 & -5 \\ 5 & -2 \end{bmatrix}$, then $A^{-1} = \dfrac{1}{-24 + 25}\begin{bmatrix} -2 & 5 \\ -5 & 12 \end{bmatrix} = \begin{bmatrix} -2 & 5 \\ -5 & 12 \end{bmatrix}$, and so

$\begin{bmatrix} x \\ y \end{bmatrix} = \begin{bmatrix} -2 & 5 \\ -5 & 12 \end{bmatrix}\begin{bmatrix} 10 \\ 17 \end{bmatrix} = \begin{bmatrix} 65 \\ 154 \end{bmatrix}$. Therefore, the solution is $(65, 154)$.

77. $\begin{bmatrix} 2 & 1 & 5 \\ 1 & 2 & 2 \\ 1 & 0 & 3 \end{bmatrix} \begin{bmatrix} x \\ y \\ z \end{bmatrix} = \begin{bmatrix} \frac{1}{3} \\ \frac{1}{4} \\ \frac{1}{6} \end{bmatrix}$. Let $A = \begin{bmatrix} 2 & 1 & 5 \\ 1 & 2 & 2 \\ 1 & 0 & 3 \end{bmatrix}$. Then $\begin{bmatrix} 2 & 1 & 5 & 1 & 0 & 0 \\ 1 & 2 & 2 & 0 & 1 & 0 \\ 1 & 0 & 3 & 0 & 0 & 1 \end{bmatrix} \xrightarrow{R_1 \leftrightarrow R_2} \begin{bmatrix} 1 & 2 & 2 & 0 & 1 & 0 \\ 2 & 1 & 5 & 1 & 0 & 0 \\ 1 & 0 & 3 & 0 & 0 & 1 \end{bmatrix}$

$\xrightarrow[R_3 - R_1 \to R_3]{R_2 - 2R_1 \to R_2} \begin{bmatrix} 1 & 2 & 2 & 0 & 1 & 0 \\ 0 & -3 & 1 & 1 & -2 & 0 \\ 0 & -2 & 1 & 0 & -1 & 1 \end{bmatrix} \xrightarrow{R_2 - 2R_3 \to R_2} \begin{bmatrix} 1 & 2 & 2 & 0 & 1 & 0 \\ 0 & 1 & -1 & 1 & 0 & -2 \\ 0 & -2 & 1 & 0 & -1 & 1 \end{bmatrix} \xrightarrow[R_3 \to R_3 + 2R_2]{R_1 - 2R_2 \to R_1}$

$\begin{bmatrix} 1 & 0 & 4 & -2 & 1 & 4 \\ 0 & 1 & -1 & 1 & 0 & -2 \\ 0 & 0 & -1 & 2 & -1 & -3 \end{bmatrix} \xrightarrow{-R_3} \begin{bmatrix} 1 & 0 & 4 & -2 & 1 & 4 \\ 0 & 1 & -1 & 1 & 0 & -2 \\ 0 & 0 & 1 & -2 & 1 & 3 \end{bmatrix} \xrightarrow[R_2 + R_3 \to R_2]{R_1 - 4R_3 \to R_1} \begin{bmatrix} 1 & 0 & 0 & 6 & -3 & -8 \\ 0 & 1 & 0 & -1 & 1 & 1 \\ 0 & 0 & 1 & -2 & 1 & 3 \end{bmatrix}$.

Hence, $A^{-1} = \begin{bmatrix} 6 & -3 & -8 \\ -1 & 1 & 1 \\ -2 & 1 & 3 \end{bmatrix}$ and $\begin{bmatrix} x \\ y \\ z \end{bmatrix} = \begin{bmatrix} 6 & -3 & -8 \\ -1 & 1 & 1 \\ -2 & 1 & 3 \end{bmatrix} \begin{bmatrix} \frac{1}{3} \\ \frac{1}{4} \\ \frac{1}{6} \end{bmatrix} = \begin{bmatrix} -\frac{1}{12} \\ \frac{1}{12} \\ \frac{1}{12} \end{bmatrix}$, and so the solution is

$\left(-\frac{1}{12}, \frac{1}{12}, \frac{1}{12} \right)$.

79. $|D| = \begin{vmatrix} 2 & 7 \\ 6 & 16 \end{vmatrix} = 32 - 42 = -10$, $|D_x| = \begin{vmatrix} 13 & 7 \\ 30 & 16 \end{vmatrix} = 208 - 210 = -2$, and $|D_y| = \begin{vmatrix} 2 & 13 \\ 6 & 30 \end{vmatrix} = 60 - 78 = -18$.

Therefore, $x = \frac{-2}{-10} = \frac{1}{5}$ and $y = \frac{-18}{-10} = \frac{9}{5}$, and so the solution is $\left(\frac{1}{5}, \frac{9}{5} \right)$.

81. $|D| = \begin{vmatrix} 2 & -1 & 5 \\ -1 & 7 & 0 \\ 5 & 4 & 3 \end{vmatrix} = 5 \begin{vmatrix} -1 & 7 \\ 5 & 4 \end{vmatrix} + 3 \begin{vmatrix} 2 & -1 \\ -1 & 7 \end{vmatrix} = -195 + 39 = -156$,

$|D_x| = \begin{vmatrix} 0 & -1 & 5 \\ 9 & 7 & 0 \\ -9 & 4 & 3 \end{vmatrix} = 5 \begin{vmatrix} 9 & 7 \\ -9 & 4 \end{vmatrix} + 3 \begin{vmatrix} 0 & -1 \\ 9 & 7 \end{vmatrix} = 495 + 27 = 522$,

$|D_y| = \begin{vmatrix} 2 & 0 & 5 \\ -1 & 9 & 0 \\ 5 & -9 & 3 \end{vmatrix} = 5 \begin{vmatrix} -1 & 9 \\ 5 & -9 \end{vmatrix} + 3 \begin{vmatrix} 2 & 0 \\ -1 & 9 \end{vmatrix} = -180 + 54 = -126$, and

$|D_z| = \begin{vmatrix} 2 & -1 & 0 \\ -1 & 7 & 9 \\ 5 & 4 & -9 \end{vmatrix} = -9 \begin{vmatrix} 2 & -1 \\ 5 & 4 \end{vmatrix} - 9 \begin{vmatrix} 2 & -1 \\ -1 & 7 \end{vmatrix} = -117 - 117 = -234$.

Therefore, $x = \frac{522}{-156} = -\frac{87}{26}$, $y = \frac{-126}{-156} = \frac{21}{26}$, and $z = \frac{-234}{-156} = \frac{3}{2}$, and so the solution is $\left(-\frac{87}{26}, \frac{21}{26}, \frac{3}{2} \right)$.

83. The area is $\pm \frac{1}{2} \begin{vmatrix} -1 & 3 & 1 \\ 3 & 1 & 1 \\ -2 & -2 & 1 \end{vmatrix} = \pm \frac{1}{2} \left(\begin{vmatrix} 3 & 1 \\ -2 & -2 \end{vmatrix} - \begin{vmatrix} -1 & 3 \\ -2 & -2 \end{vmatrix} + \begin{vmatrix} -1 & 3 \\ 3 & 1 \end{vmatrix} \right) = \pm \frac{1}{2} (-4 - 8 - 10) = 11$.

85. $\dfrac{3x+1}{x^2-2x-15} = \dfrac{3x+1}{(x-5)(x+3)} = \dfrac{A}{x-5} + \dfrac{B}{x+3}$. Thus, $3x+1 = A(x+3) + B(x-5) = x(A+B) + (3A-5B)$,

and so $\begin{cases} A+B=3 \\ 3A-5B=1 \end{cases} \Leftrightarrow \begin{cases} -3A-3B=-9 \\ 3A-5B=1 \end{cases}$ Adding, we have $-8B=-8 \Leftrightarrow B=1$, and $A=2$. Hence,

$\dfrac{3x+1}{x^2-2x-15} = \dfrac{2}{x-5} + \dfrac{1}{x+3}$.

87. $\dfrac{2x-4}{x(x-1)^2} = \dfrac{A}{x} + \dfrac{B}{x-1} + \dfrac{C}{(x-1)^2}$. Then $2x-4 = A(x-1)^2 + Bx(x-1) + Cx = Ax^2 - 2Ax + A + Bx^2 - Bx + Cx = x^2(A+B) + x(-2A-B+C) + A$. So $A=-4$, $-4+B=0 \Leftrightarrow B=4$, and $8-4+C=2 \Leftrightarrow C=-2$.

Therefore, $\dfrac{2x-4}{x(x-1)^2} = -\dfrac{4}{x} + \dfrac{4}{x-1} - \dfrac{2}{(x-1)^2}$.

89. $\dfrac{2x-1}{x^3+x} = \dfrac{2x-1}{x(x^2+1)} = \dfrac{A}{x} + \dfrac{Bx+C}{x^2+1}$. Then $2x-1 = A(x^2+1) + (Bx+C)x = Ax^2 + A + Bx^2 + Cx = (A+B)x^2 + Cx + A$. So $A=-1$, $C=2$, and $A+B=0$ gives us $B=1$. Thus $\dfrac{2x-1}{x^3+x} = -\dfrac{1}{x} + \dfrac{x+2}{x^2+1}$.

91. $\begin{cases} 2x+3y=7 \\ x-2y=0 \end{cases}$ By inspection of the graph, it appears that $(2,1)$ is the solution to the system. We check this in both

equations to verify that it is the solution. $2(2) + 3(1) = 4+3 = 7$ and $2 - 2(1) = 2-2 = 0$. Since both equations are satisfied, the solution is indeed $(2,1)$.

93. $\begin{cases} x^2+y=2 \\ x^2-3x-y=0 \end{cases}$ By inspection of the graph, it appears that $(2,-2)$ is a solution to the system, but is difficult

to get accurate values for the other point. Adding the equations, we get $2x^2 - 3x = 2 \Leftrightarrow 2x^2 - 3x - 2 = 0 \Leftrightarrow (2x+1)(x-2) = 0$. So $2x+1 = 0 \Leftrightarrow x = -\frac{1}{2}$ or $x=2$. If $x = -\frac{1}{2}$, then $\left(-\frac{1}{2}\right)^2 + y = 2 \Leftrightarrow y = \frac{7}{4}$. If $x=2$, then $2^2 + y = 2 \Leftrightarrow y = -2$. Thus, the solutions are $\left(-\frac{1}{2}, \frac{7}{4}\right)$ and $(2,-2)$.

95. The boundary is a solid curve, so we have the inequality $x + y^2 \le 4$. We take the test point $(0,0)$ and verify that it satisfies the inequality: $0 + 0^2 \le 4$.

97. $3x + y \le 6$

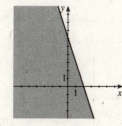

99. $x^2 + y^2 > 9$

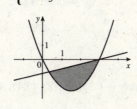

101. $\begin{cases} y \ge x^2 - 3x \\ y \le \frac{1}{3}x - 1 \end{cases}$

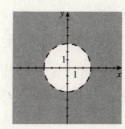

103. $\begin{cases} x+y \ge 2 \\ y-x \le 2 \\ x \le 3 \end{cases}$

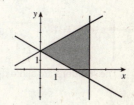

105. $\begin{cases} x^2+y^2 < 9 \\ x+y < 0 \end{cases}$ The vertices occur where $y = -x$. By substitution,

$x^2 + x^2 = 9 \Leftrightarrow x = \pm\frac{3}{\sqrt{2}}$, and so $y = \mp\frac{3}{\sqrt{2}}$. Therefore, the vertices are

$\left(\frac{3}{\sqrt{2}}, -\frac{3}{\sqrt{2}}\right)$ and $\left(-\frac{3}{\sqrt{2}}, \frac{3}{\sqrt{2}}\right)$ and the solution set is bounded.

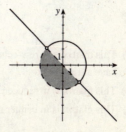

107. $\begin{cases} x \geq 0, y \geq 0 \\ x + 2y \leq 12 \\ y \leq x + 4 \end{cases}$ The intersection points are $(-4, 0)$, $(0, 4)$, $\left(\frac{4}{3}, \frac{16}{3}\right)$, $(0, 6)$,

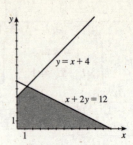

$(0, 0)$, and $(12, 0)$. Since the points $(-4, 0)$ and $(0, 6)$ are not in the solution set,

the vertices are $(0, 4)$, $\left(\frac{4}{3}, \frac{16}{3}\right)$, $(12, 0)$, and $(0, 0)$. The solution set is bounded.

109. $\begin{cases} -x + y + z = a \\ x - y + z = b \\ x + y - z = c \end{cases} \Leftrightarrow \begin{cases} -x + y + z = a \\ \quad\quad 2z = a + b \\ \quad 2y \quad\ = a + c \end{cases}$ Thus, $y = \dfrac{a+c}{2}$, $z = \dfrac{a+b}{2}$, and $-x + \dfrac{a+c}{2} + \dfrac{a+b}{2} = a \Leftrightarrow$

$x = \dfrac{b+c}{2}$. The solution is $\left(\dfrac{b+c}{2}, \dfrac{a+c}{2}, \dfrac{a+b}{2}\right)$.

111. Solving the second equation for y, we have $y = kx$. Substituting for y in the first equation gives us

$x + kx = 12 \Leftrightarrow (1 + k) x = 12 \Leftrightarrow x = \dfrac{12}{k+1}$. Substituting for y in the third equation gives us $kx - x = 2k \Leftrightarrow$

$(k - 1) x = 2k \Leftrightarrow x = \dfrac{2k}{k-1}$. These points of intersection are the same when the x-values are equal. Thus, $\dfrac{12}{k+1} = \dfrac{2k}{k-1}$

$\Leftrightarrow 12 (k - 1) = 2k (k + 1) \Leftrightarrow 12k - 12 = 2k^2 + 2k \Leftrightarrow 0 = 2k^2 - 10k + 12 = 2\left(k^2 - 5k + 6\right) = 2 (k - 3) (k + 2)$.

Hence, $k = 2$ or $k = 3$.

CHAPTER 10 TEST

1. (a) The system is linear.

(b) $\begin{cases} x + 3y = 7 \\ 5x + 2y = -4 \end{cases}$ Multiplying the first equation by -5 and then adding gives $-13y = -39 \Leftrightarrow y = 3$. So

$x + 3 (3) = 7 \Leftrightarrow x = -2$. Thus, the solution is $(-2, 3)$.

3. $\begin{cases} x - 2y = 1 \\ y = x^3 - 2x^2 \end{cases}$

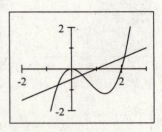

The solutions are approximately $(-0.55, -0.78)$, $(0.43, -0.29)$, and $(2.12, 0.56)$.

5. (a) This matrix is in row-echelon form, but not in reduced row-echelon form, since the leading 1 in the second row does not
have a 0 above it.

(b) This matrix is in reduced row-echelon form.

(c) This matrix is in neither row-echelon form nor reduced row-echelon form.

7. $\begin{cases} x + 3y - z = 0 \\ 3x + 4y - 2z = -1 \\ -x + 2y = 1 \end{cases}$ has the matrix representation $\begin{bmatrix} 1 & 3 & -1 & 0 \\ 3 & 4 & -2 & -1 \\ -1 & 2 & 0 & 1 \end{bmatrix}$ $\xrightarrow[R_3 + R_1 \to R_3]{R_2 - 3R_1 \to R_2}$ $\begin{bmatrix} 1 & 3 & -1 & 0 \\ 0 & -5 & 1 & -1 \\ 0 & 5 & -1 & 1 \end{bmatrix}$

$\xrightarrow{R_3 - R_2 \to R_3}$ $\begin{bmatrix} 1 & 3 & -1 & 0 \\ 0 & -5 & 1 & -1 \\ 0 & 0 & 0 & 0 \end{bmatrix}$ $\xrightarrow{-\frac{1}{5}R_2}$ $\begin{bmatrix} 1 & 3 & -1 & 0 \\ 0 & 1 & -\frac{1}{5} & \frac{1}{5} \\ 0 & 0 & 0 & 0 \end{bmatrix}$ $\xrightarrow{R_1 - 3R_2 \to R_1}$ $\begin{bmatrix} 1 & 0 & -\frac{2}{5} & -\frac{3}{5} \\ 0 & 1 & -\frac{1}{5} & \frac{1}{5} \\ 0 & 0 & 0 & 0 \end{bmatrix}$.

Since this system is dependent, let $z = t$. Then $y - \frac{1}{5}t = \frac{1}{5} \Leftrightarrow y = \frac{1}{5}t + \frac{1}{5}$ and $x - \frac{2}{5}t = -\frac{3}{5} \Leftrightarrow x = \frac{2}{5}t - \frac{3}{5}$. Thus, the

solution is $\left(\frac{2}{5}t - \frac{3}{5}, \frac{1}{5}t + \frac{1}{5}, t \right)$.

9. $A = \begin{bmatrix} 2 & 3 \\ 2 & 4 \end{bmatrix}$, $B = \begin{bmatrix} 2 & 4 \\ -1 & 1 \\ 3 & 0 \end{bmatrix}$, and $C = \begin{bmatrix} 1 & 0 & 4 \\ -1 & 1 & 2 \\ 0 & 1 & 3 \end{bmatrix}$.

(a) $A + B$ is undefined because A is 2×2 and B is 3×2, so they have incompatible dimensions.

(b) AB is undefined because A is 2×2 and B is 3×2, so they have incompatible dimensions.

(c) $BA - 3B = \begin{bmatrix} 2 & 4 \\ -1 & 1 \\ 3 & 0 \end{bmatrix} \begin{bmatrix} 2 & 3 \\ 2 & 4 \end{bmatrix} - 3 \begin{bmatrix} 2 & 4 \\ -1 & 1 \\ 3 & 0 \end{bmatrix} = \begin{bmatrix} 12 & 22 \\ 0 & 1 \\ 6 & 9 \end{bmatrix} - \begin{bmatrix} 6 & 12 \\ -3 & 3 \\ 9 & 0 \end{bmatrix} = \begin{bmatrix} 6 & 10 \\ 3 & -2 \\ -3 & 9 \end{bmatrix}$

(d) $CBA = \begin{bmatrix} 1 & 0 & 4 \\ -1 & 1 & 2 \\ 0 & 1 & 3 \end{bmatrix} \begin{bmatrix} 2 & 4 \\ -1 & 1 \\ 3 & 0 \end{bmatrix} \begin{bmatrix} 2 & 3 \\ 2 & 4 \end{bmatrix} = \begin{bmatrix} 14 & 4 \\ 3 & -3 \\ 8 & 1 \end{bmatrix} \begin{bmatrix} 2 & 3 \\ 2 & 4 \end{bmatrix} = \begin{bmatrix} 36 & 58 \\ 0 & -3 \\ 18 & 28 \end{bmatrix}$

(e) $A = \begin{bmatrix} 2 & 3 \\ 2 & 4 \end{bmatrix} \Leftrightarrow A^{-1} = \frac{1}{8 - 6} \begin{bmatrix} 4 & -3 \\ -2 & 2 \end{bmatrix} = \begin{bmatrix} 2 & -\frac{3}{2} \\ -1 & 1 \end{bmatrix}$

(f) B^{-1} does not exist because B is not a square matrix.   **(g)** $\det(B)$ is not defined because B is not a square matrix.

(h) $\det(C) = \begin{vmatrix} 1 & 0 & 4 \\ -1 & 1 & 2 \\ 0 & 1 & 3 \end{vmatrix} = 1 \begin{vmatrix} 1 & 2 \\ 1 & 3 \end{vmatrix} + 4 \begin{vmatrix} -1 & 1 \\ 0 & 1 \end{vmatrix} = 1 - 4 = -3$

11. $|A| = \begin{vmatrix} 1 & 4 & 1 \\ 0 & 2 & 0 \\ 1 & 0 & 1 \end{vmatrix} = 2 \begin{vmatrix} 1 & 1 \\ 1 & 1 \end{vmatrix} = 0$, $|B| = \begin{vmatrix} 1 & 4 & 0 \\ 0 & 2 & 0 \\ -3 & 0 & 1 \end{vmatrix} = 2 \begin{vmatrix} 1 & 0 \\ -3 & 1 \end{vmatrix} = 2$. Since $|A| = 0$, A does not have an inverse, and

since $|B| \neq 0$, B does have an inverse. $\begin{bmatrix} 1 & 4 & 0 & 1 & 0 & 0 \\ 0 & 2 & 0 & 0 & 1 & 0 \\ -3 & 0 & 1 & 0 & 0 & 1 \end{bmatrix}$ $\xrightarrow{R_3 + 3R_1 \to R_3}$ $\begin{bmatrix} 1 & 4 & 0 & 1 & 0 & 0 \\ 0 & 2 & 0 & 0 & 1 & 0 \\ 0 & 12 & 1 & 3 & 0 & 1 \end{bmatrix}$ $\xrightarrow[R_3 - 6R_2 \to R_3]{R_1 - 2R_2 \to R_1}$

$\begin{bmatrix} 1 & 0 & 0 & 1 & -2 & 0 \\ 0 & 2 & 0 & 0 & 1 & 0 \\ 0 & 0 & 1 & 3 & -6 & 1 \end{bmatrix}$ $\xrightarrow{\frac{1}{2}R_2}$ $\begin{bmatrix} 1 & 0 & 0 & 1 & -2 & 0 \\ 0 & 1 & 0 & 0 & \frac{1}{2} & 0 \\ 0 & 0 & 1 & 3 & -6 & 1 \end{bmatrix}$. Therefore, $B^{-1} = \begin{bmatrix} 1 & -2 & 0 \\ 0 & \frac{1}{2} & 0 \\ 3 & -6 & 1 \end{bmatrix}$.

13. (a) $\dfrac{4x-1}{(x-1)^2\,(x+2)} = \dfrac{A}{x-1} + \dfrac{B}{(x-1)^2} + \dfrac{C}{x+2}$. Thus,

$$4x-1 = A\,(x-1)\,(x+2) + B\,(x+2) + C\,(x-1)^2 = A\left(x^2+x-2\right) + B\,(x+2) + C\left(x^2-2x+1\right)$$

$$= (A+C)\,x^2 + (A+B-2C)\,x + (-2A+2B+C)$$

which leads to the system of equations

$$\begin{cases} A \qquad\;\; + C = 0 \\ A + B - 2C = 4 \\ -2A + 2B + \; C = -1 \end{cases} \Leftrightarrow \begin{cases} A \qquad + C = 0 \\ B - 3C = 4 \\ 2B + 3C = -1 \end{cases} \Leftrightarrow \begin{cases} A \quad + C = 0 \\ B - 3C = 4 \\ 9C = -9 \end{cases}$$

Therefore, $9C = -9 \Leftrightarrow C = -1$, $B - 3\,(-1) = 4 \Leftrightarrow B = 1$, and $A + (-1) = 0 \Leftrightarrow A = 1$. Therefore,

$$\dfrac{4x-1}{(x-1)^2\,(x+2)} = \dfrac{1}{x-1} + \dfrac{1}{(x-1)^2} - \dfrac{1}{x+2}.$$

(b) $\dfrac{2x-3}{x^3+3x} = \dfrac{2x-3}{x\,(x^2+3)} = \dfrac{A}{x} + \dfrac{Bx+C}{x^2+3}$. Then

$$2x-3 = A\left(x^2+3\right) + (Bx+C)\,x = Ax^2 + 3A + Bx^2 + Cx = (A+B)\,x^2 + Cx + 3A.$$

So $3A = -3 \Leftrightarrow A = -1$, $C = 2$ and $A + B = 0$ gives us $B = 1$. Thus $\dfrac{2x-3}{x^3+x} = -\dfrac{1}{x} + \dfrac{x+2}{x^2+3}$.

FOCUS ON MODELING   Linear Programming

1.

Vertex	$M = 200 - x - y$
$(0, 2)$	$200 - (0) - (2) = 198$
$(0, 5)$	$200 - (0) - (5) = 195$
$(4, 0)$	$200 - (4) - (0) = 196$

Thus, the maximum value is 198 and the minimum value is 195.

3. $\begin{cases} x \geq 0,\, y \geq 0 \\ 2x + y \leq 10 \\ 2x + 4y \leq 28 \end{cases}$ The objective function is $P = 140 - x + 3y$. From the graph, the vertices are $(0, 0)$, $(5, 0)$, $(2, 6)$, and $(0, 7)$.

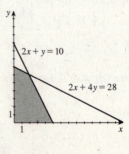

Vertex	$P = 140 - x + 3y$
$(0, 0)$	$140 - (0) + 3\,(0) = 140$
$(5, 0)$	$140 - (5) + 3\,(0) = 135$
$(2, 6)$	$140 - (2) + 3\,(6) = 156$
$(0, 7)$	$140 - (0) + 3\,(7) = 161$

Thus the maximum value is 161, and the minimum value is 135.

5. Let t be the number of tables made daily and c be the number of chairs made daily. Then the data given can be summarized by the following table:

	Tables t	Chairs c	Available time
Carpentry	2 h	3 h	108 h
Finishing	1 h	$\frac{1}{2}$ h	20 h
Profit	$35	$20	

Thus we wish to maximize the total profit $P = 35t + 20c$ subject to the constraints $\begin{cases} 2t + 3c \le 108 \\ t + \frac{1}{2}c \le 20 \\ t \ge 0, c \ge 0 \end{cases}$

From the graph, the vertices occur at $(0, 0)$, $(20, 0)$, $(0, 36)$, and $(3, 34)$.

Vertex	$P = 35t + 20c$
$(0, 0)$	$35\,(0) + 20\,(0) = 0$
$(20, 0)$	$35\,(20) + 20\,(0) = 700$
$(0, 36)$	$35\,(0) + 20\,(36) = 720$
$(3, 34)$	$35\,(3) + 20\,(34) = 785$

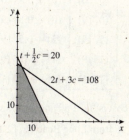

Hence, 3 tables and 34 chairs should be produced daily for a maximum profit of $785.

7. Let x be the number of crates of oranges and y the number of crates of grapefruit. Then the data given can be summarized by the following table:

	Oranges	Grapefruit	Available
Volume	4 ft^3	6 ft^3	300 ft^3
Weight	80 lb	100 lb	5600 lb
Profit	$2.50	$4.00	

In addition, $x \ge y$. Thus we wish to maximize the total profit $P = 2.5x + 4y$ subject to the constraints

$$\begin{cases} x \ge 0, y \ge 0, x \ge y \\ 4x + 6y \le 300 \\ 80x + 100y \le 5600 \end{cases}$$

From the graph, the vertices occur at $(0, 0)$, $(30, 30)$, $(45, 20)$, and $(70, 0)$.

Vertex	$P = 2.5x + 4y$
$(0, 0)$	$2.5\,(0) + 4\,(0) = 0$
$(30, 30)$	$2.5\,(30) + 4\,(30) = 195$
$(45, 20)$	$2.5\,(45) + 4\,(20) = 192.5$
$(70, 0)$	$2.5\,(70) + 4\,(0) = 175$

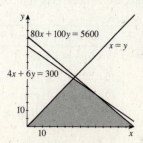

Thus, she should carry 30 crates of oranges and 30 crates of grapefruit for a maximum profit of $195.

9. Let x be the number of stereo sets shipped from Long Beach to Santa Monica and y the number of stereo sets shipped from Long Beach to El Toro. Thus, $15 - x$ sets must be shipped to Santa Monica from Pasadena and $19 - y$ sets to El Toro from Pasadena. Thus, $x \geq 0$, $y \geq 0$, $15 - x \geq 0$, $19 - y \geq 0$, $x + y \leq 24$, and $(15 - x) + (19 - y) \leq 18$. Simplifying, we get

the constraints $\begin{cases} x \geq 0, y \geq 0 \\ x \leq 15, y \leq 19 \\ x + y \leq 24 \\ x + y \geq 16 \end{cases}$

The objective function is the cost $C = 5x + 6y + 4(15 - x) + 5.5(19 - y) = x + 0.5y + 164.5$, which we wish to minimize. From the graph, the vertices occur at $(0, 16)$, $(0, 19)$, $(5, 19)$, $(15, 9)$, and $(15, 1)$.

Vertex	$C = x + 0.5y + 164.5$
$(0, 16)$	$(0) + 0.5(16) + 164.5 = 172.5$
$(0, 19)$	$(0) + 0.5(19) + 164.5 = 174$
$(5, 19)$	$(5) + 0.5(19) + 164.5 = 179$
$(15, 9)$	$(15) + 0.5(9) + 164.5 = 184$
$(15, 1)$	$(15) + 0.5(1) + 164.5 = 180$

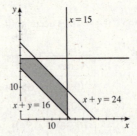

The minimum cost is $172.50 and occurs when $x = 0$ and $y = 16$. Hence, no stereo should be shipped from Long Beach to Santa Monica, 16 from Long Beach to El Toro, 15 from Pasadena to Santa Monica, and 3 from Pasadena to El Toro.

11. Let x be the number of bags of standard mixtures and y be the number of bags of deluxe mixtures. Then the data can be summarized by the following table:

	Standard	Deluxe	Available
Cashews	100 g	150 g	15 kg
Peanuts	200 g	50 g	20 kg
Selling price	$1.95	$2.20	

Thus the total revenue, which we want to maximize, is given by $R = 1.95x + 2.25y$. We have the constraints

$\begin{cases} x \geq 0, y \geq 0, x \geq y \\ 0.1x + 0.15y \leq 15 \\ 0.2x + 0.05y \leq 20 \end{cases}$ $\Leftrightarrow$ $\begin{cases} x \geq 0, y \geq 0, x \geq y \\ 10x + 15y \leq 1500 \\ 20x + 5y \leq 2000 \end{cases}$

From the graph, the vertices occur at $(0, 0)$, $(60, 60)$, $(90, 40)$, and $(100, 0)$.

Vertex	$R = 1.95x + 2.25y$
$(0, 0)$	$1.96(0) + 2.25(0) = 0$
$(60, 60)$	$1.95(60) + 2.25(60) = 252$
$(90, 40)$	$1.95(90) + 2.25(40) = 265.5$
$(100, 0)$	$1.95(100) + 2.25(0) = 195$

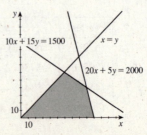

Hence, he should pack 90 bags of standard and 40 bags of deluxe mixture for a maximum revenue of $265.50.

13. Let x be the amount in municipal bonds and y the amount in bank certificates, both in dollars. Then $12000 - x - y$ is the amount in high-risk bonds. So our constraints can be stated as

$$\begin{cases} x \geq 0, y \geq 0, x \geq 3y \\ 12{,}000 - x - y \geq 0 \\ 12{,}000 - x - y \leq 2000 \end{cases} \Leftrightarrow \begin{cases} x \geq 0, y \geq 0, x \geq 3y \\ x + y \leq 12{,}000 \\ x + y \geq 10{,}000 \end{cases}$$

From the graph, the vertices occur at $(7500, 2500)$, $(10000, 0)$, $(12000, 0)$, and $(9000, 3000)$. The objective function is $P = 0.07x + 0.08y + 0.12(12000 - x - y) = 1440 - 0.05x - 0.04y$, which we wish to maximize.

Vertex	$P = 1440 - 0.05x - 0.04y$
$(7500, 2500)$	$1440 - 0.05(7500) - 0.04(2500) = 965$
$(10000, 0)$	$1440 - 0.05(10{,}000) - 0.04(0) = 940$
$(12000, 0)$	$1440 - 0.05(12{,}000) - 0.04(0) = 840$
$(9000, 3000)$	$1440 - 0.05(9000) - 0.04(3000) = 870$

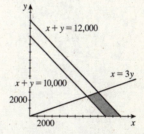

Hence, she should invest $7500 in municipal bonds, $2500 in bank certificates, and the remaining $2000 in high-risk bonds for a maximum yield of $965.

15. Let g be the number of games published and e be the number of educational programs published. Then the number of utility programs published is $36 - g - e$. Hence we wish to maximize profit, $P = 5000g + 8000e + 6000(36 - g - e) = 216{,}000 - 1000g + 2000e$, subject to the constraints

$$\begin{cases} g \geq 4, e \geq 0 \\ 36 - g - e \geq 0 \\ 36 - g - e \leq 2e \end{cases} \Leftrightarrow \begin{cases} g \geq 4, e \geq 0 \\ g + e \leq 36 \\ g + 3e \geq 36. \end{cases}$$

From the graph, the vertices are at $\left(4, \frac{32}{3}\right)$, $(4, 32)$, and $(36, 0)$. The objective function is $P = 216{,}000 - 1000g + 2000e$.

Vertex	$P = 216{,}000 - 1000g + 2000e$
$\left(4, \frac{32}{3}\right)$	$216{,}000 - 1000(4) + 2000\left(\frac{32}{3}\right) = 233{,}333.33$
$(4, 32)$	$216{,}000 - 1000(4) + 2000(32) = 276{,}000$
$(36, 0)$	$216{,}000 - 1000(36) + 2000(0) = 180{,}000$

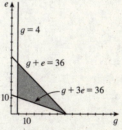

So, they should publish 4 games, 32 educational programs, and no utility program for a maximum profit of $276,000 annually.

11 CONIC SECTIONS

11.1 PARABOLAS

1. A parabola is the set of all points in the plane equidistant from a fixed point called the *focus* and a fixed line called the *directrix* of the parabola.

3. The graph of the equation $y^2 = 4px$ is a parabola with focus $F(p, 0)$ and directrix $x = -p$. So the graph of $y^2 = 12x$ is a parabola with focus $F(3, 0)$ and directrix $x = -3$.

5. $y^2 = 2x$ is Graph III, which opens to the right and is not as wide as the graph for Exercise 5.

7. $x^2 = -6y$ is Graph II, which opens downward and is narrower than the graph for Exercise 6.

9. $y^2 - 8x = 0$ is Graph VI, which opens to the right and is wider than the graph for Exercise 1.

11. $x^2 = 9y$. Then $4p = 9 \Leftrightarrow p = \frac{9}{4}$. The focus is $\left(0, \frac{9}{4}\right)$, the directrix is $y = -\frac{9}{4}$, and the focal diameter is 9.

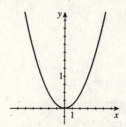

13. $y^2 = 4x$. Then $4p = 4 \Leftrightarrow p = 1$. The focus is $(1, 0)$, the directrix is $x = -1$, and the focal diameter is 4.

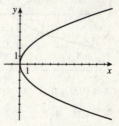

15. $y = 5x^2 \Leftrightarrow x^2 = \frac{1}{5}y$. Then $4p = \frac{1}{5} \Leftrightarrow p = \frac{1}{20}$. The focus is $\left(0, \frac{1}{20}\right)$, the directrix is $y = -\frac{1}{20}$, and the focal diameter is $\frac{1}{5}$.

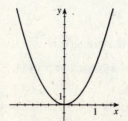

17. $x = -8y^2 \Leftrightarrow y^2 = -\frac{1}{8}x$. Then $4p = -\frac{1}{8} \Leftrightarrow p = -\frac{1}{32}$. The focus is $\left(-\frac{1}{32}, 0\right)$, the directrix is $x = \frac{1}{32}$, and the focal diameter is $\frac{1}{8}$.

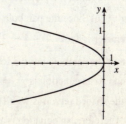

19. $x^2 + 6y = 0 \Leftrightarrow x^2 = -6y$. Then $4p = -6 \Leftrightarrow p = -\frac{3}{2}$.   **21.** $5x + 3y^2 = 0 \Leftrightarrow y^2 = -\frac{5}{3}x$. Then $4p = -\frac{5}{3} \Leftrightarrow$

The focus is $\left(0, -\frac{3}{2}\right)$, the directrix is $y = \frac{3}{2}$, and the focal $p = -\frac{5}{12}$. The focus is $\left(-\frac{5}{12}, 0\right)$, the directrix is $x = \frac{5}{12}$,

diameter is 6.   and the focal diameter is $\frac{5}{3}$.

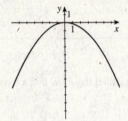   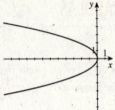

23. $x^2 = 16y$ **25.** $y^2 = -\frac{1}{3}x$

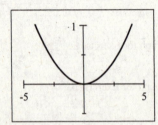   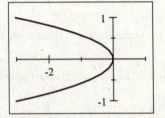

27. $4x + y^2 = 0$

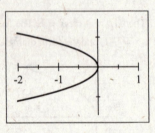

29. Since the focus is $(0, 2)$, $p = 2 \Leftrightarrow 4p = 8$. Hence, an equation of the parabola is $x^2 = 8y$.

31. Since the focus is $(-8, 0)$, $p = -8 \Leftrightarrow 4p = -32$. Hence, an equation of the parabola is $y^2 = -32x$.

33. Since the directrix is $x = 2$, $p = -2 \Leftrightarrow 4p = -8$. Hence, an equation of the parabola is $y^2 = -8x$.

35. Since the directrix is $y = -10$, $p = 10 \Leftrightarrow 4p = 40$. Hence, an equation of the parabola is $x^2 = 40y$.

37. The focus is on the positive x-axis, so the parabola opens horizontally with $2p = 2 \Leftrightarrow 4p = 4$. So an equation of the parabola is $y^2 = 4x$.

39. Since the parabola opens upward with focus 5 units from the vertex, the focus is $(5, 0)$. So $p = 5 \Leftrightarrow 4p = 20$. Thus an equation of the parabola is $x^2 = 20y$.

41. $p = 2 \Leftrightarrow 4p = 8$. Since the parabola opens upward, its equation is $x^2 = 8y$.

43. $p = 4 \Leftrightarrow 4p = 16$. Since the parabola opens to the left, its equation is $y^2 = -16x$.

45. The focal diameter is $4p = \frac{3}{2} + \frac{3}{2} = 3$. Since the parabola opens to the left, its equation is $y^2 = -3x$.

47. The equation of the parabola has the form $y^2 = 4px$. Since the parabola passes through the point $(4, -2)$, $(-2)^2 = 4p(4)$ $\Leftrightarrow 4p = 1$, and so an equation is $y^2 = x$.

49. The area of the shaded region is width $\times$ height $= 4p \cdot p = 8$, and so $p^2 = 2 \Leftrightarrow p = -\sqrt{2}$ (because the parabola opens downward). Therefore, an equation is $x^2 = 4py = -4\sqrt{2}y \Leftrightarrow x^2 = -4\sqrt{2}y$.

51. (a) A parabola with directrix $y = -p$ has equation $x^2 = 4py$. If the directrix

is $y = \frac{1}{2}$, then $p = -\frac{1}{2}$, so an equation is $x^2 = 4\left(-\frac{1}{2}\right)y \Leftrightarrow x^2 = -2y$.

If the directrix is $y = 1$, then $p = -1$, so an equation is $x^2 = 4(-1)y \Leftrightarrow$

$x^2 = -4y$. If the directrix is $y = 4$, then $p = -4$, so an equation is

$x^2 = 4(-4)y \Leftrightarrow x^2 = -16y$. If the directrix is $y = 8$, then $p = -8$, so

an equation is $x^2 = 4(-8)y \Leftrightarrow x^2 = -32y$.

(b)

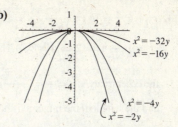

As the directrix moves further from
the vertex, the parabolas get flatter.

53. (a) Since the focal diameter is 12 cm, $4p = 12$. Hence, the parabola has equation $y^2 = 12x$.

  (b) At a point 20 cm horizontally from the vertex, the parabola passes through the point $(20, y)$, and hence from part (a),
$y^2 = 12(20) \Leftrightarrow y^2 = 240 \Leftrightarrow y = \pm 4\sqrt{15}$. Thus, $|CD| = 8\sqrt{15} \approx 31$ cm.

55. With the vertex at the origin, the top of one tower will be at the point $(300, 150)$. Inserting this point into the equation
$x^2 = 4py$ gives $(300)^2 = 4p(150) \Leftrightarrow 90000 = 600p \Leftrightarrow p = 150$. So an equation of the parabolic part of the cables is
$x^2 = 4(150)y \Leftrightarrow x^2 = 600y$.

57. Many answers are possible: satellite dish TV antennas, sound surveillance equipment, solar collectors for hot water heating
or electricity generation, bridge pillars, etc.

11.2 ELLIPSES

1. An ellipse is the set of all points in the plane for which the *sum* of the distances from two fixed points F_1 and F_2 is constant.
The points F_1 and F_2 are called the *foci* of the ellipse.

3. The graph of the equation $\dfrac{x^2}{b^2} + \dfrac{y^2}{a^2} = 1$ with $a > b > 0$ is an ellipse with vertices $(0, a)$ and $(0, -a)$ and foci $(0, \pm c)$,

where $c = \sqrt{a^2 - b^2}$. So the graph of $\dfrac{x^2}{4^2} + \dfrac{y^2}{5^2} = 1$ is an ellipse with vertices $(0, 5)$ and $(0, -5)$ and foci $(0, 3)$ and

$(0, -3)$.

5. $\dfrac{x^2}{16} + \dfrac{y^2}{4} = 1$ is Graph II. The major axis is horizontal and the vertices are $(\pm 4, 0)$.

7. $4x^2 + y^2 = 4$ is Graph I. The major axis is vertical and the vertices are $(0, \pm 2)$.

9. $\dfrac{x^2}{25} + \dfrac{y^2}{9} = 1$. This ellipse has $a = 5$, $b = 3$, and so $c^2 = a^2 - b^2 = 16 \Leftrightarrow c = 4$.

The vertices are $(\pm 5, 0)$, the foci are $(\pm 4, 0)$, the eccentricity is $e = \dfrac{c}{a} = \dfrac{4}{5} = 0.8$,

the length of the major axis is $2a = 10$, and the length of the minor axis is $2b = 6$.

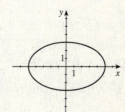

11. $9x^2 + 4y^2 = 36 \Leftrightarrow \dfrac{x^2}{4} + \dfrac{y^2}{9} = 1$. This ellipse has $a = 3$, $b = 2$, and so

$c^2 = 9 - 4 = 5 \Leftrightarrow c = \sqrt{5}$. The vertices are $(0, \pm 3)$, the foci are $\left(0, \pm\sqrt{5}\right)$, the

eccentricity is $e = \dfrac{c}{a} = \dfrac{\sqrt{5}}{3}$, the length of the major axis is $2a = 6$, and the length

of the minor axis is $2b = 4$.

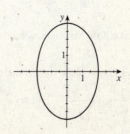

13. $x^2 + 4y^2 = 16 \Leftrightarrow \dfrac{x^2}{16} + \dfrac{y^2}{4} = 1$. This ellipse has $a = 4$, $b = 2$, and so

$c^2 = 16 - 4 = 12 \Leftrightarrow c = 2\sqrt{3}$. The vertices are $(\pm 4, 0)$, the foci are $\left(\pm 2\sqrt{3}, 0\right)$,

the eccentricity is $e = \dfrac{c}{a} = \dfrac{2\sqrt{3}}{4} = \dfrac{\sqrt{3}}{2}$, the length of the major axis is $2a = 8$,

and the length of the minor axis is $2b = 4$.

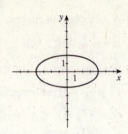

15. $2x^2 + y^2 = 3 \Leftrightarrow \dfrac{x^2}{\frac{3}{2}} + \dfrac{y^2}{3} = 1$. This ellipse has $a = \sqrt{3}$, $b = \sqrt{\frac{3}{2}}$, and so

$c^2 = 3 - \frac{3}{2} = \frac{3}{2} \Leftrightarrow c = \sqrt{\frac{3}{2}} = \dfrac{\sqrt{6}}{2}$. The vertices are $\left(0, \pm\sqrt{3}\right)$, the foci are

$\left(0, \pm\frac{\sqrt{6}}{2}\right)$, the eccentricity is $e = \dfrac{c}{a} = \dfrac{\frac{\sqrt{6}}{2}}{\sqrt{3}} = \dfrac{\sqrt{2}}{2}$, the length of the major axis is

$2a = 2\sqrt{3}$, and the length of the minor axis is $2b = 2 \cdot \dfrac{\sqrt{6}}{2} = \sqrt{6}$.

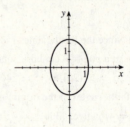

17. $x^2 + 4y^2 = 1 \Leftrightarrow \dfrac{x^2}{1} + \dfrac{y^2}{\frac{1}{4}} = 1$. This ellipse has $a = 1$, $b = \frac{1}{2}$, and so

$c^2 = 1 - \frac{1}{4} = \frac{3}{4} \Leftrightarrow c = \dfrac{\sqrt{3}}{2}$. The vertices are $(\pm 1, 0)$, the foci are $\left(\pm\frac{\sqrt{3}}{2}, 0\right)$, the

eccentricity is $e = \dfrac{c}{a} = \dfrac{\sqrt{3}/2}{1} = \dfrac{\sqrt{3}}{2}$, the length of the major axis is $2a = 2$, and

the length of the minor axis is $2b = 1$.

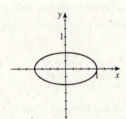

19. $\frac{1}{2}x^2 + \frac{1}{8}y^2 = \frac{1}{4} \Leftrightarrow 2x^2 + \frac{1}{2}y^2 = 1 \Leftrightarrow \dfrac{x^2}{\frac{1}{2}} + \dfrac{y^2}{2} = 1$. This ellipse has $a = \sqrt{2}$,

$b = \dfrac{1}{\sqrt{2}}$, and so $c^2 = 2 - \frac{1}{2} = \frac{3}{2} \Leftrightarrow c = \sqrt{\frac{3}{2}} = \dfrac{\sqrt{6}}{2}$. The vertices are $\left(0, \pm\sqrt{2}\right)$,

the foci are $\left(0, \pm\frac{\sqrt{6}}{2}\right)$, the eccentricity is $e = \dfrac{c}{a} = \dfrac{\frac{\sqrt{6}}{2}}{\sqrt{2}} = \dfrac{\sqrt{3}}{2}$, the length of the

major axis is $2a = 2\sqrt{2}$, and length of the minor axis is $2b = \sqrt{2}$.

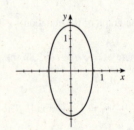

21. $y^2 = 1 - 2x^2 \Leftrightarrow 2x^2 + y^2 = 1 \Leftrightarrow \dfrac{x^2}{\frac{1}{2}} + \dfrac{y^2}{1} = 1$. This ellipse has $a = 1$, $b = \dfrac{\sqrt{2}}{2}$,

and so $c^2 = 1 - \frac{1}{2} = \frac{1}{2} \Leftrightarrow c = \dfrac{\sqrt{2}}{2}$. The vertices are $(0, \pm 1)$, the foci are

$\left(0, \pm\frac{\sqrt{2}}{2}\right)$, the eccentricity is $e = \dfrac{c}{a} = \dfrac{1/\sqrt{2}}{1} = \dfrac{\sqrt{2}}{2}$, the length of the major axis is

$2a = 2$, and the length of the minor axis is $2b = \sqrt{2}$.

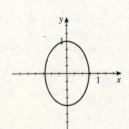

23. This ellipse has a horizontal major axis with $a = 5$ and $b = 4$, so an equation is $\dfrac{x^2}{(5)^2} + \dfrac{y^2}{(4)^2} = 1 \Leftrightarrow \dfrac{x^2}{25} + \dfrac{y^2}{16} = 1$.

25. This ellipse has a vertical major axis with $c = 2$ and $b = 2$. So $a^2 = c^2 + b^2 = 2^2 + 2^2 = 8 \Leftrightarrow a = 2\sqrt{2}$. So an equation

is $\dfrac{x^2}{(2)^2} + \dfrac{y^2}{\left(2\sqrt{2}\right)^2} = 1 \Leftrightarrow \dfrac{x^2}{4} + \dfrac{y^2}{8} = 1$.

27. This ellipse has a horizontal major axis with $a = 16$, so an equation of the ellipse is of the form $\dfrac{x^2}{16^2} + \dfrac{y^2}{b^2} = 1$. Substituting

the point $(8, 6)$ into the equation, we get $\dfrac{64}{256} + \dfrac{36}{b^2} = 1 \Leftrightarrow \dfrac{36}{b^2} = 1 - \dfrac{1}{4} \Leftrightarrow \dfrac{36}{b^2} = \dfrac{3}{4} \Leftrightarrow b^2 = \dfrac{4\,(36)}{3} = 48$. Thus, an

equation of the ellipse is $\dfrac{x^2}{256} + \dfrac{y^2}{48} = 1$.

29. $\dfrac{x^2}{25} + \dfrac{y^2}{20} = 1 \Leftrightarrow \dfrac{y^2}{20} = 1 - \dfrac{x^2}{25} \Leftrightarrow y^2 = 20 - \dfrac{4x^2}{5} \Rightarrow$ **31.** $6x^2 + y^2 = 36 \Leftrightarrow y^2 = 36 - 6x^2 \Rightarrow y = \pm\sqrt{36 - 6x^2}$.

$y = \pm\sqrt{20 - \dfrac{4x^2}{5}}$.

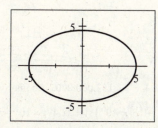

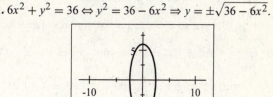

33. The foci are $(\pm 4, 0)$, and the vertices are $(\pm 5, 0)$. Thus, $c = 4$ and $a = 5$, and so $b^2 = 25 - 16 = 9$. Therefore, an equation

of the ellipse is $\dfrac{x^2}{25} + \dfrac{y^2}{9} = 1$.

35. The length of the major axis is $2a = 4 \Leftrightarrow a = 2$, the length of the minor axis is $2b = 2 \Leftrightarrow b = 1$, and the foci are on the

y-axis. Therefore, an equation of the ellipse is $x^2 + \dfrac{y^2}{4} = 1$.

37. The foci are $(0, \pm 2)$, and the length of the minor axis is $2b = 6 \Leftrightarrow b = 3$. Thus, $a^2 = 4 + 9 = 13$. Since the foci are on the

y-axis, an equation is $\dfrac{x^2}{9} + \dfrac{y^2}{13} = 1$.

39. The endpoints of the major axis are $(\pm 10, 0) \Leftrightarrow a = 10$, and the distance between the foci is $2c = 6 \Leftrightarrow c = 3$. Therefore,

$b^2 = 100 - 9 = 91$, and so an equation of the ellipse is $\dfrac{x^2}{100} + \dfrac{y^2}{91} = 1$.

41. The length of the major axis is 10, so $2a = 10 \Leftrightarrow a = 5$, and the foci are on the x-axis, so the form of the equation is

$\dfrac{x^2}{25} + \dfrac{y^2}{b^2} = 1$. Since the ellipse passes through $\left(\sqrt{5}, 2\right)$, we know that $\dfrac{\left(\sqrt{5}\right)^2}{25} + \dfrac{(2)^2}{b^2} = 1 \Leftrightarrow \dfrac{5}{25} + \dfrac{4}{b^2} = 1 \Leftrightarrow \dfrac{4}{b^2} = \dfrac{4}{5} \Leftrightarrow$

$b^2 = 5$, and so an equation is $\dfrac{x^2}{25} + \dfrac{y^2}{5} = 1$.

43. Since the foci are $(\pm 1.5, 0)$, we have $c = \dfrac{3}{2}$. Since the eccentricity is $0.8 = \dfrac{c}{a}$, we have $a = \dfrac{\frac{3}{2}}{\frac{4}{5}} = \dfrac{15}{8}$, and so

$b^2 = \dfrac{225}{64} - \dfrac{9}{4} = \dfrac{225 - 16 \cdot 9}{64} = \dfrac{81}{64}$. Therefore, an equation of the ellipse is $\dfrac{x^2}{(15/8)^2} + \dfrac{y^2}{81/64} = 1 \Leftrightarrow \dfrac{64x^2}{225} + \dfrac{64y^2}{81} = 1$.

45. $\begin{cases} 4x^2 + y^2 = 4 \\ 4x^2 + 9y^2 = 36 \end{cases}$ Subtracting the first equation from the second gives

$8y^2 = 32 \Leftrightarrow y^2 = 4 \Leftrightarrow y = \pm 2$. Substituting $y = \pm 2$ in the first equation gives

$4x^2 + (\pm 2)^2 = 4 \Leftrightarrow x = 0$, and so the points of intersection are $(0, \pm 2)$.

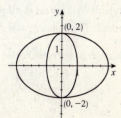

47. $\begin{cases} 100x^2 + 25y^2 = 100 \\ x^2 + \dfrac{y^2}{9} = 1 \end{cases}$ Dividing the first equation by 100 gives $x^2 + \dfrac{y^2}{4} = 1$.

Subtracting this equation from the second equation gives $\dfrac{y^2}{9} - \dfrac{y^2}{4} = 0 \Leftrightarrow$

$\left(\dfrac{1}{9} - \dfrac{1}{4}\right) y^2 = 0 \Leftrightarrow y = 0$. Substituting $y = 0$ in the second equation gives

$x^2 + (0)^2 = 1 \Leftrightarrow x = \pm 1$, and so the points of intersection are $(\pm 1, 0)$.

49. (a) $x^2 + ky^2 = 100 \Leftrightarrow ky^2 = 100 - x^2 \Leftrightarrow y = \pm \dfrac{1}{k}\sqrt{100 - x^2}$. For the top

half, we graph $y = \dfrac{1}{k}\sqrt{100 - x^2}$ for $k = 4, 10, 25,$ and 50.

(b) This family of ellipses have common major axes and vertices, and the
eccentricity increases as k increases.

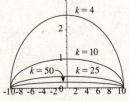

51. Using the perihelion, $a - c = 147{,}000{,}000$, while using the aphelion, $a + c = 153{,}000{,}000$. Adding, we have

$2a = 300{,}000{,}000 \Leftrightarrow a = 150{,}000{,}000$. So $b^2 = a^2 - c^2 = \left(150 \times 10^6\right)^2 - \left(3 \times 10^6\right)^2 = 22{,}491 \times 10^{12} = 2.2491 \times 10^{16}$.

Thus, an equation of the orbit is $\dfrac{x^2}{2.2500 \times 10^{16}} + \dfrac{y^2}{2.2491 \times 10^{16}} = 1$.

53. Using the perilune, $a - c = 1075 + 68 = 1143$, and using the apolune, $a + c = 1075 + 195 = 1270$. Adding, we get

$2a = 2413 \Leftrightarrow a = 1206.5$. So $c = 1270 - 1206.5 \Leftrightarrow c = 63.5$. Therefore, $b^2 = (1206.5)^2 - (63.5)^2 = 1{,}451{,}610$. Since

$a^2 \approx 1{,}455{,}642$, an equation of Apollo 11's orbit is $\dfrac{x^2}{1{,}455{,}642} + \dfrac{y^2}{1{,}451{,}610} = 1$.

55. From the diagram, $a = 40$ and $b = 20$, and so an equation of the ellipse whose top half is the window is $\dfrac{x^2}{1600} + \dfrac{y^2}{400} = 1$.

Since the ellipse passes through the point $(25, h)$, by substituting, we have $\dfrac{25^2}{1600} + \dfrac{h^2}{400} = 1 \Leftrightarrow 625 + 4y^2 = 1600 \Leftrightarrow$

$y = \dfrac{\sqrt{975}}{2} = \dfrac{5\sqrt{39}}{2} \approx 15.61$ in. Therefore, the window is approximately 15.6 inches high at the specified point.

57. We start with the flashlight perpendicular to the wall; this shape is a circle. As the angle of elevation increases, the shape of
the light changes to an ellipse. When the flashlight is angled so that the outer edge of the light cone is parallel to the wall,
the shape of the light is a parabola. Finally, as the angle of elevation increases further, the shape of the light is hyperbolic.

59. The shape drawn on the paper is almost, but not quite, an ellipse. For example, when the bottle has radius 1 unit and the
compass legs are set 1 unit apart, then it can be shown that an equation of the resulting curve is $1 + y^2 = 2\cos x$. The
graph of this curve differs very slightly from the ellipse with the same major and minor axis. This example shows that in
mathematics, things are not always as they appear to be.

11.3 HYPERBOLAS

1. A hyperbola is the set of all points in the plane for which the *difference* of the distances from two fixed point F_1 and F_2 is
constant. The points F_1 and F_2 are called the *foci* of the hyperbola.

3. The graph of the equation $\dfrac{y^2}{a^2} - \dfrac{x^2}{b^2} = 1$ with $a > 0, b > 0$ is a hyperbola with vertices $(0, a)$ and $(0, -a)$ and foci $(0, \pm c)$,

where $c = \sqrt{a^2 + b^2}$. So the graph of $\dfrac{y^2}{4^2} - \dfrac{x^2}{3^2} = 1$ is a hyperbola with vertices $(0, 4)$ and $(0, -4)$ and foci $(0, 5)$ and

$(0, -5)$.

5. $\dfrac{x^2}{4} - y^2 = 1$ is Graph III, which opens horizontally and has vertices at $(\pm 2, 0)$.

7. $16y^2 - x^2 = 144$ is Graph II, which pens vertically and has vertices at $(0, \pm 3)$.

9. The hyperbola $\dfrac{x^2}{4} - \dfrac{y^2}{16} = 1$ has $a = 2$, $b = 4$, and $c^2 = 16 + 4 \Rightarrow c = 2\sqrt{5}$. The vertices are $(\pm 2, 0)$, the foci are $\left(\pm 2\sqrt{5}, 0\right)$, and the asymptotes are $y = \pm\frac{4}{2}x \Leftrightarrow y = \pm 2x$.

11. The hyperbola $\dfrac{y^2}{1} - \dfrac{x^2}{25} = 1$ has $a = 1$, $b = 5$, and $c^2 = 1 + 25 = 26 \Rightarrow c = \sqrt{26}$. The vertices are $(0, \pm 1)$, the foci are $\left(0, \pm\sqrt{26}\right)$, and the asymptotes are $y = \pm\frac{1}{5}x$.

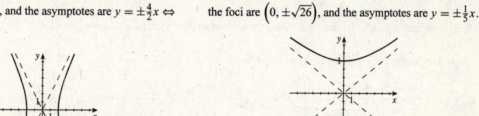

13. The hyperbola $x^2 - y^2 = 1$ has $a = 1$, $b = 1$, and $c^2 = 1 + 1 = 2 \Rightarrow c = \sqrt{2}$. The vertices are $(\pm 1, 0)$, the foci are $\left(\pm\sqrt{2}, 0\right)$, and the asymptotes are $y = \pm x$.

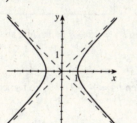

15. The hyperbola $25y^2 - 9x^2 = 225 \Leftrightarrow \dfrac{y^2}{9} - \dfrac{x^2}{25} = 1$ has $a = 3$, $b = 5$, and $c^2 = 25 + 9 = 34 \Rightarrow c = \sqrt{34}$. The vertices are $(0, \pm 3)$, the foci are $\left(0, \pm\sqrt{34}\right)$, and the asymptotes are $y = \pm\frac{3}{5}x$.

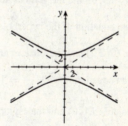

17. The hyperbola $x^2 - 4y^2 - 8 = 0 \Leftrightarrow \dfrac{x^2}{8} - \dfrac{y^2}{2} = 1$ has $a = \sqrt{8}$, $b = \sqrt{2}$, and $c^2 = 8 + 2 = 10 \Rightarrow c = \sqrt{10}$. The vertices are $\left(\pm 2\sqrt{2}, 0\right)$, the foci are $\left(\pm\sqrt{10}, 0\right)$, and the asymptotes are $y = \pm\dfrac{\sqrt{2}}{\sqrt{8}}x = \pm\frac{1}{2}x$.

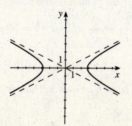

19. The hyperbola $4y^2 - x^2 = 1 \Leftrightarrow \dfrac{y^2}{\frac{1}{4}} - x^2 = 1$ has $a = \frac{1}{2}$, $b = 1$, and $c^2 = \frac{1}{4} + 1 = \frac{5}{4} \Rightarrow c = \dfrac{\sqrt{5}}{2}$. The vertices are $\left(0, \pm\frac{1}{2}\right)$, the foci are $\left(0, \pm\dfrac{\sqrt{5}}{2}\right)$, and the asymptotes are $y = \pm\dfrac{1/2}{1}x = \pm\frac{1}{2}x$.

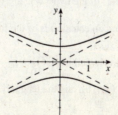

21. From the graph, the foci are $(\pm 4, 0)$, and the vertices are $(\pm 2, 0)$, so $c = 4$ and $a = 2$. Thus, $b^2 = 16 - 4 = 12$, and since

the vertices are on the x-axis, an equation of the hyperbola is $\dfrac{x^2}{4} - \dfrac{y^2}{12} = 1$.

23. From the graph, the vertices are $(0, \pm 4)$, the foci are on the y-axis, and the hyperbola passes through the point $(3, -5)$. So

the equation is of the form $\dfrac{y^2}{16} - \dfrac{x^2}{b^2} = 1$. Substituting the point $(3, -5)$, we have $\dfrac{(-5)^2}{16} - \dfrac{(3)^2}{b^2} = 1 \Leftrightarrow \dfrac{25}{16} - 1 = \dfrac{9}{b^2} \Leftrightarrow$

$\dfrac{9}{16} = \dfrac{9}{b^2} \Leftrightarrow b^2 = 16$. Thus, an equation of the hyperbola is $\dfrac{y^2}{16} - \dfrac{x^2}{16} = 1$.

25. The vertices are $(\pm 3, 0)$, so $a = 3$. Since the asymptotes are $y = \pm \frac{1}{2}x = \pm \dfrac{b}{a}x$, we have $\dfrac{b}{3} = \dfrac{1}{2} \Leftrightarrow b = \dfrac{3}{2}$. Since the

vertices are on the x-axis, an equation is $\dfrac{x^2}{3^2} - \dfrac{y^2}{(3/2)^2} = 1 \Leftrightarrow \dfrac{x^2}{9} - \dfrac{4y^2}{9} = 1$.

27. $x^2 - 2y^2 = 8 \Leftrightarrow 2y^2 = x^2 - 8 \Leftrightarrow y^2 = \frac{1}{2}x^2 - 4 \Rightarrow$

$y = \pm\sqrt{\frac{1}{2}x^2 - 4}$

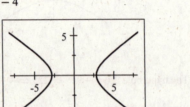

29. $\dfrac{y^2}{2} - \dfrac{x^2}{6} = 1 \Leftrightarrow \dfrac{y^2}{2} = \dfrac{x^2}{6} + 1 \Leftrightarrow y^2 = \dfrac{x^2}{3} + 2 \Rightarrow$

$y = \pm\sqrt{\dfrac{x^2}{3} + 2}$

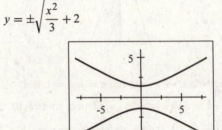

31. The foci are $(\pm 5, 0)$ and the vertices are $(\pm 3, 0)$, so $c = 5$ and $a = 3$. Then $b^2 = 25 - 9 = 16$, and since the vertices are on

the x-axis, an equation of the hyperbola is $\dfrac{x^2}{9} - \dfrac{y^2}{16} = 1$.

33. The foci are $(0, \pm 2)$ and the vertices are $(0, \pm 1)$, so $c = 2$ and $a = 1$. Then $b^2 = 4 - 1 = 3$, and since the vertices are on

the y-axis, an equation is $y^2 - \dfrac{x^2}{3} = 1$.

35. The vertices are $(\pm 1, 0)$ and the asymptotes are $y = \pm 5x$, so $a = 1$. The asymptotes are $y = \pm \dfrac{b}{a}x$, so $\dfrac{b}{1} = 5 \Leftrightarrow b = 5$.

Therefore, an equation of the hyperbola is $x^2 - \dfrac{y^2}{25} = 1$.

37. The foci are $(0, \pm 8)$, and the asymptotes are $y = \pm \frac{1}{2}x$, so $c = 8$. The asymptotes are $y = \pm \dfrac{a}{b}x$, so $\dfrac{a}{b} = \dfrac{1}{2}$ and $b = 2a$.

Since $a^2 + b^2 = c^2 = 64$, we have $a^2 + 4a^2 = 64 \Leftrightarrow a^2 = \frac{64}{5}$ and $b^2 = 4a^2 = \frac{256}{5}$. Thus, an equation of the hyperbola is

$\dfrac{y^2}{64/5} - \dfrac{x^2}{256/5} = 1 \Leftrightarrow \dfrac{5y^2}{64} - \dfrac{5x^2}{256} = 1$.

39. The asymptotes of the hyperbola are $y = \pm x$, so $b = a$. Since the hyperbola passes through the point $(5, 3)$, its foci are on

the x-axis, and an equation is of the form, $\dfrac{x^2}{a^2} - \dfrac{y^2}{a^2} = 1$, so it follows that $\dfrac{25}{a^2} - \dfrac{9}{a^2} = 1 \Leftrightarrow a^2 = 16 = b^2$. Therefore, an

equation of the hyperbola is $\dfrac{x^2}{16} - \dfrac{y^2}{16} = 1$.

41. The foci are $(\pm 5, 0)$, and the length of the transverse axis is 6, so $c = 5$ and $2a = 6 \Leftrightarrow a = 3$. Thus, $b^2 = 25 - 9 = 16$, and

an equation is $\dfrac{x^2}{9} - \dfrac{y^2}{16} = 1$.

43. (a) The hyperbola $x^2 - y^2 = 5 \Leftrightarrow \dfrac{x^2}{5} - \dfrac{y^2}{5} = 1$ has $a = \sqrt{5}$ and $b = \sqrt{5}$. Thus, the asymptotes are $y = \pm x$, and their slopes are $m_1 = 1$ and $m_2 = -1$. Since $m_1 \cdot m_2 = -1$, the asymptotes are perpendicular.

(b) Since the asymptotes are perpendicular, they must have slopes ± 1, so $a = b$. Therefore, $c^2 = 2a^2 \Leftrightarrow a^2 = \dfrac{c^2}{2}$, and

since the vertices are on the x-axis, an equation is $\dfrac{x^2}{\frac{1}{2}c^2} - \dfrac{y^2}{\frac{1}{2}c^2} = 1 \Leftrightarrow x^2 - y^2 = \dfrac{c^2}{2}$.

45. $\sqrt{(x+c)^2 + y^2} - \sqrt{(x-c)^2 + y^2} = \pm 2a$.   Let us consider the positive case only.   Then

$\sqrt{(x+c)^2 + y^2} = 2a + \sqrt{(x-c)^2 + y^2}$, and squaring both sides gives $x^2 + 2cx + c^2 + y^2 = 4a^2 +$

$4a\sqrt{(x-c)^2 + y^2} + x^2 - 2cx + c^2 + y^2 \Leftrightarrow 4a\sqrt{(x-c)^2 + y^2} = 4cx - 4a^2$. Dividing by 4 and squaring both sides

gives $a^2\left(x^2 - 2cx + c^2 + y^2\right) = c^2x^2 - 2a^2cx + a^4 \Leftrightarrow a^2x^2 - 2a^2cx + a^2c^2 + a^2y^2 = c^2x^2 - 2a^2cx + a^4$

$\Leftrightarrow a^2x^2 + a^2c^2 + a^2y^2 = c^2x^2 + a^4$. Rearranging the order, we have $c^2x^2 - a^2x^2 - a^2y^2 = a^2c^2 - a^4 \Leftrightarrow$

$\left(c^2 - a^2\right)x^2 - a^2y^2 = a^2\left(c^2 - a^2\right)$. The negative case gives the same result.

47. (a) From the equation, we have $a^2 = k$ and $b^2 = 16 - k$. Thus, $c^2 = a^2 + b^2 = k + 16 - k = 16 \Rightarrow c = \pm 4$. Thus the foci of the family of hyperbolas are $(0, \pm 4)$.

(b) $\dfrac{y^2}{k} - \dfrac{x^2}{16 - k} = 1 \Leftrightarrow y^2 = k\left(1 + \dfrac{x^2}{16 - k}\right) \Rightarrow y = \pm\sqrt{k + \dfrac{kx^2}{16 - k}}$. For

the top branch, we graph $y = \sqrt{k + \dfrac{kx^2}{16 - k}}$, $k = 1, 4, 8, 12$. As k

increases, the asymptotes get steeper and the vertices move further apart.

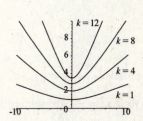

49. Since the asymptotes are perpendicular, $a = b$. Also, since the sun is a focus and the closest distance is 2×10^9, it follows

that $c - a = 2 \times 10^9$. Now $c^2 = a^2 + b^2 = 2a^2$, and so $c = \sqrt{2}a$. Thus, $\sqrt{2}a - a = 2 \times 10^9 \Rightarrow a = \dfrac{2 \times 10^9}{\sqrt{2} - 1}$ and

$a^2 = b^2 = \dfrac{4 \times 10^{18}}{3 - 2\sqrt{2}} \approx 2.3 \times 10^{19}$. Therefore, an equation of the hyperbola is $\dfrac{x^2}{2.3 \times 10^{19}} - \dfrac{y^2}{2.3 \times 10^{19}} = 1 \Leftrightarrow$

$x^2 - y^2 = 2.3 \times 10^{19}$.

51. Some possible answers are: as cross-sections of nuclear power plant cooling towers, or as reflectors for camouflaging the location of secret installations.

11.4   SHIFTED CONICS

1. (a) If we replace x by $x - 3$ the graph of the equation is shifted to the *right* by 3 units. If we replace x by $x + 3$ the graph is shifted to the *left* by 3 units.

(b) If we replace y by $y - 1$ the graph of the equation is shifted *upward* by 1 unit. If we replace y by $y + 1$ the graph is shifted *downward* by 1 unit.

3. $\dfrac{x^2}{5^2} + \dfrac{y^2}{4^2} = 1$, from left to right: vertex $(-5, 0)$, focus $(-3, 0)$, focus $(3, 0)$, vertex $(5, 0)$. $\dfrac{(x-3)^2}{5^2} + \dfrac{(y-1)^2}{4^2} = 1$, from left to right: vertex $(-2, 1)$, focus $(0, 1)$, focus $(6, 1)$, vertex $(8, 1)$.

5. The ellipse $\dfrac{(x-2)^2}{9} + \dfrac{(y-1)^2}{4} = 1$ is obtained from the ellipse $\dfrac{x^2}{9} + \dfrac{y^2}{4} = 1$

by shifting it 2 units to the right and 1 unit upward. So $a = 3$, $b = 2$, and

$c = \sqrt{9-4} = \sqrt{5}$. The center is $(2, 1)$, the foci are $\left(2 \pm \sqrt{5}, 1\right)$, the vertices are

$(2 \pm 3, 1) = (-1, 1)$ and $(5, 1)$, the length of the major axis is $2a = 6$, and the

length of the minor axis is $2b = 4$.

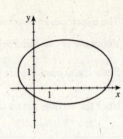

7. The ellipse $\dfrac{x^2}{9} + \dfrac{(y+5)^2}{25} = 1$ is obtained from the ellipse $\dfrac{x^2}{9} + \dfrac{y^2}{25} = 1$ by

shifting it 5 units downward. So $a = 5$, $b = 3$, and $c = \sqrt{25-9} = 4$. The center

is $(0, -5)$, the foci are $(0, -5 \pm 4) = (0, -9)$ and $(0, -1)$, the vertices are

$(0, -5 \pm 5) = (0, -10)$ and $(0, 0)$, the length of the major axis is $2a = 10$, and the

length of the minor axis is $2b = 6$.

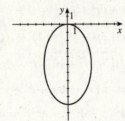

9. The parabola $(x-3)^2 = 8(y+1)$ is obtained from the parabola $x^2 = 8y$ by

shifting it 3 units to the right and 1 unit down. So $4p = 8 \Leftrightarrow p = 2$. The vertex is

$(3, -1)$, the focus is $(3, -1+2) = (3, 1)$, and the directrix is $y = -1 - 2 = -3$.

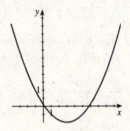

11. The parabola $-4\left(x + \tfrac{1}{2}\right)^2 = y \Leftrightarrow \left(x + \tfrac{1}{2}\right)^2 = -\tfrac{1}{4}y$ is obtained from the

parabola $x^2 = -\tfrac{1}{4}y$ by shifting it $\tfrac{1}{2}$ unit to the left. So $4p = -\tfrac{1}{4} \Leftrightarrow p = -\tfrac{1}{16}$.

The vertex is $\left(-\tfrac{1}{2}, 0\right)$, the focus is $\left(-\tfrac{1}{2}, 0 - \tfrac{1}{16}\right) = \left(-\tfrac{1}{2}, -\tfrac{1}{16}\right)$, and the

directrix is $y = 0 + \tfrac{1}{16} = \tfrac{1}{16}$.

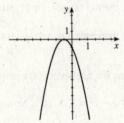

13. The hyperbola $\dfrac{(x+1)^2}{9} - \dfrac{(y-3)^2}{16} = 1$ is obtained from the hyperbola

$\dfrac{x^2}{9} - \dfrac{y^2}{16} = 1$ by shifting it 1 unit to the left and 3 units up. So $a = 3$, $b = 4$, and

$c = \sqrt{9+16} = 5$. The center is $(-1, 3)$, the foci are $(-1 \pm 5, 3) = (-6, 3)$ and

$(4, 3)$, the vertices are $(-1 \pm 3, 3) = (-4, 3)$ and $(2, 3)$, and the asymptotes are

$(y - 3) = \pm\tfrac{4}{3}(x+1) \Leftrightarrow y = \pm\tfrac{4}{3}(x+1) + 3 \Leftrightarrow 3y = 4x + 13$ and

$3y = -4x + 5$.

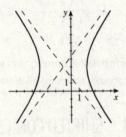

15. The hyperbola $y^2 - \dfrac{(x+1)^2}{4} = 1$ is obtained from the hyperbola $y^2 - \dfrac{x^2}{4} = 1$ by

shifting it 1 unit to the left. So $a = 1$, $b = 2$, and $c = \sqrt{1+4} = \sqrt{5}$. The center is

$(-1, 0)$, the foci are $\left(-1, \pm\sqrt{5}\right) = \left(-1, -\sqrt{5}\right)$ and $\left(-1, \sqrt{5}\right)$, the vertices are

$(-1, \pm 1) = (-1, -1)$ and $(-1, 1)$, and the asymptotes are $y = \pm\tfrac{1}{2}(x+1) \Leftrightarrow$

$y = \tfrac{1}{2}(x+1)$ and $y = -\tfrac{1}{2}(x+1)$.

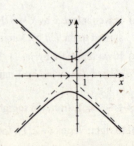

17. This is a parabola that opens down with its vertex at $(0, 4)$, so its equation is of the form $x^2 = a(y - 4)$. Since $(1, 0)$ is a point on this parabola, we have $(1)^2 = a(0 - 4) \Leftrightarrow 1 = -4a \Leftrightarrow a = -\frac{1}{4}$. Thus, an equation is $x^2 = -\frac{1}{4}(y - 4)$.

19. This is an ellipse with the major axis parallel to the x-axis, with one vertex at $(0, 0)$, the other vertex at $(10, 0)$, and one focus at $(8, 0)$. The center is at $\left(\frac{0+10}{2}, 0\right) = (5, 0)$, $a = 5$, and $c = 3$ (the distance from one focus to the center). So $b^2 = a^2 - c^2 = 25 - 9 = 16$. Thus, an equation is $\dfrac{(x-5)^2}{25} + \dfrac{y^2}{16} = 1$.

21. This is a hyperbola with center $(0, 1)$ and vertices $(0, 0)$ and $(0, 2)$. Since a is the distance form the center to a vertex, we have $a = 1$. The slope of the given asymptote is 1, so $\dfrac{a}{b} = 1 \Leftrightarrow b = 1$. Thus, an equation of the hyperbola is $(y - 1)^2 - x^2 = 1$.

23. $y^2 = 4(x + 2y) \Leftrightarrow y^2 - 8y = 4x \Leftrightarrow y^2 - 8y + 16 = 4x + 16 \Leftrightarrow$ 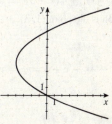 $(y - 4)^2 = 4(x + 4)$. This is a parabola with $4p = 4 \Leftrightarrow p = 1$. The vertex is $(-4, 4)$, the focus is $(-4 + 1, 4) = (-3, 4)$, and the directrix is $x = -4 - 1 = -5$.

25. $x^2 - 4y^2 - 2x + 16y = 20 \Leftrightarrow (x^2 - 2x + 1) - 4(y^2 - 4y + 4) = 20 + 1 - 16 \Leftrightarrow$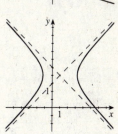

$(x - 1)^2 - 4(y - 2)^2 = 5 \Leftrightarrow \dfrac{(x-1)^2}{5} - \dfrac{(y-2)^2}{\frac{5}{4}} = 1$. This is a hyperbola with

$a = \sqrt{5}$, $b = \frac{1}{2}\sqrt{5}$, and $c = \sqrt{5 + \frac{5}{4}} = \frac{5}{2}$. The center is $(1, 2)$, the foci are

$\left(1 \pm \frac{5}{2}, 2\right) = \left(-\frac{3}{2}, 2\right)$ and $\left(\frac{7}{2}, 2\right)$, the vertices are $\left(1 \pm \sqrt{5}, 2\right)$, and the

asymptotes are $y - 2 = \pm \frac{1}{2}(x - 1) \Leftrightarrow y = \pm \frac{1}{2}(x - 1) + 2 \Leftrightarrow y = \frac{1}{2}x + \frac{3}{2}$ and

$y = -\frac{1}{2}x + \frac{5}{2}$.

27. $4x^2 + 25y^2 - 24x + 250y + 561 = 0 \Leftrightarrow$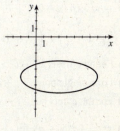

$4\left(x^2 - 6x + 9\right) + 25\left(y^2 + 10y + 25\right) = -561 + 36 + 625 \Leftrightarrow$

$4(x - 3)^2 + 25(y + 5)^2 = 100 \Leftrightarrow \dfrac{(x-3)^2}{25} + \dfrac{(y+5)^2}{4} = 1$. This is an ellipse

with $a = 5$, $b = 2$, and $c = \sqrt{25 - 4} = \sqrt{21}$. The center is $(3, -5)$, the foci are

$\left(3 \pm \sqrt{21}, -5\right)$, the vertices are $(3 \pm 5, -5) = (-2, -5)$ and $(8, -5)$, the length

of the major axis is $2a = 10$, and the length of the minor axis is $2b = 4$.

29. $16x^2 - 9y^2 - 96x + 288 = 0 \Leftrightarrow 16\left(x^2 - 6x\right) - 9y^2 + 288 = 0 \Leftrightarrow$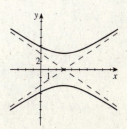

$16\left(x^2 - 6x + 9\right) - 9y^2 = 144 - 288 \Leftrightarrow 16(x - 3)^2 - 9y^2 = -144 \Leftrightarrow$

$\dfrac{y^2}{16} - \dfrac{(x-3)^2}{9} = 1$. This is a hyperbola with $a = 4$, $b = 3$, and

$c = \sqrt{16 + 9} = 5$. The center is $(3, 0)$, the foci are $(3, \pm 5)$, the vertices are

$(3, \pm 4)$, and the asymptotes are $y = \pm \frac{4}{3}(x - 3) \Leftrightarrow y = \frac{4}{3}x - 4$ and $y = 4 - \frac{4}{3}x$.

31. $x^2 + 16 = 4\left(y^2 + 2x\right) \Leftrightarrow x^2 - 8x - 4y^2 + 16 = 0 \Leftrightarrow$

$\left(x^2 - 8x + 16\right) - 4y^2 = -16 + 16 \Leftrightarrow 4y^2 = (x - 4)^2 \Leftrightarrow y = \pm\frac{1}{2}(x - 4).$

Thus, the conic is degenerate, and its graph is the pair of lines $y = \frac{1}{2}(x - 4)$ and

$y = -\frac{1}{2}(x - 4).$

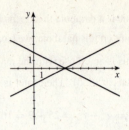

33. $3x^2 + 4y^2 - 6x - 24y + 39 = 0 \Leftrightarrow 3\left(x^2 - 2x\right) + 4\left(y^2 - 6y\right) = -39 \Leftrightarrow$

$3\left(x^2 - 2x + 1\right) + 4\left(y^2 - 6y + 9\right) = -39 + 3 + 36 \Leftrightarrow$

$3(x - 1)^2 + 4(y - 3)^2 = 0 \Leftrightarrow x = 1$ and $y = 3$. This is a degenerate conic whose

graph is the point $(1, 3)$.

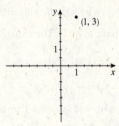

35. $2x^2 - 4x + y + 5 = 0 \Leftrightarrow y = -2x^2 + 4x - 5.$

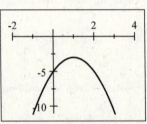

37. $9x^2 + 36 = y^2 + 36x + 6y \Leftrightarrow x^2 - 36x + 36 = y^2 + 6y \Leftrightarrow$

$9x^2 - 36x + 45 = y^2 + 6y + 9 \Leftrightarrow 9\left(x^2 - 4x + 5\right) = (y + 3)^2 \Leftrightarrow$

$y + 3 = \pm\sqrt{9\left(x^2 - 4x + 5\right)} \Leftrightarrow y = -3 \pm 3\sqrt{x^2 - 4x + 5}$

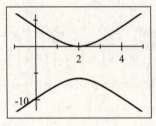

39. $4x^2 + y^2 + 4(x - 2y) + F = 0 \Leftrightarrow 4\left(x^2 + x\right) + \left(y^2 - 8y\right) = -F \Leftrightarrow$

$4\left(x^2 + x + \frac{1}{4}\right) + \left(y^2 - 8y + 16\right) = 16 + 1 - F \Leftrightarrow 4\left(x + \frac{1}{2}\right)^2 + (y - 1)^2 = 17 - F$

(a) For an ellipse, $17 - F > 0 \Leftrightarrow F < 17.$

(b) For a single point, $17 - F = 0 \Leftrightarrow F = 17.$

(c) For the empty set, $17 - F < 0 \Leftrightarrow F > 17.$

41. (a) $x^2 = 4p(y + p)$, for

$p = -2, -\frac{3}{2}, -1, -\frac{1}{2}, \frac{1}{2}, 1, \frac{3}{2}, 2.$

(b) The graph of $x^2 = 4p(y + p)$ is obtained by shifting the graph of $x^2 = 4py$ vertically $-p$ units so that the vertex is at $(0, -p)$. The focus of $x^2 = 4py$ is at $(0, p)$, so this point is also shifted $-p$ units vertically to the point $(0, p - p) = (0, 0)$. Thus, the focus is located at the origin.

(c) The parabolas become narrower as the vertex moves toward the origin.

43. Since the height of the satellite above the earth varies between 140 and 440, the length of the major axis is $2a = 140 + 2\,(3960) + 440 = 8500 \Leftrightarrow a = 4250$. Since the center of the earth is at one focus, we have

$a - c = \text{(earth radius)} + 140 = 3960 + 140 = 4100 \Leftrightarrow$

$c = a - 4100 = 4250 - 4100 = 150$. Thus, the center of the ellipse is $(-150, 0)$.

So $b^2 = a^2 - c^2 = 4250^2 - 150^2 = 18{,}062{,}500 - 22500 = 18{,}040{,}000$. Hence,

an equation is $\dfrac{(x+150)^2}{18{,}062{,}500} + \dfrac{y^2}{18{,}040{,}000} = 1$.

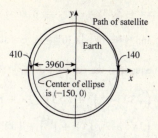

11.5 ROTATION OF AXES

1. If the x- and y-axes are rotated through an acute angle ϕ to produce the new X- and Y-axes, then the xy-coordinates (x, y) and the XY-coordinates (X, Y) of a point P in the plane are related by the formulas $x = X \cos \phi - Y \sin \phi$, $y = X \sin \phi + Y \cos \phi$, $X = x \cos \phi + y \sin \phi$, and $Y = -x \sin \phi + y \cos \phi$.

3. $(x, y) = (1, 1)$, $\phi = 45°$. Then $X = x \cos \phi + y \sin \phi = 1 \cdot \dfrac{1}{\sqrt{2}} + 1 \cdot \dfrac{1}{\sqrt{2}} = \sqrt{2}$ and

$Y = -x \sin \phi + y \cos \phi = -1 \cdot \dfrac{1}{\sqrt{2}} + 1 \cdot \dfrac{1}{\sqrt{2}} = 0$. Therefore, the XY-coordinates of the given point are $(X, Y) = \left(\sqrt{2}, 0\right)$.

5. $(x, y) = \left(3, -\sqrt{3}\right)$, $\phi = 60°$. Then $X = x \cos \phi + y \sin \phi = 3 \cdot \dfrac{1}{2} - \sqrt{3} \cdot \dfrac{\sqrt{3}}{2} = 0$ and

$Y = -x \sin \phi + y \cos \phi = -3 \cdot \dfrac{\sqrt{3}}{2} - \sqrt{3} \cdot \dfrac{1}{2} = -2\sqrt{3}$. Therefore, the XY-coordinates of the given point are $(X, Y) = \left(0, -2\sqrt{3}\right)$.

7. $(x, y) = (0, 2)$, $\phi = 55°$. Then $X = x \cos \phi + y \sin \phi = 0 \cos 55° + 2 \sin 55° \approx 1.6383$ and $Y = -x \sin \phi + y \cos \phi = -0 \sin 55° + 2 \cos 55° \approx 1.1472$. Therefore, the XY-coordinates of the given point are approximately $(X, Y) = (1.6383, 1.1472)$.

9. $x^2 - 3y^2 = 4$, $\phi = 60°$. Then $x = X \cos 60° - Y \sin 60° = \dfrac{1}{2}X - \dfrac{\sqrt{3}}{2}Y$ and $y = X \sin 60° + Y \cos 60° = \dfrac{\sqrt{3}}{2}X + \dfrac{1}{2}Y$.

Substituting these values into the equation, we get $\left(\dfrac{1}{2}X - \dfrac{\sqrt{3}}{2}Y\right)^2 - 3\left(\dfrac{\sqrt{3}}{2}X + \dfrac{1}{2}Y\right)^2 = 4 \Leftrightarrow$

$\dfrac{X^2}{4} - \dfrac{\sqrt{3}XY}{2} + \dfrac{3Y^2}{4} - 3\left(\dfrac{3X^2}{4} + \dfrac{\sqrt{3}XY}{2} + \dfrac{Y^2}{4}\right) = 4 \Leftrightarrow \dfrac{X^2}{4} - \dfrac{9}{4}X^2 + \dfrac{3Y^2}{4} - \dfrac{3Y^2}{4} - \dfrac{\sqrt{3}XY}{2} - \dfrac{3\sqrt{3}XY}{2} = 4 \Leftrightarrow$

$-2X^2 - 2\sqrt{3}XY = 4 \Leftrightarrow X^2 + \sqrt{3}XY = -2$.

11. $x^2 - y^2 = 2y$, $\phi = \cos^{-1}\left(\dfrac{3}{5}\right)$. So $\cos \phi = \dfrac{3}{5}$ and $\sin \phi = \dfrac{4}{5}$. Then

$(X \cos \phi - Y \sin \phi)^2 - (X \sin \phi + Y \cos \phi)^2 = 2(X \sin \phi + Y \cos \phi) \Leftrightarrow \left(\dfrac{3}{5}X - \dfrac{4}{5}Y\right)^2 - \left(\dfrac{4}{5}X + \dfrac{3}{5}Y\right)^2 = 2\left(\dfrac{4}{5}X + \dfrac{3}{5}Y\right)$

$\Leftrightarrow \dfrac{9X^2}{25} - \dfrac{24XY}{25} + \dfrac{16Y^2}{25} - \dfrac{16X^2}{25} - \dfrac{24XY}{25} - \dfrac{9Y^2}{25} = \dfrac{8X}{5} + \dfrac{6Y}{5} \Leftrightarrow -\dfrac{7X^2}{25} - \dfrac{48XY}{25} + \dfrac{7Y^2}{25} - \dfrac{8X}{5} - \dfrac{6Y}{5} = 0 \Leftrightarrow$

$7Y^2 - 48XY - 7X^2 - 40X - 30Y = 0$.

13. $x^2 + 2\sqrt{3}xy - y^2 = 4$, $\phi = 30°$. Then $x = X\cos 30° - Y\sin 30° = \frac{\sqrt{3}}{2}X - \frac{1}{2}Y = \frac{1}{2}\left(\sqrt{3}X - Y\right)$

and $y = X\sin 30° + Y\cos 30° = \frac{1}{2}X + \frac{\sqrt{3}}{2}Y = \frac{1}{2}\left(X + \sqrt{3}Y\right)$. Substituting these values into the

equation, we get $\left[\frac{1}{2}\left(\sqrt{3}X - Y\right)\right]^2 + 2\sqrt{3}\left[\frac{1}{2}\left(\sqrt{3}X - Y\right)\right]\left[\frac{1}{2}\left(X + \sqrt{3}Y\right)\right] - \left[\frac{1}{2}\left(X + \sqrt{3}Y\right)\right]^2 = 4 \Leftrightarrow$

$\left(\sqrt{3}X - Y\right)^2 + 2\sqrt{3}\left(\sqrt{3}X - Y\right)\left(X + \sqrt{3}Y\right) - \left(X + \sqrt{3}Y\right)^2 = 16 \Leftrightarrow$

$\left(3X^2 - 2\sqrt{3}XY + Y^2\right) + \left(6X^2 + 4\sqrt{3}XY - 6Y^2\right) - \left(X^2 + 2\sqrt{3}XY + 3Y^2\right) = 16 \Leftrightarrow 8X^2 - 8Y^2 = 16 \Leftrightarrow$

$\dfrac{X^2}{2} - \dfrac{Y^2}{2} = 1$. This is a hyperbola.

15. (a) $xy = 8 \Leftrightarrow 0x^2 + xy + 0y^2 = 8$. So $A = 0$, $B = 1$, and $C = 0$, and so the
discriminant is $B^2 - 4AC = 1^2 - 4(0)(0) = 1$. Since the discriminant is
positive, the equation represents a hyperbola.

(b) $\cot 2\phi = \dfrac{A - C}{B} = 0 \Rightarrow 2\phi = 90° \Leftrightarrow \phi = 45°$. Therefore,

$x = \frac{\sqrt{2}}{2}X - \frac{\sqrt{2}}{2}Y$ and $y = \frac{\sqrt{2}}{2}X + \frac{\sqrt{2}}{2}Y$. After substitution, the original

equation becomes $\left(\frac{\sqrt{2}}{2}X - \frac{\sqrt{2}}{2}Y\right)\left(\frac{\sqrt{2}}{2}X + \frac{\sqrt{2}}{2}Y\right) = 8 \Leftrightarrow$

$\dfrac{(X - Y)(X + Y)}{2} = 8 \Leftrightarrow \dfrac{X^2}{16} - \dfrac{Y^2}{16} = 1$. This is a hyperbola with $a = 4$, $b = 4$, and $c = 4\sqrt{2}$. Hence, the vertices

are $V(\pm 4, 0)$ and the foci are $F\left(\pm 4\sqrt{2}, 0\right)$.

(c)

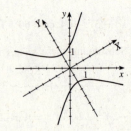

17. (a) $x^2 + 2\sqrt{3}xy - y^2 + 2 = 0$. So $A = 1$, $B = 2\sqrt{3}$, and $C = -1$, and so the
discriminant is $B^2 - 4AC = \left(2\sqrt{3}\right)^2 - 4(1)(-1) > 0$. Since the
discriminant is positive, the equation represents a hyperbola.

(b) $\cot 2\phi = \dfrac{A - C}{B} = \dfrac{1 + 1}{2\sqrt{3}} = \dfrac{1}{\sqrt{3}} \Rightarrow 2\phi = 60° \Leftrightarrow \phi = 30°$. Therefore,

$x = \frac{\sqrt{3}}{2}X - \frac{1}{2}Y$ and $y = \frac{1}{2}X + \frac{\sqrt{3}}{2}Y$. After substitution, the original

equation becomes

$\left(\frac{\sqrt{3}}{2}X - \frac{1}{2}Y\right)^2 + 2\sqrt{3}\left(\frac{\sqrt{3}}{2}X - \frac{1}{2}Y\right)\left(\frac{1}{2}X + \frac{\sqrt{3}}{2}Y\right) - \left(\frac{1}{2}X + \frac{\sqrt{3}}{2}Y\right)^2 + 2 = 0 \Leftrightarrow$

$\frac{3}{4}X^2 - \frac{\sqrt{3}}{2}XY + \frac{1}{4}Y^2 + \frac{\sqrt{3}}{2}\left(\sqrt{3}X^2 + 2XY - \sqrt{3}Y^2\right) - \frac{1}{4}X^2 - \frac{\sqrt{3}}{2}XY - \frac{3}{4}Y^2 + 2 = 0 \Leftrightarrow$

$X^2\left(\frac{3}{4} + \frac{3}{2} - \frac{1}{4}\right) + XY\left(-\frac{\sqrt{3}}{2} + \sqrt{3} - \frac{\sqrt{3}}{2}\right) + Y^2\left(\frac{1}{4} - \frac{3}{2} - \frac{3}{4}\right) = -2 \Leftrightarrow 2X^2 - 2Y^2 = -2 \Leftrightarrow Y^2 - X^2 = 1$.

(c)

19. (a) $11x^2 - 24xy + 4y^2 + 20 = 0$. So $A = 11$, $B = -24$, and $C = 4$, and so the discriminant is $B^2 - 4AC = (-24)^2 - 4\,(11)\,(4) > 0$. Since the discriminant is positive, the equation represents a hyperbola.

(c)

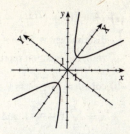

(b) $\cot 2\phi = \dfrac{A - C}{B} = \dfrac{11 - 4}{-24} = -\dfrac{7}{24} \Rightarrow \cos 2\phi = -\dfrac{7}{25}$. Therefore,

$\cos\phi = \sqrt{\dfrac{1+(-7/25)}{2}} = \dfrac{3}{5}$ and $\sin\phi = \sqrt{\dfrac{1-(-7/25)}{2}} = \dfrac{4}{5}$. Hence,

$x = \dfrac{3X}{5} - \dfrac{4}{5}Y$ and $y = \dfrac{4}{5}X + \dfrac{3}{5}Y$. After substitution, the original equation becomes

Since $\cos 2\phi = -\dfrac{7}{25}$, we have $2\phi \approx 106.26°$, so $\phi \approx 53°$.

$11\left(\dfrac{3}{5}X - \dfrac{4}{5}Y\right)^2 - 24\left(\dfrac{3}{5}X - \dfrac{4}{5}Y\right)\left(\dfrac{4}{5}X + \dfrac{3}{5}Y\right) + 4\left(\dfrac{4}{5}X + \dfrac{3}{5}Y\right)^2 + 20 = 0 \Leftrightarrow$

$\dfrac{11}{25}\left(9X^2 - 24XY + 16Y^2\right) - \dfrac{24}{25}\left(12X^2 - 7XY - 12Y^2\right) + \dfrac{4}{25}\left(16X^2 + 24XY + 9Y^2\right) + 20 = 0 \Leftrightarrow$

$X^2\,(99 - 288 + 64) + XY\,(-264 + 168 + 96) + Y^2\,(176 + 288 + 36) = -500 \Leftrightarrow -125X^2 + 500Y^2 = -500 \Leftrightarrow$

$\dfrac{1}{4}X^2 - Y^2 = 1$.

21. (a) $\sqrt{3}x^2 + 3xy = 3$. So $A = \sqrt{3}$, $B = 3$, and $C = 0$, and so the discriminant is $B^2 - 4AC = (3)^2 - 4\left(\sqrt{3}\right)(0) = 9$. Since the discriminant is positive, the equation represents a hyperbola.

(c)

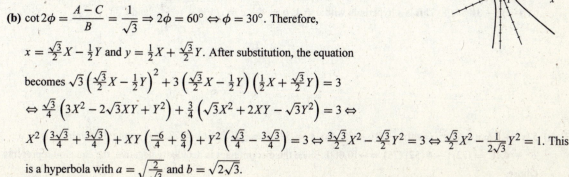

(b) $\cot 2\phi = \dfrac{A - C}{B} = \dfrac{1}{\sqrt{3}} \Rightarrow 2\phi = 60° \Leftrightarrow \phi = 30°$. Therefore,

$x = \dfrac{\sqrt{3}}{2}X - \dfrac{1}{2}Y$ and $y = \dfrac{1}{2}X + \dfrac{\sqrt{3}}{2}Y$. After substitution, the equation

becomes $\sqrt{3}\left(\dfrac{\sqrt{3}}{2}X - \dfrac{1}{2}Y\right)^2 + 3\left(\dfrac{\sqrt{3}}{2}X - \dfrac{1}{2}Y\right)\left(\dfrac{1}{2}X + \dfrac{\sqrt{3}}{2}Y\right) = 3$

$\Leftrightarrow \dfrac{\sqrt{3}}{4}\left(3X^2 - 2\sqrt{3}XY + Y^2\right) + \dfrac{3}{4}\left(\sqrt{3}X^2 + 2XY - \sqrt{3}Y^2\right) = 3 \Leftrightarrow$

$X^2\left(\dfrac{3\sqrt{3}}{4} + \dfrac{3\sqrt{3}}{4}\right) + XY\left(\dfrac{-6}{4} + \dfrac{6}{4}\right) + Y^2\left(\dfrac{\sqrt{3}}{4} - \dfrac{3\sqrt{3}}{4}\right) = 3 \Leftrightarrow \dfrac{3\sqrt{3}}{2}X^2 - \dfrac{\sqrt{3}}{2}Y^2 = 3 \Leftrightarrow \dfrac{\sqrt{3}}{2}X^2 - \dfrac{1}{2\sqrt{3}}Y^2 = 1$. This

is a hyperbola with $a = \sqrt{\dfrac{2}{\sqrt{3}}}$ and $b = \sqrt{2\sqrt{3}}$.

23. (a) $x^2 + 2xy + y^2 + x - y = 0$. So $A = 1$, $B = 2$, and $C = 1$, and so the discriminant is $B^2 - 4AC = 2^2 - 4(1)(1) = 0$. Since the discriminant is zero, the equation represents a parabola.

(c)

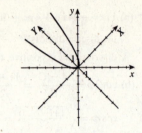

(b) $\cot 2\phi = \dfrac{A - C}{B} = 0 \Rightarrow 2\phi = 90° \Leftrightarrow \phi = 45°$. Therefore,

$x = \frac{\sqrt{2}}{2}X - \frac{\sqrt{2}}{2}Y$ and $y = \frac{\sqrt{2}}{2}X + \frac{\sqrt{2}}{2}Y$. After substitution, the original equation becomes

$$\left(\tfrac{\sqrt{2}}{2}X - \tfrac{\sqrt{2}}{2}Y\right)^2 + 2\left(\tfrac{\sqrt{2}}{2}X - \tfrac{\sqrt{2}}{2}Y\right)\left(\tfrac{\sqrt{2}}{2}X + \tfrac{\sqrt{2}}{2}Y\right)$$

$$+ \left(\tfrac{\sqrt{2}}{2}X + \tfrac{\sqrt{2}}{2}Y\right)^2 + \left(\tfrac{\sqrt{2}}{2}X - \tfrac{\sqrt{2}}{2}Y\right) - \left(\tfrac{\sqrt{2}}{2}X + \tfrac{\sqrt{2}}{2}Y\right) = 0 \Leftrightarrow$$

$\frac{1}{2}X^2 - XY + \frac{1}{2}Y^2 + X^2 - Y^2 + \frac{1}{2}X^2 + XY + Y^2 - \sqrt{2}Y = 0 \Leftrightarrow 2X^2 - \sqrt{2}Y = 0 \Leftrightarrow X^2 = \frac{\sqrt{2}}{2}Y$. This is a

parabola with $4p = \frac{1}{\sqrt{2}}$ and hence the focus is $F\left(0, \frac{1}{4\sqrt{2}}\right)$.

25. (a) $2\sqrt{3}x^2 - 6xy + \sqrt{3}x + 3y = 0$. So $A = 2\sqrt{3}$, $B = -6$, and $C = 0$, and so the discriminant is $B^2 - 4AC = (-6)^2 - 4\left(2\sqrt{3}\right)(0) = 36$. Since the discriminant is positive, the equation represents a hyperbola.

(c)

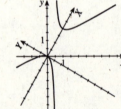

(b) $\cot 2\phi = \dfrac{A - C}{B} = \dfrac{2\sqrt{3}}{-6} = -\dfrac{1}{\sqrt{3}} \Rightarrow 2\phi = 120° \Leftrightarrow \phi = 60°$. Therefore,

$x = \frac{1}{2}X - \frac{\sqrt{3}}{2}Y$ and $y = \frac{\sqrt{3}}{2}X + \frac{1}{2}Y$, and substituting gives

$$2\sqrt{3}\left(\tfrac{1}{2}X - \tfrac{\sqrt{3}}{2}Y\right)^2 - 6\left(\tfrac{1}{2}X - \tfrac{\sqrt{3}}{2}Y\right)\left(\tfrac{\sqrt{3}}{2}X + \tfrac{1}{2}Y\right) + \sqrt{3}\left(\tfrac{1}{2}X - \tfrac{\sqrt{3}}{2}Y\right) + 3\left(\tfrac{\sqrt{3}}{2}X + \tfrac{1}{2}Y\right) = 0 \Leftrightarrow$$

$$\tfrac{\sqrt{3}}{2}\left(X^2 - 2\sqrt{3}XY + 3Y^2\right) - \tfrac{3}{2}\left(\sqrt{3}X^2 - 2XY - \sqrt{3}Y^2\right) + \tfrac{\sqrt{3}}{2}\left(X - \sqrt{3}Y\right) + \tfrac{3}{2}\left(\sqrt{3}X + Y\right) = 0 \Leftrightarrow$$

$$X^2\left(\tfrac{\sqrt{3}}{2} - \tfrac{3\sqrt{3}}{2}\right) + X\left(\tfrac{\sqrt{3}}{2} + \tfrac{3\sqrt{3}}{2}\right) + XY(-3 + 3) + Y^2\left(\tfrac{3\sqrt{3}}{2} + \tfrac{3\sqrt{3}}{2}\right) + Y\left(-\tfrac{3}{2} + \tfrac{3}{2}\right) = 0 \Leftrightarrow$$

$$-\sqrt{3}X^2 + 2\sqrt{3}X + 3\sqrt{3}Y^2 = 0 \Leftrightarrow -X^2 + 2X + 3Y^2 = 0 \Leftrightarrow 3Y^2 - \left(X^2 - 2X + 1\right) = -1 \Leftrightarrow$$

$(X - 1)^2 - 3Y^2 = 1$. This is a hyperbola with $a = 1$, $b = \frac{\sqrt{3}}{3}$, $c = \sqrt{1 + \frac{1}{3}} = \frac{2}{\sqrt{3}}$, and $C(1, 0)$.

27. (a) $52x^2 + 72xy + 73y^2 = 40x - 30y + 75$. So $A = 52$, $B = 72$, and $C = 73$, and so the discriminant is $B^2 - 4AC = (72)^2 - 4(52)(73) = -10{,}000$. Since the discriminant is decidedly negative, the equation represents an ellipse.

(b) $\cot 2\phi = \dfrac{A - C}{B} = \dfrac{52 - 73}{72} = -\dfrac{7}{24}$. Therefore, as in Exercise 19(b), we get $\cos\phi = \frac{3}{5}$, $\sin\phi = \frac{4}{5}$, and

$x = \frac{3}{5}X - \frac{4}{5}Y$, $y = \frac{4}{5}X + \frac{3}{5}Y$. By substitution,

$$52\left(\tfrac{3}{5}X - \tfrac{4}{5}Y\right)^2 + 72\left(\tfrac{3}{5}X - \tfrac{4}{5}Y\right)\left(\tfrac{4}{5}X + \tfrac{3}{5}Y\right) + 73\left(\tfrac{4}{5}X + \tfrac{3}{5}Y\right)^2$$

$$= 40\left(\tfrac{3}{5}X - \tfrac{4}{5}Y\right) - 30\left(\tfrac{4}{5}X + \tfrac{3}{5}Y\right) + 75 \Leftrightarrow$$

$$\tfrac{52}{25}\left(9X^2 - 24XY + 16Y^2\right) + \tfrac{72}{25}\left(12X^2 - 7XY - 12Y^2\right) + \tfrac{73}{25}\left(16X^2 + 24XY + 9Y^2\right)$$

$$= 24X - 32Y - 24X - 18Y + 75 \Leftrightarrow$$

$468X^2 + 832Y^2 + 864X^2 - 864Y^2 + 1168X^2 + 657Y^2 = -1250Y + 1875 \Leftrightarrow 2500X^2 + 625Y^2 + 1250Y = 1875 \Leftrightarrow$

$100X^2 + 25Y^2 + 50Y = 75 \Leftrightarrow X^2 + \frac{1}{4}(Y + 1)^2 = 1$. This is an ellipse with $a = 2$, $b = 1$, $c = \sqrt{4 - 1} = \sqrt{3}$, and center $C\,(0, -1)$.

(c)

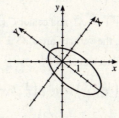

Since $\cos 2\phi = -\frac{7}{25}$, we have $2\phi = \cos^{-1}\left(-\frac{7}{25}\right) \approx 106.26°$ and so $\phi \approx 53°$.

29. (a) The discriminant is $B^2 - 4AC = (-4)^2 + 4\,(2)\,(2) = 0$. Since the discriminant is 0, the equation represents a parabola.

(b) $2x^2 - 4xy + 2y^2 - 5x - 5 = 0 \Leftrightarrow 2y^2 - 4xy = -2x^2 + 5x + 5 \Leftrightarrow$

$2\left(y^2 - 2xy\right) = -2x^2 + 5x + 5 \Leftrightarrow$

$2\left(y^2 - 2xy + x^2\right) = -2x^2 + 5x + 5 + 2x^2 \Leftrightarrow 2\,(y - x)^2 = 5x + 5 \Leftrightarrow$

$(y - x)^2 = \frac{5}{2}x + \frac{5}{2} \Rightarrow y - x = \pm\sqrt{\frac{5}{2}x + \frac{5}{2}} \Leftrightarrow y = x \pm\sqrt{\frac{5}{2}x + \frac{5}{2}}$

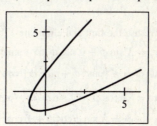

31. (a) The discriminant is $B^2 - 4AC = 10^2 + 4\,(6)\,(3) = 28 > 0$. Since the discriminant is positive, the equation represents a hyperbola.

(b) $6x^2 + 10xy + 3y^2 - 6y = 36 \Leftrightarrow 3y^2 + 10xy - 6y = 36 - 6x^2 \Leftrightarrow$

$3y^2 + 2\,(5x - 3)\,y = 36 - 6x^2 \Leftrightarrow y^2 + 2\left(\tfrac{5}{3}x - 1\right)y = 12 - 2x^2 \Leftrightarrow$

$y^2 + 2\left(\tfrac{5}{3}x - 1\right)y + \left(\tfrac{5}{3}x - 1\right)^2 = \left(\tfrac{5}{3}x - 1\right)^2 + 12 - 2x^2 \Leftrightarrow$

$\left[y + \left(\tfrac{5}{3}x - 1\right)\right]^2 = \tfrac{25}{9}x^2 - \tfrac{10}{3}x + 1 + 12 - 2x^2 \Leftrightarrow$

$\left[y + \left(\tfrac{5}{3}x - 1\right)\right]^2 = \tfrac{7}{9}x^2 - \tfrac{10}{3}x + 13 \Leftrightarrow$

$y + \left(\tfrac{5}{3}x - 1\right) = \pm\sqrt{\tfrac{7}{9}x^2 - \tfrac{10}{3}x + 13} \Leftrightarrow$

$y = -\tfrac{5}{3}x + 1 \pm\sqrt{\tfrac{7}{9}x^2 - \tfrac{10}{3}x + 13}$

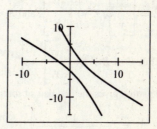

33. (a) $7x^2 + 48xy - 7y^2 - 200x - 150y + 600 = 0$. Then $A = 7$, $B = 48$, and $C = -7$, and so the discriminant is $B^2 - 4AC = (48)^2 - 4(7)(7) > 0$. Since the discriminant is positive, the equation represents a hyperbola. We now find the equation in terms of XY-coordinates. We have $\cot 2\phi = \dfrac{A - C}{B} = \dfrac{7}{24} \Rightarrow \cos\phi = \frac{4}{5}$ and $\sin\phi = \frac{3}{5}$.

Therefore, $x = \frac{4}{5}X - \frac{3}{5}Y$ and $y = \frac{3}{5}X + \frac{4}{5}Y$, and substitution gives

$$7\left(\tfrac{4}{5}X - \tfrac{3}{5}Y\right)^2 + 48\left(\tfrac{4}{5}X - \tfrac{3}{5}Y\right)\left(\tfrac{3}{5}X + \tfrac{4}{5}Y\right) - 7\left(\tfrac{3}{5}X + \tfrac{4}{5}Y\right)^2 - 200\left(\tfrac{4}{5}X - \tfrac{3}{5}Y\right) - 150\left(\tfrac{3}{5}X + \tfrac{4}{5}Y\right) + 600 = 0$$

$$\Leftrightarrow \tfrac{7}{25}\left(16X^2 - 24XY + 9Y^2\right) + \tfrac{48}{25}\left(12X^2 + 7XY - 12Y^2\right) - \tfrac{7}{25}\left(9X^2 + 24XY + 16Y^2\right)$$
$$- 160X + 120Y - 90X - 120Y + 600 = 0$$

$$\Leftrightarrow 112X^2 - 168XY + 63Y^2 + 576X^2 + 336XY - 576Y^2 - 63X^2 - 168XY - 112Y^2 - 6250X + 15{,}000 = 0 \Leftrightarrow$$

$25X^2 - 25Y^2 - 250X + 600 = 0 \Leftrightarrow 25\left(X^2 - 10X + 25\right) - 25Y^2 = -600 + 625 \Leftrightarrow (X - 5)^2 - Y^2 = 1$. This is a hyperbola with $a = 1$, $b = 1$, $c = \sqrt{1 + 1} = \sqrt{2}$, and center $C(5, 0)$.

(b) In the XY-plane, the center is $C(5, 0)$, the vertices are $V(5 \pm 1, 0) = V_1(4, 0)$ and $V_2(6, 0)$, and the foci are $F\left(5 \pm \sqrt{2}, 0\right)$. In the xy-plane, the center is $C\left(\frac{4}{5} \cdot 5 - \frac{3}{5} \cdot 0, \frac{3}{5} \cdot 5 + \frac{4}{5} \cdot 0\right) = C(4, 3)$, the vertices are $V_1\left(\frac{4}{5} \cdot 4 - \frac{3}{5} \cdot 0, \frac{3}{5} \cdot 4 + \frac{4}{5} \cdot 0\right) = V_1\left(\frac{16}{5}, \frac{12}{5}\right)$ and $V_2\left(\frac{4}{5} \cdot 6 - \frac{3}{5} \cdot 0, \frac{3}{5} \cdot 6 + \frac{4}{5} \cdot 0\right) = V_2\left(\frac{24}{5}, \frac{18}{5}\right)$, and the foci are $F_1\left(4 + \frac{4}{5}\sqrt{2}, 3 + \frac{3}{5}\sqrt{2}\right)$ and $F_2\left(4 - \frac{4}{5}\sqrt{2}, 3 - \frac{3}{5}\sqrt{2}\right)$.

(c) In the XY-plane, the equations of the asymptotes are $Y = X - 5$ and $Y = -X + 5$. In the xy-plane, these equations become $-x \cdot \frac{3}{5} + y \cdot \frac{4}{5} = x \cdot \frac{4}{5} + y \cdot \frac{3}{5} - 5 \Leftrightarrow 7x - y - 25 = 0$. Similarly, $-x \cdot \frac{3}{5} + y \cdot \frac{4}{5} = -x \cdot \frac{4}{5} - y \cdot \frac{3}{5} + 5 \Leftrightarrow x + 7y - 25 = 0$.

35. We use the hint and eliminate Y by adding: $x = X \cos\phi - Y \sin\phi \Leftrightarrow x \cos\phi = X \cos^2\phi - Y \sin\phi\cos\phi$ and $y = X \sin\phi + Y \cos\phi \Leftrightarrow y \sin\phi = X \sin^2\phi + Y \sin\phi\cos\phi$, and adding these two equations gives $x \cos\phi + y \sin\phi = X\left(\cos^2\phi + \sin^2\phi\right) \Leftrightarrow x \cos\phi + y \sin\phi = X$. In a similar manner, we eliminate X by subtracting: $x = X \cos\phi - Y \sin\phi \Leftrightarrow -x \sin\phi = -X \cos\phi\sin\phi + Y \sin^2\phi$ and $y = X \sin\phi + Y \cos\phi \Leftrightarrow y \cos\phi = X \sin\phi\cos\phi + Y \cos^2\phi$, so $-x \sin\phi + y \cos\phi = Y\left(\cos^2\phi + \sin^2\phi\right) \Leftrightarrow -x \sin\phi + y \cos\phi = Y$. Thus, $X = x \cos\phi + y \sin\phi$ and $Y = -x \sin\phi + y \cos\phi$.

37. $Z = \begin{bmatrix} x \\ y \end{bmatrix}$, $Z' = \begin{bmatrix} X \\ Y \end{bmatrix}$, and $R = \begin{bmatrix} \cos\phi & -\sin\phi \\ \sin\phi & \cos\phi \end{bmatrix}$.

Thus $Z = RZ' \Leftrightarrow \begin{bmatrix} x \\ y \end{bmatrix} = \begin{bmatrix} \cos\phi & -\sin\phi \\ \sin\phi & \cos\phi \end{bmatrix}\begin{bmatrix} X \\ Y \end{bmatrix} = \begin{bmatrix} X\cos\phi - Y\sin\phi Y \\ X\sin\phi + Y\cos\phi \end{bmatrix}$. Equating the entries in this matrix equation gives the first pair of rotation of axes formulas. Now

$$R^{-1} = \frac{1}{\cos^2\phi + \sin^2\phi}\begin{bmatrix} \cos\phi & \sin\phi \\ -\sin\phi & \cos\phi \end{bmatrix} = \begin{bmatrix} \cos\phi & \sin\phi \\ -\sin\phi & \cos\phi \end{bmatrix}$$ and so $Z' = R^{-1}Z \Leftrightarrow$

$$\begin{bmatrix} X \\ Y \end{bmatrix} = \begin{bmatrix} \cos\phi & \sin\phi \\ -\sin\phi & \cos\phi \end{bmatrix}\begin{bmatrix} x \\ y \end{bmatrix} = \begin{bmatrix} x\cos\phi + y\sin\phi \\ -x\sin\phi + y\cos\phi \end{bmatrix}$$. Equating the entries in this matrix equation gives the second pair of rotation of axes formulas.

39. Let P be the point (x_1, y_1) and Q be the point (x_2, y_2) and let $P'(X_1, Y_1)$ and $Q'(X_2, Y_2)$ be the images of P and Q under the rotation of ϕ. So $X_1 = x_1 \cos \phi + y_1 \sin \phi$, $Y_1 = -x_1 \sin \phi + y_1 \cos \phi$, $X_2 = x_2 \cos \phi + y_2 \sin \phi$, and $Y_2 = -x_2 \sin \phi + y_2 \cos \phi$. Thus $d(P', Q') = \sqrt{(X_2 - X_1)^2 + (Y_2 - Y_1)^2}$, where

$$
\begin{aligned}
(X_2 - X_1)^2 &= \left[(x_2 \cos \phi + y_2 \sin \phi) - (x_1 \cos \phi + y_1 \sin \phi)\right]^2 = \left[(x_2 - x_1) \cos \phi + (y_2 - y_1) \sin \phi\right]^2 \\
&= (x_2 - x_1)^2 \cos^2 \phi + (x_2 - x_1)(y_2 - y_1) \sin \phi \cos \phi + (y_2 - y_1)^2 \sin^2 \phi
\end{aligned}
$$

and

$$
\begin{aligned}
(Y_2 - Y_1)^2 &= \left[(-x_2 \sin \phi + y_2 \cos \phi) - (-x_1 \sin \phi + y_1 \cos \phi)\right]^2 = \left[-(x_2 - x_1) \sin \phi + (y_2 - y_1) \cos \phi\right]^2 \\
&= (x_2 - x_1)^2 \sin^2 \phi - (x_2 - x_1)(y_2 - y_1) \sin \phi \cos \phi + (y_2 - y_1)^2 \cos^2 \phi
\end{aligned}
$$

So

$$
\begin{aligned}
(X_2 - X_1)^2 + (Y_2 - Y_1)^2 &= (x_2 - x_1)^2 \cos^2 \phi + (x_2 - x_1)(y_2 - y_1) \sin \phi \cos \phi + (y_2 - y_1)^2 \sin^2 \phi \\
&\quad + (x_2 - x_1)^2 \sin^2 \phi - (x_2 - x_1)(y_2 - y_1) \sin \phi \cos \phi + (y_2 - y_1)^2 \cos^2 \phi \\
&= (x_2 - x_1)^2 \cos^2 \phi + (y_2 - y_1)^2 \sin^2 \phi + (x_2 - x_1)^2 \sin^2 \phi + (y_2 - y_1)^2 \cos^2 \phi \\
&= (x_2 - x_1)^2 \left(\cos^2 \phi + \sin^2 \phi\right) + (y_2 - y_1)^2 \left(\sin^2 \phi + \cos^2 \phi\right) = (x_2 - x_1)^2 + (y_2 - y_1)^2
\end{aligned}
$$

Putting these equations together gives $d(P', Q') = \sqrt{(X_2 - X_1)^2 + (Y_2 - Y_1)^2} = \sqrt{(x_2 - x_1)^2 + (y_2 - y_1)^2} = d(P, Q)$.

11.6 POLAR EQUATIONS OF CONICS

1. All conics can be described geometrically using a fixed point F called the *focus* and a fixed line ℓ called the *directrix*. For a fixed positive number e the set of all points P satisfying $\dfrac{\text{distance from } P \text{ to } F}{\text{distance from } P \text{ to } \ell} = e$ is a *conic section*. If $e = 1$ the conic is a *parabola*, if $e < 1$ the conic is an *ellipse*, and if $e > 1$ the conic is a *hyperbola*. The number e is called the *eccentricity* of the conic.

3. Substituting $e = \frac{2}{3}$ and $d = 3$ into the general equation of a conic with vertical directrix, we get $r = \dfrac{\frac{2}{3} \cdot 3}{1 + \frac{2}{3} \cos \theta} \Leftrightarrow$

$r = \dfrac{6}{3 + 2 \cos \theta}$.

5. Substituting $e = 1$ and $d = 2$ into the general equation of a conic with horizontal directrix, we get $r = \dfrac{1 \cdot 2}{1 + \sin \theta} \Leftrightarrow$

$r = \dfrac{2}{1 + \sin \theta}$.

7. $r = 5 \sec \theta \Leftrightarrow r \cos \theta = 5 \Leftrightarrow x = 5$. So $d = 5$ and $e = 4$ gives $r = \dfrac{4 \cdot 5}{1 + 4 \cos \theta} \Leftrightarrow r = \dfrac{20}{1 + 4 \cos \theta}$.

9. Since this is a parabola whose focus is at the origin and vertex at $(5, \pi/2)$, the directrix must be $y = 10$. So $d = 10$ and $e = 1$ gives $r = \dfrac{1 \cdot 10}{1 + \sin \theta} = \dfrac{10}{1 + \sin \theta}$.

11. $r = \dfrac{6}{1 + \cos \theta}$ is Graph II. The eccentricity is 1, so this is a parabola. When $\theta = 0$, we have $r = 3$ and when $\theta = \frac{\pi}{2}$, we have $r = 6$.

13. $r = \dfrac{3}{1 - 2 \sin \theta}$ is Graph VI. $e = 2$, so this is a hyperbola. When $\theta = 0$, $r = 3$, and when $\theta = \pi$, $r = 3$.

15. $r = \dfrac{12}{3 + 2\sin\theta}$ is Graph IV. $r = \dfrac{4}{1 + \frac{2}{3}\sin\theta}$, so $e = \frac{2}{3}$ and this is an ellipse. When $\theta = 0$, $r = 4$, and when $\theta = \pi$, $r = 4$.

17. (a) The equation $r = \dfrac{4}{1 - \sin\theta}$ has $e = 1$ and $d = 4$, so it represents a parabola.

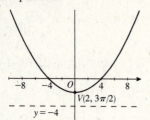

(b) Because the equation is of the form $r = \dfrac{ed}{1 - e\sin\theta}$, the directrix is parallel to the polar axis and has equation $y = -4$. The vertex is $\left(2, \frac{3\pi}{2}\right)$.

19. (a) The equation $r = \dfrac{5}{3 + 3\cos\theta} = \dfrac{\frac{5}{3}}{1 + \cos\theta}$ has $e = 1$ and $d = \frac{5}{3}$, so it represents a parabola.

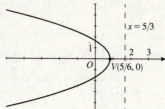

(b) Because the equation is of the form $r = \dfrac{ed}{1 + e\cos\theta}$, the directrix is parallel to the polar axis and has equation $x = d = \frac{5}{3}$. The vertex is $\left(\frac{5}{6}, 0\right)$.

21. (a) The equation $r = \dfrac{4}{2 - \cos\theta} = \dfrac{2}{1 - \frac{1}{2}\cos\theta}$ has $e = \frac{1}{2} < 1$, so it represents an ellipse.

(b) Because the equation is of the form $r = \dfrac{ed}{1 - e\cos\theta}$ with $d = 4$, the directrix is vertical and has equation $x = -4$. Thus, the vertices are $V_1\,(4, 0)$ and $V_2\left(\frac{4}{3}, \pi\right)$.

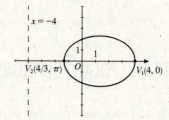

(c) The length of the major axis is $2a = |V_1 V_2| = 4 + \frac{4}{3} = \frac{16}{3}$ and the center is at the midpoint of $V_1 V_2$, $\left(\frac{4}{3}, 0\right)$. The minor axis has length $2b$ where

$$b^2 = a^2 - c^2 = a^2 - (ae)^2 = \left(\frac{8}{3}\right)^2 - \left(\frac{8}{3} \cdot \frac{1}{2}\right)^2 = \frac{16}{3}, \text{ so}$$

$$2b = 2 \cdot \sqrt{\frac{16}{3}} = \frac{8\sqrt{3}}{3} \approx 4.62.$$

23. (a) The equation $r = \dfrac{12}{4 + 3\sin\theta} = \dfrac{3}{1 + \frac{3}{4}\sin\theta}$ has $e = \frac{3}{4} < 1$, so it represents an ellipse.

(b) Because the equation is of the form $r = \dfrac{ed}{1 + e\sin\theta}$ with $d = 4$, the directrix is horizontal and has equation $y = 4$. Thus, the vertices are $V_1\left(\frac{12}{7}, \frac{\pi}{2}\right)$ and $V_2\left(12, \frac{3\pi}{2}\right)$.

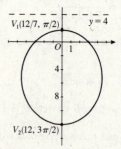

(c) The length of the major axis is $2a = |V_1 V_2| = \frac{12}{7} + 12 = \frac{96}{7}$ and the center is at the midpoint of $V_1 V_2$, $\left(\frac{36}{7}, \frac{3\pi}{2}\right)$. The minor axis has length $2b$ where

$$b^2 = a^2 - c^2 = a^2 - (ae)^2 = \left(\frac{48}{7}\right)^2 - \left(\frac{48}{7} \cdot \frac{3}{4}\right)^2 = \frac{144}{7}, \text{ so}$$

$$2b = 2 \cdot \sqrt{\frac{144}{7}} = \frac{24\sqrt{7}}{7} \approx 9.07.$$

25. (a) The equation $r = \dfrac{8}{1 + 2\cos\theta}$ has $e = 2 > 1$, so it represents a hyperbola.

(b) Because the equation has the form $r = \dfrac{ed}{1 + \cos\theta}$ with $d = 4$, the transverse axis is horizontal and the directrix has equation $x = 4$. The vertices are $V_1\left(\dfrac{8}{3}, 0\right)$ and $V_2\left(-8, \pi\right) = (8, 0)$.

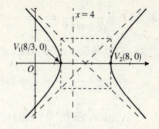

(c) The center is the midpoint of $V_1 V_2$, $\left(\dfrac{16}{3}, 0\right)$. To sketch the central box and the asymptotes, we find a and b. The length of the transverse axis is $2a = \dfrac{16}{3}$, and so

$a = \dfrac{8}{3}$, and $b^2 = c^2 - a^2 = (ae)^2 - a^2 = \left(\dfrac{8}{3} \cdot 2\right)^2 - \left(\dfrac{8}{3}\right)^2 = \dfrac{64}{3}$, so

$b = \sqrt{\dfrac{64}{3}} = \dfrac{8\sqrt{3}}{3} \approx 4.62$.

27. (a) The equation $r = \dfrac{20}{2 - 3\sin\theta} = \dfrac{10}{1 - \frac{3}{2}\sin\theta}$ has $e = \dfrac{3}{2} > 1$, so it represents a hyperbola.

(b) Because the equation has the form $r = \dfrac{ed}{1 + \cos\theta}$ with $d = \dfrac{20}{3}$, the transverse axis is vertical and the directrix has equation $y = -\dfrac{20}{3}$. The vertices are

$V_1\left(-20, \dfrac{\pi}{2}\right) = \left(20, \dfrac{3\pi}{2}\right)$ and $V_2\left(4, \dfrac{3\pi}{2}\right)$.

(c) The center is the midpoint of $V_1 V_2$, $\left(12, \dfrac{3\pi}{2}\right)$. To sketch the central box and the asymptotes, we find a and b. The length of the transverse axis is $2a = 16$, and so

$a = 8$, and $b^2 = c^2 - a^2 = (ae)^2 - a^2 = \left(8 \cdot \dfrac{3}{2}\right)^2 - 8^2 = 80$, so

$b = \sqrt{80} = 4\sqrt{5} \approx 8.94$.

29. (a) $r = \dfrac{4}{1 + 3\cos\theta} \Rightarrow e = 3$, so the conic is a hyperbola.

(b) The vertices occur where $\theta = 0$ and $\theta = \pi$. Now $\theta = 0 \Rightarrow r = \dfrac{4}{1 + 3\cos 0} = 1$,

and $\theta = \pi \Rightarrow r = \dfrac{4}{1 + 3\cos\pi} = \dfrac{4}{-2} = -2$. Thus the vertices are $(1, 0)$ and $(-2, \pi)$.

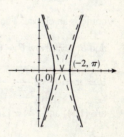

31. (a) $r = \dfrac{2}{1 - \cos\theta} \Rightarrow e = 1$, so the conic is a parabola.

(b) Substituting $\theta = \pi$, we have $r = \dfrac{2}{1 - \cos\pi} = \dfrac{2}{2} = 1$. Thus the vertex is $(1, \pi)$.

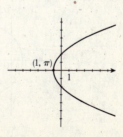

33. (a) $r = \dfrac{6}{2 + \sin\theta} \Leftrightarrow r = \dfrac{\frac{1}{2} \cdot 6}{1 + \frac{1}{2}\sin\theta} \Rightarrow e = \frac{1}{2}$, so the conic is an ellipse.

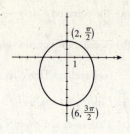

(b) The vertices occur where $\theta = \frac{\pi}{2}$ and $\theta = \frac{3\pi}{2}$. Now $\theta = \frac{\pi}{2} \Rightarrow$

$r = \dfrac{6}{2 + \sin\frac{\pi}{2}} = \dfrac{6}{3} = 2$ and $\theta = \frac{3\pi}{2} \Rightarrow r = \dfrac{6}{2 + \sin\frac{3\pi}{2}} = \dfrac{6}{1} = 6$. Thus, the

vertices are $\left(2, \frac{\pi}{2}\right)$ and $\left(6, \frac{3\pi}{2}\right)$.

35. (a) $r = \dfrac{7}{2 - 5\sin\theta} \Leftrightarrow r = \dfrac{\frac{7}{2}}{1 - \frac{5}{2}\sin\theta} \Rightarrow e = \frac{5}{2}$, so the conic is a hyperbola.

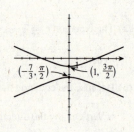

(b) The vertices occur where $\theta = \frac{\pi}{2}$ and $\theta = \frac{3\pi}{2}$. $r = \dfrac{7}{2 - 5\sin\frac{\pi}{2}} = \dfrac{7}{-3} = -\dfrac{7}{3}$ and

$\theta = \frac{3\pi}{2} \Rightarrow r = \dfrac{7}{2 - 5\sin\frac{3\pi}{2}} = \dfrac{7}{7} = 1$. Thus, the vertices are $\left(-\frac{7}{3}, \frac{\pi}{2}\right)$ and

$\left(1, \frac{3\pi}{2}\right)$.

37. (a) $r = \dfrac{1}{4 - 3\cos\theta} = \dfrac{\frac{1}{4}}{1 - \frac{3}{4}\cos\theta} \Rightarrow e = \frac{3}{4}$, so the

conic is an ellipse. The vertices occur where $\theta = 0$

and $\theta = \pi$. Now $\theta = 0 \Rightarrow r = \dfrac{1}{4 - 3\cos 0} = 1$ and

$\theta = \pi \Rightarrow r = \dfrac{1}{4 - 3\cos\pi} = \frac{1}{7}$. Thus, the vertices

are $(1, 0)$ and $\left(\frac{1}{7}, \pi\right)$. We have $d = \frac{1}{3}$, so the

directrix is $x = -\frac{1}{3}$.

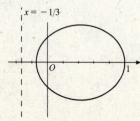

(b) If the ellipse is rotated through $\frac{\pi}{3}$, the equation of

the resulting conic is $r = \dfrac{1}{4 - 3\cos\left(\theta - \frac{\pi}{3}\right)}$.

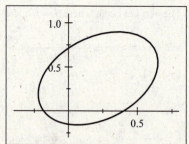

39. (a) $r = \dfrac{2}{1+\sin\theta} \Rightarrow e = 1$, so the conic is a parabola.

Substituting $\theta = \frac{\pi}{2}$, we have $r = t\dfrac{2}{1+\sin\frac{\pi}{2}} = 1$,

so the vertex is $\left(1, \frac{\pi}{2}\right)$. Because $d = 2$, the directrix

is $y = 2$.

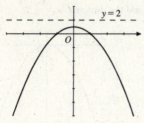

(b) If the ellipse is rotated through $\theta = -\frac{\pi}{4}$, the equation of the resulting conic is

$$r = \dfrac{2}{1+\sin\left(\theta + \frac{\pi}{4}\right)}.$$

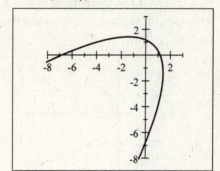

41. The ellipse is nearly circular when e is close to 0 and becomes more elongated as $e \to 1^-$. At $e = 1$, the curve becomes a parabola.

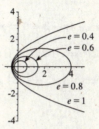

43. (a) Since the polar form of an ellipse with directrix $x = -d$ is $r = \dfrac{ed}{1-e\cos\theta}$ we need to show that $ed = a\left(1-e^2\right)$.

From the proof of the Equivalent Description of Conics we have $a^2 = \dfrac{e^2 d^2}{\left(1-e^2\right)^2}$. Since the conic is an ellipse, $e < 1$

and so the quantities a, d, and $\left(1-e^2\right)$ are all positive. Thus we can take the square roots of both sides and maintain

equality. Thus $a^2 = \dfrac{e^2 d^2}{\left(1-e^2\right)^2} \Leftrightarrow a = \dfrac{ed}{1-e^2} \Leftrightarrow ed = a\left(1-e^2\right)$. As a result, $r = \dfrac{ed}{1-e\cos\theta} \Leftrightarrow r = \dfrac{a\left(1-e^2\right)}{1-e\cos\theta}$.

(b) Since $2a = 2.99 \times 10^8$ we have $a = 1.495 \times 10^8$, so a polar equation for the earth's orbit (using $e \approx 0.017$) is

$$r = \dfrac{1.495 \times 10^8 \left[1 - (0.017)^2\right]}{1 - 0.017\cos\theta} \approx \dfrac{1.49 \times 10^8}{1 - 0.017\cos\theta}.$$

45. From Exercise 44, we know that at perihelion $r = 4.43 \times 10^9 = a\left(1-e\right)$ and at aphelion $r = 7.37 \times 10^9 = a\left(1+e\right)$.

Dividing these equations gives $\dfrac{7.37 \times 10^9}{4.43 \times 10^9} = \dfrac{a\left(1+e\right)}{a\left(1-e\right)} \Leftrightarrow 1.664 = \dfrac{1+e}{1-e} \Leftrightarrow 1.664\left(1-e\right) = 1+e \Leftrightarrow$

$1.664 - 1 = e + 1.664 \Leftrightarrow 0.664 = 2.664e \Leftrightarrow e = \dfrac{0.664}{2.664} \approx 0.25$.

47. The r-coordinate of the satellite will be its distance from the focus (the center of the earth). From the r-coordinate we can easily calculate the height of the satellite.

CHAPTER 11 REVIEW

1. $y^2 = 4x$. This is a parabola with $4p = 4 \Leftrightarrow p = 1$. The vertex is $(0, 0)$, the focus is $(1, 0)$, and the directrix is $x = -1$.

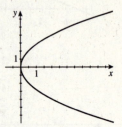

3. $x^2 + 8y = 0 \Leftrightarrow x^2 = -8y$. This is a parabola with $4p = -8 \Leftrightarrow p = -2$. The vertex is $(0, 0)$, the focus is $(0, -2)$, and the directrix is $y = 2$.

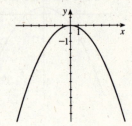

5. $x - y^2 + 4y - 2 = 0 \Leftrightarrow x - \left(y^2 - 4y + 4\right) - 2 = -4 \Leftrightarrow x - (y - 2)^2 = -2 \Leftrightarrow$ $(y - 2)^2 = x + 2$. This is a parabola with $4p = 1 \Leftrightarrow p = \frac{1}{4}$. The vertex is $(-2, 2)$, the focus is $\left(-2 + \frac{1}{4}, 2\right) = \left(-\frac{7}{4}, 2\right)$, and the directrix is $x = -2 - \frac{1}{4} = -\frac{9}{4}$.

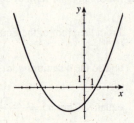

7. $\frac{1}{2}x^2 + 2x = 2y + 4 \Leftrightarrow x^2 + 4x = 4y + 8 \Leftrightarrow x^2 + 4x + 4 = 4y + 8 \Leftrightarrow$ $(x + 2)^2 = 4(y + 3)$. This is a parabola with $4p = 4 \Leftrightarrow p = 1$. The vertex is $(-2, -3)$, the focus is $(-2, -3 + 1) = (-2, -2)$, and the directrix is $y = -3 - 1 = -4$.

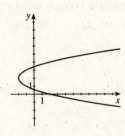

9. $\frac{x^2}{9} + \frac{y^2}{25} = 1$. This is an ellipse with $a = 5$, $b = 3$, and $c = \sqrt{25 - 9} = 4$. The center is $(0, 0)$, the vertices are $(0, \pm 5)$, the foci are $(0, \pm 4)$, the length of the major axis is $2a = 10$, and the length of the minor axis is $2b = 6$.

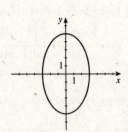

11. $x^2 + 4y^2 = 16 \Leftrightarrow \frac{x^2}{16} + \frac{y^2}{4} = 1$. This is an ellipse with $a = 4$, $b = 2$, and $c = \sqrt{16 - 4} = 2\sqrt{3}$. The center is $(0, 0)$, the vertices are $(\pm 4, 0)$, the foci are $\left(\pm 2\sqrt{3}, 0\right)$, the length of the major axis is $2a = 8$, and the length of the minor axis is $2b = 4$.

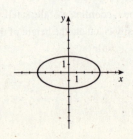

13. $\dfrac{(x-3)^2}{9} + \dfrac{y^2}{16} = 1$. This is an ellipse with $a = 4$, $b = 3$, and $c = \sqrt{16-9} = \sqrt{7}$.

The center is $(3, 0)$, the vertices are $(3, \pm 4)$, the foci are $\left(3, \pm\sqrt{7}\right)$, the length of

the major axis is $2a = 8$, and the length of the minor axis is $2b = 6$.

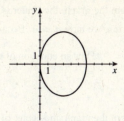

15. $4x^2 + 9y^2 = 36y \Leftrightarrow 4x^2 + 9\left(y^2 - 4y + 4\right) = 36 \Leftrightarrow 4x^2 + 9(y-2)^2 = 36 \Leftrightarrow$

$\dfrac{x^2}{9} + \dfrac{(y-2)^2}{4} = 1$. This is an ellipse with $a = 3$, $b = 2$, and $c = \sqrt{9-4} = \sqrt{5}$.

The center is $(0, 2)$, the vertices are $(\pm 3, 2)$, the foci are $\left(\pm\sqrt{5}, 2\right)$, the length of

the major axis is $2a = 6$, and the length of the minor axis is $2b = 4$.

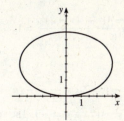

17. $-\dfrac{x^2}{9} + \dfrac{y^2}{16} = 1 \Leftrightarrow \dfrac{y^2}{16} - \dfrac{x^2}{9} = 0$. This is a hyperbola with $a = 4$, $b = 3$, and

$c = \sqrt{16+9} = \sqrt{25} = 5$. The center is $(0, 0)$, the vertices are $(0, \pm 4)$, the foci are

$(0, \pm 5)$, and the asymptotes are $y = \pm\frac{4}{3}x$.

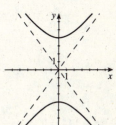

19. $x^2 - 2y^2 = 16 \Leftrightarrow \dfrac{x^2}{16} - \dfrac{y^2}{8} = 1$. This is a hyperbola with $a = 4$, $b = 2\sqrt{2}$, and

$c = \sqrt{16+8} = \sqrt{24} = 2\sqrt{6}$. The center is $(0, 0)$, the vertices are $(\pm 4, 0)$, the foci

are $\left(\pm 2\sqrt{6}, 0\right)$, and the asymptotes are $y = \pm\frac{2\sqrt{2}}{4}x \Leftrightarrow y = \pm\frac{1}{\sqrt{2}}x$.

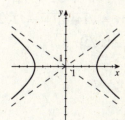

21. $\dfrac{(x+4)^2}{16} - \dfrac{y^2}{16} = 1$. This is a hyperbola with $a = 4$, $b = 4$ and

$c = \sqrt{16+16} = 4\sqrt{2}$. The center is $(-4, 0)$, the vertices are $(-4 \pm 4, 0)$ which

are $(-8, 0)$ and $(0, 0)$, the foci are $\left(-4 \pm 4\sqrt{2}, 0\right)$, and the asymptotes are

$y = \pm(x+4)$.

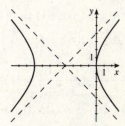

23. $9y^2 + 18y = x^2 + 6x + 18 \Leftrightarrow 9\left(y^2 + 2y + 1\right) = \left(x^2 + 6x + 9\right) + 9 - 9 + 18$

$\Leftrightarrow 9(y+1)^2 - (x+3)^2 = 18 \Leftrightarrow \dfrac{(y+1)^2}{2} - \dfrac{(x+3)^2}{18} = 1$. This is a hyperbola

with $a = \sqrt{2}$, $b = 3\sqrt{2}$, and $c = \sqrt{2+18} = 2\sqrt{5}$. The center is $(-3, -1)$, the

vertices are $\left(-3, -1 \pm \sqrt{2}\right)$, the foci are $\left(-3, -1 \pm 2\sqrt{5}\right)$, and the asymptotes

are $y + 1 = \pm\frac{1}{3}(x+3) \Leftrightarrow y = \frac{1}{3}x$ and $y = -\frac{1}{3}x - 2$.

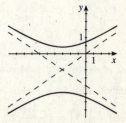

25. This is a parabola that opens to the right with its vertex at $(0, 0)$ and the focus at $(2, 0)$. So $p = 2$, and the equation is

$y^2 = 4(2)x \Leftrightarrow y^2 = 8x$.

27. From the graph, the center is $(0, 0)$, and the vertices are $(0, -4)$ and $(0, 4)$. Since a is the distance from the center to a vertex, we have $a = 4$. Because one focus is $(0, 5)$, we have $c = 5$, and since $c^2 = a^2 + b^2$, we have $25 = 16 + b^2 \Leftrightarrow b^2 = 9$. Thus an equation of the hyperbola is $\dfrac{y^2}{16} - \dfrac{x^2}{9} = 1$.

29. From the graph, the center of the ellipse is $(4, 2)$, and so $a = 4$ and $b = 2$. The equation is $\dfrac{(x - 4)^2}{4^2} + \dfrac{(y - 2)^2}{2^2} = 1 \Leftrightarrow$

$\dfrac{(x - 4)^2}{16} + \dfrac{(y - 2)^2}{4} = 1$.

31. $\dfrac{x^2}{12} + y = 1 \Leftrightarrow \dfrac{x^2}{12} = -(y - 1) \Leftrightarrow x^2 = -12 \, (y - 1)$. This is a parabola with $4p = -12 \Leftrightarrow p = -3$. The vertex is $(0, 1)$ and the focus is $(0, 1 - 3) = (0, -2)$.

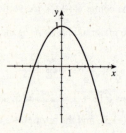

33. $x^2 - y^2 + 144 = 0 \Leftrightarrow \dfrac{y^2}{144} - \dfrac{x^2}{144} = 1$. This is a hyperbola with $a = 12$, $b = 12$, and $c = \sqrt{144 + 144} = 12\sqrt{2}$. The vertices are $(0, \pm 12)$ and the foci are $\left(0, \pm 12\sqrt{2}\right)$.

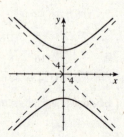

35. $4x^2 + y^2 = 8 \, (x + y) \Leftrightarrow 4 \left(x^2 - 2x\right) + \left(y^2 - 8y\right) = 0 \Leftrightarrow$

$4 \left(x^2 - 2x + 1\right) + \left(y^2 - 8y + 16\right) = 4 + 16 \Leftrightarrow 4 \, (x - 1)^2 + (y - 4)^2 = 20 \Leftrightarrow$

$\dfrac{(x - 1)^2}{5} + \dfrac{(y - 4)^2}{20} = 1$. This is an ellipse with $a = 2\sqrt{5}$, $b = \sqrt{5}$, and

$c = \sqrt{20 - 5} = \sqrt{15}$. The vertices are $\left(1, 4 \pm 2\sqrt{5}\right)$ and the foci are

$\left(1, 4 \pm \sqrt{15}\right)$.

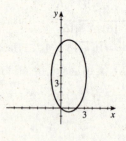

37. $x = y^2 - 16y \Leftrightarrow x + 64 = y^2 - 16y + 64 \Leftrightarrow (y - 8)^2 = x + 64$. This is a parabola with $4p = 1 \Leftrightarrow p = \frac{1}{4}$. The vertex is $(-64, 8)$ and the focus is

$\left(-64 + \frac{1}{4}, 8\right) = \left(-\frac{255}{4}, 8\right)$.

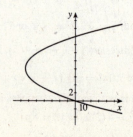

39. $2x^2 - 12x + y^2 + 6y + 26 = 0 \Leftrightarrow 2\left(x^2 - 6x\right) + \left(y^2 + 6y\right) = -26 \Leftrightarrow$

$2\left(x^2 - 6x + 9\right) + \left(y^2 + 6y + 9\right) = -26 + 18 + 9 \Leftrightarrow 2\left(x - 3\right)^2 + \left(y + 3\right)^2 = 1$

$\Leftrightarrow \dfrac{(x - 3)^2}{\frac{1}{2}} + (y + 3)^2 = 1$. This is an ellipse with $a = 1$, $b = \frac{\sqrt{2}}{2}$, and

$c = \sqrt{1 - \frac{1}{2}} = \frac{\sqrt{2}}{2}$. The vertices are $(3, -3 \pm 1) = (3, -4)$ and $(3, -2)$, and the

foci are $\left(3, -3 \pm \frac{\sqrt{2}}{2}\right)$.

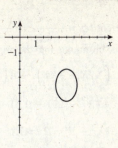

41. $9x^2 + 8y^2 - 15x + 8y + 27 = 0 \Leftrightarrow 9\left(x^2 - \frac{5}{3}x + \frac{25}{36}\right) + 8\left(y^2 + y + \frac{1}{4}\right) = -27 + \frac{25}{4} + 2 \Leftrightarrow 9\left(x - \frac{5}{6}\right)^2 + 8\left(y + \frac{1}{2}\right)^2 = -\frac{75}{4}$.
However, since the left-hand side of the equation is greater than or equal to 0, there is no point that satisfies this equation.
The graph is empty.

43. The parabola has focus $(0, 1)$ and directrix $y = -1$. Therefore, $p = 1$ and so $4p = 4$. Since the focus is on the y-axis and the vertex is $(0, 0)$, an equation of the parabola is $x^2 = 4y$.

45. The hyperbola has vertices $(0, \pm 2)$ and asymptotes $y = \pm\frac{1}{2}x$. Therefore, $a = 2$, and the foci are on the y-axis. Since the

slopes of the asymptotes are $\pm\frac{1}{2} = \pm\dfrac{a}{b} \Leftrightarrow b = 2a = 4$, an equation of the hyperbola is $\dfrac{y^2}{4} - \dfrac{x^2}{16} = 1$.

47. The ellipse has foci $F_1\,(1, 1)$ and $F_2\,(1, 3)$, and one vertex is on the x-axis. Thus, $2c = 3 - 1 = 2 \Leftrightarrow c = 1$, and so the

center of the ellipse is $C\,(1, 2)$. Also, since one vertex is on the x-axis, $a = 2 - 0 = 2$, and thus $b^2 = 4 - 1 = 3$. So an

equation of the ellipse is $\dfrac{(x - 1)^2}{3} + \dfrac{(y - 2)^2}{4} = 1$.

49. The ellipse has vertices $V_1\,(7, 12)$ and $V_2\,(7, -8)$ and passes through the point $P\,(1, 8)$. Thus, $2a = 12 - (-8) = 20 \Leftrightarrow$

$a = 10$, and the center is $\left(7, \dfrac{-8 + 12}{2}\right) = (7, 2)$. Thus an equation of the ellipse has the form $\dfrac{(x - 7)^2}{b^2} + \dfrac{(y - 2)^2}{100} = 1$.

Since the point $P\,(1, 8)$ is on the ellipse, $\dfrac{(1 - 7)^2}{b^2} + \dfrac{(8 - 2)^2}{100} = 1 \Leftrightarrow 3600 + 36b^2 = 100b^2 \Leftrightarrow 64b^2 = 3600 \Leftrightarrow b^2 = \frac{225}{4}$.

Therefore, an equation of the ellipse is $\dfrac{(x - 7)^2}{225/4} + \dfrac{(y - 2)^2}{100} = 1 \Leftrightarrow \dfrac{4\,(x - 7)^2}{225} + \dfrac{(y - 2)^2}{100} = 1$.

51. The length of the major axis is $2a = 186{,}000{,}000 \Leftrightarrow a = 93{,}000{,}000$. The eccentricity is $e = c/a = 0.017$, and so
$c = 0.017\,(93{,}000{,}000) = 1{,}581{,}000$.

(a) The earth is closest to the sun when the distance is $a - c = 93{,}000{,}000 - 1{,}581{,}000 = 91{,}419{,}000$.

(b) The earth is furthest from the sun when the distance is $a + c = 93{,}000{,}000 + 1{,}581{,}000 = 94{,}581{,}000$.

53. (a) The graphs of $\dfrac{x^2}{16 + k^2} + \dfrac{y^2}{k^2} = 1$ for $k = 1$, 2, 4, and 8 are shown in

the figure.

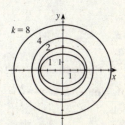

(b) $c^2 = \left(16 + k^2\right) - k^2 = 16 \Rightarrow c = \pm 4$. Since the center is

$(0, 0)$, the foci of each of the ellipses are $(\pm 4, 0)$.

55. (a) $x^2 + 4xy + y^2 = 1$. Then $A = 1$, $B = 4$, and $C = 1$, so the discriminant is $4^2 - 4\,(1)\,(1) = 12$. Since the discriminant
is positive, the equation represents a hyperbola.

(b) $\cot 2\phi = \dfrac{A - C}{B} = \dfrac{1 - 1}{4} = 0 \Rightarrow 2\phi = 90° \Leftrightarrow \phi = 45°$. Therefore, $x = \frac{\sqrt{2}}{2}X - \frac{\sqrt{2}}{2}Y$ and $y = \frac{\sqrt{2}}{2}X + \frac{\sqrt{2}}{2}Y$.
Substituting into the original equation gives

$$\left(\tfrac{\sqrt{2}}{2}X - \tfrac{\sqrt{2}}{2}Y\right)^2 + 4\left(\tfrac{\sqrt{2}}{2}X - \tfrac{\sqrt{2}}{2}Y\right)\left(\tfrac{\sqrt{2}}{2}X + \tfrac{\sqrt{2}}{2}Y\right) + \left(\tfrac{\sqrt{2}}{2}X + \tfrac{\sqrt{2}}{2}Y\right)^2 = 1 \Leftrightarrow$$

$$\tfrac{1}{2}\left(X^2 - 2XY + Y^2\right) + 2\left(X^2 + XY - XY - Y^2\right) + \tfrac{1}{2}\left(X^2 + 2XY + Y^2\right) = 1 \Leftrightarrow \quad \textbf{(c)}$$

$3X^2 - Y^2 = 1 \Leftrightarrow 3X^2 - Y^2 = 1$. This is a hyperbola with $a = \frac{1}{\sqrt{3}}$, $b = 1$, and

$c = \sqrt{\frac{1}{3} + 1} = \frac{2}{\sqrt{3}}$. Therefore, the hyperbola has vertices $V\left(\pm\frac{1}{\sqrt{3}}, 0\right)$ and foci

$F\left(\pm\frac{2}{\sqrt{3}}, 0\right)$, in XY-coordinates.

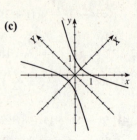

57. (a) $7x^2 - 6\sqrt{3}xy + 13y^2 - 4\sqrt{3}x - 4y = 0$. Then $A = 7$, $B = -6\sqrt{3}$, and $C = 13$, so the discriminant is

$\left(-6\sqrt{3}\right)^2 - 4(7)(13) = -256$. Since the discriminant is negative, the equation represents an ellipse.

(b) $\cot 2\phi = \dfrac{A - C}{B} = \dfrac{7 - 13}{-6\sqrt{3}} = \dfrac{1}{\sqrt{3}} \Rightarrow 2\phi = 60° \Leftrightarrow \phi = 30°$. Therefore, $x = \frac{\sqrt{3}}{2}X - \frac{1}{2}Y$ and $y = \frac{1}{2}X + \frac{\sqrt{3}}{2}Y$.
Substituting into the original equation gives

$$7\left(\tfrac{\sqrt{3}}{2}X - \tfrac{1}{2}Y\right)^2 - 6\sqrt{3}\left(\tfrac{\sqrt{3}}{2}X - \tfrac{1}{2}Y\right)\left(\tfrac{1}{2}X + \tfrac{\sqrt{3}}{2}Y\right)$$

$$+ 13\left(\tfrac{1}{2}X + \tfrac{\sqrt{3}}{2}Y\right)^2 - 4\sqrt{3}\left(\tfrac{\sqrt{3}}{2}X - \tfrac{1}{2}Y\right) - 4\left(\tfrac{1}{2}X + \tfrac{\sqrt{3}}{2}Y\right) = 0 \Leftrightarrow$$

$$\tfrac{7}{4}\left(3X^2 - 2\sqrt{3}XY + Y^2\right) - \tfrac{3\sqrt{3}}{2}\left(\sqrt{3}X^2 + 3XY - XY - \sqrt{3}Y^2\right)$$

$$+ \tfrac{13}{4}\left(X^2 + 2\sqrt{3}XY + 3Y^2\right) - 6X + 2\sqrt{3}Y - 2X - 2\sqrt{3}Y = 0 \Leftrightarrow$$

$X^2\left(\tfrac{21}{4} - \tfrac{9}{2} + \tfrac{13}{4}\right) - 8X + Y^2\left(\tfrac{7}{4} + \tfrac{9}{2} + \tfrac{39}{4}\right) = 0 \Leftrightarrow 4X^2 - 8X + 16Y^2 = 0 \Leftrightarrow \quad \textbf{(c)}$

$4\left(X^2 - 2X + 1\right) + 16Y^2 = 4 \Leftrightarrow (X - 1)^2 + 4Y^2 = 1$. This ellipse has $a = 1$,

$b = \frac{1}{2}$, and $c = \sqrt{1 - \frac{1}{4}} = \frac{1}{2}\sqrt{3}$. Therefore, the vertices are

$V\left(1 \pm 1, 0\right) = V_1\left(0, 0\right)$ and $V_2\left(2, 0\right)$ and the foci are $F\left(1 \pm \frac{1}{2}\sqrt{3}, 0\right)$.

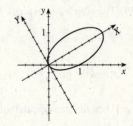

59. $5x^2 + 3y^2 = 60 \Leftrightarrow 3y^2 = 60 - 5x^2 \Leftrightarrow y^2 = 20 - \frac{5}{3}x^2$. This conic is an ellipse.

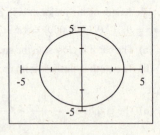

61. $6x + y^2 - 12y = 30 \Leftrightarrow y^2 - 12y = 30 - 6x \Leftrightarrow y^2 - 12y + 36 = 66 - 6x \Leftrightarrow$
$(y - 6)^2 = 66 - 6x \Leftrightarrow y - 6 = \pm\sqrt{66 - 6x} \Leftrightarrow y = 6 \pm \sqrt{66 - 6x}$. This conic is
a parabola.

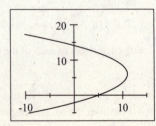

63. (a) $r = \dfrac{1}{1 - \cos\theta} \Rightarrow e = 1$. Therefore, this is a parabola.

(b)

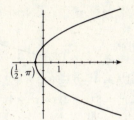

65. (a) $r = \dfrac{4}{1 + 2\sin\theta} \Rightarrow e = 2$. Therefore, this is a hyperbola.

(b)

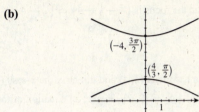

CHAPTER 11 TEST

1. $x^2 = -12y$. This is a parabola with $4p = -12 \Leftrightarrow p = -3$. The focus is $(0, -3)$ and the directrix is $y = 3$.

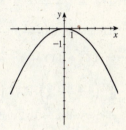

3. $\dfrac{y^2}{9} - \dfrac{x^2}{16} = 1$. This is a hyperbola with $a = 3$, $b = 4$, and $c = \sqrt{9 + 16} = 5$. The vertices are $(0, \pm 3)$, the foci are $(0, \pm 5)$, and the asymptotes are $y = \pm\frac{3}{4}x$.

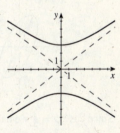

5. This is an ellipse tangent to the x-axis at $(0, 0)$ and with one vertex at the point $(4, 3)$. The center is $(0, 3)$, and $a = 4$ and $b = 3$. Thus the equation is $\dfrac{x^2}{16} + \dfrac{(y - 3)^2}{9} = 1$.

7. $16x^2 + 36y^2 - 96x + 36y + 9 = 0 \Leftrightarrow 16\left(x^2 - 6x\right) + 36\left(y^2 + y\right) = -9 \Leftrightarrow$

$16\left(x^2 - 6x + 9\right) + 36\left(y^2 + y + \frac{1}{4}\right) = -9 + 144 + 9 \Leftrightarrow$

$16(x - 3)^2 + 36\left(y + \frac{1}{2}\right)^2 = 144 \Leftrightarrow \dfrac{(x - 3)^2}{9} + \dfrac{\left(y + \frac{1}{2}\right)^2}{4} = 1$. This is an

ellipse with $a = 3$, $b = 2$, and $c = \sqrt{9 - 4} = \sqrt{5}$. The center is $\left(3, -\frac{1}{2}\right)$, the

vertices are $\left(3 \pm 3, -\frac{1}{2}\right) = \left(0, -\frac{1}{2}\right)$ and $\left(6, -\frac{1}{2}\right)$, and the foci are

$(h \pm c, k) = \left(3 \pm \sqrt{5}, -\frac{1}{2}\right) = \left(3 + \sqrt{5}, -\frac{1}{2}\right)$ and $\left(3 - \sqrt{5}, -\frac{1}{2}\right)$.

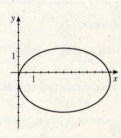

9. $2x + y^2 + 8y + 8 = 0 \Leftrightarrow y^2 + 8y + 16 = -2x - 8 + 16 \Leftrightarrow$

$(y + 4)^2 = -2(x - 4)$. This is a parabola with $4p = -2 \Leftrightarrow p = -\frac{1}{2}$. The vertex

is $(4, -4)$ and the focus is $\left(4 - \frac{1}{2}, -4\right) = \left(\frac{7}{2}, -4\right)$.

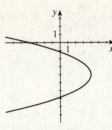

11. The parabola has focus $(2, 4)$ and directrix the x-axis ($y = 0$). Therefore, $2p = 4 - 0 = 4 \Leftrightarrow p = 2 \Leftrightarrow 4p = 8$, and the

vertex is $(2, 4 - p) = (2, 2)$. Hence, an equation of the parabola is $(x - 2)^2 = 8(y - 2) \Leftrightarrow x^2 - 4x + 4 = 8y - 16 \Leftrightarrow$

$x^2 - 4x - 8y + 20 = 0$.

13. **(a)** $5x^2 + 4xy + 2y^2 = 18$. Then $A = 5$, $B = 4$, and $C = 2$, so the discriminant is $(4)^2 - 4(5)(2) = -24$. Since the

discriminant is negative, the equation represents an ellipse.

(b) $\cot 2\phi = \dfrac{A - C}{B} = \dfrac{5 - 2}{4} = \dfrac{3}{4}$. Thus, $\cos 2\phi = \frac{3}{5}$ and so $\cos \phi = \sqrt{\dfrac{1 + (3/5)}{2}} = \dfrac{2\sqrt{5}}{5}$,

$\sin \phi = \sqrt{\dfrac{1 - (3/5)}{2}} = \dfrac{\sqrt{5}}{5}$. It follows that $x = \dfrac{2\sqrt{5}}{5}X - \dfrac{\sqrt{5}}{5}Y$ and $y = \dfrac{\sqrt{5}}{5}X + \dfrac{2\sqrt{5}}{5}Y$. By

substitution, $5\left(\dfrac{2\sqrt{5}}{5}X - \dfrac{\sqrt{5}}{5}Y\right)^2 + 4\left(\dfrac{2\sqrt{5}}{5}X - \dfrac{\sqrt{5}}{5}Y\right)\left(\dfrac{\sqrt{5}}{5}X + \dfrac{2\sqrt{5}}{5}Y\right) + 2\left(\dfrac{\sqrt{5}}{5}X + \dfrac{2\sqrt{5}}{5}Y\right)^2 = 18$

$\Leftrightarrow 4X^2 - 4XY + Y^2 + \dfrac{4}{5}\left(2X^2 + 4XY - XY - 2Y^2\right) + \dfrac{2}{5}\left(X^2 + 4XY + 4Y^2\right) = 18 \Leftrightarrow$

$X^2\left(4 + \dfrac{8}{5} + \dfrac{2}{5}\right) + XY\left(-4 + \dfrac{12}{5} + \dfrac{8}{5}\right) + Y^2\left(1 - \dfrac{8}{5} + \dfrac{4}{5}\right) = 18 \Leftrightarrow 6X^2 + Y^2 = 18 \Leftrightarrow \dfrac{X^2}{3} + \dfrac{Y^2}{18} = 1$. This is an

ellipse with $a = 3\sqrt{2}$ and $b = \sqrt{3}$.

(c)

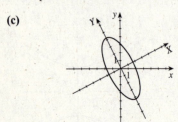

Since $\cos 2\phi = \frac{3}{5}$ we have

$2\phi = \cos^{-1}\frac{3}{5} \approx 53.13°$, so $\phi \approx 27°$.

(d) In XY-coordinates, the vertices are $V\left(0, \pm 3\sqrt{2}\right)$. Therefore, in

xy-coordinates, the vertices are $x = -\dfrac{3\sqrt{2}}{\sqrt{5}}$ and $y = \dfrac{6\sqrt{2}}{\sqrt{5}} \Rightarrow$

$V_1\left(-\dfrac{3\sqrt{2}}{\sqrt{5}}, \dfrac{6\sqrt{2}}{\sqrt{5}}\right)$, and $x = \dfrac{3\sqrt{2}}{\sqrt{5}}$ and $y = -\dfrac{6\sqrt{2}}{\sqrt{5}} \Rightarrow$

$V_2\left(\dfrac{3\sqrt{2}}{\sqrt{5}}, \dfrac{-6\sqrt{2}}{\sqrt{5}}\right)$.

FOCUS ON MODELING   Conics in Architecture

1. Answers will vary.

5. **(a)** The tangent line passes though the point $\left(a, a^2\right)$, so an equation is $y - a^2 = m(x - a)$.

(b) Because the tangent line intersects the parabola at only the one point $\left(a, a^2\right)$, the system $\begin{cases} y - a^2 = m(x - a) \\ y = x^2 \end{cases}$ has

only one solution, namely $x = a$, $y = a^2$.

(c) $\begin{cases} y - a^2 = m(x - a) \\ y = x^2 \end{cases} \Leftrightarrow \begin{cases} y = a^2 + m(x - a) \\ y = x^2 \end{cases} \Leftrightarrow a^2 + m(x - a) = x^2 \Leftrightarrow x^2 - mx + am - a^2 = 0$. This quadratic

has discriminant $(-m)^2 - 4(1)\left(am - a^2\right) = m^2 - 4am + 4a^2 = (m - 2a)^2$. Setting this equal to 0, we find $m = 2a$.

(d) An equation of the tangent line is $y - a^2 = 2a\,(x - a) \Leftrightarrow y = a^2 + 2ax - 2a^2 \Leftrightarrow y = 2ax - a^2$.

CUMULATIVE REVIEW TEST: CHAPTERS 10 and 11

1. **(a)** Because some variables are raised to the power 2, the system $\begin{cases} x^2 + y^2 = 4y \\ x^2 - 2y = 0 \end{cases}$ is nonlinear.

   (b) Subtracting the second equation from the first, we obtain

   $y^2 + 2y = 4y \Leftrightarrow y^2 - 2y = 0 \Leftrightarrow y\,(y - 2) = 0 \Leftrightarrow y = 0$ or 2.

   If $y = 0$, then the first equation gives $x^2 + 0 = 0 \Leftrightarrow x = 0$, and

   if $y = 2$, then the first equation gives $x^2 + 2^2 = 4 \cdot 2 \Leftrightarrow x^2 = 4$

   $\Leftrightarrow x = \pm 2$. Therefore, the three possible solutions are $(0, 0)$,

   $(-2, 2)$, and $(2, 2)$. We can verify that all three satisfy the

   second equation as well.

   (c) Completing the square in y in the first equation, we have

   $x^2 + y^2 = 4y \Leftrightarrow x^2 + (y - 2)^2 = 4$, a circle with radius 2

   centered at $(0, 2)$. The second equation can be written as

   $y = \frac{1}{2}x^2$, a parabola opening upward.

   (d), (e) Taking $(0, 1)$ as a test point, we find that

   $0^2 + 1^2 \le 4\,(1)$ and $0^2 - 2\,(1) \le 0$, so

   the region containing this point satisfies

   both inequalities.

   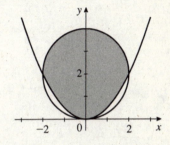

3. Let X, Y, and Z represent the number of fish caught by each of Xavier, Yolanda, and Zachary. "Yolanda catches as many fish as Xavier and Zachary put together" means that $Y = X + Z$. "Zachary catches 2 more fish than Xavier" means that $Z = X + 2$. "The total catch for all three people is 20 fish," so $X + Y + Z = 20$. From the first and third equations, we have $2Y = 20$, so $Y = 10$. From the first and second equations, we have $Y = X + (X + 2) = 2X + 2$, so $X = 4$. Finally, the third equation gives $Z = 20 - 10 - 4 = 6$. Therefore, Xavier caught 4 fish, Yolanda caught 10, and Zachary caught 6.

5. **(a)** A matrix equation equivalent to the system $\begin{cases} 5x - 3y = 5 \\ 6x - 4y = 0 \end{cases}$ is $AX = B$, where $A = \begin{bmatrix} 5 & -3 \\ 6 & -4 \end{bmatrix}$, $X = \begin{bmatrix} x \\ y \end{bmatrix}$, and

   $B = \begin{bmatrix} 5 \\ 0 \end{bmatrix}$.

   (b) Using the rule for finding the inverse of a 2×2 matrix, we get $A^{-1} = \dfrac{1}{5\,(-4) - (-3)\,6} \begin{bmatrix} -4 & -(-3) \\ -6 & 5 \end{bmatrix} = \begin{bmatrix} 2 & -\frac{3}{2} \\ 3 & -\frac{5}{2} \end{bmatrix}$.

   (c) $AX = B \Leftrightarrow A^{-1}AX = A^{-1}B \Leftrightarrow X = A^{-1}B = -\dfrac{1}{2} \begin{bmatrix} -4 & 3 \\ -6 & 5 \end{bmatrix} \begin{bmatrix} 5 \\ 0 \end{bmatrix} = -\dfrac{1}{2} \begin{bmatrix} -4\,(5) + 3\,(0) \\ -6\,(5) + 5\,(0) \end{bmatrix} = \begin{bmatrix} 10 \\ 15 \end{bmatrix}$.

   Thus, $x = 10$ and $y = 15$.

   (d) For this system we have $|D| = |A| = -2$, $|D_x| = \begin{vmatrix} 5 & -3 \\ 0 & -4 \end{vmatrix} = -20$, and $|D_y| = \begin{vmatrix} 5 & 5 \\ 6 & 0 \end{vmatrix} = -30$. Thus, the solution is

   $x = \dfrac{|D_x|}{|D|} = \dfrac{-20}{-2} = 10$ and $y = \dfrac{|D_y|}{|D|} = \dfrac{-30}{-2} = 15$, as in part (c).

7. The parabola with vertex at the origin and focus $F\,(0, 3)$ opens upward and has equation $x^2 = 4py$, where $p = 3$ is the y-coordinate of the focus. Thus, an equation is $x^2 = 12y$.

9. The center of the hyperbola lies halfway between its vertices; that is, at $\left(\frac{1+9}{2}, 0\right) = (5, 0)$. Because its transverse axis has length 8, we have $a = 4$. The foci lie 1 unit to the left and right of the center, so $c = 5$, and thus $b^2 = c^2 - a^2 = 3$. Therefore, an equation is $\dfrac{(x-5)^2}{16} - \dfrac{y^2}{9} = 1$.

12 SEQUENCES AND SERIES

12.1 SEQUENCES AND SUMMATION NOTATION

1. A sequence is a function whose domain is *the natural numbers*.

3. $a_n = n + 1$. Then $a_1 = 1 + 1 = 2$, $a_2 = 2 + 1 = 3$, $a_3 = 3 + 1 = 4$, $a_4 = 4 + 1 = 5$, and $a_{100} = 100 + 1 = 101$.

5. $a_n = \dfrac{1}{n+1}$. Then $a_1 = \dfrac{1}{1+1} = \dfrac{1}{2}$, $a_2 = \dfrac{1}{2+1} = \dfrac{1}{3}$, $a_3 = \dfrac{1}{3+1} = \dfrac{1}{4}$, $a_4 = \dfrac{1}{4+1} = \dfrac{1}{5}$, and $a_{100} = \dfrac{1}{100+1} = \dfrac{1}{101}$.

7. $a_n = \dfrac{(-1)^n}{n^2}$. Then $a_1 = \dfrac{(-1)^1}{1^2} = -1$, $a_2 = \dfrac{(-1)^2}{2^2} = \dfrac{1}{4}$, $a_3 = \dfrac{(-1)^3}{3^2} = -\dfrac{1}{9}$, $a_4 = \dfrac{(-1)^4}{4^2} = \dfrac{1}{16}$, and

$a_{100} = \dfrac{(-1)^{100}}{100^2} = \dfrac{1}{10,000}$.

9. $a_n = 1 + (-1)^n$. Then $a_1 = 1 + (-1)^1 = 0$, $a_2 = 1 + (-1)^2 = 2$, $a_3 = 1 + (-1)^3 = 0$, $a_4 = 1 + (-1)^4 = 2$, and $a_{100} = 1 + (-1)^{100} = 2$.

11. $a_n = n^n$. Then $a_1 = 1^1 = 1$, $a_2 = 2^2 = 4$, $a_3 = 3^3 = 27$, $a_4 = 4^4 = 256$, and $a_{100} = 100^{100} = 10^{200}$.

13. $a_n = 2\left(a_{n-1} - 2\right)$ and $a_1 = 3$. Then $a_2 = 2\left[(3) - 2\right] = 2$, $a_3 = 2\left[(2) - 2\right] = 0$, $a_4 = 2\left[(0) - 2\right] = -4$, and $a_5 = 2\left[(-4) - 2\right] = -12$.

15. $a_n = 2a_{n-1} + 1$ and $a_1 = 1$. Then $a_2 = 2\,(1) + 1 = 3$, $a_3 = 2 \cdot (3) + 1 = 7$, $a_4 = 2\,(7) + 1 = 15$, and $a_5 = 2\,(15) + 1 = 31$.

17. $a_n = a_{n-1} + a_{n-2}$, $a_1 = 1$, and $a_2 = 2$. Then $a_3 = 2 + 1 = 3$, $a_4 = 3 + 2 = 5$, and $a_5 = 5 + 3 = 8$.

19. (a) $a_1 = 7$, $a_2 = 11$, $a_3 = 15$, $a_4 = 19$, $a_5 = 23$,
$a_6 = 27$, $a_7 = 31$, $a_8 = 35$, $a_9 = 39$, $a_{10} = 43$

(b)

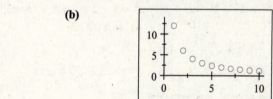

21. (a) $a_1 = \dfrac{12}{1} = 12$, $a_2 = \dfrac{12}{2} = 6$, $a_3 = \dfrac{12}{3} = 4$,
$a_4 = \dfrac{12}{4} = 3$, $a_5 = \dfrac{12}{5}$, $a_6 = \dfrac{12}{6} = 2$, $a_7 = \dfrac{12}{7}$,
$a_8 = \dfrac{12}{8} = \dfrac{3}{2}$, $a_9 = \dfrac{12}{9} = \dfrac{4}{3}$, $a_{10} = \dfrac{12}{10} = \dfrac{6}{5}$

(b)

23. (a) $a_1 = 2$, $a_2 = 0.5$, $a_3 = 2$, $a_4 = 0.5$, $a_5 = 2$,
$a_6 = 0.5$, $a_7 = 2$, $a_8 = 0.5$, $a_9 = 2$, $a_{10} = 0.5$

b.

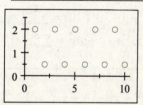

25. $2, 4, 8, 16, \ldots$. All are powers of 2, so $a_1 = 2$, $a_2 = 2^2$, $a_3 = 2^3$, $a_4 = 2^4, \ldots$ Thus $a_n = 2^n$.

27. $1, 4, 7, 10, \ldots$. The difference between any two consecutive terms is 3, so $a_1 = 3\,(1) - 2$, $a_2 = 3\,(2) - 2$, $a_3 = 3\,(3) - 2$,
$a_4 = 3\,(4) - 2, \ldots$ Thus $a_n = 3n - 2$.

29. $1, \frac{3}{4}, \frac{5}{9}, \frac{7}{16}, \frac{9}{25}, \ldots$. We consider the numerator separately from the denominator. The numerators of the terms differ

by 2, and the denominators are perfect squares. So $a_1 = \dfrac{2\,(1) - 1}{1^2}$, $a_2 = \dfrac{2\,(2) - 1}{2^2}$, $a_3 = \dfrac{2\,(3) - 1}{3^2}$, $a_4 = \dfrac{2\,(4) - 1}{4^2}$,

$a_5 = \dfrac{2\,(5) - 1}{5^2}, \ldots$ Thus $a_n = \dfrac{2n - 1}{n^2}$.

401

31. $0, 2, 0, 2, 0, 2, \ldots$. These terms alternate between 0 and 2. So $a_1 = 1 - 1$, $a_2 = 1 + 1$, $a_3 = 1 - 1$, $a_4 = 1 + 1$, $a_5 = 1 - 1$, $a_6 = 1 + 1, \ldots$ Thus $a_n = 1 + (-1)^n$.

33. $a_1 = 1$, $a_2 = 3$, $a_3 = 5$, $a_4 = 7, \ldots$. Therefore, $a_n = 2n - 1$. So $S_1 = 1$, $S_2 = 1 + 3 = 4$, $S_3 = 1 + 3 + 5+ = 9$, $S_4 = 1 + 3 + 5 + 7 = 16$, $S_5 = 1 + 3 + 5 + 7 + 9 = 25$, and $S_6 = 1 + 3 + 5 + 7 + 9 + 11 = 36$.

35. $a_1 = \frac{1}{3}$, $a_2 = \frac{1}{3^2}$, $a_3 = \frac{1}{3^3}$, $a_4 = \frac{1}{3^4}, \ldots$. Therefore, $a_n = \frac{1}{3^n}$. So $S_1 = \frac{1}{3}$, $S_2 = \frac{1}{3} + \frac{1}{3^2} = \frac{4}{9}$, $S_3 = \frac{1}{3} + \frac{1}{3^2} + \frac{1}{3^3} = \frac{13}{27}$, $S_4 = \frac{1}{3} + \frac{1}{3^2} + \frac{1}{3^3} + \frac{1}{3^4} = \frac{40}{81}$, and $S_5 = \frac{1}{3} + \frac{1}{3^2} + \frac{1}{3^3} + \frac{1}{3^4} + \frac{1}{3^5} = \frac{121}{243}$, $S_6 = \frac{1}{3} + \frac{1}{3^2} + \frac{1}{3^3} + \frac{1}{3^4} + \frac{1}{3^5} + \frac{1}{3^6} = \frac{364}{729}$.

37. $a_n = \frac{2}{3^n}$. So $S_1 = \frac{2}{3}$, $S_2 = \frac{2}{3} + \frac{2}{3^2} = \frac{8}{9}$, $S_3 = \frac{2}{3} + \frac{2}{3^2} + \frac{2}{3^3} = \frac{26}{27}$, and $S_4 = \frac{2}{3} + \frac{2}{3^2} + \frac{2}{3^3} + \frac{2}{3^4} = \frac{80}{81}$. Therefore, $S_n = \dfrac{3^n - 1}{3^n}$.

39. $a_n = \sqrt{n} - \sqrt{n+1}$. So $S_1 = \sqrt{1} - \sqrt{2} = 1 - \sqrt{2}$, $S_2 = \left(\sqrt{1} - \sqrt{2}\right) + \left(\sqrt{2} - \sqrt{3}\right) = 1 + \left(-\sqrt{2} + \sqrt{2}\right) - \sqrt{3} = 1 - \sqrt{3}$,

$S_3 = \left(\sqrt{1} - \sqrt{2}\right) + \left(\sqrt{2} - \sqrt{3}\right) + \left(\sqrt{3} - \sqrt{4}\right) = 1 + \left(-\sqrt{2} + \sqrt{2}\right) + \left(-\sqrt{3} + \sqrt{3}\right) - \sqrt{4} = 1 - \sqrt{4}$,

$S_4 = \left(\sqrt{1} - \sqrt{2}\right) + \left(\sqrt{2} - \sqrt{3}\right) + \left(\sqrt{3} - \sqrt{4}\right) + \left(\sqrt{4} - \sqrt{5}\right)$

$= 1 + \left(-\sqrt{2} + \sqrt{2}\right) + \left(-\sqrt{3} + \sqrt{3}\right) + \left(-\sqrt{4} + \sqrt{4}\right) - \sqrt{5} = 1 - \sqrt{5}$

Therefore,

$S_n = \left(\sqrt{1} - \sqrt{2}\right) + \left(\sqrt{2} - \sqrt{3}\right) + \cdots + \left(\sqrt{n} - \sqrt{n+1}\right)$

$= 1 + \left(-\sqrt{2} + \sqrt{2}\right) + \left(-\sqrt{3} + \sqrt{3}\right) + \cdots + \left(-\sqrt{n} + \sqrt{n}\right) - \sqrt{n+1} = 1 - \sqrt{n+1}$

41. $\sum_{k=1}^{4} k = 1 + 2 + 3 + 4 = 10$

43. $\sum_{k=1}^{3} \frac{1}{k} = 1 + \frac{1}{2} + \frac{1}{3} = \frac{6}{6} + \frac{3}{6} + \frac{2}{6} = \frac{11}{6}$

45. $\sum_{i=1}^{8} \left[1 + (-1)^i\right] = 0 + 2 + 0 + 2 + 0 + 2 + 0 + 2 = 8$

47. $\sum_{k=1}^{5} 2^{k-1} = 2^0 + 2^1 + 2^2 + 2^3 + 2^4 = 1 + 2 + 4 + 8 + 16 = 31$

49. 385 **51.** 46,438 **53.** 22

55. $\sum_{k=1}^{5} \sqrt{k} = \sqrt{1} + \sqrt{2} + \sqrt{3} + \sqrt{4} + \sqrt{5}$

57. $\sum_{k=0}^{6} \sqrt{k+4} = \sqrt{4} + \sqrt{5} + \sqrt{6} + \sqrt{7} + \sqrt{8} + \sqrt{9} + \sqrt{10}$

59. $\sum_{k=3}^{100} x^k = x^3 + x^4 + x^5 + \cdots + x^{100}$

61. $1 + 2 + 3 + 4 + \cdots + 100 = \sum_{k=1}^{100} k$

63. $1^2 + 2^2 + 3^2 + \cdots + 10^2 = \sum_{k=1}^{10} k^2$

65. $\dfrac{1}{1 \cdot 2} + \dfrac{1}{2 \cdot 3} + \dfrac{1}{3 \cdot 4} + \cdots + \dfrac{1}{999 \cdot 1000} = \sum_{k=1}^{999} \dfrac{1}{k(k+1)}$

67. $1 + x + x^2 + x^3 + \cdots + x^{100} = \sum_{k=0}^{100} x^k$

69. $\sqrt{2}, \sqrt{2\sqrt{2}}, \sqrt{2\sqrt{2\sqrt{2}}}, \sqrt{2\sqrt{2\sqrt{2\sqrt{2}}}}, \ldots$. We simplify each term in an attempt to determine a formula for a_n. So $a_1 = 2^{1/2}$, $a_2 = \sqrt{2 \cdot 2^{1/2}} = \sqrt{2^{3/2}} = 2^{3/4}$, $a_3 = \sqrt{2 \cdot 2^{3/4}} = \sqrt{2^{7/4}} = 2^{7/8}$, $a_4 = \sqrt{2 \cdot 2^{7/8}} = \sqrt{2^{15/8}} = 2^{15/16}, \ldots$. Thus $a_n = 2^{(2^n - 1)/2^n}$.

71. (a) $A_1 = \$2004$, $A_2 = \$2008.01$, $A_3 = \$2012.02$, $A_4 = \$2016.05$, $A_5 = \$2020.08$, $A_6 = \$2024.12$

(b) Since 3 years is 36 months, we get $A_{36} = \$2149.16$.

73. (a) $P_1 = 35{,}700$, $P_2 = 36{,}414$, $P_3 = 37{,}142$, $P_4 = 37{,}885$, $P_5 = 38{,}643$

(b) Since 2014 is 10 years after 2004, $P_{10} = 42{,}665$.

75. (a) The number of catfish at the end of the month, P_n, is the population at the start of the month, P_{n-1}, plus the increase in population, $0.08P_{n-1}$, minus the 300 catfish harvested. Thus $P_n = P_{n-1} + 0.08P_{n-1} - 300 \Leftrightarrow P_n = 1.08P_{n-1} - 300$.

(b) $P_1 = 5100$, $P_2 = 5208$, $P_3 = 5325$, $P_4 = 5451$, $P_5 = 5587$, $P_6 = 5734$, $P_7 = 5892$, $P_8 = 6064$, $P_9 = 6249$, $P_{10} = 6449$, $P_{11} = 6665$, $P_{12} = 6898$. Thus there should be 6898 catfish in the pond at the end of 12 months.

77. (a) Let S_n be his salary in the nth year. Then $S_1 = \$30{,}000$. Since his salary increase by 2000 each year, $S_n = S_{n-1} + 2000$. Thus $S_1 = \$30{,}000$ and $S_n = S_{n-1} + 2000$.

(b) $S_5 = S_4 + 2000 = (S_3 + 2000) + 2000 = (S_2 + 2000) + 4000 = (S_1 + 2000) + 6000 = \$38{,}000$.

79. Let F_n be the number of pairs of rabbits in the nth month. Clearly $F_1 = F_2 = 1$. In the nth month each pair that is two or more months old (that is, F_{n-2} pairs) will add a pair of offspring to the F_{n-1} pairs already present. Thus $F_n = F_{n-1} + F_{n-2}$. So F_n is the Fibonacci sequence.

81. $a_{n+1} = \begin{cases} \dfrac{a_n}{2} & \text{if } a_n \text{ is even} \\ 3a_n + 1 & \text{if } a_n \text{ is odd} \end{cases}$. With $a_1 = 11$, we have $a_2 = 34$, $a_3 = 17$, $a_4 = 52$, $a_5 = 26$, $a_6 = 13$, $a_7 = 40$, $a_8 = 20$, $a_9 = 10$, $a_{10} = 5$, $a_{11} = 16$, $a_{12} = 8$, $a_{13} = 4$, $a_{14} = 2$, $a_{15} = 1$, $a_{16} = 4$, $a_{17} = 2$, $a_{18} = 1$, $\ldots$ (with 4, 2, 1 repeating). So $a_{3n+1} = 4$, $a_{3n+2} = 2$, and $a_{3n} = 1$, for $n \geq 5$. With $a_1 = 25$, we have $a_2 = 76$, $a_3 = 38$, $a_4 = 19$, $a_5 = 58$, $a_6 = 29$, $a_7 = 88$, $a_8 = 44$, $a_9 = 22$, $a_{10} = 11$, $a_{11} = 34$, $a_{12} = 17$, $a_{13} = 52$, $a_{14} = 26$, $a_{15} = 13$, $a_{16} = 40$, $a_{17} = 20$, $a_{18} = 10$, $a_{19} = 5$, $a_{20} = 16$, $a_{21} = 8$, $a_{22} = 4$, $a_{23} = 2$, $a_{24} = 1$, $a_{25} = 4$, $a_{26} = 2$, $a_{27} = 1$, $\ldots$ (with 4, 2, 1 repeating). So $a_{3n+1} = 4$, $a_{3n+2} = 2$, and $a_{3n+3} = 1$ for $n \geq 7$.
We conjecture that the sequence will always return to the numbers 4, 2, 1 repeating.

12.2 ARITHMETIC SEQUENCES

1. An arithmetic sequence is sequence where the *difference* between successive terms is constant.

3. True. The nth partial sum of an arithmetic sequence is the average of the first and last terms times n.

5. (a) $a_1 = 5 + 2(1 - 1) = 5$,
$a_2 = 5 + 2(2 - 1) = 5 + 2 = 7$,
$a_3 = 5 + 2(3 - 1) = 5 + 4 = 9$,
$a_4 = 5 + 2(4 - 1) = 5 + 6 = 11$,
$a_5 = 5 + 2(5 - 1) = 5 + 8 = 13$

(b) The common difference is 2.

(c)

7. (a) $a_1 = \frac{5}{2} - (1 - 1) = \frac{5}{2}$, $a_2 = \frac{5}{2} - (2 - 1) = \frac{3}{2}$,
$a_3 = \frac{5}{2} - (3 - 1) = \frac{1}{2}$, $a_4 = \frac{5}{2} - (4 - 1) = -\frac{1}{2}$,
$a_5 = \frac{5}{2} - (5 - 1) = -\frac{3}{2}$

(b) The common difference is -1.

(c)

9. $a = 3$, $d = 5$, $a_n = a + d(n - 1) = 3 + 5(n - 1)$. So $a_{10} = 3 + 5(10 - 1) = 48$.

11. $a = \frac{5}{2}$, $d = -\frac{1}{2}$, $a_n = a + d(n - 1) = \frac{5}{2} - \frac{1}{2}(n - 1)$. So $a_{10} = \frac{5}{2} - \frac{1}{2}(10 - 1) = -2$.

13. $a_4 - a_3 = 14 - 11 = 3$, $a_3 - a_2 = 11 - 8 = 3$, $a_2 - a_1 = 8 - 5 = 3$. This sequence is arithmetic with common difference 3.

15. Since $a_2 - a_1 = 4 - 2 = 2$ and $a_4 - a_3 = 16 - 8 = 8$, the terms of the sequence do not have a common difference. This sequence is not arithmetic.

17. $a_4 - a_3 = -\frac{3}{2} - 0 = -\frac{3}{2}$, $a_3 - a_2 = 0 - \frac{3}{2} = -\frac{3}{2}$, $a_2 - a_1 = \frac{3}{2} - 3 = -\frac{3}{2}$. This sequence is arithmetic with common difference $-\frac{3}{2}$.

19. $a_4 - a_3 = 7.7 - 6.0 = 1.7$, $a_3 - a_2 = 6.0 - 4.3 = 1.7$, $4. - a_1 = 4.3 - 2.6 = 1.7$. This sequence is arithmetic with common difference 1.7.

21. $a_1 = 4 + 7(1) = 11$, $a_2 = 4 + 7(2) = 18$, $a_3 = 4 + 7(3) = 25$, $a_4 = 4 + 7(4) = 32$, $a_5 = 4 + 7(5) = 39$. This sequence is arithmetic, the common difference is $d = 7$ and $a_n = 4 + 7n = 4 + 7n - 7 + 7 = 11 + 7(n - 1)$.

23. $a_1 = \frac{1}{1 + 2(1)} = \frac{1}{3}$, $a_2 = \frac{1}{1 + 2(2)} = \frac{1}{5}$, $a_3 = \frac{1}{1 + 2(3)} = \frac{1}{7}$, $a_4 = \frac{1}{1 + 2(4)} = \frac{1}{9}$, $a_5 = \frac{1}{1 + 2(5)} = \frac{1}{11}$. Since $a_4 - a_3 = \frac{1}{9} - \frac{1}{7} = -\frac{2}{63}$ and $a_3 - a_2 = \frac{1}{7} - \frac{1}{5} = -\frac{2}{21}$, the terms of the sequence do not have a common difference. This sequence is not arithmetic.

25. $a_1 = 6(1) - 10 = -4$, $a_2 = 6(2) - 10 = 2$, $a_3 = 6(3) - 10 = 8$, $a_4 = 6(4) - 10 = 14$, $a_5 = 6(5) - 10 = 20$. This sequence is arithmetic, the common difference is $d = 6$ and $a_n = 6n - 10 = 6n - 6 + 6 - 10 = -4 + 6(n - 1)$.

27. 2, 5, 8, 11, …. Then $d = a_2 - a_1 = 5 - 2 = 3$, $a_5 = a_4 + 3 = 11 + 3 = 14$, $a_n = 2 + 3(n - 1)$, and $a_{100} = 2 + 3(99) = 299$.

29. 4, 9, 14, 19, …. Then $d = a_2 - a_1 = 9 - 4 = 5$, $a_5 = a_4 + 5 = 19 + 5 = 24$, $a_n = 4 + 5(n - 1)$, and $a_{100} = 4 + 5(99) = 499$.

31. $-12, -8, -4, 0, \ldots$. Then $d = a_2 - a_1 = -8 - (-12) = 4$, $a_5 = a_4 + 4 = 0 + 4 = 4$, $a_n = -12 + 4(n - 1)$, and $a_{100} = -12 + 4(99) = 384$.

33. 25, 26.5, 28, 29.5, …. Then $d = a_2 - a_1 = 26.5 - 25 = 1.5$, $a_5 = a_4 + 1.5 = 29.5 + 1.5 = 31$, $a_n = 25 + 1.5(n - 1)$, $a_{100} = 25 + 1.5(99) = 173.5$.

35. $2, 2 + s, 2 + 2s, 2 + 3s, \ldots$. Then $d = a_2 - a_1 = 2 + s - 2 = s$, $a_5 = a_4 + s = 2 + 3s + s = 2 + 4s$, $a_n = 2 + (n - 1)s$, and $a_{100} = 2 + 99s$.

37. $a_{10} = \frac{55}{2}$, $a_2 = \frac{7}{2}$, and $a_n = a + d(n - 1)$. Then $a_2 = a + d = \frac{7}{2} \Leftrightarrow d = \frac{7}{2} - a$. Substituting into $a_{10} = a + 9d = \frac{55}{2}$ gives $a + 9\left(\frac{7}{2} - a\right) = \frac{55}{2} \Leftrightarrow a = \frac{1}{2}$. Thus, the first term is $a_1 = \frac{1}{2}$.

39. $a_{100} = 98$ and $d = 2$. Note that $a_{100} = a + 99d = a + 99(2) = a + 198$. Since $a_{100} = 98$, we have $a + 198 = a_{100} = 98 \Leftrightarrow a = -100$. Hence, $a_1 = -100$, $a_2 = -100 + 2 = -98$, and $a_3 = -100 + 4 = -96$.

41. The arithmetic sequence is 1, 4, 7, …. So $d = 4 - 1 = 3$ and $a_n = 1 + 3(n - 1)$. Then $a_n = 88 \Leftrightarrow 1 + 3(n - 1) = 88 \Leftrightarrow 3(n - 1) = 87 \Leftrightarrow n - 1 = 29 \Leftrightarrow n = 30$. So 88 is the 30th term.

43. $a = 1$, $d = 2$, $n = 10$. Then $S_{10} = \frac{10}{2}[2a + (10 - 1)d] = \frac{10}{2}[2 \cdot 1 + 9 \cdot 2] = 100$.

45. $a = 4$, $d = 2$, $n = 20$. Then $S_{20} = \frac{20}{2}[2a + (20 - 1)d] = \frac{20}{2}[2 \cdot 4 + 19 \cdot 2] = 460$.

47. $a_1 = 55$, $d = 12$, $n = 10$. Then $S_{10} = \frac{10}{2}[2a + (10 - 1)d] = \frac{10}{2}[2 \cdot 55 + 9 \cdot 12] = 1090$.

49. $1 + 5 + 9 + \cdots + 401$ is a partial sum of an arithmetic series, where $a = 1$ and $d = 5 - 1 = 4$. The last term is $401 = a_n = 1 + 4(n - 1)$, so $n - 1 = 100 \Leftrightarrow n = 101$. So the partial sum is $S_{101} = \frac{101}{2}(1 + 401) = 101 \cdot 201 = 20{,}301$.

51. $0.7 + 2.7 + 4.7 + \cdots + 56.7$ is a partial sum of an arithmetic series, where $a = 0.7$ and $d = 2.7 - 0.7 = 2$. The last term is $56.7 = a_n = 0.7 + 2(n - 1) \Leftrightarrow 28 = n - 1 \Leftrightarrow n = 29$. So the partial sum is $S_{29} = \frac{29}{2}(0.7 + 56.7) = 832.3$.

53. $\sum_{k=0}^{10}(3 + 0.25k)$ is a partial sum of an arithmetic series where $a = 3 + 0.25 \cdot 0 = 3$ and $d = 0.25$. The last term is $a_{11} = 3 + 0.25 \cdot 10 = 5.5$. So the partial sum is $S_{11} = \frac{11}{2}(3 + 5.5) = 46.75$.

55. Let x denote the length of the side between the length of the other two sides. Then the lengths of the three sides of the triangle are $x - a$, x, and $x + a$, for some $a > 0$. Since $x + a$ is the longest side, it is the hypotenuse, and by the Pythagorean Theorem, we know that $(x - a)^2 + x^2 = (x + a)^2 \Leftrightarrow x^2 - 2ax + a^2 + x^2 = x^2 + 2ax + a^2 \Leftrightarrow x^2 - 4ax = 0 \Leftrightarrow x(x - 4a) = 0 \Rightarrow x = 4a$ ($x = 0$ is not a possible solution). Thus, the lengths of the three sides are $x - a = 4a - a = 3a$, $x = 4a$, and $x + a = 4a + a = 5a$. The lengths $3a$, $4a$, $5a$ are proportional to 3, 4, 5, and so the triangle is similar to a 3-4-5 triangle.

57. The sequence $1, \frac{3}{5}, \frac{3}{7}, \frac{1}{3}, \ldots$ is harmonic if $1, \frac{5}{3}, \frac{7}{3}, 3, \ldots$ forms an arithmetic sequence. Since $\frac{5}{3} - 1 = \frac{7}{3} - \frac{5}{3} = 3 - \frac{7}{3} = \frac{2}{3}$, the sequence of reciprocals is arithmetic and thus the original sequence is harmonic.

59. We have an arithmetic sequence with $a = 5$ and $d = 2$. We seek n such that $2700 = S_n = \frac{n}{2}\left[2a + (n-1)d\right]$. Solving for n, we have $2700 = \frac{n}{2}\left[10 + 2(n-1)\right] \Leftrightarrow 5400 = 10n + 2n^2 - 2n \Leftrightarrow n^2 + 4n - 2700 = 0 \Leftrightarrow (n-50)(n+54) = 0 \Leftrightarrow n = 50$ or $n = -54$. Since n is a positive integer, 50 terms of the sequence must be added to get 2700.

61. The diminishing values of the computer form an arithmetic sequence with $a_1 = 12{,}500$ and common difference $d = -1875$. Thus the value of the computer after 6 years is $a_7 = 12{,}500 + (7-1)(-1875) = \$1250$.

63. The increasing values of the man's salary form an arithmetic sequence with $a_1 = 30{,}000$ and common difference $d = 2300$. Then his total earnings for a ten-year period are $S_{10} = \frac{10}{2}\left[2(30{,}000) + 9(2300)\right] = 403{,}500$. Thus his total earnings for the 10 year period are $\$403{,}500$.

65. The number of seats in the nth row is given by the nth term of an arithmetic sequence with $a_1 = 15$ and common difference $d = 3$. We need to find n such that $S_n = 870$. So we solve $870 = S_n = \frac{n}{2}\left[2(15) + (n-1)3\right]$ for n. We have $870 = \frac{n}{2}(27 + 3n) \Leftrightarrow 1740 = 3n^2 + 27n \Leftrightarrow 3n^2 + 27n - 1740 = 0 \Leftrightarrow n^2 + 9n - 580 = 0 \Leftrightarrow (x-20)(x+29) = 0 \Rightarrow n = 20$ or $n = -29$. Since the number of rows is positive, the theater must have 20 rows.

67. The number of gifts on the 12th day is $1 + 2 + 3 + 4 + \cdots + 12$. Since $a_2 - a_1 = a_3 - a_2 = a_4 - a_3 = \cdots = 1$, the number of gifts on the 12th day is the partial sum of an arithmetic sequence with $a = 1$ and $d = 1$. So the sum is
$$S_{12} = 12\left(\frac{1+12}{2}\right) = 6 \cdot 13 = 78.$$

12.3 GEOMETRIC SEQUENCES

1. A geometric sequence is a sequence where the *ratio* between successive terms is constant.

3. True. If we know the first and second terms of a geometric sequence then we can find all other terms.

5. (a) $a_1 = 5(2)^0 = 5$, $a_2 = 5(2)^1 = 10$, $a_3 = 5(2)^2 = 20$, $a_4 = 5(2)^3 = 40$, $a_5 = 5(2)^4 = 80$

(b) The common ratio is 2.

(c)

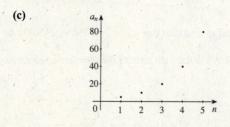

7. (a) $a_1 = \frac{5}{2}\left(-\frac{1}{2}\right)^0 = \frac{5}{2}$, $a_2 = \frac{5}{2}\left(-\frac{1}{2}\right)^1 = -\frac{5}{4}$, $a_3 = \frac{5}{2}\left(-\frac{1}{2}\right)^2 = \frac{5}{8}$, $a_4 = \frac{5}{2}\left(-\frac{1}{2}\right)^3 = -\frac{5}{16}$, $a_5 = \frac{5}{2}\left(-\frac{1}{2}\right)^4 = \frac{5}{32}$

(b) The common ratio is $-\frac{1}{2}$.

(c)

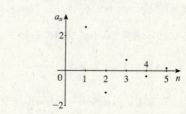

9. $a = 3$, $r = 5$. So $a_n = ar^{n-1} = 3(5)^{n-1}$ and $a_4 = 3 \cdot 5^3 = 375$.

11. $a = \frac{5}{2}$, $r = -\frac{1}{2}$. So $a_n = ar^{n-1} = \frac{5}{2}\left(-\frac{1}{2}\right)^{n-1}$ and $a_4 = \frac{5}{2} \cdot \left(-\frac{1}{2}\right)^3 = -\frac{5}{16}$.

13. $\frac{a_2}{a_1} = \frac{4}{2} = 2$, $\frac{a_3}{a_2} = \frac{8}{4} = 2$, $\frac{a_4}{a_3} = \frac{16}{8} = 2$. Since these ratios are the same, the sequence is geometric with the common ratio 2.

15. $\frac{a_2}{a_1} = \frac{3/2}{3} = \frac{1}{2}, \frac{a_3}{a_2} = \frac{3/4}{3/2} = \frac{1}{2}, \frac{a_4}{a_3} = \frac{3/8}{3/4} = \frac{1}{2}$. Since these ratios are the same, the sequence is geometric with the common ratio $\frac{1}{2}$.

17. $\frac{a_2}{a_1} = \frac{1/3}{1/2} = \frac{2}{3}, \frac{a_4}{a_3} = \frac{1/5}{1/4} = \frac{4}{5}$. Since these ratios are not the same, this is not a geometric sequence.

19. $\frac{a_2}{a_1} = \frac{1.1}{1.0} = 1.1, \frac{a_3}{a_2} = \frac{1.21}{1.1} = 1.1, \frac{a_4}{a_3} = \frac{1.331}{1.21} = 1.1$. Since these ratios are the same, the sequence is geometric with the common ratio 1.1.

21. $a_1 = 2\,(3)^1 = 6, a_2 = 2\,(3)^2 = 18, a_3 = 2\,(3)^3 = 54, a_4 = 2\,(3)^4 = 162, a_5 = 2\,(3)^5 = 486$. This sequence is geometric, the common ratio is $r = 3$ and $a_n = a_1 r^{n-1} = 6\,(3)^{n-1}$.

23. $a_1 = \frac{1}{4}, a_2 = \frac{1}{4^2} = \frac{1}{16}, a_3 = \frac{1}{4^3} = \frac{1}{64}, a_4 = \frac{1}{4^4} = \frac{1}{256}, a_5 = \frac{1}{4^5} = \frac{1}{1024}$. This sequence is geometric, the common ratio is $r = \frac{1}{4}$ and $a_n = a_1 r^{n-1} = \frac{1}{4}\left(\frac{1}{4}\right)^{n-1}$.

25. Since $\ln a^b = b \ln a$, we have $a_1 = \ln\left(5^0\right) = \ln 1 = 0, a_2 = \ln\left(5^1\right) = \ln 5, a_3 = \ln\left(5^2\right) = 2\ln 5, a_4 = \ln\left(5^3\right) = 3\ln 5$, $a_5 = \ln\left(5^4\right) = 4\ln 5$. Since $a_1 = 0$ and $a_2 \neq 0$, this sequence is not geometric.

27. $2, 6, 18, 54, \ldots$. Then $r = \frac{a_2}{a_1} = \frac{6}{2} = 3, a_5 = a_4 \cdot 3 = 54\,(3) = 162$, and $a_n = 2 \cdot 3^{n-1}$.

29. $0.3, -0.09, 0.027, -0.0081, \ldots$. Then $r = \frac{a_2}{a_1} = \frac{-0.09}{0.3} = -0.3, a_5 = a_4 \cdot (-0.3) = -0.0081\,(-0.3) = 0.00243$, and $a_n = 0.3\,(-0.3)^{n-1}$.

31. $144, -12, 1, -\frac{1}{12}, \ldots$. Then $r = \frac{a_2}{a_1} = \frac{-12}{144} = -\frac{1}{12}, a_5 = a_4 \cdot \left(-\frac{1}{12}\right) = -\frac{1}{12}\left(-\frac{1}{12}\right) = \frac{1}{144}, a_n = 144\left(-\frac{1}{12}\right)^{n-1}$.

33. $3, 3^{5/3}, 3^{7/3}, 27, \ldots$. Then $r = \frac{a_2}{a_1} = \frac{3^{5/3}}{3} = 3^{2/3}, a_5 = a_4 \cdot \left(3^{2/3}\right) = 27 \cdot 3^{2/3} = 3^{11/3}$, and $a_n = 3\left(3^{2/3}\right)^{n-1} = 3 \cdot 3^{(2n-2)/3} = 3^{(2n+1)/3}$.

35. $1, s^{2/7}, s^{4/7}, s^{6/7}, \ldots$. Then $r = \frac{a_2}{a_1} = \frac{s^{2/7}}{1} = s^{2/7}, a_5 = a_4 \cdot s^{2/7} = s^{6/7} \cdot s^{2/7} = s^{8/7}$, and $a_n = \left(s^{2/7}\right)^{n-1} = s^{(2n-2)/7}$.

37. $a_1 = 8, a_2 = 4$. Thus $r = \frac{a_2}{a_1} = \frac{4}{8} = \frac{1}{2}$ and $a_5 = a_1 r^{5-1} = 8\left(\frac{1}{2}\right)^4 = \frac{8}{16} = \frac{1}{2}$.

39. $r = \frac{2}{5}, a_4 = \frac{5}{2}$. Since $r = \frac{a_4}{a_3}$, we have $a_3 = \frac{a_4}{r} = \frac{5/2}{2/5} = \frac{25}{4}$.

41. The geometric sequence is $2, 6, 18, \ldots$. Thus $r = \frac{a_2}{a_1} = \frac{6}{2} = 3$. We need to find n so that $a_n = 2 \cdot 3^{n-1} = 118{,}098 \Leftrightarrow$ $3^{n-1} = 59{,}049 \Leftrightarrow n - 1 = \log_3 59{,}049 = 10 \Leftrightarrow n = 11$. Therefore, $118{,}098$ is the 11th term of the geometric sequence.

43. $a = 5, r = 2, n = 6$. Then $S_6 = 5 \frac{1 - 2^6}{1 - 2} = (-5)\,(-63) = 315$.

45. $a_3 = 28, a_6 = 224, n = 6$. So $\frac{a_6}{a_3} = \frac{ar^5}{ar^2} = r^3$. So we have $r^3 = \frac{a_6}{a_3} = \frac{224}{28} = 8$, and hence $r = 2$. Since $a_3 = a \cdot r^2$, we get $a = \frac{a_3}{r^2} = \frac{28}{2^2} = 7$. So $S_6 = 7 \frac{1 - 2^6}{1 - 2} = (-7)\,(-63) = 441$.

47. $1 + 3 + 9 + \cdots + 2187$ is a partial sum of a geometric sequence, where $a = 1$ and $r = \frac{a_2}{a_1} = \frac{3}{1} = 3$. Then the last term is $2187 = a_n = 1 \cdot 3^{n-1} \Leftrightarrow n - 1 = \log_3 2187 = 7 \Leftrightarrow n = 8$. So the partial sum is $S_8 = (1)\frac{1 - 3^8}{1 - 3} = 3280$.

49. $\sum_{k=0}^{10} 3 \left(\frac{1}{2}\right)^k$ is a partial sum of a geometric sequence, where $a = 3$, $r = \frac{1}{2}$, and $n = 11$. So the partial sum is

$$S_{11} = (3) \frac{1 - \left(\frac{1}{2}\right)^{11}}{1 - \left(\frac{1}{2}\right)} = 6\left[1 - \left(\frac{1}{2}\right)^{11}\right] = \frac{6141}{1024} \approx 5.997070313.$$

51. $1 + \frac{1}{3} + \frac{1}{9} + \frac{1}{27} + \cdots$ is an infinite geometric series with $a = 1$ and $r = \frac{1}{3}$. Therefore, it is convergent with sum

$$S = \frac{a}{1-r} = \frac{1}{1 - \left(\frac{1}{3}\right)} = \frac{3}{2}.$$

53. $1 - \frac{1}{3} + \frac{1}{9} - \frac{1}{27} + \cdots$ is an infinite geometric series with $a = 1$ and $r = -\frac{1}{3}$. Therefore, it is convergent with sum

$$S = \frac{a}{1-r} = \frac{1}{1 - \left(-\frac{1}{3}\right)} = \frac{3}{4}.$$

55. $1 + \frac{3}{2} + \left(\frac{3}{2}\right)^2 + \left(\frac{3}{2}\right)^3 + \cdots$ is an infinite geometric series with $a = 1$ and $r = \frac{3}{2} > 1$. Therefore, the series diverges.

57. $3 - \frac{3}{2} + \frac{3}{4} - \frac{3}{8} + \cdots$ is an infinite geometric series with $a = 3$ and $r = -\frac{1}{2}$. Therefore, it is convergent with sum

$$S = \frac{3}{1 - \left(-\frac{1}{2}\right)} = 2.$$

59. $3 - 3\,(1.1) + 3\,(1.1)^2 - 3\,(1.1)^3 + \cdots$ is an infinite geometric series with $a = 3$ and $r = 1.1 > 1$. Therefore, the series diverges.

61. $\frac{1}{\sqrt{2}} + \frac{1}{2} + \frac{1}{2\sqrt{2}} + \frac{1}{4} + \cdots$ is an infinite geometric series with $a = \frac{1}{\sqrt{2}}$ and $r = \frac{1}{\sqrt{2}}$. Therefore, the sum of the series is

$$S = \frac{\frac{1}{\sqrt{2}}}{1 - \frac{1}{\sqrt{2}}} = \frac{1}{\sqrt{2} - 1} = \sqrt{2} + 1.$$

63. $0.777\ldots = \frac{7}{10} + \frac{7}{100} + \frac{7}{1000} + \cdots$ is an infinite geometric series with $a = \frac{7}{10}$ and $r = \frac{1}{10}$. Thus

$$0.777\ldots = \frac{a}{1-r} = \frac{\frac{7}{10}}{1 - \frac{1}{10}} = \frac{7}{9}.$$

65. $0.030303\ldots = \frac{3}{100} + \frac{3}{10,000} + \frac{3}{1,000,000} + \cdots$ is an infinite geometric series with $a = \frac{3}{100}$ and $r = \frac{1}{100}$. Thus

$$0.030303\ldots = \frac{a}{1-r} = \frac{\frac{3}{100}}{1 - \frac{1}{100}} = \frac{3}{99} = \frac{1}{33}.$$

67. $0.\overline{112} = 0.112112112\ldots = \frac{112}{1000} + \frac{112}{1,000,000} + \frac{112}{1,000,000,000} + \cdots$ is an infinite geometric series with $a = \frac{112}{1000}$ and

$r = \frac{1}{1000}$. Thus $0.112112112\ldots = \frac{a}{1-r} = \frac{\frac{112}{1000}}{1 - \frac{1}{1000}} = \frac{112}{999}$.

69. Since we have 5 terms, let us denote $a_1 = 5$ and $a_5 = 80$. Also, $\frac{a_5}{a_1} = r^4$ because the sequence is geometric, and so

$r^4 = \frac{80}{5} = 16 \Leftrightarrow r = \pm 2$. If $r = 2$, the three geometric means are $a_2 = 10$, $a_3 = 20$, and $a_4 = 40$. (If $r = -2$, the three geometric means are $a_2 = -10$, $a_3 = 20$, and $a_4 = -40$, but these are not between 5 and 80.)

71. (a) The value at the end of the year is equal to the value at beginning less the depreciation, so

$V_n = V_{n-1} - 0.2V_{n-1} = 0.8V_{n-1}$ with $V_1 = 160,000$. Thus $V_n = 160,000 \cdot 0.8^{n-1}$.

(b) $V_n < 100,000 \Leftrightarrow 0.8^{n-1} \cdot 160,000 < 100,000 \Leftrightarrow 0.8^{n-1} < 0.625 \Leftrightarrow (n-1)\log 0.8 < \log 0.625 \Leftrightarrow$

$n - 1 > \frac{\log 0.625}{\log 0.8} = 2.11$. Thus it will depreciate to below \$100,000 during the fourth year.

73. Since the ball is dropped from a height of 80 feet, $a = 80$. Also since the ball rebounds three-fourths of the distance fallen, $r = \frac{3}{4}$. So on the nth bounce, the ball attains a height of $a_n = 80\left(\frac{3}{4}\right)^n$. Hence, on the fifth bounce, the ball goes $a_5 = 80\left(\frac{3}{4}\right)^5 = \frac{80 \cdot 243}{1024} \approx 19$ ft high.

75. Let a_n be the amount of water remaining at the nth stage. We start with 5 gallons, so $a = 5$. When 1 gallon (that is, $\frac{1}{5}$ of the mixture) is removed, $\frac{4}{5}$ of the mixture (and hence $\frac{4}{5}$ of the water in the mixture) remains. Thus, $a_1 = 5 \cdot \frac{4}{5}$, $a_2 = 5 \cdot \frac{4}{5} \cdot \frac{4}{5}, \ldots$, and in general, $a_n = 5\left(\frac{4}{5}\right)^n$. The amount of water remaining after 3 repetitions is $a_3 = 5\left(\frac{4}{5}\right)^3 = \frac{64}{25}$, and after 5 repetitions it is $a_5 = 5\left(\frac{4}{5}\right)^5 = \frac{1024}{625}$.

77. Let a_n be the height the ball reaches on the nth bounce. From the given information, a_n is the geometric sequence $a_n = 9 \cdot \left(\frac{1}{3}\right)^n$. (Notice that the ball hits the ground for the fifth time after the fourth bounce.)

(a) $a_0 = 9$, $a_1 = 9 \cdot \frac{1}{3} = 3$, $a_2 = 9 \cdot \left(\frac{1}{3}\right)^2 = 1$, $a_3 = 9 \cdot \left(\frac{1}{3}\right)^3 = \frac{1}{3}$, and $a_4 = 9 \cdot \left(\frac{1}{3}\right)^4 = \frac{1}{9}$. The total distance traveled is

$a_0 + 2a_1 + 2a_2 + 2a_3 + 2a_4 = 9 + 2 \cdot 3 + 2 \cdot 1 + 2 \cdot \frac{1}{3} + 2 \cdot \frac{1}{9} = \frac{161}{9} = 17\frac{8}{9}$ ft.

(b) The total distance traveled at the instant the ball hits the ground for the nth time is

$$
\begin{aligned}
D_n &= 9 + 2 \cdot 9 \cdot \frac{1}{3} + 2 \cdot 9 \cdot \left(\frac{1}{3}\right)^2 + 2 \cdot 9 \cdot \left(\frac{1}{3}\right)^3 + 2 \cdot 9 \cdot \left(\frac{1}{3}\right)^4 + \cdots + 2 \cdot 9 \cdot \left(\frac{1}{3}\right)^{n-1} \\
&= 2\left[9 + 9 \cdot \frac{1}{3} + 9 \cdot \left(\frac{1}{3}\right)^2 + 9 \cdot \left(\frac{1}{3}\right)^3 + 9 \cdot \left(\frac{1}{3}\right)^4 + \cdots + 9 \cdot \left(\frac{1}{3}\right)^{n-1}\right] - 9 \\
&= 2\left[9 \cdot \frac{1 - \left(\frac{1}{3}\right)^n}{1 - \frac{1}{3}}\right] - 9 = 27\left[1 - \left(\frac{1}{3}\right)^n\right] - 9 = 18 - \left(\frac{1}{3}\right)^{n-3}
\end{aligned}
$$

79. Let $a_1 = 1$ be the man with 7 wives. Also, let $a_2 = 7$ (the wives), $a_3 = 7a_2 = 7^2$ (the sacks), $a_4 = 7a_3 = 7^3$ (the cats), and $a_5 = 7a_4 = 7^4$ (the kits). The total is $a_1 + a_2 + a_3 + a_4 + a_5 = 1 + 7 + 7^2 + 7^3 + 7^4$, which is a partial sum of a geometric sequence with $a = 1$ and $r = 7$. Thus, the number in the party is $S_5 = 1 \cdot \dfrac{1 - 7^5}{1 - 7} = 2801$.

81. Let a_n be the height the ball reaches on the nth bounce. We have $a_0 = 1$ and $a_n = \frac{1}{2}a_{n-1}$. Since the total distance d traveled includes the bounce up as well and the distance down, we have

$$
\begin{aligned}
d &= a_0 + 2 \cdot a_1 + 2 \cdot a_2 + \cdots = 1 + 2\left(\frac{1}{2}\right) + 2\left(\frac{1}{2}\right)^2 + 2\left(\frac{1}{2}\right)^3 + 2\left(\frac{1}{2}\right)^4 + \cdots \\
&= 1 + 1 + \frac{1}{2} + \left(\frac{1}{2}\right)^2 + \left(\frac{1}{2}\right)^3 + \cdots = 1 + \sum_{i=0}^{\infty}\left(\frac{1}{2}\right)^i = 1 + \frac{1}{1 - \frac{1}{2}} = 3
\end{aligned}
$$

Thus the total distance traveled is about 3 m.

83. (a) If a square has side x, then by the Pythagorean Theorem the length of the side of the square formed by joining the midpoints is, $\sqrt{\left(\frac{x}{2}\right)^2 + \left(\frac{x}{2}\right)^2} = \sqrt{\frac{x^2}{4} + \frac{x^2}{4}} = \frac{x}{\sqrt{2}}$. In our case, $x = 1$ and the side of the first inscribed square is $\frac{1}{\sqrt{2}}$, the side of the second inscribed square is $\frac{1}{\sqrt{2}} \cdot \frac{1}{\sqrt{2}} = \left(\frac{1}{\sqrt{2}}\right)^2$, the side of the third inscribed square is $\left(\frac{1}{\sqrt{2}}\right)^3$, and so on. Since this pattern continues, the total area of all the squares is

$$
A = 1^2 + \left(\frac{1}{\sqrt{2}}\right)^2 + \left(\frac{1}{\sqrt{2}}\right)^4 + \left(\frac{1}{\sqrt{2}}\right)^6 + \cdots = 1 + \frac{1}{2} + \left(\frac{1}{2}\right)^2 + \left(\frac{1}{2}\right)^3 + \cdots = \frac{1}{1 - \frac{1}{2}} = 2.
$$

(b) As in part (a), the sides of the squares are $1, \frac{1}{\sqrt{2}}, \left(\frac{1}{\sqrt{2}}\right)^2, \left(\frac{1}{\sqrt{2}}\right)^3, \dots$. Thus the sum of the perimeters is

$$S = 4 \cdot 1 + 4 \cdot \frac{1}{\sqrt{2}} + 4 \cdot \left(\frac{1}{\sqrt{2}}\right)^2 + 4 \cdot \left(\frac{1}{\sqrt{2}}\right)^3 + \cdots, \text{ which is an infinite geometric series with } a = 4 \text{ and } r = \frac{1}{\sqrt{2}}. \text{ Thus}$$

the sum of the perimeters is $S = \dfrac{4}{1 - \frac{1}{\sqrt{2}}} = \dfrac{4\sqrt{2}}{\sqrt{2}-1} = \dfrac{4\sqrt{2}}{\sqrt{2}-1} \cdot \dfrac{\sqrt{2}+1}{\sqrt{2}+1} = \dfrac{4 \cdot 2 + 4\sqrt{2}}{2-1} = 8 + 4\sqrt{2}.$

85. Let a_n denote the area colored blue at nth stage. Since only the middle squares are colored blue,

$a_n = \frac{1}{9} \times$ (area remaining yellow at the $(n-1)$th stage). Also, the area remaining yellow at the nth stage is $\frac{8}{9}$ of the area

remaining yellow at the preceding stage. So $a_1 = \frac{1}{9}, a_2 = \frac{1}{9}\left(\frac{8}{9}\right), a_3 = \frac{1}{9}\left(\frac{8}{9}\right)^2, a_4 = \frac{1}{9}\left(\frac{8}{9}\right)^3, \dots$. Thus the total area

colored blue $A = \frac{1}{9} + \frac{1}{9}\left(\frac{8}{9}\right) + \frac{1}{9}\left(\frac{8}{9}\right)^2 + \frac{1}{9}\left(\frac{8}{9}\right)^3 + \cdots$ is an infinite geometric series with $a = \frac{1}{9}$ and $r = \frac{8}{9}$. So the total

area is $A = \dfrac{\frac{1}{9}}{1 - \frac{8}{9}} = 1.$

87. Let $a_1, a_2, a_3, \dots$ be a geometric sequence with common ratio r. Thus $a_2 = a_1 r, a_3 = a_1 \cdot r^2, \dots, a_n = a_1 \cdot r^{n-1}$. Hence,

$$\frac{1}{a_2} = \frac{1}{a_1 \cdot r} = \frac{1}{a_1} \cdot \frac{1}{r}, \; \frac{1}{a_3} = \frac{1}{a_1 \cdot r^2} = \frac{1}{a_1} \cdot \frac{1}{r^2} = \frac{1}{a_1}\left(\frac{1}{r}\right)^2, \dots \; \frac{1}{a_n} = \frac{1}{a_1 \cdot r^{n-1}} = \frac{1}{a_1} \cdot \frac{1}{r^{n-1}} = \frac{1}{a_1}\left(\frac{1}{r}\right)^{n-1}, \text{ and so}$$

$\dfrac{1}{a_1}, \dfrac{1}{a_2}, \dfrac{1}{a_3}, \dots$ is a geometric sequence with common ratio $\dfrac{1}{r}$.

89. Since $a_1, a_2, a_3, \dots$ is an arithmetic sequence with common difference d, the terms can be expressed as $a_2 = a_1 + d$,

$a_3 = a_1 + 2d, \dots, a_n = a_1 + (n-1)d$. So $10^{a_2} = 10^{a_1+d} = 10^{a_1} \cdot 10^d$, $10^{a_3} = 10^{a_1+2d} = 10^{a_1} \cdot \left(10^d\right)^2, \dots,$

$10^{a_n} = 10^{a_1+(n-1)d} = 10^{a_1} \cdot \left(10^d\right)^{n-1}$, and so $10^{a_1}, 10^{a_2}, 10^{a_3}, \dots$ is a geometric sequence with common ratio $r = 10^d$.

12.4   MATHEMATICS OF FINANCE

1. An annuity is a sum of money that is paid in regular equal payments. The *amount* of an annuity is the sum of all the individual payments together with all the interest.

3. $n = 10, R = \$1000, i = 0.06$. So $A_f = R\dfrac{(1+i)^n - 1}{i} = 1000\dfrac{(1+0.06)^{10} - 1}{0.06} = \$13{,}180.79.$

5. $n = 20, R = \$5000, i = 0.12$. So $A_f = R\dfrac{(1+i)^n - 1}{i} = 5000\dfrac{(1+0.12)^{20} - 1}{0.12} = \$360{,}262.21.$

7. $n = 16, R = \$300, i = \dfrac{0.08}{4} = 0.02$. So $A_f = R\dfrac{(1+i)^n - 1}{i} = 300\dfrac{(1+0.02)^{16} - 1}{0.02} = \$5{,}591.79.$

9. $A_f = 5000, n = 4 \cdot 2 = 8, i = \dfrac{0.10}{4} = 0.025$. So $R = \dfrac{iA_f}{(1+i)^n - 1} = \dfrac{(0.025)(5000)}{(1.025)^8 - 1} = \$572.34.$

11. $R = 1000, n = 20, i = \dfrac{0.09}{2} = 0.045$. So $A_p = R\dfrac{1 - (1+i)^{-n}}{i} = 1000\dfrac{1 - (1.045)^{-20}}{0.045} = \$13{,}007.94.$

13. $R = \$200, n = 20, i = \dfrac{0.09}{2} = 0.045$. So $A_p = R\dfrac{1 - (1+i)^{-n}}{i} = (200)\dfrac{1 - (1+0.045)^{-20}}{0.045} = \$2601.59.$

15. $A_p = \$12{,}000, i = \dfrac{0.105}{12} = 0.00875, n = 48$. Then $R = \dfrac{iA_p}{1 - (1+i)^{-n}} = \dfrac{(0.00875)(12000)}{1 - (1+0.00875)^{-48}} = \$307.24.$

17. $A_p = \$100{,}000, i = \dfrac{0.08}{12} \approx 0.006667, n = 360$. Then $R = \dfrac{iA_p}{1 - (1+i)^{-n}} = \dfrac{(0.006667)(100{,}000)}{1 - (1+0.006667)^{-360}} = \$733.76.$

Therefore, the total amount paid on this loan over the 30 year period is $(360)(733.76) = \$264{,}153.60.$

19. $R = 3500$, $n = 12\,(30) = 360$, $i = \dfrac{0.06}{12} = 0.005$. So $A_p = R\dfrac{1 - (1 + i)^{-n}}{i} = 3500\dfrac{1 - (1.005)^{-360}}{0.005} = \$583{,}770.65$.

Therefore, Dr. Gupta can afford a loan of $583,770.65.

21. $R = 220$, $n = 12\,(3) = 36$, $i = \dfrac{0.08}{12} \approx 0.00667$.   The amount borrowed is

$A_p = R\dfrac{1 - (1 + i)^{-n}}{i} = 220\dfrac{1 - (1.00667)^{-36}}{0.00667} = \$7{,}020.60$.   So she purchased the car for

$7,020.60 + \$2000 = \$9020.60$.

23. $A_p = 100{,}000$, $n = 360$, $i = \dfrac{0.0975}{12} = 0.008125$.

(a) $R = \dfrac{i A_p}{1 - (1 + i)^{-n}} = \dfrac{(0.008125)\,(100{,}000)}{1 - (1 + 0.008125)^{-360}} = \$859.15$.

(b) The total amount that will be paid over the 30 year period is $(360)\,(859.15) = \$309{,}294.00$.

(c) $R = \$859.15$, $i = \dfrac{0.0975}{12} = 0.008125$, $n = 360$. So $A_f = 859.15\dfrac{(1 + 0.008125)^{360} - 1}{0.008125} = \$1{,}841{,}519.29$.

25. $A_p = \$640$, $R = \$32$, $n = 24$. We want to solve the equation $R = \dfrac{i A_p}{1 - (1 + i)^n}$

for the interest rate i. Let x be the interest rate, then $i = \dfrac{x}{12}$. So we can express R

as a function of x by $R\,(x) = \dfrac{\dfrac{x}{12} \cdot 640}{1 - \left(1 + \dfrac{x}{12}\right)^{-24}}$. We graph $R\,(x)$ and $y = 32$ in

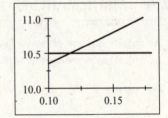

the rectangle $[0.12, 0.22] \times [30, 34]$. The x-coordinate of the intersection is about
0.1816, which corresponds to an interest rate of 18.16%.

27. $A_p = \$189.99$, $R = \$10.50$, $n = 20$. We want to solve the equation

$R = \dfrac{i A_p}{1 - (1 + i)^n}$ for the interest rate i. Let x be the interest rate, then $i = \dfrac{x}{12}$. So

we can express R as a function of x by $R\,(x) = \dfrac{\dfrac{x}{12} \cdot 189.99}{1 - \left(1 + \dfrac{x}{12}\right)^{-20}}$. We graph

$R\,(x)$ and $y = 10.50$ in the rectangle $[0.10, 0.18] \times [10, 11]$. The x-coordinate of
the intersection is about 0.1168, which corresponds to an interest rate of 11.68%.

29. (a) The present value of the kth payment is $PV = R\,(1 + i)^{-k} = \dfrac{R}{(1 + i)^k}$. The present value of an annuity is the sum of

the present values of each of the payments of R dollars, as shown in the time line.

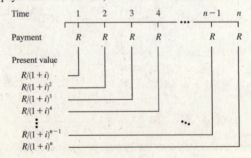

(b)
$$A_p = \frac{R}{1+i} + \frac{R}{(1+i)^2} + \frac{R}{(1+i)^3} + \cdots + \frac{R}{(1+i)^n}$$

$$= \frac{R}{1+i} + \left(\frac{R}{1+i}\right)\left(\frac{1}{1+i}\right) + \left(\frac{R}{1+i}\right)\left(\frac{1}{1+i}\right)^2 + \cdots + \left(\frac{R}{1+i}\right)\left(\frac{1}{1+i}\right)^{n-1}$$

This is a geometric series with $a = \dfrac{R}{1+i}$ and $r = \dfrac{1}{1+i}$. Since $S_n = a\dfrac{1-r^n}{1-r}$, we have

$$A_p = \left(\frac{R}{1+i}\right)\frac{1-\left[\frac{1}{(1+i)}\right]^n}{1-\left(\frac{1}{1+i}\right)} = R\frac{1-(1+i)^{-n}}{(1+i)\left[1-\left(\frac{1}{1+i}\right)\right]} = R\frac{1-(1+i)^{-n}}{(1+i)-1} = R\frac{1-(1+i)^{-n}}{i}.$$

31. (a) Using the hint, we calculate the present value of the remaining 240 payments with $R = 724.17$, $i = 0.0075$, and

$n = 240$. Since $A_p = R\dfrac{1-(1+i)^{-n}}{i} = (724.17)\dfrac{1-(1.0075)^{-240}}{0.0075} = 80{,}487.84$, they still owe \$80,487.84 on their

mortgage.

(b) On their next payment, $0.0075\,(80{,}487.84) = \$603.66$ is interest and $\$724.17 - 603.66 = \$120.51$ goes toward the
principal.

12.5   MATHEMATICAL INDUCTION

1. Mathematical induction is a method of proving that a statement $P(n)$ is true for all *natural* numbers n. In Step 1 we prove
that $P(1)$ is true.

3. Let $P(n)$ denote the statement $2 + 4 + 6 + \cdots + 2n = n(n+1)$.

Step 1: $P(1)$ is the statement that $2 = 1(1+1)$, which is true.

Step 2: Assume that $P(k)$ is true; that is, $2 + 4 + 6 + \cdots + 2k = k(k+1)$. We want to use this to show that $P(k+1)$ is
true. Now

$$2 + 4 + 6 + \cdots + 2k + 2(k+1) = k(k+1) + 2(k+1) \qquad \text{induction hypothesis}$$
$$= (k+1)(k+2) = (k+1)[(k+1)+1]$$

Thus, $P(k+1)$ follows from $P(k)$. So by the Principle of Mathematical Induction, $P(n)$ is true for all n.

5. Let $P(n)$ denote the statement $5 + 8 + 11 + \cdots + (3n+2) = \dfrac{n(3n+7)}{2}$.

Step 1: We need to show that $P(1)$ is true. But $P(1)$ says that $5 = \dfrac{1\cdot(3\cdot 1 + 7)}{2}$, which is true.

Step 2: Assume that $P(k)$ is true; that is, $5 + 8 + 11 + \cdots + (3k+2) = \dfrac{k(3k+7)}{2}$. We want to use this to show that

$P(k+1)$ is true. Now

$$5 + 8 + 11 + \cdots + (3k+2) + [3(k+1)+2] = \frac{k(3k+7)}{2} + (3k+5) \qquad \text{induction hypothesis}$$
$$= \frac{3k^2+7k}{2} + \frac{6k+10}{2} = \frac{3k^2+13k+10}{2}$$
$$= \frac{(3k+10)(k+1)}{2} = \frac{(k+1)[3(k+1)+7]}{2}$$

Thus, $P(k+1)$ follows from $P(k)$. So by the Principle of Mathematical Induction, $P(n)$ is true for all n.

7. Let $P(n)$ denote the statement $1 \cdot 2 + 2 \cdot 3 + 3 \cdot 4 + \cdots + n(n+1) = \dfrac{n(n+1)(n+2)}{3}$.

   Step 1: $P(1)$ is the statement that $1 \cdot 2 = \dfrac{1 \cdot (1+1) \cdot (1+2)}{3}$, which is true.

   Step 2: Assume that $P(k)$ is true; that is, $1 \cdot 2 + 2 \cdot 3 + 3 \cdot 4 + \cdots + k(k+1) = \dfrac{k(k+1)(k+2)}{3}$. We want to use this to show that $P(k+1)$ is true. Now

   $1 \cdot 2 + 2 \cdot 3 + 3 \cdot 4 + \cdots + k(k+1) + (k+1)[(k+1)+1]$

   $$= \frac{k(k+1)(k+2)}{3} + (k+1)(k+2) \qquad \text{induction hypothesis}$$

   $$= \frac{k(k+1)(k+2)}{3} + \frac{3(k+1)(k+2)}{3} = \frac{(k+1)(k+2)(k+3)}{3}$$

   Thus, $P(k+1)$ follows from $P(k)$. So by the Principle of Mathematical Induction, $P(n)$ is true for all n.

9. Let $P(n)$ denote the statement $1^3 + 2^3 + 3^3 + \cdots + n^3 = \dfrac{n^2(n+1)^2}{4}$.

   Step 1: $P(1)$ is the statement that $1^3 = \dfrac{1^2 \cdot (1+1)^2}{4}$, which is clearly true.

   Step 2: Assume that $P(k)$ is true; that is, $1^3 + 2^3 + 3^3 + \cdots + k^3 = \dfrac{k^2(k+1)^2}{4}$. We want to use this to show that $P(k+1)$ is true. Now

   $$1^3 + 2^3 + 3^3 + \cdots + k^3 + (k+1)^3 = \frac{k^2(k+1)^2}{4} + (k+1)^3 \qquad \text{induction hypothesis}$$

   $$= \frac{(k+1)^2 \left[k^2 + 4(k+1)\right]}{4} = \frac{(k+1)^2 \left[k^2 + 4k + 4\right]}{4}$$

   $$= \frac{(k+1)^2(k+2)^2}{4} = \frac{(k+1)^2[(k+1)+1]^2}{4}$$

   Thus, $P(k+1)$ follows from $P(k)$. So by the Principle of Mathematical Induction, $P(n)$ is true for all n.

11. Let $P(n)$ denote the statement $2^3 + 4^3 + 6^3 + \cdots + (2n)^3 = 2n^2(n+1)^2$.

    Step 1: $P(1)$ is true since $2^3 = 2(1)^2(1+1)^2 = 2 \cdot 4 = 8$.

    Step 2: Assume that $P(k)$ is true; that is, $2^3 + 4^3 + 6^3 + \cdots + (2k)^3 = 2k^2(k+1)^2$. We want to use this to show that $P(k+1)$ is true. Now

    $$2^3 + 4^3 + 6^3 + \cdots + (2k)^3 + [2(k+1)]^3 = 2k^2(k+1)^2 + [2(k+1)]^3 \qquad \text{induction hypothesis}$$

    $$= 2k^2(k+1)^2 + 8(k+1)(k+1)^2 = (k+1)^2\left(2k^2 + 8k + 8\right)$$

    $$= 2(k+1)^2(k+2)^2 = 2(k+1)^2[(k+1)+1]^2$$

    Thus, $P(k+1)$ follows from $P(k)$. So by the Principle of Mathematical Induction, $P(n)$ is true for all n.

13. Let $P(n)$ denote the statement $1 \cdot 2 + 2 \cdot 2^2 + 3 \cdot 2^3 + 4 \cdot 2^4 + \cdots + n \cdot 2^n = 2\left[1 + (n-1)2^n\right]$.

Step 1: $P(1)$ is the statement that $1 \cdot 2 = 2[1+0]$, which is clearly true.

Step 2: Assume that $P(k)$ is true; that is, $1 \cdot 2 + 2 \cdot 2^2 + 3 \cdot 2^3 + 4 \cdot 2^4 + \cdots + k \cdot 2^k = 2\left[1 + (k-1)2^k\right]$. We want to use this to show that $P(k+1)$ is true. Now

$1 \cdot 2 + 2 \cdot 2^2 + 3 \cdot 2^3 + 4 \cdot 2^4 + \cdots + k \cdot 2^k + (k+1) \cdot 2^{(k+1)}$

$$= 2\left[1 + (k-1)2^k\right] + (k+1) \cdot 2^{k+1} \qquad \text{induction hypothesis}$$

$$= 2\left[1 + (k-1) \cdot 2^k + (k+1) \cdot 2^k\right] = 2\left[1 + 2k \cdot 2^k\right]$$

$$= 2\left[1 + k \cdot 2^{k+1}\right] = 2\left\{1 + [(k+1) - 1]2^{k+1}\right\}$$

Thus $P(k+1)$ follows from $P(k)$. So by the Principle of Mathematical Induction, $P(n)$ is true for all n.

15. Let $P(n)$ denote the statement $n^2 + n$ is divisible by 2.

Step 1: $P(1)$ is the statement that $1^2 + 1 = 2$ is divisible by 2, which is clearly true.

Step 2: Assume that $P(k)$ is true; that is, $k^2 + k$ is divisible by 2. Now

$(k+1)^2 + (k+1) = k^2 + 2k + 1 + k + 1 = \left(k^2 + k\right) + 2k + 2 = \left(k^2 + k\right) + 2(k+1)$. By the induction hypothesis,

$k^2 + k$ is divisible by 2, and clearly $2(k+1)$ is divisible by 2. Thus, the sum is divisible by 2, so $P(k+1)$ is true. Therefore, $P(k+1)$ follows from $P(k)$. So by the Principle of Mathematical Induction, $P(n)$ is true for all n.

17. Let $P(n)$ denote the statement that $n^2 - n + 41$ is odd.

Step 1: $P(1)$ is the statement that $1^2 - 1 + 41 = 41$ is odd, which is clearly true.

Step 2: Assume that $P(k)$ is true; that is, $k^2 - k + 41$ is odd. We want to use this to show that $P(k+1)$ is true. Now,

$(k+1)^2 - (k+1) + 41 = k^2 + 2k + 1 - k - 1 + 41 = \left(k^2 - k + 41\right) + 2k$, which is also odd because $k^2 - k + 41$ is

odd by the induction hypothesis, $2k$ is always even, and an odd number plus an even number is always odd. Therefore, $P(k+1)$ follows from $P(k)$. So by the Principle of Mathematical Induction, $P(n)$ is true for all n.

19. Let $P(n)$ denote the statement that $8^n - 3^n$ is divisible by 5.

Step 1: $P(1)$ is the statement that $8^1 - 3^1 = 5$ is divisible by 5, which is clearly true.

Step 2: Assume that $P(k)$ is true; that is, $8^k - 3^k$ is divisible by 5. We want to use this to show that $P(k+1)$ is true. Now,

$8^{k+1} - 3^{k+1} = 8 \cdot 8^k - 3 \cdot 3^k = 8 \cdot 8^k - (8-5) \cdot 3^k = 8 \cdot \left(8^k - 3^k\right) + 5 \cdot 3^k$, which is divisible by 5 because $8^k - 3^k$

is divisible by 5 by our induction hypothesis, and $5 \cdot 3^k$ is divisible by 5. Thus $P(k+1)$ follows from $P(k)$. So by the Principle of Mathematical Induction, $P(n)$ is true for all n.

21. Let $P(n)$ denote the statement $n < 2^n$.

Step 1: $P(1)$ is the statement that $1 < 2^1 = 2$, which is clearly true.

Step 2: Assume that $P(k)$ is true; that is, $k < 2^k$. We want to use this to show that $P(k+1)$ is true. Adding 1 to both sides of $P(k)$ we have $k+1 < 2^k + 1$. Since $1 < 2^k$ for $k \geq 1$, we have $2^k + 1 < 2^k + 2^k = 2 \cdot 2^k = 2^{k+1}$. Thus $k+1 < 2^{k+1}$, which is exactly $P(k+1)$. Therefore, $P(k+1)$ follows from $P(k)$. So by the Principle of Mathematical Induction, $P(n)$ is true for all n.

23. Let $P(n)$ denote the statement $(1+x)^n \geq 1 + nx$, if $x > -1$.

Step 1: $P(1)$ is the statement that $(1+x)^1 \geq 1 + 1x$, which is clearly true.

Step 2: Assume that $P(k)$ is true; that is, $(1+x)^k \geq 1 + kx$. Now, $(1+x)^{k+1} = (1+x)(1+x)^k \geq (1+x)(1+kx)$, by the induction hypothesis. Since $(1+x)(1+kx) = 1 + (k+1)x + kx^2 \geq 1 + (k+1)x$ (since $kx^2 \geq 0$), we have $(1+x)^{k+1} \geq 1 + (k+1)x$, which is $P(k+1)$. Thus $P(k+1)$ follows from $P(k)$. So the Principle of Mathematical Induction, $P(n)$ is true for all n.

25. Let $P(n)$ be the statement that $a_n = 5 \cdot 3^{n-1}$.

Step 1: $P(1)$ is the statement that $a_1 = 5 \cdot 3^0 = 5$, which is true.

Step 2: Assume that $P(k)$ is true; that is, $a_k = 5 \cdot 3^{k-1}$. We want to use this to show that $P(k+1)$ is true. Now, $a_{k+1} = 3a_k = 3 \cdot \left(5 \cdot 3^{k-1}\right)$, by the induction hypothesis. Therefore, $a_{k+1} = 3 \cdot \left(5 \cdot 3^{k-1}\right) = 5 \cdot 3^k$, which is exactly $P(k+1)$. Thus, $P(k+1)$ follows from $P(k)$. So by the Principle of Mathematical Induction, $P(n)$ is true for all n.

27. Let $P(n)$ be the statement that $x - y$ is a factor of $x^n - y^n$ for all natural numbers n.

Step 1: $P(1)$ is the statement that $x - y$ is a factor of $x^1 - y^1$, which is clearly true.

Step 2: Assume that $P(k)$ is true; that is, $x - y$ is a factor of $x^k - y^k$. We want to use this to show that $P(k+1)$ is true. Now, $x^{k+1} - y^{k+1} = x^{k+1} - x^k y + x^k y - y^{k+1} = x^k(x-y) + \left(x^k - y^k\right)y$, for which $x - y$ is a factor because $x - y$ is a factor of $x^k(x-y)$, and $x - y$ is a factor of $\left(x^k - y^k\right)y$, by the induction hypothesis. Thus $P(k+1)$ follows from $P(k)$. So by the Principle of Mathematical Induction, $P(n)$ is true for all n.

29. Let $P(n)$ denote the statement that F_{3n} is even for all natural numbers n.

Step 1: $P(1)$ is the statement that F_3 is even. Since $F_3 = F_2 + F_1 = 1 + 1 = 2$, this statement is true.

Step 2: Assume that $P(k)$ is true; that is, F_{3k} is even. We want to use this to show that $P(k+1)$ is true. Now, $F_{3(k+1)} = F_{3k+3} = F_{3k+2} + F_{3k+1} = F_{3k+1} + F_{3k} + F_{3k+1} = F_{3k} + 2 \cdot F_{3k+1}$, which is even because F_{3k} is even by the induction hypothesis, and $2 \cdot F_{3k+1}$ is even. Thus $P(k+1)$ follows from $P(k)$. So by the Principle of Mathematical Induction, $P(n)$ is true for all n.

31. Let $P(n)$ denote the statement that $F_1^2 + F_2^2 + F_3^2 + \cdots + F_n^2 = F_n \cdot F_{n+1}$.

Step 1: $P(1)$ is the statement that $F_1^2 = F_1 \cdot F_2$ or $1^2 = 1 \cdot 1$, which is true.

Step 2: Assume that $P(k)$ is true, that is, $F_1^2 + F_2^2 + F_3^2 + \cdots + F_k^2 = F_k \cdot F_{k+1}$. We want to use this to show that $P(k+1)$ is true. Now

$$
\begin{aligned}
F_1^2 + F_2^2 + F_3^2 + \cdots + F_k^2 + F_{k+1}^2 &= F_k \cdot F_{k+1} + F_{k+1}^2 && \text{induction hypothesis} \\
&= F_{k+1}\left(F_k + F_{k+1}\right) \\
&= F_{k+1} \cdot F_{k+2} && \text{by definition of the Fibonacci sequence}
\end{aligned}
$$

Thus $P(k+1)$ follows from $P(k)$. So by the Principle of Mathematical Induction, $P(n)$ is true for all n.

33. Let $P(n)$ denote the statement $\begin{bmatrix} 1 & 1 \\ 1 & 0 \end{bmatrix}^n = \begin{bmatrix} F_{n+1} & F_n \\ F_n & F_{n-1} \end{bmatrix}$.

Step 1: Since $\begin{bmatrix} 1 & 1 \\ 1 & 0 \end{bmatrix}^2 = \begin{bmatrix} 1 & 1 \\ 1 & 0 \end{bmatrix}\begin{bmatrix} 1 & 1 \\ 1 & 0 \end{bmatrix} = \begin{bmatrix} 2 & 1 \\ 1 & 1 \end{bmatrix} = \begin{bmatrix} F_3 & F_2 \\ F_2 & F_1 \end{bmatrix}$, it follows that $P(2)$ is true.

Step 2: Assume that $P(k)$ is true; that is, $\begin{bmatrix} 1 & 1 \\ 1 & 0 \end{bmatrix}^k = \begin{bmatrix} F_{k+1} & F_k \\ F_k & F_{k-1} \end{bmatrix}$. We show that $P(k+1)$ follows from this. Now,

$$\begin{bmatrix} 1 & 1 \\ 1 & 0 \end{bmatrix}^{k+1} = \begin{bmatrix} 1 & 1 \\ 1 & 0 \end{bmatrix}^k \begin{bmatrix} 1 & 1 \\ 1 & 0 \end{bmatrix} = \begin{bmatrix} F_{k+1} & F_k \\ F_k & F_{k-1} \end{bmatrix}\begin{bmatrix} 1 & 1 \\ 1 & 0 \end{bmatrix} \qquad \text{induction hypothesis}$$

$$= \begin{bmatrix} F_{k+1} + F_k & F_{k+1} \\ F_k + F_{k-1} & F_k \end{bmatrix} = \begin{bmatrix} F_{k+2} & F_{k+1} \\ F_{k+1} & F_k \end{bmatrix} \qquad \text{by definition of the Fibonacci sequence}$$

Thus $P(k+1)$ follows from $P(k)$. So by the Principle of Mathematical Induction, $P(n)$ is true for all $n \geq 2$.

35. Since $F_1 = 1$, $F_2 = 1$, $F_3 = 2$, $F_4 = 3$, $F_5 = 5$, $F_6 = 8$, $F_7 = 13, \ldots$ our conjecture is that $F_n \geq n$, for all $n \geq 5$. Let $P(n)$ denote the statement that $F_n \geq n$.

Step 1: $P(5)$ is the statement that $F_5 = 5 \geq 5$, which is clearly true.

Step 2: Assume that $P(k)$ is true; that is, $F_k \geq k$, for some $k \geq 5$. We want to use this to show that $P(k+1)$ is true. Now, $F_{k+1} = F_k + F_{k-1} \geq k + F_{k-1}$ (by the induction hypothesis)$\geq k + 1$ (because $F_{k-1} \geq 1$). Thus $P(k+1)$ follows from $P(k)$. So by the Principle of Mathematical Induction, $P(n)$ is true for all $n \geq 5$.

37. (a) $P(n) = n^2 - n + 11$ is prime for all n. This is false as the case for $n = 11$ demonstrates: $P(11) = 11^2 - 11 + 11 = 121$, which is not prime since $11^2 = 121$.

(b) $n^2 > n$, for all $n \geq 2$. This is true. Let $P(n)$ denote the statement that $n^2 > n$.

Step 1: $P(2)$ is the statement that $2^2 = 4 > 2$, which is clearly true.

Step 2: Assume that $P(k)$ is true; that is, $k^2 > k$. We want to use this to show that $P(k+1)$ is true. Now $(k+1)^2 = k^2 + 2k + 1$. Using the induction hypothesis (to replace k^2), we have $k^2 + 2k + 1 > k + 2k + 1 = 3k + 1 > k + 1$, since $k \geq 2$. Therefore, $(k+1)^2 > k + 1$, which is exactly $P(k+1)$. Thus $P(k+1)$ follows from $P(k)$. So by the Principle of Mathematical Induction, $P(n)$ is true for all n.

(c) $2^{2n+1} + 1$ is divisible by 3, for all $n \geq 1$. This is true. Let $P(n)$ denote the statement that $2^{2n+1} + 1$ is divisible by 3.

Step 1: $P(1)$ is the statement that $2^3 + 1 = 9$ is divisible by 3, which is clearly true.

Step 2: Assume that $P(k)$ is true; that is, $2^{2k+1} + 1$ is divisible by 3. We want to use this to show that $P(k+1)$ is true. Now, $2^{2(k+1)+1} + 1 = 2^{2k+3} + 1 = 4 \cdot 2^{2k+1} + 1 = (3+1)2^{2k+1} + 1 = 3 \cdot 2^{2k+1} + \left(2^{2k+1} + 1\right)$, which is divisible by 3 since $2^{2k+1} + 1$ is divisible by 3 by the induction hypothesis, and $3 \cdot 2^{2k+1}$ is clearly divisible by 3. Thus $P(k+1)$ follows from $P(k)$. So by the Principle of Mathematical Induction, $P(n)$ is true for all n.

(d) The statement $n^3 \geq (n+1)^2$ for all $n \geq 2$ is false. The statement fails when $n = 2$: $2^3 = 8 < (2+1)^2 = 9$.

(e) $n^3 - n$ is divisible by 3, for all $n \geq 2$. This is true. Let $P(n)$ denote the statement that $n^3 - n$ is divisible by 3.

Step 1: $P(2)$ is the statement that $2^3 - 2 = 6$ is divisible by 3, which is clearly true.

Step 2: Assume that $P(k)$ is true; that is, $k^3 - k$ is divisible by 3. We want to use this to show that $P(k+1)$ is true. Now

$$(k+1)^3 - (k+1) = k^3 + 3k^2 + 3k + 1 - (k+1) = k^3 + 3k^2 + 2k = k^3 - k + 3k^2 + 2k + k = \left(k^3 - k\right) + 3\left(k^2 + k\right).$$

The term $k^3 - k$ is divisible by 3 by our induction hypothesis, and the term $3\left(k^2 + k\right)$ is clearly divisible by 3. Thus $(k+1)^3 - (k+1)$ is divisible by 3, which is exactly $P(k+1)$. So by the Principle of Mathematical Induction, $P(n)$ is true for all n.

(f) $n^3 - 6n^2 + 11n$ is divisible by 6, for all $n \geq 1$. This is true. Let $P(n)$ denote the statement that $n^3 - 6n^2 + 11n$ is divisible by 6.

Step 1: $P(1)$ is the statement that $(1)^3 - 6(1)^2 + 11(1) = 6$ is divisible by 6, which is clearly true.

Step 2: Assume that $P(k)$ is true; that is, $k^3 - 6k^2 + 11k$ is divisible by 6. We show that $P(k+1)$ is then also true. Now

$$\begin{aligned}(k+1)^3 - 6(k+1)^2 + 11(k+1) &= k^3 + 3k^2 + 3k + 1 - 6k^2 - 12k - 6 + 11k + 11 \\ &= k^3 - 3k^2 + 2k + 6 = k^3 - 6k^2 + 11k + \left(3k^2 - 9k + 6\right) \\ &= \left(k^3 - 6k^2 + 11k\right) + 3\left(k^2 - 3k + 2\right) = \left(k^3 - 6k^2 + 11k\right) + 3(k-1)(k-2)\end{aligned}$$

In this last expression, the first term is divisible by 6 by our induction hypothesis. The second term is also divisible by 6. To see this, notice that $k-1$ and $k-2$ are consecutive natural numbers, and so one of them must be even (divisible by 2). Since 3 also appears in this second term, it follows that this term is divisible by 2 and 3 and so is divisible by 6. Thus $P(k+1)$ follows from $P(k)$. So by the Principle of Mathematical Induction, $P(n)$ is true for all n.

12.6 THE BINOMIAL THEOREM

1. An algebraic expression of the form $a + b$, which consists of a sum of two terms, is called a *binomial*.

3. The binomial coefficients can be calculated directly using the formula $\dbinom{n}{k} = \dfrac{n!}{k!\,(n-k)!}$. So $\dbinom{4}{3} = \dfrac{4!}{3!\,1!} = 4$.

5. $(x+y)^6 = x^6 + 6x^5 y + 15x^4 y^2 + 20x^3 y^3 + 15x^2 y^4 + 6xy^5 + y^6$

7. $\left(x + \dfrac{1}{x}\right)^4 = x^4 + 4x^3 \cdot \dfrac{1}{x} + 6x^2 \left(\dfrac{1}{x}\right)^2 + 4x \left(\dfrac{1}{x}\right)^3 + \left(\dfrac{1}{x}\right)^4 = x^4 + 4x^2 + 6 + \dfrac{4}{x^2} + \dfrac{1}{x^4}$

9. $(x-1)^5 = x^5 - 5x^4 + 10x^3 - 10x^2 + 5x - 1$

11. $\left(x^2 y - 1\right)^5 = \left(x^2 y\right)^5 - 5\left(x^2 y\right)^4 + 10\left(x^2 y\right)^3 - 10\left(x^2 y\right)^2 + 5x^2 y - 1 = x^{10} y^5 - 5x^8 y^4 + 10x^6 y^3 - 10x^4 y^2 + 5x^2 y - 1$

13. $(2x - 3y)^3 = (2x)^3 - 3(2x)^2\, 3y + 3 \cdot 2x\,(3y)^2 - (3y)^3 = 8x^3 - 36x^2 y + 54xy^2 - 27y^3$

15. $\left(\dfrac{1}{x} - \sqrt{x}\right)^5 = \left(\dfrac{1}{x}\right)^5 - 5\left(\dfrac{1}{x}\right)^4 \sqrt{x} + 10\left(\dfrac{1}{x}\right)^3 x - 10\left(\dfrac{1}{x}\right)^2 x\sqrt{x} + 5\left(\dfrac{1}{x}\right)x^2 - x^2\sqrt{x}$

$$= \dfrac{1}{x^5} - \dfrac{5}{x^{7/2}} + \dfrac{10}{x^2} - \dfrac{10}{x^{1/2}} + 5x - x^{5/2}$$

17. $\dbinom{6}{4} = \dfrac{6!}{4!\,2!} = \dfrac{6 \cdot 5 \cdot 4!}{2 \cdot 1 \cdot 4!} = 15$

19. $\dbinom{100}{98} = \dfrac{100!}{98!\,2!} = \dfrac{100 \cdot 99 \cdot 98!}{98! \cdot 2 \cdot 1} = 4950$

21. $\dbinom{3}{1}\dbinom{4}{2} = \dfrac{3!}{1!\,2!}\dfrac{4!}{2!\,2!} = \dfrac{3 \cdot 2! \cdot 4 \cdot 3 \cdot 2!}{1 \cdot 2! \cdot 2 \cdot 1 \cdot 2!} = 18$

23. $\binom{5}{0} + \binom{5}{1} + \binom{5}{2} + \binom{5}{3} + \binom{5}{4} + \binom{5}{5} = (1+1)^5 = 2^5 = 32$

25. $(x+2y)^4 = \binom{4}{0}x^4 + \binom{4}{1}x^3 \cdot 2y + \binom{4}{2}x^2 \cdot 4y^2 + \binom{4}{3}x \cdot 8y^3 + \binom{4}{4}16y^4 = x^4 + 8x^3y + 24x^2y^2 + 32xy^3 + 16y^4$

27. $\left(1 + \dfrac{1}{x}\right)^6 = \binom{6}{0}1^6 + \binom{6}{1}1^5\left(\dfrac{1}{x}\right) + \binom{6}{2}1^4\left(\dfrac{1}{x}\right)^2 + \binom{6}{3}1^3\left(\dfrac{1}{x}\right)^3 + \binom{6}{4}1^2\left(\dfrac{1}{x}\right)^4 + \binom{6}{5}1\left(\dfrac{1}{x}\right)^5 + \binom{6}{6}\left(\dfrac{1}{x}\right)^6$

$\qquad = 1 + \dfrac{6}{x} + \dfrac{15}{x^2} + \dfrac{20}{x^3} + \dfrac{15}{x^4} + \dfrac{6}{x^5} + \dfrac{1}{x^6}$

29. The first three terms in the expansion of $(x+2y)^{20}$ are $\binom{20}{0}x^{20} = x^{20}$, $\binom{20}{1}x^{19} \cdot 2y = 40x^{19}y$, and $\binom{20}{2}x^{18} \cdot (2y)^2 = 760x^{18}y^2$.

31. The last two terms in the expansion of $\left(a^{2/3} + a^{1/3}\right)^{25}$ are $\binom{25}{24}a^{2/3} \cdot \left(a^{1/3}\right)^{24} = 25a^{26/3}$, and $\binom{25}{25}a^{25/3} = a^{25/3}$.

33. The middle term in the expansion of $\left(x^2 + 1\right)^{18}$ occurs when both terms are raised to the 9th power. So this term is $\binom{18}{9}\left(x^2\right)^9 1^9 = 48{,}620x^{18}$.

35. The 24th term in the expansion of $(a+b)^{25}$ is $\binom{25}{23}a^2b^{23} = 300a^2b^{23}$.

37. The 100th term in the expansion of $(1+y)^{100}$ is $\binom{100}{99}1^1 \cdot y^{99} = 100y^{99}$.

39. The term that contains x^4 in the expansion of $(x+2y)^{10}$ has exponent $r = 4$. So this term is $\binom{10}{4}x^4 \cdot (2y)^{10-4} = 13{,}440x^4y^6$.

41. The rth term is $\binom{12}{r}a^r\left(b^2\right)^{12-r} = \binom{12}{r}a^rb^{24-2r}$. Thus the term that contains b^8 occurs where $24 - 2r = 8 \Leftrightarrow r = 8$. So the term is $\binom{12}{8}a^8b^8 = 495a^8b^8$.

43. $x^4 + 4x^3y + 6x^2y^2 + 4xy^3 + y^4 = (x+y)^4$

45. $8a^3 + 12a^2b + 6ab^2 + b^3 = \binom{3}{0}(2a)^3 + \binom{3}{1}(2a)^2 b + \binom{3}{2}2ab^2 + \binom{3}{3}b^3 = (2a+b)^3$

47. $\dfrac{(x+h)^3 - x^3}{h} = \dfrac{x^3 + 3x^2h + 3xh^2 + h^3 - x^3}{h} = \dfrac{3x^2h + 3xh^2 + h^3}{h} = \dfrac{h\left(3x^2 + 3xh + h^2\right)}{h} = 3x^2 + 3xh + h^2$

49. $(1.01)^{100} = (1 + 0.01)^{100}$. Now the first term in the expansion is $\binom{100}{0}1^{100} = 1$, the second term is $\binom{100}{1}1^{99}(0.01) = 1$, and the third term is $\binom{100}{2}1^{98}(0.01)^2 = 0.495$. Now each term is nonnegative, so $(1.01)^{100} = (1 + 0.01)^{100} > 1 + 1 + .0.495 > 2$. Thus $(1.01)^{100} > 2$.

51. $\binom{n}{1} = \dfrac{n!}{1!\,(n-1)!} = \dfrac{n\,(n-1)!}{1\,(n-1)!} = \dfrac{n}{1} = n$. $\binom{n}{n-1} = \dfrac{n!}{(n-1)!\,1!} = \dfrac{n\,(n-1)!}{(n-1)!\,1} = n$. Therefore,

$\binom{n}{1} = \binom{n}{n-1} = n$.

53. (a) $\binom{n}{r-1} + \binom{n}{r} = \dfrac{n!}{(r-1)!\,[n-(r-1)]!} + \dfrac{n!}{r!\,(n-r)!}$.

(b) $\dfrac{n!}{(r-1)!\,[n-(r-1)]!} + \dfrac{n!}{r!\,(n-r)!} = \dfrac{r \cdot n!}{r \cdot (r-1)!\,(n-r+1)!} + \dfrac{(n-r+1) \cdot n!}{r!\,(n-r+1)\,(n-r)!}$

$\qquad = \dfrac{r \cdot n!}{r!\,(n-r+1)!} + \dfrac{(n-r+1) \cdot n!}{r!\,(n-r+1)!}$

Thus a common denominator is $r!\,(n-r+1)!$.

(c) Therefore, using the results of parts (a) and (b),

$\binom{n}{r-1} + \binom{n}{r} = \dfrac{n!}{(r-1)!\,[n-(r-1)]!} + \dfrac{n!}{r!\,(n-r)!} = \dfrac{r \cdot n!}{r!\,(n-r+1)!} + \dfrac{(n-r+1) \cdot n!}{r!\,(n-r+1)!}$

$\qquad = \dfrac{r \cdot n! + (n-r+1) \cdot n!}{r!\,(n-r+1)!} = \dfrac{n!\,(r+n-r+1)}{r!\,(n-r+1)!} = \dfrac{n!\,(n+1)}{r!\,(n+1-r)!} = \dfrac{(n+1)!}{r!\,(n+1-r)!} = \binom{n+1}{r}$

55. By the Binomial Theorem, the volume of a cube of side $x + 2$ inches is

$(x + 2)^3 = \binom{3}{0}x^3 + \binom{3}{1}x^2(2) + \binom{3}{2}x(2)^2 + \binom{3}{3}2^3 = x^3 + 3 \cdot 2x^2 + 3 \cdot 4x + 8 = x^3 + 6x^2 + 12x + 8.$ The volume

of a cube of side x inches is x^3, so the difference in volumes is $x^3 + 6x^2 + 12x + 8 - x^3 = 6x^2 + 12x + 8$ cubic inches.

57. Notice that $(100!)^{101} = (100!)^{100} \cdot 100!$ and $(101!)^{100} = (101 \cdot 100!)^{100} = 101^{100} \cdot (100!)^{100}$. Now

$100! = 1 \cdot 2 \cdot 3 \cdot 4 \cdots 99 \cdot 100$ and $101^{100} = 101 \cdot 101 \cdot 101 \cdots 101$. Thus each of these last two expressions consists of

100 factors multiplied together, and since each factor in the product for 101^{100} is larger than each factor in the product for

$100!$, it follows that $100! < 101^{100}$. Thus $(100!)^{100} \cdot 100! < (100!)^{100} \cdot 101^{100}$. So $(100!)^{101} < (101!)^{100}$.

59. $0 = 0^n = (-1 + 1)^n = \binom{n}{0}(-1)^0(1)^n + \binom{n}{1}(-1)^1(1)^{n-1} + \binom{n}{2}(-1)^2(1)^{n-2} + \cdots + \binom{n}{n}(-1)^n(1)^0$

$= \binom{n}{0} - \binom{n}{1} + \binom{n}{2} - \cdots + (-1)^k\binom{n}{k} + \cdots + (-1)^n\binom{n}{n}$

CHAPTER 12 REVIEW

1. $a_n = \dfrac{n^2}{n+1}$. Then $a_1 = \dfrac{1^2}{1+1} = \dfrac{1}{2}$, $a_2 = \dfrac{2^2}{2+1} = \dfrac{4}{3}$, $a_3 = \dfrac{3^2}{3+1} = \dfrac{9}{4}$, $a_4 = \dfrac{4^2}{4+1} = \dfrac{16}{5}$, and $a_{10} = \dfrac{10^2}{10+1} = \dfrac{100}{11}$.

3. $a_n = \dfrac{(-1)^n + 1}{n^3}$. Then $a_1 = \dfrac{(-1)^1 + 1}{1^3} = 0$, $a_2 = \dfrac{(-1)^2 + 1}{2^3} = \dfrac{2}{8} = \dfrac{1}{4}$, $a_3 = \dfrac{(-1)^3 + 1}{3^3} = 0$,

$a_4 = \dfrac{(-1)^4 + 1}{4^3} = \dfrac{2}{64} = \dfrac{1}{32}$, and $a_{10} = \dfrac{(-1)^{10} + 1}{10^3} = \dfrac{1}{500}$.

5. $a_n = \dfrac{(2n)!}{2^n n!}$. Then $a_1 = \dfrac{(2 \cdot 1)!}{2^1 \cdot 1!} = 1$, $a_2 = \dfrac{(2 \cdot 2)!}{2^2 \cdot 2!} = 3$, $a_3 = \dfrac{(2 \cdot 3)!}{2^3 \cdot 3!} = \dfrac{6 \cdot 5 \cdot 4}{8} = 15$,

$a_4 = \dfrac{(2 \cdot 4)!}{2^4 \cdot 4!} = \dfrac{8 \cdot 7 \cdot 6 \cdot 5}{16} = 105$, and $a_{10} = \dfrac{(2 \cdot 10)!}{2^{10} \cdot 10!} = 654,729,075$.

7. $a_n = a_{n-1} + 2n - 1$ and $a_1 = 1$. Then $a_2 = a_1 + 4 - 1 = 4$, $a_3 = a_2 + 6 - 1 = 9$, $a_4 = a_3 + 8 - 1 = 16$,

$a_5 = a_4 + 10 - 1 = 25$, $a_6 = a_5 + 12 - 1 = 36$, and $a_7 = a_6 + 14 - 1 = 49$.

9. $a_n = a_{n-1} + 2a_{n-2}$, $a_1 = 1$ and $a_2 = 3$. Then $a_3 = a_2 + 2a_1 = 5$, $a_4 = a_3 + 2a_2 = 11$, $a_5 = a_4 + 2a_3 = 21$,

$a_6 = a_5 + 2a_4 = 43$, and $a_7 = a_6 + 2a_5 = 85$.

11. (a) $a_1 = 2(1) + 5 = 7$, $a_2 = 2(2) + 5 = 9$,

$a_3 = 2(3) + 5 = 11$, $a_4 = 2(4) + 5 = 13$,

$a_5 = 2(5) + 5 = 15$

(b)

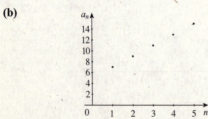

(c) $S_5 = 7 + 9 + 11 + 13 + 15 = 55$

(d) This sequence is arithmetic with common difference 2.

13. (a) $a_1 = \dfrac{3^1}{2^2} = \dfrac{3}{4}$, $a_2 = \dfrac{3^2}{2^3} = \dfrac{9}{8}$, $a_3 = \dfrac{3^3}{2^4} = \dfrac{27}{16}$,

$a_4 = \dfrac{3^4}{2^5} = \dfrac{81}{32}$, $a_5 = \dfrac{3^5}{2^6} = \dfrac{243}{64}$

(b)

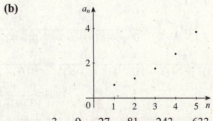

(c) $S_5 = \dfrac{3}{4} + \dfrac{9}{8} + \dfrac{27}{16} + \dfrac{81}{32} + \dfrac{243}{64} = \dfrac{633}{64}$

(d) This sequence is geometric with common ratio $\frac{3}{2}$.

15. $5, 5.5, 6, 6.5, \ldots$. Since $5.5 - 5 = 6 - 5.5 = 6.5 - 6 = 0.5$, this is an arithmetic sequence with $a_1 = 5$ and $d = 0.5$. Then $a_5 = a_4 + 0.5 = 7$.

17. $t-3, t-2, t-1, t, \ldots$. Since $(t-2)-(t-3) = (t-1)-(t-2) = t-(t-1) = 1$, this is an arithmetic sequence with $a_1 = t-3$ and $d = 1$. Then $a_5 = a_4 + 1 = t + 1$.

19. $t^3, t^2, t, 1, \ldots$. Since $\dfrac{t^2}{t^3} = \dfrac{t}{t^2} = \dfrac{1}{t}$, this is a geometric sequence with $a_1 = t^3$ and $r = \dfrac{1}{t}$. Then $a_5 = a_4 \cdot r = 1 \cdot \dfrac{1}{t} = \dfrac{1}{t}$.

21. $\frac{3}{4}, \frac{1}{2}, \frac{1}{3}, \frac{2}{9}, \ldots$. Since $\dfrac{\frac{1}{2}}{\frac{3}{4}} = \dfrac{\frac{1}{3}}{\frac{1}{2}} = \dfrac{\frac{2}{9}}{\frac{1}{3}} = \frac{2}{3}$, this is a geometric sequence with $a_1 = \frac{3}{4}$ and $r = \frac{2}{3}$. Then

$a_5 = a_4 \cdot r = \frac{2}{9} \cdot \frac{2}{3} = \frac{4}{27}$.

23. $3, 6i, -12, -24i, \ldots$. Since $\dfrac{6i}{3} = 2i$, $\dfrac{-12}{6i} = \dfrac{-2}{i} = \dfrac{-2i}{i^2} = 2i$, $\dfrac{-24i}{-12} = 2i$, this is a geometric sequence with common

ratio $r = 2i$.

25. $a_6 = 17 = a + 5d$ and $a_4 = 11 = a + 3d$. Then, $a_6 - a_4 = 17 - 11 \Leftrightarrow (a+5d) - (a+3d) = 6 \Leftrightarrow 6 = 2d \Leftrightarrow d = 3$. Substituting into $11 = a + 3d$ gives $11 = a + 3 \cdot 3$, and so $a = 2$. Thus $a_2 = a + (2-1)d = 2 + 3 = 5$.

27. $a_3 = 9$ and $r = \frac{3}{2}$. Then $a_5 = a_3 \cdot r^2 = 9 \cdot \left(\frac{3}{2}\right)^2 = \frac{81}{4}$.

29. (a) $A_n = 32{,}000 \cdot 1.05^{n-1}$

(b) $A_1 = \$32{,}000$, $A_2 = 32{,}000 \cdot 1.05^1 = \$33{,}600$, $A_3 = 32{,}000 \cdot 1.05^2 = \$35{,}280$, $A_4 = 32{,}000 \cdot 1.05^3 = \$37{,}044$,
$A_5 = 32{,}000 \cdot 1.05^4 = \$38{,}896.20$, $A_6 = 32{,}000 \cdot 1.05^5 = \$40{,}841.01$, $A_7 = 32{,}000 \cdot 1.05^6 = \$42{,}883.06$,
$A_8 = 32{,}000 \cdot 1.05^7 = \$45{,}027.21$

31. Let a_n be the number of bacteria in the dish at the end of $5n$ seconds. So $a_0 = 3$, $a_1 = 3 \cdot 2$, $a_2 = 3 \cdot 2^2$, $a_3 = 3 \cdot 2^3, \ldots$. Then, clearly, a_n is a geometric sequence with $r = 2$ and $a = 3$. Thus at the end of $60 = 5\,(12)$ seconds, the number of bacteria is $a_{12} = 3 \cdot 2^{12} = 12{,}288$.

33. Suppose that the common ratio in the sequence $a_1, a_2, a_3, \ldots$ is r. Also, suppose that the common ratio in the sequence $b_1, b_2, b_3, \ldots$ is s. Then $a_n = a_1 r^{n-1}$ and $b_n = b_1 s^{n-1}$, $n = 1, 2, 3, \ldots$. Thus $a_n b_n = a_1 r^{n-1} \cdot b_1 s^{n-1} = (a_1 b_1)(rs)^{n-1}$. So the sequence $a_1 b_1, a_2 b_2, a_3 b_3, \ldots$ is geometric with first term $a_1 b_1$ and common ratio rs.

35. (a) $6, x, 12, \ldots$ is arithmetic if $x - 6 = 12 - x \Leftrightarrow 2x = 18 \Leftrightarrow x = 9$.

(b) $6, x, 12, \ldots$ is geometric if $\dfrac{x}{6} = \dfrac{12}{x} \Leftrightarrow x^2 = 72 \Leftrightarrow x = \pm 6\sqrt{2}$.

37. $\sum_{k=3}^{6} (k+1)^2 = (3+1)^2 + (4+1)^2 + (5+1)^2 + (6+1)^2 = 16 + 25 + 36 + 49 = 126$

39. $\sum_{k=1}^{6} (k+1) 2^{k-1} = 2 \cdot 2^0 + 3 \cdot 2^1 + 4 \cdot 2^2 + 5 \cdot 2^3 + 6 \cdot 2^4 + 7 \cdot 2^5 = 2 + 6 + 16 + 40 + 96 + 224 = 384$

41. $\sum_{k=1}^{10} (k-1)^2 = 0^2 + 1^2 + 2^2 + 3^2 + 4^2 + 5^2 + 6^2 + 7^2 + 8^2 + 9^2$

43. $\sum_{k=1}^{50} \dfrac{3^k}{2^{k+1}} = \dfrac{3}{2^2} + \dfrac{3^2}{2^3} + \dfrac{3^3}{2^4} + \dfrac{3^4}{2^5} + \cdots + \dfrac{3^{49}}{2^{50}} + \dfrac{3^{50}}{2^{51}}$

45. $3 + 6 + 9 + 12 + \cdots + 99 = 3\,(1) + 3\,(2) + 3\,(3) + \cdots + 3\,(33) = \sum_{k=1}^{33} 3k$

47. $1 \cdot 2^3 + 2 \cdot 2^4 + 3 \cdot 2^5 + 4 \cdot 2^6 + \cdots + 100 \cdot 2^{102}$

$$= (1)\,2^{(1)+2} + (2)\,2^{(2)+2} + (3)\,2^{(3)+2} + (4)\,2^{(4)+2} + \cdots + (100)\,2^{(100)+2}$$
$$= \sum_{k=1}^{100} k \cdot 2^{k+2}$$

49. $1 + 0.9 + (0.9)^2 + \cdots + (0.9)^5$ is a geometric series with $a = 1$ and $r = \dfrac{0.9}{1} = 0.9$. Thus, the sum of the series is

$S_6 = \dfrac{1 - (0.9)^6}{1 - 0.9} = \dfrac{1 - 0.531441}{0.1} = 4.68559$.

51. $\sqrt{5} + 2\sqrt{5} + 3\sqrt{5} + \cdots + 100\sqrt{5}$ is an arithmetic series with $a = \sqrt{5}$ and $d = \sqrt{5}$. Then $100\sqrt{5} = a_n = \sqrt{5} + \sqrt{5}\,(n-1)$
$\Leftrightarrow n = 100$. So the sum is $S_{100} = \dfrac{100}{2}\left(\sqrt{5} + 100\sqrt{5}\right) = 50\left(101\sqrt{5}\right) = 5050\sqrt{5}$.

53. $\sum_{n=0}^{6} 3 \cdot (-4)^n$ is a geometric series with $a = 3$, $r = -4$, and $n = 7$. Therefore, the sum of the series is

$$S_7 = 3 \cdot \frac{1 - (-4)^7}{1 - (-4)} = \frac{3}{5} \left(1 + 4^7\right) = 9831.$$

55. $1 - \frac{2}{5} + \frac{4}{25} - \frac{8}{125} + \cdots$ is a geometric series with $a = 1$ and $r = -\frac{2}{5}$. Therefore, it is convergent with sum

$$S = \frac{a}{1 - r} = \frac{1}{1 - \left(-\frac{2}{5}\right)} = \frac{5}{7}.$$

57. $5 - 5(1.01) + 5(1.01)^2 - 5(1.01)^3 + \cdots$ is an infinite geometric series with $a = 5$ and $r = 1.01$. Because $|r| = 1.01 > 1$, the series diverges.

59. $-1 + \frac{9}{8} - \left(\frac{9}{8}\right)^2 + \left(\frac{9}{8}\right)^3 - \cdots$ is an infinite geometric series with $a = -1$ and $r = -\frac{9}{8}$. Because $|r| = \frac{9}{8} > 1$, the series diverges.

61. We have an arithmetic sequence with $a = 7$ and $d = 3$. Then

$$S_n = 325 = \frac{n}{2}\left[2a + (n-1)d\right] = \frac{n}{2}\left[14 + 3(n-1)\right] = \frac{n}{2}(11 + 3n) \Leftrightarrow 650 = 3n^2 + 11n \Leftrightarrow (3n + 50)(n - 13) = 0 \Leftrightarrow$$

$n = 13$ (because $n = -\frac{50}{3}$ is inadmissible). Thus, 13 terms must be added.

63. This is a geometric sequence with $a = 2$ and $r = 2$. Then $S_{15} = 2 \cdot \dfrac{1 - 2^{15}}{1 - 2} = 2\left(2^{15} - 1\right) = 65{,}534$, and so the total number of ancestors is 65,534.

65. $A = 10{,}000$, $i = 0.03$, and $n = 4$. Thus, $10{,}000 = R \dfrac{(1.03)^4 - 1}{0.03} \Leftrightarrow R = \dfrac{10{,}000 \cdot 0.03}{(1.03)^4 - 1} = \$2390.27$.

67. Let $P(n)$ denote the statement that $1 + 4 + 7 + \cdots + (3n - 2) = \dfrac{n(3n-1)}{2}$.

Step 1: $P(1)$ is the statement that $1 = \dfrac{1[3(1) - 1]}{2} = \dfrac{1 \cdot 2}{2}$, which is true.

Step 2: Assume that $P(k)$ is true; that is, $1 + 4 + 7 + \cdots + (3k - 2) = \dfrac{k(3k-1)}{2}$. We want to use this to show that $P(k+1)$ is true. Now

$$
\begin{aligned}
1 + 4 + 7 + 10 + \cdots + (3k - 2) + [3(k+1) - 2] &= \frac{k(3k-1)}{2} + 3k + 1 \qquad\qquad \text{induction hypothesis} \\
&= \frac{k(3k-1)}{2} + \frac{6k + 2}{2} = \frac{3k^2 - k + 6k + 2}{2} \\
&= \frac{3k^2 + 5k + 2}{2} = \frac{(k+1)(3k+2)}{2} \\
&= \frac{(k+1)[3(k+1) - 1]}{2}
\end{aligned}
$$

Thus, $P(k+1)$ follows from $P(k)$. So by the Principle of Mathematical Induction, $P(n)$ is true for all n.

69. Let $P(n)$ denote the statement that $\left(1+\frac{1}{1}\right)\left(1+\frac{1}{2}\right)\left(1+\frac{1}{3}\right)\cdots\left(1+\frac{1}{n}\right)=n+1$.

Step 1: $P(1)$ is the statement that $1+\frac{1}{1}=1+1$, which is clearly true.

Step 2: Assume that $P(k)$ is true; that is, $\left(1+\frac{1}{1}\right)\left(1+\frac{1}{2}\right)\left(1+\frac{1}{3}\right)\cdots\left(1+\frac{1}{k}\right)=k+1$. We want to use this to

show that $P(k+1)$ is true. Now

$$\left(1+\frac{1}{1}\right)\left(1+\frac{1}{2}\right)\left(1+\frac{1}{3}\right)\cdots\left(1+\frac{1}{k}\right)\left(1+\frac{1}{k+1}\right)$$

$$=\left[\left(1+\frac{1}{1}\right)\left(1+\frac{1}{2}\right)\left(1+\frac{1}{3}\right)\cdots\left(1+\frac{1}{k}\right)\right]\left(1+\frac{1}{k+1}\right)$$

$$=(k+1)\left(1+\frac{1}{k+1}\right) \qquad\qquad \text{induction hypothesis}$$

$$=(k+1)+1$$

Thus, $P(k+1)$ follows from $P(k)$. So by the Principle of Mathematical Induction, $P(n)$ is true for all n.

71. $a_{n+1}=3a_n+4$ and $a_1=4$. Let $P(n)$ denote the statement that $a_n=2\cdot3^n-2$.

Step 1: $P(1)$ is the statement that $a_1=2\cdot3^1-2=4$, which is clearly true.

Step 2: Assume that $P(k)$ is true; that is, $a_k=2\cdot3^k-2$. We want to use this to show that $P(k+1)$ is true. Now

$$a_{k+1}=3a_k+4 \qquad\qquad \text{definition of } a_{k+1}$$

$$=3\left(2\cdot3^k-2\right)+4 \qquad\qquad \text{induction hypothesis}$$

$$=2\cdot3^{k+1}-6+4=2\cdot3^{k+1}-2$$

Thus $P(k+1)$ follows from $P(k)$. So by the Principle of Mathematical Induction, $P(n)$ is true for all n.

73. $\dbinom{5}{2}\dbinom{5}{3}=\frac{5!}{2!\,3!}\cdot\frac{5!}{3!\,2!}=\frac{5\cdot4}{2}\cdot\frac{5\cdot4}{2}=10\cdot10=100$

75. $\sum_{k=0}^{5}\binom{5}{k}=\binom{5}{0}+\binom{5}{1}+\binom{5}{2}+\binom{5}{3}+\binom{5}{4}+\binom{5}{5}=2\left(\frac{5!}{0!\,5!}+\frac{5!}{1!\,4!}+\frac{5!}{2!\,3!}\right)=2(1+5+10)=32$

77. $(A-B)^3=\binom{3}{0}A^3-\binom{3}{1}A^2B+\binom{3}{2}AB^2-\binom{3}{3}B^3=A^3-3A^2B+3AB^2-B^3$

79. $\left(1-x^2\right)^6=\binom{6}{0}1^6-\binom{6}{1}1^5x^2+\binom{6}{2}1^4x^4-\binom{6}{3}1^3x^6+\binom{6}{4}1^2x^8-\binom{6}{5}x^{10}+\binom{6}{6}x^{12}$

$$=1-6x^2+15x^4-20x^6+15x^8-6x^{10}+x^{12}$$

81. The 20th term is $\binom{22}{19}a^3b^{19}=1540a^3b^{19}$.

83. The rth term in the expansion of $(A+3B)^{10}$ is $\binom{10}{r}A^r(3B)^{10-r}$. The term that contains A^6 occurs when $r=6$. Thus, the

term is $\binom{10}{6}A^6(3B)^4=210A^681B^4=17{,}010A^6B^4$.

CHAPTER 12 TEST

1. $a_n = 2n^2 - n \Rightarrow a_1 = 1, a_2 = 6, a_3 = 15, a_4 = 28, a_5 = 45, a_6 = 66$, and $S_6 = 1 + 6 + 15 + 28 + 45 + 66 = 161$.

3. (a) The common difference is $d = 5 - 2 = 3$.

    (b) $a_n = 2 + (n - 1)3$

    (c) $a_{35} = 2 + 3(35 - 1) = 104$

5. (a) $a_1 = 25$, $a_4 = \frac{1}{5}$. Then $r^3 = \dfrac{\frac{1}{5}}{25} = \dfrac{1}{125} \Leftrightarrow r = \frac{1}{5}$, so $a_5 = ra_4 = \frac{1}{25}$.

    (b) $S_8 = 25 \dfrac{1 - \left(\frac{1}{5}\right)^8}{1 - \frac{1}{5}} = \dfrac{5^8 - 1}{12{,}500} = \dfrac{97{,}656}{3125}$

7. Let the common ratio for the geometric series $a_1, a_2, a_3, \ldots$ be r, so that $a_n = a_1 r^{n-1}$, $n = 1, 2, 3, \ldots$. Then $a_n^2 = \left(a_1 r^{n-1}\right)^2 = \left(a_1^2\right)\left(r^2\right)^{n-1}$. Therefore, the sequence $a_1^2, a_2^2, a_3^2, \ldots$ is geometric with common ratio r^2.

9. (a) The geometric sum $\frac{1}{3} + \frac{2}{3^2} + \frac{2^2}{3^3} + \frac{2^3}{3^4} + \cdots + \frac{2^9}{3^{10}}$ has $a = \frac{1}{3}$, $r = \frac{2}{3}$, and $n = 10$. So

$$S_{10} = \frac{1}{3} \cdot \frac{1 - (2/3)^{10}}{1 - (2/3)} = \frac{1}{3} \cdot 3\left(1 - \frac{1024}{59{,}049}\right) = \frac{58{,}025}{59{,}049}.$$

    (b) The infinite geometric series $1 + \frac{1}{2^{1/2}} + \frac{1}{2} + \frac{1}{2^{3/2}} + \cdots$ has $a = 1$ and $r = 2^{-1/2} = \frac{1}{\sqrt{2}}$. Thus,

$$S = \frac{1}{1 - 1/\sqrt{2}} = \frac{\sqrt{2}}{\sqrt{2} - 1} = \frac{\sqrt{2}}{\sqrt{2} - 1} \cdot \frac{\sqrt{2} + 1}{\sqrt{2} + 1} = 2 + \sqrt{2}.$$

11. $\left(2x + y^2\right)^5 = \binom{5}{0}(2x)^5 + \binom{5}{1}(2x)^4 y^2 + \binom{5}{2}(2x)^3 \left(y^2\right)^2 + \binom{5}{3}(2x)^2 \left(y^2\right)^3 + \binom{5}{4}(2x)\left(y^2\right)^4 + \binom{5}{5}\left(y^2\right)^5$

$$= 32x^5 + 80x^4 y^2 + 80x^3 y^4 + 40x^2 y^6 + 10xy^8 + y^{10}$$

13. (a) Each week he gains 24% in weight, that is, $0.24a_n$. Thus, $a_{n+1} = a_n + 0.24a_n = 1.24a_n$ for $n \geq 1$. a_0 is given to be 0.85 lb. Then $a_0 = 0.85$, $a_1 = 1.24(0.85)$, $a_2 = 1.24(1.24(0.85)) = 1.24^2(0.85)$, $a_3 = 1.24\left(1.24^2(0.85)\right) = 1.24^3(0.85)$, and so on. So we can see that $a_n = 0.85(1.24)^n$.

    (b) $a_6 = 1.24a_5 = 1.24(1.24a_4) = \cdots = 1.24^6 a_0 = 1.24^6(0.85) \approx 3.1$ lb

    (c) The sequence $a_1, a_2, a_3, \ldots$ is geometric with common ratio 1.24.

FOCUS ON MODELING   Modeling with Recursive Sequences

1. (a) Since there are 365 days in a year, the interest earned per day is $\dfrac{0.0365}{365} = 0.0001$. Thus the amount in the account at the end of the nth day is $A_n = 1.0001 A_{n-1}$ with $A_0 = \$275{,}000$.

    (b) $A_0 = \$275{,}000$, $A_1 = 1.0001 A_0 = 1.0001 \cdot 275{,}000 = \$275{,}027.50$,

    $A_2 = 1.0001 A_1 = 1.0001(1.0001 A_0) = 1.0001^2 A_0 = \$275{,}055.00$,

    $A_3 = 1.0001 A_2 = 1.0001^3 A_0 = \$275{,}082.51$, $A_4 = 1.0001^4 A_0 = \$275{,}110.02$, $A_5 = 1.0001^5 A_0 = \$275{,}137.53$,

    $A_6 = 1.0001^6 A_0 = \$275{,}165.04$, $A_7 = 1.0001^7 A_0 = \$275{,}192.56$

    (c) $A_n = 1.0001^n \cdot 275{,}000$

3. (a) Since there are 12 months in a year, the interest earned per day is $\dfrac{0.03}{12} = 0.0025$. Thus the amount in the account at the end of the nth month is $A_n = 1.0025 A_{n-1} + 100$ with $A_0 = \$100$.

(b) $A_0 = \$100$, $A_1 = 1.0025 A_0 + 100 = 1.0025 \cdot 100 + 100 = \$200.25$,

$A_2 = 1.0025 A_1 + 100 = 1.0025\,(1.0025 \cdot 100 + 100) + 100 = 1.0025^2 \cdot 100 + 1.0025 \cdot 100 + 100 = \$300.75$,

$A_3 = 1.0025 A_2 + 100 = 1.0025 \left(1.0025^2 \cdot 100 + 1.0025 \cdot 100 + 100\right) + 100$

$= 1.0025^3 \cdot 100 + 1.0025^2 \cdot 100 + 1.0025 \cdot 100 + 100 = \$401.50$,

$A_4 = 1.0025 A_3 + 100 = 1.0025 \left(1.0025^3 \cdot 100 + 1.0025^2 \cdot 100 + 1.0025 \cdot 100 + 100\right) + 100$

$= 1.0025^4 \cdot 100 + 1.0025^3 \cdot 100 + 1.0025^2 \cdot 100 + 1.0025 \cdot 100 + 100 = \$502.51$

(c) $A_n = 1.0025^n \cdot 100 + \cdots + 1.0025^2 \cdot 100 + 1.0025 \cdot 100 + 100$, the partial sum of a geometric series, so

$$A_n = 100 \cdot \frac{1 - 1.0025^{n+1}}{1 - 1.0025} = 100 \cdot \frac{1.0025^{n+1} - 1}{0.0025}.$$

(d) Since 5 years is 60 months, we have $A_{60} = 100 \cdot \dfrac{1.0025^{61} - 1}{0.0025} \approx \$6{,}580.83$.

5. (a) The amount A_n of pollutants in the lake in the nth year is 30% of the amount from the preceding year ($0.30 A_{n-1}$) plus the amount discharged that year (2400 tons). Thus $A_n = 0.30 A_{n-1} + 2400$.

(b) $A_0 = 2400$, $A_1 = 0.30\,(2400) + 2400 = 3120$,

$A_2 = 0.30\,[0.30\,(2400) + 2400] + 2400 = 0.30^2\,(2400) + 2400\,(2400) + 2400 = 3336$,

$A_3 = 0.30 \left[0.30^2\,(2400) + 2400\,(2400) + 2400\right] + 2400$

$= 0.03^3\,(2400) + 0.30^2\,(2400) + 2400\,(2400) + 2400 = 3400.8$,

$A_4 = 0.30 \left[0.03^3\,(2400) + 0.30^2\,(2400) + 2400\,(2400) + 2400\right] + 2400$

$= 0.03^4\,(2400) + 0.03^3\,(2400) + 0.30^2\,(2400) + 2400\,(2400) + 2400 = 3420.2$

(c) A_n is the partial sum of a geometric series, so

$$A_n = 2400 \cdot \frac{1 - 0.30^{n+1}}{1 - 0.30} = 2400 \cdot \frac{1 - 0.30^{n+1}}{0.70}$$

$$\approx 3428.6 \left(1 - 0.30^{n+1}\right)$$

(e)

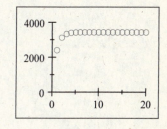

(d) $A_6 = 2400 \cdot \dfrac{1 - 0.30^7}{0.70} = 3427.8$ tons. The sum of a geometric

series, is $A = 2400 \cdot \dfrac{1}{0.70} = 3428.6$ tons.

7. (a) In the nth year since Victoria's initial deposit the amount V_n in her CD is the amount from the preceding year (V_{n-1}), plus the 5% interest earned on that amount ($0.05 V_{n-1}$), plus \$500 times the number of years since her initial deposit ($500n$). Thus $V_n = 1.05 V_{n-1} + 500n$.

(b) Ursula's savings surpass Victoria's savings in the 35th year.

13 LIMITS: A PREVIEW OF CALCULUS

13.1 FINDING LIMITS NUMERICALLY AND GRAPHICALLY

1. When we write $\lim\limits_{x \to a} f(x) = L$ then, roughly speaking, the values of $f(x)$ get closer and closer to the number L as the values of x get closer and closer to a. To determine $\lim\limits_{x \to 5} \dfrac{x-5}{x-5}$ we try values for x closer and closer to 5 and find that the limit is 1.

3.

x	4.9	4.99	4.999	5.001	5.01	5.1
$\dfrac{x^2-25}{x-5}$	9.9	9.99	9.999	10.001	10.01	10.1

The limit appears to be 10.

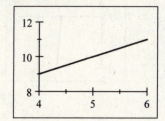

5.

x	3.9	3.99	3.999	4.001	4.01	4.1
$f(x)$	0.25158	0.25016	0.25002	0.24998	0.24984	0.24845

$\lim\limits_{x \to 4} \dfrac{\sqrt{x}-2}{x-4} = 0.25$

7.

x	0.9	0.99	0.999	1.001	1.01	1.1
$f(x)$	0.36900	0.33669	0.33367	0.3330	0.33002	0.30211

$\lim\limits_{x \to 1} \dfrac{x-1}{x^3-1} = 0.33333$

9.

x	± 1	± 0.5	± 0.1	± 0.05	± 0.01
$f(x)$	0.84147	0.95885	0.99833	0.99958	0.99998

$\lim\limits_{x \to 0} \dfrac{\sin x}{x} = 1$

11.

x	$f(x)$
-3.9	-1.11111
-3.99	-1.01010
-3.99	-1.00100
-4.001	-0.99900
-4.01	-0.99010
-4.1	-0.90909

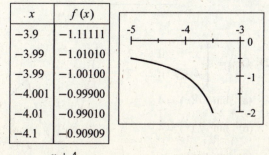

$\lim\limits_{x \to -4} \dfrac{x+4}{x^2+7x+12} = -1$

13.

x	$f(x)$
-0.1	0.44619
-0.01	0.50396
$-0.0.01$	0.51013
0.001	0.51152
0.01	0.51779
0.1	0.58496

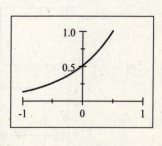

$\lim\limits_{x \to 0} \dfrac{5^x - 3^x}{x} = 0.5108$

425

15.

x	$f(x)$
0.9	0.50878
0.99	0.50084
0.999	0.50008
1.001	0.49992
1.01	0.49917
1.1	0.49206

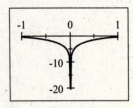

$$\lim_{x \to 1}\left(\frac{1}{\ln x} - \frac{1}{x-1}\right) = 0.5000$$

17. (a) $\lim_{x \to 1^-} f(x) = 2$

(b) $\lim_{x \to 1^+} f(x) = 3$

(c) $\lim_{x \to 1} f(x)$ does not exist;

$\lim_{x \to 1^-} f(x) \neq \lim_{x \to 1^+} f(x)$.

(d) $\lim_{x \to 5} f(x) = 4$

(e) $f(5)$ is not defined.

Note: In the first printing of the text, there was an error in the graph. The open circle at $(5, 4)$ was missing.

19. (a) $\lim_{t \to 0^-} g(t) = -1$

(b) $\lim_{t \to 0^+} g(t) = -2$

(c) $\lim_{t \to 0} g(t)$ does not exist; $\lim_{t \to 0^-} g(t) \neq \lim_{t \to 0^+} g(t)$.

(d) $\lim_{t \to 2^-} g(t) = 2$

(e) $\lim_{t \to 2^+} g(t) = 0$

(f) $\lim_{t \to 2} g(t)$ does not exist; $\lim_{t \to 2^-} g(t) \neq \lim_{t \to 2^+} g(t)$.

(g) $g(2) = 1$

(h) $\lim_{t \to 4} g(t) = 3$

21. $\lim_{x \to 1} \dfrac{x^3 + x^2 + 3x - 5}{2x^2 - 5x + 3} = -8.00$

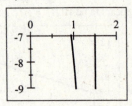

23. $\lim_{x \to 0} \ln\left(\sin^2 x\right)$ does not exist.

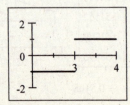

25. $\lim_{x \to 0} \cos\dfrac{1}{x}$ does not exist.

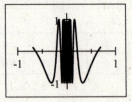

27. $\lim_{x \to 3} \dfrac{|x - 3|}{x - 3}$ does not exist.

29. $f(x) = \begin{cases} x^2 & \text{if } x \leq 2 \\ 6 - x & \text{if } x > 2 \end{cases}$

(a) $\lim_{x \to 2^-} f(x) = 4$

(b) $\lim_{x \to 2^+} f(x) = 4$

(c) $\lim_{x \to 2} f(x) = 4$

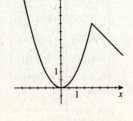

31. $f(x) = \begin{cases} -x+3 & \text{if } x < -1 \\ 3 & \text{if } x \geq -1 \end{cases}$

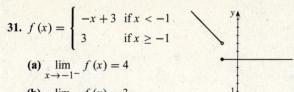

(a) $\lim\limits_{x \to -1^-} f(x) = 4$

(b) $\lim\limits_{x \to -1^+} f(x) = 3$

(c) $\lim\limits_{x \to -1} f(x)$ does not exist.

33. There are an infinite number of functions satisfying the given conditions. The graph of one such function is shown.

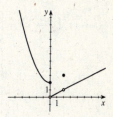

13.2 FINDING LIMITS ALGEBRAICALLY

1. If $\lim\limits_{x \to a} f(x)$ and $\lim\limits_{x \to a} g(x)$ exist, then $\lim\limits_{x \to a} [f(x) + g(x)] = \lim\limits_{x \to a} f(x) + \lim\limits_{x \to a} g(x)$ and

$\lim\limits_{x \to a} [f(x) g(x)] = \lim\limits_{x \to a} f(x) \cdot \lim\limits_{x \to a} g(x)$. The limit of a sum is the *sum* of the limits and the limit of a product is the *product* of the limits.

3. (a) $\lim\limits_{x \to a} [f(x) + h(x)] = \lim\limits_{x \to a} f(x) + \lim\limits_{x \to a} h(x) = -3 + 8 = 5$

(b) $\lim\limits_{x \to a} [f(x)]^2 = \left[\lim\limits_{x \to a} f(x)\right]^2 = (-3)^2 = 9$

(c) $\lim\limits_{x \to a} \sqrt[3]{h(x)} = \sqrt[3]{\lim\limits_{x \to a} h(x)} = \sqrt[3]{8} = 2$

(d) $\lim\limits_{x \to a} \dfrac{1}{f(x)} = \dfrac{1}{\lim\limits_{x \to a} f(x)} = -\dfrac{1}{3}$

(e) $\lim\limits_{x \to a} \dfrac{f(x)}{h(x)} = \dfrac{\lim\limits_{x \to a} f(x)}{\lim\limits_{x \to a} h(x)} = -\dfrac{3}{8}$

(f) $\lim\limits_{x \to a} \dfrac{g(x)}{f(x)} = \dfrac{\lim\limits_{x \to a} g(x)}{\lim\limits_{x \to a} f(x)} = \dfrac{0}{-3} = 0$

(g) $\lim\limits_{x \to a} \dfrac{f(x)}{g(x)}$ does not exist since $\lim\limits_{x \to a} g(x) = 0$.

(h) $\lim\limits_{x \to a} \dfrac{2f(x)}{h(x) - f(x)} = \dfrac{\lim\limits_{x \to a} [2f(x)]}{\lim\limits_{x \to a} [h(x) - f(x)]} = \dfrac{2 \cdot \left[\lim\limits_{x \to a} f(x)\right]}{\left[\lim\limits_{x \to a} h(x)\right] - \left[\lim\limits_{x \to a} f(x)\right]} = \dfrac{2(-3)}{8 - (-3)} = -\dfrac{6}{11}$

5. $\lim\limits_{x \to 4} \left(5x^2 - 2x + 3\right) = 5 \lim\limits_{x \to 4} x^2 - 2 \lim\limits_{x \to 4} x + \lim\limits_{x \to 4} 3$ Laws 1, 2, and 3

$\qquad\qquad\qquad = 5(4)^2 - 2(4) + 3$ Special Limits 1, 2, and 3

$\qquad\qquad\qquad = 80 - 8 + 3 = 75$

7. $\displaystyle\lim_{x\to-1}\frac{x-2}{x^2+4x-3} = \frac{\displaystyle\lim_{x\to-1} x-2}{\displaystyle\lim_{x\to-1} x^2+4x-3}$ Law 5

$\displaystyle = \frac{\displaystyle\lim_{x\to-1} x - \lim_{x\to-1} 2}{\displaystyle\lim_{x\to-1} x^2 + 4\lim_{x\to-1} x - \lim_{x\to-1} 3}$ Laws 1, 2, and 3

$\displaystyle = \frac{(-1)-2}{(-1)^2+4(-1)-3}$ Special Limits 1, 2, and 3

$\displaystyle = \frac{-3}{-6} = \frac{1}{2}$

9. $\displaystyle\lim_{t\to-2}(t+1)^9\left(t^2-1\right) = \left[\lim_{t\to-2}(t+1)^9\right]\left[\lim_{t\to-2} t^2-1\right]$ Law 4

$\displaystyle = \left[\lim_{t\to-2} t+1\right]^9\left[\lim_{t\to-2} t^2-1\right]$ Law 6

$\displaystyle = \left[\lim_{t\to-2} t + \lim_{t\to-2} 1\right]^9\left[\lim_{t\to-2} t^2 - \lim_{t\to-2} 1\right]$ Laws 1, 2, and 3

$\displaystyle = [(-2)+1]^9\left[(-2)^2-1\right]$ Special Limits 1, 2, and 3

$\displaystyle = (-1)^9(4-1) = (-1)(3) = -3$

11. $\displaystyle\lim_{x\to2}\frac{x^2+x-6}{x-2} = \lim_{x\to2}\frac{(x-2)(x+3)}{x-2} = \lim_{x\to2} x+3 = 5$

13. $\displaystyle\lim_{x\to2}\frac{x^2-x+6}{x+2} = \frac{(2)^2-(2)+6}{(2)+2} = \frac{8}{4} = 2$

15. $\displaystyle\lim_{t\to-3}\frac{t^2-9}{2t^2+7t+3} = \lim_{t\to-3}\frac{(t-3)(t+3)}{(t+3)(2t+1)} = \lim_{t\to-3}\frac{t-3}{2t+1} = \frac{(-3)-3}{2(-3)+1} = \frac{-6}{-5} = \frac{6}{5}$

17. $\displaystyle\lim_{h\to0}\frac{(2+h)^3-8}{h} = \lim_{h\to0}\frac{8+12h+6h^2+h^3-8}{h} = \lim_{h\to0}\frac{12h+6h^2+h^3}{h}$

$\displaystyle = \lim_{h\to0}\frac{h\left(12+6h+h^2\right)}{h} = \lim_{h\to0} 12+6h+h^2 = 12$

19. $\displaystyle\lim_{x\to7}\frac{\sqrt{x+2}-3}{x-7} = \lim_{x\to7}\frac{\left(\sqrt{x+2}-3\right)\left(\sqrt{x+2}+3\right)}{(x-7)\left(\sqrt{x+2}+3\right)} = \lim_{x\to7}\frac{x+2-9}{(x-7)\left(\sqrt{x+2}+3\right)}$

$\displaystyle = \lim_{x\to7}\frac{x-7}{(x-7)\sqrt{x+2}+3} = \lim_{x\to7}\frac{1}{\sqrt{x+2}+3} = \frac{1}{\sqrt{7+2}+3} = \frac{1}{6}$

21. $\displaystyle\lim_{x\to-4}\frac{\dfrac{1}{4}+\dfrac{1}{x}}{4+x} = \lim_{x\to-4}\frac{\left(\dfrac{1}{4}+\dfrac{1}{x}\right)4x}{(4+x)4x} = \lim_{x\to-4}\frac{x+4}{(4+x)4x} = \lim_{x\to-4}\frac{1}{4x} = \frac{1}{4(-4)} = -\frac{1}{16}$

23. $\displaystyle\lim_{x\to1}\frac{x^2-1}{\sqrt{x}-1} = \lim_{x\to1}\frac{(x-1)(x+1)\left(\sqrt{x}+1\right)}{\left(\sqrt{x}-1\right)\left(\sqrt{x}+1\right)} = \lim_{x\to1}\frac{(x-1)(x+1)\left(\sqrt{x}+1\right)}{x-1}$

$\displaystyle = \lim_{x\to1}(x+1)\left(\sqrt{x}+1\right) = [(1)+1]\left[\sqrt{1}+1\right] = 2\cdot2 = 4$

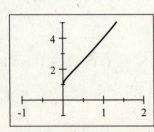

25. $\lim\limits_{x \to -1} \dfrac{x^2 - x - 2}{x^3 - x} = \lim\limits_{x \to -1} \dfrac{(x-2)(x+1)}{x(x-1)(x+1)} = \lim\limits_{x \to -1} \dfrac{x-2}{x(x-1)}$

$\qquad = \dfrac{(-1)-2}{(-1)[(-1)-1]} = -\dfrac{3}{2}$

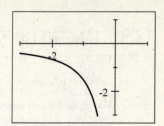

27. (a)

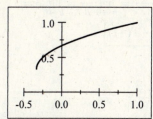

From the graph,

$\lim\limits_{x \to 0} \dfrac{x}{\sqrt{1+3x}-1} \approx 0.67$

(b)

x	-0.1	-0.01	-0.001	0.001	0.01	0.1
$f(x)$	0.61222	0.66163	0.66617	0.66717	0.67183	0.71339

(c) $\lim\limits_{x \to 0} \dfrac{x}{\sqrt{1+3x}-1} = \lim\limits_{x \to 0} \dfrac{x\left(\sqrt{1+3x}+1\right)}{\left(\sqrt{1+3x}-1\right)\left(\sqrt{1+3x}+1\right)}$

$\qquad = \lim\limits_{x \to 0} \dfrac{x\left(\sqrt{1+3x}+1\right)}{(1+3x)-1} = \lim\limits_{x \to 0} \dfrac{x\left(\sqrt{1+3x}+1\right)}{3x}$

$\qquad = \lim\limits_{x \to 0} \dfrac{\sqrt{1+3x}+1}{3} = \dfrac{\sqrt{1+3(0)}+1}{3} = \dfrac{2}{3}$

29. $\lim\limits_{x \to -4^-} |x+4| = \lim\limits_{x \to -4^-} -(x+4) = 0$ and $\lim\limits_{x \to -4^+} |x+4| = \lim\limits_{x \to -4^+} x+4 = 0$. Therefore $\lim\limits_{x \to -4^+} |x+4| = 0$.

31. $\lim\limits_{x \to 2^-} \dfrac{|x-2|}{x-2} = \lim\limits_{x \to 2^-} \dfrac{-(x-2)}{x-2} = -1$ and $\lim\limits_{x \to 2^+} \dfrac{|x-2|}{x-2} = \lim\limits_{x \to 2^+} \dfrac{x-2}{x-2} = 1$. So $\lim\limits_{x \to 2} \dfrac{|x-2|}{x-2}$ does not exist.

33. $\lim\limits_{x \to 0^-} \left(\dfrac{1}{x} - \dfrac{1}{|x|} \right) = \lim\limits_{x \to 0^-} \left(\dfrac{1}{x} - \dfrac{1}{-x} \right) = \lim\limits_{x \to 0^-} \dfrac{2}{x}$ which does not exist.

35. (a) $\lim\limits_{x \to 2^-} f(x) = \lim\limits_{x \to 2^-} (x-1) = 1$ and

$\lim\limits_{x \to 2^+} f(x) = \lim\limits_{x \to 2^+} \left(x^2 - 4x + 6 \right) = 2$.

(b) Since $\lim\limits_{x \to 2^-} f(x) \neq \lim\limits_{x \to 2^+} f(x)$, $\lim\limits_{x \to 2} f(x)$ does not exist.

(c)

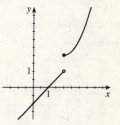

37. (a) The equation is true only for $x \neq 2$. That is, $x = 2$ is in the domain of $f(x) = x + 3$ but not in the domain of

$g(x) = \dfrac{x^2 + x - 6}{x - 2}$.

(b) When we take the limit we have $x \neq 2$.

39. (a) Many answers are possible. For example, let $f(x) = \dfrac{x+4}{x-2}$ and $g(x) = \dfrac{x-8}{x-2}$. Then neither $\lim\limits_{x \to 2} f(x)$ nor $\lim\limits_{x \to 2} g(x)$

exist, but $\lim\limits_{x \to 2} f(x) + g(x) = \lim\limits_{x \to 2} \left(\dfrac{x+4}{x-2} + \dfrac{x-8}{x-2} \right) = \lim\limits_{x \to 2} \dfrac{2x-4}{x-2} = \lim\limits_{x \to 2} \dfrac{2(x-2)}{x-2} = \lim\limits_{x \to 2} 2 = 2$.

(b) Many answers are possible. For example let $f(x) = \dfrac{|x-2|}{x-2}$ and $g(x) = \dfrac{x-2}{|x-2|}$. Then neither $\lim\limits_{x \to 2} f(x)$ nor

$\lim\limits_{x \to 2} g(x)$ exist, but $\lim\limits_{x \to 2} \left[f(x) g(x) \right] = \lim\limits_{x \to 2} \left(\dfrac{|x-2|}{x-2} \cdot \dfrac{x-2}{|x-2|} \right) = \lim\limits_{x \to 2} 1 = 1$.

13.3   TANGENT LINES AND DERIVATIVES

1. The derivative of a function f at a number a is $f'(a) = \lim\limits_{h \to 0} \dfrac{f(a+h) - f(a)}{h}$ if the limit exists. The derivative $f'(a)$ is

the *slope* of the tangent line to the curve $y = f(x)$ at the point $(a, f(a))$.

3. $m = \lim\limits_{h \to 0} \dfrac{f(1+h) - f(1)}{h} = \lim\limits_{h \to 0} \dfrac{[3(1+h)+4] - [3(1)+4]}{h} = \lim\limits_{h \to 0} \dfrac{7 + 3h - 7}{h} = \lim\limits_{h \to 0} \dfrac{3h}{h} = \lim\limits_{h \to 0} 3 = 3$

5. $m = \lim\limits_{h \to 0} \dfrac{f(-1+h) - f(-1)}{h} = \lim\limits_{h \to 0} \dfrac{\left[4(-1+h)^2 - 3(-1+h)\right] - \left[4(-1)^2 - 3(-1)\right]}{h}$

$= \lim\limits_{h \to 0} \dfrac{4 - 8h + 4h^2 + 3 - 3h - 7}{h} = \lim\limits_{h \to 0^-} \dfrac{-11h + 4h^2}{h} = \lim\limits_{h \to 0} \dfrac{h(-11 + 4h)}{h}$

$= \lim\limits_{h \to 0} (-11 + 4h) = -11$

7. $m = \lim\limits_{h \to 0} \dfrac{f(2+h) - f(2)}{h} = \lim\limits_{h \to 0} \dfrac{\left[2(2+h)^3\right] - \left[2(2)^3\right]}{h} = \lim\limits_{h \to 0} \dfrac{16 + 24h + 12h^2 + 2h^3 - 16}{h}$

$= \lim\limits_{h \to 0} \dfrac{24h + 12h^2 + 2h^3}{h} = \lim\limits_{h \to 0} \dfrac{h\left(24 + 12h + 2h^2\right)}{h} = \lim\limits_{h \to 0} \left(24 + 12h + 2h^2\right) = 24$

9. $m = \lim\limits_{h \to 0} \dfrac{\left[(-1+h) + (-1+h)^2\right] - \left[(-1) + (-1)^2\right]}{h} = \lim\limits_{h \to 0} \dfrac{-1 + h + 1 - 2h + h^2}{h}$

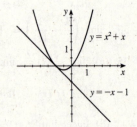

$= \lim\limits_{h \to 0} \dfrac{-h + h^2}{h} = \lim\limits_{h \to 0} \dfrac{h(-1 + h)}{h} = \lim\limits_{h \to 0} (-1 + h) = -1$

Thus an equation of the tangent line at $(-1, 0)$ is $y - 0 = -1(x + 1) \Leftrightarrow$

$y = -x - 1$.

11. $m = \lim\limits_{h \to 0} \dfrac{\dfrac{2+h}{2+h-1} - \dfrac{2}{2-1}}{h} = \lim\limits_{h \to 0} \dfrac{\left(\dfrac{2+h}{1+h} - 2\right)}{h} \cdot \dfrac{(1+h)}{(1+h)}$

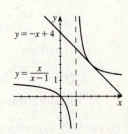

$= \lim\limits_{h \to 0} \dfrac{2 + h - 2 - 2h}{h(1+h)} = \lim\limits_{h \to 0} \dfrac{-h}{h(1+h)} = \lim\limits_{h \to 0} \dfrac{-1}{1+h} = -1$

Thus an equation of the tangent line at $(2, 2)$ is $y - 2 = -1(x - 2) \Leftrightarrow y = -x + 4$.

13. $m = \lim\limits_{h \to 0} \dfrac{\sqrt{(1+h)+3} - \sqrt{1+3}}{h} = \lim\limits_{h \to 0} \dfrac{\sqrt{h+4} - 2}{h} = \lim\limits_{h \to 0} \dfrac{(\sqrt{h+4} - 2)(\sqrt{h+4} + 2)}{h(\sqrt{h+4} + 2)}$

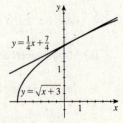

$= \lim\limits_{h \to 0} \dfrac{h + 4 - 4}{h(\sqrt{h+4} + 2)} = \lim\limits_{h \to 0} \dfrac{h}{h(\sqrt{h+4} + 2)} = \lim\limits_{h \to 0} \dfrac{1}{\sqrt{h+4} + 2} = \dfrac{1}{4}$

Thus an equation of the tangent line at $(1, 2)$ is $y - 2 = \frac{1}{4}(x - 1) \Leftrightarrow y = \frac{1}{4}x + \frac{7}{4}$.

15. $f'(2) = \lim\limits_{h \to 0} \dfrac{f(2+h) - f(2)}{h} = \lim\limits_{h \to 0} \dfrac{\left[1 - 3(2+h)^2\right] - \left[1 - 3(2)^2\right]}{h} = \lim\limits_{h \to 0} \dfrac{1 - 12 - 12h - 3h^2 + 11}{h}$

$= \lim\limits_{h \to 0} \dfrac{-12h - 3h^2}{h} = \lim\limits_{h \to 0} \dfrac{h(-12 - 3h)}{h} = \lim\limits_{h \to 0} (-12 - 3h) = -12$

17. $g'(1) = \lim_{h\to 0} \dfrac{g(1+h) - g(1)}{h} = \lim_{h\to 0} \dfrac{(1+h)^4 - (1)^4}{h} = \lim_{h\to 0} \dfrac{4h + 6h^2 + 4h^3 + h^4}{h} = \lim_{h\to 0} \dfrac{h\left(4 + 6h + 4h^2 + h^3\right)}{h}$

$\qquad = \lim_{h\to 0} \left(4 + 6h + 4h^2 + h^3\right) = 4$

19. $F'(4) = \lim_{h\to 0} \dfrac{F(4+h) - F(4)}{h} = \lim_{h\to 0} \dfrac{\left[\dfrac{1}{\sqrt{4+h}}\right] - \left[\dfrac{1}{\sqrt{4}}\right]}{h} = \lim_{h\to 0} \dfrac{\dfrac{1}{\sqrt{4+h}} - \dfrac{1}{2}}{h}$

$\qquad = \lim_{h\to 0} \dfrac{\left(\dfrac{1}{\sqrt{4+h}} - \dfrac{1}{2}\right) 2\sqrt{4+h}}{h \cdot 2\sqrt{4+h}} = \lim_{h\to 0} \dfrac{2 - \sqrt{4+h}}{2h\sqrt{4+h}} = \lim_{h\to 0} \dfrac{(2 - \sqrt{4+h})(2 + \sqrt{4+h})}{2h\sqrt{4+h}(2 + \sqrt{4+h})}$

$\qquad = \lim_{h\to 0} \dfrac{4 - (4+h)}{2h\sqrt{4+h}(2 + \sqrt{4+h})} = \lim_{h\to 0} \dfrac{-h}{2h\sqrt{4+h}(2 + \sqrt{4+h})} = \lim_{h\to 0} \dfrac{-1}{2\sqrt{4+h}(2 + \sqrt{4+h})} = -\dfrac{1}{16}$

21. $f'(a) = \lim_{x\to a} \dfrac{f(x) - f(a)}{x - a} = \lim_{x\to a} \dfrac{\left[x^2 + 2x\right] - \left[a^2 + 2a\right]}{x - a} = \lim_{x\to a} \dfrac{x^2 + 2x - a^2 - 2a}{x - a} = \lim_{x\to a} \dfrac{x^2 - a^2 + 2x - 2a}{x - a}$

$\qquad = \lim_{x\to a} \dfrac{(x - a)(x + a) + 2(x - a)}{x - a} = \lim_{x\to a} \dfrac{(x - a)[(x + a) + 2]}{x - a} = \lim_{x\to a}(x + a + 2) = 2a + 2$

23. $f'(a) = \lim_{x\to a} \dfrac{f(x) - f(a)}{x - a} = \lim_{x\to a} \dfrac{\left[\dfrac{x}{x+1}\right] - \left[\dfrac{a}{a+1}\right]}{x - a} = \lim_{x\to a} \dfrac{\left(\left[\dfrac{x}{x+1}\right] - \left[\dfrac{a}{a+1}\right]\right)(a+1)(x+1)}{(x - a)(a+1)(x+1)}$

$\qquad = \lim_{x\to a} \dfrac{x(a+1) - a(x+1)}{(x-a)(a+1)(x+1)} = \lim_{x\to a} \dfrac{ax + x - ax - a}{(x-a)(a+1)(x+1)} = \lim_{x\to a} \dfrac{x - a}{(x-a)(a+1)(x+1)}$

$\qquad = \lim_{x\to a} \dfrac{1}{(a+1)(x+1)} = \dfrac{1}{(a+1)^2}$

25. (a) $f'(a) = \lim_{x\to a} \dfrac{f(x) - f(a)}{x - a} = \lim_{x\to a} \dfrac{\left[x^3 - 2x + 4\right] - \left[a^3 - 2a + 4\right]}{x - a} = \lim_{x\to a} \dfrac{x^3 - 2x + 4 - a^3 + 2a - 4}{x - a}$

$\qquad = \lim_{x\to a} \dfrac{x^3 - a^3 - 2x + 2a}{x - a} = \lim_{x\to a} \dfrac{(x - a)\left(x^2 + ax + a^2\right) - 2(x - a)}{x - a}$

$\qquad = \lim_{x\to a} \dfrac{(x - a)\left(x^2 + ax + a^2 - 2\right)}{x - a} = \lim_{x\to a} \left(x^2 + ax + a^2 - 2\right) = 3a^2 - 2$

(b) $f(0) = 0^3 - 2(0) + 4 = 4$ and $f'(0) = 3(0)^2 - 2 = -2$. So the equation of line tangent to $f(x)$ at $x = 0$

is $y - 4 = -2(x - 0) \Leftrightarrow y = -2x + 4$. $f(1) = 1^3 - 2(1) + 4 = 3$ and $f'(1) = 3(1)^2 - 2 = 1$. So an

equation of the line tangent to $f(x)$ at $x = 1$ is $y - 3 = 1(x - 1) \Leftrightarrow y = x + 2$. $f(2) = 2^3 - 2(2) + 4 = 8$ and

$f'(2) = 3(2)^2 - 2 = 10$, so an equation of the line tangent to $f(x)$ at $x = 2$ is $y - 8 = 10(x - 2) \Leftrightarrow y = 10x - 12$.

(c)

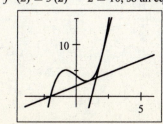

27. Let $s(t) = 40t - 16t^2$. Then

$$v(2) = \lim_{t \to 2} \frac{\left(40t - 16t^2\right) - 16}{t - 2} = \lim_{t \to 2} \frac{-16t^2 + 40t - 16}{t - 2} = \lim_{t \to 2} \frac{-8\left(2t^2 - 5t + 2\right)}{t - 2}$$

$$= \lim_{t \to 2} \frac{-8(t - 2)(2t - 1)}{t - 2} = -8 \lim_{t \to 2} (2t - 1) = -8(3) = -24$$

Thus, the instantaneous velocity when $t = 2$ is -24 ft/s.

29. $v(a) = \lim\limits_{h \to 0} \dfrac{s(a + h) - s(a)}{h} = \lim\limits_{h \to 0} \dfrac{\left[4(a + h)^3 + 6(a + h) + 2\right] - \left[4a^3 + 6a + 2\right]}{h}$

$$= \lim_{h \to 0} \frac{4a^3 + 12a^2h + 12ah^2 + 4h^3 + 6a + 6h + 2 - 4a^3 - 6a - 2}{h}$$

$$= \lim_{h \to 0} \frac{12a^2h + 12ah^2 + 4h^3 + 6h}{h} = \lim_{h \to 0} \frac{h\left(12a^2 + 12ah + 4h^2 + 6\right)}{h}$$

$$= \lim_{h \to 0} \left(12a^2 + 12ah + 4h^2 + 6\right) = 12a^2 + 6 \text{ m/s}$$

So $v(1) = 12(1)^2 + 6 = 18$ m/s, $v(2) = 12(2)^2 + 6 = 54$ m/s, and $v(3) = 12(3)^2 + 6 = 114$ m/s.

31. The slope of the tangent (that is, the rate of change in temperature with respect to time) at $t = 1$ h seems to be about
$$\frac{147 - 106}{90 - 30} \approx 0.7° \text{ F/min.}$$

33. (a) For the time interval $[10, 15]$ the average rate is $\dfrac{250 - 444}{15 - 10} = -38.8$ gal/min. For the time interval $[15, 20]$ the average

rate is $\dfrac{111 - 250}{20 - 15} = -27.8$ gal/min.

(b) $V'(15) \approx \dfrac{-38.8 - 27.8}{2} = -33.3$ gal/min.

35. $g'(0)$ is the only negative value. The slope at $x = 4$ is smaller than the slope at $x = 2$ and both are smaller than the slope at $x = -2$. Thus, $g'(0) < 0 < g'(4) < g'(2) < g'(-2)$.

13.4   LIMITS AT INFINITY; LIMITS OF SEQUENCES

1. If f is a function defined on some interval (a, ∞), then $\lim\limits_{x \to \infty} f(x) = L$ means that the values of $f(x)$ can be made

arbitrarily close to L by taking x sufficiently large. In this case, the line $y = L$ is called a *horizontal asymptote* of the curve

$y = f(x)$. For example, $\lim\limits_{x \to \infty} \dfrac{1}{x} = 0$ and the line $y = 0$ is a horizontal asymptote.

3. (a) $\lim\limits_{x \to \infty} f(x) = -1$ and $\lim\limits_{x \to -\infty} f(x) = 2$.   **5.** $\lim\limits_{x \to \infty} \dfrac{6}{x} = 0$

(b) Horizontal asymptotes: $y = -1$ and $y = 2$.

7. $\lim\limits_{x \to \infty} \dfrac{2x + 1}{5x - 1} = \lim\limits_{x \to \infty} \dfrac{2 + \dfrac{1}{x}}{5 - \dfrac{1}{x}} = \dfrac{\lim\limits_{x \to \infty} 2 + \lim\limits_{x \to \infty} \dfrac{1}{x}}{\lim\limits_{x \to \infty} 5 - \lim\limits_{x \to \infty} \dfrac{1}{x}} = \dfrac{2 + 0}{5 - 0} = \dfrac{2}{5}$

9. $\lim\limits_{x \to \infty} \dfrac{4x^2 + 1}{2 + 3x^2} = \lim\limits_{x \to \infty} \dfrac{4 + \dfrac{1}{x^2}}{\dfrac{2}{x^2} + 3} = \dfrac{\lim\limits_{x \to \infty} 4 + \lim\limits_{x \to \infty} \dfrac{1}{x^2}}{\lim\limits_{x \to \infty} \dfrac{2}{x^2} + \lim\limits_{x \to \infty} 3} = \dfrac{4 + 0}{0 + 3} = \dfrac{4}{3}$

11. $\displaystyle\lim_{t\to\infty} \frac{8t^3 + t}{(2t-1)(2t^2+1)} = \lim_{t\to\infty} \frac{8 + \dfrac{1}{t^2}}{\left(\dfrac{2t-1}{t}\right)\left(\dfrac{2t^2+1}{t^2}\right)} = \lim_{t\to\infty} \frac{8 + \dfrac{1}{t^2}}{\left(2 - \dfrac{1}{t}\right)\left(2 + \dfrac{1}{t^2}\right)}$

$$= \frac{\displaystyle\lim_{t\to\infty} 8 + \lim_{t\to\infty} \frac{1}{t^2}}{\left[\displaystyle\lim_{t\to\infty}\left(2 - \frac{1}{t}\right)\right]\left[\displaystyle\lim_{t\to\infty}\left(2 + \frac{1}{t^2}\right)\right]} = \frac{\displaystyle\lim_{t\to\infty} 8 + \lim_{t\to\infty} \frac{1}{t^2}}{\left[\displaystyle\lim_{t\to\infty} 2 - \lim_{t\to\infty}\frac{1}{t}\right]\left[\displaystyle\lim_{t\to\infty} 2 + \lim_{t\to\infty}\frac{1}{t^2}\right]}$$

$$= \frac{8 + 0}{(2-0)(2+0)} = 2$$

13. $\displaystyle\lim_{x\to\infty} \frac{x^4}{1 - x^2 + x^3} = \lim_{x\to\infty} \frac{1}{\dfrac{1}{x^4} - \dfrac{1}{x^2} + \dfrac{1}{x}} = \frac{\displaystyle\lim_{x\to\infty} 1}{\displaystyle\lim_{x\to\infty}\frac{1}{x^4} - \lim_{x\to\infty}\frac{1}{x^2} + \lim_{x\to\infty}\frac{1}{x}} = \frac{1}{0}$, so the limit does not exist.

15. $\displaystyle\lim_{x\to-\infty}\left(\frac{x-1}{x+1} + 6\right) = \lim_{x\to-\infty}\left(\frac{1 - \dfrac{1}{x}}{1 + \dfrac{1}{x}} + 6\right) = \lim_{x\to-\infty}\frac{1 - \dfrac{1}{x}}{1 + \dfrac{1}{x}} + \lim_{x\to-\infty} 6 = \frac{\displaystyle\lim_{x\to-\infty} 1 - \lim_{x\to-\infty}\frac{1}{x}}{\displaystyle\lim_{x\to-\infty} 1 + \lim_{x\to-\infty}\frac{1}{x}} + \lim_{x\to-\infty} 6$

$$= \frac{1-0}{1+0} + 6 = 7$$

17. Because of the periodic nature of the cosine function, the values of $\cos x$ oscillate between -1 and 1 and do not approach any definite number. Therefore, $\displaystyle\lim_{x\to\infty} \cos x$ does not exist.

19.

x	$\dfrac{\sqrt{x^2 + 4x}}{4x + 1}$
-10	-0.1986
-100	-0.2456
-1000	-0.2496

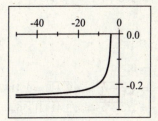

From the table it appears that $\displaystyle\lim_{x\to-\infty} \frac{\sqrt{x^2 + 4x}}{4x + 1} = -0.25$.

21.

x	x^5/e^x
10	4.5399
100	3.720×10^{-34}
1000	0

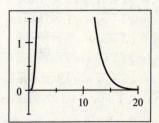

From the table it appears that $\displaystyle\lim_{x\to\infty} \frac{x^5}{e^x} = 0$.

23. $\displaystyle\lim_{n\to\infty} a_n = \lim_{n\to\infty} \frac{1+n}{n + n^2} = \lim_{n\to\infty} \frac{\dfrac{1}{n^2} + \dfrac{1}{n}}{\dfrac{1}{n} + 1} = \frac{\displaystyle\lim_{n\to\infty}\frac{1}{n^2} + \lim_{n\to\infty}\frac{1}{n}}{\displaystyle\lim_{n\to\infty}\frac{1}{n} + \lim_{n\to\infty} 1} = \frac{0+0}{0+1} = 0$

25. $a_n = \dfrac{n^2}{n+1}$ is divergent since $\displaystyle\lim_{n\to\infty} a_n = \lim_{n\to\infty} \frac{n^2}{n+1} = \lim_{n\to\infty} \frac{1}{\dfrac{1}{n} + \dfrac{1}{n^2}} = \frac{\displaystyle\lim_{n\to\infty} 1}{\displaystyle\lim_{n\to\infty}\frac{1}{n} + \lim_{n\to\infty}\frac{1}{n^2}} = \frac{1}{0+0}$ which is not

defined.

27. $\displaystyle\lim_{n\to\infty} a_n = \lim_{n\to\infty} \frac{1}{3^n} = 0$

29. $a_n = \sin(n\pi/2)$, because of the periodic nature of the sine function, the terms of this sequence repeat the sequence 1, 0, -1, 0 infinitely often and so they don't approach any definite number. Therefore, $\lim\limits_{n\to\infty} a_n$ does not exist, and the sequence is divergent.

31. $\lim\limits_{n\to\infty} a_n = \lim\limits_{n\to\infty} \dfrac{3}{n^2}\left[\dfrac{n(n+1)}{2}\right] = \lim\limits_{n\to\infty} \dfrac{3}{2}\left(\dfrac{n}{n}\right)\left(\dfrac{n+1}{n}\right) = \lim\limits_{n\to\infty} \dfrac{3}{2}\,(1)\left(1+\dfrac{1}{n}\right)$

$\qquad = \lim\limits_{n\to\infty} \dfrac{3}{2} \cdot \lim\limits_{n\to\infty} 1 \cdot \lim\limits_{n\to\infty}\left(1+\dfrac{1}{n}\right) = \frac{3}{2} \cdot 1 \cdot 1 = \frac{3}{2}$

33. $\lim\limits_{n\to\infty} a_n = \lim\limits_{n\to\infty} \dfrac{24}{n^3}\left[\dfrac{n(n+1)(2n+1)}{6}\right] = \lim\limits_{n\to\infty} 4\left(\dfrac{n}{n}\right)\left(\dfrac{n+1}{n}\right)\left(\dfrac{2n+1}{n}\right)$

$\qquad = \lim\limits_{n\to\infty} 4\,(1)\left(1+\dfrac{1}{n}\right)\left(2+\dfrac{1}{n}\right) = 4\lim\limits_{n\to\infty}\left(1+\dfrac{1}{n}\right) \cdot \lim\limits_{n\to\infty}\left(2+\dfrac{1}{n}\right) = 4 \cdot 1 \cdot 2 = 8$

35. (a) $C(t) = \dfrac{\text{amount of salt}}{\text{volume of water}} = \dfrac{30 \cdot 25t}{5000 + 25t} = \dfrac{30t}{200 + t}$.

$\quad$ **(b)** $\lim\limits_{t\to\infty} \dfrac{30t}{200+t} = \lim\limits_{t\to\infty} \dfrac{30}{\dfrac{200}{t}+1} = \dfrac{\lim\limits_{t\to\infty} 30}{\lim\limits_{t\to\infty}\dfrac{200}{t} + \lim\limits_{t\to\infty} 1} = \dfrac{30}{0+1} = 30 \text{ g/L}.$

37. (a) $a_1 = 0$

$\qquad a_2 = \sqrt{2+a_1} = \sqrt{2+0} = 1.41421356$

$\qquad a_3 = \sqrt{2+a_2} = \sqrt{3.41421356} = 1.84775907$

$\qquad a_4 = \sqrt{2+a_3} = \sqrt{3.84775907} = 1.96157056$

$\qquad a_5 = \sqrt{2+a_4} = \sqrt{3.96157056} = 1.99036945$

$\qquad a_6 = \sqrt{2+a_5} = \sqrt{3.99036945} = 1.99759091$

$\qquad a_7 = \sqrt{2+a_6} = \sqrt{3.99759091} = 1.99939764$

$\qquad a_8 = \sqrt{2+a_7} = \sqrt{3.99939764} = 1.99984940$

$\qquad a_9 = \sqrt{2+a_8} = \sqrt{3.99984940} = 1.99996235$

$\qquad a_{10} = \sqrt{2+a_9} = \sqrt{3.99996235} = 1.99999059$

From the table, it appears that $\lim\limits_{n\to\infty} a_n = 2$.

(b) Let $\lim\limits_{n\to\infty} a_n = L$. Then we can also have $\lim\limits_{n\to\infty} a_{n=1} = L$. Thus, we have

$$L = \lim\limits_{n\to\infty} a_{n+1} = \lim\limits_{n\to\infty} \sqrt{2+a_n}$$

$$= \sqrt{\lim\limits_{n\to\infty}(2+a_n)} = \sqrt{\lim\limits_{n\to\infty} 2 + \lim\limits_{n\to\infty} a_n}$$

$$= \sqrt{2+L}$$

So $L = \sqrt{2+L} \Rightarrow L^2 = 2 + L \Leftrightarrow L^2 - L - 2 = 0 \Leftrightarrow (L-2)(L+1) = 0 \Leftrightarrow L = 2$ or $L = -1$. But $L \geq 0$, so $L \neq -1$. Therefore, $\lim\limits_{n\to\infty} a_n = 2$.

13.5   AREAS

1. To find the area under the graph of f we first approximate the area by *rectangles*. The area of the approximation with four rectangles is $R_4 = f(x_1)(x_1 - a) + f(x_2)(x_2 - x_1) + f(x_3)(x_3 - x_2) + f(b)(b - x_3)$.

3. (a) Since f is *increasing*, we can obtain a *lower* estimate by using *left* endpoints. We are instructed to use five rectangles, so $n = 5$.

$$L_5 = \sum_{i=1}^{5} f(x_{i-1})\,\Delta x \quad [\Delta x = \tfrac{10-0}{5} = 2]$$
$$= f(x_0)\cdot 2 + f(x_1)\cdot 2 + f(x_2)\cdot 2 + f(x_3)\cdot 2 + f(x_4)\cdot 2$$
$$= 2[f(0) + f(2) + f(4) + f(6) + f(8)]$$
$$\approx 2(1 + 3 + 4.3 + 5.4 + 6.3) = 2(20) = 40$$

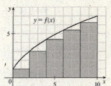

Since f is *increasing*, we can obtain an *upper* estimate by using *right* endpoints.

$$R_5 = \sum_{i=1}^{5} f(x_i)\,\Delta x$$
$$= 2[f(x_1) + f(x_2) + f(x_3) + f(x_4) + f(x_5)]$$
$$= 2[f(2) + f(4) + f(6) + f(8) + f(10)]$$
$$\approx 2(3 + 4.3 + 5.4 + 6.3 + 7) = 2(26) = 52$$

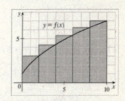

(b) $L_{10} = \sum_{i=1}^{10} f(x_{i-1})\,\Delta x \quad [\Delta x = \tfrac{10-0}{10} = 1]$
$$= 1[f(x_0) + f(x_1) + \cdots + f(x_9)]$$
$$= f(0) + f(1) + \cdots + f(9)$$
$$\approx 1 + 2.1 + 3 + 3.7 + 4.3 + 4.9 + 5.4 + 5.8 + 6.3 + 6.7$$
$$= 43.2$$

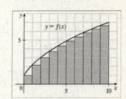

$$R_{10} = \sum_{i=1}^{10} f(x_i)\,\Delta x = f(1) + f(2) + \cdots + f(10)$$
$$= L_{10} + 1\cdot f(10) - 1\cdot f(0) \quad \begin{bmatrix} \text{add rightmost rectangle,} \\ \text{subtract leftmost} \end{bmatrix}$$
$$= 43.2 + 7 - 1 = 49.2$$

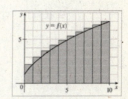

5. $\Delta x = \tfrac{2-0}{4} = \tfrac{1}{2}$, so $R_4 = \sum_{i=1}^{4} f(x_i)\,\Delta x = f\left(\tfrac{1}{2}\right)\cdot\tfrac{1}{2} + f(1)\cdot\tfrac{1}{2} + f\left(\tfrac{3}{2}\right)\cdot\tfrac{1}{2} + f(2)\cdot\tfrac{1}{2} = \tfrac{1}{2}\left(\tfrac{9}{4} + \tfrac{10}{4} + \tfrac{11}{4} + \tfrac{12}{4}\right) = \tfrac{42}{8} = \tfrac{21}{4}$

7. $\Delta x = \tfrac{7-1}{6} = 1$, so

$$R_6 = \sum_{i=1}^{6} f(x_i)\,\Delta x = f(2)\cdot 1 + f(3)\cdot 1 + f(4)\cdot 1 + f(5)\cdot 1 + f(6)\cdot 1 + f(7)\cdot 1 = 2 + \tfrac{4}{3} + 1 + \tfrac{4}{5} + \tfrac{4}{6} + \tfrac{4}{7} = \tfrac{223}{35} \approx 6.37$$

9. (a) $R_4 = \sum_{i=1}^{4} f(x_i)\,\Delta x \quad (\Delta x = \tfrac{5-1}{4} = 1)$
$$= f(x_1)\cdot 1 + f(x_2)\cdot 1 + f(x_3)\cdot 1 + f(x_4)\cdot 1$$
$$= f(2) + f(3) + f(4) + f(5)$$
$$= \tfrac{1}{2} + \tfrac{1}{3} + \tfrac{1}{4} + \tfrac{1}{5} = \tfrac{77}{60} = 1.28\overline{3}$$

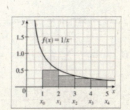

Since f is decreasing on $[1, 5]$, an underestimate is obtained by the right endpoint approximation, R_4.

(b) $L_4 = \sum\limits_{i=1}^{4} f\left(x_{i-1}\right) \Delta x$

$= f\left(1\right) + f\left(2\right) + f\left(3\right) + f\left(4\right)$

$= 1 + \frac{1}{2} + \frac{1}{3} + \frac{1}{4} = \frac{25}{12} = 2.08\overline{3}$

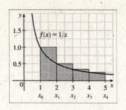

L_4 is an overestimate. Alternatively, we could just add the area of the leftmost rectangle and subtract the area of the rightmost; that is,

$L_4 = R_4 + f\left(1\right) \cdot 1 - f\left(5\right) \cdot 1.$

11. (a) $f\left(x\right) = 1 + x^2$ and $\Delta x = \dfrac{2 - \left(-1\right)}{3} = 1$, so

$R_3 = f\left(0\right) \cdot 1 + f\left(1\right) \cdot 1 + f\left(2\right) \cdot 1 = 1\left[1 + 2 + 5\right] = 8.$

If $\Delta x = \dfrac{2 - \left(-1\right)}{6} = 0.5$ then

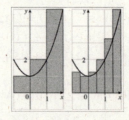

$R_6 = 0.5\left[f\left(-0.5\right) + f\left(0\right) + f\left(0.5\right) + f\left(1\right) + f\left(1.5\right) + f\left(2\right)\right]$

$= 0.5\left(1.25 + 1 + 1.25 + 2 + 3.25 + 5\right) = 0.5\left(13.75\right) = 6.875$

(b) $L_3 = f\left(-1\right) \cdot 1 + f\left(0\right) \cdot 1 + f\left(1\right) \cdot 1 = 1\left[2 + 1 + 2\right] = 5$

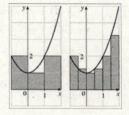

$L_6 = 0.5\left[f\left(-1\right) + f\left(-0.5\right) + f\left(0\right) + f\left(0.5\right) + f\left(1\right) + f\left(1.5\right)\right]$

$= 0.5\left(2 + 1.25 + 1 + 1.25 + 2 + 3.25\right)$

$= 0.5\left(10.75\right) = 5.375$

13. $f\left(x\right) = 3x$ and $\Delta x = \dfrac{5 - 0}{n} = \dfrac{5}{n}$ and $x_k = \dfrac{5k}{n}$, so

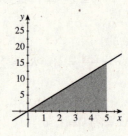

$A = \lim\limits_{n \to \infty} \sum\limits_{k=1}^{n} f\left(x_k\right) \Delta x = \lim\limits_{n \to \infty} \sum\limits_{k=1}^{n} 3\left(\dfrac{5k}{n}\right) \cdot \dfrac{5}{n}$

$= \lim\limits_{n \to \infty} \sum\limits_{k=1}^{n} \dfrac{75k}{n^2} = \lim\limits_{n \to \infty} \dfrac{75}{n^2} \sum\limits_{k=1}^{n} k = \lim\limits_{n \to \infty} \dfrac{75}{n^2} \cdot \dfrac{n\left(n+1\right)}{2}$

$= \lim\limits_{n \to \infty} \dfrac{75\left(n+1\right)}{2n} = \lim\limits_{n \to \infty} \dfrac{75}{2}\left(1 + \dfrac{1}{n}\right) = \dfrac{75}{2}\left(1 + 0\right) = \dfrac{75}{2}$

Using geometry, the area of the triangle is $\frac{1}{2}\left(5\right)\left(15\right) = \frac{75}{2}$.

15. $f\left(x\right) = 3x^2$, $a = 0$, $b = 2$, $\Delta x = \dfrac{2 - 0}{n} = \dfrac{2}{n}$, and $x_k = \dfrac{2k}{n}$, so

$A = \lim\limits_{n \to \infty} \sum\limits_{k=1}^{n} f\left(x_k\right) \Delta x = \lim\limits_{n \to \infty} \sum\limits_{k=1}^{n} 3\left(\dfrac{2k}{n}\right)^2 \cdot \dfrac{2}{n} = \lim\limits_{n \to \infty} \sum\limits_{k=1}^{n} \dfrac{24k^2}{n^3} = \lim\limits_{n \to \infty} \dfrac{24}{n^3} \cdot \sum\limits_{k=1}^{n} k^2$

$= \lim\limits_{n \to \infty} \dfrac{24}{n^3} \cdot \dfrac{n\left(n+1\right)\left(2n+1\right)}{6} = \lim\limits_{n \to \infty} \dfrac{4\left(2n^2 + 3n + 1\right)}{n^2} = \lim\limits_{n \to \infty} 4\left(2 + \dfrac{3}{n} + \dfrac{1}{n^2}\right)$

$= 4\left(2 + 0 + 0\right) = 8$

17. $f(x) = x^3 + 2$, $a = 0$, $b = 5$, $\Delta x = \dfrac{5-0}{n} = \dfrac{5}{n}$, and $x_k = \dfrac{5k}{n}$, so

$$A = \lim_{n\to\infty} \sum_{k=1}^{n} f(x_k)\,\Delta x = \lim_{n\to\infty} \sum_{k=1}^{n} \left[\left(\frac{5k}{n}\right)^3 + 2\right] \cdot \frac{5}{n} = \lim_{n\to\infty} \sum_{k=1}^{n} \left[\frac{625k^3}{n^4} + \frac{10}{n}\right]$$

$$= \lim_{n\to\infty} \left[\frac{625}{n^4} \cdot \sum_{k=1}^{n} k^3 + \frac{10}{n} \cdot \sum_{k=1}^{n} 1\right] = \lim_{n\to\infty} \left(\frac{625}{n^4} \cdot \frac{n^2(n+1)^2}{4} + \frac{10}{n} \cdot n\right)$$

$$= \lim_{n\to\infty} \left[\frac{625\left(n^2 + 2n + 1\right)}{4n^2} + 10\right] = \lim_{n\to\infty} \left[\frac{625}{4}\left(1 + \frac{2}{n} + \frac{1}{n^2}\right) + 10\right]$$

$$= \tfrac{625}{4}(1 + 0 + 0) + 10 = \tfrac{665}{4} = 166.25$$

19. $f(x) = x + 6x^2$, $a = 1$, $b = 4$, $\Delta x = \dfrac{4-1}{n} = \dfrac{3}{n}$, and $x_k = 1 + \dfrac{3k}{n}$, so

$$A = \lim_{n\to\infty} \sum_{k=1}^{n} f(x_k)\,\Delta x = \lim_{n\to\infty} \sum_{k=1}^{n} \left[\left(1 + \frac{3k}{n}\right) + 6\left(1 + \frac{3k}{n}\right)^2\right] \cdot \frac{3}{n}$$

$$= \lim_{n\to\infty} \sum_{k=1}^{n} \left[\frac{3}{n} + \frac{9k}{n^2} + \frac{18}{n} + \frac{108k}{n^2} + \frac{162k^2}{n^3}\right]$$

$$= \lim_{n\to\infty} \left[\frac{21}{n} \sum_{k=1}^{n} 1 + \frac{117}{n^2} \sum_{k=1}^{n} k + \frac{162}{n^3} \sum_{k=1}^{n} k^2\right]$$

$$= \lim_{n\to\infty} \left(\frac{21}{n} \cdot n + \frac{117}{n^2} \cdot \frac{n(n+1)}{2} + \frac{162}{n^3} \cdot \frac{n(n+1)(2n+1)}{6}\right)$$

$$= \lim_{n\to\infty} \left(21 + \tfrac{117}{2} \cdot \frac{(n+1)}{n} + 27 \cdot \frac{2n^2 + 3n + 1}{n^2}\right)$$

$$= \lim_{n\to\infty} \left[21 + \tfrac{117}{2}\left(1 + \frac{1}{n}\right) + 27\left(2 + \frac{3}{n} + \frac{1}{n^2}\right)\right] = 21 + \tfrac{117}{2} + 54 = 133.5$$

21. (a) $R_{10} \approx 161.2656$, $R_{20} \approx 147.9666$, and $R_{100} \approx 137.8067$.

(b) (i) For $f(x) = \sin x$ on $[0, \pi]$, $R_{100} \approx 1.99984$.

(ii) For $f(x) = e^{-x^2}$ on $[-1, 1]$, $R_{100} \approx 1.49360$.

CHAPTER 13 REVIEW

1.

x	$f(x)$
1.9	1.11111
1.99	1.01010
1.999	1.00100
2.001	0.99990
2.01	0.99100
2.1	0.90909

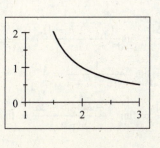

$$\lim_{x\to 2} \frac{x-2}{x^2 - 3x + 2} = 1$$

3.

t	$f(t)$
-0.1	0.66967
-0.01	0.69075
-0.001	0.69291
0.001	0.69339
0.01	0.69556
0.1	0.71773

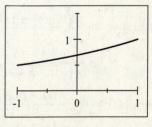

$$\lim_{x\to 0} \frac{2^x - 1}{x} = 0.693$$

5.

x	$f(x)$
1.5	-0.34657
1.1	-1.15129
1.05	-1.49787
1.01	-2.30258
1.001	-3.45388
1.0001	-4.60517

$\lim\limits_{x\to 1^+} \ln\sqrt{x-1}$ does not exist.

7. (a) $\lim\limits_{x\to 2^+} f(x)$ does not exist; f does not approach a finite value.

(b) $\lim\limits_{x\to -3^+} f(x) = 2.4$

(c) $\lim\limits_{x\to -3^-} f(x) = 2.4$

(d) $\lim\limits_{x\to -3} f(x) = 2.4$

(e) $\lim\limits_{x\to 4} f(x) = 0.5$

(f) $\lim\limits_{x\to\infty} f(x) = 1$

(g) $\lim\limits_{x\to -\infty} f(x) = 2$

(h) $\lim\limits_{x\to 0} f(x)\, f(x) = \left(\lim\limits_{x\to 0} f(x)\right)\left(\lim\limits_{x\to 0} f(x)\right)$
$$= 0\cdot 0 = 0$$

9. $\lim\limits_{x\to 2} \dfrac{x+1}{x-3} = \dfrac{2+1}{2-3} = -3$

11. $\lim\limits_{x\to 3} \dfrac{x^2+x-12}{x-3} = \lim\limits_{x\to 3} \dfrac{(x-3)(x+4)}{x-3} = \lim\limits_{x\to 3}(x+4) = 3+4 = 7$

13. $\lim\limits_{u\to 0} \dfrac{(u+1)^2-1}{u} = \lim\limits_{u\to 0} \dfrac{u^2+2u+1-1}{u} = \lim\limits_{u\to 0} \dfrac{u^2+2u}{u} = \lim\limits_{u\to 0}(u+2) = 0+2 = 2$

15. $\lim\limits_{x\to 3^-} \dfrac{x-3}{|x-3|} = \lim\limits_{x\to 3^-} \dfrac{x-3}{-(x-3)} = \lim\limits_{x\to 3^-} -1 = -1$

17. $\lim\limits_{x\to\infty} \dfrac{2x}{x-4} = \lim\limits_{x\to\infty} \dfrac{2}{1-\dfrac{4}{x}} = \dfrac{2}{1-0} = 2$

19. $\lim\limits_{x\to\infty} \cos^2 x$ does not exist since cosine squared alternates between 0 and 1.

21. $f'(4) = \lim\limits_{h\to 0} \dfrac{f(4+h)-f(4)}{h} = \lim\limits_{h\to 0} \dfrac{[3(4+h)-5]-[3(4)-5]}{h} = \lim\limits_{h\to 0} \dfrac{12+3h-5-12+5}{h}$

$= \lim\limits_{h\to 0} \dfrac{3h}{h} = \lim\limits_{h\to 0} 3 = 3$

23. $f'(16) = \lim\limits_{h\to 0} \dfrac{f(16+h)-f(16)}{h} = \lim\limits_{h\to 0} \dfrac{\sqrt{16+h}-\sqrt{16}}{h} = \lim\limits_{h\to 0} \dfrac{\left(\sqrt{16+h}-4\right)}{h}\cdot\dfrac{\left(\sqrt{16+h}+4\right)}{\left(\sqrt{16+h}+4\right)}$

$= \lim\limits_{h\to 0} \dfrac{16+h-16}{h\left(\sqrt{16+h}+4\right)} = \lim\limits_{h\to 0} \dfrac{h}{h\left(\sqrt{16+h}+4\right)} = \lim\limits_{h\to 0} \dfrac{1}{\sqrt{16+h}+4} = \dfrac{1}{8}$

25. (a) $f'(a) = \lim\limits_{x\to a} \dfrac{f(x)-f(a)}{x-a} = \lim\limits_{x\to a} \dfrac{[6-2x]-[6-2a]}{x-a} = \lim\limits_{x\to a} \dfrac{6-2x-6+2a}{x-a} = \lim\limits_{x\to a} \dfrac{-2x+2a}{x-a}$

$= \lim\limits_{x\to a} \dfrac{-2(x-a)}{x-a} = \lim\limits_{x\to a} -2 = -2$

(b) $f'(2) = -2$ and $f'(-2) = -2$.

27. (a) $f'(a) = \lim\limits_{x \to a} \dfrac{f(x) - f(a)}{x - a} = \lim\limits_{x \to a} \dfrac{\sqrt{x+6} - \sqrt{a+6}}{x - a} = \lim\limits_{x \to a} \dfrac{\left(\sqrt{x+6} - \sqrt{a+6}\right)\left(\sqrt{x+6} + \sqrt{a+6}\right)}{(x-a)\left(\sqrt{x+6} + \sqrt{a+6}\right)}$

$= \lim\limits_{x \to a} \dfrac{x + 6 - a - 6}{(x-a)\left(\sqrt{x+6} + \sqrt{a+6}\right)} = \lim\limits_{x \to a} \dfrac{x - a}{(x-a)\left(\sqrt{x+6} + \sqrt{a+6}\right)}$

$= \lim\limits_{x \to a} \dfrac{1}{\sqrt{x+6} + \sqrt{a+6}} = \dfrac{1}{2\sqrt{a+6}}$

(b) $f'(2) = \dfrac{1}{2\sqrt{2+6}} = \dfrac{1}{4\sqrt{2}} = \dfrac{\sqrt{2}}{8}$ and $f'(-2) = \dfrac{1}{2\sqrt{-2+6}} = \tfrac{1}{4}$.

29. $m = \lim\limits_{h \to 0} \dfrac{f(1+h) - f(1)}{h} = \lim\limits_{h \to 0} \dfrac{\left[4(1+h) - (1+h)^2\right] - \left[4(1) - (1)^2\right]}{h}$

$= \lim\limits_{h \to 0} \dfrac{4 + 4h - 1 - 2h - h^2 - 3}{h} = \lim\limits_{h \to 0} \dfrac{2h - h^2}{h} = \lim\limits_{h \to 0} \dfrac{h(2-h)}{h} = \lim\limits_{h \to 0}(2 - h) = 2$

So an equation of the line tangent to $y = 4x - x^2$ at $(1, 3)$ is $y - 3 = 2(x - 1) \Leftrightarrow y = 2x + 1$.

31. $m = \lim\limits_{h \to 0} \dfrac{f(3+h) - f(3)}{h} = \lim\limits_{h \to 0} \dfrac{[2(3+h)] - [2(3)]}{h} = \lim\limits_{h \to 0} \dfrac{6 + 2h - 6}{h} = \lim\limits_{h \to 0} \dfrac{2h}{h} = \lim\limits_{h \to 0} 2 = 2$.

So an equation of the line tangent to $y = 2x$ at $(3, 6)$ is $y - 6 = 2(x - 3) \Leftrightarrow y = 2x$.

33. $m = \lim\limits_{h \to 0} \dfrac{f(2+h) - f(2)}{h} = \lim\limits_{h \to 0} \dfrac{\dfrac{1}{2+h} - \dfrac{1}{2}}{h} = \lim\limits_{h \to 0} \dfrac{\left(\dfrac{1}{2+h} - \dfrac{1}{2}\right)}{h} \cdot \dfrac{2(2+h)}{2(2+h)}$

$= \lim\limits_{h \to 0} \dfrac{2 - (2+h)}{2h(2+h)} = \lim\limits_{h \to 0} \dfrac{-h}{2h(2+h)} = \lim\limits_{h \to 0} \dfrac{-1}{2(2+h)} = -\dfrac{1}{4}$

So an equation of the line tangent to $y = \dfrac{1}{x}$ at $\left(2, \tfrac{1}{2}\right)$ is $y - \tfrac{1}{2} = -\tfrac{1}{4}(x - 2) \Leftrightarrow y = -\tfrac{1}{4}x + 1$.

35. (a) $v(2) = \lim\limits_{t \to 2} \dfrac{h(t) - h(2)}{t - 2} = \lim\limits_{t \to 2} \dfrac{\left[640 - 16t^2\right] - \left[640 - 16(2)^2\right]}{t - 2} = \lim\limits_{t \to 2} \dfrac{640 - 16t^2 - 640 + 64}{t - 2}$

$= \lim\limits_{t \to 2} \dfrac{-16t^2 + 64}{t - 2} = \lim\limits_{t \to 2} \dfrac{-16(t+2)(t-2)}{t - 2} = \lim\limits_{t \to 2}[-16(t + 2)] = -64 \text{ ft/s}$

(b) $v(a) = \lim\limits_{t \to a} \dfrac{h(t) - h(a)}{t - a} = \lim\limits_{t \to a} \dfrac{\left[640 - 16t^2\right] - \left[640 - 16a^2\right]}{t - a} = \lim\limits_{t \to a} \dfrac{640 - 16t^2 - 640 + 16a^2}{t - a}$

$= \lim\limits_{t \to a} \dfrac{-16t^2 + 16a^2}{t - a} = \lim\limits_{t \to a} \dfrac{-16(t+a)(t-a)}{t - a} = \lim\limits_{t \to a} \dfrac{58h - 1.66ah - 0.83h^2}{h}$

$= \lim\limits_{t \to a} \dfrac{h(58 - 1.66a - 0.83h)}{h} = \lim\limits_{t \to a}[-16(t + a)] = -32a \text{ ft/s}$

(c) $h(t) = 0$ when $640 - 16t^2 = 0 \Leftrightarrow -16\left(t^2 - 40\right) = 0 \Leftrightarrow t^2 = 40 \Rightarrow t = \sqrt{40} \approx 6.32 \text{ s}$ (since $t \geq 0$).

(d) $v(6.32) = -32(6.32) = -202.4 \text{ ft/s}$.

37. $\lim\limits_{n \to \infty} a_n = \lim\limits_{n \to \infty} \dfrac{n}{5n + 1} = \lim\limits_{n \to \infty} \dfrac{1}{5 + \dfrac{1}{n}} = \dfrac{1}{5 + 0} = \dfrac{1}{5}$

39. $\lim\limits_{n \to \infty} a_n = \lim\limits_{n \to \infty} \left(\dfrac{n(n+1)}{2n^2}\right) = \lim\limits_{n \to \infty} \dfrac{1}{2}\left(\dfrac{n}{n}\right)\left(\dfrac{n+1}{n}\right) = \lim\limits_{n \to \infty} \dfrac{1}{2}(1)\left(1 + \dfrac{1}{n}\right)$

$= \lim\limits_{n \to \infty} \dfrac{1}{2} \cdot \lim\limits_{n \to \infty} 1 \cdot \lim\limits_{n \to \infty} \left(1 + \dfrac{1}{n}\right) = \dfrac{1}{2} \cdot 1 \cdot 1 = \dfrac{1}{2}$

41. The sequence $a_n = \cos\left(\dfrac{n\pi}{2}\right)$ is the sequence $0, -1, 0, 1, 0, -1, 0, 1, 0, -1, 0, 1, \ldots$ which does not go to any one value.

Thus this sequence does not converge.

43. $f(x) = \sqrt{x}$ and $\Delta x = 0.5$, so

$$R_6 = \sum_{i=1}^{6} f(x_i)\, \Delta x = f(x_1) \cdot 0.5 + f(x_2) \cdot 0.5 + f(x_3) \cdot 0.5 + f(x_4) \cdot 0.5 + f(x_5) \cdot 0.5 + f(x_6) \cdot 0.5$$

$$= 0.5 \left[f(0.5) + f(1) + f(1.5) + f(2) + f(2.5) + f(3) \right]$$

$$\approx 0.5\,(0.7071 + 1 + 1.2247 + 1.4142 + 1.5811 + 1.7320) = 0.5\,(7.6591) = 3.8296$$

45. $f(x) = 2x + 3$, $a = 0$, $b = 2$, $\Delta x = \dfrac{2-0}{n} = \dfrac{2}{n}$, and $x_k = \dfrac{2k}{n}$, so

$$A = \lim_{n \to \infty} \sum_{k=1}^{n} f(x_k)\, \Delta x = \lim_{n \to \infty} \sum_{k=1}^{n} \left[2\left(\frac{2k}{n} \right) + 3 \right] \cdot \frac{2}{n} = \lim_{n \to \infty} \sum_{k=1}^{n} \left[\frac{8k}{n^2} + \frac{6}{n} \right] = \lim_{n \to \infty} \left[\frac{8}{n^2} \cdot \sum_{k=1}^{n} k + \frac{6}{n} \cdot \sum_{k=1}^{n} 1 \right]$$

$$= \lim_{n \to \infty} \left(\frac{8}{n^2} \cdot \frac{n(n+1)}{2} + \frac{6}{n} \cdot n \right) = \lim_{n \to \infty} \left[\frac{4(n+1)}{n} + 6 \right] = \lim_{n \to \infty} \left[4\left(1 + \frac{1}{n} \right) + 6 \right] = 4(1+0) + 6 = 10$$

47. $f(x) = x^2 - x$, $a = 1$, $b = 2$, $\Delta x = \dfrac{2-1}{n} = \dfrac{1}{n}$, and $x_k = 1 + \dfrac{k}{n}$, so

$$A = \lim_{n \to \infty} \sum_{k=1}^{n} f(x_k)\, \Delta x = \lim_{n \to \infty} \sum_{k=1}^{n} \left[\left(1 + \frac{k}{n} \right)^2 - \left(1 + \frac{k}{n} \right) \right] \cdot \frac{1}{n}$$

$$= \lim_{n \to \infty} \sum_{k=1}^{n} \left[1 + \frac{2k}{n} + \frac{k^2}{n^2} - 1 - \frac{k}{n} \right] \cdot \frac{1}{n} = \lim_{n \to \infty} \sum_{k=1}^{n} \left[\frac{k}{n^2} + \frac{k^2}{n^3} \right] = \lim_{n \to \infty} \left[\sum_{k=1}^{n} \frac{k}{n^2} + \sum_{k=1}^{n} \frac{k^2}{n^3} \right]$$

$$= \lim_{n \to \infty} \left[\frac{1}{n^2} \sum_{k=1}^{n} k + \frac{1}{n^3} \sum_{k=1}^{n} k^2 \right] = \lim_{n \to \infty} \left[\frac{1}{n^2} \cdot \frac{n(n+1)}{2} + \frac{1}{n^3} \cdot \frac{n(n+1)(2n+1)}{6} \right]$$

$$= \lim_{n \to \infty} \left(\frac{1}{2} \cdot \frac{(n+1)}{n} + \frac{1}{6} \cdot \frac{2n^2 + 3n + 1}{n^2} \right) = \lim_{n \to \infty} \left[\frac{1}{2}\left(1 + \frac{1}{n} \right) + \frac{1}{6}\left(2 + \frac{3}{n} + \frac{1}{n^2} \right) \right]$$

$$= \frac{1}{2}(1+0) + \frac{1}{6}(2+0+0) = \frac{5}{6} = 0.8\overline{3}$$

CHAPTER 13 TEST

1. (a)

x	$f(x)$
-0.1	0.50335
-0.01	0.50003
-0.001	0.50000

x	$f(x)$
0.001	0.50000
0.01	0.50003
0.1	0.50335

$$\lim_{x \to 0} \frac{x}{\sin 2x} = 0.5$$

(b)

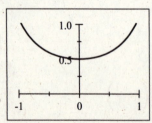

3. (a) $\displaystyle \lim_{x \to 2} \frac{x^2 + 2x - 8}{x - 2} = \lim_{x \to 2} \frac{(x-2)(x+4)}{x-2} = \lim_{x \to 2} (x+4) = 2 + 4 = 6$

(b) $\displaystyle \lim_{x \to 2} \frac{x^2 - 2x - 8}{x + 2} = \frac{2^2 - 2(2) - 8}{2 + 2} = -2$

(c) $\lim\limits_{x \to 2} \dfrac{1}{x-2}$ does not exist because $\left|\dfrac{1}{x-2}\right|$ grows without bound as $x \to 2$.

(d) $\lim\limits_{x \to 2^-} \dfrac{x-2}{|x-2|} = \lim\limits_{x \to 2^-} \dfrac{x-2}{-(x-2)} = \lim\limits_{x \to 2^-} -1 = -1$ and $\lim\limits_{x \to 2^+} \dfrac{x-2}{|x-2|} = \lim\limits_{x \to 2^+} \dfrac{x-2}{x-2} = \lim\limits_{x \to 2^+} 1 = 1$. Since

$\lim\limits_{x \to 2^-} \dfrac{x-2}{|x-2|} \neq \lim\limits_{x \to 2^+} \dfrac{x-2}{|x-2|}$, $\lim\limits_{x \to 2} \dfrac{x-2}{|x-2|}$ does not exist.

(e) $\lim\limits_{z \to 4} \dfrac{\sqrt{z}-2}{z-4} = \lim\limits_{z \to 4} \dfrac{\sqrt{x}-2}{\left(\sqrt{x}+2\right)\left(\sqrt{x}-2\right)} = \lim\limits_{z \to 4} \dfrac{1}{\sqrt{x}+2} = \dfrac{1}{\sqrt{4}+2} = \dfrac{1}{4}$

(f) $\lim\limits_{x \to \infty} \dfrac{2x^2-4}{x^2+x} = \lim\limits_{x \to \infty} \dfrac{2-4/x^2}{1-1/x} = \dfrac{2-0}{1-0} = 2$

5. $m = \lim\limits_{h \to 0} \dfrac{f(9+h)-f(9)}{h} = \lim\limits_{h \to 0} \dfrac{\sqrt{9+h}-\sqrt{9}}{h} = \lim\limits_{h \to 0} \dfrac{\sqrt{9+h}-3}{h}$

$= \lim\limits_{h \to 0} \dfrac{\left(\sqrt{9+h}-3\right)}{h} \cdot \dfrac{\left(\sqrt{9+h}+3\right)}{\left(\sqrt{9+h}+3\right)} = \lim\limits_{h \to 0} \dfrac{(9+h)-9}{h\left(\sqrt{9+h}+3\right)} = \lim\limits_{h \to 0} \dfrac{h}{h\left(\sqrt{9+h}+3\right)}$

$= \lim\limits_{h \to 0} \dfrac{1}{\sqrt{9+h}+3} = \dfrac{1}{3+3} = \tfrac{1}{6}$

So an equation of the line tangent to $f(x) = \sqrt{x}$ at $(9,3)$ is $y - 3 = \tfrac{1}{6}(x-9) \Leftrightarrow y = \tfrac{1}{6}x + \tfrac{3}{2}$.

7. (a) $f(x) = 4 - x^2$ and $\Delta x = \dfrac{1-0}{5} = 0.2$, so

$$R_5 = \sum_{k=1}^{5} f(x_k)\,\Delta x = f(0.2) \cdot 0.2 + f(0.4) \cdot 0.2 + f(0.6) \cdot 0.2 + f(0.8) \cdot 0.2 + f(1) \cdot 0.2$$

$$= 0.2\,(3.96 + 3.84 + 3.64 + 3.36 + 3) = 0.2\,(17.8) = 3.56$$

(b) $f(x) = 4 - x^2$, $a = 0$, $b = 1$, $\Delta x = \dfrac{1-0}{n} = \dfrac{1}{n}$, and $x_k = \dfrac{k}{n}$, so

$$A = \lim_{n \to \infty} \sum_{k=1}^{n} f(x_k)\,\Delta x = \lim_{n \to \infty} \sum_{k=1}^{n} \left[4 - \left(\dfrac{k}{n}\right)^2\right] \cdot \dfrac{1}{n} = \lim_{n \to \infty} \sum_{k=1}^{n} \left[\dfrac{1}{n} - \dfrac{k^2}{n^3}\right]$$

$$= \lim_{n \to \infty} \left[\dfrac{4}{n} \cdot \sum_{k=1}^{n} 1 - \dfrac{1}{n^3} \cdot \sum_{k=1}^{n} k^2\right] = \lim_{n \to \infty} \left(\dfrac{4}{n} \cdot n - \dfrac{1}{n^3} \cdot \dfrac{n(n+1)(2n+1)}{6}\right)$$

$$= \lim_{n \to \infty} \left[4 - \dfrac{1}{6}\left(\dfrac{n+1}{n}\right)\left(\dfrac{2n+1}{n}\right)\right] = \lim_{n \to \infty} \left[4 - \dfrac{1}{6}\left(1 + \dfrac{1}{n}\right)\left(2 + \dfrac{1}{n}\right)\right]$$

$$= 4 - \tfrac{1}{6}(1+0)(2+0) = \tfrac{11}{3}$$

FOCUS ON MODELING   Interpretations of Area

1. Work $=$ force $\times$ distance, so the work done in moving the tree the kth interval is $w_k = f(x_k)\,\Delta x$, where $f(x) = 1500 + 10x - \frac{1}{2}x^2$, $a = 0$, $b = 40$, $\Delta x = \dfrac{40 - 0}{n} = \dfrac{40}{n}$, and $x_k = \dfrac{40k}{n}$. Thus,

$$
\begin{aligned}
W &= \lim_{n\to\infty} \sum_{k=1}^{n} f(x_k)\,\Delta x = \lim_{n\to\infty} \sum_{k=1}^{n} \left[1500 + 10\left(\frac{40k}{n}\right) - \frac{1}{2}\left(\frac{40k}{n}\right)^2 \right] \cdot \frac{40}{n} \\
&= \lim_{n\to\infty} \sum_{k=1}^{n} \left[\frac{60{,}000}{n} + \frac{16{,}000k}{n^2} - \frac{32{,}000k^2}{n^3} \right] = \lim_{n\to\infty} \left[\frac{60{,}000}{n} \sum_{k=1}^{n} 1 + \frac{16{,}000}{n^2} \sum_{k=1}^{n} k - \frac{32{,}000}{n^3} \sum_{k=1}^{n} k^2 \right] \\
&= \lim_{n\to\infty} \left(\frac{60{,}000}{n} \cdot n + \frac{16{,}000}{n^2} \cdot \frac{n(n+1)}{2} - \frac{32{,}000}{n^3} \cdot \frac{n(n+1)(2n+1)}{6} \right) \\
&= \lim_{n\to\infty} \left(60{,}000 + 8{,}000 \cdot \frac{n+1}{n} - \frac{16{,}000}{3} \cdot \frac{(n+1)(2n+1)}{n^2} \right) \\
&= \lim_{n\to\infty} \left[60{,}000 + 8{,}000 \cdot \left(1 + \frac{1}{n} \right) - \frac{16{,}000}{3} \cdot \left(1 + \frac{1}{n} \right)\left(2 + \frac{1}{n} \right) \right] \\
&= 60{,}000 + 8000 - \frac{32{,}000}{3} = 57{,}333.33 \text{ ft-lb}
\end{aligned}
$$

3. **(a)** The force on the kth strip is $375x_k\,\Delta x$, so to find the force exerted on the window panel we must add the forces on each of the strips, giving $\displaystyle\lim_{n\to\infty} \sum_{k=1}^{n} 375x_k\,\Delta x$.

   (b) This is the area under the graph of $p(x) = 375x$ between $x = 0$ and $x = 4$.

   (c) Using $\Delta x = \dfrac{4}{n}$, and $x_k = \dfrac{4k}{n}$,

$$
\begin{aligned}
P &= \lim_{n\to\infty} \sum_{k=1}^{n} 375x_k\,\Delta x = \lim_{n\to\infty} \sum_{k=1}^{n} 375\left(\frac{4k}{n}\right) \cdot \frac{4}{n} = \lim_{n\to\infty} \sum_{k=1}^{n} \left[\frac{6000k}{n^2} \right] = \lim_{n\to\infty} \left[\frac{6000}{n^2} \sum_{k=1}^{n} k \right] \\
&= \lim_{n\to\infty} \left(\frac{6000}{n^2} \cdot \frac{n(n+1)}{2} \right) = \lim_{n\to\infty} \left[3000\left(\frac{n+1}{n}\right) \right] = \lim_{n\to\infty} \left[3000\left(1 + \frac{1}{n}\right) \right] = 3000 \text{ lb}
\end{aligned}
$$

   (d) The area of the kth strip is $3\,\Delta x$, so the pressure on the kth strip is $187.5x_k\,\Delta x$. Now using $\Delta x = \dfrac{4}{n}$, and $x_k = \dfrac{4k}{n}$, we have

$$
\begin{aligned}
P &= \lim_{n\to\infty} \sum_{k=1}^{n} 187.5x_k\,\Delta x = \lim_{n\to\infty} \sum_{k=1}^{n} 187.5\left(\frac{4k}{n}\right) \cdot \frac{4}{n} = \lim_{n\to\infty} \sum_{k=1}^{n} \left[\frac{3000k}{n^2} \right] = \lim_{n\to\infty} \left[\frac{3000}{n^2} \sum_{k=1}^{n} k \right] \\
&= \lim_{n\to\infty} \left(\frac{3000}{n^2} \cdot \frac{n(n+1)}{2} \right) = \lim_{n\to\infty} \left[1500\left(\frac{n+1}{n}\right) \right] = \lim_{n\to\infty} \left[1500\left(1 + \frac{1}{n}\right) \right] = 1500 \text{ lb}
\end{aligned}
$$

5. (a) $D(t) = 61 + \frac{6}{5}t - \frac{1}{25}t^2$, $a = 0$, $b = 24$, $\Delta x = \dfrac{24 - 0}{n} = \dfrac{24}{n}$, and $x_k = \dfrac{24k}{n}$, so the heating capacity generated is

$$\lim_{n \to \infty} \sum_{k=1}^{n} D(t_k)\,\Delta t = \lim_{n \to \infty} \sum_{k=1}^{n} \left[61 + \frac{6}{5}\left(\frac{24k}{n}\right) - \frac{1}{25}\left(\frac{24k}{n}\right)^2 \right] \cdot \frac{24}{n} = \lim_{n \to \infty} \sum_{k=1}^{n} \left[\frac{1464}{n} + \frac{3456k}{5n^2} - \frac{13824k^2}{25n^3} \right]$$

$$= \lim_{n \to \infty} \left[\frac{1464}{n} \sum_{k=1}^{n} 1 + \frac{3456}{5n^2} \sum_{k=1}^{n} k - \frac{13824}{25n^3} \sum_{k=1}^{n} k^2 \right]$$

$$= \lim_{n \to \infty} \left(\frac{1464}{n} \cdot n + \frac{3456}{5n^2} \cdot \frac{n(n+1)}{2} - \frac{13824}{25n^3} \cdot \frac{n(n+1)(2n+1)}{6} \right)$$

$$= \lim_{n \to \infty} \left(1464 + \frac{1728}{5} \cdot \frac{n+1}{n} - \frac{2304}{25} \cdot \frac{(n+1)(2n+1)}{n^2} \right)$$

$$= \lim_{n \to \infty} \left[1464 + \frac{1728}{5} \cdot \left(1 + \frac{1}{n}\right) - \frac{2304}{25} \cdot \left(1 + \frac{1}{n}\right)\left(2 + \frac{1}{n}\right) \right]$$

$$= 1464 + \frac{1728}{5} - \frac{4608}{25} = 1625.28 \text{ heating degree-hours}$$

(b) $D(t) = 61 + \frac{6}{5}t - \frac{1}{25}t^2 = -\frac{1}{25}\left(t^2 - 30t\right) + 61 = -\frac{1}{25}\left(t^2 - 30t + 225\right) + 61 + 9 = -\frac{1}{25}(t - 15)^2 + 70$. Thus the maximum temperature during the day was $70°$.

(c) In this case, $D(t) = 50 + 5t - \frac{1}{4}t^2$, $a = 0$, $b = 24$, $\Delta x = \dfrac{24 - 0}{n} = \dfrac{24}{n}$, and $x_k = \dfrac{24k}{n}$, so the heating capacity generated is

$$\lim_{n \to \infty} \sum_{k=1}^{n} D(t_k)\,\Delta t = \lim_{n \to \infty} \sum_{k=1}^{n} \left[50 + 5\left(\frac{24k}{n}\right) - \frac{1}{4}\left(\frac{24k}{n}\right)^2 \right] \cdot \frac{24}{n} = \lim_{n \to \infty} \sum_{k=1}^{n} \left[\frac{1200}{n} + \frac{2880k}{n^2} - \frac{3456k^2}{n^3} \right]$$

$$= \lim_{n \to \infty} \left[\frac{1200}{n} \sum_{k=1}^{n} 1 + \frac{2880}{n^2} \sum_{k=1}^{n} k - \frac{3456}{n^3} \sum_{k=1}^{n} k^2 \right]$$

$$= \lim_{n \to \infty} \left(\frac{1200}{n} \cdot n + \frac{2880}{n^2} \cdot \frac{n(n+1)}{2} - \frac{3456}{n^3} \cdot \frac{n(n+1)(2n+1)}{6} \right)$$

$$= \lim_{n \to \infty} \left(1200 + 1440 \cdot \frac{n+1}{n} - 576 \cdot \frac{(n+1)(2n+1)}{n^2} \right)$$

$$= \lim_{n \to \infty} \left[1200 + 1440 \cdot \left(1 + \frac{1}{n}\right) - 576 \cdot \left(1 + \frac{1}{n}\right)\left(2 + \frac{1}{n}\right) \right]$$

$$= 1200 + 1440 - 1152 = 1488 \text{ heating degree-hours}$$

(d) $D(t) = 50 + 5t - \frac{1}{4}t^2 = -\frac{1}{4}\left(t^2 - 20t\right) + 50 = -\frac{1}{4}\left(t^2 - 20t + 100\right) + 50 + 25 = -\frac{1}{4}(t - 10)^2 + 75$. Thus the maximum temperature during the day was $75°$.

(e) The day described in part (a) was hotter because the number of heating degree-hours was greater, meaning the average temperature of that day was higher.

CUMULATIVE REVIEW TEST: CHAPTERS 12 and 13

1. (a) The sequence $\frac{1}{3}, \frac{2}{5}, \frac{3}{7}, \frac{4}{9}, \frac{5}{11}, \ldots$ has nth term $a_n = \dfrac{n}{2n+1}$. Thus, the 7th term is $a_7 = \dfrac{7}{2\cdot 7+1} = \dfrac{7}{15}$, the 20th term

is $a_{20} = \dfrac{20}{2\cdot 20+1} = \dfrac{20}{41}$, and the limit of the sequence is $\lim\limits_{n\to\infty} \dfrac{n}{2n+1} = \dfrac{1}{2}$.

(b) The sequence with $a_n = \dfrac{2n^2+1}{n^3-n+4}$ has 7th term $a_7 = \dfrac{2(7)^2+1}{7^3-7+4} = \dfrac{99}{340}$, 20th term $a_{20} = \dfrac{2(20)^2+1}{20^3-20+4} = \dfrac{801}{7984}$,

and limit $\lim\limits_{n\to\infty} \dfrac{2n^2+1}{n^3-n+4} = \lim\limits_{n\to\infty} \dfrac{\frac{1}{n^3}\left(2n^2+1\right)}{\frac{1}{n^3}\left(n^3-n+4\right)} = \lim\limits_{n\to\infty} \dfrac{\frac{2}{n}+\frac{1}{n^3}}{1-\frac{1}{n^2}+\frac{4}{n^3}} = 0.$

(c) The arithmetic sequence with initial term $a = \frac{1}{2}$ and common difference $d = 3$ has nth term

$a_n = a + (n-1)d = \frac{1}{2} + (n-1)3 = 3n - \frac{5}{2}$. Thus, the 7th term is $a_7 = 3(7) - \frac{5}{2} = \frac{37}{2}$ and the 20th term is

$a_{20} = 3(20) - \frac{5}{2} = \frac{115}{2}$. The limit is $\lim\limits_{n\to\infty} \left(3n - \frac{5}{2}\right)$, which does not exist.

(d) The geometric sequence with initial term $a = 12$ and common ratio $r = \frac{5}{6}$ has nth term $a_n = ar^{n-1} = 12\left(\frac{5}{6}\right)^{n-1}$.

Thus, $a_7 = 12\left(\frac{5}{6}\right)^6$, $a_{20} = 12\left(\frac{5}{6}\right)^{19}$, and the limit is $\lim\limits_{n\to\infty} \left[\frac{72}{5}\left(\frac{5}{6}\right)^n\right] = 0.$

(e) The sequence defined recursively by $a_1 = 0.01$ and $a_n = -2a_{n-1}$ is geometric with $a = 0.01$ and $r = -2$, so its nth

term is $a_n = ar^{n-1} = 0.01\,(-2)^{n-1}$. Thus, $a_7 = 0.01\,(-2)^6 = 0.64$, $a_{20} = 0.01\,(-2)^{19} = -5242.88$, and the limit is

$\lim\limits_{n\to\infty} \left[0.01\,(-2)^{n-1}\right]$, which does not exist.

3. We use the formula for installment buying with $A_P = 350{,}000 - 35{,}000 = 315{,}000$, $i = 0.06/12 = 0.005$, and

$n = 12 \times 15 = 180$, and find that $R = \dfrac{iA_P}{1-(1+i)^{-n}} = \dfrac{0.005\,(315{,}000)}{1-(1.005)^{-180}} \approx \$2658.15.$

5. (a) $\left(2x - \frac{1}{2}\right)^5 = \binom{5}{0}(2x)^5 + \binom{5}{1}(2x)^4\left(-\frac{1}{2}\right) + \binom{5}{2}(2x)^3\left(-\frac{1}{2}\right)^2$

$$+ \binom{5}{3}(2x)^2\left(-\frac{1}{2}\right)^3 + \binom{5}{4}(2x)\left(-\frac{1}{2}\right)^4 + \binom{5}{5}\left(-\frac{1}{2}\right)^5$$

$$= 32x^5 - 40x^4 + 20x^3 - 5x^2 + \tfrac{5}{8}x - \tfrac{1}{32}$$

(b) The general term in the expansion is $\binom{12}{12-r}(2x)^r\left(-\frac{1}{2}\right)^{12-r}$. The term that contains x^4 is the term in which $r = 4$, so

the required term is $\binom{12}{12-4}(2x)^4\left(-\frac{1}{2}\right)^{12-4} = \frac{495}{16}x^4.$

7.

x	-1	-0.1	-0.01	0.01	0.1	1
$\dfrac{1-\cos x}{x^2}$	0.45970	0.49958	0.50000	0.50000	0.49958	0.45970

$\lim\limits_{x\to 0} \dfrac{1-\cos x}{x^2} = \dfrac{1}{2}$

9. (a) $g(x) = x^3 \Rightarrow$

$g'(x) = \lim\limits_{h\to 0} \dfrac{(x+h)^3 - x^3}{h} = \lim\limits_{h\to 0} \dfrac{x^3 + 3x^2h + 3xh^2 + h^3 - x^3}{h} = \lim\limits_{h\to 0} \dfrac{3x^2h + 3xh^2 + h^3}{h}$

$= \lim\limits_{h\to 0} \left(3x^2 + 3xh + h^2\right) = 3x^2$

(b) $g'(-3) = 3(-3)^2 = 27$, $g'(0) = 3(0)^2 = 0$, and $g'(a) = 3a^2$.

(c) $g'(2) = 3(2)^2 = 12$, so an equation of the tangent line at $(2, 8)$ is $y - 8 = 12(x-2)$, or $y = 12x - 16$.